# Conceptual Issues in Evolutionary Biology

# 亅Ŀ Bradford Books

Edward C. T. Walker, Editor. Explorations in THE BIOLOGY OF LANGUAGE. 1979. The M.I.T. Work Group in the Biology of Language: Noam Chomsky, Salvador Luria, et alia.

Daniel C. Dennett. BRAINSTORMS: Philosophical Essays on Mind and Psychology. 1979.

Charles Marks. COMISSUROTOMY, CONSCIOUSNESS AND UNITY OF MIND. 1980.

John Haugeland, Editor. MIND DESIGN. 1981.

Fred I. Dretske. KNOWLEDGE AND THE FLOW OF INFORMATION. 1981.

Jerry A. Fodor. REPRESENTATIONS: Philosophical Essays on the Foundations of Cognitive Science. 1981.

Ned Block, Editor. IMAGERY. 1981.

Roger N. Shepard and Lynn A. Cooper. MENTAL IMAGES AND THEIR TRANSFORMATIONS. 1982.

Hubert L. Dreyfus, Editor, in collaboration with Harrison Hall. HUSSERL, INTENTIONALITY AND COGNITIVE SCIENCE. 1982.

John Macnamara. NAMES FOR THINGS: A Study of Human Learning. 1982.

Natalie Abrams and Michael D. Buckner, Editors. MEDICAL ETHICS; A Clinical Textbook and Reference for the Health Care Professions. 1983.

Morris Halle and G. N. Clements. PROBLEM BOOK IN PHONOLOGY: A Workbook for Introductory Courses in Linguistics and in Modern Phonology. 1983.

Irvin Rock. THE LOGIC OF PERCEPTION. 1983.

Jon Barwise and John Perry. SITUATIONS AND ATTITUDES. 1983.

Stephen Stich. FOLK PSYCHOLOGY AND COGNITIVE SCIENCE. 1983.

Jerry A. Fodor. MODULARITY OF MIND: An Essay on Faculty Psychology. 1983.

George D. Romanos. QUINE AND ANALYTIC PHILOSOPHY. 1983.

Robert Cummins. THE NATURE OF PSCYHOLOGICAL EXPLANATION. 1983.

Elliott Sober, Editor. CONCEPTUAL ISSUES IN EVOLUTIONARY BIOLOGY: A Book of Readings. 1984.

# Conceptual Issues in Evolutionary Biology

## An Anthology

ELLIOTT SOBER, EDITOR

*A BRADFORD BOOK*

THE MIT PRESS
CAMBRIDGE, MASSACHUSETTS
LONDON, ENGLAND

Third printing, 1986
Copyright © 1984 by
The Massachusetts Institute of Technology

Library of Congress Cataloging in Publication Data:
Main entry under title:

Conceptual issues in evolutionary biology.
   "A Bradford book."
   Bibliography: p.
   Includes index.
   1. Evolution–Addresses, essays, lectures. I. Sober, Elliott.
QH366.2.C64 1984      575       83–43025
ISBN 0-262-19220-9 (h)
    0-262-69084-5 (p)

Typographic design by David Horne.
Composition by Horne Associates, Inc.,
West Lebanon, New Hampshire.
This book was printed and bound
in the United States of America.

# Contents

# Preface

"Philosophy assembles definitions of concepts; science applies those concepts to the empirical world." This idea is familiar to any philosopher or scientist who has thought about the connection between the two fields. Logical positivism raised this rough and ready characterization to the level of principle. Postpositivist philosophers—especially Quine—have challenged its adequacy at the level of theory, claiming that scientific and philosophical questions lie on a continuum, there being no sharp demarcation between conceptual problems and empirical ones.[1] Although this is not the place to consider whether there is *in principle* some fundamental gap between the two kinds of questions, the contents of this anthology suggest that a blurring of boundaries has occurred at the level of practice. In this book you will find philosophers and biologists engaging in the same line of work—clarifying and connecting some of the major concepts of evolutionary theory. Although you may have a firm conviction about the usefulness of this or that article, you probably will find yourself hard pressed at times to say whether what is being done is science or philosophy.

The first section, "Guiding Ideas in Evolutionary Biology," focuses on how evolutionary theory has tried to cope with diversity and complexity. In sexual populations, individuals (except for identical twins) are genetically unique; they are subject to a host of forces, which themselves vary in space and in time. Lewontin, Mayr, and Levins describe some of the conceptual tools available for making sense of this blooming, buzzing confusion.

Darwin was not the first to think of evolution. Nor was the idea of a selection process original with him (think of Hobbes's war of all against all and the invisible-hand explanations of the Scottish economists). But Darwin was the first to synthesize these elements; the idea of evolution (including speciation)

1. W. V. O. Quine, *Word and Object,* Cambridge, MIT Press, 1960.

by a process of natural selection was his. He found Spencer's "survival of the fittest" a handy motto. The subsequent development of evolutionary theory shows that this slogan was confusing if not confused. Critics seized upon the phrase in the hope of showing that evolutionary theory is no scientific theory at all. What organisms are fit? Those that survive. What organisms will survive? Those that are fit. This is a disappointingly small circle. If evolutionary biology has no more to offer than this, it is an impoverished discipline indeed. The papers in Section 2, "Fitness," dissect the concept of fitness and evaluate the power of this line of attack.

Section 3, "The Units of Selection," concerns another problem we inherited from Darwin. He thought of natural selection as pitting organism against organism. With the possible exception of his discussion of human morality,[2] he never viewed selection as favoring some *groups* over others. This individualistic (organismic) perspective implies that selection will tend to eliminate "altruism." An altruist may benefit its group at its own expense: its group will do better than other groups, but it will do worse than its neighbors in the same group. With organismic selection in control, the result is "the survival of the selfish"; group selection, however, can produce the opposite result. Contemporary discussion of this problem was triggered by V. C. Wynne-Edwards' *Animal Dispersion in Relation to Social Behavior.*[3] He argued that altruistic characteristics are widespread in nature and are to be explained by hypotheses of group selection. George C. Williams published *Adaptation and Natural Selection* in reply.[4] He urged a return to the Darwinian fold, but with a twist. The pioneering work of Wright, Fisher, and Haldane earlier in this century had made it possible for evolutionary processes to be represented in terms of changes in gene frequencies. Williams brought this genetic perspective to bear: not the group, nor even the organism, but the single gene is the unit of selection.[5] A variety of quantitative models were subsequently developed with the aim of establishing what natural conditions would enhance the power of different sorts of selection processes. But a fundamental conceptual issue has also engaged the attention of biologists and philosophers: What does it mean to call a gene, an organism, a family, a small local population, or a species a unit of selection? This, then, is the principal focus of the articles in Section 3, "The Units of Selection."

Williams concluded his book with a plea for a systematic exploration of concepts like adaptation and function. The essays in Sections 4 and 5 attempt to fill the gap. Questions of meaning and questions of evidence are jointly addressed. What does it mean to say that a characteristic of some organism (or

2. See *The Descent of Man and Selection in Relation to Sex,* London, Murray, pp. 163–166.

3. V. C. Wynne-Edwards, *Animal Dispersion in Relation to Social Behavior,* Edinburgh, Oliver and Boyd, 1962.

4. George C. Williams, *Adaptation and Natural Selection,* Princeton, Princeton University Press, 1966.

5. Richard Dawkins provided a popularized presentation of this idea in his book *The Selfish Gene,* Oxford University Press, 1976.

group, or gene) is an adaptation for performing some task, or has the performing of that task as its function? And given some clarification of these twin concepts, how are scientists to construct and evaluate explanations that postulate adaptations and functions?

The contributors to these two sections aim for teleology without vitalism. Camouflaged moths, unlike the soot on the trees they inhabit, may be black because it is advantageous for them to be so. Their coloration has a function; it achieves some goal. Although there is a difference between moth and soot, it is not to be found in some special immaterial fluid, present in the one but absent in the other. Both the biologists and the philosophers writing here want their teleology to be appropriately physicalistic.

This constraint leaves much room for maneuver, as the lively debate in Section 5 shows. In addition, the requirement of physicalism leaves unsolved the methodological problem of how adaptationist and functionalist explanations are best deployed. Earlier themes concerning the survival of the fittest reappear here. Is the assumption that evolution is mainly or exclusively guided by natural selection "unfalsifiable" or "untestable"? What other evolutionary forces may circumscribe or counter the action of selection? Some questions in this vein were occasioned by the sociobiology debate.[6] But whatever the merits or limitations of that line of thought, the questions posed by adaptation and adaptationism are foundational.

The waning of vitalism as a viable approach to teleology opens the door for another *ism*—reductionism. Can biology be unified with the rest of science by reducing it to chemistry and then (utlimately, and in principle) to physics? Simply banishing immaterial vital fluids from the furniture of the world is not enough to resolve this problem. The concept of reduction must itself be clarified. Section 6 interweaves questions of explanation, definability, and methodological utility.[7]

The final section of the anthology addresses the species concept. It might be thought that the discovery that species evolve must definitively settle what the aims and methods of systematics should be. But the spirited and continuing controversy among *pheneticists, cladists,* and *evolutionary taxonomists,* as they are called, shows that it is possible to agree that species evolve without agreeing on much else. For example, should organisms be included in the same species (or higher taxonomic unit) by virtue of their overall similarity, propinquity of descent, or some combination of the two? And even if this question of meaning were settled, many methodological problems would still remain. We do not observe one species evolving into another. At best, we observe character distributions: we see that different species share some traits but not others. How is it

6. On which see Arthur Caplan's anthology, *The Sociobiology Debate,* New York, Harper and Row, 1978.

7. Problems of reductionism are also discussed in previous selections—for example, in essays 8, 11, and 12. Discussion of reductionism in the philosophy of psychology is extremely relevant to the biological case. The reader is urged to consult Ned Block's anthology *Readings in the Philosophy of Psychology,* Cambridge, Harvard University Press, 1981.

possible to reconstruct the historical relations of species from facts about their sameness and difference? Several of the papers reprinted in Section 7 are controversial; I regret not being able to include more. The attentive reader will want to look to the journals to see how, in each case, the other side gets in its licks.

When you write a book, you don't have to apologize for your interests. After all, it's *your* book. But anthologies are different, and the right to be idiosyncratic is less absolute. I have followed my interests in selecting material, but I have also sought advice. I especially want to thank John Beatty, Richard Boyd, Robert Brandon, Richard Burian, Marjorie Grene, David Hull, Philip Kitcher, and William Wimsatt. They will forgive me, I hope, for not taking their suggestions at every turn. I also am grateful to Harry and Betty Stanton and to Ruth Saunders for their help in planning and producing the book. Many other articles merited reprinting, but space forbade. In fact, some *seminal* articles could not be included. In publishing, as in evolution itself, design constraints prevent selection from achieving perfection. I hope this book will spur the interest of philosophers and biologists alike. May it soon be rendered obsolete!

*Elliott Sober*

*Madison, Wisconsin*

# Contributors

John Beatty, Department of Philosophy, Arizona State University, Temple, Arizona 85287.

Christopher Boorse, Department of Philosophy, University of Delaware, Newark, Delaware 19711.

Robert Brandon, Department of Philosophy, Duke University, Durham, North Carolina 27708.

Theodore J. Crovello, Department of Biology, Notre Dame University, Notre Dame, Indiana 46556.

Robert Cummins, Department of Philosophy, University of Illinois–Circle Campus, Chicago, Illinois 60680.

Richard Dawkins, Zoology–Animal Behavior Research, South Parks Road, University of Oxford, Oxford OX1 3P5, England.

Steven Farris, 41 Admiral Street, Port Jefferson Station, New York 11776.

Joe Felsenstein, Department of Genetics, University of Washington, Seattle, Washington 98195.

Stephen Jay Gould, Museum of Comparative Zoology, Harvard University, Cambridge, Massachusetts 02138.

David L. Hull, Department of Philosophy, University of Wisconsin, Milwaukee, Wisconsin 53201.

Richard Levins, School of Public Health, Harvard University, Cambridge, Massachusetts 02138.

R. C. Lewontin, Museum of Comparative Zoology, Harvard University, Cambridge, Massachusetts 02138.

Nancy Maull, Special Assistant to the President, University of Chicago, Chicago, Illinois 60637.

Ernst Mayr, Museum of Comparative Zoology, Harvard University, Cambridge, Massachusetts 02138.

Susan Mills, 945 Judson Way, Evanston, Illinois 60202.

Ernest Nagel, Department of Philosophy, Columbia University, New York, N.Y. 10027.

George Oster, Department of Biology, University of California, Berkeley, California 94720.

Alex Rosenberg, Department of Philosophy, Syracuse University, Syracuse, New York 13210.

Michael Ruse, Department of Philosophy, Guelph University, Guelph, Ontario N1G 2W1, Canada.

Kenneth Schaffner, Department of Philosophy, University of Pittsburgh, Pittsburgh, Pennsylvania 15260.

John Maynard Smith, Population Biology Group, University of Sussex, Brighton BM1 90G, England.

Robert Sokal, Department of Biology, State University of New York, Stony Brook, New York 11794.

Mary Williams, Department of Biological Sciences, University of Delaware, Newark, Delaware 18711.

Edward O. Wilson, Museum of Comparative Zoology, Harvard University, Cambridge, Massachusetts 02138.

William Wimsatt, Department of Philosophy, University of Chicago, Chicago, Illinois 60637.

Larry Wright, Department of Philosophy, University of California, Riverside, California 92521.

# I
# GUIDING IDEAS IN
# EVOLUTIONARY BIOLOGY

# 1

# The Structure of Evolutionary Genetics

## RICHARD C. LEWONTIN

WHEN DIFFERENT PEOPLE CONSIDER the work of a revolutionary like Darwin, they see different aspects of it as representing the "real" or "fundamental" element that separates it from the preexistent conformity of thought. To many, Darwinism means "evolution" and the commitment to an evolutionary world view, but historical evidence makes clear that Darwin only applied rigorously to the organic world what was already accepted as characteristic of the inorganic universe and of human culture (Lewontin, 1968). To others, especially students of the evolutionary process, Darwin's unique intellectual contribution was the idea of natural selection. For them, Darwinism is the theory that evolution occurs because, in a world of finite resources, some organisms will make more efficient use of those resources in producing their progeny and so will leave more descendants than their less efficient relatives.

Yet it is by no means certain, even now, what proportion of all evolutionary change arises from natural selec-tion. Attitudes toward the importance of random events as opposed to selective ones vary from time to time and place to place. The famous Committee on Common Problems of Genetics, Paleontology and Systematics, whose work led to the publication of Genetics, Paleontology and Evolution (Jepsen, Mayr, and Simpson, 1949) embodied within it divergent views on this question. In his article entitled "Speciation and Systematics," Mayr made clear that, in his view, the divergence between isolated populations is the result of their adaptation to different environments and that "geographical races are invariably to a lesser or greater degree also ecological races" (p. 291). But Muller's essay in the same volume puts emphasis on "cryptic genetic change" (p. 425) that is not reflected in phenotypic differentiation but that may result in sufficient genetic divergence between groups to result in speciation. For Mayr, natural selection is vital in the divergence between isolated populations, while for Muller natural selection is always primarily a cleansing agent, rejecting "inharmonious" gene combinations, and not necessarily the causative agent in the initial divergence between incipient species. Nor is this conflict of viewpoints yet resolved. During the last few years there has been a flowering of interest in evolution by purely random processes in which natural selection plays no role at all. Kimura and Ohta suggest, for example (1971) that *most* of the genetic divergence between species that is observable at the molecular level is nonselective, or, as proponents of this view term it, "nonDarwinian," although they do not deny that the evolution of obviously adaptive characters like the elephant's trunk and the camel's hump are the result of natural selection. If the empirical fact should be that most of the genetic change in species formation is indeed of this non-Darwinian sort, then where is the revolution that Darwin made?

The answer is that the essential nature of the Darwinian revolution was neither the introduction of evolutionism as a world view (since historically that is not the case) nor the emphasis on natural selection as the main force in evolution (since empirically that may not be the case), but rather the replacement of a metaphysical view of variation among organisms by a materialistic view (Lewontin, 1973). For Darwin, evolution was the conversion of the variation among individuals within an interbreeding group into variation between groups in space and time. Such a theory of evolution necessarily takes the variation between individuals as of the essence. Ernst Mayr has many times pointed out, especially in

*Animal Species and Evolution* (1963), that this emphasis on individual variation as the central reality of the living world is the mark of modern evolutionary thought and distinguishes it from the typological doctrine of previous times.

In the thought of the pre-Darwinians, the Platonic and Aristotelian notion of the "ideal" or "type" to which actual objects were imperfect approximations was a central feature. Nature, and not just living nature, was understood by the pre-Darwinians only in terms of the ideal; and the failure of individual cases to match the ideal was a measure of the imperfection of nature. Such a metaphysical construct is not without importance in science, as Newton's mechanics prove so well. The first law in the *De Motu Corporum* is that

> Every body perseveres in its state of rest, or of uniform motion in a right line, unless it is compelled to change that state by forces impressed thereon.

Yet Newton points out immediately that even "the great bodies of the planets and comets" have such perturbing forces impressed upon them and that no body perseveres indefinitely in its motion. The metaphysical introduction of ideal bodies moving in ideal paths, so essential to the proper development of physics and so constant with the habits of thought of the seventeenth century, was precisely what had to be destroyed in the creation of evolutionary biology. Darwin rejected the metaphyiscal object and replaced it with the material one. He called attention to the *actual* variation among *actual* organisms as the most essential and illuminating fact of nature. Rather than regarding the variation among members of the same species as an annoying distraction, as a shimmering of the air that distorts our view of the essential object, he made that variation the cornerstone of his theory. Let us remember that the *Origin of Species* begins with a discussion of variation under domestication.

The conflict between the real and the ideal was also important in a second realm that is relevant to evolution. What we know as the science of genetics is meant to explain two apparently antithetical observations—that organisms resemble their parents and differ from their parents. That is, genetics deals with both the problem of heredity and the problem of variation. It is, in fact, the triumph of genetics that a single theory, down to the molecular level, explains in one synthesis both the constancy of inheritance and its variation. It is the Hegelian's dream. But this synthesis was not possible until sufficient importance was attached to variation. Francis Galton attempted to construct a theory of inheritance based upon the degree of resemblance of offspring to their parents. Galton's Law of Filial Regression placed the emphasis on the fact that the offspring of extreme parents tended to "regress" back to the mean of the parental generation. The law was derived from observations of the mean height of all offspring whose parents belonged to a specific height class, but it placed no weight at all on the variation between parental pairs whose offspring were the same height, nor between sibs, nor between sibships whose parents belonged to the same height class. Galton's scheme depended entirely upon the regression of *means* on *means*.

In striking contrast, Mendel placed his emphasis on the *variations* among the offspring, rather than on any average description of them, *and derived his laws from the nature of the variations*. Thus, for Mendelism, as for Darwinism, the fact of variation and its nature are central and essential. Of course, at a second level, Mendel idealized his laws, much as Newton did, and every student knows that Mendel's $F_2$ ratios were suspiciously close to those 3 to 1 ideals. But this in no way vitiates the centrality of variation in Mendel's thought. Modern evolutionary genetics is then a union of two systems of knowledge, both of which took variation to be the essential fact of nature. It is not surprising, therefore, that the study of genetically determined variation within and between species should be the starting point of evolutionary investigation.

### A DIGRESSION INTO FORMALISM

*Dynamic sufficiency.* When we say that we have an evolutionary perspective on a system or that we are interested in the evolutionary dynamics of some phenomenon, we mean that we are interested in the change of state of some universe in time. Whether we look at the evolution of societies, languages, species, geological features, or stars, there is a formal representation that is in common to all. At some time $t$ the system is in some state $E$, and we are interested in the state of the system, $E'$, at a future time, or past time, $\tau$ time units away. We must then construct laws of transformation $T$ that will enable us to predict $E'$ given $E$. Formally, we may represent this as

$$E(t) \xrightarrow{T} E'(t + \tau)$$

The laws of transformation $T$ cannot be of an arbitrary form. Usually they contain some parameters $\Pi$, values that are not themselves a function of time or the state of the system. Second, they will contain the elapsed time $\tau$, except in the description of equilibrium systems in which no change is taking place. They may or may not refer specifically to the absolute time $t$, depending on whether the system carries in its present state some history of its past. For example, if I wish to know what proportion of a population will be alive a year from now, I need to know how long ago the individuals were born, since survival probabilities change with age. But if I wish to know what proportion of my teacups will last through the next year, I do not have to know how old they are since, to a first order of approximation, their breakage probability per year is independent of their age. Finally, and essentially, the laws of transformation must contain the present state of the system $E$ and obviously must produce as part of their output the new state $E'$. The laws of transformation are a machinery that must be built to process information about the current state $E$ and to produce, as an output, the new state $E'$. But that means that the description of the system, $E$, must be chosen in such a way that laws of transformation can indeed be constructed using it.

For example, one cannot predict the future position of a space capsule from its present position alone. No set of laws can be constructed that

will transform $E(t)$ into $E'(t + \Pi)$ if $E$ is the position of the capsule in three-dimensional space. It is necessary, in addition, to specify the present velocity of the capsule in three orthogonal directions, and its present acceleration in three orthogonal directions. If the state description is a function of those nine variables, then laws of transformation are possible and, as a result, capsules get to the moon and back. We will say that the state description with all nine variables is a *dynamically sufficient description* because, given that description, it is possible to find laws $T(E, \Pi, \tau, t)$:

$$E(t) \xrightarrow{T(E,\ \Pi,\ \tau,\ t)} E'(t + \tau)$$

The transformation of state in time has a geometrical interpretation. Let us take each variable used in specifying the state of the system as an axis in a Cartesian space. Then the state of the system at any time is represented as a point in that space, located by the projection onto the various axes. Evolution of the system is movement of the point through the space, tracing out a trajectory, and the laws of transformation are the equation whose solution is that trajectory. The axes of the space are the *state variables,* and the space they span is the *state space* of our system.

Looked at in this way, the problem of constructing an evolutionary theory is the problem of constructing a state space that will be dynamically sufficient, and a set of laws of transformation in that state space that will transform all the state variables. It is not always appreciated that the problem of theory building is a constant interaction between constructing laws and finding an appropriate set of descriptive state variables such that laws can be constructed. We cannot go out and describe the world in any old way we please and then sit back and demand that an explanatory and predictive theory be built on that description. The description may be dynamically insufficient. Such is the agony of community ecology. We do not really know what a sufficient description of a community is because we do not know what the laws of transformation are like, nor can we construct those laws until we have chosen a set of state variables. That is not to say that there is an insoluble contradiction. Rather, there is a process of trial and synthesis going on in community ecology, in which both state descriptions and laws are being fitted together.

*Tolerance limits.* In the development of a real science about a real and practical world, it is impossible and undesirable to search for an exactly sufficient description. The nature of the physical universe is such that the change of state of every part of it affects the change of state of every other part, no matter how remote. While a space capsule evolves in a nine-dimensional space for all practical purposes, an absolutely exact treatment of its motion would have to take into account the fact that it affects the motion of the earth and moon and of every other celestial body in some tiny degree. Yet no serious person would suggest that we really need to take into account the impetus given to the earth at the moment of launching. In each domain of practice we have a notion of the appropriate accuracy of prediction, and we are satisfied if our theory gives results that are somehow close enough. More exactly, corresponding to each state of a system $E$ we establish a *tolerance set $\epsilon$* of states that are similar enough to be regarded as indistinguishable. That does not mean that they *are* indistinguishable, but only that we do not care about the differences among states within a tolerance set. In practice, then, a sufficient dimensionality is one that allows us to describe the evolution of tolerance sets in time rather than the evolution of exact state descriptions.

In population ecology and evolutionary genetics, the tolerance limits remain matters of debate and choice. Ought we to be satisfied with theories that predict only that one community will have more species than another or that one population will be more polymorphic than another? If so, a very low dimensionality may be sufficient. Or do we really want to predict the number of breeding individuals of each species each year, or the gene frequencies at various loci, and if so, how accurately? Exactly the same domain of science may require quite different degrees of accuracy in different applications and thus use models of very different dimensionality. For example, population ecologists are generally satisfied to explain to one order of magnitude the increases and decreases in population size of the organisms studied, and for this purpose net fecundity and mortality are ususally sufficient. Game and fish management, however, may require prediction of population changes to an accuracy of 10 to 20 percent, and for this purpose complete age-specific mortality and fecundity schedules are required. Finally, the human demographer needs to project human population sizes to better than 1 percent accuracy, and to do so needs fecundity and mortality figures by age, sex, socioeconomic class, education, geographical location, and so on. The building of a dynamically sufficient theory of evolutionary processes will really entail the simultaneous development of theories of different dimensionalities, each appropriate to the tolerance limits acceptable in its domain of explanation.

*Empirical sufficiency.* There is a second problem of theory construction in evolution, which we may term the problem of *empirical sufficiency*. The laws of transformation contain two elements that require measurement: the state variables that make up $E$ and the parameters that make up $\Pi$. Even when a dynamically sufficent state space and a set of transformation laws have been arrived at, some of the state variables or the parameters may turn out to be, in practice, unmeasurable. Such unmeasurability is often not absolute; instead, the accuracy of measurement is low compared to the sensitivity of predictions to small perturbations in the values of the variables. To return to our space capsule example, when the capsule leaves its "parking orbit" around the earth, the smallest error in exit angle, smaller than can be controlled, will cause it to miss the moon by many miles. That is why mid-course corrections are necessary.

But empirical sufficiency is not always a matter of accuracy. It may lie much deeper. An example to which I shall return is the measurement of fitness of genotypes. One component of fitness is the probability of survival from conception to the age of reproduction. But by definition the probability of survival is an ensemble property, not the property of a single individual, who either will or will not survive. To measure this probability we then need to produce an ensemble of individuals, all of the same genotype. But if we are concerned with the alternative genotypes at a single locus, we need to randomize the rest of the genome. In sexually reproducing organisms, there is no way known to produce an ensemble of individuals that are all identical with respect to a single locus but randomized over other loci. Thus a theory of evolution that depends on the characterization of fitness of genotypes with respect to single loci is in serious trouble, trouble that cannot be cleared up by a quantitative improvement in the accuracy of measurement. The theory suffers from an epistemological paradox.

It is a remarkable feature of the sociology of science that evolutionary biologists have persistently ignored the problem of empirical sufficiency. The literature of population genetics is littered with estimates lacking standard errors and with methods for deciding between alternatives that have no sensitivity analyses or tests of hypotheses. No one has been exempt from this methodological naïveté, including myself, so that any specific discussion of the problem becomes immediately offensive to almost everyone. There are, however, a few positive landmarks in the assessment of these methodological difficulties that may be mentioned to the credit of their discoverers.

Since the invention by Muller of the C/B technique in 1928, many investigations have been made of the distribution of viabilities and fecundities of chromosomal homozygotes in various species of Drosophila. Yet it was not until 25 years later that Wallace and Madden (1953) and Dobzhansky and Spassky (1953) showed what proportion of the observed variance among chromosomal homozygotes and heterozygotes was, in fact, genetic, as opposed to experimental error. Since the error variance turned out to be between 59 and 99 percent of the total variance in different samples of heterozygotes and between 24 and 68 percent in homozygotes (Dobzhansky and Spassky, 1953, table 6), it is difficult to know what to make of studies without estimates of error.

An important parameter to be estimated for an understanding of the differentiation between populations is the effective breeding size, $N$. An elaborate theory exists for the estimation of this quantity from the allelism of lethal genes. The estimation procedure involves the reciprocal of the difference between two small numbers, a fact that in itself makes the use of the technique dubious. In an empirical test of procedure, Prout (1954) showed that in practice one cannot detect the difference between a population of 5000 and an infinite population.

Since the first experiments of L'Heritier and Teissier (1933) and of Wright and Dobzhansky (1946), adaptive values of Drosophila genotypes in population cages have been estimated from changes in genotype frequencies, but despite the use of a number of statistically sophisticated estimation techniques, standard errors of fitness estimates from population cages were not commonly calculated until Anderson, Oshima, Watanabe, Dobzhansky, and Pavlovsky did so in 1968, with rather disturbing results. Fitness estimates as disparate as 0.72 and 0.19 were not significantly different and estimates of 0.54, 0.47, and 0.59 had standard errors of 0.78, 1.15, and 1.23, respectively. Admittedly these experiments involved the estimation of nine fitness values, but this simply points up the problem of estimation for even a small number of genotypes under rigorously controlled laboratory conditions. The possibility of fitness estimation with smaller standard errors in natural populations seems remote, under the circumstances.

While dynamic sufficiency is an absolute and basic requirement for the building of an evolutionary theory, empirical sufficiency adds yet another stricture that may render a formally perfect theory useless. If one simply cannot measure the state variables or the parameters with which the theory is constructed, or if their measurement is so laden with error that no discrimination between alternative hypotheses is possible, the theory becomes a vacuous exercise in formal logic that has no points of contact with the contingent world. The theory explains nothing because it explains everything. It is my contention that a good deal of the structure of evolutionary genetics comes perilously close to being of this sort.

## POPULATION GENETICS

Population genetics sets a much more modest goal than general evolutionary theory. If we take the Darwinian view that evolution is the conversion of variation between individuals into variation between populations and species in time and space, then an essential ingredient in the study of evolution is a study of the origin and dynamics of genetic variation within populations. This study, population genetics, is *an* essential ingredient, but it is not the entire soup. While population genetics has a great deal to say about changes or stability of the frequencies of genes in populations and about the rate of divergence of gene frequencies in populations partly or wholly isolated from each other, it has contributed little to our understanding of speciation and nothing to our understanding of extinction. Yet speciation and extinction are as much aspects of evolution as is the phyletic evolution that is the subject of evolutionary genetics, strictly speaking. That is not to say that speciation and extinction are not the natural extensions of changes within populations, but only that our present theories do not deal with these processes except on the most general and nonrigorous plane.

Even the partial task set for population genetics is a tremendously ambitious program. If we could succeed in providing a description of the genetic state of populations and laws of transformation of state that were both dynamically and empirically sufficient, we would create a complete theory concerning a vastly more complex domain than any yet dealt with by physics or molecular biology.

The present structure of population genetics theory may be represented as

$$G_1 \xrightarrow{T_1} P_1 \xrightarrow{T_2} P_2 \xrightarrow{T_3} G_2 \xrightarrow{T_4} G'_1 \xrightarrow{T_1}$$

where $G_1$ and $G'_1$ represent a genetic description of population at times $t$ and $t + \Delta t$, and $P_1$, $P_2$

and $G_2$ represent phenotypic and genotypic descriptions of states during the transformation

and the laws of transformation are

$T_1$ : a set of epigenetic laws that give the distribution of phenotypes that result from the development of various genotypes in various environments

$T_2$ : the laws of mating, of migration, and of natural selection that transform the phenotypic array in a population *within* the span of a generation

$T_3$ : an immense set of epigenetic relations that allow inferences about the distribution of genotypes corresponding to the distribution of phenotypes, $P_2$

$T_4$ : the genetic rules of Mendel and Morgan that allow us to predict the array of genotypes in the next generation produced from gametogenesis and fertilization, given an array of parental genotypes

It would appear that both genotypes and phenotypes are state variables and that what population genetic theory does is to map a set of genotypes into a set of phenotypes, provide a transformation in the phenotype space, then map these new phenotypes back into genotypes, where a final transformation occurs to produce the genotypic array in the next generation. Figure 1 shows these transformations schematically. From this schema one would imagine that both phenotypic and genotypic state variables would enter into a sufficient dimensionality for the description of population evolution, and that the laws of population

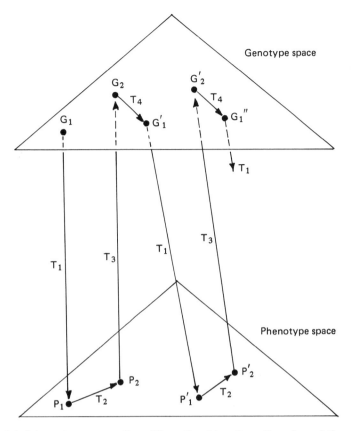

Figure 1-1. Schematic representation of the paths of transformation of population genotype from one generation to the next. G and P are the spaces of genotype description. $G_1$, $G'_1$, $G_2$, and $G'_2$ are genotypic descriptions at various points in time within successive generations. $P_1$, $P'_1$, $P_2$, and $P'_2$ are phenotypic descriptions. $T_1$, $T_2$, $T_3$, and $T_4$ are laws of transformation. Details are given in the text.

genetics would be framed in terms of both genetic and phenotypic variables. Yet an examination of population genetic theory shows a paradoxical situation in this respect. One body of theory, what we might call the "Mendelian" corpus, used almost exclusively by those interested in the genetics of natural populations, seems to be framed entirely in genetic terms. The other theoretical system, the "biometric," used almost exclusively in plant and animal breeding, appears to be framed in completely phenotypic terms. In Mendelian population genetics, for example, the expression

$$\Delta q = \frac{q(1-q)}{2} \frac{d\ln\bar{w}}{dq} \quad (1)$$

expresses the change in relative *allele frequency* $\Delta q$ of an allele at a locus after one generation, in terms of the present allele frequency $q$ and the mean fitness $\bar{w}$ of the genotypes in the population.

In biometric genetics, the prediction equation

$$\Delta P = ih^2 = i\frac{\sigma_g^2}{\sigma_p^2} \tag{2}$$

predicts the change in mean *phenotype, $\Delta P$,* in one generation in terms of $i$, the difference in phenotype between the population as a whole and the selected parents, the phenotypic variance of the character in the population, $\sigma_p^2$, the additive genetic variance, $\sigma_g^2$, and the heritability of the trait, $h^2$.

Apparently, then, we have two parallel systems of evolutionary dynamics, one operating in the space of genotypes and bypassing the phenotypic space, and another operating entirely in the phenotypic domain. But this impression is illusory and arises from a bit of sleight-of-hand in which variables are made to appear as merely parameters that need to be experimentally determined, constants that are not themselves transformed by the evolutionary process. In equation (1) these pseudoparameters are the fitnesses associated with the individual genotypes in computing the function $\bar{w}$. Fitness, however, is a function of the phenotype, not the genotype, although in special circumstances it might turn out that a one-to-one constant correspondence existed between genotype and fitness. More usually, however, relative fitnesses of phenotypes will be a function of the phenotype composition of the population as a whole, so that the fitnesses assigned to genotypes in each generation will themselves follow a law of transformation that depends upon the genotype-phenotype relations. Although, for convenience, geneticists usually use a "constant genotypic

fitness" model, this had led to a number of paradoxes that can be resolved properly only when phenotypic relations are taken fully into account. An example is the paradox of *genetic load.*

The biometric model presents even greater difficulty. The heritability $h^2$ and the genetic and phenotypic variances are themselves determined by the genetic variation in the population and undergo constant change in the course of the population's evolution. The laws of change of these variables cannot be framed without reference to the genetic determination of phenotype, including the degree of dominance of genes, the amount of interaction between them, and the relative frequencies of the alleles at the loci determining the character under selection. Thus, while equation (2) masquerades as a phenotypic law, genetic state variables must be added to provide a sufficient dimensionality. Because the transition between generations depends upon genetic laws, no sufficient description in terms of phenotypes alone is ever possible, and attempts like those of Slatkin (1970) can only succeed under extremely simplified conditions with very limited applicability. It is only fair to point out that for some purposes of plant and animal breeding, equation (2) is a sufficient predictor because the value of $h^2$ may evolve very slowly as compared with the mean phenotype, especially if a very large number of genes, each of small effect, are influencing the trait under selection. For a long-term prediction of progress under selection, or for an estimate of the eventual limit to the selection process, however, equation (2) is insufficient. *The sufficient set of state variables for describing an evolutionary process within a population must include some information about the statistical distribution of genotypic frequencies. It is for this* *reason that the empirical study of population genetics has always begun with and centered on the characterization of the genetic variation in populations.*

## REFERENCES

Anderson, W. W., C. Oshima, T. Watanabe, Th. Dobzhansky, and O. Pavlovsky, 1968, Genetics of natural populations. XXXIX. A test of the possible influence of two insecticides on the chromosomal polymorphism in *Drosophila pseudoobscura. Genetics,* 58: 423–434.

Dobzhansky, Th., and B. Spassky, 1953, Genetics of natural populations. XXI. Concealed variability in two sympatric species of Drosophila. *Genetics,* 38: 471–484.

Jepsen, G. L., E. Mayr, and G. G. Simpson, eds., 1949, *Genetics, Paleontology and Evolution,* Princeton, N.J., Princeton.

Kimura, M. and T. Ohta, 1971, *Theoretical Aspects of Population Genetics,* Princeton, N.J., Princeton.

Lewontin, R. C., 1968, The concept of evolution. In *International encyclopedia of Social Science,* ed. D. L. Sills, vol. 5, pp. 202–210. New York, Macmillan, Free Press.

Lewontin, R. C., 1973, Darwin and Mendel: The triumph of materialism. In: *The Copernican Revolution,* ed., J. Neyman. In press.

L'Heritier, Ph., and G. Teissier, 1933, Etude d'une population de Drosophiles en equilibre. C. R. Acad. Sci. 198: 770–772.

Mayr, E., 1963, *Animal Species and Evolution,* Cambridge, Mass., Harvard, Belknap Press.

Prout, T., 1954, Genetic drift in irradiated experimental populations of *Drosophila melanogaster. Genetics,* 39: 529–545.

Slatkin, M., 1970, Selection and polygenic characters. *Proc. Nat. Acad. Sci. U.S.,* 66: 87–93.

Wallace, B. and C. Madden, 1953, The frequencies of sub- and supervitals in experimental populations of *Drosophila melanogaster. Genetics,* 38: 456–470.

Wright, S. and Th. Dobzhansky, 1946, Genetics of natural populations. XII. Experimental reproduction of some of the changes caused by natural selection in certain populations of *Drosophila pseudoobscura. Genetics,* 31: 125–156.

# 2

# Typological versus Population Thinking

## ERNST MAYR

*Rather imperceptibly, a new way of thinking began to spread through biology soon after the beginning of the nineteenth century. It is now most often referred to as population thinking. What its roots were is not at all clear, but the emphasis of animal and plant breeders on the distinct properties of individuals was clearly influential. The other major influence seems to have come from systematics. Naturalists and collectors realized increasingly often that there are individual differences in collected series of animals, corresponding to the kind of differences one would find in a group of human beings. Population thinking, despite its immense importance, spread rather slowly, except in those branches of biology that deal with natural populations.*

*In systematics it became a way of life in the second half of the nineteenth century, particularly in the systematics of the better-known groups of animals, such as birds, mammals, fishes, butterflies, carabid beetles, and land snails. Collectors were urged to gather large samples at many localities, and the variation within populations was studied as assiduously as differences between localities. From systematics, population thinking spread, through the Russian school, to population genetics and to evolutionary biology. By and large it was an empirical approach with little explicit recognition of the rather revolutionary change in conceptualization on which it rested. So far as I know, the following essay, excerpted from a paper originally published in 1959, was the first presentation of the contrast between essentialist and population thinking, the first full articulation of this revolutionary change in the philosophy of biology.*

THE YEAR OF PUBLICATION OF Darwin's *Origin of Species,* 1859, is rightly considered the year in which the modern science of evolution was born. It must not be forgotten, however, that preceding this zero year of history there was a long prehistory. Yet, despite the existence in 1859 of a widespread belief in evolution, much published evidence on its course, and numerous speculations on its causation, the impact of Darwin's publication was so immense that it ushered in a completely new era.

It seems to me that the significance of the scientific contribution made by Darwin is threefold:

1. He presented an overwhelming mass of evidence demonstrating the occurence of evolution.

2. He proposed a logical and biologically well-substantiated mechanism that might account for evolutionary change, namely, natural selection. Muller (1949:459) has characterized this contribution as follows:

Darwin's theory of evolution through natural selection was undoubtedly the most revolutionary theory of all time. It surpassed even the astronomical revolution ushered in by Copernicus in the significance of its implications for our understanding of the nature of the universe and of our place and role in it. . . . Darwin's masterly marshalling of the evidence for this [the ordering effect of natural selection], and his keen-sighted development of many of its myriad facets, remains to this day an intellectual monument that is unsurpassed in the history of human thought.

3. He replaced typological thinking by population thinking.

The first two contributions of Darwin's are generally known and sufficiently stressed in the scientific literature. Equally important but almost consistently overlooked is the fact that Darwin introduced into the scientific literature a new way of thinking, "population thinking," What is this population thinking and how does it differ from typological thinking, the then prevailing mode of thinking? Typological thinking no doubt had its roots in the earliest efforts of primitive man to classify the bewildering diversity of nature into categories. The *eidos* of Plato is the formal philosophical codification of this form of thinking. According to it, there are a limited number of fixed, unchangeable "ideas" underlying the observed variability, with the *eidos* (idea) being the only thing that is fixed and real, while the observed variability has no more reality than the shadows of an object on a cave wall, as it is stated in Plato's allegory. The discontinuities between these natural "ideas" (types), it was believed, account for the frequency of gaps in nature. Most of the great philosophers of the seventeenth, eighteenth, and nineteenth centuries were influenced by the idealistic philosophy of Plato, and the thinking of this school dominated the thinking of the period. Since there is no gradation between types, gradual evolution is basically a logical impossibility for the typologist. Evolution, if it occurs at all, has to proceed in steps or jumps.

The assumptions of population thinking are diametrically opposed to those of the typologist. The populationist stresses the uniqueness of everything in the organic world. What is true for the human species—that no two individuals are alike—is equally true for all other species of animals and plants. Indeed, even the same individual changes continuously throughout its lifetime and when placed into different environments. All organisms and organic phenomena are composed of unique features and can be described collectively only in statistical terms. Individuals, or any kind of organic entities, form populations of which we can determine only the

*typologist*

arithemetic mean and the statistics of variation. Averages are merely statistical abstractions; only the individuals of which the populations are composed have reality. The ultimate conclusions of the population thinker and of the typologist are precisely the opposite. For the typologist, the type (*eidos*) is real and the variation an illusion, while for the populationist the type (average) is an abstraction and only the variation is real. No two ways of looking at nature could be more different.

The importance of clearly differentiating these two basic philosophies and concepts of nature cannot be overemphasized. Virtually every controversy in the field of evolutionary theory, and there are few fields of science with as many controversies, was a controversy between a typologist and a populationist. Let me take two topics, race and natural selection, to illustrate the great difference in interpretation that results when the two philosophies are applied to the same data.

### RACE

The typologist stresses that every representative of a race has the typical characteristics of that race and differs from all representatives of all other races by the characteristics "typical" for the given race. All racist theories are built on this foundation. Essentially, it asserts that every representative of a race conforms to the type and is separated from the representatives of any other race by a distinct gap. The populationist also recognizes races but in totally different terms. Race for him is based on the simple fact that no two individuals are the

same in sexually reproducing organisms and that consequently no two aggregates of individuals can be the same. If the average difference between two groups of individuals is sufficiently great to be recognizable on sight, we refer to such groups of individuals as different races. Race, thus described, is a universal phenomenon of nature occurring not only in man but in two thirds of all species of animals and plants.

Two points are especially important as far as the views of the populationist on race are concerned. First, he regards races as potentially overlapping population curves. For instance, the smallest individual of a large-sized race is usually smaller than the largest individual of a small-sized race. In a comparison of races the same overlap will be found for nearly all examined characters. Second, nearly every character varies to a greater or lesser extent independently of the others. Every individual will score in some traits above, in others below the average for the population. An individual that will show in all of its characters the precise mean value for the population as a whole does not exist. In other words, the ideal type does not exist.

### NATURAL SELECTION

A full comprehension of the difference between population and typological thinking is even more necessary as a basis for a meaningful discussion of the most important and most controversial evolutionary theory—namely, Darwin's theory of evolution through natural selection. For the typologist everything in nature is either "good" or "bad," "useful" or "detrimental." Natural selection is an all-or-none phenomenon. It either selects or rejects, with rejection being by far more obvious and conspicuous. Evolution to him consists of the testing of newly arisen "types." Every new type is put through a screening test and is either kept or, more probably, rejected. Evolution is defined as the preservation of superior types and the rejection of inferior ones, "survival of the fittest" as Spencer put it. Since it can be shown rather easily in any thorough analysis that natural selection does not operate in this described fashion, the typologist comes by necessity to the conclusions: (1) that natural selection does not work, and (2) that some other forces must be in operation to account for evolutionary progress.

The populationist, on the other hand, does not interpret natural selection as an all-or-none phenomenon. Every individual has thousands or tens of thousands of traits in which it may be under a given set of conditions selectively superior or inferior in comparison with the mean of the population. The greater the number of superior traits an individual has, the greater the probability that it will not only survive but also reproduce. But this is merely a probability, because under certain environmental conditions and temporary circumstances, even a "superior" individual may fail to survive or reproduce. This statistical view of natural selection permits an operational definition of "selective superiority" in terms of the contribution to the gene pool of the next generation.

**REFERENCE**

Muller, H. J., 1949, The Darwinian and modern conceptions of natural selection. *Proc. Amer. Phil. Soc.,* 93: 459–470.

# 3

# The Strategy
# of Model Building
# in Population Biology

## RICHARD LEVINS

MODERN POPULATION BIOLOGY arises from the coming together of what were previously independent clusters of more or less coherent theory. Population genetics and population ecology, the most mathematical areas of population biology, had developed with quite different assumptions and techniques, while mathematical biogeography is essentially a new field.

For population genetics, a population is specified by the frequencies of genotypes without reference to the age distribution, physiological state as a reflection of past history, or population density. A single population or species is treated at a time, and evolution is usually assumed to occur in a constant environment.

Population ecology, on the other hand, recognizes multispecies systems, describes populations in terms of their age distributions, physiological states, and densities. The environment is allowed to vary but the species are treated as genetically homogeneous, so that evolution is ignored.

But there is increasing evidence that demographic time and evolutionary time are commensurate. Thus population biology must deal simultaneously with genetic, physiological, and age heterogeneity within species of multi-species systems changing demographically and evolving under the fluctuating influences of other species in a heterogeneous environment. The problem is how to deal with such a complex system.

The naive, brute force approach would be to set up a mathematical model which is a faithful, one-to-one reflection of this complexity. This would require using perhaps 100 simultaneous partial differential equations with time lags; measuring hundreds of parameters, solving the equations to get numerical predictions, and then measuring these predictions against nature. However:

(a) there are too many parameters to measure; some are still only vaguely defined; many would require a lifetime each for their measurement.

(b) The equations are insoluble analytically and exceed the capacity of even good computers.

(c) Even if soluble, the result expressed in the form of quotients of sums of products of parameters would have no meaning for us.

Clearly we have to simplify the models in a way that preserves the essential features of the problem. The difference between legitimate and illegitimate simplifications depends not only on the reality to be described but also on the state of the science. The early pioneering work in population genetics by Haldane, Fisher, and Wright all assumed a constant environment in the models although each author was aware that environments are not constant. But the problem at hand was: Could weak natural selection account for evolutionary change? For the purposes of this problem, a selection coefficient that varies between .001 and .01 will have effects somewhere between constant selection pressures at those values, and would be an unnecessary complication. But for us today environmental heterogeneity is an essential ingredient of the problems and therefore of our mathematical models.

It is desirable, of course, to work with manageable models which maximize generality, realism, and precision toward the overlapping but not identical goals of understanding, predicting, and modifying nature. But this cannot be done. Therefore, several alternative strategies have evolved:

1. Sacrifice generality to realism and precision. This is the approach of Holling (e.g., 1959), of many fishery biologists, and of Watt (1956). These

workers can reduce the parameters to those relevant to the short-term behavior of their organism, make fairly accurate measurements, solve numerically on the computer, and end with precise, testable predictions applicable to these particular situations.

2. Sacrifice realism to generality and precision. The approach of Kerner (1957), Leigh (1965), and most physicists who enter population biology work in this tradition, which involves setting up quite general equations from which precise results may be obtained. Their equations are clearly unrealistic. For instance, they use the Volterra predator-prey systems, which omit time lags, physiological states, and the effect of a species' population density on its own rate of increase. But these workers hope that their model is analogous to assumptions of frictionless systems or perfect gases. They expect that many of the unrealistic assumptions will cancel each other, that small deviations from realism result in small deviations in the conclusions, and that, in any case, the way in which nature departs from theory will suggest where further complications will be useful. Starting with precision, they hope to increase realism.

3. Sacrifice precision to realism and generality. This approach is favored by MacArthur (1965) and myself. Since we are really concerned in the long run with qualitative rather than quantitative results (which are only important in testing hypotheses) we can resort to very flexible models, often graphical, which generally assume that functions are increasing or decreasing, convex or concave, greater or less than some value, instead of specifying the mathematical form of

of an equation. This means that the predictions we can make are also expressed as inequalities, as between tropical and temperate species, insular *versus* continental faunas, patchy *versus* uniform environments, etc.

However, even the most flexible models have artificial assumptions. There is always room to doubt whether a result depends on the essentials of a model or on the details of the simplifying assumptions. This problem does not arise in the more familiar models, such as the geographic map, where we all know that contiguity on the map implies contiguity in reality, relative distances on the map correspond to relative distances in reality, but color is arbitrary and a microscopic view of the map would only show the fibers of the paper on which it is printed. In the mathematical models of population biology, on the other hand, it is not always obvious when we are using too high a magnification.

Therefore, we attempt to treat the same problem with several alternative models each with different simplifications but with a common biological assumption. Then, if these models, despite their different assumptions, lead to similar results, we have what we can call a robust theorem that is relatively free of the details of the model. Hence, our truth is the intersection of independent lies.

### ROBUST AND NON-ROBUST THEOREMS

As an example of a robust theorem, consider the proposition that in an uncertain environment species will evolve broad niches and tend toward polymorphism. We will use three models, the fitness set of Levins

(1962), a calculus of variation argument, and one which specifies the genetic system (Levins and MacArthur, 1966).

Model I assumes:

1. For each phenotype $i$ there is a best environment $s_i$, and fitness $w$ declines with the deviation of $s_i$ from the actual environment. Although the curves $W(s-s_i)$ in nature may differ in the location of the peak at $s_i$, the height of the peak, and the rate at which fitness declines with the deviation from the optimum, our model treats all the curves as identical except for the location of the peak $s_i$.

2. The environment consists of two (easily extended to $N$) alternative facies or habitats or conditions. Thus on a graph whose axes are $W_1$ and $W_2$, the fitnesses in environments 1 and 2, each phenotype is represented by a point, as in Figure 1. The set of all available phenotypes is designated the fitness set. Since a mixed population of two phenotypes would be represented by a point on the straight line joining their points, the extended fitness set of all possible populations is the smallest convex set enclosing the fitness set. In particular, if the fitness set is convex, then population heterogeneity adds no new fitness points, whereas on a concave fitness set there are polymorphic populations represented by new points. It remains to add that if the two environments are similar compared to the rate at which fitness declines with deviation (that is, similar compared to the tolerance of an individual phenotype), the fitness set will be convex. But as the environments diverge the set becomes concave.

3. In an environment that is uniform in time but shows fine-grained

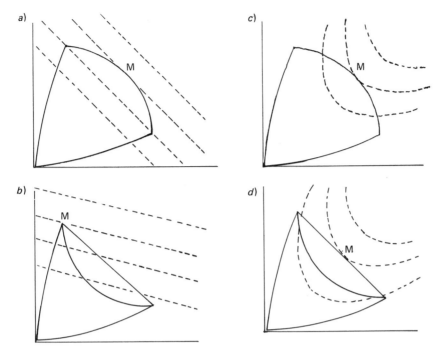

Figure 1. (a) The family of straight lines $pW_1 + (1-p)W_2 = C$ are the fitness measures for a fine-grained stable environment. Optimum fitness occurs for the phenotype which is represented by the point of tangency of these lines with the fitness set. (b) Same for a concave fitness set. (c) In an uncertain environment the optimum population is the one that maximizes $p \log W_1 + (1-p) \log W_2$. On a convex fitness set this is monomorphic unspecialized. (d) Same for a concave fitness set. Here polymorphism creates a broad niche.

heterogeneity in space, each individual is exposed to many units of environment of both kinds in the proportions $p$ to $1-p$ of their occurrence. Thus the rate of increase of the population is $pW_1 + (1-p)W_2$. If the environment is uniform in space but variable in time, the rate of increase is a product of fitnesses in successive generations, $W_1{}^p W_2{}^{1-p}$. For the two situations these alternative functions would be maximized to maximize over-all fitness. The 1962 paper did not distinguish between coarse- and fine-grained environments and therefore gave the linear expression for spatial heterogeneity in general.

The rest of the argument is given in the figure. The result is that if the environment is not very diverse (convex fitness set), the populations will all be monomorphic of a type intermediately well-adapted to both environments. If the environmental diversity exceeds the tolerance of the individual (concave fitness set), then spatial diversity results in specialization to the more common habitat, while temporal diversity results in polymorphism.

Model II does not fix the shape of the curve $W(s-s_i)$. Instead we fix the area under the curve so that $\int W(s)ds = C$. Subject to this restriction, we maximize the rate of increase, which is $\int W(s)P(s)ds$ for a fine-grained spatial heterogeneity and $\int \log W(s)P(s)ds$ for temporal heterogeneity. $P(S)$ is the frequency of environment $S$. In the first case, the optimum population would assign all its fitness to the most abundant environment while in the second case the optimum is $W(S)=CP(S)$. At optimum, the fitness is $\log (C) + \int \log P(s) \cdot P(S)ds$, or $\log (C)$ minus the uncertainty of the environment. Thus the more variable the environment, the flatter and more spread out the $W(S)$ curve and the broader the niche. This analysis does not mention polymorphism directly, since it discusses the assignment of the fitness of the whole population. But if the $P(S)$ curve is broader than the maximum breadth attainable by individual phenotypes, polymorphism will be optimal.

These two models differ in several ways. While the first allows only discretely different environments, the second permits a continuum. While the fitness set specifies how different environments are by showing the relation between fitness in both environments for each phenotype, the second treats each environment as totally different, so that fitness assigned to one contributes nothing to survival in any other. Therefore, that they coincide in their major results adds to the robustness of the theorem. Both models are similar in that they use optimization arguments and ignore the genetic system. We did not assert that evolution will, in fact, establish the optimum population but only the weaker expectation that populations will differ in the direction of their optima. Even this is not obvious. Therefore, in model III we examine a simple genetic model with one locus and two alleles. The graph in Figure 2 has, as before, two axes which represent fitnesses in environments 1 and 2. The points $A_1 A_1$, $A_1 A_2$, $A_2 A_2$ are the fitness points of the three possible genotypes at that locus. The rules of genetic segregation restrict the possible populations to points on the curve joining the two homozygous points and bending halfway toward the heterozygote's point.

We already know from Fisher that for rather general conditions, natural selection will move toward gene frequencies which maximize the log fitness averaged over all individuals. In a fine-grained environment, this means that selection maximizes log $[p W_1 + (1-p)W_2]$, which is the same as maximizing $p W_1 + (1-p)W_2$. But as the environment becomes more coarse-grained, each individual is exposed to fewer units of environment until, in the limit, each one lives either in environment 1 or in environment 2 for the relevant parts of his life. Thus in a fine-grained environment, heterogeneity appears as an average, in a coarse-grained environment as alternatives, and hence uncertainty. Here selection maximizes: $p \log W_1 + (1-p) \log W_2$ or $W_1^p W_2^{1-p}$. We note the following from the figures.

1. In a fine-grained environment average superiority of the heterozygote is necessary for polymorphism. This has nothing to do with the "mixed strategy" polymorphism of previous arguments.

2. If most of the environment is of type 1, there can be no polymorphism. Only when $p$ is closer to 0.5 than the slope of the curve $A_1 A_1$, $A_2 A_2$ will the population become unspecialized.

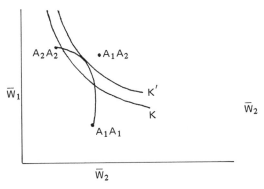

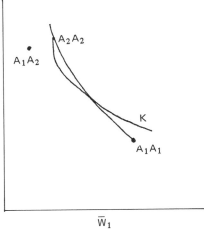

Figure 2. (*a*) Selection in an environment with average heterosis. Possible populations are represented by points on the curve $A_1A_1$, $A_2A_2$. The curves K, K' are samples from an infinite family of curves each connecting points with equal fineness. In a fine-grained, stable environment, the K-curves would be linear and governed by the equation $pW_1 + (1-p)W_2 = K$ for values of $p$ between zero and one. When these K lines are steeper (or flatter,) than the slope of the tangent to the $A_1A_1$, $A_2A_2$ curve at the point $A_1A_1$ (or $A_2A_2$ resp.) then specialization will replace polymorphism.

Figure 2. (*b*) Selection in an uncertain (coarse-grained) environment without average heterosis. The points of tangency of the K curve to the curve $A_1A_1$, $A_2A_2$ are the favored gene frequencies, one specialized and one polymorphic.

3. In a coarse-grained environment the same holds—a sufficient heterogeneity is required to broaden the niche. But now polymorphism is an optimum strategy in the sense that a population of all heterozygotes would not be optimal.

4. As the two environments become more similar, $W_1$ approaches $W_2$ for each genotype. Then the optimum is a single genotype, and polymorphism can only come about as an imposition of the facts of segregation when the heterozygote is superior.

Other work on the joint evolution of habitat selection and niche breadth, on the role of productivity of the environment, and on food-getting procedures all converge to support the theorem that environmental uncertainty leads to increased niche breadth while certain but diverse environments lead to specialization.

As an example of a non-robust theorem, consider the proposition that a high intrinsic rate of increase leads to a smaller average population (productivity as opposed to biomass). This result can be derived from the logistic equation for population growth

$$dx/dt = rx\,(K-x)/K$$

where $x$ is the population size, $r$ is the intrinsic rate of increase, and $K$, the carrying capacity or saturation level, is an environmental variable. Leigh (1965) showed that it also holds for a Volterra system of many prey- and predator-species without a saturation level $K$. However both results have a common explanation in the model in which high $r$ not only increases the rate of approach toward $K$ from below but also speeds the crash toward $K$ from above. This latter property was certainly not in-

tended in the definition of $r$, and
may or may not be true depending on how resources are used. Furthermore, even in the simple logistic,
the result can be reversed if we
add a term $-px$ to the right hand
side to indicate extraneous predation. Thus the theorem, although
interesting, is fragile and cannot
be asserted as a biological fact.
We may be dealing here with a
case of examining a map under
the microscope.

## SUFFICIENT PARAMETERS

The thousand or so variables of our
original equations can be reduced to
manageable proportions by a process
of abstraction whereby many terms
enter into consideration only by way
of a reduced number of higher-level
entities. Thus, all the physiological
interactions of genes in a genotype
enter the models of population genetics only as part of "fitness." The
great diversity in populations appears
mostly as "additive genetic variance"
and "total genetic variance." The
multiplicity of species interactions is
grouped in the vague notions of the
ecological niche, niche overlap, niche
breadth, and competition coefficients.
It is an essential ingredient in the concept of levels of phenomena that there
exists a set of what, by analogy with
the sufficient statistic, we can call
sufficient parameters defined on a
given level (say community) which are
very much fewer than the number of
parameters on the lower level and
which among them contain most of
the important information about
events on that level. This is by no
means equivalent to asserting that
community properties are additive

or that these sufficient parameters are
independent.

Sometimes, the sufficient parameters arise directly from the mathematics and may lack obvious intuitive
meaning. Thus, Kerner discovered a
conservation law for predator-prey
systems. But what is conserved is not
anything obvious like energy or momentum. It is a complicated function
of the species densities which may
acquire meaning for us with further
study. Similarly, working with cellular
metabolism and starting like Kerner
with a physics background, Goodwin
(1963) found an invariant which he
refers to metaphorically as a biological
"temperature."

In other cases, the sufficient parameters are formalizations of previously
held but vague properties such as
niche breadth. We would like some
measure of niche breadth which reflects the spread of a species' fitness
over a range of environments. Thus
the measure should have the following
properties: if a species utilizes $N$ resources equally, it should have niche
breadth of $N$. If it uses two resources
unequally, the niche breadth measure
should lie between 1 and 2. If two
populations which have equal niche
breadths that do not overlap are
merged, their joint niche breadth
should be the sum of their separate
breadths. It may be less if they overlap but never more. Two measures
satisfy these requirements:

$$\log B = - \Sigma p \log p$$

where $p$ is the measure of relative
abundance of the species on a given
resource or in a given habitat, and

$$1/B = \Sigma p^2$$

Neither one is the "true" measure in

TABLE 1

SEASONAL NICHE BREADTH OF SOME PUERTO RICAN DROSOPHILA

| Species | niche breadth during 1962 | |
| | Measure I | Measure II |
| --- | --- | --- |
| D. melanogaster | 14.4 | 10.5 |
| D. latifasciaeformis | 15.5 | 15.7 |
| D. dunni | 11.0 | 7.6 |
| D. tristriata | 6.9 | 5.7 |
| D. ananassae | 11.2 | 8.6 |
| D. repleta | 5.5 | 4.2 |
| D. nebulosa | 6.5 | 6.2 |
| D. paramediostriata | 11.2 | 7.9 |
| X4 (tripunctata group) | 7.2 | 6.0 |
| X6 (tripunctata group) | 13.9 | 12.0 |

The data are based on 21 collections, so that the maximum niche breadth would be 21. Method I is $\log B = -\Sigma \, p \log p$ and method II is $1/B = \Sigma \, p^2$, where $p$ is the proportion of the given species taken in each collection.

the sense that one can decide between proposed alternative structures for the hemoglobin molecule. Both are defined by us to meet heuristic criteria. The final choice of an appropriate measure of niche breadth will depend on convenience, on some new criteria which may arise, and on the extent to which the measures lead to biological predictions based on niche breadth. Meanwhile, we should use both measures in presenting ecological data so that they may be compared and studied together. In Table 1 we show some sample niche breadth measures from our study of Puerto Rican Drosophila populations.

The sufficient parameters may arise from the combination of results of more limited studies. In our robust theorem on niche breadth we found that temporal variation, patchiness of the environment, productivity of the habitat, and mode of hunting could all have similar effects, and they did this by way

of their contribution to the uncertainty of the environment. Thus uncertainty emerges as a sufficient parameter.

The sufficient parameter is a many-to-one transformation of lower-level phenomena. Therein lies its power and utility, but also a new source of imprecision. The many-to-one nature of "uncertainty" prevents us from going backward. If either temporal variation or patchiness or low productivity leads to uncertainty, the consequences of uncertainty alone cannot tell us whether the environment is variable or patchy or unproductive. Therefore we have lost information. It becomes necessary to supplement our theorem with some subordinate models which explain how to go from "uncertainty" to the components of the environment and biology of the species in question. Thus general models have three kinds of imprecision:

(1) they omit factors which have small effects or which have

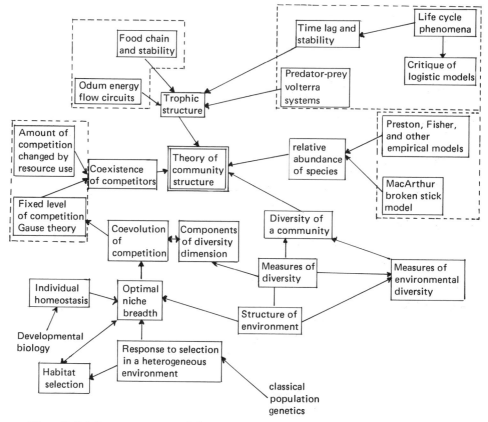

Figure 3. Relations among some of the components in a theory of the structure of an ecological community. Broken lines enclose alternative equivalent models.

large effects but only in rare cases;

(2) they are vague about the exact form of mathematical functions in order to stress qualitative properties;

(3) the many-to-one property of sufficient parameters destroys information about lower level events.

Hence, the general models are necessary but not sufficient for understanding nature. For understanding is not achieved by generality alone, but by a relation between the general and the particular.

## CLUSTERS OF MODELS

A mathematical model is neither an hypothesis nor a theory. Unlike the scientific hypothesis, a model is not verifiable directly by experiment. For all models are both true and false. Almost any plausible proposed relation among aspects of nature is likely to be true in the sense that it occurs (although rarely and slightly). Yet all models leave out a lot and are in that sense inadequate. The validation of a model is not that it is "true" but that it generates good testable hypotheses relevant to important problems. A model may be discarded in favor of a more powerful one, but it usually is simply outgrown when the live issues are no longer those for which it was designed.

Unlike the theory, models are restricted by technical considerations to a few components at a time, even in systems that are complex. Thus a satisfactory theory is usually a cluster of models. These models are related to one another in several ways: as coordinate alternative models for the same set of phenomena, they jointly produce robust theorems; as complementary models they can cope with different aspects of the same problem and give complementary as well as overlapping results; as hierarchically arranged "nested" models, each provides an interpretation of the sufficient parameters of the next higher level where they are taken as given. In Figure 3 we show schematically the relations among some of the models in the theory of community structure.

The multiplicity of models is imposed by the contradictory demands of a complex, heterogeneous nature and a mind that can cope with only a few variables at a time; by the contradictory desiderata of generality, realism, and percision; by the need to understand and also to control; even by the opposing esthetic standards which emphasize the stark simplicity and power of a general theorem as against the richness and the diversity of living nature. These conflicts are irreconcilable. Therefore, the alternative approaches even of contending schools are part of a larger mixed strategy. But the conflict is about method, not nature, for the individual models, while they are essential for understanding reality, should not be confused with that reality itself.

## REFERENCES

Goodwin, B. C., 1963, *Temporal Organization in Cells,* Academic Press.

Holling, C. S., 1959, The components of predation as revealed by a study of small-mammal predation of the European pine sawfly. *Canadian Entomologist,* 91 (5): 293-320.

Kerner, E. H., 1957, A statistical mechanics of interacting biological species. *Bull. Mat. Biophys.,* 19: 121-146.

Leigh, Egbert, 1965, On the relation between productivity, biomass, diversity, and stability of a community. *PNAS,* 53 (4): 777-783.

Levins, R., 1962, Theory of fitness in a heterogeneous environment, I. The fitness set and adaptive function. *Am. Nat.,* 96 (891): 361-373.

Levins, R., and R. H. MacArthur, 1966, The maintenance of genetic polymorphism in a spatially heterogeneous environment: variations on a theme by Howard Levene. *Am. Nat.,* 100: 585-589.

MacArthur, R. H., and R. Levins, 1965, Competition, habitat selection, and character displacement in a patchy environment. *PNAS,* 51 (3): 1207-1210.

Watt, Kenneth E. F., 1956, The choice and solution of mathematical models for predicting and maximizing the yield of a fishery. *J. Fisheries Res. Bd. of Canada,* 13, 613-645.

# II
# FITNESS

# 4

# Darwin's Untimely Burial

STEPHEN JAY GOULD

IN ONE OF THE NUMEROUS MOVIE versions of *A Christmas Carol,* Ebenezer Scrooge encounters a dignified gentleman sitting on a landing, as he mounts the steps to visit his dying partner, Jacob Marley, "Are you the doctor?" Scrooge inquires. "No," replies the man, "I'm the undertaker; our is a very competitive business." The cutthroat world of intellectuals must rank a close second, and few events attract more notice than a proclamation that popular ideas have died. Darwin's theory of natural selection has been a perennial candidate for burial. Tom Bethell held the most recent wake in a piece called "Darwin's Mistake" (*Harper's,* February 1976): "Darwin's theory, I believe, is on the verge of collapse. . . . Natural selection was quietly abandoned, even by his most ardent supporters, some years ago." News to me, and I, although I wear the Darwinian label with some pride, am not among the most ardent defenders of natural selection. I recall Mark Twain's famous response to a premature obitu-

ary: "The reports of my death are greatly exaggerated."

Bethell's argument has a curious ring for most practicing scientists. We are always ready to watch a theory fall under the impact of new data, but we do not expect a great and influential theory to collapse from a logical error in its formulation. Virtually every empirical scientist has a touch of the Philistine. Scientists tend to ignore academic philosophy as an empty pursuit. Surely, any intelligent person can think straight by intuition. Yet Bethell cites no data at all in sealing the coffin of natural selection, only an error in Darwin's reasoning: "Darwin made a mistake sufficiently serous to undermine his theory. And that mistake has only recently been recognized as such. . . . At one point in his argument, Darwin was misled."

Although I will try to refute Bethell, I also deplore the unwillingness of scientists to explore seriously the logical structure of arguments. Much of what passes for evolutionary theory is as vacuous as Bethell claims. Many great theories are held together by chains of dubious metaphor and analogy. Bethell has correctly identified the hogwash surrounding evolutionary theory. But we differ in one fundamental way: for Bethell, Darwinian theory is rotten to the core; I find a pearl of great price at the center.

Natural selection is the central concept of Darwinian theory—the fittest survive and spread their favored traits through populations. Natural selection is defined by Spencer's phrase "survival of the fittest," but what does this famous bit of jargon really mean? Who are the fittest? and how is "fitness" defined? We often read that fitness involves no more than "differential reproductive success"—the production of more surviving offspring than other competing members of the population. Whoa! cries Bethell, as many others have before him. This formulation defines fitness in terms of survival only. The crucial phrase of natural selection means no more than "the survival of those who survive"—a vacuous tautology. (A tautology is a phrase—like "my father is a man"—containing no information in the predicate ("a man") not inherent in the subject ("my father"). Tautologies are fine as definitions, but not as testable scientific statements—there can be nothing to test in a statement true by definition.)

But how could Darwin have made such a monumental, two-bit mistake? Even his severest critics have never accused him of crass stupidity. Obviously, Darwin must have tried to define fitness differently—to find a criterion for fitness independent of mere survival. Darwin did propose an independent criterion, but Bethell argues quite correctly that he relied

upon analogy to establish it, a dangerous and slippery strategy. One might think that the first chapter of such a revolutionary book as *Origin of Species* would deal with cosmic questions and general concerns. It doesn't. It's about pigeons. Darwin devotes most of his first forty pages to "artificial selection" of favored traits by animal breeders. For here an independent criterion surely operates. The pigeon fancier knows what he wants. The fittest are not defined by their survival. They are, rather, allowed to survive because they possess desired traits.

The principle of natural selection depends upon the validity of analogy with artificial selection. We must be able, like the pigeon fancier, to identify the fittest beforehand, not only by their subsequent survival. But nature is not an animal breeder; no preordained purpose regulates the history of life. In nature, any traits possessed by survivors must be counted as "more evolved"; in artificial selection, "superior" traits are defined before breeding even begins. Later evolutionists, Bethell argues, recognized the failure of Darwin's analogy and redefined "fitness" as mere survival. But they did not realize that they had undermined the logical structure of Darwin's central postulate. Nature provides no independent criterion of fitness: thus, natural selection is tautological.

Bethell then moves to two important corollaries of his major argument. First, if fitness only means survival, then how can natural selection be a "creative" force, as Darwinians insist. Natural selection can only tell us how "a given type of animal became more numerous"; it cannot explain "how one type of animal gradually changed into another." Secondly, why were Darwin and other eminent Victorians so sure that mindless nature could be compared with conscious selection by breeders? Bethell argues that the cultural climate of triumphant industrial capitalism had defined any change as inherently progressive. Mere survival in nature could only be for the good: "It is beginning to look as though what Darwin really discovered was nothing more than the Victorian propensity to believe in progress."

I believe that Darwin was right and that Bethell and his colleagues are mistaken: criteria of fitness independent of survival can be applied to nature and have been used consistently by evolutionists. But let me first admit that Bethell's criticism applies to much of the technical literature in evolutionary theory, especially to the abstract mathematical treatments that consider evolution only as an alteration in numbers, not as a change in quality. These studies do assess fitness only in terms of differential survival. What else can be done with abstract models that trace the relative successes of hypothetical genes A and B in populations that exist only on computer tape? Nature, however, is not limited by the calculations of theoretical geneticists. In nature, A's "superiority" over B will be *expressed* as differential survival, but it is not *defined* by it—or, at least, it better not be so defined, least Bethell et al. triumph and Darwin surrender.

My defense of Darwin is not startling, novel or profound. I merely assert that Darwin was justified in analogizing natural selection with animal breeding. In artificial selection, a breeder's desire represents a "change of environment" for a population. In this new environment, certain traits are superior a priori (they survive and spread by our breeder's choice, but this is a *result* of their fitness, not a definition of it). In nature, Darwinian evolution is also a response to changing environments. Now, the key point: certain morphological, physiological, and behavioral traits should be superior a priori as designs for living in new environments. These traits confer fitness by an engineer's criterion of good design, not by the empirical fact of their survival and spread. It got colder before the woolly mammoth evolved its shaggy coat.

Why does this issue agitate evolutionists so much? OK, Darwin was right: superior design in changed environments is an independent criterion of fitness. So what. Did anyone ever seriously propose that the poorly designed shall triumph? Yes, in fact, many did. In Darwin's day, many rival evolutionary theories asserted that the fittest (best designed) must perish. One popular notion—the theory of racial life cycles—was championed by a former inhabitant of the office I now occupy, the great American paleontologist Alpheus Hyatt. Hyatt claimed that evolutionary lineages, like individuals, had cycles of youth, maturity, old age, and death (extinction). Decline and extinction are programmed into the history of species. As maturity yields to old age, the best-designed individuals die and the hobbled, deformed creatures of phyletic senility take over. Another anti-Darwinian notion, the theory of orthogenesis, held that certain trends, once initiated, could not be halted, even though they must lead to

*[Handwritten marginal notes: "Gould ↓" beside left column; "But by 'good' do you mean 'promotes survival'?" beside right column]*

extinction caused by increasingly inferior design. Many nineteenth-century evolutionists (perhaps a majority) held that Irish elks became extinct because they could not halt their evolutionary increase in antler size; thus, they died—caught in trees or bowed (literally) in the mire. Likewise, the demise of saber-toothed "tigers" was often attributed to canine teeth grown so long that the poor cats couldn't open their jaws wide enough to use them.

Thus, it is not true, as Bethell claims, that any traits possessed by survivors must be designated as fitter. "Survival of the fittest" is not a tautology. It is also not the only imaginable or reasonable reading of the evolutionary record. It is testable. It had rivals that failed under the weight of contrary evidence and changing attitudes about the nature of life. It has rivals that may succeed, at least in limiting its scope.

If I am right, how can Bethell claim, "Darwin, I suggest, is in the process of being discarded, but perhaps in deference to the venerable old gentleman, resting comfortably in Westminster Abbey next to Sir Isaac Newton, it is being done as discreetly and gently as possible with a minimum of publicity." I'm afraid I must say that Bethell has not been quite fair in his report of prevailing opinion. He cites the gadflies C. H. Waddington an H. J. Muller as though they epitomized a consensus. He never mentions the leading selectionists of our present generation—E. O. Wilson or D. Janzen, for example. And he quotes the architects of neo-Darwinism—Dobzhansky, Simpson, Mayr, and J. Huxley—only to ridicule their metaphors on the "creativity" of natural selec-

tion. (I am not claiming that Darwinism should be cherished because it is still popular; I am enough of a gadfly to believe that uncriticized consensus is a sure sign of impending trouble. I merely report that, for better or for worse, Darwinism is alive and thriving, despite Bethell's obituary.)

But why was natural selection compared to a composer by Dobzhansky; to a poet by Simpson; to a sculptor by Mayr; and to, of all people, Mr. Shakespeare by Julian Huxley? I won't defend the choice of metaphors, but I will uphold the intent, namely, to illustrate the essence of Darwinism—the creativity of natural selection. Natural selection has a place in all anti-Darwinian theories that I know. It is cast in a negative role as an executioner, a headsman for the unfit (while the fit arise by such non-Darwinian mechanisms as the inheritance of acquired characters or direct induction of favorable variation by the environment). The essence of Darwinism lies in its claim that natural selection creates the fit. Variation is ubiquitous and random in direction. It supplies the raw material only. Natural selection directs the course of evolutionary change. It preserves favorable variants and builds fitness gradually. In fact, since artists fashion their creations from the raw material of notes, words, and stone, the metaphors do not strike me as inappropriate. Since Bethell does not accept a criterion of fitness independent of mere survival, he can hardly grant a creative role to natural selection.

According to Bethell, Darwin's concept of natural selection as a creative force can be no more than an illusion encouraged by the social and political climate of his times. In the throes of Victorian optimism in imperial Britain, change seemed to be inherently progressive; why not equate survival in nature with increasing fitness in the nontautological sense of improved design.

I am a strong advocate of the general argument that "truth" as preached by scientists often turns out to be no more than prejudice inspired by prevailing social and political beliefs. I have devoted several essays to this theme because I believe that it helps to demystify the practice of science by showing its similarity to all creative human activity. But the truth of a general argument does not validate any specific application, and I maintain that Bethell's application is badly misinformed.

Darwin did two very separate things: he convinced the scientific world that evolution had occurred and he proposed the theory of natural selection as its mechanism. I am quite willing to admit that the common equation of evolution with progress made Darwin's first claim more palatable to his contemporaries. But Darwin failed in his second quest during his own lifetime. The theory of natural selection did not triumph until the 1940s. Its Victorian unpopularity, in my view, lay primarily in its denial of general progress as inherent in the workings of evolution. Natural selection is a theory of *local* adaptation to changing environments. It proposes no perfecting principles, no guarantee of general improvement; in short, no reason for general approbation in a political climate favoring innate progress in nature.

Darwin's independent criterion of fitness is, indeed, "improved design," but not "improved" in the cosmic sense that contemporary Britain favored. To Darwin, improved meant only "better designed for an immediate, local environment." Local environments change constantly: they get colder or hotter, wetter or drier, more grassy or more forested. Evolution by natural selection is no more than a tracking of these changing environments by differential preservation of organisms better designed to live in them: hair on a mammoth is not progressive in any cosmic sense. Natural selection can produce a trend that tempts us to think of more general progress—increase in brain size does characterize the evolution of group after group of mammals. But big brains have their uses in local environments; they do not mark intrinsic trends to higher states. And Darwin delighted in showing that local adaptation often produced "degeneration" in design—anatomical simplification in parasites, for example.

If natural selection is not a doctrine of progress, then its popularity cannot reflect the politics that Bethell invokes. If the theory of natural selection contains an independent criterion of fitness, then it is not tautological. I maintain, perhaps naïvely, that its current, unabated popularity must have something to do with its success in explaining the admittedly imperfect information we now possess about evolution. I rather suspect that we'll have Charles Darwin to kick around for some time.

# 5

# The Propensity
# Interpretation of Fitness

## SUSAN MILLS AND JOHN BEATTY

*The concept of "fitness" is a notion of central importance to evolutionary theory. Yet the interpretation of this concept and its role in explanations of evolutionary phenomena have remained obscure. We provide a propensity interpretation of fitness, which we argue captures the intended reference of this term as it is used by evolutionary theorists. Using the propensity interpretation of fitness, we provide a Hempelian reconstruction of explanations of evolutionary phenomena, and we show why charges of circularity which have been leveled against explanations in evolutionary theory are mistaken. Finally, we provide a definition of natural selection which follows from the propensity interpretation of fitness, and which handles all the types of selection discussed by biologists, thus improving on extant definitions.*

## 1. INTRODUCTION

The testability and logical status of evolutionary theory have been brought into question by numerous authors in recent years (e.g., Manser 1965, Smart 1963, Popper 1974). Many of the claims that evolutionary theory is not testable, that it parades tautologies in the guise of empirical claims, and that its explanations are circular, resulted from misunderstandings which have since been rebutted (e.g., by Ruse 1969, 1973 and Williams 1970, 1973a, 1973b). Yet despite the skilled rejoinders which have been given to most of these charges, the controversy continues to flourish, and has even

We wish to thank Professor Michael Ruse, for initially drawing our attention to the problems of the logical status of evolutionary theory, and for insightful criticisms of an early draft of this paper. We are heavily indebted to Alberto Coffa for providing us with explications both of propensities and of the nature of explanation, and for innumerable criticisms and suggestions. Ron Giere also suggested that the propensity interpretation was a little more complex than we originally suspected. However, we claim complete originality for our mistakes.

found its way beyond philosophical and biological circles and into the pages of *Harper's Magazine*. In the spring of 1976, journalist Tom Bethell reported to the unsuspecting public that:

> Darwin's theory . . . is on the verge of collapse. In his famous book, *On the Origin of Species* . . . Darwin made a mistake sufficiently serious to undermine his theory. The machinery of evolution that he supposedly discovered has been challenged, and it is beginning to look as though what he really discovered was nothing more than the Victorian propensity to believe in progress. [1976, p. 72].

Those familiar with the details of evolutionary theory, and with the history of this controversy, will rightfully feel no sympathy with such challenges, and may wonder whether it is worth bothering with them. But the fact is that there is a major problem in the foundations of evolutionary theory which remains unsolved, and which continues to give life to the debate. The definition of fitness remains in dispute, and the role of appeals to fitness in biologists' explanations is a mystery. This is a problem which ought to concern biologists and philosophers of science, quite independent of the vicissitudes of the controversy which it perpetuates.

Biologists agree on how to *measure* fitness, and they routinely appeal to fitness in their explanations, attributing the relative predominance of certain traits to the relative fitness of those traits. However, these explanations can and have been criticized on the grounds that, given the definitions of fitness offered by most biologists, these explanations are no more than re-descriptions of the phenomena to be explained (e.g., Popper, 1974, Manser 1965, Smart 1963). Philosophers have proposed new treatments of fitness designed to avoid these charges of explanatory circularity (e.g. Hull 1974 and Williams 1973). Unfortunately, none of these interpretations succeeds in avoiding the charges, while providing a definition *useful* to evolutionary theory.

Thus it is high time that an analysis of fitness is provided which reveals the empirical content implicit in evolutionary biologist's explanations. To this end, we propose and defend the *propensity interpretation* of fitness. We argue that the propensity interpretation captures the intended reference of "fitness" as biologists use the term. Furthermore, using this interpretation, we show how references to fitness play a crucial role in explanations in evolutionary theory, and we provide a Hempelian reconstruction of such explanations which reveals the precise nature of this role. We answer the charges of explanatory circularity leveled against evolutionary theory by showing how these charges arise from mistaken interpretations of fitness.

The concepts of fitness and natural selection are closely linked, since it is through the process of natural selection that the fittest gain predominance, according to the theory of evolution. Thus it is not surprising to find misinterpretations of fitness paralleled by misunderstandings of natural selection. The propensity analysis suggests a definition of "selection" which (unlike previously proposed definitions) accords with all the diverse types of selection dealt with by biologists.

But before proceeding with the positive analyses just promised, we consider the charge of explanatory circularity which arises from the lack of a satisfactory interpretation of fitness, and the reasons for the inadequacy of the replies so far offered in answer to the charge.

## 2. THE CHARGE OF CIRCULARITY

According to the most frequently cited definitions of "fitness," that term refers to the *actual* number of offspring left by an individual or type relative to the actual contribution of some reference individual or type. For instance, Waddington (1968, p. 19) suggests that the fittest individuals are those which are "most effective in leaving gametes to the next generation." According to Lerner (1958), "the individuals who have more offspring are fitter in the Darwinian sense." Grant (1977, p. 66) construes fitness as "a measure of reproductive success." And Crow and Kimura (1970, p. 5) regard fitness "as a measure of both survival and reproduction ..." (see also Dobzhansky 1970, p. 101–102; Wilson 1975, p. 585; Mettler and Gregg 1969, p. 93).

These definitions of "fitness" in terms of actual survival and reproductive success are straightforward and initially intuitively satisfying. However, such definitions lead to justifiable charges that certain explanations invoking fitness differences are circular. The explanations in question are those which point to fitness differences between alternate types in a population in order to account for (1) differences

in the average offspring contributions of those phenotypes, and (2) changes in the proportions of the types over time (i.e., evolutionary changes). Where fitness is defined in terms of survival and reproductive success, to say that type *A* is fitter than type *B* is just to say that type *A* is leaving a higher average number of offspring than type *B*. Clearly, we cannot say that the difference in fitness of *A* and *B explains* the difference in actual offspring contribution of *A* and *B*, when fitness is defined in terms of actual reproductive success. Yet evolutionary biologists seem to think that type frequency changes (i.e., evolutionary changes) can be *explained* by invoking the relative fitnesses of the type concerned. For instance, Kettlewell (1955 and 1956) hypothesized that fitness differences were the cause of frequency changes of dark- and light-colored pepper moths in industrial areas of England. And he devised experiments to determine whether the frequency changes were correlated with fitness differences. Several philosophers have pointed to the apparent circularity involved in these explanations. Manser (1965) describes Kettlewell's account of the frequency differences in terms of fitness differences as "... only a description in slightly theory-laden terms which gives the illusion of an explanation in the full scientific sense" (1965, p. 27).

The whole idea of setting up empirical investigations to determine whether fitness differences are correlated with actual descendant contribution differences seems absurd, given the above definitions of "fitness." If this type of charge is coupled with the assumption that the only testable claims of evolutionary theory are of this variety, (i.e., tests of whether individuals identified as "the fittest" are most reproductively successful), it appears that evolutionary theory is not testable. As Bethell puts it, "If only there were some way of identifying the fittest beforehand, without always having to wait and see which ones survive, Darwin's theory would be testable rather than tautological" (1976, p. 75).

However, as Ruse (1969) and Williams (1973a and b) have made clear, this latter charge is mistaken. Evolutionary theory embodies many testable claims. To take but one of many examples cited by Williams, Darwinian evolutionary theory predicted the existence of *transitional forms* intermediate between ancestral and descendant species. The saltationist (creationist) view of the origin of species which was accepted at the time when Darwin wrote on *The Origin of Species* predicted no such plethora of intermediate forms. Ruse has called attention to the predictions concerning distributions of types in populations which can be made on the basis of the Hardy-Weinberg law (1973, p. 36).

While these replies are well taken, they fail to clarify the role of fitness ascriptions in evolutionary theory. We agree with Williams and Ruse that evolutionary theory does make testable claims, and that many of these claims can be seen to be testable without providing an analysis of the role of fitness ascriptions. Nevertheless, some claims of evolutionary theory cannot be shown to be empirical without clarifying the role of "fitness." Moreover, our understanding of other straightforwardly empirical claims of evolutionary theory will be enhanced by an explication of "fitness" in these claims.

## 3. WHAT FITNESS IS NOT

There are two questions to be clarified in defining fitness: What sorts of entities does this predicate apply to, and what does it predicate of these entities? Both these questions have received disparate answers from various biologists and philosophers. Fitness has been claimed to apply to types (e.g., Dobzhansky 1970, pp. 101–102; Crow and Kimura 1970) as well as individuals (Lerner 1958, Waddington 1968, p. 19). As will become apparent in the course of the positive analysis, the question what sorts of entities "fitness" applies to should not be given a univocal answer. Fitness may be predicated of individual organisms, and (in a somewhat different sense) of phenotypes and genotypes. In this section we will only consider the question which one is predicating of individuals and types in ascribing to them a fitness value, according to the various proposals under scrutiny.

Before moving on to alternatives to the definition of "fitness" in terms of actual survival and reproductive success, we need to consider the acceptability of this definition, independent of the criticism that it leads to explanatory circularity. This criticism alone is obviously not sufficient to show that the interpretation is incorrect. For proponents of this definition can reply that fitness is actual reproductive success, since that is the way biologists use the term, and there is no other feasible definition. That references to fitness lead to explanatory circularity just shows that fitness has no explanatory role to play in evolutionary theory. In fact, Bethell (1976, p. 75) makes this claim, and even maintains that biologists have abandoned references to fitness in their accounts of evolutionary phenomena. This is a scandalous claim.[1] A survey of evolutionary journals like *American Naturalist* and *Evolution* reveals that fitness ascriptions still play a major role in explanations of evolutionary phenomena. Indeed, the current literature on evolutionary theory reveals that the notion of fitness is of tremendous concern. Rather than abandoning the notion, modern evolutionary biologists have chosen to refine and extend it. Levins (1968) has raised the problem of fitness in changing environments. Thoday (1953) has pointed to the distinction between short-term and long-term fitness. An analysis, and evidence of, "variable fitness" or "frequency dependent fitness" was given by Kojima (1971). The effects of "overdominance with regard to fitness" on the maintenance of polymorphisms continues to be studied. And one very promising model of sociobiological evolution has been developed via an extension of traditional notions of fitness (the new notion is one of "inclusive fitness" (Cf. Hamilton 1964). As we will argue below, biologists are well advised *not* to abandon references to fitness, for such references play a crucial role in explanations of evolutionary phenomena.

Fortunately, we do have grounds quite independent of the issue of explanatory circularity for deeming inadequate definitions of "fitness" in terms of actual survival and reproductive success. For, such definitions conflict with biologists' usage of the term, as is demonstrated by the following considerations. Surely two organisms which are genetically and phenotypically identical, and which inhabit the same environment, should be given the same fitness value. Yet where fitness is defined in terms of actual number of offspring left, two such organisms may receive radically different fitness values, if it happens that one of them succeeds in reproducing while the other does not. Scriven (1959) invites us to imagine a case in which two identical twins are standing together in the forest. As it happens, one of them is struck by lightning, and the other is spared. The latter goes on to reproduce while the former leaves no offspring. Surely in this case there is no difference between the two organisms which accounts for their difference in reproductive success. Yet on the traditional definition of "fitness," the lucky twin is *far* fitter. Most undesirably, such a definition commits us to calling the intuitively less fit of two organisms the fitter if it happens that this organism leaves the greater number of offspring of the two.[2]

Nor can these counterintuitive results be avoided by shifting the reference of fitness from individual organisms to groups. For, precisely as was the case with individuals, the intuitive less fit subgroup of a population may by chance come to predominate.

---

1. Bethell may have been misled by the fact that evolutionary biologists recognize mechanisms of evolutionary change other than fitness differences (e.g., drift). Nevertheless, there is no question that fitness differences have been and still are considered effective in producing evolutionary changes.

2. The counter-intuitiveness of the traditional definition is also suggested by the following

For example, an earthquake or forest fire may destroy individuals irrespective of any traits they possess. In such a case we do not wish to be committed to attributing the highest fitness values to whichever subgroup is left.

Since an organism's traits are obviously important in determining its fitness, it is tempting to suggest that fitness be defined independently of survival and reproduction, as some function of traits. Hull (1974) hints at the desirability of such a definition. This suggestion derives *prima facie* support from the fact that given such a definition, explanations of differential offspring contribution which appeal to differences in fitness are noncircular. However, no one has seriously proposed such a definition, and it is easy to see why. The features of organisms which contribute to their survival and reproductive success are endlessly varied and context dependent. What do the fittest germ, the fittest geranium, and the fittest chimpanzee have in common? It cannot be any concretely characterized physical property, given that one and the same physical trait can be helpful in one environment and harmful in another. This is not to say that it is impossible that some as yet unsuspected (no doubt abstractly characterized) feature of organisms may be found which correlates with reproductive success. Rather, it is to say that we

need not and should not, wait for the discovery of such a feature in order to define "fitness."

So far, we have seen that we cannot define fitness simply in terms of survival and reproductive success. But neither can we define fitness entirely independently of any reference to survival and reproduction. An ingenious alternative to either of these approaches has been offered by Williams (1970, 1973a and b). She suggests that we regard "fitness" as a primitive term of evolutionary theory, and that we therefore refuse to define it. As she points out, in the formal axiomatization of a theory, it is not possible that all terms be explicitly defined, on pain of circularity. However, the fact that we cannot formally define all the terms of a theory *within* the framework of the theory does not prevent us from stepping outside the theory and explaining the meaning of the term in a broader linguistic framework.[3] Such an explication need not amount to anything as restrictive as an operational definition or an explicit definition making the term eliminable without loss from the theory. Rather, such an explication should allow us to understand what sort of property fitness is, its relation to natural selection, and the role of references to fitness in evolutionary theorists' explanations. Thus our criticism of Williams is not that she is wrong about fitness but that she does not go

hypothetical case. Imagine two butterflies of the same species, which are phenotypically identical except that one (C) has color markings which camouflage it from its species' chief predator, while the second (N) does not have such markings and is hence more conspicuous. If N nevertheless happens to leave more offspring than C, we are committed on the definition of fitness under consideration to conclude that (1) both butterflies had the same degree of fitness before reaching maturity (i.e., zero fitness) and (2) in the end, N is fitter, since it left more offspring than C.

3. Gary Hardegree suggested this to us in a conversation.

far enough. We believe that a more thorough explication is possible through the *propensity* interpretation of fitness.[4]

## 4. PROPENSITY ANALYSIS OF FITNESS

Levins (1968) has remarked that "Fitness enters population biology as a vague heuristic notion, rich in metaphor but poor in precision." No doubt this is accurate as a characterization of the unclarity surrounding the role of fitness in evolutionary theory, even among biologists who use the term. But such unclarity is compatible with the fact that fitness plays an essential explanatory role in evolutionary theory. It is to the task of increasing the precision of the concept of fitness as well as making explicit its explanatory role that we now turn.

We have already seen that fitness is somehow connected with success at survival and reproduction, although it cannot be defined in terms of actual survival and reproductive success. Why have evolutionary biologists continued to confuse fitness with actual descendant contribution? We believe that the confusion involves a misidentification of the *post facto* survival and reproductive success of an organism with the *ability* of an organism to survive and reproduce. We believe that "fitness" refers to the ability. Actual offspring contribution, on the other hand, is a sometimes reliable—sometimes unreliable—indicator of that ability. In the hypothetical cases above, actual descendant contribution is clearly an unreliable indicator of descendant contribution capability. The identical twins are equally *capable* of leaving offspring. And the camouflaged butterfly is more *capable* of leaving offspring than is the non-camouflaged butterfly.

Thus we suggest that fitness be regarded as a complex *dispositional* property of organisms. Roughly speaking, the fitness of an organism is its *propensity* to survive and reproduce in a particularly specified environment and population. A great deal more will have to be added before the substance of this interpretation becomes clear. But before launching into details, let us note a few general features of this proposal.

First, if we take fitness to be a dispositional property of organisms, we can immediately see how references to fitness can be explanatory.[5] The fitness of an organism explains its success at survival and reproduction in a particular environment in the same way that the solubility of

4. As we recently learned, Mary Williams supports the propensity interpretation and has independently worked toward an application of this interpretation.

5. Where fitness is defined as a propensity we can also squeeze the empirical content out of the phrase "survival of the fittest" (i.e., the claim that the fittest survive), which has frequently been claimed to be tautological (e.g., by Bethell 1976. Popper 1974, and Smart 1963). Just as the claim that "the soluble (substance) is dissolving" is an empirical claim, so the claim that "those which could gain predominance in a particular environment are in fact gaining predominance" is an empirical claim. In short, to claim that a dispositional property is manifesting itself is to make an empirical claim. Such a claim suggests that the conditions usually known to trigger the manifestation are present, and no factors are present to override this manifestation. It seems plausible to interpret "the survival of the fittest" as a loose way of claiming that the organisms which are leaving most offspring are also the most fit. That this is a plausible interpretation of Darwin's use of the phrase is also suggested by

a substance explains the fact that it has dissolved in a particular liquid. When we say that an entity has a propensity (disposition, tendency, capability) to behave in a particular way, we mean that certain physical properties of the entity determine, or are causally relevant to, the particular behavior whenever the entity is subjected to appropriate "triggering conditions." For instance, the propensity of salt to dissolve in water (the "water-solubility" of salt) consists in (i.e., "water solubility" *refers to*) its ionic crystalline character, which causes salt to dissolve whenever the appropriate triggering condition—immersion in water—is met. Likewise, the fitness of an organism consists in its having traits which condition its production of offspring in a given environment. For instance, the dark coloration of pepper moths in sooted, industrial areas of England effectively camouflages the moths from predators, enabling them to survive longer and leave more offspring. Thus, melanism is one of many physical properties which constitute the fitness, or reproductive propensity, of pepper moths in polluted areas (in

the same sense that the ionic crystalline character of salt constitutes its propensity to dissolve in water).

The appropriate triggering conditions for the realization of offspring contribution dispositions include particular environmental conditions. We do not say that melanic moths are equally fit in polluted and unpolluted environments, any more than we claim that salt is as soluble in water as it is in mercury or swiss cheese.[6]

In addition to the triggering conditions which cause a disposition to be manifested, we must, in explaining or predicting the manifestation of a disposition, consider whether any factors other than the relevant triggering conditions were present to interfere with the manifestation. When we say that salt has dissolved in water because it is soluble in water, we assume the absence of disturbing factors, such as the salt's having been coated in plastic before immersion. Likewise, when we explain an organism's (or type's) offspring contribution by referring to its degree of fitness, we assume, for instance, that environmental catastrophes (e.g., atomic holocausts, forest fires, etc.) and

---

Darwin's concern (in *The Origin of Species*) to demonstrate that conditions favoring natural selection are widely in effect. But it should be emphasized that nothing hinges on providing such an interpetation for "the survival of the fittest." This catch-phrase is not an important feature of evolutionary theory, in spite of the controversy its alleged tautological status has generated.

6. As this discussion suggests, an organism's fitness is not only a function of the organism's traits, but also of characteristics of the organism's environment. Actually, this function may be even more complicated. For evolutionary biologists have also noted that the fitness of an individual may depend upon the characteristics of the population to which it belongs. For instance, there is evidence of "frequency dependent selection" in several species of Drosophila (Kojima 1971). This kind of selection is said to occur whenever the fitness of a type depends upon the frequency of the type. Some types appear to be fitter, and are selected for, when they are rare. Thus, fitness is relative to environmental and population characteristics. And consequently, the appropriate triggering conditions for the realization of descendant contribution dispositions include envrionmental and population structure conditions.

human intervention have not inter-
fered with the manifestation of offspring contribution dispositions. In
general, we want to rule out the
occurrence of any environmental conditions which separate successful from
unsuccessful reproducers without regard to physical differences between
them.

Now let us fill in some of the details of this proposal. First, we must
clarify the view of propensities we
are presupposing. In our view, propensities are dispositions of *individual
objects*.[7] It is each hungry rat which
has a tendency or propensity to
move in the maze in a certain way;
not the class of hungry rats. Classes
—abstract objects, in general—do not
have dispositions, tendencies, or propensities in any orthodox sense of the
term. This aspect of propensities in
general is also a feature of the (unexplicated) notion of fitness employed by biologists. Evolutionary
biologists often speak of fitness as if
it were a phenotypic trait—i.e., a
property of individuals. For instance,
Wallace (1963, p. 633) remarks,
"That instances of overdominance
exist, especially in relation to a trait
as complex as fitness, is generally
conceded."

However, evolutionary biologists
also employ a notion of fitness which
refers to *types* (e.g., Dobzhansky
1970, pp. 101–102). Fitness cannot
be a propensity in this case, although,
as we will argue, it is a derivative of
individual fitness propensities. We will
therefore introduce two definitions of
"fitness": Fitness$_1$ of individual organisms and fitness$_2$ of types.

## 4.1. FITNESS$_1$: FITNESS OF INDIVIDUAL ORGANISMS

A paradigm case of a propensity is
a subatomic particle's propensity to
decay in a certain period of time.
Whether a particle decays during
some time interval is a qualitative,
nonrepeatable property of that particle's event history. It might initially
be thought that "propensity to reproduce" is also a qualitative nonrepeatable property of an organism:
either it reproduces during its lifetime
or it does not. However, the property
of organisms which is of interest to
the evolutionary biologist is not the
organism's propensity to reproduce or
not to reproduce, but rather the
*quantity* of offspring which the organism has the propensity to contribute. For the evolutionary biologist
is interested in explaining proportions
of types in populations, and from this
point of view an organism which
leaves one offspring is much more
similar to an organism which leaves
no offspring than it is to an organism
which leaves 100 offspring. Thus
when we speak of "reproductive propensity," this should be understood as
a quantitative propensity like that of
a lump of radioactive material (considered as an individual) to emit particles over time, rather than as a
"yes-no" propensity, like that of an

7. Given that propensities apply to individual objects (rather than chance set-ups or
sequencies of trials), we also take them to be ontologically real—not merely epistemic
properties. Our view is similar to Mellor's (for a good review of the views on propensities,
cf. Kyburg, 1974), but it most closely follows Coffa's analysis (1977, and his unpublished
dissertation, "Foundations of Inductive Explanation").

individual particle to decay or not decay during some time interval.

It may have struck the reader that given this quantitative understanding of "propensity to reproduce," there are many such propensities. There is an organism's propensity to leave zero offspring, its propensity to leave 1 offspring, 2 offspring, ... $n$ offspring (during its lifetime). Determinists might claim that there is a unique number of offspring which an organism is determined to leave (i.e., with propensity 1) in a given environment. For nondeterminists, however, things are more complicated. Organisms may have propensities of different strengths to leave various numbers of offspring. The standard dispositions philosophers talk about are tendencies of objects to instantiate certain properties invariably under appropriate circumstances. But besides such "deterministic" dispositions, there are the tendencies of objects to produce one or another of a distribution of outcomes with predetermined frequency. As Coffa (1977) argues, it seems just as legitimate to suppose there are such nondeterministic, "probabilistic" causes as to posit deterministic dispositions.[8]

If we could assume that there were a unique number of offspring which any organism is determined to produce (i.e., which the organism has propensity 1 to produce), then the fitness$_1$ of an organism could be valued simply as the number of offspring which that organism is disposed to produce. But since it is quite possible that organisms may have a range or distribution of reproductive propensities, as was suggested above, we derive fitness$_1$ values taking these various propensities into consideration.

Unfortunately, we also cannot simply choose the number of offspring which an organism has the *highest* propensity to leave—i.e., the mode of the distribution. For in the first place, an organism may not have a *high* propensity to leave any particular number of offspring. And in the second place, there may not be one number of offspring which corresponds to the mode of distribution. For example, an organism might have a .5 propensity to leave 10 offspring and a .5 propensity to leave 20 offspring. And finally, even if there is a number of offspring which an organism has a significantly higher propensity to leave than any other number of offspring, we must take into account

8. While an organism has a number of different propensities to leave $n$ offspring, for different values of $n$, we do not have the additional complication that an organism has a number of different propensities to leave a particular number of offspring, $n$. An object has many different *relative probabilities* to manifest a given property, depending on the reference class in which it is placed. (In practice, choice of reference classes is dependent on our knowledge of the statistically relevant features of the situation). But an object's *propensity* to manifest a certain property is a function of all the causally relevant features of the situation, independent of our knowledge or ignorance of these factors. The totality of causally relevant features determines the unique correct reference class, and thus the unique strength of the propensity to manifest the property in question. (Thus it cannot be the case that an object has more than one propensity to manifest a particular property in a particular situation.)

the remainder of the distribution of reproductive propensities as well. For example, an organism with a .7 propensity to leave 5 offspring and .3 propensity to leave 50 offspring is very different from an organism with a .7 propensity to leave 5 offspring and a .3 propensity to leave no offspring, even though each has the propensity to leave 5 offspring as its highest reproductive propensity.[9]

In lieu of these considerations, one might suggest that the $fitness_1$ of an organism be valued in terms of the entire distribution of its reproductive propensities. The simplest way to do this is to assign distributions as values. For example, the $fitness_1$ of an imaginary organism $x$ might be the following distribution.

number of offspring
  1  2  3  4  5  6  7  8  9  10
propensity
.05 .05 .05 .2  .3  .2 .05 .05 .05

However, our intuitions fail us in regard to the comparison of such distributions. How can we determine whether one organism is fitter than another, on the basis of their distributions alone? For instance, is $x$ fitter or less fit than $y$ and $z$, whose distributions (below) differ from x's?

number of offspring
    1  2 3 4 5 6 7 8 9  10
$y$           1.0
$z$    .5    .3          .2

In order to avoid the uncertainties inherent in this method of valuation, and still take into account all an organism's reproductive propensities, we suggest that $fitness_1$ values reflect an organism's *expected number* of offspring. The expected value of an event is the weighted sum of the values of its possible outcomes, where the appropriate weights are the probabilities of the various outcomes. As regards $fitness_1$, the event in question is an individual's total offspring contribution. The possible outcomes $0_1$, $0_2$, ... $0_n$ are contributions of different numbers of offspring. Values $(1, 2, ... n)$ of the outcomes correspond to the number of offspring left. And the weighting probability for each outcome $0_1$ is just the organism's propensity to contribute $i$ offspring. Thus the imaginary organisms $x$, $y$, and $z$ above all have the same expected number of offspring, or fitness value, of 5.

We propose, then, that "individual fitness" or "$fitness_1$" be defined as follows:

The *fitness*$_1$ of an organism $x$ in environment $E$ equals $n = _{df} n$ is the expected number of descendants which $x$ will leave in $E$.[10]

It may have occurred to the reader that the fitness values assigned to organisms are not literally propensity values, since they do not range from

---

9. It might initially be thought that these examples are highly artificial, since there are no such "bimodal" organisms. But some organisms tend to have offspring in litters and swarms. For such organisms, their offspring contribution propensities will cluster around multiples of numbers typical of the litter or hatching size.

10. A note of clarification is in order concerning our definition of "fitness." It is not clear whether "expected descendant contribution" refers to expected offspring contribution, or expected second generation descendant contribution, or expected 100th generation de-

0 to 1. But this does not militate against our saying that the fitness of an organism is a complex of its various reproductive propensities. Consider for comparison another dispositional property of organisms: their intelligence. If everyone could agree that a particular intelligence test really measured intelligence, then an organism's intelligence could be defined as the expected score on this test. (We would not value intelligence as the score actually obtained in a particular taking of the test, for reasons precisely analogous to those which militate against definitions of fitness in terms of actual numbers of organisms left. Intelligence is a competence or capacity of organisms, rather than simply a measure concept). Obviously, intelligence would not be valued as the strength of the propensity to obtain a *particular* score. Similarly, it is the expected number of offspring which determines an organism's fitness values, not the strength of the propensity to leave a particular number of offspring.

### 4.2. FITNESS: FITNESS OF TYPES

Having defined fitness$_1$, we are in a position to define the fitness$_2$ of types. As will become apparent in what follows, it is the fitness of types which figures primarily in explanations of microevolutionary change.

Intuitively, the fitness of a type (genotype or phenotype) reflects the contribution of a particular gene or trait to the expected descendant contribution (i.e., the fitness$_1$) of possessors of the gene or trait. Differences in the contributions of alternate genes or traits would be easy to detect in populations of individuals that were phenotypically identical except in regard to the trait or gene in question. In reality, though, individuals differ with regard to many traits, so that the contribution of one or another trait to fitness$_1$ is not so straightforward. In fact, the notion of any simple, absolute contribution is quite meaningless. For a trait acts in conjunction with many other traits in influencing the survival and reproductive success of its possessors. Thus its contribution to different organisms will depend upon the different traits it is associated with in those organisms.

Yet, in order to explain the evolution and/or persistence of a gene or its phenotypic manifestation in a temporally extended population, we would like to show that possessors of the

---

scendant contribution. The problem can be illustrated as follows. One kind of individual may contribute a large number of offspring which are all very well adapted to the environment into which they are born, but cannot adapt to environmental changes. As a result, an individual of this type contributes a large number of offspring at time $t$, but because of an environmental change at $t + \Delta t$, these offspring in turn leave very few offspring, so that the original individual actually has very few second- or third-generation descendants. On the other hand, individuals of an alternate type may leave fewer offspring, yet these offspring may be very adaptable to environmental changes. Thus, although an individual of the latter type contributes a lower average number of offspring at time $t$, that individual may have a greater descendant contribution at $t + \Delta t$. Which individual is fitter? We suggest differentiating between long-term fitness and short-term fitness—or between first generation fitness, second generation fitness, ... $n$-generation fitness. Thus the latter type is fitter in the long term, while the former is fitter in the short term.

gene or trait were generally better able to survive and reproduce than possessors of alternate traits or genes. (By "alternate genes" we mean alternate alleles, or alternate genes at the same locus of the chromosome. "Alternate traits" are phenotypic manifestations of alternate genes.) In other words, we want to invoke the *average* fitness$_1$ of the members of each of the types under consideration. Let us refer to average fitness$_1$ as "fitness$_2$." Given some information about the fitness$_2$ of each of a set of alternate types in a population, and given some information about the mechanisms of inheritance involved, we can predict and explain the evolutionary fate of the genes or traits which correspond to the alternate types. For instance, if we knew that possessors of a homozygous-based trait were able to contribute a higher average number of offspring than possessors of any of the alternate traits present in the population, we would have good grounds for predicting the eventual predominance of the trait in the population.

As the above discussion suggests, we actually invoke *relative* fitness$_2$ values in predictions and explanations of the evolutionary fate of genes and traits. That is, we need to know whether members of a particular type have a *higher* or *lower* average fitness$_1$ in order to predict the fate of the type. In order to capture this notion, and to accomodate biologists' extensive references to "relative fitness" or "Darwinian fitness," we introduce "relative fitness$_2$." Given a set of specified alternate types, there will be a type which is fittest in the fitness$_2$ sense (i.e., has highest average fitness$_1$, designated "Max Fit-

ness$_2$, we define relative fitness$_2$ as follows:

The relative fitness$_2$ of type $X$ in $E$ $=_{df}$ the fitness$_2$ of $X$ in $E$/Max fitness$^2$ in $E$.

The role of relative fitness$_2$ ascriptions in evolutionary explanations has been acknowledged (for instance by Williams' "condition 3" in her 1976 analysis of functional explanations). Yet very little attention has been paid to the establishment of these ascriptions. Perhaps we should say a few words about these claims, for it might be supposed that the *only* way in which fitness$_2$ ascriptions can be derived is through measurements of actual average offspring contributions of types. If this were the case, even though "fitness$_2$" is not *defined* in terms of such measures (so that explanations employing fitness$_2$ ascriptions to explain actual offspring contribution differences would not be formally circular), claims concerning the influence of fitness$_2$ differences upon offspring contribution could not be *tested*. This would obviously be disastrous for our analysis.

Evolutionary biologists frequently *derive* relative fitness claims from optimality models (e.g., Cody 1966); this is basically an engineering design problem. It involves determining, solely on the basis of design considerations, which of a set of specified alternate phenotypes maximizes expected descendant contribution. The solution to such a problem is only optimal relative to the other *specified* alternatives (there may be an unspecified, more optimal solution). Thus, optimality models provide some insight into the relative fitness of members of alternate types.

The theorems derived from optimality models can be confirmed by measurements of actual descendant contribution. Such measures can also be used to generate fitness$_2$ ascriptions. Given evidence that descendant contribution was affected primarily or solely by individual propensities for descendant contribution, we can infer that descendant contribution measurements are indicative of individual or type fitness.

### 4.3 EXPLAINING MICROEVOLUTIONARY PHENOMENA

Having elaborated the notions of fitness$_1$ and relative fitness$_2$, we hope to show how these concepts function in explanations of evolutionary phenomena. Perhaps the clearest means of showing this is to work through an example of such an explanation. The example we are going to consider involves a change in the proportion of the two alleles at a single chromosomal locus, and a change in the frequency of genotypes associated with this locus, in a large population of organisms. In this population, at the locus in question, there are two alleles, $A$ and $a$. $A$ is fully dominant over $a$, so that $AA$ and $Aa$ individuals are phenotypically indistinguishable with respect to the trait determined by this locus. This trait is the "natural gun" trait. All individuals which are either homozygous ($AA$) or heterozygous ($Aa$) at this locus have a natural gun, whereas the unfortunate individuals of genotype $aa$ have no gun. Let us suppose that for many generations this population has lived

in peace in an environment $E$, in which no ammunition is available. (Were the terminology not in question, we would say that there had been no "selective pressure" for or against the natural gun trait.) However, at generation $n$, environment $E$ changes to environment $E'$, by the introduction of ammunition usable by the individuals with natural guns. At generation $n$, the proportion of $A$ alleles is .5 and the proportion of $a$ alleles is .5, with the genotypes distributed as follows:

$$AA: .25 \ Aa: .50 \ aa: .25$$

What we want to explain is that in generation $n + 1$, the new frequency of genotypes is as follows:

$$AA: .29 \ Aa: .57 \ aa: .14$$

Let us suppose that the large size of this population makes such a change in frequency extremely improbably ($p = .001$) on the basis of chance.

We need two pieces of information concerning this population in order to explain the change in frequency. We need to know (1) the relative fitness$_2$ of the natural gun and non-natural gun types, and (2) whether any conditions obtain which would interfere with the actualization of the descendant contribution propensities which the relative fitness$_2$ valuations reflect. As was noted above, the fact that an organism does not survive and reproduce in an environment in which periodic cataclysms occur is no indication of its fitness (any more than the failure of salt to dissolve in water when coated with plastic would count against its solubility).

The latter qualification, stating that no factors other than fitness$_2$

differences were responsible for descendant contribution, corresponds to the "extremal clause" which, as Coffa (1977, p. 194) has made clear, is a component in the specification of most scientific laws. Such clauses state that no physical properties or events relevant to the occurrence of the outcome described in the law (other than those specified in the initial conditions) are present to interfere with that outcome. In stating scientific laws, the assumption is often tacitly made that no such disturbing factors are present. But as Coffa has pointed out, it is important to make this assumption explicit in an extremal clause, for no scientific law can be falsified by an instance in which the event predicted by the law fails to occur unless the extremal clause is satisfied. Thus our ability to fill in the details of the extremal clause will determine our ability to distinguish between contexts which count as genuine falsifications of a law and contexts which do not. That evolutionary theorists are fairly specific about the types of conditions which interfere with selection is an indication in favor of the testability of claims about fitness. As noted above, the influence of fitness upon offspring contribution is disturbed by any factors which separate successful from unsuccessful reproducers without regard to *physical* differences between them. In addition, certain other evolutionary factors, such as mutation, migration, and departures from panmixia, may disturb the systematic influence of fitness differences between types upon proportions of those types in subsequent generations.

Let us suppose that we do know the relative fitnesses$_2$ of the natural gun and non-natural gun types, and let us suppose the natural selection conditions are present (i.e., nothing is interfering with the manifestations of the fitness propensities). This information together with the relevant laws of inheritance will allow us to predict (and explain) the frequencies of types in generation $n + 1$. We need not detail the principles of inheritance which allow this computation here (since they are available in any genetics text) other than to note that the Hardy-Weinberg Law allows us to compute the relative frequencies of types in a population, given information about the heritability of the types in question together with hypotheses about fitness$_2$ differences.

In light of these considerations, we construct the promised schema as follows:

1. In $E'$, in generation $n$, the distribution of genotypes is:
   $AA: .25$  $Aa: .50$  $aa: .25$
2. $(x) (AAx \supset tx)$ & $(x) (Aax \supset tx)$ & $(x)(aax \supset -tx)$
3. In $E'$, the relative fitness$_2$ of type $t$ is 1.0.
4. In $E'$, the relative fitness$_2$ of type not-$t$ is 0.5.
5. For any three distinct genotypes $X, Y, Z$ (generated from a single locus), if the proportions of $X$, $Y$, $Z$ in generation $n$ are $P$, $Q$ and $R$, respectively, and if the relative fitnesses$_2$ of genotypes $X$, $Y$, and $Z$ are $F(Y)$ and $F(Z)$, respectively, then the proportion of $X$ in generation $n + 1$ is:
   $P \cdot F(X)/[P \cdot F(X) + Q \cdot F(Y) + R \cdot F(Z)]$
6. $EC(E)$

7. Given the size of population $P$, the probability that the obtained frequencies were due to chance is less than .001.

$$=.99$$

In $E'$ at generation $n + 1$ the frequency of genotypes is:
$AA$: .29    $Aa$: .57    $aa$: .14

This explanation is of the inductive-statistical variety, with the strength of the connection between explanans and explanandum determined, as indicated in premise (7), by the size of the population. Premise (1) is, obviously, a statement of the initial conditions. Premise (2) allows us to determine which genotypes determine each phenotype: all individuals with genotype $AA$ or $Aa$ have trait $t$, and all individuals of genotype $aa$ lack trait $t$. Premises (3) and (4) indicate the relative fitness$_2$ of types $t$ and not-$t$ in environment $E$. Premise (5) is the above-mentioned consequence of the Hardy-Weinberg Law which allows computation of the expected frequencies in generation $n + 1$, given information about reproductive rates at generation $n$, together with information about initial frequences of individuals of each genotype at generation $n$. Premise (6) affirms that the extremal clause ($EC$) was satisfied—i.e., that the "natural selection conditions" were present for the environment ($E$) in question. Thus we can infer that propensities to contribute descendants will be reflected in actual reproductive rates. Each genotype receives the relative fitness$_2$ associated with the phenotype it determines, as indicated in premise (2). Thus by substitution of the values provided in premises (3) and (4) in formula

(5) (i.e., $X = AA$, $F(X) = 1.0$, $P = .25$; $Y = Aa$, $F(Y) = 1.0$, $Q = .50$, . . . etc.) we can obtain the values which appear in the explanandum.

To summarize, from knowledge of (1) initial frequencies of genotypes in generation $n$, (2) the relative fitness$_2$ of those genotypes, and (3) the fact that the extremal clause was satisfied, we can infer what the frequencies of genotypes will be in generation $n + 1$.

Of course, in this admittedly artificial example it was presumed that the appropriate relative fitness$_2$ values were known. This suggests that we somehow investigated reproductive *capabilities*, and not just reproductive differences. We must emphasize, however, that actual reproductive differences may be regarded as measures of differences in reproductive capability as long *as the measured differences are statistically significant*. This is the means of fitness determination in many, if not most, evolutionary investigations. But this must not mislead the reader into identifying fitness with actual reproductive contribution. For *statistically significant* differences would not be required to establish fitness differences in this case. Rather, statistically significant differences are required to establish that certain variables (fitness differences, in this case) are causally connected with other variables (in this case, differences in offspring contribution). Statistically significant differences are thus quite appropriate measures for fitness differences, given the propensity interpretation of fitness.

Having explained the role of statistical significance in measuring fitness differences, we can now consider a more realistic example of the role of fitness in population biology. Certainly one of the greatest controversies in the history of population genetics concerns the differences in fitness of heterozygotes and homozygotes. The importance of the controversy lies in the fact that if heterozygotes are generally fitter than homozygotes, then breeding groups will retain a greater amount of genetic variation than if homozygotes were generally superior in fitness. And the amount of variation present in a population is of considerable importance to the evolutionary fate of the population. (For instance, greater variation provides some "flexibility" in the sense that a genetically variable population has more alternatives for adapting to changing environmental conditions.) Theodosius Dobzhansky, a principal protagonist in this controversy, maintained that heterozygotes at many loci were fitter than homozygotes at the same loci, and he and his collaborators gathered a good deal of statistically significant data to support this contention.

For instance, in one article, it was reported that members of the species *Drosophila pseudoobscura* which were heterozygous in regard to the structure of their third chromosome were more viable than the flies which were homozygous. Dobzhansky et al. correlated viability differences (note that *viability* differences are dispositional property differences) with fitness differences, and they performed a statistical analysis on their data, in order to conclude that:

> Heterosis [heterozygote superiority in fitness] has . . . developed during the experiment, as indicated by the attainment of equilibrium and by a study of the viability of the flies derived from the cage. Both tests gave statistically significant results. [1951, p. 263]

Again, statistical significance would be of no concern if fitness were identified straightforwardly with offspring contribution. Statistical significance is important, however, if fitness is identified with phenotypic properties causally connected with offspring contribution.

As these examples demonstrate, fitness ascriptions play not only a legitimate, but a crucial role in explanations of evolutionary change. While biologists have not been able to justify their usage of the concept of "fitness," their concept has nevertheless been consistent and appropriate. Philosphers have accused biologists of giving circular explanations of evolutionary phenomena because they have taken into account only the definitions of fitness that biologists explicitly cite, and they have not looked for the interpretation implicit in biologists' usage.

### 4.4 A PROPENSITY ANALYSIS OF NATURAL SELECTION

One consequence of our propensity interpretation of fitness is that the analysis also points to an improved definition of "natural selection." As was noted earlier, the concepts of fitness and natural selection are inextricably bound—so much so that misinterpretations of fitness are reflected in misinterpretations of natural selection.

Thus, according to one of the more

popular interpretations of natural selection, that process occurs whenever two or more individuals leave different numbers of offspring, or whenever two or more types leave different average numbers of offspring. For example, Crow and Kimura (1970) stipulate that "Selection occurs when one genotype leaves a different number of progeny than another" (p. 173). Insofar as it is correct to say that the *fittest* are *selected,* this definition of "selection" clearly reflects a definition of "fitness" in terms of actual descendant contribution.

But surely these definitions (see also Wallace, 1963, p. 160; Wilson, 1975, p. 489) do not adequately delimit the reference of "natural selection." For evolutionary biologists do not refer to just any case of differential offspring contribution as "natural selection." For instance, if predatory birds were to kill light and dark-colored moths indiscriminately and yet by chance killed more light than dark ones, we would not attribute the differential offspring contribution of light and dark moths to natural selection. But if the dark coloration acted as camouflage, enabling the dark moths to escape predation and leave more offspring, we would attribute the resulting differential offspring contribution to the action of natural selection. For only in the latter case are differences in offspring contribution due to differences in offspring contribution dispositions.

Thus Kettlewell (1955, 1956) did not presume to have demonstrated the occurrence of natural selection simply by pointing out the dramatic increase in frequency of dark-colored

pepper moths within industrial areas of England. In order to demonstrate that selection (vs. chance fluctuations, migration, etc.) had accounted for the change, Kettlewell had to provide evidence that the dark-colored moths were better able to survive and reproduce in the sooted forests of these regions. Nor did Cain and Sheppard (1950, 1954) and Ford (1964) consider differential contribution to be a sufficient demonstration of natural selection in their celebrated accounts of the influence of selection on geographical distribution. In order to support the hypothesis that natural selection had affected the geographic distribution of various color and banding-pattern traits of snails of the species *Cepaea nemoralis,* these men argued that the colors and band-patterns peculiar to an area were correlated with the background color and uniformity of that area. More precisely, yellow snails were predominant in green areas; red and brown snails were predominant in beechwoods "with their red litter and numerous exposures of blackish soil . . ." (Ford 1964, p. 153); and unbanded shells were predominant in more uniform environments. These traits effectively camouflaged their possessors from the sight of predators (Ford, 1964, p. 155), thus *enabling* suitably marked snails to contribute more offspring than the unsuitably marked snails.

In each of these cases, selection is construed as involving more than just differential perpetuation. Rather, selection involves differential perpetuation caused by differential reproductive capabilities. So, just as we amended traditional definitions of "fitness" to take into account descendant contribution propensities, we must also amend traditional definitions of "selection" so as to emphasize the role of differential descendant contribution of descendants, but a differential contribution *caused* by differential propensities to contribute. On the basis of these considerations, let us define "individual selection" and "type selection" as follows:

> Natural selection is occurring in population $P$ in environment $E$ with regard to organisms $x, y, z$ (members of $P$) = $_{df}$ $x, y, z$ differ in their descendant contribution dispositions in $E$, and these differences are manifested in $E$ in $P$.

> Natural selection is occurring in population $P$ in environment $E$ with regard to types $X, Y, Z$ (included in $P$) = $_{df}$ members of $X, Y, Z$ types differ in their average descendant contribution dispositions in $E$, and these differences are manifested in $E$ in $P$.

We know from our previous analysis that when organisms leave numbers of offspring which reflect their reproductive propensities (i.e., when reproductive propensities are manifesting themselves) in a particular environment, this implies that no factors are interfering with the manifestation of these propensities. (cf. our remarks on extremal clauses above). Put more positively, we have grounds for believing that, for example, no cataclysms, cases of human intervention, etc., are occurring. Of course, the occurence of natural selection is not precluded by the

incidence of such factors. Fitter individuals might leave more offspring than less fit individuals (on account of their fitness differences), even though nondiscriminating factors are operating to minimize the reproductive effects of fitness differences. In other words, the incidence of nondiscriminating factors will not necessarily override the effects of fitness differences. Thus we do not have to rule out the occurrence of nonselective factors in our definition of "natural selection." But in explanations (such as our Hempelian schema above) of the precise evolutionary effects of selection, we must take these nonselective factors into account.

## 5. CONCLUSION

A science may well progress even though its practitioners are unable to account for aspects of its foundations in any illuminating way. We believe that this has been the case with evolutionary theory, but that the *propensity* analysis of fitness which we have described captures the implicit content in biologists' usage of the term. The propensity interpretation allows us to reconstruct explanations of microevolutionary phenomena in such a way that these explanations appear to be entirely respectable and noncircular. By their form, and by inspection of the premises and conclusion, such explanations appear to satisfy Hempelian adequacy requirements for explanations, and even appear to incorporate recent modifications of the Hempelian model for inductive explanations (Coffa 1974). We chose an example of microevolutionary change, since we wanted the least complicated instance possible in order to illuminate the form of explanations utilizing fitness ascriptions. We know of no reason to believe that a similar reconstruction could not be given for the case of macroevolutionary change.[11]

11. A great deal more needs to be done by way of clarifying the concepts of fitness and natural selection, given the many uses biologists make of these concepts. But we believe that the broad analyses we have given provide an adequate framework within which further distinctions and clarifications can be made. For example, within the categories of $fitness_1$ and relative $fitness_2$, distinctions can be drawn between short- and long-term fitness, by distinguishing between propensities to leave descendants in the short run (in the next few generations) *vs.* propensities to leave descendants in the long run (cf. footnote 7).

The propensity interpretation also lends itself to the much-discussed notion of "frequency dependent fitness," wherein the fitness of a type differs according to the frequency of the type. Certain cases of mimicry have been explained via reference to frequency dependent fitness. For instance, it has been suggested that the mimetic resemblance of a prey species to a distasteful model may enhance the survival of the mimics so long as they are rare, because individual predators most readily learn to avoid the distasteful type (and hence the mimic) when the model is more common than the mimic. Surely the survival *ability* of the mimics, and not just their survival rates, are enhanced by the scarcity of their type.

The sociobiological notion of "inclusive fitness" also seems susceptible to a propensity analysis. Biologists have invoked this notion in order to explain the evolution of certain altruistic traits. The idea (very simply) is that some of the organisms benefiting from an altruistic action may be genetically related to the altruistic actor, and may therefore share the behavioral trait which led to the action (if the trait is genetically based). Thus, although an altruistic action may decrease the $fitness_1$ of the actor, it may increase the $fitness_2$ of the altruistic trait. As a result, the trait may come to predominate within the population. "Inclusive fitness" values have been proposed as appropriate indicators of the evolutionary fate of altruistic traits. These values take into account not only the effect of altruistic actions upon the fitness of the actors, but also the probability that the action will benefit genetic relatives, and the extent of the benefit to relatives (cf. Hamilton 1964). Our colleague Greg Robischon is currently considering a propensity interpretation of inclusive fitness.

## REFERENCES

Bethell, T., 1976, Darwin's mistake. *Harper's Magazine,* 70–75.

Cain, A. J., and P. M. Sheppard, 1950, Selection in the polymorphic land snail *Cepala nemoralis. Heredity,* 4: 275–294.

Cain, A. J., and P. M. Sheppard, 1954, Natural selection in Cepaea. *Genetics,* 39: 89–116.

Cody, M., 1966, A general theory of clutch size. *Evolution,* 20: 174–184.

Coffa, J. A., 1974, Hempel's ambiguity. *Synthese,* 28: 141–163.

Coffa, J. A., 1977, Probabilities: reasonable or true? *Philosophy of Science,* 43: 186–198.

Crow, J. F., and Kimura, M., 1970, *An Introduction to Population Genetics Theory,* New York, Harper and Row.

Dobzhansky, T., 1970, *Genetics of the Evolutionary Process,* New York, Columbia University Press.

Dobzhansky, T., and Levene, H., 1951, Development of heterosis through natural selection in experimental populations of Drosophila pseudoobscura. *American Naturalist,* 85: 246–264.

Ford, E. B., 1964, *Ecological Genetics,* New York, Wiley.

Grant, V., 1977, *Organismic Evolution,* San Francisco, W. H. Freeman.

Hamilton, W. D., 1964, The genetical evolution of social behavior, 1. *Journal of Theoretical Biology,* 7: 1–16.

Hull, D., 1974, *Philosophy of Biological Theory,* Englewood Cliffs, New Jersey, Prentice Hall.

Kettlewell, H. B. D., 1955, Selection experiments on industrial melanism in the Lepidoptera. *Heredity,* 9: 323–342.

Kettlewell, H. B. D., 1956, Further selection experiments on industrial melanism in the Lepidoptera. *Heredity,* 10: 287–301.

Kojima, K., 1971, Is there a constant fitness value for a given genotype? *Evolution,* 25: 281–285.

Kyburg, H., 1974, Propensities and probabilities. *British Journal for the Philosophy of Science,* 25, 4: 358–374.

Lerner, I. M., 1958, *The Genetic Basis of Selection,* New York, Wiley.

Levins, R., 1968, *Evolution in Changing Environments,* Princeton, Princeton University Press.

Levins, R., 1970, Fitness and optimization. *Mathematical Topics in Population Genetics.* New York, Springer Verlag.

Manser, A. R., 1965, The concept of evolution. *Philosophy,* XL: 18–34.

Mettler, L. E., and Gregg, T. G., 1969, *Population Genetics and Evolution,* Englewood Cliffs, New Jersey, Prentice Hall.

Popper, K., 1974, Intellectual autobiography. *The Philosophy of Karl Popper* (Shilpp, ed.). LaSalle, Illinois, Open Court.

Ruse, M., 1969, Confirmation and falsification of theories of evolution. *Scientia,* 1–29.

Ruse, M., 1973, *The Philosophy of Biology,* London, Hutchinson.

Scriven, M., 1959, Explanation and prediction in evolutionary theory. *Science,* 130: 477–482.

Smart, J. J. C., 1963, *Philosophy and Scientific Realism,* London, Routledge and Kegan Paul.

Thoday, J. M., 1953, Components of fitness. *Symposium of the Society for Experimental Biology,* 7: 96–113.

Waddington, C. H., 1968, The basic ideas of biology. *Towards a Theoretical Biology,* vol. 1, Chicago, Aldine.

Wallace, B., 1963, Further Data on the overdominance of induced mutations. *Genetics,* 48: 633–651.

Wallace, B., 1968, *Topics in Population Genetics,* New York, W. W. Norton.

Williams, M. B., 1970, Deducing the consequences of evolution: a mathematical model. *Journal of Theoretical Biology,* 29: 343–385.

Williams, M. B., 1973a, The logical status of natural selection and other evolutionary controversies: resolution by axiomatization. *Methodological Unity of Science* (Bunge, ed.). Dordrecht, Holland, Reidel.

Williams, M. B., 1973b. Falsifiable predictions of evolutionary theory. *Philosophy of Science,* 40: 518–537.

Williams, M. B., 1976, The logical structure of functional explanations in biology. *Proceedings of the Philosophy of Science Association 1976,* 37–46. East Lansing, Philosophy of Science Asso.

Wilson, E. O. 1975, *Sociobiology,* Cambridge, Massachusetts, Harvard University Press.

Wright, S., 1955, Classification of the factors of evolution. *Cold Spring Harbor Symposia on Quantitative Biology,* 20: 16–24.

# 6

# Adaptation and Evolutionary Theory

## ROBERT BRANDON

> There is virtually universal disagreement among students of evolution as to the meaning of adaptation.
>
> —Lewontin, 1957

> Much of past and current disagreement on adaptation centers about the definition of the concept and its application to particular examples: these arguments would lessen greatly if precise definitions for adaptations were available.
>
> —Bock and von Wahlert, 1965

> The development of a predictive theory [of evolution] depends on being able to specify when a population is in better or worse evolutionary state. For this purpose an objective definition of adaptedness is necessary.
>
> —Slobodkin, 1968

THE CONCEPTION OF ADAPTATION was not introduced into biology in 1859. Rather, what Darwin did was to offer a radically new type of explanation of adaptations, and in so doing he altered the conception. As the above quotes indicate, we have not in the last century sufficiently delimited this conception, and it is important to do so.

In this paper I will analyze and, I hope, clarify one aspect of the conception of adaptation. One of my aims is a theoretically adequate definition of relative adaptedness. As we will see, such analysis cannot be divorced from an analysis of the structure of evolutionary theory. My other major aim is to expose this

I owe a debt of gratitude to all those who read earlier verisons of this paper and helped me improve it. Where possible I have tried to footnote contributions. Here I want to give special thanks to Ernst Mayr and Paul Ziff, whose comments and criticisms have had pervasive effects on the evolution of this paper.

structure, to show how it differs from the standard philosophical models of scientific theories, and to defend this differentiating feature (and, hence, to show the inadequacy of certain views about the structure of scientific theories which purport to be complete).

A note on defining is needed. Definitions are often thought to be of two kinds, descriptive and stipulative. (See, for example, Hempel, 1966, chapter 7.) Descriptive definitions simply describe the meaning of terms already in use; stipulative definitions assign, by stipulation, special meaning to a term (either newly coined or previously existing). According to this view the project of defining a term is either purely descriptive or purely stipulative. This view is mistaken. The project at hand calls for neither pure linguistic analysis nor pure stipulation; it is much more complex. Briefly, I examine the conceptual network of evolutionary biology. I find that according to evolutionary theory there is a biological property, adaptedness, which some organisms have more of than others. Those having more of it, or those better adapted, tend to leave more offspring. And this is the mechanism of evolution. The project calls for conceptual analysis, but such analysis is sterile unless it is coupled with an examination of the biological property that is the object of the conception. Any definition that fails to fit the conceptual network must be rejected, as must any that fails to apply to the property. The project calls for an element of stipulation, but our stipulatory freedom is con-

strained both by the theoretical and conceptual requirements and, one hopes, by the real world.

A note on the restricted scope of this paper: Biologists talk about the adaptedness of individual organisms and of populations. Selection occurs at the level of individuals and, presumably, at higher levels. That is, there is intrapopulational selection and interpopulational selection. It is vital that we keep these levels separate and that we see the relation between selection and adaptation.[1] Selection at the level of individual organisms has as its cause differences in individual adaptedness and its effect is adaptions for individual organisms. I will follow standard practice in calling selection at this level natural selection. Any benefit to the population from natural selection is purely fortuitous. One must distinguish between a group of adapted organisms and an adapted group of organisms. For instance, a herd of fleet gazelles is not necessarily a fleet herd of gazelles. Similarly, group selection will have as its cause differences in group adaptedness and as its effect group adaptations. The theory of group selection is clear; its occurrence in nature is controversial. One could speak of an abstract theory of evolution which covers natural selection, group selection, and even the selection of tin cans in junk yards. But most of the interesting problems don't arise at that level of generality. In this paper I will be primarily concerned with natural selection—that is, with intraspecific intraenvironmental selection, and therefore, with the adaptedness

1. G. C. Williams (1966) does an excellent job of clarifying these matters. Also see Lewontin (1970).

of individual organisms, not with the adaptedness of populations.

Let me illustrate the confusion that results from the failure to relate adaptedness to the proper level of selection. One of the more prominent definitions of relative adaptedness is due to Thoday.[2] Basically, it says: *a* is better adapted than *b* if and only if *a* is more likely than *b* to have offspring surviving $10^8$ (or some other large number) years from now. Either the long-range probability of offspring corresponds to the short-range probability of offspring or it does not. ("Corresponds" means *a*'s long-range probability of offspring is greater than *b*'s long-range probability of offspring if and only if *a*'s short-range probability of offspring is greater than *b*'s short-range probability of offspring.) If it does correspond, we should stick to the more easily measurable short-range probability. If not, then, since natural selection is not foresighted—that is, it operates only on the differential adaptedness of present organisms to present environments—the long-range probability of offspring is irrelevant to natural selection.

Why has Thoday's definition been so favorably received? Because the long-range probability of descendants is important to selection at or above the species level. For instance, one plausible explanation of the predominance of sexual reproduction over asexual modes of reproduction is that the long-range chances of

survival are greater for populations having sex (see Maynard Smith, 1975, pp. 185ff). But if one is interested in selection at the population level, the relevant notion of adaptedness would be that which applies to populations. Until recently even biologists have failed to distinguish intra- and interpopulational selection. Thoday's definition, not being selection relative, lends itself to this confusion. To keep matters as clear as possible, we will be concerned only with natural selection and with that notion of adaptation which properly relates to it.

## 1. THE ROLE OF THE CONCEPT OF RELATIVE ADAPTEDNESS IN EVOLUTIONARY THEORY

The following three statements are crucial components of the Darwinian (or neo-Darwinian) theory of evolution.[3]

(1) Variation: There is (significant) variation in morphological, physiological, and behavioral traits among members of a species.

(2) Heredity: Some traits are heritable so that individuals resemble their relations more than they resemble unrelated individuals and, in particular, offspring resemble their parents.

(3) Differential Fitness: Different variants (or different types of organisms) leave different num-

2. Thoday (1953, 1958). Actually he uses the word "fitness" not "adaptedness," but I think he is like most biologists in using the words interchangeably.

3. This characterization of evolutionary theory is adopted from Lewontin (1977). For less satisfactory versions see Lewontin (1968) and (1970). For a more historical and fuller sketch of the major components of the theory see Mayr (1977).

bers of offspring in immediate or remote generations.

When the conditions described above are satisfied, organic evolution occurs. A thorough examination of the history of our awareness of these conditions would be interesting and worthwhile but will not be attempted here (see Mayr, 1977). Suffice it to say that in Darwin's time each was a nontrivial statement. In what follows we will examine them predominantly from our own point of view.

Ignoring the parenthetical "significant," (1) cannot help being true. The uniqueness of complex material systems is now taken for granted; and so we expect variation among individuals of a species. Their similarity needs explaining, not their variation. (1) becomes less empty from our point of view when "significant" is added. What sort of variation is significant? That which can lead to adaptive evolutionary changes. Though the world is such that individuals must be unique, the recognition of this fact is of fairly recent origin and is necessary for an evolutionary world view.

Unlike (1), (2) is not trivial. There is no metaphysical necessity in offspring resembling their parents. (2) can now be derived from our modern theories of genetics; in Darwin's time it was an observation common

to naturalists and animal breeders. Darwin's theories of heredity were notoriously muddled, but fortunately a correct theory of genetics is not a prerequisite for a Darwinian theory of evolution (see Mayr, 1977, p. 325). What is important to note is that given that there is variation, (1), and that some of the traits which vary are heritable, (2), it follows that the variation within a species tends to be preserved. (Of course this tendency can be counterbalanced by other factors.)

When (3) holds, when there are differences in reproductive rates, it follows from (1) and (2) that the variation status quo is disrupted— that is, there are changes in the patterns of variation within the species. For our purposes we can count such changes as evolution. (For a fuller explication of the concept of evolution see Brandon, 1978.) Thus when (1)–(3) hold, evolution occurs.

We have seen that (1) is in a sense trivial and requires no explanation. We have also seen that (2) is nontrivial and is to be explained by modern theories of genetics, but that this explanation is not essential to Darwinian theory. In contrast, the distinguishing feature of a Darwinian theory of evolution is its explanation of (3).[4] The focus of this paper is the conception used for such explanations.

4. Perhaps one should not speak of *the* distinguishing feature of Darwinian theory. One should recognize that evolutionary theory is not a monolithic whole. For instance, theories of speciation are quite distinct from the part of Darwinian theory on which we are focusing —namely, the theory of evolution within a species by natural selection. Apropos the history of the subject it is useful to distinguish four subtheories or four parts of Darwin's theory (pointed out to me by Ernst Mayr): (a) Evolution at all; (b) Gradual evolution; (c) Evolution by common descent; and (d) Evolution by natural selection. Nevertheless, both from an historical and a contemporary perspective the most salient feature of a Darwinian theory of evolution is its explanation of evolution by natural selection.

The distinguishing feature of a Darwinian theory of evolution is explaining evolutionary change by a theory of natural selection. Of course, that is not the only possible sort of explanation of evolution. In his own time Darwin convinced the majority of the scientific community that evolution has and does occur, but hardly anyone bought his natural-selection explanation of it. (For an excellent source book on the reception of Darwin's theory see Hull, 1973.) The alternatives of Darwin's day—for example, divine intervention and the unfolding of some predetermined plan, are no longer scientifically acceptable. But there is one present day alternative we should consider.

It is not unsurprising that in finite populations of unique individuals some variants leave more offspring than others. We would expect such differences in reproductive success simply from chance. And if there are chance differences in reproductive success between two types of organisms (or similarity classes of organisms), we expect one type ultimately to predominate by what statisticians call random walk. If we can explain (3), and so the occurrence of evolution, in terms of chance, is the hypothesis of natural selection necessary?

It is becoming the received view in the philosophy of science that hypotheses are not evaluated in isolation but rather in comparison with rival hypotheses. This view is, I think, for the most part correct but not entirely so; some hypotheses we reject as unacceptable without comparison with specific alternatives. Unacceptable hypotheses are those that violate deeper-seated beliefs, theories or metaphysics. Similarly some forms of explanation are unacceptable in that no investigation into the particular phenomenon is required to reject them. We reject them without considering any particular alternative explanation simply because we believe there must be a better alternative. For example, accepting Darwinian theory, we reject the explanation that bees make honey in order to provide food for bears and without examining bees, bears, or honey. (An acceptable form of explanation is not one which is necessarily correct or even accepted; it simply is one which is not unacceptable.)

The theory of evolution by chance or by random walk has been developed in recent years and is often called the theory of non-Darwinian evolution, or better, the neutrality theory of evolution (see King and Jukes, 1969). We cannot give it the discussion it deserves, but it is worth pointing out that explanation in terms of chance is an acceptable form of explaining short-term evolutionary change but not any interesting sort of long-term evolutionary change. (The truth of this hinges on what counts as interesting. I will not try to delimit interesting long-term evolutionary change; suffice it to say that any seemingly directed change is interesting.)

The neutrality theory supposes that certain alternative alleles (and therefore certain protein molecules coded by them) are functionally equivalent—that is, selectively *neutral*. Given this supposition, the neutrality theory predicts (and so is able to explain) the sorts of changes in frequencies of these alleles expected by a process of random sampling in different situations. As Ayala (1974) points out, these predictions differ both qualitatively and quantitatively from those given by the selectionist theory. (Ayala presents data on different species of *Drosophila* which tend to corroborate the natural selectionist hypothesis and refute the neutrality hypothesis.) Whether evolution by random walk is a common or rare phenomenon, we cannot reject a priori a chance-explanation of short-term evolutionary change.

The situation is different for interesting long-term evolutionary phenomena. Of course we do not directly observe long-term evolutionary change. Presumably any complex feature of an organism is the product of long-term evolutionary change. On the one hand some complex features of organisms, such as the eye of a human, are so obviously useful to their possessor that we cannot believe that this usefulness plays no part in explaining their existence. That is, given Darwinian theory and the obvious usefulness of sight, we have a better alternative to the chance-explanation. On the other hand, there are features whose usefulness is unclear for which we still reject chance-explanations, because of their high degree of complexity and constancy. Complexity and constancy are not made likely on the hypothesis of evolution by random sampling. A good example is lateral lines in fish. This organ is structurally complex and shows a structural constancy within taxa, yet until recently it was not known how the lateral line was useful to its possessor. In this case the rejection of a chance-explanation was good policy; studies eventually showed

that the lateral line is a sense organ of audition. (This example is taken from G. C. Williams, 1966, pp. 10–11).

One can contrast the lateral line in fish with the tailless condition of Manx cats. This feature is not even constant within the species, and a nonexistent tail is hardly complex. (Actually what is relevant concerning complexity is that the historical process leading from tailed to tailless is most probably not complex.) Furthermore legend has it that Manx cats originated on the Isle of Man in what would be a small isolated population, thus increasing the probable role of chance. The tailless condition of Manx cats may have evolved by natural selection but for all we know the best explanation of it is the explanation in terms of chance.

It is important to keep in mind the possibility of evolution by random walk, for it is important that Darwinian explanations be testably different (at least in principle) from chance-explanations. What is the Darwinian explanation of (3)? The conventional wisdom is that Darwin explained (3) by his postulate of the "struggle for existence" (or in Spencer's words, which Darwin later used, the "survival of the fittest") and that this explanation, or this discovery of the mechanism of evolution, was Darwin's greatest contribution.

How does the "struggle for existence" or the "survival of the fittest" explain (3)? Following current practice, let us define the *reproductive success* or the *Darwinian fitness* of an organism in terms of its actual genetic contribution to the next generation. I will not try to make this definition precise and complete. The genetic contribution to the next generation can usefully be identified with the number of sufficiently similar offspring when "sufficiently similar" is sufficiently explicated. This would disallow, for example, sterile offspring from counting toward Darwinian fitness. There are two options: either let the Darwinian fitness of an individual equal its actual number of sufficiently similar offspring or let the Darwinian fitness of an individual equal the mean number of sufficiently similar offspring of members of the similarity class to which it belongs. In either case Darwinian fitness is defined in terms of numbers of *actual* offspring. I should point out that most biologists use the words "fitness" and "adaptedness" interchangeably. In this paper "fitness" will only be used to refer to Darwinian fitness. Adaptedness, as we will see, cannot be identified with Darwinian fitness. (3) says that Darwinian fitness is correlated with certain morphological, physiological, or behavioral traits. Why is there this correlation? Why is there differential fitness? Darwin's answer, which he arrived at after reading Malthus' *Essay on Population*,[5] was that since in each generation more individuals are produced than can survive to reproduce, there is a struggle for existence. In this "struggle" (which in its broadest sense is a struggle of the organism with its environment not just with other individuals; see Darwin, 1859, p. 62), certain traits will render an organism *better adapted* to its environment than conspecifics with certain other

traits. The better adapted individuals will tend to be fitter (that is, produce more offspring) than the less well adapted. Why are those who happen to be the fittest in fact the fittest? The Darwinian answer is: They are (for the most part) better adapted to their environment.

What does this explanation presuppose? It seems to presuppose the following as a law of nature:

(D) If *a* is better adapted than *b* in environment *E,* then (probably) *a* will have more (sufficiently similar) offspring than *b* in *E.*

Certainly if (D) is a true law, the Darwinian explanation is acceptable. Darwin seems to presuppose (D), but it is not to be found stated explicitly in the *Origin.* Nor is it to be found in modern evolutionary works. But if one examines work in modern evolutionary biology—the theorizing done, the inferences made, the explanations offered—one finds that (D) or something like (D) is required as the foundation of evolutionary theory. I take it that this conclusion will be so uncontroversial that it need not be further supported by examining examples of evolutionary reasoning. But later in this paper we will give some examples to show how (D) is to employed.

Philosophers of science talk about laws more often than they display actual examples of them. In particular, many people have discussed whether the "survival of the fittest" is a tautology without displaying some-

---

5. Malthus (1798). It seems that Malthus was more of a coagulant than a catalyst for Darwin's ideas on this matter: See Hull (1973), pp. 344, 345, and Mayr (1977).

thing other than that phrase which might be a tautology. (As for example J. J. C. Smart 1963, p. 59.) The phrase itself, not being a declarative sentence, could not be a tautology. An exception is Mary Williams,[6] who has attempted to give a "precise, concise, and testable" version of the phrase, and so has attempted to give a precise, concise, and testable version of the fundamental law of evolutionary theory.

Williams defines the clan of a set $\beta$ as the members of $\beta$ plus all their descendants. On a phylogenetic tree the clan of $\beta$ would be those nodes which are in $\beta$ plus all nodes after them which are on a branch which passes through one of the original nodes. A subclan is either a whole clan or a clan with one or more branches removed. A Darwinian subclan is a subclan which is held together by cohesive forces so that it acts as a unit with respect to selection (this crucial concept is not defined by Williams; she takes it as primitive). Informally, Williams' version of the fundamental law of evolutionary theory states that for any subclan $D_1$ of any Darwinian subclan D,

If $D_1$ is superior in fitness to the rest of D for sufficiently many generations . . . then the proportion of $D_1$ in D will increase during these generations. [1970, p. 362]

(D) is a "law"[7] about properties of individual organisms; Williams' version is a law about properties of sets of organisms. Which is fundamental? Some properties of sets (notable exceptions being set-theoretic properties like cardinality) are a function of the properties of the sets' members. In particular, as Williams herself points out (1973, p. 528), the fitness of a clan is to be identified with the average fitness of the members of the clan. Thus the property of individuals (or more precisely the property of individuals in some environment—what we will call adaptedness, what Williams calls fitness—is fundamental. Likewise (D) is fundamental in that Williams' law can be derived from it and the laws of population genetics but not vice versa. Perhaps the only way of testing (D) is to apply it to fairly large populations and so to test something like Williams' law, but this does not change our conclusion. (D) is required as the foundation of evolutionary theory.

## 2. FOUR DESIDERATA OF DEFINITIONS OF RELATIVE ADAPTEDNESS

We have seen the role the relational concept of adaptedness is to play in a Darwinian theory of evolution: It is the explanatory concept in what I have called the fundamental law of evolutionary theory. Philosophers have not been able to come up with a set

---

6. Another exception is Michael Ruse (1971). He has attacked the problem from an historical perspective and has tried to show that what Darwin said on natural selection was not tautological.

7. In speaking of (D) as a "law" I could continue to put "law" in scare-quotes in order not to prejudge its status, but I will not. We will, in due course carefully evaluate its status.

of necessary and jointly sufficient conditions for scientific lawhood, but there is wide agreement on some necessary conditions. In particular laws of the empirical sciences are to be empirically testable universal statements. It is also desirable, whether or not definitionally necessary, that laws be empirically correct or at least nearly true. One cannot just look at the surface logic of a statement in order to determine whether it is a scientific law (as done in Ruse, 1975). To determine whether (D) is a scientific law, we will have to look deeply into the conception of adaptation. My strategy is to try to construct a definition of relative adaptedness that makes (D) a respectable scientific law. In this section I will argue that from any definition (construction, explication) of this concept we would want the following: (a) independence from actual reproductive values; (b) generality; (c) epistemological applicability; and (d) empirical correctness. After arguing for the above desiderata I will show how current definitions fail to satisfy all four, and then I will produce a general argument showing that no explication of the concept will satisfy all four desiderata. In the final section I will attempt to draw the ramifications of this result.

(a) *Independence.* The relational concept of adaptation is to explain differential fitness. To do so (D) must not be a tautology. Clearly if (D) is to be a scientific law rather than a tautology, the relational concept of adaptation cannot be defined in terms of actual reproductive values. That is, we cannot define it as follows:

> *a* is better adapted than *b* in *E* if and only if *a* has more offspring than *b* in *E*.

Most biologists treat "fitness" and "adaptedness" as synonymous, and many define relational fitness in just this way. (See Stern, 1970, p. 47, where he quotes Simpson, Waddington, Lerner, and Mayr[8] to this effect. Stern approves of this definition.) They thus deprive evolutionary theory of its explanatory power.

To avoid turning (D) into a tautology, it seems we must also avoid defining relative adaptedness in terms of probable reproductive values. That is, the following definition also seems to render (D) a tautology:

> *a* is better adapted than *b* in *E* if and only if *a* will probably have more offspring than *b* in *E*.

(See Munson, 1971, p. 211, for a definition of this form; but he substitutes survival for reproductive values.) Actually, things are not as simple as they seem to be. Whether the above definition makes (D) tautological depends on the interpretation of probability being used. More will be said about this, but for the moment we may conclude the obvious: If the relational concept of adaptation is to play its explanatory role in evolutionary theory, it must be defined so that (D) does not become a tautology. We will call this requirement the condition of independence from actual reproductive values.

(b) *Generality.* As stated earlier, we are primarily interested in intraspecific selection; hence for the set of

8. Mayr, it seems, was quoted out of context. See Mayr, (1963), pp. 182–184.

ordered triples $<x, y, z>$ which satisfy "$x$ is better adapted than $y$ in $z$," the first two members of those triples will be members of the same species. In other, less formal, words we are interested in what it is for one alligator to be better adapted than another alligator to their particular environment but not in what it is for one elephant to be better adapted than one swallow to their environment (since they are not in direct reproductive competition with each other; see Ghiselin, 1974). But we do expect one and the same explication or definition of relative adaptedness to apply to ants, birds, and elephants. That is, we want (D) to be a general law that applies to the whole biosphere.

Suppose for some precursors of modern giraffes it was true that one was better adapted than another to their environment if and only if it was taller than the other. (Suppose this only for the sake of this discussion. Even within a given species it is doubtful that any single-dimensional analysis of adaptedness will be adequate.) It won't do to define relative adaptedness in terms of relative height, because even though such a definition may truly apply to some giraffe precursors, it will not apply to most other plants and animals. Such a definition would make (D) a true law of giraffe precursors but make it false or inapplicable to other plants and animals. If (D) is to be a general law, our definition of relative adaptedness must meet what we will call the condition of generality—that is, it must apply to all plants and animals.

(c) *Epistemological Applicability.*

One way of stating this requirement is to say that our definition of relative adaptedness must render (D) testable. However, I prefer to stress another side of what is perhaps the same coin and say that our definition of relative adaptedness should tell us something about how (D) is to be applied to particular cases. I choose this stress because I think testing (D) is a pipe dream, whereas applying it to explain certain phenomena should not be. (Such thoughts are in consonance with Scriven, 1959, and Mayr, 1961.)

One sometimes hears talk of adaptedness as a "close correlation with the environment." We could define relative adaptedness as follows:

> $a$ is better adapted than $b$ in $E$ if and only if $a$ is more closely correlated than $b$ to $E$.

This is a good example of a definition which fails epistemological applicability. Without further information we have no idea how it applies to particular organisms, simply because we have no idea what it means. Consider the following:

> $a$ is better adapted than $b$ in $E$ if and only if God prefers $a$ to $b$ in $E$.

At least for those theistically inclined there is no problem of meaning here. But this definition is clearly useless, since we have no way of knowing which organisms God favors.

The definition discussed above in terms of relative height is a good example of a definition which meets the requirement of epistemological applicability. We know what it is for one organism to be taller than another. Unfortunately this definition lacks generality (or if general, then it is empirically incorrect).

To say a definition is epistemologically applicable does not imply that there is an easy mechanical test for its application. Perhaps a paradigm for an epistemologically applicable definition of relative adaptedness is the definition in terms of actual reproductive values (which explains its popularity). But if we try to apply it to two female Pacific salmon in the sea, we are faced with real difficulties. We would have to try to follow them up river to their spawning ground. And if we managed to do that and if they both managed to make it, we would be faced with the task of counting numerous eggs dispersed in the water. And then we would have to follow each egg's progress to sexual maturity or to death.

But these practical difficulties need not matter. What matters is that theoretically we know what it is for *a* to be better adapted than *b* in *E,* and that for at least some cases we can apply it and so test (D); and in those cases where we cannot test (D), we have a good explanation of why we cannot. Thus by requiring epistemological applicability, I do not mean to require an operational definition; theoretical applicability is enough.

(d) *Empirical Correctness.* I hardly

need to argue that we want our definition of relative adaptedness to be empirically correct, but I do need to say something about what it is for our definition to be empirically correct and how we go about determining its correctness.

There may be many features of organisms, such as strength, beauty, or even longevity, which, we will be disappointed to find out, are not invariably selected. In fact, quite often there is no selection for higher fecundity.[9] The best adapted may not always be the strongest or the most beautiful or even the most prolific. But natural selection, rather than personal or collective taste, must be the ultimate criterion against which we test our explication of adaptedness.[10] If we define natural selection in terms of relative adaptedness (as we will; see below, p. 71) then those selected will by definition be the better adapted. Yet it does not follow that those organisms with higher reproductive values will by definition be better adapted. If it did then (D) would be tautologous. We must allow that some instances of differential reproduction are not instances of natural selection.

If natural selection is to be defined in terms of relative adaptedness, how can we use it to test the empirical correctness of our definition of relative adaptedness? Suppose for a

9. As shown by Lack (1954). This must be quite surprising to those with only a superficial understanding of evolution. For example, Popper (1972, p. 271) thinks it is "one of the countless difficulties of Darwin's theory" that natural selection should do anything other than increase fecundity. The explanation is really quite simple: Increased fecundity often results in a decreased number of offspring surviving in the next generation. See Williams (1966, chapter 6) for discussion.

10. To some unfamiliar with the problem of adaptation, this may not be obvious. Rather than reargue the generally accepted, I refer the reader to Stern (1970), which is a good introduction to the relevant literature.

certain species of organisms we pick out 2 similarity classes of members of this species, A and B. (For our purposes these classes should be formed on the basis of the functional or epigenetical similarity of the genotypes of the members; see Brandon, 1978.) Suppose further that by our definition of relative adaptedness, all members of A are better adapted than any member of B to their mutual environment. Our theory of natural selection, of which (D) is a major component, tells us that in statistically large populations (where chance differences in fitness are canceled out), A's will have a higher average reproductive rate than B's. If repeated observations (either in the lab or in the field) show that A's do, in fact, outreproduce B's, our definition of relative adaptedness fits these facts of natural selection and so is corroborated; if not, it is on its way to being falsified (of course no one observation would falsify it).

It should be clear that any definition that fails to satisfy the condition of independence from actual reproductive values will fail to be testable in the way described above. Yet it is important to note that once we accept some theory of adaptedness—that is, some theory of what it is for an organism to be adapted to its environment, we can criticize a definition failing (a) as empirically incorrect. In fact, as we will see, on any decent theory of adaptedness

any definition failing (a) will also fail (d).[11] We want our definition of relative adaptedness to fit the facts of natural selection. We cannot accept a definition that renders (D) false.

To summarize, our strategy is to construct a definition of relative adaptedness that makes (D) a respectable scientific law (from the received point of view of philosophy of science). Requirement (a) is that (D) cannot be a tautology. Requirement (b) is that (D) must be general—that is, universally applicable throughout the biosphere. Requirement (c) is that (D) not be so vague or so obscure that we have no idea how to apply it to particular cases—or that (D) be testable. And requirement (d) is that (D) must not be false—or more precisely, that (D) must be nontautologously true.

### 3. CURRENT DEFINITIONS AND THE POSSIBILITY OF SATISFYING THE FOUR DESIDERATA

Let us now examine current approaches to the problem of defining relative adaptedness in the light of the four desiderata discussed above. As I said earlier, the simplest approach is perhaps the most popular: *a* is better adapted than *b* in *E* if and only if *a* has more offspring than *b* in *E*. Besides making (D) a tautology and so stripping the concept of its explanatory power, this approach totally ignores the fact that natural

---

11. See below, pp. 70–72. Of course, one might wonder how a definition could fail both (a) and (d), or how a tautology could be empirically incorrect. It can be in just this sense; *given* an adequate theory of adaptedness, we have a notion of adaptedness which differs from any notion failing (a) (that is, any notion which identifies adaptedness with actual reproductive success). These two notions will not be extensionally equivalent. So, from the standpoint of our theory, the definition which fails (a) will also fail (d).

selection is a statistical phenomenon. Differential fitness may be correlated with certain differences in traits, but the correlation is not expected to be perfect. For example, in a certain population of moths, darker-winged individuals may on average produce more offspring than lighter-winged individuals, but this certainly does not imply that for every pair of moths the darker-winged one will have a greater number of offspring than the lighter winged one. Appreciating that natural selection is a statistical rather than a deterministic process has led some theorists to suggest a more sophisticated approach to our problem (see Mayr, 1963, pp. 182–184).

This more sophisticated approach would define relative adaptedness in terms of the statistical probability of reproductive success. How is this probability to be determined? Suppose we separate the members of a population (of moths, for example) into similarity classes formed on the basis of the functional or epigenetical similarity of their genotypes. To fix ideas let us say that we form two such classes and that the members of one are all darker winged than any of the members of the other (this difference being the result of genetic differences between members of the two classes). Further suppose that these classes are epistemically homogeneous with respect to reproduction—that is, no other division of this class of moths that we can make (based on our knowledge) will be statistically more relevant to reproduction, except divisions based tautologically on actual reproduction. We can now determine the probability of reproductive success of any individual as a simple function of the

average reproductive success of the members of the similarity class to which it belongs. And so the reproductive success of the individual is statistically determined by the functional properties of its genotype.

This approach, which we will call the statistical approach, fits some existing paradigms of statistical explanation (see Salmon, 1970), but, as I will show, it fails not only desideratum (a) but also (d). The statistical approach is most closely related to the frequentist interpretation of probability, which identifies the probability of an event with its relative frequency "in the long run". The leading proponents of this interpretation have been Richard von Mises and Hans Reichenbach. In what follows I am only criticizing the application of this interpretation of probability to defining relative adaptedness. This, of course, does not constitute a general criticism of that interpretation. In the next section I will suggest a definition using a rival conception of probability.

Since the statistical approach uses actual reproductive values, its empirical correctness cannot be tested by prediction and observation, it can only be tested against certain general theoretical principles. Consider the following case. Four dogs are on an island—two German shepherds, one of each sex, and two basset hounds, one of each sex. Both bitches go into heat, basset mounts basset and German shepherd mounts German shepherd. While copulating, the shepherds are fatally struck by lightning. The bassets, on the other hand, raise a nice family. Are the bassets therefore better adapted to the island environment than the sheperds? To put the question another way, do we count this differential reproduction as natural selection?

Biologists usually define natural selection simply as differential reproduction (of genes, genotypes, or phenotypes). But this is due to carelessness, not lack of understanding. Most biologists would agree that the above case is not an instance of natural selection but rather a case of chance differences in fitness. (Not that it could not be natural selection, but nothing in the story indicates that it is. We can elaborate the story in ways that make it clear that it is not a case of natural selection. For instance, the only food source for dogs on our island might be animals whose size and ferocity would make it relatively easier for the larger shepherds to eat than the bassets. Furthermore, lightning might be a rare phenomenon and indifferent between bassets and shepherds.) How then shall we characterize natural selection? The concept must be defined in terms of the as yet undefined notion of adaptedness. Natural selection is not just differential reproduction but rather differential reproduction due to the adaptive superiority of those who leave more offspring.

Even without a definition of relative adaptedness, we can be confident that cases like the basset-shepherd case are not instances of natural selection. Given that natural selection is a statistical phenomenon, it should not be surprising that in small populations Darwinian fitness is not always correlated with adaptedness. Yet the statistical approach to defining relative adaptedness cannot recognize this. According to our story the basset-shepherd case is unique; no

such population of dogs has ever been or will ever be on this island or on any sufficiently similar island. Thus our four dogs exhaust the data available for the statistical approach. So according to the statistical approach the bassets are better adapted to the island environment than the shepherds. Yet by ecological analysis, in which we determine what it takes for a dog to survive and reproduce on our island, we conclude that the shepherds are better adapted to the island than the bassets. This conflict raises questions concerning the empirical correctness of the statistical approach.

If the basset-shepherd case were just an ad hoc counterexample dreamed up to refute the statistical approach, perhaps we should ignore it. But statistically small populations are not uncommon in nature, and they are of considerable evolutionary significance (especially for speciation by what Mayr calls the *founder principle;* see Mettler and Gregg, 1969, pp. 130–135; and Mayr, 1963). When applied to small populations, the statistical approach will quite predictably conflict with our best analyses of the organism-environment relation; hence, we are led to conclude that this approach, which renders (D) a tautology, is also empirically incorrect. (It should be clear that defining the relative adaptedness of an individual in terms of *its* actual reproductive success is likewise empirically incorrect.)

Let me criticize the statistical approach in a slightly different way to show the connection between its empirical incorrectness and its explanatory failure. The role in evolutionary theory of the relational concept of adaptation is to explain differential fitness. The question is: Why are those features which happen to be highly correlated with reproductive success in fact highly correlated with reproductive success? The Darwinian answer is: Organisms having these features are (for the most part) better adapted to their environment than their conspecifics lacking them. This higher degree of adaptedness causes the fitter organisms to be fitter and is the explanation of their higher fitness. The idea behind the statistical approach to defining relative adaptedness is that high statistical correlations between certain features and Darwinian fitness will indeed be causal connections and so will explain differential fitness. Yet we have seen that there are conceptually clear-cut types of cases (involving small populations) where the high statistical correlation is not a causal connection (in any interesting sense) and so cannot be used to explain differential fitness. In our basset-shepherd case certain distinctively basset features (such as shortness and color of coat) are perfectly correlated with fitness. Yet our bassets are fitter than our shepherds not because they are shorter or are a certain color, but because the shepherds were in the wrong place at the wrong time. In our case it's not that shepherds are characteristically in the wrong place at the wrong time but just that they happened to be once. Because the population is small, once is enough, and an essentially random process has radically altered our island population of dogs. Here differential fitness is explained (some might worry over how this is an explanation—I can't concern myself with that here) in terms of a chance process and small

population size. Thus if evolutionists are to explain what they want to explain, if they are to have the sort of explanatory theory they want, some other approach to the problem of adaptation is needed.

Early in this paper we were led to distinguish adaptedness from Darwinian fitness. As we have seen from the basset-shepherd example, in small populations the two do not always coincide. Are there other types of cases where the two do not coincide? I can think of only three candidates for such cases: cases of artificial selection, cases of domestication, such as in modern man, where selection seems to have been relaxed, and cases of sexual selection. But none of these types are ones where the correlation of fitness and adaptedness should not be expected, and it is important to see why this is so. I will focus my attention on artificial selection; what is said about it can easily be applied to the other two types of cases by analogy.

Artificial selection quite often results in organisms which could not survive in their 'natural' habitat. Organisms which under "natural" conditions would be the fittest are prevented from breeding, while other organisms, less fit under "natural" conditions, are allowed to breed. By such a process we end with chickens without feathers, dogs so small they can fit in your hand and fruit flies with legs where they should have antennae. Such cases, it could be claimed, are clear cases where Darwinian fitness does not coincide with adaptedness. But how could one argue for this claim?

Suppose we are following the relative frequency of a segregating genetic entity, say a chromosome inversion in a population of fruit flies. We divide this population into two genetically identical subpopulations, leave one subpopulation in its original habitat, and move the other to some new and different habitat. After a few generations we observe that the frequency of this chromosome inversion has changed in the moved population (while remaining the same in the control population). Are we to conclude that this change in frequency is the result of some divergence between fitness and adaptedness, since some flies which would have been less fit in the original evironment have had a higher relative fitness in the new environment? Obviously not. Whatever adaptedness is, it has something to do with the organism-environment relation. With a change in environment a change in relative adaptedness is not unexpected. Man is often thought of as the zenith of evolution, yet he can hardly get by in his fishy ancestors' environment.

Artifical selection is just a human-induced change in environment. I presume that it is true that a fly with leglike antennae would not be as well adapted to his ancestral home as many of his more normal relatives. But is he not much better adapted than his normal relatives to the laboratory where the experimenter is selecting for an extra set of legs? In this environment he is much better able to survive and reproduce than his more normal colleagues. The flies are living and breeding in the laboratory; what their relative adaptedness in the wild would be is irrelevant to an assessment of their relative adaptedness in the lab.

To argue that in cases of artificial selection fitness and adaptedness do not coincide is clearly to ignore the environment in which the selection is taking place; in particular it is to ignore the experimenter's or breeder's part in this environment. But that is no more justified than ignoring the part of predators in the prey's environment and is a bit of anthropocentrism. To objective biologists, experimenters and breeders are no different from those English birds who for hundreds of years have steadfastly selected against (*i.e.* eaten) moths not cryptically colored.

Thus artificial selection is just a type of natural selection. This point will have a crucial role to play in an argument later in this paper, so I should make it clear that it is not a quibble over words. How would we reply to one who says that by "natural selection" he means all cases of selection excluding those involving man? To this we should reply that the concept he has defined is not as useful for theoretical purposes as the more inclusive concept we have defined. He can try to use words however he wants, but he can't justify an anthropocentric point of view toward the concept of adaptedness.

We have seen that the simplest approach to defining relative adaptedness, which does so in terms of actual reproductive values, and the more sophisticated statistical approach, fail both desiderata (a) and (d). This failure, especially the failure to meet (a), is fairly apparent and is presumably due to the neglect of theorists to formulate desiderata concerning the concept of relative adaptation. However, there is the novel approach by Walter Bock and Gerd von Wahlert (1965) which might be taken as an attempt to meet (a)–(d); at least it does not obviously fail them.

Bock and von Wahlert argue that a measure of adaptedness should be expressed in terms of energy requirements. First, they point out that the energy available to an organism at any given time (from both internal and external sources) is limited and that there is interindividual variation in the amount of energy available to organisms (as well as intraindividual variation over the lifespan of an individual). Next, they point out that for an organism to maintain the proper relation to its environment (that is, to stay alive), it must expend energy. The amount of energy expended will vary, depending, for example, on whether the organism is resting or escaping predation. Since an organism must expend energy to live and reproduce and since its available energy is limited, it is advantageous, they argue, for the organism to minimize the amount of energy required to maintain successfully its ecological niche (p. 287). Thus the following definition is suggested by their work:

> *a* is better adapted than *b* in *E* if and only if *a* requires less energy to maintain successfully its niche in *E* than does *b*.

There are a number of problems with this definition. First, we must ask whether it really meets requirement (a). Stern (1970, p. 48) suggests that it does not. He asks what it means to maintain a niche *successfully*. He quotes Bock and von Wahlert as follows: "The relative factor of survival or the relative number of progeny left which is usual when comparing the adaptedness of individuals is accounted for by the relative nature of the term 'successful'" (Bock and von Wahlert, p. 287). This, according to Stern, "is tantamount to admitting that their criterion is really subservient to reproduction, and that success in adaptation is still to be measured by more conventional means. That a niche will be maintained more successfully if less energy is required is clearly only an unsupported conclusion, not a matter of definition" (Stern, p. 48). But here Stern misses the point. Bock and von Wahlert clearly assert that "unsupported conclusion." They say, "The less energy used, the more successfully . . . the niche will be maintained" (p. 287). If they are right, differences in fitness can and will be explained in terms of differences in energy requirements. It remains for us to ask whether they are right.

We may not be able to answer this question. Although their definition of relative adaptedness *seems* to be applicable—that is, it seems to satisfy desideratum (c)—it may not be. We can turn to Bock and von Wahlert for suggestions on how their definition is to be applied to particular cases. Unfortunately, they do not discuss intraspecific comparisons; but from their discussion of comparing the energy requirements of sparrows versus woodpeckers for clinging to vertical surfaces, we can reconstruct how they would make such a comparison (see Bock and von Wahlert, pp. 287 ff.). They would determine the amount of energy expended in clinging to a vertical surface by measuring the amount of oxygen consumed. Thus for two woodpeckers they would determine which is better adapted to clinging to vertical surfaces by measuring their oxygen consumption while clinging to some surface. One would be better adapted than the other if it used less oxygen than the other. Recall that we want to explain differential reproductive success. One could test the hypothesis that if one woodpecker requires less energy than another to cling to a vertical surface, it (probably) will have more offspring than the other. But it is not likely to be true. Even for woodpeckers there is more to life than hanging on trees. What seems to be needed is a determination of all the activities necessary for survival and reproduction in a particular environment. We would then compare the relative adaptedness of two organisms by comparing their energy requirements for these activities. But would not these activities have to be weighted according to their importance? How would they be weighted? And isn't it possible, and doesn't it happen frequently, that one organism can bypass some "necessary activity" because of some difference from his conspecifics in morphology, physiology, or behavior? These questions lead me to believe that the Bock and von Wahlert definition is, in fact, not epistemologically applicable—that is, it fails (c)—but I will not pursue this further. Rather, let us grant for the sake of argument that it is applicable and ask whether it is empirically correct.

I have already outlined how to test the empirical correctness of a definition of relative adaptedness (see above, pp. 70-71). In brief, we take paradigmatic cases of natural selection and see if the definition fits the case. In the well known case of melanism in English moths, we would check to see if darker-winged moths required on average less energy than lighter-winged moths. I have raised doubts whether the Bock and von Wahlert definition is so testable, and since I can't overcome the problems raised for its testability, I can't subject it to this case-study type of test. But if it is testable (or epistemologically applicable), I will argue that it can be shown to be empirically incorrect.

Suppose we have in our laboratory a population of genetically diverse individuals whose diversity is phenotypically expressed in an easily recognizable manner. By Bock and von Wahlert's definition, some variants are better adapted than others. I, as a perverse Popperian, prevent the so-called "better adapted" from breeding while allowing the so-called "less well adapted" to breed. I do this in a large population over a number of generations. Since artificial selection is just a type of natural selection, we have here a case of natural selection which does not fit Bock and von Wahlert's definition. If more falsifying cases are wanted, we can produce them. And so, it seems, if Bock and von Wahlert's definition is epistemologically applicable, it is not empirically correct. Clearly this argument applies not only to the Bock and von Wahlert definition but to all definitions which meet desiderata (a)–(c).

This argument is not conclusive. When we begin to select for the so-called "less well adapted," we change the environment of the organisms. It is open for the theorist whose definition we are criticizing to claim that our change of environment has reversed his estimations of adaptedness,

adaptedness being environment relative. This doesn't deter us; again we try to refute the implications of the definition. But what if our most perverse efforts fail to contradict the proposed definition? Here I think we must conclude that empirical correctness has been purchased at the price of epistemological applicability. (Consider how one would try to defend the Bock and von Wahlert definition against such countercases.) That is, the definition has become so vague and malleable as to make (D) unfalsifiable. My claim is that for any proposed definition of relative adaptedness satisfying desiderata (a)–(c), I can produce cases showing that it fails (d)–that is, is empirically incorrect–and that to resist falsification by artificial selection is to give up (c) (that is, to cease being epistemologically applicable or testable). To exhaustively prove this would be to take every possible definition of relative adaptedness and produce the relevant countercases. It is not surprising that I can't do this. But I do hope my argument is convincing.

I'm sure some will feel that this argument from artificial selection is a cheap victory. If we could find a definition of relative adaptedness that truly applied to all organisms in "natural" environments, wouldn't we be justified in ignoring counterexamples produced by artificial selection? That is a difficult theoretical question, but we can say this: such a definition would represent a tremendous advance in our knowledge of ecology and would be welcomed. But artificial selection is as much a natural phenomenon as predation, starvation, mate selection, etc. The argument from artificial selection should, if

nothing else, decrease the plausibility of the possibility of such a definition. Naturalists are well aware that natural selection is an opportunistic process, often leading to evolutionary dead ends and extinction. Are not some "natural" cases of selection just as bizarre as our concocted cases?

The point emphasized in the argument from artificial selection is this: The environments in which organisms find themselves competing are radically different from each other, and at least practically speaking there is no way to specify all possible environments. Thus there is conflict between desiderata (c) and (d). To make (D) testable is to expose (D) to falsification from some radically new ecological situation. And to protect (D) from such falsification is to make it so general that it ceases to be applicable. This point should be accepted even by those who fail to subsume artificial selection under natural selection. Since there are good reasons to doubt that any definition of relative adaptedness will satisfy (a)–(d), the question should be: is there any reason to suppose such a definition possible? I've found none.

## 4. A SUGGESTED DEFINITION

The attempt has been to construct a definition of relative adaptedness that renders (D) an explanatory law. Accepting the received view of philosophy of science, I pointed out that for (D) to be an explanatory law it must be nontautologous, general, testable, and true. I argued that for (D) to be such the definition of relative adaptedness must satisfy desiderata (a)–(d). Finally I showed that no definition of relative adaptedness can satisfy (a)–(d). In the light of these conclusions, I will now suggest what I take to be the best possible definition of relative adaptedness.

Recall our desiderata. Apparently we will have to give up at least one of them. We should retain (a) and (d); tautologies and false statements explain little (one should note that giving up (a) would also entail giving up (d)). As we will see, there is a trade-off between desiderata (b) and (c), and my suggested definition will, in a sense, preserve both.

First I will suggest a nontechnical definition of relative adaptedness and then a more technical version. The nontechnical version follows:

> (RA) *a* is better adapted than *b* in *E* if and only if *a* is better able to survive and reproduce in *E* than is *b*.

This definition avoids tautology—that is, it is independent of actual reproductive values. (We can confidently assert that a particular Mercedes-Benz 450 SEL is *able* to do 150 mph while knowing that it never has and never will go that fast.) It is also a general definition and it is empirically correct (insofar as this makes sense; at least it is not empirically incorrect). But how are we to apply it to particular situations? I think it is clear that as it stands, (RA) is not epistemologically applicable. So this suggested definition has the effect of preserving (a), (b), and (d) at the expense of (c), and given that we cannot have all four, (RA)'s obvious failure of (c) is a virtue. It is an unpretentious definition; it wears its epistemological inapplicability on its sleeve.

We can construct a more technical (and more pretentious) definition.

Earlier I criticized what I called the statistical approach to defining relative adaptedness. This approach identified adaptedness with the statistical mean of observed reproductive rates. As pointed out then, it is not too distorting to call the interpretation of probability used in this approach the frequentist interpretation. There are other interpretations of probability. Some, for instance the logical and subjective interpretations (associated with Carnap and de Finetti respectively), are here irrelevant. But the approach (best expounded by Hacking, 1965; also see Popper, 1959) according to which probabilities are deduced from theory rather than identified with observed frequencies is relevant.

In discussing the basset-shepherd case, I said that observed reproductive rates can conflict with estimations of adaptedness based on ecological analysis. Suppose our ecological theories to be so well developed that for any given environment and organism we could deduce the distribution of probabilities of the number of offspring left by that organism (in the next generation). That is, from our theories we deduce for each organism *O* and environment *E* a range of possible numbers of sufficiently similar offspring, $Q_1{}^{OE}, Q_2{}^{OE}, \ldots, Q_n{}^{OE}$ and for each $Q_i{}^{OE}$ our theory associates a number $P(Q_i{}^{OE})$ which is the probability (or chance or propensity) of *O* leaving $Q_i$ sufficiently similar offspring in *E*. Given all this, we define the adaptedness *O* in *E* (symbolized as $A(O, E)$ as follows:

$$A(O,E) = \Sigma P(Q_i{}^{OE})Q_i{}^{OE}$$

That is, the adaptedness of *O* in *E*

equals the expected value of its genetic contribution to the next generation. (The units of value are arbitrary. All that matters here are the ordinal relations among the numbers associated with each pair $<O,E>$. Outside of this context the numbers have no significance.) Our new more exacting definition of relative adaptedness, (RA'), is as follows:[12]

> (RA') $a$ is better adapted than $b$ in $E$ if and only if $A(a,E)>A(b,E)$.

Two things should be clear: First, (RA') only makes sense for intraspecific intra-environmental comparisons. Second, (RA') is a step in the right direction only on the proper interpretation of probability.

Before evaluating (RA') I should say something about its basic presupposition: that from detailed ecological analysis we can give good estimates of the probabilities of reproductive success of organisms in environments independent of observations of their actual reproductive success. For example, given the characteristics of a certain island environment and the particular characteristics of some basset hounds and German shepherds, such theories should be able to predict the relative reproductive success of each even without any relevant statistics. Clearly such predictions are falsifiable (as falsifiable as any statistical hypothesis), but do we have any reason to expect them to be success-

ful? There are few, if any, outstanding examples of such success in the corpus of biological science. On the other hand, there seems to be no theoretical obstacle to successful predictions of this sort.

The informal definition of relative adaptedness suggested above, (RA), satisfied desiderata (a), (b), and (d) but not (c). How does (RA') fare on our desiderata? Given the proper interpretation of probability, it satisfies (a). According to this interpretation the probability of reproductive success (or expected genetic contribution to future generations) is some biological property of the organism and its environment (just as the probability of heads for a coin is a physical property of the coin and the tossing device). The organism in its environment has this property even if it is struck by lightning prior to leaving any offspring (just as the chance of heads may be one half for a coin even if it is unique and is melted before it is ever tossed). Thus (RA') is independent of actual reproductive values. The occurence of "probably" in (D) may be confusing, but (RA') does not turn (D) into a tautology.[13] (RA') clearly satisfies (b); that is, it is general. Like (RA), (RA') is not empirically incorrect and so we will say it satisfies (d)—that is, it is empirically correct. Although the failure of (RA') to satisfy (c) may not be be as apparent as the failure of (RA), it also fails to be epistemologically applicable. If there were a single all-encom-

---

12. The move to this sort of definition was suggested to me by Hilary Putnam.

13. (D) becomes something like an instance of what Hacking calls the Law of Likelihood and is analogous to the following: If the chance of heads for coin $a$ is one half and the chance of heads for $b$ is one quarter, then (probably) when both coins are tossed a small number of times, $a$ will land on heads more that $b$ will.

passing theory of adaptedness from which we could derive the adaptedness (as defined above) of any organism in any environment, then (RA) would be epistemologically applicable. But as I've argued, no such theory is possible. (I presented Bock and von Wahlert's theory as an attempt at such completeness.)

How is the suggested definition useful? It is useful as what we might call a schematic definition. It is neither applicable nor testable but particular instances of it are. What do I mean by an instance of (RA')? Formally, in an instance of (RA') we fix the value of the environment parameter $E$ and limit the range of the individual variables 'a' and 'b' to a particular population of organisms living in $E$. Such an instantiation would represent a hypothesis concerning what it takes for certain types of organisms to survive and reproduce in a certain type of environment. Good hypotheses of this kind can result only from detailed ecological analysis. (Where "ecological" is used in a broad, perhaps too broad, sense, I would include in such analysis the study of the sorts of genetic variation that occur and are likely to occur in the relevant organisms and the study of the phenotypic effects of this variation.)

For a simplified example suppose that the only variation in a certain population of moths is in wing color. These moths all rest on dark-colored tree trunks during the day. Birds prey on the moths by sight in daytime. We analyze this simplified solution as follows: The darker the wing color, the closer it is to the color of the tree trunks. Moths whose wings are col-

ored most like the tree trunks are least likely to be eaten by birds. Moths less likely to be eaten are more likely to leave offspring. Thus we instantiate (RA') as follows:

> Moth $a$ is better adapted than $b$ in (our specified) $E$ if and only if $a$'s wings are darker colored than $b$'s (in $E$).

(I am primarily interested in illustrating certain logical points, but I don't want to appear to take an overly naive and sanguine view toward the sort of ecological analysis necessary for complex organisms in complex environments. Lewontin (1977) discusses some of the problems involved. Suffice it to say that although successful ecological analysis is difficult, it does not seem impossible.)

With a schematic definition of relative adaptedness, (D) becomes a schematic law, and with an instantiation of (RA') we get an instantiation of (D). For our moths (D) says:

> If $a$ is darker winged than $b$ (in $E$), then (probably) $a$ will have more offspring than $b$ (in $E$).

Such an instantiation of (D) is clearly testable (in fact it has been tested; see Kettlewell, 1955 and 1956). Moreover it does what we want it to do: it explains differential reproduction and so explains evolution by natural selection (as in this instance we explain the evolution of industrial melanism in certain species of English moths).

To summarize: I have suggested that we give up epistemological applicability and adopt a schematic definition of relative adaptedness,

(RA'). This correlatively makes (D) schematic and so not testable. When we instantiate (RA'), we give up generality for applicability. Likewise, instances of (D) become testable and explanatory but not general.

## 5. THE STRUCTURE OF EVOLUTIONARY THEORY

(D) is the fundamental law of evolutionary theory. What sort of foundation is (D) for a scientific theory? Critics have often maintained that evolutionary theory rests on a tautology. As I hope I have made clear, (D) is not a tautology. But I have shown that no definition of relative adaptedness can render (D) nontautological, general, testable, and true. (D) as a schematic law is not testable; instantiations of (D) are not general. This may not be so bad. If disconfirming an instantiation of (D) disconfirms (D), then (D) may be a respectable law. But this relation between (D) and its instances does not hold. That is, no amount of falsification of instances of (D) even begins to falsify (or disconfirm) (D).

Consider the instantiation of (D) concerning moths. If through experiments and observations it proved to be false, our response would be and should be that we have incorrectly analyzed the ecological situation. Perhaps the birds prey on these moths by using heat-sensing devices, thus making color variation irrelevant (unless that variation is correlated with variation in heat irradiation). We reanalzye the situation and test our new hypothesis. If the falsification of one instance of (D) doesn't begin to cast doubt on (D), will large numbers of falsifications change matters? If, as

is the case, some instances of (D) have proved successful, even large numbers of falsifications of instances of (D) will not cast doubt on (D). If no instance of (D) ever succeeded, we would doubt the usefulness of (D), but even this would not lead us to say (D) is false. In our world (where some instances of (D) have successfully explained and predicted certain phenomena), no set of test results could falsify (D). Thus (D) is unfalsifiable.

With this in mind, and given that through informative instantiations of (RA') we get testable and explanatory instances of (D), one might question the status of the schematic (RA') and (D). Neither meets our philosophical expectations, so why should they be granted any status in our expurgated science? To answer this question we must consider some of the aims of scientific inquiry and some of the criteria by which theories are judged. Perhaps the distinguishing feature between science and myth is that science, unlike myth, aims at testable explanations. So theories and laws are judged according to their (in-principle-) testability. Instantiations of (D) fare well on this criterion; (D) itself does not. But scientific inquiry also aims at the systematic unification of broad bodies of diverse phenomena. Without (D) there is no theory of evolution, there are only low-level theories about the evolution of certain organisms in certain environments. (And at present there are very few of those.) With (D), Darwinian theory is possible.

I have not simply presented a case where philosophy of science is at variance with actual science. Rather, I have presented a case where two philosophical principles conflict. There is, as I have shown, a trade-off between desiderata (b) and (c), and so a conflict between testability and systematic unification. I have suggested adopting (RA') and so treating (D) as a schematic law as the best possible solution to this dilemma.

## 6. SUMMARY

The conception of adaptation has been one of the most troublesome and yet one of the most important concepts in the biological sciences. I hope that this paper has cleared up much of that trouble. We have constructed an adequate definition of relative adaptedness. Our analysis of the conception of relative adapted-ness went hand in hand, as it had to, with an analysis of the structure of evolutionary theory. We found that Darwinian evolutionary theory has as its foundation what I called a schematic law; thus its structure does not fit any existing philosophical paradigms for scientific theories. Heretofore, schematic definitions and schematic laws have not been recognized or investigated by philosophers of science.

In constructing a definition of relative adaptedness, we posited the biological property of adaptedness. In this paper I said much about what this property is and what it is not. But its particular ontological status has not been discussed and remains somewhat mysterious.

## REFERENCES

Ayala, F. J., 1974, Biological evolution: natural selection or random walk? *Am. Scient.*, 62: 692–701.

Brandon, R. N., 1978, Evolution. *Philosophy of Science*, 45: 96–109.

Bock, W. J. and von Wahlert, G., 1965, Adaptation and the form-function complex. *Evolution*, 19: 269–299.

Darwin, C., 1859, *On the Origin of the Species*, London, John Murray.

Dobzhansky, T., Adaptedness and fitness. *Population Biology and Evolution*. R. C. Lewontin ed., Syracuse, Syracuse University Press, pp. 109–121.

Ghiselin, M. T., 1974, A radical solution to the species problem. *Systematic Zoology*, 23: 536–544.

Hacking, I., 1965, *Logic of Statistical Inference*, Cambridge, Cambridge University Press.

Hempel, C. G., 1966, *Philosophy of Natural Science*, Englewood Cliffs, Prentice-Hall.

Hull, D. L., 1973, *Darwin and his Critics*, Harvard University Press.

Kettlewell, H. B. D., 1955, Selection experiments on industrial melanism in the lepidoptera. *Heredity*, 9: 323–342.

——1956, Further selection experiments on industrial melanism in the lepidoptera. *Heredity*, 10: 287–301.

King, J. L. and Jukes, T. H., 1969, Non-Darwinian evolution. *Science*, 164: 788–798.

Lack, D., 1954, The evolution of reproductive rates. *Evolution as a Process*. J. S. Huxley, A. C. Hardy and E. B. Ford eds., London, Allen & Unwin, pp. 143–156.

Lewontin, R. C., 1957, The adaptations of populations to varying environments. *Symposium of Quantitative Biology*, 22: 395–408.

——1968, The concept of evolution. *International Encyclopedia of the Social Sciences*, New York, Macmillan, pp. 202–210.

——1970, The units of selection. *Annual Review of Systematics and Ecology*, 1: 1–18.

——1977. Adattamento. *Encyclopedia Eiaudi*, Torino, Italy.

Malthus, T. R., 1798, *An Essay on the Principle of Population,* London.

Mayr, E., 1961, Cause and effect in biology. *Science,* 134: 1501–1506.

——1963, *Animal Species and Evolution,* Cambridge, Harvard University Press.

——1977, Darwin and natural selection. *Am. Scient.,* 65: 321–327.

Mettler, L. E. and Gregg, T. G., 1969, *Population Genetics and Evolution,* Englewood Cliffs, Prentice-Hall.

Munson, R., 1971, Biological adaptation. *Philosophy of Science,* 38: 200–215.

Popper, K. R., 1959, The propensity interpretation of probability. *Br. J. Phil.,* 10: 25–42.

——1972, *Objective Knowledge,* Oxford, Oxford University Press.

Ruse, M., 1971, Natural selection in *The Origin of Species. Stud. Hist. Phil. Sci.,* 1: 311–351.

——1975, Charles Darwin's theory of evolution: an analysis. *J. Hist. Biol.,* 8: 219–241.

Salmon, W., 1970, Statistical explanation. *The Nature and Function of Scientific Theories* (Pittsburgh Studies in the Philosophy of Science, IV), R. G. Colodny, ed., pp. 173–321.

Scriven, M., 1959, Explanation and prediction in evolutionary theory. *Science,* 130: 477–482.

Slobodkin, L. B., 1968, Toward a predictive theory of evolution. *Population Biology and Evolution.* Syracuse, Syracuse University Press, pp. 187–205.

Smart, J. J. C., 1963, *Philosophy and Scientific Realism,* London, Routledge and Kegan Paul.

Smith, J. Maynard, 1975, *The Theory of Evolution* 3rd edn. Middlesex, Penguin Books Ltd.

Stern, J. T., 1970, The meaning of "adaptation" and its relation to the phenomenon of natural selection. *Evolutionary Biology,* 4: 39–66.

Thoday, J. M., 1953, Components of fitness. *Symposium of the Society for Experimental Biology,* 7: 96–113.

——1958, Natural selection and biological process. *A Century of Darwin,* S. A. Barnett ed., London, Heinemann, pp. 313–333.

Williams, G. C., 1966, *Adaptation and Natural Selection,* Princeton, Princeton University Press.

Williams, M. B., 1970, Deducing the consequences of evolution. *Journal of Theoretical Biology,* 29: 343–385.

——1973, Falsifiable predictions of evolutionary theory. *Philosophy of Science,* 40: 518–537.

# 7

# The Logical Status
# of Natural Selection and
# Other Evolutionary Controversies

∿∿∿∿∿∿∿∿∿∿∿∿∿∿∿∿∿∿∿∿

## MARY WILLIAMS

*This paper shows that the theory of natural selection is not tautological and that the belief that it is stems from formulations of the theory which attempt to defy logical impossibility by defining all the words used in the theory. The correct method is to treat some of them as primitives characterized by axioms. The axiomatization used to resolve this controversy also resolves controversies about the units of evolution and the target of selection. The concatenation of the axiomatization, revealing the fundamental structure of the theory, and the controversies arising in the preaxiomatic theory gives insight into the structure underlying controversies about other preaxiomatic theories.*

THE PHRASE "SURVIVAL" OF THE fittest" is a source of nagging embarrassment to evolutionists. On the one hand, it is clearly ambiguous, purportedly tautological, and (so far) impossible to translate into a statement that is nontautological and nonambiguous, and also captures the essence of natural selection; but, on the other hand, it is so evocative of the essence of the principle of natural selection that even those who believe that it is tautological and therefore completely meaningless find themselves using it, albeit with many *caveats,* when teaching novices. The arguments about this phrase and the many suggested replacements are reflected in Lerner's comment: "with all of the knowledge now acquired on the process of selection in nature, its logical status in evolution is still uncertain and undoubtedly controversial" (12, p. 180). This controversy is settled by axiomatization of the theory of natural selection (given in 21) but it will not disappear until the fallacy underlying the belief that it is tautologous is exposed. The purpose of this paper is to expose that fallacy and to provide a nonambiguous nontautological translation of "survival of the fittest."

To give an understanding of the term that is the referent of this

1. This work was supported by NSF Grant GU 1590.

phrase, it is necessary (and, with the aid of the axiomatization, easy) to resolve the controversies about the units of evolution and the primary target of selection. The paper therefore resolves these controversies and exposes the fallacy which underlies them.

## 1. DESCRIPTION OF THE AXIOMATIZATION

The axiomatization presented in (21) is a naive axiomatization, closer in style to the axiomatizations of Euclid and Newton than to the formal axiomatizations of the Russell-Whitehead school. In this section I will present the main points of the axiomatization even more informally; because definition of the technical terms used in the later axioms would require an inordinate amount of space, the later axioms will be stated in informal "translations" which do not render their full meaning. These translations should be sufficient, however, to provide the necessary insights into the phrase "survival of the fittest."

There are two sets of axioms. The first set delineates properties of the set $B$ of reproducing organisms on which natural selection works. The primitive terms introduced in this first set are *biological entity* and $\succ$. $\succ$ is an asymmetric, irreflexive, and nontransitive relation between biological entities and should be read "is a parent of." Some possible interpretations for biological entity are: organism, gene, chromosome, and population.

*Axiom B1:* For any $b_1$ in $B$, $\sim (b_1 \succ b_1)$. (I.e., $\succ$ is irreflexive.)

The following defines another relation, $\rhd$, or "is an ancestor of."

*Definition B1:* $b_1 \rhd b_2$ if and only if $b_1 \succ b_2$ or there exists a finite nonempty set of biological entities, $b_3$, $b_4, \ldots b_k$, such that $b_1 \succ b_3 \succ b_4 \succ \ldots \succ b_k \succ b_2$.

*Axiom B2:* For any $b_1$ and $b_2$ in $B$, if $b_1 \rhd b_2$ then $\sim (b_2 \rhd b_1)$. (i.e., $\rhd$ is asymmetric.)

A set $B$ with a relation $\succ$ satisfying these axioms[1] will be called a *biocosm*. From these axioms it can be proved that $\rhd$ is irreflexive and transitive and that $\succ$ is asymmetric. $\succ$ is not transitive, although in special cases $b_1 \succ b_2$, $b_2 \succ b_3$ may all be true; e.g., let $b_1$ be Jocasta, $b_2$ be Oedipus, and $b_3$ be Antigone.

The primary purpose for stating these axioms is to enable the concepts of *clan* and *subclan* to be rigorously defined. A clan is a temporally extended (over many generations) set of related organisms. (In this description I will, for clarity, usually use the "organism" interpretation of "biological entity.")

*Translation of Definition B7:* The clan of a set $S$ is the set containing $S$ and all of its descendants.

A crucial biological phenomenon occurs when a part of a clan becomes isolated from the rest of the clan (perhaps by an earthquake) and is subjected for many generations to different selective pressures until ultimately the organisms in one part

---

1. Axiom B1 is really a theorem, since it can be proved from Definition B1 and Axiom B2.

of the clan are so different from their contemporaries in the other part that they would be described as different species; this is a situation of great interest for evolutionary theory, so it is clearly important to have a word for some particular kinds of parts of a clan. The first one I introduce is a *subclan,* which, roughly speaking, is either a whole clan or a clan with one or several branches removed; a subclan is an organized chunk of a clan, not just a set of isolated descendants.

*Translation of Definition B9:* $C_1$ is a subclan of $C$ if and only if $C_1$ is contained in $C$ and for every organism $b_1$ in $C_1$ there exists a lineage (or line of descent) from an organism in the first generation of $C$ to $b_1$ such that every biological entity in that lineage is in $C_1$.

*Translation of Definition B10:* $C_1$ is the subclan of $C$ derived from $C_1(j)$ if and only if $C_1$ contains all of the descendants of $C_1(j)$ and all the ancestors of $C_1(j)$ which are in $C$.

Now I can introduce the second set of axioms. This set uses two new primitive terms, *Darwinian subclan* and *fitness.* The following quotation from (21) gives an intuitive introduction to the meaning of the concept of Darwinian *subclan:*

Using an imperfect but intuitively useful analogy, we may think of selection as a force which is pushing a subclan in a certain direction (e.g. toward longer necks, or toward an optimum proportion of different forms in a polymorphic population). The direction and strength of this force is always an average over the differing forces on different individuals of the sub-clan, but, as long as cohesive forces (e.g., interbreeding, same environment) hold the subclan together, it is meaningful to speak of selection pushing the subclan as a whole in some direction. If, on the other hand, the subclan consists of two parts which are isolated from each other and subject to different selective forces (e.g. one favoring longer necks and the other favoring shorter necks), it is not very meaningful to speak of selection pushing this subclan in one direction; it is pushing it in two opposite directions and the average of the two directions would be a mathematical abstraction which would only serve to conceal the interesting biological phenomenon. Clearly I need a term to denote a subclan which is held together by cohesive forces so that it acts as a unit with respect to selection; I shall call such a subclan a *Darwinian subclan.*

Darwinian subclan is a primitive term but, unlike the other primitive terms, it does not correspond closely to any familiar intuitive concept. A Darwinian subclan may be an entire clan from which no descendants have been separated; it may be an interbreeding population; it may be an entire species. It cannot be a species which is splitting into two species; such a species will consist of two Darwinian subclans.

Although a Darwinian subclan is a set of biological entities, it cannot be defined solely in terms of properties definable within the biocosm; it is determined both by properties of the parent relation among biological entities and by properties of the relationship between a set of biological entities and their environment.

Fitness is introduced as a real valued function on the set of biological entities; $\varphi(b_1)$ is a real number expressing the fitness of the biological entity $b_1$ in the environment in which it lives. The following quotation from (21) is an intuitive introduction to the concept of fitness:

> ... fitness is a measure of the quality of the relationship between an organism and its environment, where the environment of an organism is the set of all external factors which have influenced it during its life ... This relationship is determined by such factors as fertility, ability to get food, ability to avoid dangers, etc.

Fitter organisms have a better *chance* of surviving long enough to leave descendants, but a fitter organism does not necessarily leave more descendants than its less fit brother.

*Translation of Axiom D1:* Every Darwinian subclan is a subclan of a clan in some biocosm.

*Translation of Axiom D2:* There is an upper limit to the number of organisms in any generation of a Darwinian subclan.

(This limit is important in inducing the "severe struggle for life" which Darwin noted.)

*Translation of Axiom D3:* For each organism, $b_1$, there is a positive real number, $\varphi(b_1)$, which describes its fitness in its environment.

Before giving the next axiom we must introduce another term denoting a particular type of subclan. Suppose that at a particular point in time the Darwinian subclan $D$ contains a subset of organisms with a hereditary trait that gives them a selective advantage over their contemporaries, and suppose that this trait continues to give a selective advantage for many generations. Then members of this subset will, on average, have more offspring than their contemporaries, and if we consider the sub-subclan derived from this set of organisms (call it $D_1$), then in the offspring generation the proportion of the members of $D$ which are in $D_1$ will be larger than it was in the parent generation. Similarly, since this trait continues to give a selective advantage, the proportion of $D_1$ will continue to increase in subsequent generations. It is by this increase in the proportion of organisms with particular traits that characteristics of populations (and species) are changed over time. It is the increase of the fitter sub-subclan which causes descent with adaptive modification. This sub-subclan is the referent of 'survival of the fittest'. It will be called a *subcland.*[2]

*Translation of Definition D2:* Let $D_1(j)$ be a set of biological entities in the $j$th generation of the Darwinian subclan $D$. Then $D_1$ is the subcland derived from $D_1(j)$ if and only if $D_1 = D \cap C_1$, where $C_1$ is the subclan derived from $D_1(j)$.

*Translation of Axiom D4:* Consider a subcland $D_1$ of $D$. If $D_1$ is superior in fitness to the rest of $D$ for sufficiently many generations (where how many is 'sufficiently many' is determined by how much superior $D_1$ is

---

2. *Biocosm, biological entity,* ⧐, ▷, *subclan, Darwinian subclan,* and *subcland* are all terms that I have coined.

and how large $D_1$ is), then the proportion of $D_1$ in $D$ will increase during these generations.

The final axiom asserts the existence of sufficiently hereditary fitness differences.

*Translation of Axiom D5:* In every generation $m$ of a Darwinian subclan $D$ which is not on the verge of extinction, there is a subcland $D_1$ such that: $D_1$ is superior to the rest of $D$ for long enough to ensure that $D_1$ will increase relative to $D$; and as long as $D$ contains biological entities that are not in $D_1$, $D_1$ retains sufficient superiority to ensure further increases relative to $D$.

The full formal translation of the law of the survival of the fittest is contained in these axioms, and in particular in Axiom D4. This axiom could be called the 'survival of the fittest' axiom; but an even better informal descriptive phrase would be the 'expansion of the fitter subcland' axiom.

## 2. FITNESS

Let us first examine the attempts to translate "fittest." The fundamental cause of the problems associated with these attempts is the acceptance, by essentially all biologists, of the metaphysical doctrine that all words used in a scientific theory should be defined. It is these problems which have been central in creating the belief that the theory of natural selection is circular and is, therefore, a linguistic epiphenomenon which can be shown by close examination to have no real

meaning. (In this controversy the theory is usually accused of being tautological, but "tautological" is always used in these accusations in the sense of "circular and therefore vacuous." Since there are other meanings of "tautological" which are scientifically respectable, I shall use the term "circular.") In this paper I am concerned not so much with proving that the theory is not circular as with exposing the fallacious reasoning which led to the accusations of circularity.[3]

The best introduction to the dilemma is to look through the eyes of Ernst Mayr (15, p. 182), one of the most powerful opponents of the idea that Darwin's reasoning is circular:

> Darwin ... has therefore been accused of tautological (circular) reasoning: 'What will survive? The fittest. What are the fittest? Those that survive.' To say that this is the essence of natural selection is nonsense! To be sure, those individuals that have the most offspring are by definition (Lerner, 1959) the fittest ones. However, this fitness is determined (statistically) by their genetic constitution ... A superior genotype has a greater probability of leaving offspring than has an inferior one. Natural selection, simply, is the differential perpetuation of genotypes.

In the last sentence Mayr is simply repeating the usual contemporary description of natural selection. (The

3. There are two ways of showing that a theory is not circular. The first is by axiomatizing the theory; this I have done in (21). The second is by exhibiting falsifiable predictions of the theory; this I have done in (22).

virtually complete acceptance by biologists of this description is due to the fact that it does capture the essence of *one* of the Darwinian insights, and possibly also to the compelling, though fallacious, argument: without differential perpetuation natural selection could not occur; therefore natural selection *is* differential perpetuation.) But the context shows that Mayr does not really believe that natural selection is *nothing more than* the differential perpetuation of genotypes; he believes that it is the differential perpetuation of *superior* genotypes. It is essential to his (and Darwin's) conception of natural selection that over the long run the adaptively superior genotypes have more offspring; without some such adjective the important Darwinian insight that the environment has something to do with which genotypes have more offspring is lost. But "superior," "adaptively superior," "adaptively complex" (Maynard Smith, 18), "greater ability of phenotypes to obtain representation in the next generation" (Bossert and Wilson, 2), "adaptedness" (Dobzhansky, 5), etc. are all simply disguised ways of saying "fitter." How do we know that a superior genotype has a greater probability of leaving offspring? What is superior? In this context "superior" has the same intuitive meaning as "fitter" had to Darwin. Therefore, if we accept Lerner's definition of fitness as a true rendition of the meaning of fitness in the Darwinian phrase, we should define "superior" in the same way; the

superior genotypes are the genotypes that have more offspring. Once this is done, of course, "differential perpetuation of superior genotypes" is just as circular as "survival of the fittest."

The real source of the dilemma is that this definition is *not* a true rendition of Darwin's meaning.[4] Lerner's unfortunately all too true comment (12, p. 176) about this definition is

> If there is one thing upon which the most factious partisans of various currents of evolutionary thought agree, it is that fitness of an individual, in the context of the natural selection principle, can mean *only* the extent to which the organism is represented by descendants in succeeding generations. [italics mine]

This definition of fitness makes no reference to the fact that fitness is a property of the relationship between the organism and its environment; it states by omission that the environment is irrelevant to fitness. It is not surprising that the acceptance of this definition has led to statements that the theory is vacuous, for the fact that fitness is related to the environment is absolutely essential for the Darwinian insight. With this definition Darwinian theory is reduced to "differential perpetuation of genotypes," that is, to a theory which implies that the properties (e.g., morphological characters) of the organisms will change over the generations but gives no indication that this

---

4. The situation is made more confusing by the fact that the concept denoted by this definition is *called* "Darwinian fitness"! I hope that the users of this term follow Humpty Dumpty's example and pay it extra when it comes round for its wages Saturday night.

change is systematically related to the environment. (It would be possible for the geneticists to retain the Darwinian insight by stating a new law that differential perpetuation is systematically related to the environment. But the statement of such a law would inevitably require a theoretical term with the same objectionable qualities that "fitness" has.) Some sternly ascetic geneticists actually do restrict themselves to this reduced version of the theory of natural selection; most evolutionists, however, while using this definition of "fitness," sneak in a new term (superior, adaptedness, etc.) to take the place of "fitness" in order to retain as a part of the theory the important Darwinian insight that changes in the properties of species are systematically related to the environment. The controversy, then, is between those who, relying on the deep intuitive knowledge that Darwinian theory is far from being vacuous, save the Darwinian insight by reintroducing a term equivalent to fitness without explicitly defining the new term, and those who, relying on logical reasoning from the accepted definition of fitness, throw away the Darwinian insight. This definition of fitness has put both parties to the controversy into untenable positions; the only way to resolve the controversy is to throw out this definition of fitness.

Both parties to the controversy would probably respond to this statement by asserting that they cannot throw out this definition until a suitable replacement is available. And this reveals the deeper source of the controversy: the fallacious doctrine that all words used in a scientific theory can (at least in principle) and should

be defined. (In practice this means that definitions are not insisted on for those theoretical terms which, like "organism," are close enough to experience that scientists rarely have difficulty deciding what they mean in specific situations, while definitions are demanded for those theoretical terms which, like "fitness," are so abstract that scientists frequently must struggle to find an adequate interpretation.) Now it is a simple logical fact that it is impossible to define all words of a theory in terms of other words of the theory without introducing a circularity into the theory. Since the doctrine in question would not allow a definition in terms of words from another theory which themselves ultimately depend on undefined terms, no noncircular theoretical definition of fitness will be acceptable.

The scientist hopes to avoid this problem by using operational definitions for his basic words and defining the rest in terms of them. Theories which contain only terms denoting concepts which are close to experience appear to be stated in simple, operationally defined (or, at least, definable) terms, and it is this appearance which gives rise to the belief that any theory can be so stated. (Also, of course, this belief comes from the reasoning that since the theoretical terms have their ultimate origin in experience, they must be stable in terms of experience. But this is the same as reasoning that since scientific laws have their ultimate origin in observation statements, they must be stable as some conjunction of observation statements. The problem of induction is clearly as much a problem with regard to general concepts as with regard to general laws.) As Hempel has shown (9, p. 129), there are difficulties which prevent theoretically satisfactory operational definitions of even simple concepts like "length," but these are not the difficulties that ambush attempts to define "fitness." Scientists are generally willing to accept as operational any definition stated in terms of (potentially) directly observable results of manipulations, where what is to count as directly observable is decided not by philosophical analysis but by consensus.[5] The difficulty that ambushes fitness is that abstract terms which are not close to experience (i.e., to direct observation) are not amenable to definition in terms of direct observation. As Einstein pointed out, theories which have been rigorously stated meet this difficulty by "the application of complicated logical processes in order to reach conclusions from the premises that can be confronted with observation", (6, p. 5), though even this philosophically acceptable technique causes "an almost irresistible feeling of aversion [to arise] in people who are inexperienced in epistemological analysis" (ibid.). Since evolutionary theory had not been stated sufficiently rigorously to allow the application of complicated logical processes, this way of validating the concept of fitness was not available. Therefore the evolutionists could not, for the full Darwinian meaning of the concept, satisfy the demands of this metaphysical doctrine by using either direct or indirect operational criteria.

Faced with this impasse, the evolutionist defined fitness in terms of its most important known property; he defined "the fittest" as "those that survive". This was disastrous for two reasons: (1) Because the law of the survival of the fittest had not itself been explicitly stated, the definition enshrined an extremely impoverished version of this important property; (2) when an axiom is used as a definition of one of its terms, it is no longer an axiom; the theory remaining after that axiom has been removed will obviously be less powerful, and, when the axiom that was removed expressed the most important insight of the theory, its removal makes the theory virtually meaningless. (Of course, in a formally stated theory it can be seen that *calling* an axiom a definition does not make it one; it will be a creative definition, and therefore really an axiom. But in an intuitively stated theory once you convert the axiom into a definition you are forced to admit that it is purely circular and gives no power to the theory; no one can save you by proving it is a creative definition.) Thus this definition of fitness weakened the theory both by replacing the intuitive definition, which recognized that fitness is related to the environment, with a definition that ignores this relationship and by removing from the theory its most important law. It is this replacement of an important law by an (apparently) circular definition that is responsible for the belief that Darwinian theory is circular.

5. Hull's discussion of operationism in (10) provides a useful view of what has been accepted by various groups of biologists as operational.

The belief that all (nonobvious) terms must be defined is common among scientists. Since the most important explicitly known property of an abstract term is the property stated in the deepest law containing the term, it will always be very tempting to respond to the demand for a definition of the term by using that property as a definition. It is probably because of this that we so frequently see theories accused of being circular on the ground that their most important insights are merely definitions, or of being meaningless on the ground that their most central concepts cannot be defined. Recall, for example, the *reductio ad absurdum:* What is intelligence? Intelligence is what is measured by intelligence tests. Or consider the accusations in the ecological literature that the exclusion principle of Gause, which says that two species cannot coexist in the same niche, is circular because this is the defining property of the niche. Anyone familiar with the "soft" sciences can give more examples. But an example from a "hard" science will strengthen our understanding of the theoretical structure which underlies these problematic definitions. Consider the following quotations from Newton's *Principia* (17, pp. 2, 13):

Definition IV: An impressed force is an action exerted upon a body, in order to change its state, either of rest, or of uniform motion in a right line.

Law I: Every body continues in its state of rest, or of uniform motion in a right line, unless it is compelled to change that state by forces impressed upon it.

Let us subject the concept of force to the catechism used in the quotation from Mayr for the concept of fitness. "What will change the state of rest or of uniform motion of a body? An impressed force. What is an impressed force? Something that will change the state of rest or of uniform motion of a body." Clearly Newton's theory of mechanics is just as circular as Darwin's theory of natural selection. (More so, in fact, since Darwin at least didn't give that definition of fitness.)

Fitness is a theoretical term which cannot be explicitly defined. That such terms exist in any axiomatized theory is well known; they are the primitive terms of the theory. That such terms must exist also in the intuitively stated' precursor of an axiomatized theory is obvious. In any deep theory some of the terms will be so abstract that the lack of adequate definitions will be painfully obvious. But until the theory is rigorously stated, it is impossible to validate experimentally such an abstract concept. If it is a primitive term, only an axiomatization of the theory will allow a full and explicit statement of its meaning in the theory, for its meaning in the theory is completely given by the statements that the theory makes about it. After the theory has been axiomatized, the various successful interpretations of the primitive term give the most comprehensive possible statement about its meaning in the real world. Thus only an axiomatization can provide a final resolution of the problems stemming from the need to specify the meanings of abstract concepts. But during the preaxiomatic stages of a theory it is important simply to recognize that these problems are caused by a fallacious metaphysical doctrine and do not indicate that the theory is worthless; in fact, they indicate that the theory is deep.

### 3. THE REFERENT

The major source of the controversy surrounding the phrase "survival of the fittest' is that dealt with in the preceding section. But to understand the referent of this phrase we must first use the axiomatization to resolve controversies arising from the following questions: What is the fundamental unit of selection? What is the primary target of selection? Both spring from the fact that our analytic metaphysics leads us to assume (incorrectly) that there is *one* fundamental entity which is the object of selection and which is such that all results of selection can, and should, be described with reference only to that fundmental entity. (For example, in the argument about whether the target of selection is the gene or the organism, the argument is about which is the fundamental entity with reference to which the basic theory of selection should be stated and all other entities which undergo selection should be described.) This assumption has led to confusion, because the world that the evolutionist is trying to describe contains several important independent fundamental entities.

All of these different fundamental entities are physically parts of one another. With respect to the theory of natural selection, they fall into two categories: (1) those that cannot be defined in terms of one another because they correspond to different primitive terms of the theory; and (2) those that are independent of one another because they appear in different models of the theory.

(1) *Different Ontological Levels.* My primary plan in this section is to discuss the problem on a relatively intuitive biological level in order to give insight into the way things look at an intuitive (preaxiomatic) stage when the underlying theoretical structure is such that one of the entities being discussed is a collection of entities but cannot be defined as a set of these entities. Since the assertion that the entities under discussion are of this kind is a very strong assertion, I will first point out that the axiomatization makes a rigorous proof of this assertion possible. Because this assertion is equivalent to the assertion that *Darwinian subclan* is independent of the remaining primitive terms of the theory, it can be proved by using Padoa's Principle.[6]

First let us try to get some insight into why the theory needs two types of entity. Natural selection changes the characteristics of populations by depriving individuals of offspring (either by ensuring an untimely death for the individual or by interfering with the reproductive process). Thus it acts on individuals to change the characteristics of populations. This is

---

6. Padoa's principle states that to prove that a given primitive term, *D,* is independent of the other primitives it is sufficient to find two different interpretations of the axioms in which *D* has different interpretations while the other primitives have the same interpretations. I have a sketch of the necessary proof, but because it relies on an understanding of some rather complex biological situations, it would be inappropriate to present it here.

a strange kind of force, since the forces we usually deal with change the characteristics of the objects they act on; for example, an impressed force changes the state of motion of the billiard ball it acts on. Selection is a force which changes hereditary characteristics, but this change does not take place in individuals. Selection cannot change the inherited characteristics of an individual; it may, so to speak, punish the individual for his bad characteristics by preventing him from having offspring; but it is powerless to change his characteristics once he is there. (The insight that selection could change the characteristics of a population without changing the characteristics of any individual was one of Darwin's most valuable contributions.) It is clear that, while the forces of mechanics could be described with reference to a single type of entity which is both acted upon and changed, the forces of selection must be described with reference to two types of entity, one of which is acted on and the other of which is changed. (This is partly an artifact of our human-sized viewpoint; we could, theoretically, completely ignore the existence of individuals and consider the killing of an individual as an action on the population of which that individual is a component, just as we consider a knife wound as an action on the organism rather than as an action on the cells that were actually killed. But, because we *do* have a human-sized viewpoint and *do* see the individual deaths rather than the wound in the population, it is necessary for the theory to deal with the relationship between the actions on the individual and the

results on the population.) This is the reason that in the axiomatization there are two primitive terms referring to types of entities; the biological entity is the thing selection acts on and the Darwinian subclan is the thing whose characteristics are changed.

Let us consider now the effect that this indefinability has had on the definitions used in biology, remembering that biologists have simply assumed that since a species is a collection of organisms, it can be defined in terms of the properties of the organisms it contains. Since a species is a Darwinian subclan and an organism is a biological entity, we would expect that any attempt to define species in terms of organisms would be doomed to failure. And it is exactly this failure that Beckner has analyzed so beautifully in (1), where he shows that in the definitions used in systematics the definiens is typically a set of neither severally necessary nor jointly sufficient properties of the entities contained in the taxa being defined. From the point of view of normal logic such definitions are clearly not legitimate definitions. (Beckner concluded that the concept of definition should be expanded to include these polytypic definitions. Although, as I have argued elsewhere (22), definitions in the usual logical form are possible once the theory has been sufficiently rigorously stated to make clear what the defining properties are properties of, polytypic definitions probably do have a legitimate role in the pre-axiomatic stages of theories.)

And finally let us look at how these two types of entities appear in arguments about the fundamental units of selection. Consider the following statements:

"The primary focus of evolution by natural selection is the individual." (13, p. 7)
"Selection is primarily concerned with genotypes." (12, p. 178)
"Mendelian populations, rather than individuals, are the units of natural selection." (4, p. 79)
"The species are the real units of evolution." (15, p. 621)

The first two of these statements are concerned with fundamental entities corresponding to the biological entity—the entity which is acted on. The second two statements are concerned with fundamental entities corresponding to the Darwinian subclan—the entities which are changed. These statements are not contradictory: the first two are appropriate in discussions on micro-evolution (i.e., the processes occurring in the short term which will, after many short terms, result in significant evolutionary change), while the last two are appropriate in discussions of macro-evolution (i.e., significant changes in characteristics of populations). These statements were, in fact, used in the appropriate contexts, but, as the phrases "the *real* units" and "rather than individuals" indicate, the assumption that there is *one* fundamental entity (and that that one is the smallest of the entities under consideration) forces biologists to fight for the appropriate recognition of the Darwinian subclan-type entity.

(2) *Different Levels of Selection.* It has long been recognized that natural selection operates on the levels of the gene, the chromosome, the gamete, the organism, and the population, but it is usually assumed that one of these levels is primary and the others subsidiary. Thus:

Asserting the primacy of the organism:

"To consider genes independent units is . . . meaningless from the evolutionary viewpoint because the individual as a whole . . . , not individual genes, is the target of selection." (16, p. 162; see also 8, p. 230; 3, p. 173; 13, p. 7; etc.)

Asserting the primacy of the gene:

"In its ultimate essence the theory of natural selection deals with a cybernetic abstraction, the gene, and a statistical abstraction, mean phenotypic fitness." (20, p. 33)

Asserting the primacy of the chromosome over the gene:

"The selection of the chromosome as a whole is the overriding determiner of allelic frequencies." (17, p. 725)

Asserting the primacy, in at least some circumstance, of the population:

"Unless extinction of populations, species, and higher taxa occurs randomly, group selection occurs . . . [for the evolution of dispersal] it seems to be important." (19, p. 596)

Such assertions of primacy are sometimes assertions that selection on a particular level is the primary form determining the evolution of certain types of phenomena, but in other cases (e.g., the first and second assertions given above) they represent a real clash between scientists each of whom thinks that he is speaking of *the* fundamental level of selection in terms of which all other levels of selection should, and ultimately, will, be expressed.

The question is: What is the relationship between the different levels of selection? There are two important ways of answering this question. One is to answer it not as a question about the theory of natural selection but as a question about the relationship between genes, chromosomes, gametes, organisms, and populations; this answer would state that all phenomena at each level are caused by phenomena at the lowest level. (At present, of course, this is merely an assertion of a philosophical commitment to reductionism; we are nowhere near being able to specify the chain of causal mechanisms.) But when we are trying to decide whether statements about selection at the gene level *should* be expressed in terms of selection at the organism level, it is more important to ask it as a question about the theory of natural selection. Does the process at each of these different levels actually follow the laws of the theory, or are the laws of the theory truly applicable only to one level? Formally, this question is asking whether: (1) the different levels of selection are different models of the theory; or (2) only one of the levels is a model of the theory. By substituting gene, chromosome, gamete, organism, and population for *biological entity* in the axioms, we find that these different

levels of selection are different models of the theory of natural selection. The levels of selection, being different models of one theory, are analogs of one another. One level of selection may be more suitable than the other for studying a particular phenomenon, and the organism level may seem *primus inter pares* because we are more familiar with the phenomena for which it is most suitable, but *no* level of selection has absolute primacy over the others.

It has been difficult to identify the referent of "survival of the fittest" because it appears to have different forms in the different levels of selection. To understand this difficulty let us consider what subclands look like in two different models. With the organism interpretation a subcland is the set of all organisms which are in the Darwinian subclan and are descended from (or ancestral to—but we can ignore that part at present) a particular founder set of organisms; for the present discussion we can consider the founder set to be a set of contemporaneous organisms with a particular advantageous hereditary trait; some of the offspring of the founder organisms will have this trait and some will not; it is natural when discussing selection on this trait to ignore the ones that do not have it. But this should not be done: biological traits are affected by many different genes and even if a particular offspring does not have the trait it may have many of the genes which positively affect the trait; thus the fact that it does not have the trait is not an indication that its success in leaving descendants is irrelevant to the selection for the trait. By including all descendants, we may include some which are irrelevant, but this causes less difficulty than the assumption that the expression of the trait is all that counts. (For some traits, those that are controlled by a single gene, this argument does not hold. But these are traits whose selective fate should be analyzed with the gene interpretation.) Notice that one cannot define the subcland simply in terms of the morphological properties of its members.

With the gene interpretation, the Darwinian subclan frequently used in discussions of selection is the set of all alleles occupying a particular chromosomal locus. (An "allele" is "one of two or more alternate forms of a gene occupying the same locus on a chromosome." E.g., the gene for hemoglobin has several alleles: a normal hemoglobin allele, a sickle cell anemia allele, a thallassemia allele, etc.; each of these alleles has a different molecular structure.)"The sickle allele" denotes the set of all molecules (or, rather, of all portions of DNA molecules) with a particular molecular structure. In this usage the sickle allele is, barring mutations, exactly the same as the subcland derived from the set of all sickle alleles existing in A.D. 1800. Thus to say that selection is increasing the frequency of the sickle allele is the same as to say that selection is increasing the frequency of this subcland. Note how much easier it is to visualize the subcland on the gene level than on the organism level; a definition in terms of the molecular structure[7] of the gene includes virtually all members of the subcland, while a definition in terms of the morphological structure of the organisms includes only scattered portions of the subcland. This is why "differential perpetuation of genotypes," which expresses the referent in the terminology of the gene level, is used even by organism-level biologists.

We saw, in the section on fitness, that the fallacious metaphysical doctrine about definitions led many biologists into the inconsistent position of advocating that all terms be defined while at the same time sneaking in undefined terms. Similarly the fallacious metaphysical doctrine that there is *one* fundamental level has led many biologists into the inconsistent position of asserting that the organism level is *the only* level while working solely on the gene level. (Although evolutionary geneticists work as if the gene level could be studied independently; it is very difficult to find one who disagrees with the statement that the organism level is the primary focus of selection. But, as Mayr frequently points out (e.g., in 14), if this statement is true, then most of the work in evolutionary genetics is worthless.) The axiomatization has shown us that in both cases the inconsistency lies not within the theory but between generally accepted metaphysical assumptions and the theory.

---

7. It can be seen from this analysis that those who believe the organism level to be the fundamental one need not be antireductionists. They are not denying that laws may some day be found which are expressed in terms of the lowest level and from which all higher level phenomena can be deduced. They are denying that the *Darwinian* laws are applicable on the lower levels.

## 4. CONCLUSION

I have shown in this paper that the controversies about the logical status of evolutionary theory and about the fundamental entity of the theory arise from the following two fallacious metaphysical doctrines: (1) All words used in a scientific theory should be defined. (2) When a theory deals with entities which are physically parts of another, *one* of those entities is the fundamental entity of the theory and all results of the theory can, and should, be expressed in terms of that fundamental entity.

The first of these doctrines is fallacious because it is inconsistent with our logical system; it is logically impossible to define all words used. The attempt to defy this logical impossibility led to definitions which created the impression that the theory of natural selection is tautological.

The second of these doctrines is fallacious because it denies the possibility that a theory may have two primitive terms denoting entities which are such that one is contained within the other, and because it denies the possibility that a theory may have two distinct, but physically related models in the real world.

The axiomatization, and in particular Axiom D4, provides a nontautological, nonambiguous translation of "survival of the fittest." A brief phrase which captures the essence of this translation is "expansion of the fitter subcland."

## REFERENCES

1. Beckner, M., 1968, *The Biological Way of Thought*, Berkeley, University of California Press.
2. Bossert, W. H. and Wilson, E. O., 1971, *A Primer of Population Biology*, Stanford, Conn., Sinauer Associates, Inc.
3. Crow, J. F. and Kimura, M., 1970, *An Introduction to Population Genetics Theory*, New York, Harper and Row.
4. Dobzhansky, T., 1937, *Genetics and the Origin of Species*, New York, Columbia University Press.
5. ——1968, On some fundamental concepts of Darwinian biology. *Evolutionary Biology*, vol. 2 (ed. by T. Dobzhansky, M. K. Hecht, and W. C. Steere). New York, Appleton-Century-Crofts.
6. Einstein, A., 1950, On the generalized theory. *Scientific American* (April 1950), Reprint 109; San Francisco, W. H. Freeman and Company.
7. Franklin, I. and Lewontin, R. C., 1970, Is the gene the unit of selection? *Genetics*, 65: 707–734.
8. Grant, V., 1963, *The Origin of Adaptations*, New York, Columbia University Press.
9. Hempel, C. G., 1965, *Aspects of Scientific Explanation*, New York, The Free Press.
10. Hull, D. L., 1968, The operational imperative: sense and nonsense in operationism. *Systematic Zoology*, 17: 438–457.
11. ——1974, *Philosophy of Biological Science*, Englewood Cliffs, N.J., Prentice Hall.
12. Lerner, I. M., 1959, The concept of natural selection: a centennial view. *Proceedings of the American Philosophical Society*, 103: 173–182.
13. Lewontin, R. C., 1970, The units of selection. *Annual Review of Ecology and Systematics*, vol. 1 (ed. by R. F. Johnston, P. W. Frank, and C. D. Michener), Palo Alto, California, Annual Reviews Inc.
14. Mayr, E., 1959, Where are we? *Cold Spring Harbor Symposia on Quantitative*

*Biology,* vol. 24. Cold Spring Harbor, New York, The Biological Laboratory.

15. ——1963, *Animal Species and Evolution,* Boston, Harvard University Press.

16. ——1970, *Populations, Species, and Evolution,* Boston, Harvard University Press.

17. Newton, I., 1686, *Principia* (Quotations from Motte's translation revised by Cajori), Berkeley, University of California Press, 1966.

18. Smith, M. M., 1969, The status of neo-Darwinism. *Towards a Theoretical Biology,* ed. by C. H. Waddington, Chicago, Aldine.

19. Van Valen, Leigh, 1971, Group selection and the evolution of diversity. *Evolution,* 25: 591–598.

20. Williams, G. C., 1966, *Adaptation and Natural Selection,* Princeton, Princeton University Press.

21. Williams, M. B., 1970, Deducing the consequences of evolution: a mathematical model. *J. Theoret. Biol.,* 29: 343–385.

22. ——Falsifiable predictions of evolutionary theory. *Philosophy of Science,* 40: 518–37.

# 8

# The Supervenience
# of Biological Concepts

## ALEXANDER ROSENBERG

*In this paper the concept of supervenience is employed to explain the relationship between fitness as employed in the theory of natural selection and population biology and the physical, behavioral, and ecological properties of organisms that are the subjects of lower-level theories in the life sciences. The aim of this analysis is to account simultaneously for the fact that the theory of natural selection is a synthetic body of empirical claims, and for the fact that it continues to be misconstrued, even by biologists, for a tautological system. The notion of supervenience is then employed to provide a new statement of the relation of Mendelian predicates to molecular ones in order to provide for the commensurability and potential reducibility of Mendelian to molecular genetics in a way that circumvents the theoretical complications which appear to stand in the way of such a reduction.*

SOME OF THE MOST EXASPERATING problems in the philosophy of biology involve not so much the substantiation but the perspicuous exposition of important and controversial claims. Among these problems paramount seem to be the precise expression of the reductionist thesis that Mendelian genetics, and/or its successor, transmission genetics, are reducible, in a sense to be specified, to the theory of molecular genetics. Less controversial but equally important is the problem of giving a general account of the notion of fitness that is not merely consistent with, but also explains, the nontautological character of the theory of natural selection—that is, explains both why it is nontautological, and why it is so often mistaken even by biologists for an empirically empty theory unconnected to other theories in natural science. In the present paper I hope to express the first thesis and explain the second by appeal to the concept of "supervenience." I

For useful comments on earlier versions of this paper I owe thanks to David Hull, Michael Ruse, and William Wimsatt. None of these persons should, however, be assumed to agree with my claims.

begin with the notion of fitness, and its role in evolutionary theory.

I

While there is no misconception in the standard textbooks of the philosophy of biology about the logical status of the theory of natural selection, there remains among biologists a continuing conviction that the theory is untestable, and the accounts offered in the philosophical texts do not clearly disabuse this impression,[1] for they do not provide a general characterization of "fitness" which distinguishes its operational connection to rates of reproduction from its conceptual connection to such rates. If differences of fitness are defined by reference to differences in future rates of reproduction, then fitness cannot be appealed to in explanation of such rates, and the theory in which it figures as the cause of these rates will rightly be judged vacuous. On the other hand, if fitness is defined by appeal to the rates of reproduction of ancestor organisms, then although the theory may remain testable, the notion of fitness as an explanatory variable drops out. Thus consider the following semiformal characterization of fitness offered by Mary Williams:

As a possible operational definition of [the fitness of organism, b], $\phi(b)$, I might suggest the following. Let $v_1(b,k)$ be the sum over all the $k$-ancestors of $b$ of the number of reproducing off-spring of each. Then let $v_2(b,k)$ be the number of $k$ ancestors of $b$. Then $v_3(b,k) = v_1(b,k)/v_2(b,k)$ is an estimate of the average fitness of the $k$-ancestors of $b$. Now let the operational definition of $\phi(b)$ be:

$$\left\{ \phi(b) = \sum_{k=1}^{n} (.05)^k v_3(b,k), \right.$$

where $n$ is the number of generations for which data are available. Then about 0.5 of the fitness of $b$ is estimated by the "average fitness" of its parents; about 0.25 of the fitness of $b$ is estimated by the average fitness of its grandparents; etc. . . . $(0, 5)^k$ is a factor which adjusts the importance to be attached to more remote generations.[2] (16, p. 359)

Fitness is characterized by this "operational definition" just in terms of the reproduction rates of prior members of an organism's line of descent. If fitness is then called upon to explain the present organism's rate of reproduction and the rates of its

---

1. Both Michael Ruse (13) and David Hull (6) argue that the theory is a body of synthetic propositions. Yet despite the uniform clarity of the rest of these books, both discussions of this issue are curiously tortured, difficult for students to follow, and unconvincing. Cf. pp. 66–69 and pp. 38–41 respectively. One recent example of a biologist arguing vigorously for the tautological character of the theory of natural selection is R. H. Peters (12). Characteristic of his claims is the following: "the 'theory of evolution' does not make predictions . . . but is instead a logical formula which can be used only to classify empiricisms and to show the relationships which such a classification implies."

2. I have no argument with this operational definition just as long as its status as a definition is not taken seriously. Although Williams calls this operational characterization a definition, none of her important results in this undeservedly neglected paper turn on giving it this status.

descendants, then, in effect, explanations in terms of fitness are but covert appeals to prior rates of reproduction to explain future rates; the notion of "fitness" thus becomes a *facon de parler* for such prior rates, and has no explanatory standing of its own. Although this is a result which in a way Craig's theorem[3] has independently assured us of, it is clear that in evolutionary theory fitness is not simply a short-hand way of describing past and future rates of reproduction, nor is the theory of evolution committed to the claim that future rates of reproduction are dependent on nothing but past rates of reproduction. Nevertheless, determinations of such prospective and retrospective rates provide the only uniform measure of fitness generally available, and this is what leads to the mistaken view that the theory of natural selection is vacuous, or that the notion of fitness is at best doubtful as a full-fledged theoretical, explanatory term.[4]

To see that fitness is a "real" property of organisms and not merely an eliminable construct, consider two organisms with the genetic identity of twins. Suppose, despite present limitations, that one is synthesized from organic material on the basis of a complete genetic map of the other. In this case the synthesized entity will have no ancestors, and the measure of fitness given above for it will be zero or at any rate inapplicable. Since the two organisms are of exactly the same character in physical properties, it must follow that they have the same level of fitness. To deny this is to deny that fitness is a physical property of organisms. Now, destroy one organism (either one) and permit the other to reproduce. Since under these circumstances their rates of reproduction will obviously be different, it follows that differences in fitness cannot be identical to differences in reproduction rates, for *ex hypothesi* the two items were identical in levels of fitness. So, it seems, two organisms can have differing rates of reproduction, both prospectively and retrospectively, yet have the same level of fitness.

What is more, two organisms can have quite different physical properties and yet share exactly the same level of fitness in respect to a given environment. For example, a bird and a squirrel can occupy roughly the same environment, have exactly the same prospective and retrospective reproduction rates, and yet differ greatly in their anatomical, physiological, behavioral, and environmentally related properties. But how can fitness be a purely physical property of organisms and not simply a measure of their rates of reproduction if two such different organisms can have exactly the same level of fitness? By now it should be clear that this sort of case, like the previous one, is possible because "fitness" is a functionally characterized concept;[5] that is, levels

3. Cf. William Craig (2) for an informal treatment of his theorem.

4. Ruse (13, pp. 48 ff.), for example, refuses to credit "fitness" with this status, arguing that the theoretical terms of evolutionary theory are those of population genetics. For an assessment of this view and an alternative account of the relations between evolutionary theories and theories of heredity, see Rosenberg, "The Interaction of Evolutionary and Genetic Theory" in (17).

5. For an introduction to this notion cf. J. A. Fodor (5), especially chapter 3.

of fitness are the causal consequences and causal antecedents of heterogeneous classes of natural phenomena. Thus, for example, two organisms can have the level of fitness because one avoids its predators by camouflage and the other by flight, or one endures severe weather by virtue of its thick coat of fur, and the other by migration, etc. In other words, fitness among animals is interconnected with a vast number of different physical properties and environmental conditions; so vast a number of functional concommitants does fitness have that it would be impossible to specify even a small proportion of the nomological connections between a given level of fitness and all of the different properties and relations of organisms that could give rise to it, and to which it can give rise. The relations between these items and any given level of fitness are "many-one," and "one-many." These facts explain why fitness functions as it does in respect to prospective and retrospective levels of reproduction, and at the same time why there is no practicable way of measuring fitness differences between differing species (or even different organisms) other than appealing to reproduction rates. It also explains the difficulty of providing a systematic reduction of the theory of natural selection to any other theory. In fact this reduction would involve the explanation of natural selection by appeal to every theory that governs the particular physical and environmental properties of every species whose evolution natural selection explains. Both of these considerations make manifest the *force* of the conclusion that the theory of natural selection is somehow methodologically suspect: its key term cannot *in fact* and in detail be cashed in by appeal to any other theories which it is practically possible to develop or systematically deploy, or which are strategically worth pursuing. On the other hand, these considerations also help show how to state the character of the concept of fitness precisely, in a way that blunts the *force* of such conclusions, and renders perspicuous the logical status of the theory of natural selection.

## II

The concept of "fitness" is *supervenient* on the manifest properties of organisms, their anatomical, physiological, behavioral, and environmentally relative properties. And this fact alone explains the simultaneous explanatory power and empirical recalcitrance of the concept of fitness.

The concept of "supervenience" has its current locus in the work of Donald Davidson, who argues that mentalistic concepts and general statements are not reducible to physicalistic ones but are dependent upon them in the sense that "there cannot be two events alike in all physical respects but differing in some mental respect, or that an object cannot alter in some mental respect without altering in some physical respect." Davidson goes on to say that "Dependence or supervenience of this kind does not entail reducibility through law or definition" (4, p. 88). This is presumably because such dependence or supervenience does not entail any finitely expressible reducibility of mentalistic concepts and claims to physicalistic ones, since the biconditionals connecting mentalistic and physicalistic concepts could be indefinitely long and/or expressible at a level of complication that would make them systematically otiose for full explanation and practicable prediction of mental events by appeal to the known occurrence of physical events. Notwithstanding this impossibility, if Davidson's thesis of the supervenience of the mental on the physical could be substantiated, then, together with a few further (and less controversial) assumptions, the physicalist and materialist would be able to infer the claim which they have traditionally made about mental events—viz, they are nothing but physical events. Or at any rate, so Davidson avers. Whether or not he is correct in these claims about physicalism, it is possible to formalize his characterization of supervenience in a way that reveals the actual relation of the concept of fitness to other biological predicates that characterize species and organisms, and thus to show the exact relationship between the theory of natural selection and the manifest anatomical, physiological, and behavioral theories couched in these other predicates, which explain and predict events in the lives of species and individuals.

Before proceeding to expound the sense in which fitness is supervenient on what I have called the manifest properties and predicates of organisms, I should at least sketch out what sorts of properties these are. By the manifest properties of organisms I mean those properties, dispositions, and abilities which organisms have in virtue of their anatomical, physiological character, and the interaction of this character with the organism's environment. (The properties reflecting this interaction are hereafter called "ecologically relative" properties.) These properties are the ones to which appeal must be made in any account of the organism's behavior, size, strength, speed, longevity, period of sexual maturity, probability of avoiding predators or capturing prey, litter size and frequency, method of feeding young, etc. In other words, the properties of organisms on which fitness is supervenient are those which figure in nomological generalizations (but not necessary truths) that govern an organism's number of reproductive opportunities, and rate of successful reproduction. They are, in short, the causes of its rate of reproduction. Naturally, the instantiation of these properties causally connected to organisms' rates of reproduction will themselves be explained by anatomical, physiological, behavioral, and ecological theories.

The formalization of supervenience employed here is the application of one due to Jaegwon Kim.[6] Consider the notion of fitness offered by Williams and described above. This functor can

6. Kim (8), read at the Oberlin Colloquium, 1977. Forthcoming in its preceedings. My debts to this paper are by no means limited to its account of supervenience. In fact, the whole perspective adopted here with respect to the relations between theories involved in reduction and the bearing of causal and mereological determinism on this relation are inspired by this paper. However, no agreement on Kim's part with any of the views expressed, or even its particular account of supervenience, should be assumed. The reader is referred to Kim's paper for a fuller account of supervenience and a discussion of its role in ethics, aesthetics, and the philosophy of mind in general.

take on a continuum of values from 0 to 1. In effect, these values constitute a denumerable infinity of fitness predicates. Call this set of predicates $F$. Now consider the set of all the anatomical, physiological, behavioral and ecologically relative properties which it is physically possible for an organism to manifest. Call this set $P$. Let $P^*$ be the set of all properties constructable from $P$ by any combination of conjunctions and disjunctions (and their negations) of finite or infinite length. Among the members of $P^*$ will be properties which are exhaustive of $P$, in that an organism that manifests one of these exhaustive properties will be fully and completely characterized, both positively and negatively. These exhaustive members of $P^*$ will be mutually exclusive properties of organisms as well. Call this set of exhaustive mutually exclusive predicates constructable from $P$, $P_e$. Now the set of properties $F$, the properties of having various fitness levels, is supervenient on the set of properties $P$, just in case if two organisms share the same member of $P^*$, the set of all possible anatomical, physiological, behavioral, and ecologically relative properties, then they share the same property in $F$, i.e. the same level of fitness. An organism's level of fitness, $\phi_i$ is thus not dependent on or identical with its past or future rates of reproduction, but is a matter of the organism's having a particular set of properties in $P$. If an organism $o$ has fitness level $\phi_i$, then there is a set of properties in $P$ such that $o$ has this set of properties, and anything else with the same set of properties has fitness level $\phi_i$. Although fitness is thus determined by features unmentioned in the theory of natural selection, it cannot yet be said to be reduced or reducible to these features. The members of $P$ realized by $o$ are only individually sufficient conditions for particular levels of fitness; they may not be necessary and sufficient; but biconditionals, stating necessary and sufficient conditions, are required for reduction. However, such biconditionals will be at least in principle constructible from members of $P_e$, the set of exhaustive and exclusive properties constructed from $P$, if the members of $P$ are finite in number. To see this, suppose that organism $o$ has fitness level $\phi_i$. Then, since fitness levels supervene on properties in $P$, there is some member of $P_e$ which $o$ manifests, say, $P_{eo}$, and if any other organism instantiates $P_{eo}$, then it, too, has fitness level $\phi_i$. But this is tantamount to concluding that each member of $P_e$, the set of exhaustive and exclusive properties constructable from $P$, is sufficient for some level of fitness. If the set of members of $P_e$ is finite, then a finitely long enumeration of the members of $P_e$, each of which is sufficient for some particular level of fitness, will provide both necessary as well as sufficient conditions for each level of fitness. Of course if the number of members of $P_e$ is very large or infinite, then such biconditionals will be either practically or logically impossible to construct. But the important point is that any particular level of fitness is a function solely of the manifest properties of organisms, and the function in question is that of supervenience. This explains how fitness can be *nothing more than* having a certain combination of anatomical, physiological, and ecologically relative properties, even though no set of such properties may be stateably necessary and sufficient for a given level of fitness; even though differing organisms in different habitats, with differing prospective and retrospective reproduction rates may (conceivably) have identical levels of fitness, and even though we may be unable to cash any particular level of fitness in for a specific set of such manifest properties of organisms. All these features of the notion of fitness, which its supervenience on manifest properties of organisms brings out, are the very features of a theoretical term in science that has great explanatory potential, just because it is not exhausted by, and cannot be translated into, observable properties.

### III

Consider how fitness, thus understood, figures in the process of natural selection as a causal variable nomologically connected with, but conceptually independent of, rates of reproduction. A line of descent is characterized by a certain set of hereditary, physical, and behavioral properties. Together with the environment in which the members of this descent line live, these properties determine a level of fitness for each member, measured on a scale between zero and one. If the environment changes, physical and behavioral properties remaining constant, then the fitness number will also change; if mutation changes physical or be-

havioral properties, environment remaining unchanged, fitness number will also change. In part, the fitness number is a measure of the "fit" between organism and environment. This fit at any given generation causally determines the number of reproductive opportunities at that generation, and since the fit is partly hereditary, it is also nomologically connected to the number of such opportunities at previous and future generations. This is why rates of reproduction provide an estimate of the fitness number of a line of descent, even though the direction of causation is from fitness to rates of reproduction, and not vice versa. But since the environmental conditions and the physical and behavioral properties of organisms are so complex, so varied, and so difficult to separate out and quantify as determinants of fitness, fitness can at best be shown to be supervenient on these properties, and we must employ rates of reproduction in order to measure fitness.

It is only by appeal to the supervenience of fitness on physical, behavioral, and ecologically relative properties that we can account for the employment of fitness to explain differences in reproduction rates, and can account for the *intra-* and *inter-*species comparisons of fitness that biologists appeal to as the mechanism of successful competition, predator-prey relations, biogeography, niche theory, and other aspects of selection and evolution.[7] A tighter connection

7. For an account of the theoretical and computational uses to which biologists put the notion of fitness, cf. Bossart and Wilson (1). For a more sophisticated text in theoretical biology employing the notion of fitness in a way that literally cries out for the treatment offered here, see R. Levins (9) and (10). The graphs and equations there employed involve the comparison of intra- and interspecies differences in fitness to explain reproductive

to rates of reproduction deprives fitness of its explanatory force. On the other hand a tighter connection than that of supervenience between fitness and the manifest traits of organisms would deprive the notion of its systematic employment in the explanation of how differing organisms in similar or different environments can both survive and compete successfully against other organisms, and how differing organisms can supplant one another in the same environment. Only on the assumption of supervenience can different combinations of manifest and ecologically relative properties constitute the very same level of fitness, and the same manifest or ecologically relative properties constitute differing levels of fitness (*modulo* differences in ecologically relative or manifest properties). Yet it remains entirely consistent with this flexibility in the relation between manifest properties of the organism and its level of fitness that any organism's *particular* level of fitness *at a given time* consists in, is identical to, nothing more than the organism's physiological, anatomical, and behavioral properties and the environment in which it finds itself.

Thus, although "fitness" is a theoretical term, at the level of the theory of natural selection, employed to explain differences in rates of

reproduction, levels of fitness of particular organisms are in turn explained by appeal to theories about their physical and behavioral properties, and the relation of these properties to the organism's environment. This enables us also to see more exactly the relation between the theory of evolution and those other biological, behavioral, and physical theories which, together with the genetic account of heredity, *explain* the theory of evolution. The theory of natural selection rests on these theories, in *something like* the conventional reductionistic picture. Since fitness consists in properties of organisms whose instantiation is deducible from principles of population genetics (and its successors), the leading principles of (at least some formal expositions of) the theory of natural selection should follow from these theories alone. We can sketch out this reproduction in connection with the following axiomatization[8] of theory of natural selection. The axiomatic system consists of four axioms. I shall state each axiom and indicate how the reduction of the axiom to laws in other theories can proceed.

*Axiom 1.* There is an upper limit to the number of organisms in any generation of

---

differences in a way that is unintelligible except on an analysis of fitness like the one offered here. In effect the present account offers a general framework for Levins' notion of sets of "sufficient parameters" of fitness, although he treats these parameters as approximate measures of fitness and not definitions of it. I owe thanks to William Wimsatt for bringing this notion to my attention.

8. This axiomatization is adapted from Williams (16). The adaptation involves simplification and deletion of certain features of Williams' original treatment, in the interests of brevity and simplicity. For a more detailed discussion of the original axiomatization, and an explanation of the adaptation effected here, see Rosenberg in (17).

an interbreeding population.

This axiom finds its ultimate reduction in considerations from thermodynamics and needs no special biological underpinning.

> *Axiom 2.* Each organism has a certain amount of fitness with respect to its particular environment.

Since each organism has a certain proportion of the properties, dispositions, and abilities which are the causal determinants of its number of off-spring, and since fitness is supervenient on these properties, this axiom can in principle be shown to follow from the theories which account for why particular organisms have certain proportions of these properties. But because fitness is supervenient on these properties, different organisms can have the same level of fitness in a given environment, and the same organisms can have different levels of fitness in different environments. Since fitness is determined by properties causally responsible for reproductive opportunities and successes, and since genetic theory (which is separate and distinct from evolutionary theory)[9] assures us that some of these properties are hereditary, it should follow that

> *Axiom 3.* If *D* is a physically or behaviorally homogeneous subclass of an interbreeding population and *D* is superior in fitness to the rest of the population for

sufficiently many generations, then the proportion of *D* in the total population will increase.

The fourth axiom asserts that the antecedent of *axiom 3* is fulfilled. It is the assertion that fitness is sufficiently hereditary to make for this secular increase in the proportion of *D* within the entire population.

> *Axiom 4.* In every generation of an interbreeding population (not on the verge of extinction), there is a subclass, the members of *D*, such that *D* is superior to the rest of the population for long enough to ensure that *D* will increase relative to the population, and will retain sufficient superiority to continue to increase, unless it comes to constitute all the living members of the whole interbreeding population at some time.

This axiom should follow from the existence claims of other theories to the effect that among interbreeding populations there are classes that differentially instantiate the properties causally connected to reproduction, and that some of these properties are heritable.

The question of whether such an account captures all the leading ideas of the theory of natural selection (as its orginator, Williams, claims) is a difficult one to answer definitively. Similarly, filling out the reductionist

9. For a detailed argument to this effect see Roseberg in (17).

program here envisioned may be difficult in the extreme. And because the failure of any particular attempt to fill it out in no way undercuts our confidence in the theory of natural selection, it is easy, if mistaken, to infer that the theory is a vacuous one. But once it is seen that the program is at least in principle susceptible of completion, the exercise need not be carried out. For merely seeing that the theory of natural selection can be shown to rest on such plainly empirical foundations is all that is required to understand both its power and its conceptual status as an extremely broad-gauged but nevertheless entirely contingent general theory.

Recognizing the supervenience of fitness on what I have called the manifest properties of organisms (or at least those which are causally connected to reproductive opportunities and successes) is especially important because it allows us to render consistent two significant considerations: first, the belief that the theory of natural selection is a scientific theory conceptually of a piece with other hypothetico-deductive systems in natural science, embodying concepts and predicates that characterize wholly physical events, states, and conditions; and second, the admission that these concepts and predicates cannot be reduced or exhaustively defined, by any means we are ever likely to have at hand, to concepts and predicates of more fundamental theories in biology, chemistry or physics. It is important to render this belief and this admission consistent because

arguments against the scientific standing of a theory often proceed by citing the impossibility of reducing the terms of the theory to those of undeniably scientific theories, and go on to suggest that the unreducible terms of the suspect theory are methologically illegitimate. It is equally important to render the belief in question consistent with the admission that reduction is impossible, because this same impossibility of reduction is frequently appealed to in order to substantiate irrationalist claims about the incommensurability of scientific theories and skeptical doubts about versions of the thesis of the unity of the sciences.

The supervenience of fitness on manifest properties of organisms enables us to say that (1) the state of a particular organism's manifesting a given level of fitness at a given place and time is identical with, is nothing but, the state of its exemplifying a certain set of manifest properties, even though (2) in general the property of having that level of fitness is not identical with or extensionally equivalent to the complex property of having those particular manifest properties. Thus a particular exemplification of fitness is subject to just the (qualitatively and numerically) same causal and mereological forces as the instantiation of a particular set of manifest properties is, even though two distinct instantiations of the same level of fitness are each likely to be identical with instantiations of different sets of manifest properties.[10] Thus, even

10. Here I have implicitly employed a criterion of event identity due to Davidson (3): Events are identical if their causes and effects are identical. Regardless of whether this

though the terms of the theory of natural selection may not be reducable to those of more fundamental theories, these terms are employed in the theory to describe and explain the very same events, their cause and effects, as the terms of more fundamental theories; and if the more fundamental theories provide more fundamental (if less manageable) explanations of the same events, then in and of itself the failure to provide reduction functions for the terms of the theory of evolution cannot cast doubt either on the scientific legitimacy of this theory or its consistency with the thesis of the unity of science.

## IV

The flexibility which supervenience affords in the account of the nature of fitness, and the consequences of this nature for the interaction of the theory of evolution and other theories, also enables us to state the claims of the reductionist with respect to population genetics and molecular genetics in a brief but precise way and, moreover, in a way that blunts complaints that the lack of manageable reduction functions between the genes of the former theory and the molecular material of the latter theory demonstrates either the nonreducibility, or the incommensurability of these theories or the numerical diversity of genes and strings of DNA.

The chief complaint lodged against the reduction of population genetics —and in particular its transmission genetic variant—to a theory about the behavior of DNA molecules has been clearly expressed by David Hull:

> One does not have to look very deeply into the relations between Mendelian and molecular genetics to discover how naive the [reductionist's] expectations actually are. Even if all gross phenotypic traits are translated into molecularly characterized traits the relation between Mendelian and molecular predicate terms express prohibitively complex, many-many relations. Phenomena characterized by a single Mendelian predicate can be produced by several different types of molecular mechanisms. Hence any possible reduction will be complex. Conversely, the same types of molecular mechanism can produce phenomena that must be characterized by different Mendelian predicates. Hence, reduction is impossible. [6, p. 61]

Elsewhere, Hull has stated that if the reduction of transmission genetics to molecular genetics is to be possible, "the one situation which is precluded ... is a single molecular entity or mechanism being identified with two or more Mendelian entities, properties, or relations (7, p. 571). And, of course, in defense of his denial of the possibility of identifying transmission genes with strings of DNA, Hull goes on to cite cases in which a single type of molecular item can result in differing phenotypic effects, and must therefore be identified with

---

criterion gives the meaning of event-identity claims, it is, in fact, the most useful practical guide to such identities.

at least two different types of transmission genes. If this sort of identification cannot be effected, then, he infers,
there is little hope of actually deducing the laws of transmission genetics
from those of molecular genetics, and
the consequences will be grave either
for the post-positivist notion of reduction or the empiricist commitment
to the commensurability of theories.
Defenders of the prospects for reduction between these theories have responded by admitting the "many-
many" relations between Mendelian
predicates and molecular ones, but
denying that this makes for the
impossibility of identifying types of
transmission genes and molecular
items. That is, although the property
of being a Mendelian gene of type $M$
may be realized by several different
kinds of molecular structures and
combinations—and the property of
having molecular structure of type
$S$ may be realized in several different
types of Mendelian genes—nevertheless, a particular item of type $M$
may be numerically identical with an
item of type $S$. This claim is typically
made out by noting that the property
of being a Mendelian gene is a function of a (gross) phenotype for which
the items code, and that in different
molecular environments the same type
of molecular item can code for different phenotypes. Given a molecular
environment, the phenotype for
which a string of DNA will code is
determined thereby, and the type
of Mendelian gene which that string
is identical to is also determined. It
follows that the particular molecular

item, of type $S$, say, is identical *in
this case* with a Mendelian gene, of
type $M$.[11] Hull has derided this strategy, describing it as the suggestion
that "if a reduction function becomes
too complicated, then get rid of the
complexity by some vague reference
to differences in the environment."
His point is that showing the identity
of two events by showing that their
causal connections in other particular
events are identical is not enough to
show that any predicate of one theory
that explains or describes the event
is identical with or extensionally
equivalent to any particular single
predicate of another theory which
explains or describes the same event.
He writes:

> It may well be true that a particular
> molecular mechanism of some kind
> or other can be found for each
> instance of a particular pattern of
> Mendelian genetics on a case by
> case basis, but more than that is
> required for reduction. Reduction
> functions relate *kinds* of entities
> and predicates, not particulars. The
> problem is to discover natural kinds
> in the two theories which can be
> related systematically in reduction
> functions. (7, p. 638)

The only answer reductionists have
thus far offered against this conclusion is the claim that the strategies
of particular identifications may well
not be as numerous as the actual
number of identifications, and that
there are probably a large but manageable number of generalizations to

---

11. Ruse (14) argues to this effect in his detailed discussion of Hull's views. This paper
and Schaffner's cited below contain a wealth of detail useful to anyone interested in the
relations between the theories in question.

which we can appeal in effecting identifications of Mendelian and molecular items (14). Not only is this a weak answer, but it concedes Hull's crucial claim that the reduction of population genetics to molecular genetics turns on the provision of biconditionals between Mendelian "natural kinds" and molecular "natural kinds." Whether each and every event describable in Mendelian terms can also be described in molecular terms, and thus shown to be the same event under different descriptions, is an empirical matter. And no enumeration of actual successes (enlightening as they may be) in the provision of such examples will forestall complaints like Hull's. Moreover, no particular failure to show this sort of identity will ever give the reductionist serious pause. For, armed with his commitment to causal and mereological determinism, he will simply treat any such failure as a temporary practical limitation on the progress of science. Thus the issue between Hull and the reductionist comes to a standoff. In the face of Hull's complaint of "Ad hoc-ary" and "vague references to differences in environment," the reductionist needs a new way of stating his claims, a way that is precise and faces or circumvents the problem that Hull may be correct in his denial that we can provide manageably long biconditionals between molecular and Mendelian predicates. The formalization of supervenience sketched above provides such a way. It enables the reductionist to have his reduction—at least in its ontological aspects—without having to provide all or any of the reduction-functions—the biconditionals between molecular and Mendelian items—that Hull required. And, if the terms of population genetics are supervenient on those of molecular genetics, then the truth of the laws of the latter theory can be explained on the basis of the laws of the former theory, via appeal to the numerical identity of every Mendelian event with a molecular event, even though no biconditionals between the terms of these theories are stated. In short, supervenience enables the reductionist to appeal to his conviction that particular molecular and Mendelian events are identical while avoiding Hull's objection.

## V

Call the set of all Mendelian predicates $M$, and the set of all properties constructable from $M$, $M^*$. Call the parallel sets of predicates of molecular genetics, $L$ and $L^*$. The set of Mendelian predicates, $M$, is supervenient on the set of molecular predicates, $L$, if and only if any two items that both have the same properties in the set $L^*$, also have the same properties in $M^*$. Thus, on the assumption, if an item exemplifies a certain set of Mendelian properties, then there is a set of molecular properties and relations that it also exemplifies, and if any other item also exemplifies the same set of molecular properties and relations, it manifests exactly the same Mendelian properties as well. Of course, if the number of predicates in either or both of the sets $M$ and $L$ is infinite, then the first set may be supervenient on the second without any finitely long biconditionals between members of $M$ and $L$ being stateable. That is, it may be impossible, in Hull's terms, "to discover natural kinds in the two theories which can be related systematically in reduction functions." Nevertheless, if $M$ is supervenient on $L$, it is only natural to infer that a particular item's having some Mendelian property is identical to its being in some molecular state or other. This is the kind of ontological reduction, without theoretical reduction, that supervenience permits. Of course, if the number of predicates in $M$ or $L$ is finite, even though extremely large, the supervenience of $M$ on $L$ will entail at least the in-principle possibility of providing the relevant biconditionals; although the larger the number of predicates in $M$ or $L$ the more impractical and useless such biconditionals will be. This may explain why, as Hull notes, "no geneticists to [Hull's] knowledge are attempting to derive the principles of transmission genetics from those of molecular genetics" (6, p. 44, n.1); and it may explain it in a way wholly consistent with the reductionist's identity claims. The possibility in principle of producing the relevant biconditionals is thus tantamount to the claim that molecular items can be characterized by molecular predicates which are exhaustive and mutually exclusive of one another, because they are constructed from conjunctions and/or disjunctions that include every predicate in $L$ or its complement. Let us call this set of predicates $L_e$. If there are such finitely expressible predicates, then each item characterized in purely Mendelian terms exemplifies one and only one member of $L_e$. And in virtue of the supervenience of Mendelian on molecular predicates, every item that manifests the Mendelian property in question manifests the same member of $L_e$. Of course, even if we had any assurance that the number of predicates in the supervening set was finite, so that $L_e$ is finite or recursively statable, we would have no assurance that we can in fact state any of the members of $L_e$, or the biconditionals whose existence they assure us of.

One of the special advantages of the notion of supervenience is that it enables us to account for the reduction of one theory to another even when we know that the first theory is false, or at any rate incompatible with the other. In the last few sections, I have been referring to nonmolecular genetics sometimes as Mendelian theory, sometimes as transmission theory, and sometimes ambiguously as population genetics. Transmission genetics represents a successor to Mendelian genetics, more sophisticated in detail and able to account for phenomena that disconfirm its predecessor (like linkage and cross-over). In his attempt to show that Mendelian genetics can be reduced by deduction to molecular genetics, even though the theories presumably differ in truth value. Schaffner introduces the further requirements on traditional logical empiricist reduction, that the reduced theory be corrected and that there be a strong analogy between the reduced and the corrected theory.[12] Hull in turn disputes the existence of such analogy between Mendelian and transmission genetics, and consequently

12. Kenneth Schaffner (15) and in connection with Hull's views, "Reductionism in Biology: Prospects and Problems," in Michalos and Cohen (11, pp. 627–646).

the deductive reduction of Mendelian to molecular genetics (6, p. 43). The dispute about whether there is an analogy here and if so, whether it is strong enough, is bedeviled by the vagueness of the concept of analogy. Indeed this vagueness makes it difficult to expect the kind of agreement desired about putative cases of reduction by deduction. Supervenience enables us to circumvent this issue and thus Hull's objection. For the supervenience of Mendelian predicates on transmission predicates, and of both of these sets of predicates on molecular ones enables us to identify states, conditions, events, and objects characterized in any of these terms with one another, without committing ourselves to the deducibility of Mendelian or transmission laws from the laws of bio-chemistry. Indeed, because the relation between the terms of these theories is that of supervenience we should expect that laws expressed in the supervenient terms are not deducible from those expressed in the supervening terms, since the bi-conditionals required for this deduction are not available, and may not even exist. Moreover, if the general statements of Mendelian or transmission theory are in fact vitiated by exceptions, this suggests that the terms that figure in these statements are not "natural kinds" after all, for it is characteristic of a "natural kind" that all its members share the same causal consequents and antecedents. If we assume that the terms of biochemistry do describe "natural kinds" then the failure to provide the bi-conditions that Hull demands is inevitable. But the kind of reduction that supervenience allows of does not require that both sets of terms

describe "natural kinds" and thus undercuts Hull's demand that we "discover natural kinds in the two theories which can be related systematically in reduction functions."

VI

In both my discussion of the supervenience of fitness on manifest properties of organisms and of the supervenience of population genetic predicates on molecular ones, I have been assuming that if these terms are related by supervenience, then a particular characterized in one of these terms will be identical to one characterized in the other terms (provided the spatio-temporal locations of the particulars are identical). While this assumption may be deemed reasonable in the case of fitness, it might be considered quite question-begging by a serious opponent of reductionism in biology. I agree, the assumption is question-begging, in the sense that the issue here is not whether the items differently characterized are numerically identical or not. Whether they are or not is an empirical issue, which I believe is being settled in favor of the reductionist. The question at issue is not whether genes are numerically identical to strings of DNA, but how to express this identity in a way that is consistent with the complexities of the actual relations between the two theories in question.

I have tried to express this identity, subject to the constraint just mentioned, by in effect substituting *ontological* reduction for *theoretical* reduction. That is I have shown how it is possible to argue that the referents of the two theories are identical even when the predicates of the theories are not stably identifiable. To this extent my conclusions should be sympathetic to a philosopher like Hull who believes that the relation between population genetics and molecular genetics is a paradigm case of reduction, and cites the difficulties in effecting this reduction as an objection to the traditional post-positivist account of reduction. *Mutatis mutandis,* a reductionist like Schaffner will take no satisfaction in my result if his chief objective in examining the relation between these theories is to substantiate the traditional account of reduction by deduction. For supervenience erects at least a practical obstacle, and at most a logical obstacle to the satisfaction of the traditional criterion of connectability between these theories. But the really important objective of any defender of the traditional view of reduction is to demonstrate that theories are commensurable, that they can be assessed on rational grounds, that their succession represents scientific progress, not merely scientific change, and that the relation between theories reflects the unity of science. Treating the terms of population genetics as supervenient on those of molecular genetics allows for the commitment to all these goals: it does so simply by permitting us to assert that despite

the potential impossibility of manageable interconnections between these terms, the ontology of these theories is just the same. So that, whether an event or object described in one theory is identical to an event or object described in another will turn not on explicitly statable relations between the theories, but on general considerations of event- and object-identity Insofar as the claim of supervenience allows for all these traditional empiricist goals, it can hardly be much of an objection that embracing the claim involves surrendering the letter, though not the spirit, of the traditional picture of reduction.

At the outset of this paper, I described one of my aims as explaining why the theory of natural selection is a contingent one, whose contingency and *appearance* of tautology *both* rest on the supervenient character of fitness. I did not describe my other aim as expounding or defending the reductionist thesis that population genetics is reducible to molecular genetics, for this matter is an empirical one, not open to philosophical argument. My aim has rather been to express the thesis of reductionism in a new way: to state it, not to prove it, and to show how this new way of stating the thesis allows for the rich complexity which the interrelation between biological theories and concepts actually manifests.

## REFERENCES

(1) Bossert, W. & Wilson, E. O., 1971, *Primer of Population Genetics*, Stamford, Sinauer.

(2) Craig, W., 1956, "Replacement of Auxiliary Expressions." *Philosophical Review*, 65: 38-55.

(3) Davidson, D., 1969, "The Individuation of Events." In Rescher, et al., ed., *Essays in Honor of Carl Hempel*, Dordrecht, Reidel, pp. 140-147.

(4) ———, 1970, "Mental Events." In Foster, et al., eds., *Experience and Theory*, Amherst, University of Massachusetts Press.

(5) Fodor, J. A., 1968, *Psychological Explanation*, New York, Random House.

(6) Hull, D., 1974, *Philosophy of Biological Science*, Englewood Cliffs, Prentice Hall.

(7) ———, "Informal Aspects of Theory Reduction." In (11).

(8) Kim, J., 1977, "Nomological incommensurables." Oberlin Colloquium.

(9) Levins, R., 1968, *Evolution in Changing Environments*, Princeton, Princeton University Press.

(10) ———, 1966, "Strategy for Model Building in Population Biology." *American Scientist*, 54: 421-431.

(11) Michalos, A., and Cohen, R., 1976, *PSA*, Dordrecht, Reidel.

(12) Peters, R. H., 1976, "Tautology in evolution and ecology." *American Naturalist*, 110: 1-12.

(13) Ruse, M., 1973, *Philosophy of Biology*, London, Hutchinson.

(14) ———, "Reduction in Genetics." In (11), pp. 633-651.

(15) Schaffner, K., 1967, "Approaches to reduction." *Philosophy of Science*, 34: 137-147.

(16) Williams, M. B., 1970, "Deducing the consequences of evolution." *Journal of Theoretical Biology*, 29: 343-385.

(17) Wilson, F., and L. W. Sumner, et al. *Pragmatism and Purpose: Essays Presented to T. A. Goudge*, Toronto, University Press, 1981.

# III
# THE UNITS OF SELECTION

# 9

# Caring Groups and Selfish Genes

## STEPHEN JAY GOULD

THE WORLD OF OBJECTS CAN BE ordered into a hierarchy of ascending levels, box within box. From atoms to molecules made of atoms, to crystals made of molecules, to minerals, rocks, the earth, the solar system, the galaxy made of stars, and the universe of galaxies. Different forces work at different levels. Rocks fall by gravity, but at the atomic and molecular level, gravity is so weak that standard calculations ignore it.

Life, too, operates at many levels, and each has its role in the evolutionary process. Consider three major levels: genes, organisms, and species. Genes are blueprints for organisms; organisms are the building blocks of species. Evolution requires variation, for natural selection cannot operate without a large set of choices. Mutation is the ultimate source of variation, and genes are the unit of variation. Individual organisms are the units of selection. But individuals do not evolve—they can only grow, reproduce, and die. Evolutionary change occurs in groups of interact-

ing organisms; species are the unit of evolution. In short, as philosopher David Hull writes, genes mutate, individuals are selected, and species evolve. Or so the orthodox, Darwinian view proclaims.

The identification of individuals as the unit of selection is a central theme in Darwin's thought. Darwin contended that the exquisite balance of nature had no "higher" cause. Evolution does not recognize the "good of the ecosystem" or even the "good of the species." Any harmony or stability is only an indirect result of individuals relentlessly pursuing their own self-interest—in modern parlance, getting more of their genes into future generations by greater reproductive success. Individuals are the unit of selection; the "struggle for existence" is a matter among individuals.

During the past fifteen years, however, challenges to Darwin's focus on individuals have sparked some lively debates among evolutionists. These challenges have come from above and below. From above,

Scottish biologist V. C. Wynne-Edwards raised orthodox hackles fifteen years ago by arguing that groups, not individuals, are units of selection, at least for the evolution of social behavior. From below, English biologist Richard Dawkins has recently raised my hackles with his claim that genes themselves are units of selection, and individuals merely their temporary receptacles.

Wynne-Edwards presented his defense of "group selection" in a long book entitled *Animal Dispersion in Relation to Social Behavior.* He began with a dilemma: Why, if individuals only struggle to maximize their reproductive success, do so many species seem to maintain their populations at a fairly constant level, well matched to the resources available? The traditional Darwinian answer invoked external constraints of food, climate, and predation: only so many can be fed, so the rest starve (or freeze or get eaten), and numbers stabilize. Wynne-Edwards, on the other hand, argued that animals regulate their own populations by gauging the restrictions of their environment and regulating their own reproduction accordingly. He recognized right away that such a theory contravened Darwin's insistence on "individual selection," for it required that many individuals limit or forgo their own reproduction for the good of their group.

Wynne-Edwards postulated that most species are divided into many more-or-less discrete groups. Some groups never evolve a way to regulate their reproduction. Within these groups, individual selection reigns supreme. In good years, populations rise and the groups flourish; in bad years, the groups cannot regulate themselves and face severe crash and even extinction. Other groups develop systems of regulation in which many individuals sacrifice their reproduction for the group's benefit (an impossibility if selection can only favor individuals that seek their own advantage). These groups survive the good and the bad. Evolution is a struggle among groups, not individuals. And groups survive if they regulate their populations by the altruistic acts of individuals. "It is necessary," Wynne-Edwards wrote, "to postulate that social organizations are capable of progressive evolution and perfection as entities in their own right."

Wynne-Edwards reinterpreted most animal behavior in this light. The environment, if you will, prints only so many tickets for reproduction. Animals then compete for tickets through elaborate systems of conventionalized rivalry. In territorial species, each parcel of land contains a ticket and animals (usually males) posture for the parcels. Losers accept gracefully and retreat to peripheral celibacy for the good of all. (Wynne-Edwards, of course, does not impute conscious intent to winners and losers. He imagines that some unconscious hormonal mechanism underlies the good grace of losers.)

In species with dominance hierarchies, tickets are alloted to the appropriate number of places, and animals compete for rank. Competition is by bluff and posture, for animals must not destroy each other by fighting like gladiators. They are, after all, only competing for tickets to benefit the group. The contest is more of a lottery than a test of skills; a distribution of the right number of tickets is far more important than who wins. "The conventionalization of rivalry and the foundation of society are one and the same thing," Wynne-Edwards proclaimed.

But how do animals know the number of tickets? Clearly, they cannot, unless they can census their own populations. In his striking hypothesis, Wynne-Edwards suggested that flocking, swarming, communal singing, and chorusing evolved through group selection as an effective device for censusing. He included "the singing of birds, the trilling of katydids, crickets and frogs, the underwater sounds of fish, and the flashing of fireflies."

Darwinians came down hard on Wynne-Edwards in the decade following his book. They pursued two strategies. First, they accepted most of Wynne-Edward's observations, but reinterpreted them as examples of individual selection. They argued, for example, that *who* wins is what dominance hierarchies and territoriality are all about. If the sex ratio between males and females is near 50:50 and if successful males monopolize several females, then not all males can breed. Everyone competes for the Darwinian prize of passing more genes along. The losers don't walk away with grace, content that their sacrifices increase the common good. They have simply been beaten; with luck, they will win on their next try. The result may be a well-regulated population, but the mechanism is individual struggle.

Virtually all Wynne-Edward's examples of apparent altruism can be rephrased as tales of individual selfishness. In many flocks of birds, for example, the first individual that

spots a predator utters a warning cry. The flock scatters but, according to group selectionists, the crier has saved his flockmates by calling attention to himself—self-destruction (or at least danger) for the good of the flock. Groups with altruist criers prevailed in evolution over all selfish, silent groups, despite the danger to individual altruists. But the debates have brought forth at least a dozen alternatives that interpret crying as beneficial for the crier. The cry may put the flock in random motion, thus befuddling the predator and making it less likely that he will catch anyone, including the crier. Or the crier may wish to retreat to safety but dares not break rank to do it alone, lest the predator detect an individual out of step. So he cries to bring the flock along with him. As the crier, he may be disadvantaged relative to flockmates (or he may not, as the first to safety), but he may still be better off than if he had kept silent and allowed the predator to take someone (perhaps himself) at random.

The second strategy against group selection reinterprets apparent acts of disinterested altruism as selfish devices to propagate genes through surviving kin—the theory of kin selection. Siblings, on average, share half their genes. If you die to save three sibs, you pass on 150 percent of yourself through their reproduction. Again, you have acted for your own evolutionary benefit, if not for your corporeal continuity. Kin selection is a form of Darwinian individual selection.

These alternatives do not disprove group selection, for they merely retell its stories in the more conventional Darwinian mode of individual selection. The dust has yet to settle on this contentious issue but a consensus (perhaps incorrect) seems to be emerging. Most evolutionists would now admit that group selection can occur in certain special situations (species made of many very discrete, socially cohesive groups in direct competition with each other.) But they regard such situations as uncommon if only because discrete groups are often kin groups, leading to a preference for kin selection as an explanation for altruism within the group.

Yet, just as individual selection emerged relatively unscarred after its battle with group selection from above, other evolutionists launched an attack from below. Genes, they argue, not individuals, are the units of selection. They begin by recasting Butler's famous aphorism that a hen is merely the egg's way of making another egg. An animal, they argue, is only DNA's way of making more DNA. Richard Dawkins has put the case most forcefully in his recent book *The Selfish Gene.* "A body," he writes, "is the genes' way of preserving the genes unaltered."

For Dawkins, evolution is a battle among genes, each seeking to make more copies of itself. Bodies are merely the places where genes aggregate for a time. Bodies are temporary receptacles, survival machines manipulated by genes and tossed away on the geological scrap heap once genes have replicated and slaked their insatiable thirst for more copies of themselves in bodies of the next generation. He writes:

We are survival machines—robot vehicles blindly programmed to preserve the selfish molecules known as genes. . . .

They swarm in huge colonies, safe inside gigantic lumbering robots . . . they are in you and me; they created us, body and mind; and their preservation is the ultimate rationale for our existence.

Dawkins explicitly abandons the Darwinian concept of individuals as units of selection: "I shall argue that the fundamental unit of selection, and therefore of self-interest, is not the species, nor the group, nor even, strictly, the individual. It is the gene, the unit of heredity." Thus we should not talk about kin selection and apparent altruism. Bodies are not the appropriate units. Genes merely try to recognize copies of themselves wherever they occur. They act only to preserve copies and make more of them. They couldn't care less which body happens to be their temporary home.

I begin my criticism by stating that I am not bothered by what strikes most people as the most outrageous component of these statements—the imputation of conscious action to genes. Dawkins knows as well as you and I do that genes do not plan and scheme; they do not act as witting agents of their own preservation. He is only perpetuating albeit more colorfully than most, a metaphorical shorthand used (perhaps unwisely) by all popular writers on evolution, including myself (although sparingly, I hope). When he says that genes strive to make more copies of themselves, he means: "selection has operated to favor genes that, by chance, varied in such a way that more copies survived in subsequent generations." The second is quite a mouthful; the first is direct and acceptable as metaphor although literally inaccurate.

Still, I find a fatal flaw in Dawkins's attack from below. No matter how much power Dawkins wishes to assign to genes, there is one thing that he cannot give them—direct visibility to natural selection. Selection simply cannot see genes and pick among them directly. It must use bodies as an intermediary. A gene is a bit of DNA hidden within a cell. Selection views bodies. It favors some bodies because they are stronger, better insulated, earlier in their sexual maturation, fiercer in combat, or more beautiful to behold.

If, in favoring a stronger body, selection acted directly upon a gene for strength, then Dawkins might be vindicated. If bodies were unambiguous maps of their genes, then battling bits of DNA would display their colors externally and selection might act upon them directly. But bodies are no such thing.

There is no gene "for" such unambiguous bits of morphology as your left kneecap or your fingernail. Bodies cannot be atomized into parts, each constructed by an individual gene. Hundreds of genes contribute to the building of most body parts and their action is channeled through a kaleidoscopic series of environmental influences: embryonic and postnatal, internal and external. Parts are not translated genes, and selection doesn't even work directly on parts. It accepts or rejects entire organisms because suites of parts, interacting in complex ways, confer advantages. The image of individual genes, plotting the course of their own survival, bears little relationship to developmental genetics as we understand it. Dawkins will need another metaphor: genes caucusing, forming alliances, showing deference for a chance to join a pact, gauging probable environments. But when you amalgamate so many genes and tie them together in hierarchical chains of actions mediated by environments, we call the resultant object a body.

Moreover, Dawkins's vision requires that genes have an influence upon bodies. Selection cannot see them unless they translate to bits of morphology, physiology, or behavior that make a difference to the success of an organism. Not only do we need a one-to-one mapping between gene and body (criticized in the last paragraph), we also need a one-to-one *adaptive* mapping. Ironically, Dawkins's theory arrived just at a time when more and more evolutionists are rejecting the panselectionist claim that all bits of the body are fashioned in the crucible of natural selection. It may be that many, if not most, genes work equally well (or at least well enough) in all their variants and that selection does not choose among them. If most genes do not present themselves for review, then they cannot be the unit of selection.

I think, in short, that the fascination generated by Dawkins's theory arises from some bad habits of Western scientific thought—from attitudes (pardon the jargon) that we call atomism, reductionism, and determinism. The idea that wholes should be understood by decomposition into "basic" units; that properties of microscopic units can generate and explain the behavior of macroscopic results; that all events and objects have definite, predictable, determined causes. These ideas have been successful in our study of simple objects, made of few components, and uninfluenced by prior history. I'm pretty sure that my stove will light when I turn it on (it did). The gas laws build up from molecules to predictable properties of larger volumes. But organisms are much more than amalgamations of genes. They have a history that matters; their parts interact in complex ways. Organisms are built by genes acting in concert, influenced by environments, translated into parts that selection sees and parts invisible to selection. Molecules that determine the properties of water are poor analogues for genes and bodies. I may not be the master of my fate, but my intuition of wholeness probably reflects a biological truth.

# 10

# Replicator Selection and the Extended Phenotype

## RICHARD DAWKINS

*Adaptations are often spoken of as "for the good of" some entity, but what is that entity? Groups and species are now rightly unfashionable, so what are we left with? The prevailing answer is Darwin's "the individual." Individuals clearly do not maximize their own survival, so the concept of fitness had to be invented. If fitness is correctly defined in Hamilton's way as "inclusive fitness," it ceases to matter whether we speak of individuals maximizing their inclusive fitness or of genes maximizing their survival. The two formulations are mutually intertranslatable. Yet some serious mistranslations are quoted from the literature, which have led their authors into actual biological error. The present paper blames the prevailing concentration on the individual for these errors, and advocates a reversion to the replicator as the proper focus of evolutionary attention. A gene is an obvious replicator, but there are others, and the general properties of replicators are discussed. Defenders of the individual as the unit of selection often point to the unity and integration of the genome as expressed phenotypically. This paper ends by attacking even this assumption, not only by a reductionist fragmentation of the phenotype, but, on the contrary, by extending it to include more than one individual. Replicators survive by virtue of their effects on the world, and these effects are not restricted to one individual body but constitute a wider "extended phenotype."*

### INTRODUCTION

Sociobiology is a name that has acquired irritating pretensions, but we shall probably have to learn to live with it. Whatever may have been E. O. Wilson's (1975) definition, the aspect of "sociobiology" which has captured the imagination of biologists (other than the minority overexcited by political misunderstanding) is a particular neo-Darwinian view of social ethology. Wilson sums this up in his first chapter, "The Morality of

This is a modified version of a lecture given in the plenary session on "Sociobiology" at the 15th International Ethological Conference, Bielefeld, 1977.

the Gene," where he identifies the central problem of sociobiology as the problem of altruism, and gives as the answer: "kinship." I would characterize the approach as the "selfish gene" approach to ethology, and I unhesitatingly name as its founding genius W. D. Hamilton. Not only did Hamilton (1964, 1971, 1972, 1975) supply the theory of inclusive fitness, bulwark of Wilson's admirable new synthesis. John Maynard Smith has told us that Hamilton (1967) was also an inspirer of the concept of the *evolutionarily stable strategy,* which has been developed by Maynard Smith et al. (1973, 1974, 1976) into another central plank of modern sociobiological theory (Dawkins 1976a). With acknowledgment to Konrad Lorenz's well known tribute to Oskar Heinroth, I would define sociobiology as the branch of ethology inspired by W. D. Hamilton. But Hamilton did not go far enough. Paradoxically, the logical conclusion to his ideas should be the eventual abandonment of his central concept of inclusive fitness. We should also move toward giving up the term "kin selection" as well as group selection and individual selection. Instead of all of these we should substitute the single term "replicator selection."

Evolutionary models, whether they call themselves group-selectionist or individual-selectionist, are fundamentally gene-selectionist. They work within the population geneticist's assumption that natural selection acts by changing the relative frequencies of alleles in gene pools. The thing which changes in evolution is the gene pool, and the things between which nature fundamentally selects are alternative alleles. But genes don't literally float free in a pool, they go around in individual bodies and are selected by virtue of their effects on individual phenotypes. A biologist can count on a chorus of approving nods if he says that, in the last analysis, selection works on "outcomes": it is the whole individual that has to survive, the whole individual who faces the cutting edge of natural selection. Superficially sensible as this sounds, it can be called in question. Selection means differential survival, and the units which survive in the long run are not individuals but *replicators* (genes or small fragments of genome). They survive by virtue of their phenotypic outcomes, to be sure, but these are best interpreted not exclusively at the individual level, but in terms of the doctrine of the *extended phenotype.* Replicator selection and the extended phenotype will be discussed in the last two sections of this paper, after some preliminary matters have been dealt with.

Individuals in their turn go around in larger units—groups and species. Some biologists have accordingly argued that intergroup selection is an important cause of adaptive evolution (Wynne-Edwards 1962). Wynne-Edwards (1977) has recently written: "The general consensus of theoretical biologists at present is that credible models cannot be devised, by which the slow march of group selection could overtake the much faster spread of selfish genes that bring gains in individual fitness. I therefore accept their opinion." But, whether or not we found models of group selection convincing, my point here is that in any case these models were always framed as special cases of *gene-selection* models. Intergroup selection and interindividual selection, and all the other levels reviewed by Lewontin (1970) and Wickler (1976), are different proximal processes whose claim to biological importance is judged on the extent to which they can be shown to correlate with what really matters—inter-allele selection. I believe it is often superfluous, and sometimes actually misleading, to discuss natural selection at these higher levels. It is usually better to go straight to the fundamental level of selection among *replicators*—single genes or fragments of genetic material which behave like long-lived units in the gene pool.

This amounts to a plea that the good example of population geneticists should be followed by those of us who want to discuss adaptation or function. We often wish to attribute "benefit" to some entity. Thus an animal may be said to show parental care "for the good of the species" or "for the good of its own fitness." The first of these is almost certainly wrong (Williams 1966), the second right if fitness is correctly defined in Hamilton's way. But, in any case, how much more compelling it is to say: "Genes which make individuals more likely to perform parental care than their alleles work for the survival of copies of themselves in the bodies of the young cared for." Or, more briefly and generally, genes work for their own benefit, using individual bodies as their agents. We substitute the easily understood notion of *survival* (gene survival) for the complex and difficult concept of *fitness* (individual fitness).

### FITNESS

In Herbert Spencer's (1864) day the fittest survived, and the "fittest" were understood in the everyday sense of the most muscular, fleetest of foot, brainiest. For Spencer, fitness was passed on because the individuals best fitted to their way of life survived to reproduce. Fitness was the capacity to survive, and survival was a prerequisite for reproductive success. It was only later that fitness started to *mean* reproductive success, and the fitness of an individual could hence, without contradiction, be said to be increased by sexually attractive characters which detracted from individual survival.

Both in Spencer's sense, and in the sense of reproductive success, fitness was attributed to *individuals*. But population geneticists developed an independent usage of the word, and they applied it not to individuals but to genotypes at a locus. This made sense, because you can count the number of occurrences of a particular genotype, say Aa, in a population, relative to its alternatives at the same locus. An equivalent count in the next generation, followed by a normalizing division sum, leads to a direct quantitative estimate of the fitness of Aa relative to, say, AA and aa. This is quite different from the idea of individual fitness. You can't count the number of times an individual occurs in a sexually reproducing population, for he only occurs once, ever. If you want to measure the "fitness" of an individual, you have to resort to something like counting the number of his fledged offspring. In the light of Hamilton's inclusive fitness concept we can now see that this is a very crude approximation. Offspring turn out to be only a special case of close genetic relatives

*Table 1*

| "Unit of Selection"  | Quantity maximized  |
|----------------------|---------------------|
| Individual           | Inclusive fitness   |
| Gene                 | Replication         |

with a high probability of sharing one's own genes.

Hamilton's rationale is best explained at the level of genes. Thus parental care and sibling care both evolved because genes for such caring behavior tend to be present in the bodies of the individuals cared for. But Hamilton expressed the idea at the level of the individual: the individual works so as to maximize his inclusive fitness. Inclusive fitness may be defined as that property of an individual organism which will appear to be maximized when what is really being maximized is gene survival. This is not his own definition, but Dr. Hamilton allows me to say that it is the ideal inclusive fitness to which his actual concept was an approximation. Table 1 shows the two equivalent ways of expressing what happens in natural selection.

For different purposes it is convenient to use sometimes the individual/inclusive-fitness formulation, sometimes the gene/replication formulation. We should become adept at translating rapidly between the two. Unfortunately, some serious mistranslations have appeared in the literature. Since, for reasons which I have given at length (Dawkins 1976a, following Williams 1966, pp. 22–25), I believe that the gene/replication formulation is to be preferred when there is any apparent conflict, I am prepared to say that these mistranslations have led their authors into actual biological

error. Readers who do not accept this preference for replicator selection rather than "individual selection" may at least agree that the following examples demonstrate confusion. As for outright error, the only disagreement should be over whether it is mine or the authors' whom I quote.

### CONFUSION

It is an important part of my case that the concept of individual fitness has proved itself to be actively misleading. It is therefore necessary that I demonstrate from the literature that people have been misled. I do not intend this in a carping or ungracious spirit. My case is against a fashionable concept, and the more distinguished the authors who have been misled, the stronger the indictment against the concept.

The ordinary everyday usage of 'fitness' is so deeply ingrained that the special neo-Darwinian meaning is hard to get used to. Here is a distinguished American ecologist writing as recently as 1960. He first quoted Waddington's (1957) definition of survival in the modern sense: ". . . survival does not, of course mean the bodily endurance of a single individual . . . That individual 'survives' best which leaves most offspring." Then the eminent ecologist goes on: "Critical data on this contention are difficult to find, and it is likely that much new investigation is needed before the point is either verified or refuted." He apparently thought that Waddington was making a statement of fact about survival, whereas Waddington was really *defining* survival in the new sense of individual fitness. No wonder this poor ecologist had such trouble grappling with mammary glands: "It would be extremely difficult to explain the evolution of the uterus and mammary glands in mammals . . . as the result of natural selection of the fittest individual." He goes on to recommend a group-selectionist interpretation. I think it would be discourteous to regard his confusion as anything but a black mark against the concept of individual fitness.

Here is another example of the trouble that can result from carelessly looking at adaptation in terms of individual benefit. It is often pointed out that some coefficients of relationship are exact while others are probabilistic. For instance the coefficient between brothers is ½, but this "is an average figure: by the luck of the meiotic draw, it is possible for particular pairs of brothers to share more or fewer genes than his. The relatedness between parent and child is always exactly ½" (Dawkins 1976a, p. 98). Gibson (1976) correctly stated this point, but then went on to draw an incorrect inference. She supposed that an adult might invest in a son rather than in a full sibling because nature might prefer "a sure thing (relatedness = 0.5 as in the case of the son) to gambling (average relatedness = 0.5 as in the case of siblings)". But only an individual could see the son as a "sure thing." From the point of view of a single gene determining parental or brotherly behavior, the son is no more a sure thing than the brother: both are gambles with 50 percent odds (Dawkins (1976b).

Fagen (1976) made a similar mistake in the course of worrying about something called the "doting grandparent problem." The number of a grandparent's genes inherited by a given grandchild is ¼, but only *on average*. Some grandchildren will inherit more than ¼ of the grandparent's genome, others less. So, the author reasoned, "grandparents should tend to detect and favour those grandchildren having a disproportionate number of grandparental genes . . . physical resemblance of grandchildren to grandparents should serve as an important releaser of doting (and is expected to lead to endless discussions of 'grandpa's chin' or 'grandma's eyes')." The fallacy is again easily seen. What matters is the replication of the gene or genes which make for doting. As Partridge and Nunney (1977) have pointed out, unless there is genetic linkage between genes for chins and genes for doting, grandpa should behave as if completely indifferent to whether any given grandchild has inherited his chin. In practice, linkage effects and uncertainty about whether an individual is a grandchild at all could lead to Fagen's being right, but if so it would be for the wrong reason. Fagen, like Gibson, was misled by the following mathematical equivalence. The coefficient of relationship between two relatives is equivalent to two things. It is the average *proportion* of the genome of one which is shared by the other. It is also the *probability* that a given gene in one will be identical by descent with one in the relative. What matters is this probability. The proportion is merely incidentally equivalent, but all too often it is what people think in terms of.

I am grateful to L. Partridge for calling my attention to the last example, and to P. J. Greene, for showing me yet a third example of the same error in a paper devoted to "exact versus probabilistic coefficients of relationship" (Barash et al. 1978). In this paper, Fagen's fallacy is repeated, but in a more stark and general form. More general because it makes the same point about relationships other than the grandparental one, and more stark because here it is not possible to save the argument by special pleading about linkage, pleiotropy, or detectability of relationship. Thus Fagen could defend grandpa's chin by pointing out that it could help grandpa to decide whether a particular child was really his grandchild at all, or by suggesting that the genes controlling facial appearance might be linked to the genes for grandparental altruism. Barash et al. have no such defense, since they were specifically concerned to emphasize the difference between exact and probabilistic coefficients of relationship, and the arguments about linkage etc. apply regardless of this distinction.

Now I want to mention a more subtle and important source of misunderstanding resulting from the individual fitness point of view. I refer to the so-called cost of meiosis. Williams (1971) has put it like this: "Suppose there were two kinds of females in a population; one produced monoploid, fertilizable eggs, and the other . . . diploid eggs . . . each with exactly the mother's genetic makeup. These parthenogenetic eggs would each contain twice as much of the mother's genotype as is present in a reduced and fertilized egg. Other things being equal, the parthenogenetic female would be twice as well represented in the next generation as the normal one. In a few generations, meiosis and sexual recombination should disappear . . . Meiosis is therefore a way in which an individual actively reduces its genetic representation in its own offspring . . . Sexual reproduction is analogous to a roulette game in which the player throws away half his chips at each spin."

It is with particular diffidence that I criticize a quotation from one of Darwin's foremost heirs. I believe William's expression of a cost of meiosis is misleading because the important question is not what happens to the whole genome of a female, but what happens to the gene or genes determining sexuality versus asexuality (Treisman and Dawkins 1976). By the way, to avoid a mistake which already appeared in the literature (Barash 1976), I must hasten to agree with Maynard Smith and Williams (1976) that this does not mean there is *no* cost of sex. Williams (1975) is right to stress that the existence of sexual reproduction really is a huge paradox, but it is not the same paradox as he originally said. A better expression of the true nature of the paradox is that of Maynard Smith (1971), but Professor Maynard Smith will agree with me that Trivers's (1976) way of explaining it is easier to understand. The true cost of sex is an economic cost resulting from the fact that fathers usually do not invest as much in their children as mothers do. Elsewhere I have gone into the nature of this cost, which I call the cost of paternal

*Table 2:* Probability that a gene on a particular type of chromosome (row titles) will be identical by descent to a gene in a relative (column titles). Male sex assumed heterogametic; if female heterogametic, reverse sex titles.

|  | *Sex* | *Chromosome* | *Brother* | *Sister* | *Father or son* | *Mother or daughter* |
|---|---|---|---|---|---|---|
| Normal Diploid | ♀ | X | ¼ | ¾ | ½ | ½ |
|  | ♂ | X | ½ | ½ | 0 | 1 |
|  |  | Y | 1 | 0 | 1 | 0 |
|  | Either | Autosome | ½ | ½ | ½ | ½ |
| Haplodiploid | ♀ | Any | ¼ | ¾ | ½ | ½ |
|  | ♂ | Any | ½ | ½ | - | 1 |

neglect (Dawkins 1978), and which might more generally be called the cost of anisogamy. All that is relevant here is that it is different from Williams's cost of meiosis. We are misled into the formulation of a cost of meiosis because, once again, of the habit of thinking about individual fitness (genome survival) rather than gene survival.

Here is an amusing little idea which would not occur to somebody who thought in terms of genome preservation rather than gene preservation. Hamilton (1972) pointed out that, as far as a gene on a X-chromosome was concerned, its probability of being shared by two siblings of the homogametic sex was ¾, not the usual ½ (see Table 2). For instance, in birds, a gene for brother to brother altruism, if it happened to be on an X-chromosome, should be favored by the same strong selection pressure as would favor a gene for sister to sister altruism in a haplodiploid hymenopterous insect. This could favor the evolution of helping at the nest by elder brothers. Hamilton modestly considered his idea too far-fetched to merit more than a paragraph, but it has recently been rediscovered and expounded at greater length (Whitney

1976), as has a Y-chromosome version of the same idea (Wickler 1977). The "green beard effect" (Dawkins 1976a, p. 96) represents an extreme of this way of thinking. All these ideas, even if they appear far-fetched in practice, are perfectly respectable in theory, and you would never think of them if you based your ideas on individual fitness rather than on gene replication.

One of the most pernicious consequences of the "individual selection" viewpoint is the notion that explanations in terms of "kin selection" are somehow unparsimonious. Zahavi (1975) says of one of his own entertaining theories: "Such an interpretation may provide an alternative to other hypotheses which assumed complicated selective mechanisms, such as group selection or kin selection which do not act directly on the individual." When he says "act directly on the individual" he must mean individual reproductive success, i.e., number of children and lineal descendants. He is implicitly using "kin" to refer to relatives *other* than offspring. Wilson (1975) incorporates this odd usage into an explicit definition, as I have criticized elsewhere (Dawkins 1976a, p. 102). Woolfenden (1975) similarly mars his discussion of Florida scrub jays helping at the nest by speaking of a "controversy about group or kin selection versus individual selection." The literature contains many efforts to explain facts in terms of "individual selection" without having to "resort" to kin selection. Of course "resort" is an entirely inappropriate verb. "Kin selection" is not a distinct kind of natural selection, to be invoked only when "individual selection" cannot explain the facts. Both kin selection and individual selection are logical consequences of gene selection. If we accept neo-Darwinian gene-selectionism, kin selection necessarily follows. There is, indeed, no need for the term "kin selection" to exist, and I suggest that we stop using it.

We round off this section, as we began it, with mammary glands. ". . . mammary glands contribute to individual fitness, the individual in this case being the kinship group" (Hull 1976). Wilson (1975) goes so far as to define kin selection as a special case of group selection. But there is no "kinship group" unless families happen to go around together—an incidental fact, not a necessary assumption. Individuals do not, in an all or none sense, either qualify or fail to qualify as kin. They have, quantitatively, a greater or less chance of containing a particular gene. If Hull *must* talk about individuals, the post-Hamiltion "individual" in his sentence is certainly not a group. It is an animal plus ½ of each of its children plus ½ of each sibling plus ¼ of each niece and grandchild plus ⅛ of each first cousin plus $\frac{1}{32}$ of each second cousin . . . Far from being a tidy, discrete group, it is more like a sort of genetical

octopus, a probabilistic amoeboid whose pseudopodia ramify and dissolve away into the common gene pool. We have reached the Darwinian equivalent of the Ptolemaic epicycles. It is time to go back to first principles. What really happens in natural selection?

## REPLICATOR SELECTION

We may define a *replicator* as any entity in the universe which interacts with its world, including other replicators, in such a way that copies of itself are made. A corollary of the definition is that at least some of these copies, in their turn, serve as replicators, so that a replicator is, at least potentially, an ancestor of an indefinitely long line of identical descendant replicators. In practice no replication process is infallible, and defects in a replicator will tend to be passed on to descendants. If a replicator exerts some power over the world, such that its nature influences the survival of itself and its copies, natural selection, and hence progressive evolution, may occur through differential replicator survival.

A DNA molecule is the obvious replicator. The mistakes which are made in its replication are the various kinds of gene mutation and also, since multicistron fragments of chromosome can qualify as replicators (see below), crossing over. The power which a gene exerts over its world is its influence on the synthesis of proteins which in turn influence the embryonic development of phenotypes. Since the gene rides inside the body whose development it influenced, its own long-term future is affected by its nature.

Bateson (1978) has criticized the view that an animal is the genes' way of making more genes by drawing an analogy which appears to reduce the idea to an absurdity. Birds build nests, and nests protect new growing birds. So you might as well say that a bird is a nest's way of making new nests! But Bateson's amusing analogy is a false one. A nest is not a true replicator because a "mutation" which occurs in the construction of a nest, for example the accidental incorporation of a pine needle instead of the usual grass, is not perpetuated in future "generations of nests." Similarly, protein molecules are not replicators, nor is messenger RNA.

A gene in the nucleus of a germ-line cell is a replicator, but a sexually reproducing individual organism is not. It does not make copies of itself. It propagates copies of its genes, but its genome is shredded to smithereens at meiosis. Because individual bodies are big things that we can watch moving about in apparently purposeful ways, we focus our attention on them. We forget the lesson of August Weismann: organisms are but the transient engines of long-term gene replication. The qualities of a good replicator may be summed up in a slogan reminiscent of the French Revolution: Longevity, Fecundity, Fidelity (Dawkins 1976a, 1978). Genes are capable of prodigious feats of fecundity and fidelity. In the form of copies of itself, a single gene may persist for a hundred million individual lifetimes. Some genes survive better than their alleles, which is what natural selection is all about. But neither individual organisms, whose copying fidelity is destroyed by meiosis, nor groups of individuals for similar reasons, deserve to be called replicators at all.

Why "replicator selection" rather than "gene selection"? One reason for preferring replicator selection is that the phrase automatically preadapts our language to cope with non-DNA-based forms of evolution such as may be encountered on other planets, and perhaps also cultural analogues of evolution (Dawkins 1976a, pp. 203–215). The term replicator should be understood to *include* genetic replicators, but not to exclude any entity in the universe which qualifies under the criteria listed above.

The other reason for avoiding "gene selection" is that we must not be forced into the position of saying that the single gene, in the narrow molecular biologists' sense of cistron, is the unit of selection. The problem of what fragments of genome should be regarded as units of selection is discussed from time to time in the mathematical genetics literature. The details are complicated and yet to be finally resolved (Lewontin 1974), but whatever conclusions the geneticists come up with are of great importance to those of us who want to talk about adaptation. Here is one opinion:

"It is clear that when permanent linkage disequilibrium is maintained in a population, the higher order interactions are important and the chromosome tends to act as a unit. The degree to which this is true in any given system is a measure of whether the gene or the chromosome is the unit of selection, or, more accurately, what parts of the the genome can be said to be acting in unison" (Slatkin 1972).

Very well, if the geneticist says the

chromosome functions as the unit of selection, so be it. The implication for whole-animal biology is that under these conditions adaptations might be interpreted as "for the good of the chromosome." This will not always be so, as Slatkin indicates, and as Templeton, Sing, and Brokaw (1976) put it:

> ... the unit of selection is a function in part of the intensity of selection: the more intense the selection, the more the whole genome tends to hold together as a unit ... Thus selection under a broad range of conditions seems to preferentially operate upon linked blocks of genes.

So! Adaptation is sometimes for the good of the linked block of genes. By using the flexible word *replicator,* we can safely say that adaptation is for the good of the replicator, and leave it open exactly how large a chunk of genetic material we are talking about. One thing we can be sure of is that, except in special circumstances like asexual reproduction, the *individual organism* is not a replicator.

It is my contention that we should reserve the phrase "unit of selection" for replicators, that is for entities which become either more or less numerous in the world as a result of selection. Replicators exert power over their world, and it so happens that, in the forms of life with which we are familiar, groups of replicators are to be found exerting this power via relatively discrete entities which we call individuals. Because these entities have a high degree of autonomy of behavior and unity of struc-

ture, we are tempted to see them as the units of selection. But, for the reasons which we have seen, at least where reproduction is sexual, this is misleading. Individual bodies are units of replicator power. They are not replicators.

It is a remarkable fact that natural selection seems to have chosen those replicators that cooperate with each other, and go around in the large collective packages which we see as individual organisms. This is a fact that needs explaining in its own right, just as the existence of sexual reproduction needs explaining in its own right. Such extreme "gregariousness" of replicators may not be true of life all over the universe, just as it probably was not true of the earliest forms of life on earth. I will not discuss it here, but I have the hunch that something like game theory may be the right way to think about interactions between replicators. Maynard Smith (1974) has the right idea, but he should increase the time he spends on replicator games rather than individual games (Dawkins 1976a, pp. 91–93).

It has to be admitted that many biologists find attempts to dethrone the individual as the "unit of selection" unsatisfactory. At one level this shows itself as a kind of gut reaction: "What you say is all very well in theory. But when I am out in the field what I actually see is individuals. I don't see a gene pool, I see animals. Each one has four legs, two eyes, and a skin round it. Each one has its own nervous system, and it behaves like a single coherent entity, as if it had a single goal, not like a sort of federal democracy of replicators." At a more profound level, no less biologists than Ernst Mayr (1963) and E. B. Ford (1975) have poured scorn on the idea of the gene, rather than the individual, as the unit of selection. Incidentally, Mayr's attack had the additional merit of provoking a splendidly spirited "defence of beanbag genetics" from J. B. S. Haldane (1964). I really do think the argument is based on a misunderstanding. I have no trouble at all in enthusiastically endorsing all Mayr's eloquently expressed views on the unity of the genome. Of course it is true that the phenotypic effect of a gene is a meaningless concept outside the context of many, or even all, of the other genes in the genome. Yet, however complex and intricate the organism may be, however much we may agree that the organism is a unit of *function,* I still think it is misleading to call it a unit of *selection.* Genes may interact, even "blend," in their effects on embryonic development, as much as you please. But they do not blend when it comes to being passed on to future generations. I am not trying to belittle the importance of the individual phenotype in evolution. I am merely trying to sort out exactly what its role is. It is the all important instrument of replicator preservation: it is *not* that which is preserved.

Hesitantly, I will go further. It may be that, even in its role as the unit of gene action, the importance of the single individual phenotype has been exaggerated. If the word phenotype is defined physiologically, it is of course true that the phenotypic expression of a gene is confined to the one body in which it sits. But if we focus our interest on adaptation, and regard the "phenotypic expression" of a gene as

the power for its own preservation which it exerts over its surroundings, we are led to extend our view of what the word "phenotype" should mean.

### THE EXTENDED PHENOTYPE

There is a hidden assumption running right through the whole idea of individual and inclusive .fitness. This is that the individual, to the extent that it behaves in the best interests of *anybody's* genes, behaves in the best interests of its own (even if this means copies of its own genes in other individuals). There are rare cases of authors departing from this assumption. For instance Alexander (1974), in his theory of parental manipulation, suggested that offspring should be expected to behave in the best interests of their parents' genes rather than their own. I have argued that Alexander's main reason for expecting this is false (Dawkins 1976a, pp. 145–149), and my verbal criticism has been confirmed in a mathematical model by Parker and Macnair (1978). But although our refutation of Alexander was justified within the framework of ordinary replicator selection theory, Alexander's idea starts to look a lot more exciting within the framework of the extended phenotype, which I am now about to lay out. I begin with an example.

In the snail *Limnaea peregra,* the direction of coiling of the shell is controlled at a single locus. It is a classic case of simple Mendelian inheritance, right-handed coiling being straightforwardly dominant to left-handed coiling. Classic and simple except in one remarkable respect: control is exerted not by the individual's own genotype, but by its mother's. As Ford (1975) puts it: "We have here simple Mendelian inheritance the expression of which is constantly delayed one generation. It was long ago suggested that this phenomenon may be a widespread one controlling the early cleavage of the embryo until its own genes can take charge."

So, a gene can find phenotypic expression not in its own body but in a body of the next generation. This is a particular example of the concept of the extended phenotype. Here the route of the influence is presumably maternal cytoplasm, and other such "maternal effects" are known. But I want to apply the idea not just to mother and child but to influences on other members of the species, members of other species, even inanimate objects. If we can do this convincingly, we shall no longer be justified in regarding an individual as a machine programmed to preserve its *own* genes. It may be programmed to preserve somebody else's genes!

For didactic reasons I use examples which extend the idea of the phenotype gradually outward in stages. Caddis larvae live in houses which they themselves build out of stones, twigs, or some other material. The form of the house is determined by the behavior of the builder, and this in turn is presumably influenced by the builder's genes. The evolution of caddis houses came about through ordinary replicator selection—gene selection. There is nothing difficult about a genetics of caddis houses. All the ordinary genetic terms, dominance, epistasis, etc., would be perfectly applicable to traits such as stone color or stick length. Each gene

exerts its influence via building behavior, of course, and before that via control of protein synthesis. When I say the stones of houses are part of the phenotypic expression of genes, all I have done is to add one, rather minor, link to the end of an already long and complicated embryonic causal chain. Strictly speaking, it is *differences* in houses that are controlled by differences in genes, but differences, in any case, are what geneticists study.

It is easy to see a caddis house as part of the phenotype of genes, because the genes ride inside that house. It is the outer fortification of the body which they helped to build for themselves. It just happens to be made of stone rather than skin. The fates of the genes that built it are bound up inside the house that they built, just as in an ordinary body made of cells rather than stone. It is also easy to imagine a genetic account of variation in bower-bird bowers. The genes for bower building do not ride inside their bower. Nevertheless, their chances of being passed on to the next generation may depend critically on the success of the bower in attracting females. The bower is part of the phenotypic expression of genes in the bird, and the success of the genes as replicators depends on their effects on the bower. So, we have seen that the phenotypic expression of a gene may extend to inanimate objects, and it may also extend outside the body in which it sits.

The genes of parasites do not "build" the body of the host, but they can manipulate it. There is a large and interesting literature on parasites which influence the behavior of the hosts in which they ride (Holmes and Bethel 1972). Sporocysts of flukes of the genus *Leucochloridium* invade the tentacles of snails where they can be seen conspicuously pulsating through the snail's skin. This tends to make birds, who are the next host in the life cycle of the fluke, bite off the tentacles, mistaking them, Wickler (1968) suggests, for insects. What is interesting here is that the flukes seem to manipulate the behavior of the snails. The normal negative phototaxis is replaced in infected snails by positive light-seeking. This probably carries them up to open sites where they are more likely to be eaten by birds, and this benefits the fluke.

I have so far used conventional "individual level" language to describe this parasitic adaptation. The individual fluke is said to manipulate the behavior of the individual snail for its own individual advantage. But now I want to rephrase it in replicator language, in this case gene language. A mutation in the fluke can be said to have phenotypic expression in the snail's body—it changes the snail's behavior. The route of this phenotypic expression is tortuous and indirect, but not more so than the normal embryological details of phenotypic expression in a gene's "own" body. We are quite accustomed to the idea that genes are selected for their distantly ramified phenotypic effects on their *own* body. I am saying that they may also be selected for their distantly ramified phenotypic effects on *other* bodies.

Now of course selection also acts on hosts to make them resist manipulation by parasites. We expect counter-adaptations on the part of snails. Let

us again move from the language of individuals to the language of replicators. Suppose a mutation arises in snails which restores negative phototaxis even in the presence of a manipulating fluke, counteracting the tendency of the fluke gene to produce positive phototaxis. Both genes are acting on the same phenotype—the snail phenotype. They are pushing it in opposite directions but, once again, this is nothing new. We are already familiar with the idea of conflict between genes *within* a single body. This is often discussed in terms of so-called "modifier" genes. Any gene may modify the phenotypic expression of any other gene in the genome. A deleterious mutation is subject not only to direct selection against itself. There may also be selection on other, modifier, genes, to reduce the phenotypic effects of the deleterious gene.

For instance, imagine a mutant gene on a mammalian Y-chromosome. The argument would work if it was an ordinary segregation distorter (Hamilton 1967), but since it is a hypothetical gene, we may dramatize its properties a little. Any individual possessing this hypothetical gene kills his own daughters and feeds them to his sons. The death of the daughters is of no consequence to the rogue Y mutant, since they never contain it. On the other hand the sons all contain it, so the rogue Y gene will tend to spread very rapidly, and it might incidentally lead to the extinction of the whole population. But suppose modifiers arise on other chromsomes. These tend to neutralize the phenotypic expression of the rogue Y gene. The modifiers are carried not only by males themselves, but also by half the females whose lives they save. Depending on circumstances, such modifiers might therefore spread through the gene pool. Hamilton (1967) has suggested that something like this may be why so few genes on Y-chromosomes seem to have any detectable phenotypic expression. From our point of view the message is this: there can be conflicts of interest between the replicators in one individual's genome, and among the weapons at the disposal of a replicator is the modification of the phenotypic expression of another replicator. Now we can return to the fluke and the snail. The conflict between fluke genes and snail genes is no different from the conflict between genes within a single individual. In both cases the genes are struggling for *power over the phenotype.* In both cases they *modify* the phenotypic expression of other genes.

Fluke genes and snail genes ride inside the same body (though not inside the same cells). But, just as bower birds do not live inside their bowers, so parasites do not have to live inside their hosts. A cuckoo nestling manipulates the behavior of its foster mother. Once again I now switch from individual language to replicator language. If a mutation arises in a cuckoo which brightens the color of its gape so that it acts as a supernormal stimulus to a foster mother, the gene may be positively selected. The change in the behavior of the foster mother is properly regarded as part of the phenotypic expression of the cuckoo gene. The parental behavior of the foster mother is under the influence of many genes. Some of them are in her own body; some of them are in the cuckoo's body. They are struggling to push her

behavior in opposite directions. If a mutant arises in the host gene pool which causes individuals to stop treating bright gapes as supernormal, such a counteradapting mutation might be selected. I would call it a *modifier* of the cuckoo gene's phenotypic effects.

What we are talking about is *power,* replicator power. Those replicators survive which exert power over their world which leads to their own survival. Phenotypic expression is the name we are giving to the power of genes over their world and their future. The power of a gene within the body in which it sits is very considerable. Direct biochemical channels of power are available to it. No wonder we have got used to the idea that the phenotypic expression of a gene comes to an end at the wall of the individual body. Indeed, this makes very good sense if we are interested in physiological mechanisms in embryology. But if we are interested in adaptation, the logical conclusion to what I have been saying is that the whole world is potentially part of the phenotypic expression of a gene. It is only in practice that the power of a gene is limited to its immediate neighborhood. Maybe we have underestimated the extended power of replicators.

The routes of power in the extended phenotype are less purely biochemical than the routes of power in the conventional local phenotype. In the extended phenotype we must look to *behavior* rather than biochemistry. The study of animal communication turns out to be a branch of extended embryology. The same may be said of relationships between parasites and hosts, predators and prey, indeed it may be said of most of ecology. Bird song is the way it is because selection has acted on the distant phenotypic effects of genes in singing males: effects on the behavior of rivals and females (Dawkins and Krebs 1978). The peacock's tail is not the terminal phenotypic expression of the peacock's genes. It is only a way-station on the route to a more distant phenotypic expression in female behavior. Genes in orchids express themselves phenotypically in the form of changes in bee behavior, which result in the successful transference of pollen grains containing those same genes.

I end with a little flight of fancy through the ways of the extended phenotype. What is it about termites that led them to evolve eusociality? They are not haplodiploid, so that good old explanation won't do. Hamilton's (1972) inbreeding theory seems plausible enough. Other theories have invoked the termites' need to congregate in order to infect themselves with symbiotic protozoa. But, for the sake of argument, let us use the protozoa in the service of a very different idea. The symbionts in a termite colony are usually an identical clone. They are very numerous, and may constitute up to a third of each individual termite's body weight (Rietschel and Rohde 1974). They would seem to be in an excellent position to manipulate their host's physiology. Who knows, perhaps it is the protozoan genes that are really running the termite nest, exerting phenotypic power over the behavior of the termites, sterilizing the workers, making them behave eusocially.

To conclude: the replicator is the unit of selection. Adaptations are for the benefit of replicators. Individuals are manifestations of the power wielded by replicators over the world in which they live. The individual body is a convenient practical unit of combined replicator power. But we must not be misled by this parochial detail. In the light of the doctrine of the extended phenotype, the conceptual barrier of the individual body wall dissolves. We see the world as a melting pot of replicators, selected for their power to manipulate the world to their own long-term advantage. Individuals and societies are by-products.

## SUMMARY

"Sociobiology," in the sense in which the word has come to be used, may be defined as the branch of ethology inspired by W. D. Hamilton. The time has come to carry his "selfish gene" revolution to its conclusion, and give up the habit of speaking of adaptation at the individual level. Group selection, kin selection, individual selection, all may be swept away and replaced by *replicator selection*. Inclusive fitness is that property of an individual organism which will appear to be maximized when what is really being maximized is gene survival. The language of individual inclusive fitness is directly interchangeable with the language of gene replication, and it pays to learn to translate rapidly between the two languages. Examples are given of mistranslations in the literature. These have led to actual biological error, and the inherent confusingness of the concept of individual fitness is blamed. All remains clear if we stick to the language of replication. Genes are not the only conceivable replicators, and some general properties of replicators are listed. Sexually reproducing individuals are definitely not replicators. Units of genetic material larger than cistrons may be. In general, adaptations should be thought of not as for the good of the species, nor as for the good of the individual, but as for the good of the replicator. The last part of the paper develops the doctrine of the *extended phenotype*. Replicators such as genes manipulate their surroundings to their own advantage.

Manifestations of such manipulation are called phenotypic. Conventionally, the phenotypic expression of a gene is considered to be limited to the individual body in whose cells it resides. If we are interested in physiological mechanisms, this makes sense. But if we are interested in adaptation, it pays to make an imaginative leap and see the phenotypic expression of a gene as extending outside the individual body wall. The study of animal communication, and most of ecology, turn out to be branches of extended embryology.

Glenys Thomson, Jane Brockmann, and Marian Dawkins helped me by offering various combinations of skepticism and constructive suggestions. Catie Rechten translated the German summary, and helpfully critized the paper itself.

## REFERENCES

Alexander, R. D., 1974, The evolution of social behavior. *Ann. Rev. Ecol. Syst.,* 5: 325–383.

Barash, D. P., 1976, What does sex really cost? *Amer. Nat.,* 110: 894–897.

Barash, D. P., W. G. Holmes and P. J. Greene, 1978, Exact versus probabilistic coefficients of relationship: Some implications for sociobiology. *Amer. Nat.,* 112: 355–363.

Bateson, P. P. G., 1978, Book review: The Selfish Gene. *Anim. Behav.,* 26: 316–318.

Dawkins, R., 1976a, *The Selfish Gene,* Oxford and New York, Oxford Univ. Press.

Dawkins, R., 1976b, Reply to Gibson. *Nature,* 264: 381.

Dawkins, R., 1978, The value-judgments of evolution. In *Animal Economics* (Dempster, M. A. H., and D. J. McFarland, eds.), London, Acad. Press.

Dawkins, R., and J. R. Krebs, 1978, Animal signals: Information or manipulation? In *Behavioural Ecology* (Krebs, J. R., and N. B. Davies, eds.), Oxford, Blackwell Scientific Publ.

Fagen, R. M., 1976, Three-generation family conflict. *Anim. Behav.,* 24: 874–879.

Ford, E. B., 1975, *Ecological Genetics,* 4th ed., London, Chapman and Hall.

Gibson, P. A., 1976, Relatedness. *Nature,* 264: 381.

Haldane, J. B. S., 1964, A defense of beanbag genetics. *Perspectives in Biology and Medicine,* pp. 343–359.

Hamilton, W. D., 1964, The genetical theory of social behaviour, I and II. *J. Theoret. Biol.* 7: 1–52.

Hamilton, W. D., 1967, Extraordinary sex ratios. *Science,* 156: 477–488.

Hamilton, W. D., 1971, Selection of selfish and altruistic behavior in some extreme models. In *Man and Beast: Comparative Social Behavior* (Eisenberg, J. F., and W. S. Dillon, eds.), Smithsonian Institution, Washington.

Hamilton, W. D., 1972, Altruism and related phenomena, mainly in social insects. *Ann. Rev. Ecol. Syst.,* 3: 193–232.

Hamilton, W. D., 1975, Innate social aptitudes of man: An approach for evolutionary genetics. In *Biosocial An-*

*thropology* (Fox, R., ed.), London, Malaby Press.

Holmes, J. C., and W. M. Bethel, 1972, Modification of intermediate host behaviour by parasites. In *Behavioural aspects of parasite transmission* (Canning, E. V., and C. A. Wright, eds.), London, Acad. Press, pp. 123–149.

Hull, D. L., 1976, Are species really individuals? *Syst. Zool.*, 25: 174–191.

Lewontin, R. C., 1970, The units of selection. *Ann. Rev. Ecol. Syst.*, 1: 1–18.

Lewontin, R. C., 1974, *The Genetic Basis of Evolutionary Change*, New York and London, Columbia Univ. Press.

Maynard Smith, J., 1971, The origin and maintenance of sex. In *Group Selection* (Williams, G. C., ed.), Chicago, Aldine, Atherton.

Maynard Smith, J., 1974, The theory of games and the evolution of animal conflict. *J. Theoret. Biol.*, 47: 209–221.

Maynard Smith, J., and G. A. Parker, 1976, The logic of asymmetric contests. *Anim. Behav.*, 24: 159–175.

Maynard Smith, J. and G. R. Price, 1973, The logic of animal conflicts. *Nature*, 246: 15–18.

Maynard Smith, J., and G. C. Williams, 1976, Reply to Barash. *Amer. Nat.*, 110: 897.

Mayr, E., 1963, *Animal Species and Evolution*, Cambridge, Mass., Harvard Univ. Press.

Parker, G. A., and M. R. Macnair, 1978, Models of parent-offspring conflict. *Animal Behav.*, 26: 97–110.

Partridge, L., and L. Nunney, 1977, Three-generation family conflict. *Anim. Behav.*, 25: 785–786.

Rietschel, P., and K. Rohde, 1974, The unicellular animals. In *Animal Life Encyclopedia* (Grzimek, B., ed.), New York, Van Nostrand, pp. 89–137.

Slatkin, M., 1972, On treating the chromosome as the unit of selection. *Genetics*, 72: 157–168.

Spencer, H., 1864, *The Principles of Bi-*

*ology*, Vol. I, London and Edinburgh, Williams and Norgate.

Templeton, A. R., C. F. Sing, and B. Brokaw, 1976, The unit of selection in *drosophila mercatorum*. I, The Interaction of selection and meiosis in parthenogenetic strains. *Genetics*, 82: 349–376.

Treisman, M., and R. Dawkins, 1976, The cost of meiosis—is there any? *J. Theoret. Biol.*, 63: 479–484.

Trivers, R. L., 1976, Sexual selection and resource-accruing abilities in *Anolis garmani*. *Evolution*, 30: 253–269.

Waddington, C. H., 1957, *The Strategy of the Genes*, London, Allen and Unwin.

Whitney, G., 1976, Genetic substrates for the initial evolution of human sociality. I, Sex chromosome mechanisms. *Amer. Nat.*, 110: 867–875.

Wickler, W., 1968, *Mimicry*, London, World Univ. Library.

Wickler, W., 1976, Evolution-oriented ethology, kin selection, and altruistic parasites. *Z. Tierpsychol.*, 42: 206–214.

Wickler, W., 1977, Sex-linked altruism. *Z. Tierpsychol.*, 43: 106–107.

Williams, G. C., 1966, *Adaptation and Natural Selection*, New Jersey, Princeton Univ. Press.

Williams, G. C., 1971, *Group Selection*, Chicago, Atherton, Aldine.

Williams, G. C., 1975, *Sex and Evolution*, New Jersey, Princeton Univ. Press.

Wilson, E. O., 1975, *Sociobiology*, Cambridge, Mass., Harvard Univ. Press.

Woolfenden, G. E., 1975, Florida scrub jay helpers at the nest. *Auk*, 92: 1–15.

Wynne-Edwards, V. C., 1962, *Animal Dispersion in Relation to Social Behaviour*, Edinburgh, Oliver and Boyd.

Wynne-Edwards, V. C., 1977, "Intrinsic population control: An introduction." Institute of Biology Symposium on Population Control by Social Behaviour, London.

Zahavi, A., 1975, Mate selection—a selection for a handicap. *J. Theoret. Biol.*, 53: 205–214.

# 11

# Reductionistic Research Strategies and Their Biases in the Units of Selection Controversy

~~~~~~~~~~~~~~~~~~~~~~~~~~~~~~~~~~~~~~~~~

## WILLIAM WIMSATT

"A hen is but an egg's way of making another egg.

——Samuel Butler

### MOTIVATING REMARKS ON GENETIC DETERMINISM

Butler's satiric comment encapsulates the reductionistic spirit that made Darwinism objectionable to many in his own day, but has fared even better as a prophetic characterization of the explanatory tenor of modern evolutionary biology. It preceded August Weismann's doctrine of the continuity of the germ plasm advanced in his inaugural lecture in 1883 by some five years. As Weismann's views became one of the anchor points of the modern "neo-Darwinian" theory of evolution, they led to many modern recapitulations and elaborations of Butler's epigram. Thus Richard Dawkins writes (1976, p. 21):

Was there to be any end to the gradual improvement in the techniques and artifices used by the replicators to insure their own con-

tinuance in the world? . . . They did not die out, for they are past masters of the survival arts. But do not look for them floating loose in the sea. . . . Now they swarm in huge colonies, safe inside gigantic lumbering robots, sealed off from the outside world, communicating with it by tortuous indirect routes, manipulating it by indirect control. They are in you and me; they created us, body and mind; and their preservation is the ultimate rationale for our existence. . . . Now they go by the name of genes, and we are their survival machines.

Thus Dawkins announces his intention to "argue that the fundamental unit of selection, and therefore of self-interest, is not the species, nor the group, nor even, strictly speaking, the individual. It is the gene, the unit of heredity" (1976, p. 12). His purple prose gives ample food for worries that the account of evolved structure and behavior in general, and social behavior in particular, will do at best "simple justice" to its complexities and smack of genetic determinism. But his conclusions are well anchored in the dominant interpretation of the modern "genetical theory of natural selection," as R. A. Fisher called his theory (Fisher, 1930), and are espoused by many major students of evolutionary biology, as exemplified in the works of G. C. Williams (1966), J. Maynard Smith (1975), and E. O. Wilson (1975). Dawkins has in fact been a clear and ingenious expositor and elaborator of this view (1976, 1978), despite his often colorful language.

The quote from Dawkins is a direct reflection of the genetic determinism espoused by many and perhaps by most evolutionary biologists today. I take this view to involve two theses, one ontological and the other dynamic about the nature of evolutionary processes:

*T1:* (Ontological thesis): Genes are the only significant units (or individuals) required for the analysis of evolutionary processes.

*T2:* (Dynamical thesis): Processes at the genetic level determine (and are the primary and ultimate) explanations for processes at all higher levels.

Genetic determinism has its origins in a misconstrual of the nature of reductionism and of reductive explanation promulgated by most philosophers and by many biologists. A correct view makes it plausible, even inevitable:

(1) that there should be a variety of significant units of selection at various level of organization, thus denying *T1;*

(2) that understanding evolutionary processes requires the invocation and analysis of causal mechanisms and nomic regularities concerning their behavior at each of these levels of organization for the explanation of phenomena at a variety of levels, including that of the individual gene and a number of higher levels, thus denying *T2;*

(3) that sociobiology, properly conceived, should be viewed as the incorporation of an evolutionary perspective into the analysis of processes at these levels (and for most aspects of human social evolution, invoking cultural, rather than biological evolution) rather than the replacement of sociology, psychology, anthropology and the other social sciences by an extended, genetically based ethology of the selfish gene; and

(4) that the apparent success and power of genetic reductionist theories derives from distortions produced by cognitive biases arising from the uncritical application of a variety of reductionistic problem solving heuristics and research strategies.

In what follows I will outline the standard philosophical account of reduction and how it relates to genetic determinism (Section 2); discuss its inadequacy in handling problems of computational complexity (Section

3); show in particular how a claim of *in principle* reduction made by G. C. Williams fails for the simplest extension to a more complex system than that of two alleles at one locus (Section 4); and then discuss at length (Sections 5 through 8) how the problem-solving heuristics actually used by reductionistically inclined scientists result in systematic distortions biasing the case against recognizing the need for invoking various higher-level units of selection. A study of these biases suggests recommendations for methodological procedures which should at least partially mitigate their effects, and will in any case serve as a warning to those who must use these heuristic procedures.

## 2. THE PHILOSOPHERS' VIEW OF REDUCTION

The view I will be discussing is generically based on the model of Nagel (1961) and has been elaborated in somewhat different representative directions by Schaffner (1967, 1969, 1974, 1976), Ruse (1973, 1976), and Causey (1972a, 1972b, 1977). It has been widely criticized by a number of authors, including Hull (1973, 1974, 1976), Nickles (1973, 1976), Dresden (1974), Darden and Maull (1976), Maull (1977), Bantz (1980), Bogaard (1979), and myself (1974, 1976a, 1976b, 1978). My (1978) is an extensive review of the literature, but probably the best self-contained systematic critical analysis of these issues is to be found in McCauley (1979, chapters 4 and 5). Schaffner's own most recent work (1979), while not explicitly discussing reduction, seems to me to be a powerful and productive move away from

his earlier position, in this direction. I mention these sources to indicate where extensive discussion of relevant issues can be found. Most of this discussion will be presupposed here.

The traditional view of reduction holds that it is the

(i) *in principle*
(ii) *deducibility* of upper-level entities, properties, theories, and laws
(iii) in terms of the properties, laws, and relations *of any degree of complexity* of entities at the lower level.

I have emphasized in this characterization the three clauses which have caused the greatest problems for this traditional view. This view has two corollaries:

*C1:* Upper-level entities are thus shown to be "nothing more than" collections of lower-level entities (and their relations).
*C2:* Upper-level laws and causal relations are illusory or are shown to be "nothing more than" a shorthand for and to be determined by lower-level laws and causal relations.

These corollaries have a direct relation to the ontological and dynamical theses of genetic determinism, for *C1* and *C2* respectively provide the reasons for holding *T1* and *T2* to be true, if the relation of the various higher-level units of selection and phenomena to those at the genetic level is one of reduction, as it has been traditionally construed. *T1* is true *because* then the various higher-level units of selection are "nothing more than" collections of genes (and their relations) in a fashion demonstrated by the deductive or definitional relations between terms, and similarly for *T2* and *C2*.

A major problem for applying the Nagelian model of reduction, one recognized by many of its defenders (see, e.g., Schaffner, 1974), is that it appears not to fit the practice of reductionistically inclined scientists. As various writers have observed, if the Nagel model of reduction as a kind of deduction or its extensions is accepted as an adequate model of reduction, there may not *be* any cases of reduction in science (see Schaffner, 1974; Hull, 1973, 1976). A standard retreat in the face of this problem has been from claiming deducibility in practice, to defending a claim of deducibility or qualifier explicitly in my characterization of reduction, since at least this modification is required to describe the actual practice of scientists.

Further problems have arisen with the characterization of the analysis of the upper level in lower-level terms as a kind of deduction. Dresden (1974), Bogaard (1979), Bantz (1976), Sklar (1973), 1976), and others have pointed to the role of approximations, which prevent this "derivation" from being characterized as a deduction. I have emphasized (1976b, pp. 685–689) how the interpretation and elaboration of the implicit *ceteris paribus* clause in purported deductions makes any supposed "translation" context-dependent in a way that undercuts the usefulness of the deductive model and falsifies corollaries *C1* and *C2*. Furthermore, not only are there problems with filling out the *ceteris paribus* clause, but, rather anomalously on this model,

scientists appear to have no interest in trying to do so.

I have argued elsewhere that "in principle" deducibility, analyzability, or translatability is best seen not as the primary structure, focus, or thesis of reductionism but as a derivative corollary of the use of identificatory hypotheses in reductive explanation. If an upper-level entity, phenomenon, etc. is *identified* with a lower-level complex of entities, properties, and relations, then Leibniz's law tells us that any property of one is a property of the other. Thus Leibniz's Law tells us that if the purported identity holds, any upper-level thing must *in principle* be analyzable in lower-level terms. These "in principle" claims thus become an important heuristic method for moving from an identity claim to specific hypotheses about heretofore unmatched properties at the upper and lower levels (see Wimsatt, 1976a, pp. 225–237, and 1976b, pp. 697–701). This heuristic use of identity claims and *in principle* arguments provides further support for the view advocated here: that *the power, limitations, and character of reductionistic approaches in science are better analyzed in terms of the reductionistic research strategies one is led to adopt than in terms of idealized deductive accounts and ontological theses derived from them.* For more on this use of identity claims, see my (1976a; 1976b; Section 3; and 1978). It will not be further elaborated here.

The use of simpler models and approximations in reductionistic modeling produces a gap between promise and performance that has interesting consequences. The metaphysical position that the reductionist defends holds that a reductionistic analysis of upper-level phenomena must exist in terms of lower-level entities, properties, and relations *of some degree of complexity*—preferably in terms of monadic properties; but if not these, then at least in terms of *some* (possibly complex and relational) properties of the lower-level entities. This is another formulation of the claim of *in principle* deducibility of reducibility that I have argued is a corollary of the use of identity claims and Leibniz's Law.

The holist, as antireductionist, is taken normally as denying this metaphysical claim, and thus to be holding the equally metaphysical (and to most people, radically implausible) claim that no analysis of whatever complexity in lower-level terms could be adequate. But, despite appearances, the in principle claim of the reductionist is seldom in dispute. *In the cases I know in population biology, in neurophysiology, and in the history of genetics, the issue between scientists who are reductionists and holists is not over the in principle possibility of an analysis in lower-level terms but on the complexity and scope of the properties and analyses required.* The more holistically inclined scientists usually argue that higher-order relational properties of the lower-level entities are required, and the reductionists argue that a given simple, lower-level model (often one using only monadic properties) is adequate. To the extent that this is true, the portrayal of the dispute between reductionist and holist as over the *in principle* claim (a portrayal favored by most philosophers, and by many scientists) is seriously in error and turns a usually serious, compre-

hensible, and important empirical dispute into a usually one-sided and poorly motivated metaphysical one.

This reading of the dispute might seem to have the apparent disadvantage of dissolving it, for both holists and reductionists now appear to be species of reductionist—"complex" reductionists or "simple" reductionists respectively. But this species of complex reductionist is still recognizably a holist. (See Wimsatt, 1978). Complex relations relate several to many lower-level entities, and require the recognition of these complexes as entities. Furthermore, if the relationships overlap in their relata, these higher-level entities become tied together in terms of still higher-level systems in ways suggested by the discussion of descriptive and interactional complexity in Wimsatt (1974). The further presence of many-one mappings between lower- and higher-level state descriptions, required even by the existence of recognizable, stable higher-level phenomena generates a kind of autonomy and independence of the dynamics and the explanations at the higher level from detailed lower-level specifications and laws. (See Wimsatt, 1980, sections 4 and 5, for the discussions of "sufficient parameters" and "robustness"; Wimsatt, 1976a, pp. 248–251 on "explanatory primacy"; and Wimsatt, 1976b, section 6, pp. 689–692, and the appendix, pp. 701–704, for the discussion of the "screening off" relation of Salmon.)

The net effect of these considerations is that the holist can get the significance and autonomy of upper-level entities, laws, and phenomena which he desires while accepting a kind of in principle (but "complex") reductionism. The arguments in what follows for the significance of higher-level units of selection are to be interpreted as espousing this kind of holism.

### 3. THE PROBLEM OF COMPUTATIONAL COMPLEXITY AND THE USE OF REDUCTIONIST RESEARCH HEURISTICS

I claimed in the last section that the claims of "in principle" deducibility or translatability are best seen as corollaries of Leibniz's Law, and thus as consequences of the use of compositional identities in reductive explanations. There is much to be learned, however, by looking at the standard explication of these claims. Most of them seem to suppose that a claim of *in principle* translatability is to be explicated in terms of effective computability or, mirroring Laplace's definition of a deterministic system, as a translation which could be produced by a sufficiently powerful computer which was given a total state description of the microlevel of the appropriate system, together with all of the microlevel laws which applied to that system. Richard Boyd (1972) has given a brilliant criticism of this possibility in general. I wish here to make only some more pragmatic criticisms (see also my 1978).

First of all, it is unclear how a practicing scientist could make use of results concerning the effective computability of any system he is studying, since all of these results would presuppose a total knowledge of the system, which he does not possess, and a theory of that system organized in a fashion unlike any of the theories which he knows. It is all right to *talk* about writing the Schrödinger wave

equation for a particular organism, but in fact physical and quantum chemists don't even do it for chemical bonding in simple molecules (see Bantz, 1980; Bogaard, 1979; and Dresden, 1974). Instead they use simple heuristic approximations even for these far simpler cases.

But suppose that one could. Does that mean that we would, or even could study systems in this way? It does not. That can be seen by studying the game of chess. Chess is a totally deterministic game. At each stage, the possible moves and their outcomes are exhaustively specifiable —and indeed relatively straightforwardly specifiable. This means that if we specify in advance how many moves we wish to allow in the game, we can in principle write down a branching tree, beginning with the initial state of the chessboard and ending with branches of nodes corresponding to all possible games of that many moves or less. (Some of the games may already have terminated in fewer moves.)

There are twenty possible opening moves (two for each of eight pawns and two knights). Suppose that this number of alternatives continues throughout the game on the average, as a geometrical mean. (This is almost certainly an understimate.) And suppose that we wish to consider games of 100 moves (fifty pairs of moves). Then we are considering on the order of $20^{100}$ possible games. This is a large number, but clearly it is a task which is effectively computable. But that is not much consolation. Consider the size of the task: $20^{100} = 2^{100} \times 10^{100}$. Since $2^{10} \approx 10^3$, then $20^{100} \approx 10^{130}$. Now for some other relevant numbers. There are about $10^{79}$ elementary particles in the universe. There have been about $10^{19}$ seconds since the big bang. And the shortest known time for a physical event is on the order of $10^{-24}$ second, the time it takes for light to traverse the diameter of an atomic nucleus. Putting this all together, we arrive at an upper estimate of the number of events in the universe to date of about $10^{122}$. Then this task, a trivial one for a universal Turing machine, is nonetheless *not* doable by the most universal computer we could imagine—the universe as a computer! It would fall 8 orders of magnitude short of having had enough *actual* states (as opposed to possible ones) to represent all of these games, and we have not raised the questions of how these games could be mapped into the states in a usable manner and how rapidly the different parts of a computer spanning $10^{10}$ light years will be able to communicate with one another! Clearly, this effectively computable (and therefore, to many logicians and mathematicians, *in principle* possible) task is, *physically speaking, in principle impossible.*

This kind of observation led Herbert Simon and others since to look for other kinds of models than the exhaustive, brute-force algorithmic approach for human problem-solving. First, for decision-making (Simon, 1957), then for proving theorems (Newell, Shaw, and Simon, 1958), and subsequently for other problems (Newell and Simon, 1961; Simon, 1966a, 1966b, 1969, 1973), Simon

espoused a "principle of bounded rationality" (Simon, 1957, pp. 198–199) which asserts that we are generally faced with problems of such complexity that we cannot solve them exactly, and therefore, if we are to get any solutions at all, we must do so by introducing various simplifying and approximative techniques. Thus was born the idea of a heuristic.[1] As I use that notion here, I take a heuristic procedure to have three important properties:

(1) By contrast with an algorithmic procedure, the correct application of a heuristic procedure does not guarantee a solution; and if it produces a solution, does not guarantee that the solution is correct. Thus valid deduction from true premises is not a heuristic procedure. Most or all inductive procedures are, however (see Shimony, 1970).

(2) The expected time, effort, and computational complexity of producing a solution using a heuristic procedure are appreciably less than those expected using an algorithmic procedure. This is indeed the reason why heuristics are used. They are a "cost-effective" way of producing a solution, and often the only physically possible way.

(3) The failures and errors produced using a heuristic are not random, but systematic. I conjecture that *any heuristic, once*

1. Actually the term "heuristic" was used earlier and perhaps introduced by G. Polya in his book *How to Solve It* in 1945. The idea of a heuristic procedure has, however, developed substantially further, and has become one of the central theoretical concepts of artificial intelligence. See, e.g., Nilsson (1971) and Winston (1977) for more recent discussion of heuristic programming.

*we understand how it works, can be made to fail.* That is, given this knowledge of the heuristic procedure, we can construct classes of problems for which it will always fail to produce an answer, or for which it will always produce the *wrong* answer. This property of systematic production of wrong answers will be called the *bias(es)* of the heuristic.

Not only can we work forward from an understanding of a heuristic to predict its biases, but we can also work backward, hypothetically, from the observation of systematic biases as data to conjecture about the heuristic which produced them; and if we can get independent evidence on the nature of the heuristics, we can propose a well-founded theory of the structure of our heuristic reasoning in these areas. This was elegantly done for the first time by Tversky and Kahneman (1974), in their analysis of fallacies of probabilistic reasoning and the cognitive heuristics which produce them. To my mind, Simon's work and that of Tversky and Kahneman have opened up a whole new set of questions, a new area of investigation of pragmatic inference in science, which should revolutionize our discipline in the next decade, and increasing numbers of workers are moving in this direction. (See, for example, the papers of Schaffner, Darden, and Bantz in Nickles, 1980.)

The notion of a heuristic has far greater implications than can be explored in this paper. In addition to its centrality in human problem-solving, it is a pivotal concept in evolutionary biology and in evolutionary epistemology. It is a central concept in evolutionary biology because any

biological adaptation meets the conditions given for a heuristic procedure. First, it is a commonplace among evolutionary biologists that adaptations, even when functioning properly, do not guarantee survival and production of offspring. Secondly, they are, however, cost-effective ways of contributing to this end. Finally, any adaptation has systematically specifiable conditions, derivable through an understanding of the adaptation, under which its employment will actually *decrease* the fitness of the organism employing it, by causing the organism to do what is, under those conditions, the wrong thing for its survival and reproduction. (This, of course, seldom happens in the organism's "normal" environment, or the adaptation would become maladaptive and be selected against.) This fact is, indeed, systematically exploited in the functional analysis of organic adaptations. It is a truism of functional inference that learning the conditions under which a system malfunctions, and how it malfunctions under those conditions, is a powerful tool for determining how it functions normally and the conditions under which it was designed to function. (See, e.g., Gregory, 1967; Lorenz, 1965; Valenstein, 1973; and Glassman, 1978, for illuminating discussion of the problems, techniques, and fallacies of functional inference under a variety of circumstances.)

The notion of a heuristic is central to evolutionary epistemology, because Campbell's notion of a "vicarious selector" (1974, 1977), which is central to his conception of a hierarchy of adaptive and selective processes spanning subcognitive, cognitive, and social levels, is that of a heuristic procedure. A vicarious selector for Campbell is a (1) substitute, (2) less costly selection procedure acting to optimize some index which is only contingently connected with the index optimized by the selection process it is substituting for. This contingent connection allows for the possibility —indeed the inevitability— of systematic error when the conditions for the contingent consilience of the substitute and primary indices are not met. An important ramification of Campbell's idea of a vicarious selector is the possibility that one heuristic may substitute for another (rather than for an algorithmic procedure) under restricted sets of conditions, and that this process may be repeated, producing a nested hierarchy of heuristics. I believe that this is an appropriate model for describing the nested or sequential structure of many approximation techniques, limiting operations, and the families of progressively more realistic models found widely in "progressive research programs," as exemplified in the development of nineteenth-century kinetic theory, early twentieth-century genetics, and in several areas of modern population genetics and evolutionary ecology. (On this last, see, e.g., Roughgarden, 1979.)

The ultimate end of this paper is to discuss some of the heuristics used in reductionistic modeling and to show how their systematic biases have given illegitimate support to a reductionist vision of evolutionary processes culminating in genetic determinism. But before I do this, it is necessary to show that, how, and why a brute force, quasi-algorithmic, reductionistic approach cannot work in evolutionary biology and population genetics, for

just such an approach has been suggested by G. C. Williams.

## 4. WILLIAMS'S "IN PRINCIPLE" REDUCTIONISM AND THE CASE OF TWO LOCI

A problem-solving heuristic which Simon (1966a and b) has called "factoring into subproblems" appears in a variety of guises in reductionistic modelling. Simon illustrates the heuristic and its advantages using the problem of finding the right combination for a combination lock. Imagine a bicycle lock with ten wheels of ten positions each. If there is only one combination which will work, one would expect to look through about half of the possible $10^{10}$ combinations on the average before finding it. On the other hand, suppose that the lock is a cheap or defective one for which one can tell individually for each wheel when it is in the right position. Then an average of 5 tries on each wheel, for a total of 50 tries would be expected to find the right combination. The advantage that accrues from being able to break the problem down into subproblems, being able to find out parts of the combination, rather than having to solve the whole problem at once, is given by the ratio of the number of alternatives which must be inspected. This is the case, $(5 \times 10^9)/(5 \times 10) = 10^8$

Similar advantages accrue for similar combinatorial reasons if problems of evolutionary dynamics can be treated in terms of the frequencies of individual alleles, with no epistatic interactions and no probabilistic associations between alleles at different loci because of linkage or assortative mating rather than in terms of the gametic or zygotic genotype fre-

TABLE 1

Sufficient dimensionality required for the prediction of evolution of a single locus
with $a$ alleles where there are $n$ segregating loci in the system

| Level of Description: | Zygotic Classes | Gametic Classes | Allele Frequencies | Allele Frequencies |
|---|---|---|---|---|
| Dimensionality: | $\dfrac{a^n(a^n+1)}{2}-1$ | $a^n-1$ | $n(a-1)$ | $(a-1)$ |
| Assumptions: | none | 1 | 1, 2 | 1, 2, 3 |
| $n$:     $a$: | | | | |
| 2     2 | 9 | 3 | 2 | 1 |
| 3     2 | 35 | 7 | 3 | 1 |
| 3     3 | 377 | 26 | 6 | 2 |
| 5     2 | 527 | 31 | 5 | 1 |
| 10     2 | 524799 | 1023 | 10 | 1 |
| 32     2 | $9.22 \times 10^{18}$ | $4.29 \times 10^9$ | 32 | 1 |

*Assumptions:*

(1) random union of gametes (no sex linkage, no assortative mating)

(2) random statistical association of genes at different loci (linkage equilibrium).

(3) no epistatic interaction (inter-locus effects are totally additive).

(Table is adapted and extended from Table 56 of Lewontin, 1974, p. 283.)

quencies required if these assumptions do not hold. Here the simplification occurs in the number of dimensions in the phase space required to adequately describe and predict evolutionary changes, and, correlatively, in the number of state variables in the equations required to describe and predict the dynamics of evolutionary change. Table 1, derived and extended from Table 56 of Lewontin (1974, p. 283), summarizes the dimensionality of the problem under different simplifying assumptions. It is worth noting that if *no* simplifying assumptions are made, even the simplest multi-locus case of two alleles at each of two loci is analytically intractable. This should not be surprising: the problem of dimensionality nine (there are nine possible genotypes,

with independently specifiable fitness parameters) is already more complicated than the three-body problem of classical mechanics. Like the three-body problem, it has been solved for a variety of special cases (see Roughgarden, 1979, chapter 8, pp. 111–133) but has not been solved in general.

In the light of this, G. C. Williams makes a claim which gives substantial hope, for it appears to promise that the problem can be treated as one of the lowest dimensionality. His view is that since the operation of any higher-level selection processes can be mathematically expressed as resulting from the operation of selection coefficients acting independently at each locus to change the frequency of individual alleles or genes, there is no need to postulate the existence of any higher-level units of selection or selection forces. This view will look most familiar to philosophers, since it bears the strongest resemblance to traditional philosophical accounts of theory reduction. Williams expresses it as follows:

Obviously it is unrealistic to believe that a gene actually exists in its own world with no complications other than abstract selection co-efficients and mutation rates. The unity of the genotype and the functional subordination of the individual genes to each other and to their surroundings would seem, at first sight, to invalidate the one-locus model of natural selection. Actually these considerations do not bear on the basic postulates of the theory. No matter how functionally dependent a gene may be, and no matter how complicated its interactions with other genes and environmental factors, it must always be true that a given gene substitution will have an arithmetic mean effect on fitness in any population. One allele can always be regarded as having a certain selection coefficient relative to another at the same locus at any given point in time. Such coefficients are numbers that can be treated algebra-ically, and conclusions inferred from one locus can be iterated over all loci. Adaptation can thus be attributed to the effect of selection acting independently at each locus. (Williams, 1966, pp. 56–57)

Williams goes on, in the next two pages, to illustrate how this algebraic manipulation can be accomplished in a simplified genetic environment of two alleles at each of two loci, and we are to imagine the extrapolation to cases of many alleles at many loci. Complicated it would be, but *in principle,* of course (we are told), it could be done, "by iterating over all loci."

This claim might appear to involve another variant of the "factoring into subproblems" heuristic which Simon has studied and written upon at length, and which he has called "the hypothesis of near decomposability" (see Simon, 1969, pp. 99 ff, in Ando et al., 1963, and further references given there; and also Wimsatt, 1974, for further discussion).

The hypothesis of near decompos-ability involves the assumption that a complex system can be decomposed into a set of subsystems such that all strong interactions are contained within subsystems' boundaries, and interactions between variables or entities in different subsystems are appreciably weaker than those relating variables or entities in the same sub-system. In this case, an approximation to the behavior of the system, in the short run, can be had by ignoring the intersystemic interactions and analyz-ing each subsystem as if it were isolated, studying only internal vari-ables in their common approach to equilibrium. Its behavior in the long run can be approximated by ignoring the intra-systemic interactions of the subsystems (assuming that there is intrasystemic equilibrium), represent-ing each subsystem by a single index. and considering the equilibration of the various subsystems with one another as a system involving the in-teraction only of these lumped index variables.

There are thus two different approximations involved in studying the short-run and the long-run behavior of the system. Each substantially reduces the complexity of the problem, if the assumptions allowing the approximation are justified.

Indeed the hypothesis of near decomposability *is* used in this way in a number of multilocus models—in particular when it is assumed: (1) that the system starts at or near linkage equilibrium (this is the condition when all genotypes occur at frequencies given by the products of the frequencies of their constituent genes, a condition equivalent to the assumption of a multilocus Hardy-Weinberg equilibrium); (2) that selection between genotypes is relatively weak (a condition that guarantees that the population never deviates far from the multilocus Hardy-Weinberg equilibrium). Indeed, these assumptions (as well as that of random association of gametes, implying no assortative mating) are made in the original model (Lewontin and White, 1960) for which the two-locus fitness surfaces, which provide below a counterexample to Williams's claim, originally were derived. Under these conditions, recombination can be neglected as a significant contributor to genotype frequencies, and the dynamics can be treated as if they were affected by segregation and selection only. This is equivalent to using the "long range" approximation in studying the behavior of a nearly decomposable system, since if the system is far from linkage equilibrium, recombination may be a far greater contributor to genotype frequency of some genotypes than either segregation or selection, and thus behaves like an intrasystemic "strong" interaction which goes relatively rapidly to equilibrium.[2] The observation that under some conditions there can be permanent and substantial linkage disequilibrium (see Lewontin, 1974; Roughgarden, 1979; and also Maynard Smith, 1978, chapter 5) is equivalent, then, to saying that the system cannot be treated as nearly decomposable.

In fact, however, Williams' claim in the above quote appears to be far stronger than a near decomposability claim and is not made on the basis of these assumptions about linkage equilibrium and random assortment of gametes produced by random mating. He claims that the problem can be solved one locus at a time and then extended to a global solution by "iterating over all loci." His claim is thus not that the genetic system is nearly decomposable, but that it is

---

2. On the most obvious reading, in which genotypes are the subsystems containing genes, which interact to produce fitness, which is a property of genotypes which affects the multiplication ratios of the genes they contain, selection and mating would be treated as an intersystemic interaction, with segregation and recombination as intrasystemic interactions. Thus the analogy is not quite exact. In this case, with mating assumed to be at random and the genotypic frequencies in linkage equilibrium, the "long range" behavior of the system involves the interaction of an intrasystemic force (segregation) and an intersystemic one (selection). This particular decomposition into subsystems (at variance with that required for easy analysis of near decomposability) is necessitated by the particular structure of Mendelian genetics, which, through mating and differential reproduction, inextricably combines inter- and intraorganismic forces. *(Continued on page 154.)*

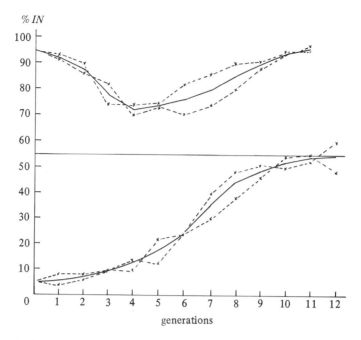

Figure 1. The frequency of an inversion, IN, in hypothetical laboratory populations. The heavy lines represent the average behavior of replicates, while X's represent individual data points. (Reprinted from Lewontin, 1974, figure 23, page 274, with permission of Columbia University Press.)

simply decomposable, like the simpler of Simon's two locks. Without all of the qualifications in Lewontin's table, this claim is simply incorrect, and can be shown to be so for the simplest case involving more than one locus— that of two alleles at each of two loci. The reasons for the failure of Williams's claim can best be seen after a discussion of this case.

A claim that evolutionary processes can be analyzed in the manner Williams suggests, as being of the lowest possible dimensionality, involves at least the claim that a deterministic theory of the change of gene frequencies at a given locus can be construed using only the frequencies of the alternative alleles of that locus. In the simplest case of two alleles at one locus, this involves saying that it is a function only of the frequency of a single gene, since if $q$ is the frequency of gene $a$, then $1-q$ must be the frequency of the other gene, $A$, because there are no other genes at that

Other ways of breaking up the system (e.g., into loci as subsystems, which might be suggested by Williams's remarks) produce similar problems: then recombination (a strong force if there is substantial linkage disequilibrium) is intersystemic, rather than intrasystemic, as it "should" be. Nonetheless, the partitioning of forces into strong and weak, characteristic of near decomposability analysis, is found here also, so there remains an important (and probably the most important) ground of analogy.

locus. (It is a function also of the fitnesses, $W_{11}$, $W_{12}$ and $W_{22}$ *of the* genotypes *AA, Aa,* and *aa,* but these are assumed to be constant parameters of the system in this discussion.)

Consider Figure 1 as a graph of gene frequency from different initial points (.05 for the bottom curve, .95 for the top curve) as it changes in successive generations. If this were the graph of an actual case (Lewontin describes it as of a "hypothetical laboratory population"), it would falsify Williams's claim. Why? Consider the topmost curve. At all points between the initial high value of gene frequency of .95 and the minimum value (of about .7, reached in generation 4) a population which is decreasing in gene frequency at that value (between generations 0 and 4) is later increasing in gene frequency at that value (in generations 5 and later). *But if gene frequency can either increase or decrease from a given value, then gene frequency* (of that gene or its allele) alone is not an adequate basis for a deterministic theory of evolutionary change.

Williams gets into trouble at this point because his claim is neither a theory of evolution in terms of gene frequencies, nor even a schematic description of the form of such a theory. His statement that "it must always be true that a given gene substitution will have an arithmetic mean effect on fitness in any population" (1966, p. 56) suggests the following procedure for evaluating

this effect on fitness: Imagine a gigantic (noninterventive) DNA sequencer that, given a population, will determine all of the genes in that population and their frequencies. Perform this genetic census at two points in time—or perhaps in each generation. For each gene, its frequency in the interval will either increase (in which case it is being selected for), decrease (it is selected against), or remain constant (it is neutral).

These data can then be used in one of two ways, They can be used to describe the evolutionary trajectory of the population in its phase space. But then this is not a *theory* of evolutionary change but a description The fitness $W_{11}$, $W_{12}$, and $W_{22}$ inferred from this are merely biological redescriptions of what is happening in successive generations[3] and may undergo arbitrary changes as the "curve-fitting" parameters that they are. Or the changes observed in one generation may be used to estimate fitness values which are then used to predict future changes. This is more of a process of trend extrapolation using an assumed model rather than a theory itself, but it is at least not totally tautological. To have a predictive tool or theory, then, Williams must intend his remarks to describe a process of trend extrapolation.

But here is where the trouble arises. The graph of Figure 1 indicates that local estimates of fitness values *cannot* be used in this way to extrapolate

---

3. The biologist's use of tautology here is looser than the philosopher's, and means roughly a relation which has no empirical content because of the way in which it is used. As such it is related to the vernacular use of tautology as in "covert tautology" rather than to the logician's sense. I will here use the term as the biologist does.

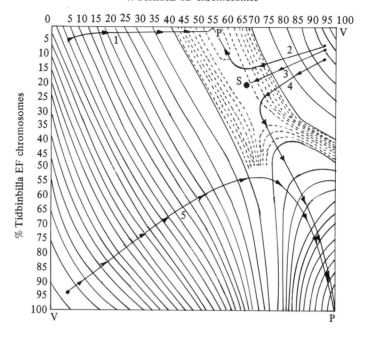

Figure 2. Projected changes in the frequency of two polymorphic inversion systems in *Moraba scurra* from different initial compositions, based on fitness estimates from nature. The trajectories, shown by arrow-marked lines, are calculated by the solution of differential equations of gene frequency change. Lines crossing the trajectories are contours of equal mean population fitness, $\bar{W}$. (Reprinted from Lewontin, 1974, figure 24, page 280, with permission of Columbia University Press.)

evolutionary trends. After all, gene *a* is apparently being selected *against* in generations 0–4, but subsequently it must be being selected *for,* as its frequency is then *increasing.* To put it more generally, local estimates of fitness or selective value are not valid globally, for other values of the frequency of that and other genes.

The reason why this is not the case becomes apparent in Figure 1. Indeed, Lewontin's hypothetical laboratory population of Figure 1 was not hypothetical at all, but a description of changes that would be expected in a field population of the grasshopper, *Moraba scurra,* whose mean Dar-

winian fitness $\bar{W}$, is given as a function of the frequency of two alleles of each of two loci in the adaptive topography of Figure 2. An adaptive topography is a plot of contours of equal mean population fitness, $W$, as a function of the gene frequencies (in this case, at two loci). Since in many simpler models (particularly where the genotypic fitnesses are constant), a population will tend to evolve in directions of increasing $\bar{W}$, the adaptive topography gives a visual means of making qualitative predictions about the direction and relative rates of local evolutionary change.

Indeed, the lower curve of Figure

1 is just trajectory 1 of Figure 2, and the upper (problematic) curve of Figure 1 is trajectory 4 of Figure 2. And the results that appeared as indeterministic in terms of the trajectory of gene frequencies of a single locus are seen to be deterministic once the frequencies of each of two loci are specified. Thus the initial points of trajectories 1 and 4 should not have been specified as .05 and .95, the frequency of a gene at the first locus, as is implicit in Figure 1, but as (.05, .05) and as (.95, .12), the frequencies of the genes at both loci as in Figure 2. Note also that trajectory 2, with initial point (.95, .07) —in which the frequency at the first locus is the same as for trajectory 4, but that at the second locus is different—also shows a violation of the deterministic assumption if only the first locus frequency is looked at when the two trajectories are compared. Trajectory 2 is not, however, by itself evidence of a violation, as is trajectory 4.

It is quite clear from this adaptive topography that what will happen in evolution is a function of the joint values of gene frequency at two loci, and no set of measurements or extrapolations looking at frequencies of just one locus at a time can provide an adequate basis for prediction. That is true in this case because of epistatic interactions between loci, which is a sufficient condition for having to go to a phase space of greater dimensionality for prediction.

Williams's proposal, then, fails in this case, in 3 ways:

(1) It does not result in a deterministic theory of evolutionary change in terms of the gene frequencies of individual loci which can be "iterated over all loci" to produce a global solution.

(2) It fails to do so because epistatic interactions among loci prevent local estimates of fitness at single loci from being projectable or extrapolatable if *gene frequencies at other loci are free to change simultaneously.* (The appearance of projectability usually arises when estimates are done locally under conditions in which there *is* no change or no significant change at other loci. But this cannot be assumed in general.) Williams in effect errs by assuming that single locus fitnesses are independent of context, when in fact they are functions of the context of other loci. Illegitimate assumptions of context-independence are a frequent error in reductionist analyses. (See Wimsatt, 1976, p. 688, and 1980 for further discussion.)

(3) In fact, Lewontin's data on *Moraba scurra* represent not gene frequency changes at single loci, but the frequencies of chromosomal inversions involving *many* loci. For reasons which I will not detail here, inversions can often act as units of selection, and Lewontin has devoted much of the earlier portions of his book to arguing that Williams's aim of measuring the fitness effects of single gene substitutions is bedeviled with a host of practical and theoretical problems. So Lewontin's one-chromosome example of

Figure 1 and the two-chromo-some counterexample of Figure 2 are already at a higher level of organization than that supposed by Williams's single locus genetic reductionism.

What goes for two loci or chromosomes, goes as well for many. In this light, Williams's remarks suggesting genetic reductionism are better seen as having more import as a kind of genetic bookkeeping than as promising a reductionistic theory of evolutionary change in terms of gene frequencies. The latter is a tempting mirage which vanishes upon closer inspection of the complexities and heuristics of the actual theory.

## 5. A GENERAL CLASS OF REDUCTIONIST RESEARCH HEURISTICS AND THEIR BIASES

The kind of mistake implicit in the reductionistic argument discussed in the preceding section seems so straightforward, yet is so pervasive and is made by so many leading practitioners of the discipline that it cries out for a deeper explanation. Any engineer knows that systematic failures in a mechanism indicate a design problem, which he then tries to locate and eliminate. This procedure is so important (and so often ignored in the traditional education of engineers) that at least one company holds seminars for the engineers of its various divisions to teach approaches and methods for doing it, as described in Moss (1979). The heuristics in our reasoning processes have similar possibilities for systematic er-

ror, and we should similarly try to analyze these failures to get an understanding of where they are likely to occur and, where possible, eliminate or moderate their effects through redesign. If redesign (through teaching different or modified heuristics) is impossible or impractical, we can at least, through an understanding of the causes of failure, be warned when they are likely to lead us into error, so that our troubleshooting efforts may be concentrated there.

I will describe here a general class of heuristics and their biases which have a common origin in the nature of reductionistic analyses, thus providing the appropriate warnings. After defining the notion of unit of selection, I will describe some of the results of Wade's review of the models of group selection (1978) which nicely illustrates several of these biases with systematic failures in the literature. I will then discuss a design modification in our heuristic which should serve to eliminate or at least moderate the effects of several of these biases.

(1) Any analysis of a system presupposes a division of the world, however tentatively, into the system being studied and its environment. This division may be made on grounds of interest, which will in turn often be determined by judgments of the scope of one's field (a molecular geneticist is unlikely, at least initially, to consider social forces as part of the subject matter of his discipline), other jurisdictional criteria, and, probably most frequently, intuitive judgments about the natural chunks and boundaries in his area.[4] Judgments

---

4. These judgments are themselves an important source of error, associated with heuristics for cutting the world up into entities, using the robustness or overdetermination of bound-

of what can be manipulated relatively independently of "outside" forces are likely to enter into any of these, and this in turn implies judgments of near decomposability or near isolatability in the individuation of systems.

(2) A reductionist adds to this a further consideration: by his description as a reductionist, he is interested in understanding the behavior of his system in terms of the interaction of its parts. This means that his *interest* at least (though not necessarily his scope of investigation) will be focused on the entities and interrelations between them *internal* to the system he is studying.

(3) The third and last constraint is to recognize the practical impossibility of generating an exhaustive, quasi-algorithmic, or exact analysis of the behavior of the system in its environment. This is an application of Simon's "principle of bounded rationality" discussed earlier. So the reductionist must start simplifying. In general, simplifying assumptions will have to be made everywhere, but given his interest in studying relations *internal* to the system, he will tend to order his list of economic priorities so as to simplify first and more extremely in his description observation, control, and analysis of the environment than in the system he is studying. After all, simplifications internal to the system face the danger of simplifying out of existence the very phenomena and mechanisms he wishes to study.

This fact alone, derived just from these three very general assumptions, is sufficient to generate and explain a wealth of heuristics and their attendant biases arising in the reductionist analysis of systems. These heuristics and biases can be classified roughly as biases of conceptualization, biases of model-building and theory construction, and biases of observation and experimental design, though any rigid classification would fail because of the interdependence and intercalation of these activities in the course of a scientific investigation.

I will here describe them only cursorily, leaving their further elaboration for other occasions. (I have already discussed item 2 in 1976a, pp. 244–245, and in 1978. Extensive discussions of items 1, 3, and 4 are in Wimsatt, 1980. Wade's work, and the work I have done so far in population biology, relates most strongly to items 4, 5, and 6, though 1, 7, 8, and 9 are also implicated.) This is possible because even a statement of the heuristic naturally suggests the pervasiveness of their use and multiplicity of their possible effects, and it is in any case necessitated by space limitations.

These heuristics and/or biases are as follows:

A. *Conceptualization:*

1. *Descriptive localization.* Describe a relational property as if it were monadic, or a lower order relational property; thus, e.g., fitness as a property of phenotype (or even of genes) rather than phenotype-environmental relation.

2. *Meaning reductionism.* As-

---

aries. These heuristics and examples of their application and misapplication are discussed at length in Wimsatt (1980), where they are particularly relevant in understanding the nature and origin of functional localization fallacies.

sume lower-level redescriptions to change meanings of scientific terms; higher-level redescriptions not. Result: philosophers (who view themselves as concerned with meaning relations) are inclined to a reductionistic bias.

3. *Interface determinism.* Assume that all that counts in analyzing the nature and behavior of a system is what comes or goes across the system-environment interface. This has two versions: (a) black-box behaviorism—all that matters about a system is how it responds to given inputs; (b) black-world perspectivalism— all that matters about the environment is what comes in across the system boundaries and how it responds to system inputs. Either can introduce reductionistic biases when conjoined with the assumption of white box analysis . . . that the order of study is from a system, with its input-output relations, to its subsystems, with theirs, and so on. The analysis of functional properties in particular is rendered incoherent and impossible by these assumptions.

B. *Model Building and Theory Construction:*

4. *Modeling localization.* Look for an intrasystemic mechanism to explain a systemic property, rather than an intersystemic one. Structural properties are regarded as more important than functional ones, and mechanisms as more important than context.

5. *Simplification.* In reductionistic model building, simplify environment before simplifying system. This strategy often legislates higher-level systems out of existence or leaves no way of describing systemic phenomena appropriately.

6. *Generalization.* When starting out to improve a simple model of system environment: focus on generalizing or elaborating the internal structure at the cost of ignoring generalizations or elaborations of the external structure. *Corollary.* If the model doesn't work, it must be because of simplifications in description of internal structure, not because of simplified descriptions of external structure.

C. *Observation and Experimental Design:*

7. *Observation.* Reductionists will tend not to monitor environmental variables, and thus will often tend not to record data necessary to detect interactional or larger scale patterns.

8. *Control.* Reductionists will tend to keep environmental variables constant, and will thus then often tend to miss dependencies of system variables on them. (*Ceteris paribus* is viewed as a qualifier on environmental variables.)

9. *Testing.* Make sure that a theory works out only locally (or only in the laboratory) rather than testing it in appropriate natural environments, or doing appropriate robustness analyses to suggest what are important environmental variables and/or parameter ranges.

These heuristics and their biases can be particularly powerful for two reasons: (1) There is, on the face of it, no way to correct for their effects —at best not in the most obvious way by producing the exact and general analysis of the behavior of the system in its environment to use as a check against the models produced. That may be all right in theory, but it won't work out in practice, as I have tried to suggest in section 3. Nonetheless, there are at least two possible corrective measures which will be discussed later. The first is robustness analysis—a term and procedure first suggested by Richard Levins in his (1966). The second, which I will call "multilevel reductionistic analysis" involves using these heuristics simultaneously at more than one level of organization—a procedure which allows discovery of errors and their correction in at least some circumstances, and which in fact implicitly followed as a species of "means-end analysis" (see Simon, 1966a and b) in the construction of inter-level theories involving compositional identities (see Wimsatt, 1976a, pp. 230–237, and 1976b, section 8).

(2) Secondly, it should be clear that *these heuristics are mutually supporting,* not only in their effective use in structuring and in solving problems, but also *in reinforcing, multiplying, and, above all, hiding the effects of their respective biases.* This effect of bias amplification is very serious and one of the biggest reasons why the effects of these biases are so hard to detect and why the proponents of extreme reductionistic positions can be so resistant to recognizing potential counterexamples to their position. Whatever can be said for theories or paradigms as self-confirming entities (and much that has been said is too excessive, and would render progressive science impossible), as much and perhaps more can be said similarly for heuristics. Indeed, I suspect that most of the blame and criticism of theories in this regard is more accurately laid at the doorstep of the heuristics used by those applying these theories and extending these paradigms.

Consider how this could work. Heuristics 1 and 4, applied in the early stage of an investigation, give apparent conceptual and theoretical reasons for locating a phenomenon of interest (say, that an organism has a given fitness in a given environment) as having causes primarily or wholly within the system under study. In the process of model-building, the environment may be simply described (e.g., as totally constant in space and in time, a frequent assumption in population genetics) that relevant variables (and the possibility of their variation) are ignored. (Bias 5), further leading to and being reinforced by tendencies not to observe variation in relevant environmental variables (Bias 7) and to make efforts to assure (or, too often, merely to assume) that they are constant as controls in the experimental analysis of the system (Bias 8). Any failures in the model are then assumed to be caused by a failure to model intrasystemic interactions in sufficient detail (Bias 6), leading to another cycle, beginning with Biases 1 and 4 applied to the properties and phenomena which were anomalous for the first model. This may result in further simplifications in the environment to offset the loss in analytical tractability arising from the increased internal complexity now assumed, or it may result in focusing in on a particular subsystem to be modeled in further detail, with much of the rest of the system now becoming part of the systematically simplified, ignored, and controlled environment. If even a part of this scenario or one like it is correct, we should not be surprised if quite remarkable failures went undetected for appreciable lengths of time. At present this remains just a hypothetical scenario, probably only one of many possible scenarios for producing this result. It would be very difficult to establish that the whole scenario, or one like it, was played out in any given case, in part because of the practice of not describing chains of hypothetical reasoning or discovery in scientific papers. Moreover, the practitioners are usually not themselves aware of the microstructure and background presuppositions of their reasoning processes, a fact which has bedeviled attempts to use protocols in which experimental subjects try to describe their reasoning processes as a basis for constructing theories of problem-solving behavior even for much simpler tasks (see Newell and Simon, 1972). Nonetheless, the scientific literature does contain suggestive

evidence of several of these heuristics in operation, and it could be hoped that future research would turn up more. A remarkable example of cumulative and systematic biases was unearthed by the work of Michael Wade on the models of group selection, which will be discussed after some preliminary discussion in the next section concerning the notion of a unit of selection.

## 6. DARWIN'S PRINCIPLES AND THE DEFINITION OF A UNIT OF SELECTION

Charles Darwin's argument in *The Origin of Species* is adumbrated[5] by R.C. Lewontin (1970, p. 1) as a scheme involving three essential principles:

1. Different individuals in a population have different morphologies, physiologies, and behaviors (*phenotypic variation*).
2. Different phenotypes have different rates of survival and reproduction in different environments (*differential fitness*).
3. There is a correlation between parents and offspring in the contribution of each to future generations (*fitness is heritable*).

Where (and while)[6] these three principles hold, evolutionary change will

5. These make no mention of the geometric rate of natural increase of organisms and the consequent inevitability of competition for resources (Malthus's observation). But this was a subsidiary argument employed by Darwin to establish the second principle—that different types of organisms had different fitnesses. Darwin needed this a priori argument because he had no direct observations of the occurrence of natural selection in nature.

6. Lewontin applies these principles on a genetic microevolutionary scale, and points out that for a population in equilibirum of gene frequencies, however temporary, conditions 2, 3, or both are not met (1970, p. 1). And obviously, if there is only a single allele

occur. Lewontin argues not only that these requirements are necessary for evolution to occur, but also that they are sufficient. They also embody what is generally regarded as Darwin's major contribution over prior evolutionists in that they specify a mechanism, natural selection, which produces this change.

Mechanism or not, these principles specify very little about the units which must meet these conditions. Although they are specified in terms of phenotypes and their properties (a form appropriate to Darwin's original theory and one to which modern evolutionists still pay lip service), Lewontin immediately applies them to genes (the units of the neo-Darwinian theory, under the impetus of Weismannism). Lewontin exploits the fact that these requirements say little about the units which must meet them, to argue that selection can operate—simultaneously and in different directions—on a variety of units (the unspecified individuals) at a number of levels of organization. In his view, he discusses selection processes at the micro- and macromolecular levels, and as operating on cell organelles, cells (in the immune system, in developmental processes, and he could have added, in cancer), gametes, individual organisms, varieties of kin groups, populations, species, and even ecological communities.

These principles give necessary conditions for an entity to act as a unit of selection, as well as necessary and sufficient conditions for evolution to occur. The three conditions must all be met by the same entity, in a way that can be summarized by saying that entities of that kind must show *heritable variance in fitness.*[7]

These conditions fail to be sufficient for the entity to be a unit of selection, however, for they guarantee only that the entity in question is either a unit of selection *or is composed of units of selection.* A further condition, which is sufficient, is given in the following definition:

A *unit of selection* is any entity for which there is heritable *context-independent* variance in fitness among entities at that level which does not appear as heritable context-independent variance in fitness (and thus, for which the variance in fitness is *context-dependent*) at any lower level of organization.

Much of population genetic theory involves the notion of additive variance in fitness. It is this quantity which, in Fisher's fundamental theorem of natural selection (Fisher, 1930) determines the rate of evolution. To say that variance in fitness is totally additive is to say that the fitness increase in a genotype is a linear function of the number of genes of a given type present in it. But this entails that the contribution to fitness of a given gene whose effect

---

at a given locus in a population (violating condition 1), no change in gene frequency (or microevolution) is possible at that locus.

7. I have analyzed these conditions and their ramifications in much greater detail in a book manuscript now in process and tentatively to be called *Reductionism, Sociobiology, and the Units of Selection.* Further excellent discussions of related issues can be found in Hull (1978a and b), Sober (1979), and, less directly, Cassidy (1978).

on fitness is totally additive is independent of the genetic background
in which it occurs, which is to say that
the variance in fitness is context-
independent. Additivity is thus a
special case of context-independence.
It is assumed for reasons of analytical
tractability, but the properties which
flow from this assumption derive
from its relation to context-independence.

One very important result follows
when this assumption holds at a given
level of organization. *If variance in
fitness is totally additive at a given
level of organization over a given
range of conditions on the environment and the system, then under
those conditions there are no higher-
level units of selection!* This is true
because fitness of any higher-level
unit is then a totally aggregative or
mass effect of the fitnesses of the
individual entities at that level of
organization. With no context-dependence of fitness, the *organization*
of these units into higher-level units
does not matter. There are no epistatic interactions to tie complexes
of these entities together as units of
selection. The higher level unit is
totally reducible in its effects to the
action of various lower-level units,

acting in a context-independent manner.[8]

It may be that this assumption (a
product of bias 4 or 5 applied at the
level of the gene to increase the analytical tractability of the model) is
one of the major reasons contributing
to the plausibility of Williams's reductionistic vision. It is clear that
once this assumption is made, it becomes plausible to attribute adaptation (and thus fitness) "to the effect
of selection acting independently at
each locus" (Williams, 1966, p. 57)
and leads naturally to regarding fitness
as a property of genes (a case of
Bias 1). It is also true that many or
most population geneticists believe
and argue (as James Crow has, in
personal conversation) that most
variance in fitness is additive—presumably, at the level of the contributions of individual genes. This is
an empirical claim and represents a
view not shared by all population
geneticists. Sewell Wright has systematically argued throughout his professional life and his magisterial
four-volume treatise that the opposite is true, that epistatic interactions
are all-pervasive and important (personal conversation; see, e.g., Wright,
1968, chapter 5, especially pp.

8. Mike Wade felt that this did not emphasize sufficiently strongly that whether an
interaction was additive or epistatic is a function of the relation of the system to the environments in which it is studied. He feels that many studies which purport to show that
variance in fitness is additive rather than epistatic suffer from looking at a restricted environment (usually in the laboratory) or range of environments, and that investigation of
the system in a wide range of environments would show that many or most of the supposedly additive interactions are in fact epistatic.

It is worth pointing out that the term "epistasis" is traditionally reserved for interactions
between genes *within a given genotype.* But the discussion here naturally suggests an extension to interactions between higher-level complexes of genes. Thus when one speaks, as
Wade does, of a *group* phenotype, it becomes natural to describe nonadditive interactions
between individuals in the group as epistatic.

71-105). Michael Wade's current research indicates the importance of epistatic interactions at the *individual* level (that is, between individuals in populations) in group selection (personal conversation; see Wade and McCauley, 1980, and McCauley and Wade, 1980). What is clearly true is that Biases 7, 8, and 9 would in general contribute substantially to failures to detect nonadditive variance if it exists because of artificially induced constancies in or ignorance of environmental conditions capable of producing nonadditive components of variance in fitness.

To summarize then, if variance in fitness at a given level is totally additive, the entities of that level are composed of units of selection, and there are no higher level units of selection. If the additive variance in fitness at that level is totally analyzable as additive variance in fitness at lower levels, then the entities at that level are composed of units of selection at these lower levels, rather than being units of selection themselves. To put it in terms of Salmon's (1971) analysis of statistical explanation, the higher-level units of selection as causal factors are then "screened off" by the lower-level units of selection. In their causal effects, they are then "nothing more than" collections of the lower-level entities, and any independent causal efficacy is illusory. This is a necessary and sufficient condition for the truth of Williams's genetic reductionism.

But in general, we would expect this partitioning of variance in fitness into additive and nonadditive components at different levels to show a number of levels—genes, gene complexes, chromosomes, individuals, even groups—at which additive variance at that level appears only as nonadditive variance at lower levels. There are units of selection at each level at which this occurs, and if it does, genetic reductionism and determinism are false.

## 7. WADE'S REVIEW OF THE MODELS OF GROUP SELECTION

For groups to act as units of selection, they must show heritable context-independent (or additive) variance in fitness[9] which is not merely a summative redescription of additive components of variance in fitness to be found at lower levels of organization. Groups must meet Darwin's three principles as well as the additional constraint of a context-independent component of fitness to so qualify.

9. The notion of additive variance has an implication that speaking of context-independent variance does not. Speaking of additive variance implies the context of a larger unit in which more than one of the smaller units which contribute to fitness will co-occur, so that their contributions will add. Thus genes which show additive variance will occur in genotypes. In the case of two alleles at one locus, this condition is met if the fitness of the heterozygote *Aa* is exactly halfway between the fitness of the two homozygotes, *AA* and *aa*. Talking about groups as units of selection may not imply a larger conspecific group whose fitness they contribute to, and in this case (and other similar cases), it is preferable to talk about *context-independent fitnesses* of groups rather than of *additive contributions to fitness* of groups.

Organisms meet Darwin's principles by having phenotypic differences (variability) which are heritable, and which have a differential effect on their survival and/or reproduction. Similarly, there may be differences of group structure and interaction (variability of group phenotype) which are transmitted to offspring groups or migrant propagules (group heritability), and which affect the rate of survival and/or reproduction of groups. The heritability of group variability may be either genetically or phenotypically transmitted. The latter results in models for cultural transmission and evolution. The former results in models for the genetic transmission of group traits and biological evolutionary models of group selection. Wade's results concern these models, in which what is inherited, to a greater or lesser degree, is the set of gene frequencies of a group's gene pool. Thus, if migration from a group occurs at random with respect to the genotypes of the individual migrants, the fact that the migrants are drawn from a given group will confer a kind of heritability or correlation between the gene frequencies of the parent population and the gene frequencies in the migrant propagule. (Indeed, this may be true to some extent even if migration from the group is not at random with respect to genotype.) This, together with a differential rate of production of migrant propagules by groups of different genetic compositions or differential rates of survival of such groups or both should make group selection a reality.

The various mathematical models of group selection surveyed by Wade all admit of the possibility of group selection. But almost all of them predict that group selection should be a significant evolutionary factor only very rarely, because they predict that group selection should have significant effects only under very special circumstances—for extreme values of parameters of the models which should be found in nature only rarely. Wade undertook an experimental test of the relative efficacy of individual and group selection—acting in concert or in opposition in laboratory populations of the flour beetle, *Tribolium.* This work (reported in Wade, 1976 and 1977) produced surprising results: group selection appeared to be a significant force in these experiments, one capable of overwhelming individual selection in the opposite direction for a wide range of parameter values, apparently contradicting the results of all of the then extant mathematical models of group selection. This led Wade to a closer analysis of these models, with results reported in Wade (1978), and described here.

All of the models surveyed made simplifying assumptions, most of them different. Five assumptions, however, were widely held in common: of the twelve models surveyed, each made at least three of these assumptions, and five of the models made all five assumptions.[10] Crucially, for present

10. Within the "traditional" models, the record was even worse. Five out of seven of the models made all five of the assumptions. The structure of the intramedic models, a newer development widely heralded as improving the case for group selection, required dropping one of the assumptions.

purposes, the five assumptions are biologically unrealistic and incorrect, and each independently has a strong negative effect on the possibility or efficacy of group selection. It is important to note that these models were advanced by a variety of different biologists, some sympathetic to and some skeptical of group selection as a significant evolutionary force. *Why then did all of them make assumptions strongly inimical to it?* Such a "coincidence", highly improbable at best, cries out for explanation. I will attempt to offer some explanations after I have presented and discussed some of the assumptions. These assumptions are given in Wade (1978, p. 103):

(1) It is assumed that the frequency of a single allele within a population can produce a significant change in the probability of survival of that population, or in the genetic contribution which the population makes to the next generation.

(2) All populations contribute migrants to a common pool, called the "migrant pool" (Levins, 1970), from which colonists are drawn at random to fill vacant habitats.

(3) The number of migrants contributed to the migrant pool by a population is often assumed to be independent of the size of the population. Thus the frequency of an allele in the migrant pool can be represented by the mean allele frequency of all contributing populations.

(4) It is assumed, often implicitly, that the variance between populations (which is a prerequisite for the operation of group selection) is created primarily by genetic drift within the populations and to a lesser extent, by sampling from the migrant pool.

(5) Group and individual selection are assumed to be operating in opposite directions with respect to the allele in question. In short, the allele is favored by selection between groups but disfavored by selection within groups.

The first assumption replicates Williams's reductionistic approach by assuming that the trait under selection at the group and individual level is a single-locus trait, rather than one with a polygenic basis. However, Wade's experiments rule out this possibility for the trait under selection, population size. In these experiments a number of replicate populations given the same selection treatment (selection for increasing size at both levels; or selection for increasing size at one level and decreasing size at the other) were assayed to determine a number of demographic parameters of the population. These demographic parameters included fecundity, fertility, body size, developmental time, and cannibalism rates for various developmental stages on other developmental stages, all factors having a known and modelable effect on population size, and on the number of different organisms of different age classes as a function of time. (The demographic effects of changes in these variables are visualized in the comparison of the number of individuals of different life stages in high-productivity and low-productivity populations as a function of time in figure 4 of McCauley and Wade, 1980.) The results of this assay showed that replicate populations drawn from the same genetic stock and given the same selection treatment, to which they responded in the same way at the level of the macroscopic trait, population size, achieved this response by different combinations of changes in the underlying demographic parameters affecting population size.[11] This means that even if each of these underlying demographic parameters is affected by a single-locus trait (which seems unlikely), *population size cannot be a single locus trait because similar values of it are produced by simultaneous independent variations in the underlying demographic variables.* Thus none of the single-locus models (nine out of the twelve) can adequately describe the selection processes in this experiment. In virtue of the increased dynamical complexity and richness of two-locus systems as compared with one-locus systems (see above, section 4, and Roughgarden, 1979, chapter 5) and the further poorly understood complexities of systems involving more than two loci (see Lewontin, 1974, chapter 6, and Maynard Smith, 1978, chapter 5), this is a very significant failure. Little if anything learned from the single-locus case is generalizable to cases involving two or more loci.

The second assumption, the analytical device of a "migrant pool," is, if anything, more serious. In this assumption, all migrants contributed by any population are thrown into a common pool (see Figure 3) from which all new founding populations are drawn. This assumption may not be unrealistic in some cases in the five intrademic selection models (it is equivalent to assuming panmixia or random mating within populations), but is surely radically unrealistic for any of the seven traditional models, where it is equivalent to assuming panmixia throughout the entire range of a species. If there is any systematic genetic differentiation throughout the range of a species, and if the average one-generation migration distance of an individual of that species is appreciably less than the dimensions of its range, then there will be mating which is assortative (or nonrandom) merely because of the limited distance a migrant can travel, thus falsifying this assumption. In general, parent populations will contribute migrants to founding populations in their immediate vicinity, and this assumption will be seriously incorrect.

The seriousness of this simplifying assumption can be better seen by exploring its analogies with the theory of blending inheritance at the individual level. In 1868 Darwin proposed his "provisional hypothesis of pangenesis in which large numbers of gemmules secreted by the various cells of an organism were combined in sexual reproduction in such a way that the characters of the offspring

---

11. Thus population size is a case of a functional or supervenient property, in the sense widely discussed in philosophy of psychology and recently in biology, by Rosenberg (1978), in that the same value of population size can be realized by a variety of underlying states or mechanisms involving lower-level variables.

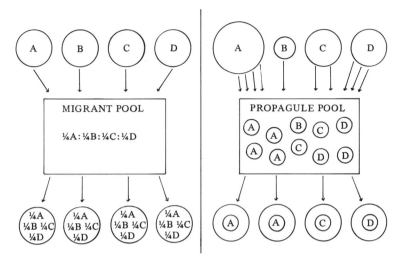

Figure 3. Diagram illustrating the differences between a migrant pool and a propagule pool. (Reprinted from Wade, 1978, figure 1, page 109, with permission of the *Quarterly Review of Biology*.)

(produced by equal contributions of gemmules from both parents) were an intermediate blend of the characters of the parents. This theory drew rapid and searching criticism (as it appeared in the fifth edition of Darwin's *Origin of Species*) by Fleeming Jenkin (1867), who pointed out that with such a "blending" mechanism of inheritance, the variation in a population would be rapidly attenuated until the population was essentially homogeneous. With no variation for selection to act upon, evolution would rapidly come to a halt.

With the rise of Mendelism (Mendelian segregation prevents significant blending in the 1-locus case, and limits its effects in the multi-locus case[12]), blending theories were rapidly forgotten, and almost without exception (one being Wallace's excellent discussion in Wallace, 1968, pp. 61–65), their characteristics and consequences are forgotten, ignored, and misunderstood.[13]

To R. A. Fisher, however, the avoidance of blending inheritance and its consequences for the loss of genetic variance was a sine qua non for the possibility of evolution. He began

12. Blending inheritance does not require fusion of the hereditary particles. The blending refers to the character traits, not to their underlying determining factors. Unconstrained random mixing of a large number of factors having an additive effect will produce the main effect of blending, a loss of genotypic variance, until Hardy-Weinberg equilibrium is achieved. For this reason a process importantly analogous to blending in the individual case can occur even with Mendelian genetics and can be significant in the loss of genotypic variance for additive multilocus traits.

13. These are explored in substantial further detail in the manuscript mentioned in note 7.

his ground-breaking treatise (Fisher, 1930, pp. 1-4) with a discussion of the character and consequences of a blending mode of inheritance. To Fisher, Mendel had clearly made the world safe for Darwinism, and he turns somersaults in a marvelously Whiggish rewriting of history to try to show that Darwin didn't really take blending inheritance too seriously.

Whatever the merit of that attempt, he makes the consequences of blending inheritance very clear. Fisher's fundamental theorem of natural selection says that a measure of the rate at which gene frequencies change is directly proportional to the additive genetic variance in fitness. No additive genetic variance in fitness means no gene frequency change, and no evolution. In those beginning pages, he derives (or actually, claims the derivability of) a formula for the rate of loss of variance resulting from mating under blending inheritance. In each generation, the variance is attenuated by a factor of $\frac{1}{2} (1 + r)$, where $r$ is the correlation between parental genotypes. Fisher expected this correlation to be small ($r$ ranges between $-1$ and $+1$, and is 0 in a randomly mating population). Thus he argued that the existing genetic variance in a population would be approximately halved in succesive generations. With no production of new variation, evolution would go as far in a given generation as it would in all successive generations, in accordance with the series $1 + \frac{1}{2} + \frac{1}{4} + \ldots$ (see Wallace, 1968).

Blending inheritance at the individual level is analogous to the assumption of the panmixia of migrants, or of a mating pool at the group level. Wade and I (Wade and Wimsatt, in preparation) have re-derived and extended Fisher's results in a manner applicable to the case of group selection. Blending inheritance at the individual level involves two parents, each making equal contributions. The group case may involve contributions of migrants from $n$ parental populations, in possibly unequal contributions, $w_1, w_2, \ldots, w_n$, which sum to 1. In a deterministic model which neglects sampling error in drawing migrants from parental populations, and in which the draw does not affect the gene frequencies in the parental population (requiring, in effect, infinite population size of both parental and migrant propagule populations with the migrants constituting a negligible proportion of the parent populations), a close analogy with Fisher's result can be derived. In this case, if the correlation between any two parental populations is constant at $r$, then the factor by which variance is attenuated in successive generations, $\alpha$ is given by equation 1:

$$(1) \quad \alpha = r + (1 - r) \, \Sigma w_i^2$$

This reduces to Fisher's derived factor of $\frac{1}{2} (1 + r)$ if there are two parents making equal contributions. With $n$ parents making equal contributions, and no correlation between them, $(r = 0)$, $\alpha = 1/n$. This is a "worst case" assumption, but it is exactly what is assumed in the "migrant pool" models! The assumption of the migrant pool with any significant number of populations guarantees that the variance goes to 0 essentially immediately.

Suppose that variance is being created anew in each generation by some mechanism or mechanisms, at a rate $V_0$. Then the equilibrium pool of variance (which will determine the rate of evolution) is that amount of variance for which variance is being lost at the same rate at which it is being created, which happens when:

(2)   $V = [1/(1 - \alpha)] V_0$

Under the migrant pool assumption, this is simply $[n/(n - 1)] V_0$, which is essentially $V_0$ for $n$ large. In effect, there is no pool of variance, but selection can only act on the variance created anew in each generation.

It can be seen by inspection of equations (1) and (2) what conditions will allow the accumulation of an appreciable amount of variance. This happens when $\alpha$ is close to 1, which is true if either there is high correlation between parent populations ($r$ close to 1), or if one population predominates in contributing to a given migrant propagule ($\Sigma w_i^2$ will be close to 1 if and only if $w_i$ is close to 1 for some $i$). In the migrant propagule model of Wade (1978), the latter condition holds, since all of a migrant population comes from a single parent population, and $\alpha = 1$ (in this deterministic model, which doesn't allow for sampling error, and thus does not exactly apply to Wade's case).[14] In more realistic models for the diffusion of migrants in a cline, the former condition would tend to hold. So Wade's models, and realistic

population models which don't make the migrant pool assumption, would be expected to allow much more rapid evolution at the group level. Plausible values of the parameters $r$ (or $r_{ij}$'s in a more general treatment) and the $w_i$'s suggest rates of evolutionary change at the group level which are easily 10 to 100 times greater than expected under comparable circumstances with the migrant pool assumption.[15]

Why the seriousness of the migrant pool assumption should have been overlooked, and the significance of how it was finally detected and analyzed, relate to the use of heuristics, and will be discussed in the closing section.

The third assumption, that parent populations contribute the same number of migrants to the migrant pool, independent of their size, in each generation, was also made for reasons of analytical tractability given by Wade (1978). Its effects, however, are far more immediately obvious, and thus it is similarly more anomalous that the assumption would ever have been made. In models where it is made, differences in reproductive output of different populations is due entirely to the populations' surviving for a different number of generations, since all alike have the same reproductive output in generations in which they are surviving.

Such an assumption would never have been made in models of selection

---

14. This conclusion must be moderated when the effect of finite population size and sampling error are taken into account. As Wade (1978, p. 110) points out, sampling error increases variance (so $\alpha$ would in effect be greater than 1) but at a cost of lowered group heritability. Some simple models taking account of sampling error have been constructed and will be discussed in our forthcoming paper.

15. See note 14, above.

William Wimsatt

TABLE 2

Ratio of intrinsic growth rates, $r$, of shorter-lived replicator $(r_b)$ to longer-lived replicator $(r_a)$ that shorter-lived replicator needs to offset shorter lifetime.[**]

| lifetime of replicator (generation) | Intrinsic growth rate (per generation) of a replicator with a lifetime of 1000 generations | | | | | |
|---|---|---|---|---|---|---|
| | 1.001 | 1.01 | 1.1 | 2 | 10 | 100 |
| 1000 | 1 | 1 | 1 | 1 | 1 | 1 |
| 100 | 1.0062 | 1.0032 | * | * | * | * |
| 10 | 1.071 | 1.067 | 1.033 | 1.001 | * | * |
| 5 | 1.148 | 1.143 | 1.101 | 1.0062 | * | * |
| 4 | 1.188 | 1.183 | 1.139 | 1.015 | * | * |
| 3 | 1.259 | 1.254 | 1.205 | 1.040 | 1.0003 | * |
| 2 | 1.413 | 1.403 | 1.352 | 1.118 | 1.005 | * |
| 1 | 1.998 | 1.990 | 1.909 | 1.500 | 1.100 | 1.010 |
| .9 | 2.158 | 2.149 | 2.062 | 1.611 | 1.141 | 1.017 |
| .75 | 2.518 | 2.507 | 2.405 | 1.863 | 1.244 | 1.042 |
| .50 | 3.997 | 3.980 | 3.816 | 2.914 | 1.733 | 1.210 |
| .25 | 15.988 | 15.921 | 15.260 | 11.485 | 5.958 | 3.001 |
| .10 | 1023.25 | 1018.92 | 976.46 | 728.44 | 345.95 | 133.18 |
| .01 | $1.27 \times 10^{30}$ | $1.26 \times 10^{30}$ | $1.21 \times 10^{30}$ | $8.97 \times 10^{29}$ | $4.04 \times 10^{29}$ | $1.30 \times 10^{29}$ |

* Added increase is less than 1 in $10^4$, Fisher's rough lower limit for selective differences to be significant.

[**]selection coefficient is $r - 1$.

$r$ short $(r_b)$ calculated from $r_{1000}$, $(r_a)$ from the relationship $r_b = \{[(r_a)^a - 1]^{b/a} + 1\}^{1/b}$ and from the approximation $r_b = [(r_a)^b + 1]^{a/b}$ when $(r_a)^a$ is greater than $1 \times 10^{100}$.

at the individual level. Darwin was acutely aware of the importance of differential reproduction as a selective force, one overshadowing in its potential intensity the effects of differential survivorship, and other evolutionists since have generally retained this awareness. It is doubly mysterious as an assumption at the group level, since presumably one of the primary ways in which increased group fitness could be manifested would be through increased production of individuals, leading to increased population size and increased migration of individuals once maximum population size is attained. One would thus expect a strong correlation between population size and migration rate, rather than the constancy assumed in these models. The association between size and reproductive rate is even more direct than it is for most cases of selection at the individual level. This assumption also has an enormous effect under some conditions, particularly those in which the average population survives for several or more (population) generations in which the average output of migrant propagules is appreciably more than one per generation. This can be seen by looking at the coefficients in Table 2, which gives the ratio in per-generation reproductive rates required of two replicators, one with very long lifetime (1000 genera- tions) and with different assumed re- productive rates given across the horizontal dimension, the shorter with a lifetime given in the vertical column and with a per-generation replication rate such that the shorter-lived replica- tor will have the same long-range replication rate as the longer-lived one. Thus the shorter-lived replicator must increase its per-generation repro- ductive rate above that of the longer- lived replicator (indicated by the amount that the ratio of reproductive rates exceeds 1) in order to offset the effect of its shorter lifetime. Comparisons in which the longer- lived replicator is not so long lived (say ten generations) can be had by dividing the appropriate coefficient of the shorter-lived replicator (say 1.0062 for comparison with a five-generation replicator if the ten-generation repli- cator is having two offspring each generation) by its appropriate co- efficient (1.0001, in this case), to get the ratio (1.0061, in this case) of per-generation reproductive rates for the five-and ten generation replicators.

What should be obvious from looking at this table is that unless one is comparing replicators which survive for a relatively short time *and* have very low reproductive rates, almost negligible differences in reproductive rate more than offset even quite substantial differences in lifetime. This is so because in this range of parameter values, changes in lifetime have a roughly linear effect on the outcome, whereas changes in repro- ductive rate have a roughly exponen-

tial effect. Thus a small increase in reproductive rate offsets a large de- crease in lifetime.[16] Given this fact, it is a bizarre and dangerous simpli- fication in a model to assume that per-generation reproductive rate (a variable to which the net reproduc- tive rate, which determines the in- tensity of selection, is very sensitive) is constant across populations, and that differences in selection are due entirely to differences in the lifetimes of different populations (a variable to which it is quite insensitive). Pre- sumably, group selection would often, even usually, act to optimize both the reproductive rate and the survival time of populations, but these considera- tions suggest that the former variable bears watching much more closely than the latter.

I will make no lengthy comments on the last two simplifications dis- covered and discussed by Wade. Clearly the fourth rules out spatial heterogeneity as a significant selective force at either the individual or group levels, an assumption which also exerts a bias against higher-level units of selection, which will be discussed on another occasion.

The fifth simplification, the as- sumption that group and individual selection are operating to change gene frequencies in opposite directions, has two interesting features.

The first is that it almost certainly has its origin not in any arguments about what would be true in nature, but in the joint action of a con- sideration of testability and a simpli- fying assumption. The consideration

16. The effects are reversed for survival times of the order of one generation or less and small reproductive rates, but survival times and reproductive rates are both inclined to be appreciably larger under a wide variety of circumstances.

of testability is that because of the complexity of interaction of fitness components and the difficulty of determining the relevant parameter values (see Lewontin, 1974), it would be helpful in determining the efficacy of group selection if we could find a trait whose presence clearly signaled the operation of group selection, because it would be selected *against* at the individual level and thus could *only* owe its presence to group selection. This does *not* mean that group selection would usually or generally tend to be opposed to individual selection in nature. Nonetheless, it was probably responsible for the concentration of analytical models on circumstances designed to investigate this condition. In the context of models of two alleles at one locus, the natural way to implement this condition is to assume that the effect of selection at one level was to increase a given gene frequency and that of selection at the other level was to decrease it. In the context of such simplified models, the move from traits to genes or genotypes is easy—all too easy as recent sociobiology has shown—but this way is fraught with error, as a longer discussion on another occasion will show (see above, note 7).

It is worth noting, however, how implausible the assumption that individual and group selection are opposed becomes once multilocus models are considered. The effects of selection may generally be described as a vector in which each component is the change in one of the stable variables (e.g., frequencies of genes, gametic, or zygotic genotypes) describing the population. *Only in a phase space of one dimension, such as that characterizing the model of two alleles at one locus, are change vectors constrained to lie in the same or in opposite directions. In spaces of any higher dimensionality, the probability that they will be identically or oppositely directed is of measure zero, and the resultant of the two vectors may similarly lie in any direction whatsoever.* Any residual plausibility of this assumption is clearly an artifact of being guided only by the simplest possible model of evolutionary change.

But then if group selection has to overcome forces of individual selection to which it must be opposed, it matters little what or how strong are the selective forces acting at the individual level in evaluating the possibility that group selection can be efficacious. This is particularly detrimental to many of Williams's (1966) arguments. In a multidimensional space, even relatively weak selective forces at the group level, when added to relatively strong forces at the individual level, can change the resultant selection vector sufficiently to cause evolution toward an adaptive peak different from that which might be achieved by individual selection acting alone. The richer dynamics and greater dimensions of a multidimensional phase space produces the possibility of a wide variety of interactions among selection forces at different levels. These interactions are surely poorly understood at this time. But we cannot hope to understand them if we don't even detect them. We fail to do so, it appears, because of biases which are almost perpetual in character.

## 8. HEURISTICS AND THEIR BIASES: SOME AMELIORATIVE REMARKS

The simplifications Wade discusses were almost all (with the possible exception of the fifth) made to improve analytical tractability. Why their biasing effects should not have been noticed is a difficult and probably multifaceted problem. I will here discuss some of the considerations which seem to me to be most salient:

(1) *Inertia.* Some assumptions have a time-honored status, in that they have been made by almost all past models. As a result, unless the model or phenomenon in question itself seems to point immediately to the need for relaxing one of these assumptions, it will be taken for granted, especially since each of these assumptions almost invariably involves advantages of increased analytical tractability. This may relate to the anchoring bias of Tversky and Kahneman (1974), but it seems equally likely that "anchoring" and "inertia" are broad phenomenological categories that cover a multitude of sins committed for a variety of reasons. Particularly well entrenched is the assumption of panmixia (equivalent to the mating pool assumption, and *sufficient to guarantee that the mode of inheritance at any higher level of organization is a form of extreme blending*). An assumption that is well entrenched in the theoretical structure is one that is widely used in, and at least apparently essential to, the derivation of many other results. These dependency relations can lead to a reluctance to give up an assumption even if it is widely known to be problematic, or even if it is generally believed to be false. The assumption

of the transitivity of preferences in decision theory, or of the transitivity of fitness in evolutionary theory, both known to be false when choice or fitness is a function of a number of variables, seems to be such a case. The assumption of panmixia is almost certainly another. A third, similarly entrenched assumption in quantitative genetics is the assumption that variance in fitness is totally additive. Both this and the assumption of panmixia are absolutely inimical to the existence or significance of higher-level units of selection.

(2) *Perceptual focus.* Given the centrality of reproductive rate in virtually all evolutionary models and in the structure of evolutionary theory, Wade's third simplifying assumption seems almost unintelligible in almost any circumstances. The only way that I can think of to rationalize it is as follows: model-building activity is performed against a background of presumed mechanisms operating in the interaction of presumed units. If the presumed units are very well entrenched in a given area, there is a strong tendency to describe and to think about even phenomena at other levels of organization in terms of these units. In traditional evolutionary theory, and even at present, the most obvious unit is the individual organism—the unit which our everyday thought and our perceptual apparatus naturally predisposes us to consider. Most of our everyday interactions (as well as those of most other organisms) that call for voluntary action are pairwise interactions with one other organism at a time, so this bias is evolutionarily well founded.

Consequently, there is a strong tendency to see and to talk about groups of organisms as *collections of individuals, rather than as unitary entities.* This is true even for colonies of social insects, whose interdependencies extend even to reproductive specialization, making the metaphor of the colony as an organism perhaps more revealing in evolutionary terms than the view of it as a collection of organisms. Hull (1978a and b) has found similar biases in his arguments that species must often be conceived of as individuals in evolutionary contexts. A quick review of Williams (1966) reveals that even in the context of discussions of group selection, groups are usually described as collections of individuals, and it is my impression that this tendency is widespread throughout the literature. But *in the context of the group selection controversy, description of an assemblage of units as a collection is a theory-laden description, since it suggests that it is an aggregate. The only time it is appropriate to describe it as an aggregate is when the fitnesses of its components are context-independent* (see section 6) *or are additive. But as we saw above, this is a sufficient condition for their not constituting a higher-level unit of selection!* Perceptually, the focus on individual organisms prevents us from at the same time seeing the groups as individuals. If we do not see the groups as individuals, then we do not see that assuming that each group contributes an equal number of migrants is equivalent to assuming that there is no variance in reproductive rate. The bias, I suspect, is a perceptual one.

Similar remarks apply to seeing panmixia as equivalent to a form of blending inheritance at the group level. Blending inheritance is traditionally viewed as applying only at the individual level. But once the group is seen as an individual—a view that emerges particularly strongly in Wade's work, even in his earliest paper (Wade, 1976, 1977)—the analogy is immediate. I saw the analogy when I read an early draft of Wade (1978) at the same time as I was teaching Darwin's blending theory and its criticisms. The phenomenon was reproducible: three months later, Ross Kiester was teaching Darwin and independently pointed out the analogy to Wade. (Seeing the analogy also requires, I suppose, more than a passing acquaintance with blending inheritance.)

The recent focus on the gene (rather than the individual) as the unit of selection has introduced its share of perceptual problems and biases. Williams (1966) and Dawkins (1976) are full of "perceptual shifts" back and forth from the genetic to the individual level, but both try to maintain the primacy of the former with consequent biases in their description of the complexities of gene interaction.

Wade (1979) has shown a marvelous case of perceptual bias in the foundational work of Hamilton on kin selection theory. In this theory, selection is seen as maximizing an individual's "inclusive fitness," in which contributions to the fitness of relatives who share genes are included, weighted by their degree of relationship. Hamilton is reported to have described this as "the gene's eye view of evolution," and in any case the description fits. Hamilton's theory involves the assignment of selection coefficients to individual genes, and makes no reference to the genotypes in which they occur, or to which genotypes mate with which to produce offspring. Wade (1979) advances one- and two-locus models in which this mimimum amount of detail is added, and in which he is able to prove an analogue of Fisher's fundamental theorem of natural selection, according to which the rate of evolutionary change is proportional to the *between-family variance in fitness.* This result shows that *in a more realistic model of kin selection, the unit of selection is not the individual gene, but the mating pair, Wade's "atomic family."* This result could not have been derived or even seen in Hamilton's model, because the simplifications of the environment of the gene were such that Hamilton had no structure in his model on which to hang the upper-level phenomena. Mating pairs do not occur in his model, and there is no way of putting them into it. This is a paradigmatic example of Bias 5 in section 5 above, but it also shows the perceptual or quasi-perceptual character of the way in which this bias operates.

(3) *Perceptual reinforcement.* A factor discussed above is the reinforcement of different biases or heuristics. When a force must be greater than a given magnitude to have an effect, several biases tending to underestimate this force may have a cumulative effect that none of them could have alone. But biases can be mutually supportive in another way, also suggested there. One bias may act in such a way as to hide the fact that another bias *is* a bias, and conversely. Detailed documentation of these cooperative effects of biasing assumptions in population genetics remains largely a

task for future analysis, but a promising one. The assumptions of panmixia and additivity of fitness variance are plausible candidates for inspection, both because of their ubiquity and because each carries with it a relatively easily intuited picture of the unstructured environment of the unit of analysis. The assumptions that fitnesses are constant in space and time, and are independent of the density or of the relative frequency of the relevant units (usually assumed to be genes) also demands closer inspection, as each is a source of potential systematic bias against recognizing higher-level units. Near decomposability assumptions are important tools which are easily misused, here as elsewhere. Some futher biases which may be relevant in this context are discussed in Wimsatt (1980), but most of the work is yet undone.

How, aside from analyzing specific cases of bias and identifying biases more generally can something be done to correct for their effects? There are at least two plausible candidate procedures:

(1) The first is what Levins (1966) has called the search for robust theorems. To counteract the biases of any given model, he suggests building families of alternative models of a given phenomenon which differ in their simplifying assumptions. The models will vary in their consequences and predictions, reflecting the variety of their assumptions, but there may be consequences which are true across all of the models. These Levins calls "robust theorems," results which are independent of the details of any particular model. Thus, says Levins, "Our truth is the intersection of independent lies" (1966, p. 126; see

also his 1968, pp. 6–8, for futher elaborations of this approach.

This method has two disadvantages; neither is fatal but both are worthy of note. The first is that we are often in situations in which we do not have even one model, much less two or more, to compare their results. Population genetics and ecology has a richness of mathematical models, unlike some other disciplines, but even here, many areas do not have models with sufficient overlap to be able to compare the results. A further contributing problem arises when the assumptions made in deriving the models are not explicit, since it is then not possible to tell how much one can trust the robustness of a result, basically for the second reason.

A second caution is necessary because it is not always possible to tell when the models are *in fact* independent. Given a number of models, if each makes a possibly well-disguised but in any case unnoticed assumption, any theorems in common may simply be a relatively direct consequence of the shared assumption or assumptions. In this case, the results will generally be assumed to be quite robust—illegitimately as it turns out. Indeed, just this seems to have happened in the group selection controversy. Until Wade's review, where the assumptions the various models had in common and their consequences were made explicit, the one result which seemed to be robust was that group selection could be efficacious only rarely and under very special circumstances, and probably the majority of evolutionary biologists still believe this to be the case.

Another approach, generally consistent with robustness analysis (see Levins, 1968, pp. 6–8) and suggested by the heuristic use of means-end analysis and identifications in interlevel reductive explanation (Wimsatt 1976a, pp. 231–237, 1976b, section 8) could be called "multilevel reductive analysis." Even if perceptual focusing may leave one blind to the biases of the nine heuristics of section 5 as applied at any one level, these same heuristics will have biases leading to *different* simplifications if they are applied to the same system at a different level of organization, *simply because the system-environment boundary has changed.* Simplified models of group selection may thus suggest particular structural features of the environment of the individual which should be included in models of individual selection, just as, e.g., an analysis of the structure of the genotype and/or mating system of the individual may suggest important internal constraints on models at the group level. Simultaneous multilevel modeling may thus eliminate the biases of proceeding at only one level, but a final caution is required: this is likely to work only to the extent that the phenomena and entities of a given level are taken seriously in their own right. Seeing them merely as an extension of another level, be it lower or higher, will merely preserve the perceptual focus of that other level, and most biases will go undetected. This is merely another expression of a view I have argued for before (Wimsatt 1976a, 1976b, 1978). Now, for pragmatic as well as for theoretical reasons, reduction in science is better seen as the attempt to understand the explanatory relations between different levels of phenomena, each of which is taken seriously in its own right, than as an unending search of firm foundations at deeper and deeper levels in which, as Roger Sperry so aptly put it (quoted in Wimsatt, 1976a), "eventually everything is explained in terms of essentially nothing."

### ACKNOWLEDGMENT

This work owes a great deal more to many people than is reflected in the footnotes: To Richard Lewontin for his leading me past the most elemental understandings (and sometimes misunderstandings) I brought to my study of population genetics as a postdoctoral fellow in his lab, and for his early advocacy of the importance of gene interactions and higher-level units of selection in evolution, which, as with many other things he has taught me, had a formative effect on my present views. To Richard Levins I owe a similar debt, for his tutelage in mathematical ecology and the analysis of complex systems, for his friendly tolerance of my early naive and principled reductionism, while he gradually taught me elements of a richer vision. To George Williams, whose book was an education to me and to many others, formulating for the first time a clear reductionistic vision and many arguments supporting that view, which I continue to respect, even though I now believe it to be mistaken. More recently, I have learned a great deal from discussions with Ross Kiester and Mike Wade about model-building and experimental design, as well as a lot of fascinating and useful biology, and Wade in particular has done a lot to shape my views on group selection. I have profited also in particular from discussions with David Hull, but also with Bob McCauley, Bill Bechtel, Bob Richardson, Elliott Sober, and Mary Jane West-Eberhardt, the participants of a Midwest Faculty Seminar on Evolution at Chicago, and seminar or conference participants at the Leonard Conference at Reno, Denison University, Ohio State, the University of Wisconsin at Madison, and the University of Colorado at Boulder, where versions of this work were presented in 1978–79.

I would like to thank Columbia University Press for permission to reprint figure 23 and 24 from Lewontin (1974), pp. 274 and 280, and to adapt table 56 from p. 283; and the *Quarterly Review of Biology* for permission to reprint figure 1, p. 109, and to quote from p. 103 of Wade (1978). Finally, all of this would have occurred substantially later or perhaps not at all without the support of the National Science Foundation (Grant SOC78–07310 research).

## REFERENCES

Ando, A., F. M. Fisher, and H. A. Simon, 1963, *Essays on the Structure of Social Science Models,* Cambridge, MIT Press.

Bantz, D. A., 1976, "Does Physics Explain Chemistry?" Ph.D. Dissertation, Committee on Conceptual Foundations of Science, Univ. of Chicago. Unpublished.

Bogaard, P. A., 1979, A reductionist's dilemma: The case of quantum chemistry. In I. Hacking and P. D. Asquith, eds., *PSA 1978,* Vol. 2, East Lansing, Michigan, Philosophy of Science Association.

Boyd, R., 1972, Determinism, laws and predictability in principle. *Philosophy of Science,* 39: 431–450.

Campbell, D. T., 1974, Evolutionary epistemology. In P. A. Schilpp, ed., *The Philosophy of Karl Popper,* Vol. 1, pp. 413–463, La Salle, Illinois, Open Court.

———, 1977, "Descriptive Epistemology: Psychological, Sociological, and Evolutionary." The William James Lectures, Harvard, Spring, 1977. Unpublished.

Cassidy, John, 1978, Philosophical aspects of the group selection controversy. *Philosophy of Science,* 45: 575–594.

Causey, R. L., 1972a, Uniform microreductions. *Synthese,* 25: 176–218.

———, 1972b, Attribute-identities in microreductions. *Journal of Philosophy,* 69: 407–422.

———, 1977, *Unity of Science,* Dordrecht, Holland, Reidel.

Cohen, R. S., C. A. Hooker, A. C. Michalos, and J. van Evra, eds., 1976, *PSA 1974,* Boston Studies in the Philosophy of Science, Vol. 32, Dordrecht, Holland, Reidel.

Darden, L., and N. Maull, 1977, Interfield theories. *Philosophy of Science,* 44: 43–64.

Dawkins, R., 1976, *The Selfish Gene,* New York, Oxford Univ. Press.

———, 1978, Replicator selection and the extended phenotype. *Zeitschrift für Tierpsychologie,* 47: 61–76.

Dresden, M., 1974, Reflections on fundamentality and complexity, in C. Enz and J. Mehra, eds., *Physical Reality and Mathematical Description,* pp. 133–166, Dordrecht, Holland, Reidel.

Fisher, R. A., 1930, *The Genetical Theory of Natural Selection,* London, Oxford Univ. Press.

Glassman, R. B., 1978, The logic of the lesion experiment and its role in the neural sciences. In S. Finger, ed., *Recovery from Brain Damage: Research and Theory,* pp. 3–31, New York, Plenum Press.

Globus, G., G. Maxwell, and I. Savodnik, eds., 1976, *Brain and Consciousness: Scientific and Philosophic Strategies,* New York, Plenum Press.

Gregory, R. L., 1967, Models and the localization of function in the central nervous system. Reprinted in C. R. Evans and A. D. J. Robertson, eds., *Key Papers: Cybernetics,* pp. 91–102, London, Butterworth.

Hull, D. L., 1973, Reduction in genetics—biology or philosophy? *Philosophy of Science,* 39: 491–498.

———, 1974, *Philosophy of Biological Science,* Englewood Cliffs, N.J., Prentice-Hall.

———, 1976, Informal aspects of theory reduction. In Cohen et al. (1976), pp. 633–652.

———, 1978a, A matter of individuality, *Philosophy of Science,* 45: 335–360.

———, 1978b, The Units of Evolution: A Metaphysical Essay, presented at the University of Aarhus symposium, "Evolution: Its Philosophical and Methodological Aspects," in August, 1978. A later version of this paper was given at the 1980 meetings of the Philosophy of Science Association, and published in the proceedings.

Jenkin, Fleeming, 1867, Review of Darwin's *Origin of Species. The North British Review,* reprinted in David L. Hull, ed., *Darwin and His Critics,* Cambridge 1973, Harvard Univ. Press.

Levins, R., 1966, The strategy of model building in population biology. *American Scientist,* 54: 421–431.

——, 1968, *Evolution in Changing Environments,* Princeton, Princeton Univ. Press.

——, Extinction. In Some mathematical questions in biology: Lectures on mathematics in the life sciences, *American Mathematical Society,* 2: 75–108.

Lewontin, R. C., 1970, The units of selection. *Annual Review of Ecology and Systematics,* 1: 1–18.

——, 1974, *The Genetic Basis of Evolutionary Change,* New York, Columbia Univ. Press.

Lewontin, R. C., and M. J. D. White, 1960, Interaction between inversion polymorphisms of two chromosome pairs in the grasshopper, *Moraba scurra. Evolution,* 14: 116–129.

Lorenz, K. Z., 1965, *Evolution and the Modification of Behavior,* Chicago, Univ. of Chicago Press.

Maull, N. L., 1977, Unifying science without reduction. *Studies in History and Philosophy of Science,* 8: 143–162.

Maull-Roth, N., 1974, "Progress in Modern Biology: An Alternative to Reduction", Ph.D. Dissertation, Committee on Conceptual Foundations of Science, Univ. of Chicago. Unpublished.

Maynard Smith, John, 1975, *The Theory of Evolution,* 3rd ed., London, Penguin.

——, 1978, *The Evolution of Sex,* London, Cambridge Univ. Press.

McCauley, D. E., and M. J. Wade, 1980, Group selection: The genetic and demographic basis for the phenotypic differentiation of small populations of *Tribolium castaneum. Evolution,* 34: 813–821.

McCauley, Robert, N., 1979, *Explanation, Cross-Scientific Study, and the Philosophy of Mind: An Examination of the Methodological Foundations of Transformational Generative Grammar,* Ph.D. Dissertation, Department of Philosophy, Univ. of Chicago.

Moss, R. Y., 1979, *Designing Reliability into Electronic Components,* Palo Alto, Calif., Hewlett-Packard.

Nagel, E., 1961, *The Structure of Science,* New York, Harcourt Brace.

Newell, A., and H. A. Simon, 1972, *Human Problem Solving,* Englewood Cliffs, N.J., Prentice-Hall.

Newell, A., C. Shaw, and H. A. Simon, 1957, Empirical explorations with the logic theory machine: A case study in heuristics. Reprinted in E. A. Feigenbaum and J. Feldman, eds., *Computers and Thought,* New York, McGraw-Hill, 1963.

Newell, A., and H. A. Simon, 1961, GPS, a program that simulates human thought. Reprinted in E. A. Feigenbaum and J. Feldman, eds., *Computers and Thought,* New York, McGraw Hill, 1963.

Nickles, T., 1973, Two concepts of intertheoretic reduction. *Journal of Philosophy,* 70: 181–201.

——, 1976, Theory generalization, problem reduction, and the unity of science. In Cohen et al. (1976), pp. 33–74.

——, ed., 1980, *Scientific Discovery,* Dordrecht, Reidel.

Nilsson, N. J., 1971, *Problem-Solving Methods in Artificial Intelligence,* New York, McGraw-Hill.

Rosenberg, A., 1978, The supervenience of biological concepts. *Philosophy of Science,* 45: 368–386.

Roughgarden, Jonathan, 1979, *Theory of Population Genetics and Evolutionary Ecology: An Introduction,* New York, Macmillan.

Ruse, M., 1973, *The Philosophy of Biology,* London, Hutchinson University Library.

——, 1976, Reduction in genetics. In Cohen, et al. (1976), pp. 633–651.

Salmon, W. C., 1971, *Statistical Explanation and Statistical Relevance,* Pittsburgh, Univ. of Pittsburgh Press.

Schaffner, K. F., 1967, Approaches to reduction, *Philosophy of Science,* 34: 137–147.

——, 1969, The Watson-Crick model and reductionism. *British Journal for the Philosophy of Science,* 20: 325–348.

——, 1974, The peripherality of reductionism in the development of molecular biology. *Journal for the History of Biology,* 7: 111–139.

Schaffner, K. F., 1976, Reduction in biology: Prospects and problems. In Cohen et al. (1976), pp. 613–632.

———, 1978, "The Logic of Discovery and Justification in the Biomedical Sciences", Department of History and Philosophy of Science, Univ. of Pittsburgh. Unpublished.

———, 1979, Theory structure in the biomedical sciences. *Journal of Medicine and Philosophy,* 3: 331–371.

Shimony, A., 1970, Statistical inference. In R. G. Colodny, ed., *The Nature and Function of Scientific Theories,* pp. 79–172, Pittsburgh, Univ. of Pittsburgh Press.

Simon, H. A., 1957, A behavioral model of rational choice. In H. A. Simon, *Models of Man,* pp. 241–256. See also introduction to Part IV, pp. 196–206, New York, Wiley.

———, 1966a, Thinking by computers. In R. G. Colodny, ed., *Mind and Cosmos,* pp. 3–21, Pittsburgh, Univ. of Pittsburgh Press.

———, 1966b, Scientific discovery and the psychology of problem solving. In R. G. Colodny, ed., *Mind and Cosmos,* pp. 22–40, Pittsburgh, Univ. of Pittsburgh Press.

———, 1969, *The Sciences of the Artificial,* Cambridge, MIT Press.

———, 1973, The structure of ill-structured problems. *Artificial Intelligence,* 4: 181–201.

Sklar, L., 1973, Statistical explanation and ergodic theory. *Philosophy of Science,* 40: 194–212.

———, 1976, Thermodynamics, statistical mechanics, and the complexity of reductions. In Cohen et al. (1976), pp. 15–32.

Sober, E., 1979, "Significant Units and the Group Selection Controversy," Department of Philosophy, University of Wisconsin at Madison. A later version of this paper was given at the 1980 meetings of the Philosoohy of Science Association and published with the proceedings.

Tversky, A., and D. Kahneman, 1974, Decision under uncertainty: Heuristics and biases. *Science,* 185: 1124–1131.

Valenstein, E., 1973, *Brain Control,* New York, Wiley.

Wade, Michael J., 1976, Group selection among laboratory populations of *Tribolium.* Proceedings of the National Academy of Sciences, U.S.A., 73: 4604–4607.

———, 1977, An experimental study of group selection. *Evolution,* 31: 134–153.

———, 1978, A critical review of the models of group selection. *Quarterly Review of Biology,* 53: 101–114.

———, 1979, The evolution of social interactions by family selection. *American Naturalist,* 113: 399–417.

Wade, Michael J., and D. E. McCauley, 1980, Group Selection: The phenotypic and genotypic differentiation of small populations. *Evolution,* 34: 799–812.

Wade, Michael J., and W. C. Wimsatt, "The Blending Theory of Group Inheritance." Forthcoming.

Wallace, B., 1968, *Topics in Population Genetics,* Norton, New York.

Williams, G. C., 1966, *Adaptation and Natural Selection,* Princeton, Princeton Univ. Press.

Wilson, E. O., 1975, *Sociobiology: The New Synthesis,* Cambridge, Harvard Univ. Press.

Wimsatt, W. C., 1974, Complexity and organization. In K. F. Schaffner and R. S. Cohen, eds., *PSA 1972,* Boston Studies in the Philosophy of Science. Vol. 20, pp. 67–86. Dordrecht, Holland, Reidel.

———, 1976a, Reductionism, levels of organization, and the mind-body problem. In Globus, et al. (1976), pp. 199–267.

———, 1976b, Reductive explanation: A functional account. In Cohen et al. (1976), pp. 671–710.

———, 1978, Reduction and reductionism. In H. Kyburg, Jr., and P. D. Asquith, eds., *Current Problems in Philosophy of Science,* Philosophy of Science Association, East Lansing, Mich.

———, 1980, Randomness and perceived-randomness in evolutionary biology. *Synthese*, 43: 287–329.

——, 1980, Robustness and functional localization: Heuristics for determining the boundaries of systems and their biases. In M. Brewer and B. Collins, eds., *Knowing and Validating in the Social Sciences: A Tribute to Donald T. Campbell*, Jossey-Bass, San Francisco.

Winston, P. H., 1977, *Artificial Intelligence*, Reading, Mass., Addison-Wesley.

Wright, Sewall, 1968, *Evolution and the Genetics of Populations*, Vol. 1, Chicago, Univ. of Chicago Press.

# 12

# Holism, Individualism, and the Units of Selection

## ELLIOTT SOBER

### 1. HOLISM AND INDIVIDUALISM

The units of selection problem, as it is discussed within evolutionary theory, recapitulates some important elements in the dispute between methodological holism and methodological individualism. Holism and individualism have for a long time occupied favored positions in the stable of old warhorses owned and operated by philosophers of social science. These particular old warhorses are thought by many to be in retirement, although there is less than universal agreement about whether holism or individualism won the battle. Part of the point I will make about group versus individual selection is that biologists would do well *not* to emulate certain aspects of the holism/individualism contro-

versy. That they have done so already is a point that I will attempt to establish. And, conversely, the substantive empirical issues involved in the units of selection controversy suggest that the holism/individualism debate within the social sciences can be reformulated in a way that makes it nontrivial and also not decidable a priori. This offers some hope that the holism/individualism dispute need not remain a dismal philosophical problem of the "dismal sciences."

Holists and individualists disagree over whether social wholes are more than the sum of their parts (see Brodbeck, 1968, for representative essays). Holists say they are and individualists say they are not. There is the *appearance* of a disagreement here. But the appearance starts to

I am very grateful to James Crow, David Hull, Richard Lewontin, and William Wimsatt. Discussions with them have been invaluable to me in developing my ideas on evolutionary theory in general and on group selection in particular. The research discussed here was supported by the John Simon Guggenheim Foundation and by the Graduate School of the University of Wisconsin, Madison. I also wish to thank the Museum of Comparative Zoology, Harvard University, for its hospitality during 1980–81.

appear illusory when one asks what each side means by "sum"; exactly what is meant when it is asserted, or denied, that the whole is more than the *sum* of its parts?

Holists are concerned to avoid the sin of *atomism*. They do not think that social entities can be understood by taking individuals in isolation from each other. To understand social wholes, they insist, one must consider individuals in their relationship to each other and to the environment. When holists assert that the whole is more than the sum of its parts, they mean that properties of the whole are not determined by the unary, non-relational properties of the parts. To me, the point that holists are making is a truism. That relational properties must be taken into account seems obvious. It looks as though this point is true not just of *social* objects, but of *any* object which has parts. One might even suspect that the principle is a priori, or as a priori as anything can be.

Do individualists seriously propose to ignore relations? Not at all, say the individualists, who insist that they should not be confused with the straw man just discussed. Individualists concede that social facts cannot be understood by taking individuals in isolation from one another. But the crucial point is that the character of the whole is fixed by the properties and relations of its parts. The whole is nothing above and beyond those interactions among individuals. Individualists will deny being atomists. What accusation might they hurl back at the holists? Holists, they say, *hypostatize* (*reify*) social wholes. Holists, according to this indictment, think that properties of social wholes

are not determined by the interactions that individuals have with one another and with the environment. Where, then, does this independent existence of social wholes come from? What is the secret added ingredient one must add to individuals, the environment, and their interactions to get social facts? Holism, thus construed, looks like old-fashioned vitalism; it isn't that some mysterious fluid must be added to matter to get life. Rather, holism is portrayed as holding that you must add some sort of occult social fluid to individuals and their interactions to get social groups.

This time it is the individualists who seem to be right. Reification is precisely what holists are up to, if they believe that the whole is not determined by its parts and their interactions with each other and the environment. Individualism does seem to be correct in claiming that properties of wholes are determined by properties of parts, in this sense.

So what happened to this dismal dispute? One *could* embrace atomism on the one hand or hypostasis on the other, and doubtless there have been social scientists who have done so, in practice if not in theory. But if one rejects both of these alternatives, there seems to be no issue left. *Yes,* in one sense, the whole *is* more than the sum of its parts, but, in another sense, *no,* it is *not*. What else is there to talk about?

I will mention three problems which retain their interest in the face of these truisms. The first is epistemological. One can agree with the truisms and still wonder what the most fruitful research strategy might be in understanding a particular social phenomenon. Even if the truisms are true, it is still an open question what facts would be most interesting to look at in trying to understand the stock market crash of 1929; this social fact may be manageable from a macroscopic perspective and completely intractable from a more individualistic point of view (see Sober, 1980, for the relevance of this distinction to the biological species concept). A second sort of question, also untouched by the truisms, concerns the amount of complexity and interaction among parts that needs to be taken into account to explain social facts. Historically, the difference between self-styled holists and self-styled individualists has often concerned this question; the issue has not been *whether* the whole is reducible to the interaction of its parts, but in what ways the determination works (Wimsatt, 1980, argues the centrality of this issue to the units of selection problem). A third question which is not addressed by the truisms can be grasped by distinguishing *type from token*.[1] What I have been discussing so far is the way in which single social events, like the stock market crash

of 1929, are the upshot of interactions among individuals. That is, I have been talking about token social facts. A rather different question concerns the nature of various *kinds* of social facts. Here, one asks not for an explanation of a particular historical event, but of the nature of a social property. What is capitalism? What is a stock market crash? These questions about types may or may not be answerable in terms of individuals and their interactions. I raise these three questions only to set them to one side.

The truisms, then, do not concern how we might best attempt to understand a particular social fact, but concern that fact's causal connections with the world of individuals. The truisms do not concern the character of social properties, like the nature of crises or production in general, but the causality of single token events, like the stock market crash of 1929. Each whole is determined by interactions among its parts. This truism would be vouchsafed if causality were transitive. I'll assume that it is.[2] The stock market crash was caused, we might suppose, by inter-

1. The argument presented in Putnam (1975) and Fodor (1976) against identifying psychological and physical properties may be applicable to the relationship between social properties and the properties of individual psychology. Just as a given psychological property may be "multiply realizable" in indefinitely many physical forms, so a given social property may have indefinitely many realizations at the level of individual psychology. For an application of this line of thinking to the relationship of biological properties like fitness to physical properties, see Rosenberg (1978).

2. The present discussion of holism and individualism and of causality assumes the truth of determinism, but this assumption is not essential to the points at issue. If quantum mechanical states at one time do not uniquely determine such states at a later time, then, on the assumption that macro-states are token/token identical with quantum mechanical states, it follows that macro-states at one time do not determine macro-states at a later time. Thus, from the point of view of causal determination, not only will facts about individuals fail to causally determine social facts; it will also be true that earlier social facts fail to causally determine later social facts. So, if both holism and individualism are construed as making claims which imply that social facts are causally determined (but disagree about what does the determining), then both are mistaken. However, there still is scope for

actions among various market conditions, which were themselves social facts. These social facts, in turn, obtained because various individuals did what they did in various physical environments. So the interactions among people caused the crash. If this is right, then it is an ill-conceived question which asks: "What caused the crash—was it the market conditions or was it the way people acted?" One should respond to this question by asking: "What are you saying *or* for?" *There is no asymmetry here.* Causality, in virtue of its transitivity, gives aid and comfort neither to the holist nor to the individualist. The causal chain just keeps rolling along.

In what follows, it will be argued that the dispute about the units of selection resembles the holism/individualism dispute, but with the concept of natural selection replacing the more general idea of causality. This replacement makes all the difference in the world, however. Holists and individualists are, or should be, driven to disgruntled agreement by considerations that are not specifically sociological and appear to be almost a priori (like the fact that causality is transitive). In contrast, group selection and individual selection hypotheses admit of no such easy resolution; this dispute, properly construed, turns out to be an interesting empirical and specifically biological one. What is

more, the lack of asymmetries just noted, when a question of social versus individual causation is broached, are replaced by asymmetries aplenty. It *is* a very real question to ask "Was a particular social characteristic caused by group or by individual selection?"

## 2. HISTORICAL AND CONCEPTUAL BACKGROUND

I just argued that holists and individualists interpret the cliché "the whole is more than the sum of its parts" in two different ways. The interpretations given a priori bias the case: holists interpret the slogan in such a way that it cannot fail to be true, whereas individualists tend to understand it in such a way that it cannot fail to be false. This same situation obtains, in much less virulent form, in the unit of selection controversy. I will address here the dispute between group and organismic selectionists, leaving to one side the issues raised concerning genic and molecular selection at one end of the spectrum, and interspecies and community selection at the other (although a brief comment will be made about this in section 6). Authors who believe that group selection has played a relatively minor role in evolution tend to use a definition of group selection that is extremely restrictive; authors who

two other issues to be raised. First, if causality does not require causal determination (as in the theories of causality of Dretske and Snyder, 1972, and of Mackie, 1974), then it still is possible for social facts to be caused by individual facts and for social facts to cause other social facts, as required above. Secondly, besides the question of the causal connections between and within levels, there is the possibility of identity relations obtaining between social and individual facts (and between macro- and micro-facts generally). This second possibility is enough to allow the holism and individualism issue to be addressed, even if the question of causality is set to one side.

attribute greater potential importance to this selective force often use a more liberal, permissive, conception of group selection. What is more, each position has fairly cogent criticisms of some of the ideas on the other side.

The philosophical focus of this paper is on determining what group selection is; the hope is that we can then use this clarification to pinpoint what distinguishes group and individual processes generally. Yet it is well to remember that biologists do not have this as their motivation for thinking about group selection, nor did they become interested in group selection as an idle conceptual exercise. The historical context for the recent incarnation of the group selection controversy is that group selection was hypothesized as an explanation of phenomena that allegedly could not be explained in any other way. In 1962, V. C. Wynne-Edwards published his book *Animal Dispersion in Relation to Social Behavior*. There he argued that certain adaptations found in nature would be counterpredicted if individual selection were the only selective force at work. Wynne-Edwards talked about the ways in which prey populations react to the approach of predators. He discussed territoriality. He devoted a great deal of attention to the idea that organisms limit their own reproduction when the population approaches the environment's carrying capacity. Each of these categories involves traits which he thought were *altruistic:* organisms possessing such traits diminished their own reproductive chances while enhancing the fitness of the group. Altruism is always at a disadvantage when compared with selfishness, as far

as individual selection is concerned. But groups of altruists may do better than groups of selfish individuals, and this, Wynne-Edwards argued, explains why altruism is such a common and stable phenomenon in nature.

Four years later, George C. Williams published his *Adaptation and Natural Selection*. Williams subjected group selection to two lines of attack. First, he looked at the alleged examples of altruism that Wynne-Edwards had discussed and argued that they could be analyzed differently. Sentinel crows issue alarm cries when a predator approaches. Wynne-Edwards saw this as an example of altruism; the sentinel places itself in peril for the good of the group. But a number of alternative, individualist construals can be offered: perhaps the sentinel's warnings benefit its own offspring more than they benefit unrelated individuals. If so, the issuing of warning cries is, like parental care, perfectly consistent with the selfish calculus of individual selection. Or perhaps, though the warning cries are heard by related and unrelated individuals alike, the acoustical properties of the cry do not expose the sentinel to increased risk; maybe predators can't localize them. Or, perhaps the cries have the effect of causing a flurry of activity in the flock, and thereby *conceal* the sentinel from the approaching predator.

The occurrences of "perhaps" above deserve notice. The empirical details of sentinel crows, and of the other phenomena that Williams and Wynne-Edwards discuss, are incompletely understood. What Williams was doing was not providing known facts which, in every instance, *refuted* Wynne-

Edwards' suggestions. Rather, he was telling an alternative story which was capable of explaining the observations from the point of view of individual selection alone. So given our paucity of details about sentinel crows and the other examples that Wynne-Edwards and Williams discuss, we might say that there are *two* possible explanations, at least, of the observations. One of them is provided by Wynne-Edwards' group selection hypothesis, the other by Williams' individual selection hypothesis. Do we have here a stand-off? According to Williams' we do not, since it is more *parsimonious,* he says to invoke individual selection alone to account for these controversial cases.

Since I have discussed the principle of parsimony and its application to the group selection controversy elsewhere (1981b), I will not go into a great deal of detail in discussing the merits of such arguments in general or of Williams' argument in particular. However, a few comments are in order. One, not uncommon, reaction to Williams' parsimony argument is to dismiss it in such a way as to imply that considerations of parsimony never count for anything. As one biologist said to me: "The fact that Williams doesn't *need* group selection has nothing to do with whether group selection exists." If the thought behind this remark were true, then Ockham's razor would be a purely aesthetic consideration, never offering us a reason for thinking that a given hypothesis is *true.*

I do not take this wholly negative view of Williams' parsimony argument, although there can be no doubt that parsimony alone does not place individual selection on a thoroughly satisfying theoretical basis. One of the limitations of parsimony arguments in general is that they do not offer us an explanation of *why* the parsimonious hypothesis is true. Maybe we ought to believe that individual selection better accounts for the phenomena that Wynne-Edwards discussed. But this does not explain *why* group selection has played so minor a role in the history of evolution. Sewall Wright (1978) discusses this point. He suggested that to place the subject on a more secure foundation, we must create quantitative models of group selection, determine under what parameter values group selection would be efficacious, and then go to nature to see when, if ever, those parameter values are satisfied. Then we would have a more substantial reason for being parsimonious; we also would have an explanation of why group selection never, or rarely, occurs.

I believe that this limitation of parsimony arguments has the curious result of revealing why they can have a rational basis. *Parsimony arguments are inductive arguments.* Induction from a sample to a containing population does not provide one with an explanation of why the containing population is as it is. Williams wanted to explain the controversial phenomena over which he and Wynne-Edwards differed in a way in which everyone agreed the noncontroversial phenomena were to be explained. What was uncontroversial (at least to the participants in this dispute) was that individual selection was the mechanism behind numerous adaptations. Williams' parsimony argument was simply the assertion that we should use old mechanisms to explain new phenomena. Since the sampled, already understood adaptations were due to individual selection, we infer that the new cases at issue are caused by the same thing.

Inductive arguments, I assume, can, if they are any good, provide us with reasons for believing their conclusions. So the issue of the quality of Williams' parsimony argument reduces to the issue of whether his inductive argument is strong. Are the known cases representative of the kinds of adaptations found in nature, or do they involve a biased sample? This and other questions are suggested by Wright's (1978) evaluation of the parsimony argument. What they show is that parsimony arguments *can* provide us with reasons, if they meet the standards of good induction. But even when they do this, they are always incomplete, from the point of view of a science which has explanation as a goal.

As I mentioned above, Williams offered a second line of argument which is supposed to count against the importance of group selection. This involved quantitative considerations based on Fisher's (1930) so-called Fundamental Theorem of Natural Selection. The upshot of these considerations (also discussed in Lewontin, 1970) is that group selection will probably play a restricted, and relatively minor, role in the history of evolution, when compared with individual selection. These quantitative arguments do not show that group selection never exists, or that it couldn't exist. As we'll see, Williams believed that at least one real case of group selection has probably been found in nature, in any case. Although these quantitative consider-

ations are separable from our goal of describing what group selection is, the definition we will arrive at has some ramifications for the quantitative question.

Wynne-Edwards' work and Williams' attack were followed by a series of theoretical papers (reviewed in Wade 1978) in which mathematical models were proposed and examined. The main result of these analyses has appeared to confirm Williams' orientation, in that the parameter values needed for group selection to have significant impact were generally found to be quite restrictive. This conclusion, however, has not gone unchallenged, in that it is arguable that the models contain several unrealistic assumptions that a priori bias the case against group selection (Wade, 1978). In my opinion, the quantitative question remains open.

As I have indicated, our interest here is not in the issue of how much of a difference group selection has made, but rather in the question what group selection is. The various arguments and approaches that have fueled the biological controversy suggest that some rather different conceptions of group selection have been at work. Before we can identify these points of divergence, however, we ought to be clear on the common conceptual structure which is not in dispute.

For natural selection to act on a set of objects, there must be *variation*—the objects must be different. Moreover, the differences between the objects must include differences in their probabilities of reproductive success—there must be *variation in fitness*. And lastly, it usually is assumed that the fitness of parents must be correlated with the fitness of offspring—there must be *heritable variation in fitness* (adapted from Lewontin, 1970). This last requirement seems to me inessential for the existence of natural selection, although it is essential if cumulative genetic evolution by means of natural selection is to take place.

A word of clarification is in order concerning how the concept of fitness will be understood here. In any model of evolutionary processes which accords a role to random drift, fitness *cannot* be defined as actual reproductive success (e.g., number of viable offspring). Since any realistic model must give drift its due, fitness is not identical with actual reproductive success. The so-called tautology of the survival of the fittest is no tautology at all; the fitter do *not* always turn out to be more successful. The natural reaction to this fact is to think of fitness as an expectation, in the mathematical sense, of reproductive success (see, e.g., Crow and Kimura, 1970, p. 178; Mills and Beatty, 1979; and Sober, 1981a, for discussion). The fitness of an object is its propensity, or disposition, to be reproductively successful. Fitness differences, thus construed, may be the causes of reproductive differences.

The conditions set out above for natural selection to act on a set of objects—namely that the objects should vary in fitness—require supplementation to avoid the following problem. Consider a set of organisms which are causally isolated from one another: they may be at opposite ends of the universe and experience entirely different kinds of environmental stress. Suppose they are different in their fitness values. Still, it would be odd to conclude that there is a selection process in which they are all involved. One solution to this problem is to require that the objects be in *competition* with each other. But, as Lewontin (1978) has argued, this is inessential; two bacterial strains may be subject to natural selection even when neither impinges on the other's access to resources. The strains, growing in an excess of nutrient broth, may have unlimited energy available, but selection may favor the strain with the faster division time. As Darwin remarked, "A plant at the edge of the desert is said to struggle for life against the drought." By implication, selection can favor the plant better suited to the desert conditions, even when the better and worse plants do not interfere with each other. To be sure, competition is a familiar way of thinking about natural selection; yet, curiously, the familiar cases of natural selection that serve as textbook examples do not involve competition. The evolution of industrial melanism and of immunity to DDT do not involve there being a common resource in short supply. Competition is a special case, not a defining characteristic, of natural selection.

A more general conception of what subsumes a set of objects under a single selection process is that there must be some common causal influence acting on the objects which affects their reproductive chances. This common influence I will call a *force*. Much latitude exists for determining whether two objects are subject to the *same* force. It may be appropriate to think of the organisms in geographically isolated local populations of the same species as all involved in a single selection process. If each experiences predation as its major environmental problem, this may suffice to say that they are exposed to the same force. If, however, some experience predation, others experience temperature fluctuation, and still others experience the disappearance of prey as the major environmental stress, it will be wrong to lump these organisms together and talk about a single selection process subsuming them all. The sameness of the forces impinging on different organisms will be determined not just by the physical characteristics of the environment, but also by the biology of the organisms involved. If a field is sprayed with one insecticide, and a second field is sprayed with a second insecticide, it *may* be perfectly correct to construe the two affected insect populations as part of the same selection process. This will be true if the physical differences in the insecticides make no difference in the way those chemicals impinge on the organisms. The idea of "sameness of force" needs to be read biologically.[3]

So far I have talked about a set of objects satisfying certain conditions. What are these objects? How should the abstract structure of these conditions be interpreted? The classical, Darwinian interpretation is that the objects are organisms that exist within the same population. Organismic or individual, selection is generally understood as this sort of within-group selection. Group selection, on the other hand, involves interpreting the structure so that the objects involved are groups. Groups differ in their capacity to contribute to the next generation. Group reproduction is here understood to require the founding of numerically distinct colonies; mere growth in size of the group isn't enough. So, to get started in considering what group selection is, imagine a set of groups which differ from one another in their expectations of reproductive success. Group selection will thereby involve selection *between groups,* whereas individual selection involves selection *within groups.*

### 3. THE ARTIFACT ARGUMENT

Can one define group selection in the way just suggested, as existing whenever there is heritable variation in the fitnesses of groups? I would say not, although some biologists have used this sort of permissive characterization. The defect of the definition is that differences in reproductive capacity that obtain between groups

---

3. It is sometimes remarked that for selection to act on a set of objects, the objects must share a "common environment." I take it that this concept of a common environment is not definable in terms of spatiotemporal proximity, but will involve the idea that some causal influence affects the objects involved. (This includes, of course, the idea that they affect each other.) Even so, the requirement of a common environment still appears to represent too stringent a demand on the concept of natural selection. For consider again two prey populations which are geographically isolated from each other; suppose that they are preyed upon by two different populations of predators. If the two prey populations are conspecific, and the two predator populations are too, it may be appropriate to view the individuals in both prey populations as participating in a single-selection process. The objects in a single-selection process must be acted on by agents which are qualitatively similar, not necessarily numerically identical.

may merely be *artifacts* of the differences in fitness that obtain between organisms. Williams (1966) again and again deploys this idea in criticizing group-selection hypotheses. That natural selection has the effect that some groups are more reproductively successful than others is not enough to show that one has group selection. Selection at lower levels of organization can have this sort of "macroscopic" upshot.

Williams' *artifact argument,* as I will call it, asserts that the mere existence of differences in group productivity or in group fitness is not enough to demonstrate that there is group selection. The crucial question is where those differences came from: are they an artifact of selection processes occurring at other levels, or are they due to selection occurring at the level of groups? The distinction being made here must be spelled out in an adequate characterization of what group selection is.[4]

This line of thinking has its counterpart in the methodological holist versus methodological individualist controversy. Suppose holists argued that their position is confirmed by the fact that some social event, like the stock market crash of 1929, was caused by the occurrence of certain market conditions which are themselves social events and states of affairs. Individualists might grant the causal claim but argue that this by no means argues in favor of holism.

After all, it still is possible, and indeed is to be expected, that individual interactions brought about those very market conditions. That social facts are causally efficacious does not show that their causal efficacy is *irreducible.* Individualists will often demand to be shown how social facts can exist without an individualist foundation. Williams, in his criticisms of group-selection hypotheses, demands to be shown cases in which the causation of group properties is not reducible to individual selection.

Let's illustrate how the artifact argument works with a simple example. Imagine a system of populations each of which is internally homogeneous with respect to height. All the individuals in group number 1 are one foot tall, all those in group number 2 are two feet tall, and so on, for six such populations. Now imagine that natural selection favors individuals who are taller over ones who are shorter. As a result group number 6 will be more reproductively successful than group number 5, and so on. Is this a case of group selection? I doubt that many biologists would want to say that it is, and I am certain that Williams' artifact argument entails that it is not. What one has here is a case in which the differential reproductive success of groups is an artifact of differences in individual fitness. Group selection isn't to be equated with heritable variation in the fitnesses of groups.[5]

---

4. In this paper I will construe William's artifact argument as favoring individual-selection hypotheses at the expense of group-selection hypotheses. This does less than full justice to Williams' considered position, in which *genic,* rather than organismic, selection is the preferred level. However, a detailed discussion of genic selection must be postponed for another occasion.

5. In the light of this argument, consider the common definition of group selection re-

One of the assumptions underlying Williams' artifact argument is that group selection and individual selection are objectively distinct forces of evolution; in the above case, the group-selection description is false while the individual selection description is true. Now it is conceivable that one might give up this assumption and view the concepts of group and individual selection as interchangeable and equivalent, the way that some positivists have viewed the relationship between Euclidean and non-Euclidean geometry. In this vein, one might hold that whenever there is heritable variation in the fitness of groups, you can say that there is group selection or not, as you please. The choice would be one of convenience, in that both descriptions would be correct. According to this view, group selection and individual selection are not related to each other the way that mutation and migration are related to each other, namely, as two objectively distinct observer-independent forces of evolution.

Actually, this "conventionalist" attitude is not merely conceivable but is suggested by one of the oldest and most influential models of group selection, that of Sewall Wright (1931). Wright postulated a system of semi-isolated local populations. In virtue of their small size, random drift has more chance to operate, so that an allele might drift to a sufficiently high frequency for individual selection to take over and then drive the trait to fixation. The population would then send out migrants who would make over

other local populations by the same process. Is this group selection? Well, it can be described as a case in which one has drift acting on *individuals* within a population, *individual* selection, and migration of *individuals*. One could, I suppose, define group selection as existing whenever these three individual-level processes occur. But then group selection is not a distinct evolutionary force, and Williams' artifact argument cannot be made.

## 4. CONTEXT SENSITIVITY OF FITNESS

How are we to strengthen our definition of group selection? What more is there to group selection beyond the fact of heritable variation in the fitness of groups? One natural suggestion is this: if the change in gene frequency one gets would not have happened if one had a single panmictic population, then one has group selection. That is, returning to our example of the populations distinguished by heights of their inhabitants, we get this result: if the differential reproductive success that obtains in this situation differs from what would have happened if there had been a single interbreeding population made up of all the individuals, then we've got group selection. The idea here, which has obvious application to Wright's model, is that group selection exists whenever *population structure* affects reproductive success. A central point of Wright's model was to show how a system of small semi-isolated local populations could undergo evolutionary changes that would be ex-

---

ported in Wade's (1978) review article: "Group selection is defined as that process of genetic change which is caused by differential extinction or proliferation of groups of organisms." Note that besides failing to distinguish group from individual selection, this definition, taken at its word, fails to distinguish group selection from drift.

tremely improbable in the kind of large panmictic population that R. A. Fisher (1930) considered.

Is this an adequate definition? Can one say that group selection exists whenever there is heritable variation in the fitness of groups, where those fitness values depend on population structure? I'll set to one side the problem of figuring out the character of the counterfactual situation we are supposed to consider; if the groups exist in different environments, what are we to imagine is the environment that the hypothetical single panmictic population occupies? This problem aside, it seems clear that the proposed definition is extremely liberal, in that it makes group selection a commonplace of evolution. It is a virtually universal fact about fitness that an organism's fitness depends on what the other organisms are like in the population. Thus, whenever one has a system of populations, one can expect individual fitnesses to depend in part on the way in which organisms are distributed into populations. In particular, one can expect the fitness values to differ from what they would be if there were a single panmictic population. Does this suffice for group selection? I would say not; group selection must involve more than the fact that fitness values are *context sensitive*.

Let me report a classic finding con-cerning this fact of context sensitivity. Levene, Pavlovsky, and Dobzhansky (1954) found that the so-called Arrowhead homozygote on the third chromosome of *Drosophila pseudoobscura* is fitter than the Chiricahua homozygote, under laboratory conditions in which just these two chromosome types competed. However, when a selection experiment was run in which Arrowhead and Chiricahua competed with the Standard chromosome type, Arrowhead turned out to be *less* fit than Chiricahua. This showed how relative fitness of a trait can depend not just on the physical environment of the population, but on what other traits are present in the population itself. Now imagine two population cages, one containing Arrowhead and Standard, the other containing Chiricahua and Standard. Fitness values in the two pair-wise competitions would be different from what would obtain if Arrowhead, Chiricahua, and Standard were all present in a single population. But that there would be a difference in no way implies that the two pair-wise competitions involve group selection. Group selection must involve more than the idea that the fitness values of organisms is influenced by the kind of groups they are in.[6]

A parallel line of argument can be traced in the holism/individualism

---

6. I take it that this point undermines the definition of group selection presented in Wimsatt (1980, p. 236): "A *unit of selection* is an entity for which there is heritable *context-independent* variance in fitness among entities at that level which does not appear as heritable context-independent variance in fitness (and thus, for which the variance in fitness is *context-dependent*) at any lower level of organization." As I understand it, this definition would imply that in our two pair-wise competitions between Arrowhead and Standard in one cage and between Chiricahua and Standard in the other, we do *not* have a case of organismic, individual selection. The reason is that the fitnesses of organisms in this situation are context dependent.

dispute. Holists sometimes argue for their position by citing cases in which the properties of individuals are influenced by the groups to which they belong. We will see later on that certain situations of this kind can be crucial for confirming holism; however, the mere fact that the properties of individuals are context sensitive does not win the day against individualism. For the individualist will simply point out that the group properties which shape the character of individuals are themselves the product of individuals and *their* interactions. Again, if one is unwilling to hypostatize groups, the individualist's assertion cannot be faulted. The point being made is simply that the context sensitivity of individual properties —the way in which they depend on group context—is perfectly consistent with individuals being the material basis for all higher-level phenomena.

## 5. ALTRUISM

So far, I have examined two rather permissive definitions of group selection and argued that they are inadequate. I argued for their inadequacy by taking seriously Williams' artifact argument. Group selection is not to be identified with heritable variation in the fitness of groups; nor should it be identified with heritable variation in the fitness of groups where organismic fitness is context sensitive. Neither of these conditions is sufficient for group selection.

In each of the two proposals, it was found that alleged cases of group selection could be analyzed

in terms of individual-level processes alone. This is a standard ploy that individualists use to combat the arguments of holists. A natural response by holists to this line of argument is to try to find some social fact which will be *counterpredicted* by the assumptions of individualism. If this can be found, then the individualists' strategy of assimilating the phenomenon into their own framework will be blocked. Although phenomena which individualists *can* accomodate may not count as "real" group processes, ones that they *cannot* accommodate will qualify as paradigm cases of what holism is all about.

This thought may lead us back to the issue which fueled the fires of the group selection dispute initially. *Altruism* is *counterpredicted* by individual selection.[7] Group selection, however, might be supposed to promote the existence of altruism, since groups which contain altruists may fare better than ones which do not. So, if one wants to argue for the efficacy of group selection, what better phenomenon to look for than cases in which altruism has emerged and remained prevalent owing to natural selection? Let us be careful here. There are two roles which the concept of altruism might be taken to play in the idea of group selection. One of them concerns how a biologist might try to discover cases of group selection: look for altruism. The other concerns what group selection *is* (not how we might find out about it); on this view, group selection must *be* selection for altruism. The

---

7. For some careful definitions of altruism, see Wilson (1980).

difference here is that between an epistemological and an ontological consideration.

The ontological thesis, that group selection must work in a direction opposite to that of individual selection, has one virtue. It has the clear implication that group selection is objectively distinct from individual-level forces like individual selection. It thereby cannot be criticized by invoking Williams' artifact argument; the typical individualist ploy of reanalyzing alleged group-level phenomena as artifacts of individual-level processes has been forestalled.

So, does group selection have to involve selection for altruism? Before considering some biological examples which show that it need not, I want to note a rather abstract oddity of this idea. It implies that if there is group selection at work on a system of populations, individual selection must be acting as well. That is, we would have here a force of evolution which could not act alone. This, by itself, is not a conclusive reason for rejecting the idea, but it does show the proposed definition to conflict with a rather plausible requirement on evolutionary forces, and, perhaps, on all forces in general. It should be possible to describe the changes a force would bring about if there were no other forces at work. That is, one might suppose that the use of the *ceteris paribus* condition is not merely *permitted* when it comes to describing a force, but is *necessary* for the adequate description of a force in order that this be possible. This corresponds to the idea that a force should be isolatable in principle. I do not propose to defend this idea of isolatability,

but merely note that it is rather standard fare in our conception of force. If group selection requires selection for altruism, this condition cannot be satisfied.

But the decisive reason for rejecting this view of group selection is more down to earth and biological. It is simply that some cases which seem clearly to be ones of group selection involve group and organismic selection acting in the same direction. A proper conception of group selection should be tailored to allow for this fact. Let me describe two examples, one of them being the sole case believed by Williams to be a real instance of group selection. Here I mean the investigation by Lewontin and Dunn (1960) of the segregator distorter *t*-allele in the house mouse *Mus musculus*.

Let me give an elementary description of how the process of segregation distortion, or meiotic drive, works (see Crow, 1979, for details). Diploid organisms are ones whose chromosomes come in pairs. In the formation of sex cells, these pairs of chromosomes separate, so that sperms and eggs contain one chromosome each from each pair—they are haploid. The normal pattern for this reduction is that 50 percent of the sex cells contain one chromosome and 50 percent the other, from each homologous pair.

But when a segregator distorter allele is present on a chromosome, it "subverts" this equality of representation and secures for itself representation of greater than 50 percent.

This is what the *t*-allele does in the house mouse. Consider the males who are heterozygote for the *t*-allele. One might expect that 50 percent of the sperm pool of this group would be made of gametes containing the *t*-allele. In fact, the representation of the segregator distorter is 85 percent. So at this level there is strong selective pressure favoring chromosomes which contain the *t*-allele. Let us call this chromosomal selection. Chromosomes having the trait are at an advantage over ones lacking it.

If we go up a level or two, from chromosomes to organisms, the *t*-allele is *not* favored by selection. Males who are homozygous for the *t*-allele are sterile. So there is strong organismic selection working against the *t*-allele.

Lewontin and Dunn combined this information about the chromosomal and organismic selection acting on the *t*-allele and derived a prediction of what the frequency of the *t*-allele should be in nature. It was wrong; the prediction erred by being too high. This suggested to them that some third force was acting against the *t*-allele. The third force was group selection. The population structure of the house mouse is one of small local demes. Whenever all the males in one of these small groups are homozygous for the *t*-allele, the entire deme becomes extinct. Females living in a group all of whose males are homozygous will have no offspring. What is more, their fitnesses (or rather the component of fitness determined by this selection process) will be 0, owing to the fact that they belong to a group of a certain kind. Females within such a group may differ in phenotype and genotype as much as you like, but such differences make no difference; their reproductive chances have been destroyed by their belonging to the kind of group they're in. Since the frequency of *t*-alleles among females in such groups will, on average, be higher than the frequency of *t*-alleles among females in groups lacking this fatal flaw, the effect of group selection will be to reduce the frequency of the *t*-allele. Notice that in this case, organismic and group selection are in the same direction; both work against the *t*-allele. So one cannot require that group selection and individual selection always be opposing forces.

The other example I want to describe in which individual and group selection work in the same direction is a group selection experiment that Michael Wade (1976) carried out on the flour beetle, *Tribolium castaneum.* Wade's experiment involved setting up and monitoring four selection processes at once. In each of them he started with 48 populations, each containing 16 beetles each. At the end of 37 days, he did his selecting. In one of the treatments, he selected for large populations; he located the population containing the largest number of adults and used it to found colonies of 16 individuals each until the population was exhausted. He then went to the next largest group and did the same, until 48 second-generation populations were founded. He repeated this process for a number of generations. The average size of populations at the end of the selection process was higher than the average size of the populations in the first generation. Here we have group selection; groups were selected in virtue of their being large.

Another group selection procedure was carried out on a second set of 48 populations. Here, Wade selected for *small* populations. The regimen was as before, and at the end of the procedure the average size was much reduced from what it had been at the start.

A third selection treatment served as a control group. Again, there were 48 canisters. At the end of 37 days, one sample of 16 individuals was drawn *from each canister* and used to found a next generation population. The only selection process that took place here was within populations. Individuals within a canister competed with each other. But, roughly speaking, each group had the same chance of representation in the next generation as any other; each contributed one and only one colony of 16 individuals. This experimental treatment involved individual selection but no group selection. What happened in the process? Under individual selection alone, the average population size declined, owing to such factors as lengthened developmental time, reduced fecundity, and increased cannibalism.[8]

Now what happened in the control treatment also happened in the two group-selection treatments just described. That is, within each canister in the group-selection treatments, individual selection was going on as well. This force, we learn from the control treatment, promoted reduction in population size. In the group-selection treatment, in which there was group selection for reduced population size, there were, in fact, *two* forces at work. Individual selection promoted reduction in population size, and group selection did the same thing. As one might expect, the magnitude of the reduction that took place in the group-selection treatment was greater than that achieved by individual selection alone in the control treatment. Two forces are better than one. Here we have the same lesson as that obtained in the *t*-allele example. Group selection and individual selection can act in the same direction; group selection does not have to be selection for altruism.[9]

## 6. GROUP SELECTION DEFINED

My purpose in discussing these two examples has not merely been to argue that altruism is inessential. I also wanted to add some data which may serve to constrain an

8. Wade's fourth treatment he calls "random selection." Each canister is assigned a number and then a canister is chosen by picking a number at random. The chosen group is then used to found colonies of 16 until it is exhausted, at which point another canister is chosen at random. This is repeated until 48 next-generation colonies are established. Although this process can be called "group selection," according to the definition of group selection used (see my footnote 5 above), it is not group selection, according to the definition to be presented in what follows. Moreover, if drift and selection are mutually distinct categories, it is hard to see how there could be such a thing as "random selection" at all.

9. It is worth pointing out that if Wade's experiment provides genuine cases of group selection, then group selection need not involve groups with complex organizational properties or ones having especially intricate forms of sociality. Although interest in group selection sparked by the issue of altruism will naturally focus on such cases, this is not a consequence of the concept of group selection itself. Wade's group selection treatments selected for group properties which are absolutely universal when there are groups at all. Of course, even though it is no problem finding groups which vary in *size*, it is not quite

adequate definition of group selection. Let me review the other requirements that a reasonable definition
should fulfill. First, the definition
should allow one to distinguish
changes in group properties due to
group selection from changes in group
properties that are due to processes
occurring at lower levels of organization. That is, an adequate definition
should take seriously Williams' artifact argument. Secondly, the definition should not have the consequence
that group selection exists whenever
fitness values are context sensitive.
Group selection does not exist simply
in virtue of the fact that the fitness
values of organisms depend on the
character of the group they are in.
And, thirdly, it should turn out that
group selection can exist in either
presence or absence of organismic
selection and can act in the same or
opposite direction from it. This last
consideration combines the rejection
of altruism as a criterion of group
selection with the earlier remarks
to the effect that group selection
should be an objectively distinct
force of evolution.

With this elaborate preamble, the
definition can be stated:

> Group selection acts on a set of
> groups if, and only if, there is a
> force impinging on those groups
> which makes it the case that for
> each group, there is some property
> of the group which determines one
> component of the fitness of every
> member of the group.

Let me try to state the intuitive idea
in a less cumbersome way. When
group selection occurs, all the organisms in the same group are bound
together by a common fate. As far
as this selective force is concerned,
they are equally fit. What determines
these identical fitness values (on the
component of fitness at issue) is their
membership in the same group. Individuals with radically different genotypes and phenotypes may have
identical fitness values because they
belong to the same group. And individuals with identical genotypes and
phenotypes may have very different
fitness values, because they belong
to different groups. Under group
selection, what is causally efficacious
in the production of reproductive
differences among organisms is membership in groups of different kinds.

Let's apply this idea to the examples discussed so far, starting with
the contrived example of populations
in which everyone has the same
height. Our definition explains why
this is not a case of group selection.
Although every individual in the same
group has the same fitness value, the
cause of this sameness is *not* common
membership in a group of a certain
kind. A simple explanation is also
available of why our two pair-wise
competitions between Arrowhead and
Standard and between Chiricahua and
Standard in adjacent population cages
was not a case of group selection.
Although the fitness of a fruit fly
depended on the kind of group it
was in, the members of the same
group were not acted on *as a unit.*
The members of the same group did
*not* have the same fitness, on any

---

so inevitable that this variation is heritable. This further requirement, as noted in section
2, is needed if the selection process is to result in cumulative evolutionary change.

component of fitness. Rather, group context served to determine the *differential* fitnesses of organisms within the group, in just the way that the environment can determine fitness differences in cases of ordinary individual selection. Group properties existed, but these failed to ramify back on the fitnesses of organisms in the appropriate way.

Our definition also explains why the two cases of group selection discussed before do really count as group selection. In the *t*-allele example, every mouse has fitness equal to 0 if it belongs to a group all of whose males are homozygous for the *t*-allele. What is crucial is that this common fitness value is caused by common membership in a group of a certain kind. The same is true of Wade's selection experiment. If there is group selection for groups of a certain size, then every individual in a group has an equal chance of finding its way into the next generation. The individuals within a population are bound together, their common fitness values determined by their common membership in a group of a certain size.

Before moving on to another example of group selection and to some further biological considerations, I want to point out a philosophically interesting feature of the definition proposed. The claim that a set of groups is subject to group selection will differ from the claim that a set of objects is subject to familiar physical forces like gravity or electromagnetism. Group selection may take endlessly many different physical forms; to say that some populations are undergoing a group selection process is not yet to say what physical

properties are causally efficacious, but rather is to say that some physical property or other is responsible for fitness values in a certain way. In contrast, claiming that a particular physical force is acting on a set of objects is a much more specific claim about the physical details; for example, to say that a physical object is in an electromagnetic field is to say that its *charge* and its *distances* from other objects play a specific kind of causal role. It is in this sense that claims about evolutionary forces can be more "abstract" than claims about physical forces. This greater degree of abstractness—this formulation of generalizations which are true of objects which differ physically from each other—is achieved in evolutionary theory by quantifying over properties. Besides the inevitable ontological commitment to numbers which any mathematical theory will involve, evolutionary theory is therby platonistic in an additional respect. This is one reason among others (discussed at greater length in Sober, 1981a) for thinking that a purely extensional ontology will be unsatisfactory for this science.

In order to give the reader some further grasp of the phenomenon that a definition of group selection is supposed to circumscribe, I want to describe another biological example which is often cited (e.g., by Lewontin, 1970) as a probable case of group selection. The empirical details to be described have biological plausibility, but they may be revised or replaced by further information. Our interest, though, is not in whether they are true, but in what the assumption of their truth tells us about what group selection is.

Here I have in mind the coevolution in Australia of the disease virus myxoma and the rabbit *Oryctolagus cuniculus*. Myxoma was introduced into Australia to cut down on the rabbit population. Two familiar epidemiological events ensued. On the one hand, rabbits became more immune to the disease; on the other, the virus became less virulent. The explanation of the latter change is that the disease is spread from rabbit to rabbit by a mosquito which only bites live rabbits. Thus an extremely virulent strain of myxoma, while it may become predominant within a single host, runs a good chance of never spreading through the rabbit population. Less virulent strains, on the other hand, while they succeed in expropriating a smaller number of the host's cells, nevertheless increase their chances of transmission.

Two, opposing, selection forces are at work here. Within each rabbit, strains of greater virulence will tend to consume a greater proportional share of the limiting resource—namely the host's own cells. So there is individual (within group) selection for increased virulence. But a virus winning this race may thereby lose another—that of spreading its genes to other rabbits. A virus population—the assemblage of different strains within a single rabbit—founds colonies, and, roughly speaking, the lower the average virulence of a population, the better the chances are that a mosquito will transport a colonizing propagule from that population to another host. Assuming that these two selection forces are the main evolutionary forces at work, that the virus declined in virulence shows that in this case the group selection force was stronger

than the force of individual selection.

Less virulent strains of myxoma are "altruists." By being less virulent, they reduce their expectation of reproductive success within the population they are in but thereby increase the group's chances of survival and reproduction by lowering the average virulence of the population. This example should correct the popular misconception that altruism must always be driven to extinction by a selection process. Evolutionary theory entails no such theorem. Rather, what this example shows is that a crucial factor in determining the evolution of a system of this kind, in which group and individual selection oppose each other, is *time*. If mosquitoes bit rabbits much more rarely, or if myxoma expropriated host cells at a much faster rate, the decline in virulence of myxoma might never have occurred.

There *is* a theorem which represents *this* general idea, however. Fisher's fundamental theorem of natural selection (1930) states that the rate of evolution under natural selection is identical to the additive genetic variance in fitness. Since the fundamental theorem has to do with the *rate* of evolution, evolution will proceed faster, the shorter the generation time of the objects involved. But since groups almost always take longer to found new colonies than the individuals within the groups take to reproduce themselves, one again has the consequence that group selection will produce smaller changes than individual selection (Crow and Kimura, 1970; Lewontin 1970). In the myxoma example, group selection was able to exert a powerful influence precisely because of the contingent facts concerning group and individual generation times.

Although I don't want to contest the correctness of applying Fisher's theorem in this case, it is important to identify a presupposition of using it in the general argument that group selection works more slowly than individual selection. It was pointed out earlier that the idea of group reproduction standardly used in discussing group selection is that of groups founding numerically distinct colonies. But there is no need to restrict our attention to this process, to the exclusion of considering the dynamics of population growth. Indeed, the definition of group selection we have arrived at is perfectly consistent with a system of groups undergoing a group selection process in which fitter groups increase in relative size. The total number of groups need not change at all. But if this kind of group selection process is considered, the argument based on Fisher's theorem cannot be made. Although individuals usually reproduce faster than their containing groups found colonies, it isn't quite so ubiquitous that individual reproduction takes place in the context of noncolonizing groups which are at their carrying capacity. This point leaves open the possibility, of course, that other broad differences between groups and organisms may be harnessed to Fisher's theorem in support of the claim that group selection is a weaker force of evolution than individual selection (see, for example, the argument of Lewontin, 1970, concerning heritability).

Before drawing a few general lessons concerning what the definition of group selection implies about general features of the concepts of fitness and selection, I want to take up an objection to the proposed definition. According to the definition, the members of the same group must have *precisely the same fitness values* (on the component of fitness at issue) if group selection is at work. But this sounds too strong; for example, group selection might exist simply by virtue of the fact that membership in groups of different kinds had some percentage effect on some other fitness parameter. For example, each member of a particular group might have its overall fitness boosted by 5 percent by belonging to a group of a certain kind. What is crucial is *uniform effect,* in some sense; identical fitness values are not required, strictly speaking.

If a property of a group drives predators away, the individuals in the group need not benefit equally. Some might have been better than others in evading predators to begin with, and so the removal of danger may represent an unequal benefit. Still, this may be a genuine case of group selection. If some groups have properties which attract predators while others have properties which repel them, a group selection process may ensue. Though the numerical increments in fitness that members of the same group obtain from the shared group property may be unequal, the fundamental causal structure of a group-selection process is still intact. The group's relation to the predator, in this case, is such that the predator reacts to the group *as a unit*. Although the numbers assigned to individuals may not transparently represent this, the biological relationship of the group to its predator tor subsumes each individual indifferently. Though fitness values within the group may differ, each individual encounters a predator to the degree that it does because of the property of the group it is in. Whether the biologist characterizes this aspect of the ecology in terms of a separate component of fitness or views it as a partial determinant of some more encompassing component is not what matters.

A number of consequences follow from our discussion concerning the concepts of fitness and selection. As soon as fitness is decoupled from actual reproductive success, it follows inevitably that one cannot read off fitness values from patterns of reproduction. That some groups reproduce more than others does not mean that the more productive groups are fitter. Nor does the fact that some species speciate and persist more than others imply that species selection is occurring, or that some species are fitter than others (see Stanley, 1975, and Gould, 1980, for discussions of species selection). Fitness and selection are both causal concepts; they describe the causes of change and not the fact that there has been differential productivity.

Perhaps a more surprising consequence of our discussion is that fitness and selection are decoupled from each other. In spite of the fact that fitness values and selection coefficients are interdefinable in mathematical models (so that, typically, $s = 1 - w$), there is an important difference between these concepts. As we saw in our simple example of a series of populations which were each internally homogeneous for height, that groups differ in

fitness does not imply that there is group selection. The groups, in this example, differed in fitness in that they had different propensities to be reproductively successful. But the cause of these fitness differences was individual, not group, selection.

Selection is a richer concept than fitness. In fact, the relation of selection to fitness is somewhat like the relation of fitness to actual reproductive success. To say that group selection occurs is to say more than simply that groups differ in fitness; it is to say *why* those fitness differences obtain. Selection is the cause of fitness differences, just as fitness differences may be the cause of differences in actual reproductive success. It follows from this that just as one cannot read off the level of selection from facts about differential productivity, one cannot read off the level of selection from facts about differential fitness. The difference between individual and group selection is *not* the same as that between within-group and between-group variance in fitness. By the same token, even if some species could be shown to have a greater tendency to speciate, this would not suffice to establish the existence of species selection. The question that remains unanswered is the causal one of why these differences in the expectations of splitting obtain.

This difference between fitness and selection is not surprising when one considers that fitness is a *disposition* while natural selection is a *force*. Although the forces at work determine certain dispositions in the objects present, the dispositions of those objects do not uniquely determine what forces are at work. Thus, to be told that two billiard balls are disposed to accelerate toward each other from their initial positions is not to say what force or forces endowed the objects with that disposition. On the other hand, to specify that the two objects generate an electromagnetic field determines one of their dispositions to move in certain ways.

Our analysis also reveals the inadequacy of two lines of argument that are sometimes offered in defense of lower-level—either organismic or genic—selectionism. It is sometimes pointed out that all of the alleged higher-level interactions which may obtain owing to population structure can be given mathematical representation in the fitness values of individual organisms or of individual genes. That is, the effects of processes at higher levels can be viewed as part of the environment of genes, and the whole selection story can be told in terms of the selection coefficients of individual genes. One criticism of this line of thinking is epistemological and, therefore, inconsequential: no one at present knows enough about any gene to define for it a selection coefficient which takes account of all this information. But this line of attack misses its mark, since the proposal does not describe what *we* as theorists can successfully codify, but makes a claim about what is going on in nature.

The fundamental flaw in this kind of argument is that it confuses the task of formulating a predictively successful mathematical apparatus with the task of accurately describing the causal structure of selection processes. It is to be granted that all of the information about higher-level selection *can* be represented in the

so-called selection coefficients of organisms or genes (Levins, 1970, 1975; Wade, 1979), but this simply does not imply that, in nature, it is individual or genic selection which is always occurring. Genes may be modeled as maximizing their fitnesses, but that leaves open the question of what causal processes propel changes in gene frequencies (Wilson, 1980). Earlier, I commented on the fact that the interdefinability of fitness values and selection coefficients should not mislead us into thinking that fitness and selection are essentially equivalent concepts. The same point applies here: it is essential not to confuse facts about the mathematics of our models with facts about the causal structure of the process modeled. It is desirable, of course, that our models be realistic. But it is a naive realism which holds that every biologically interesting distinction will be forced on us by the exigencies of mathematical modeling.

One last confusion which I hope this discussion lays to rest is that between the issue of the unit of selection and the issue of the unit of replication (see Hull, 1980, 1981, for discussion). Group selectionists do not deny that the gene is the mechanism by which biological objects pass on their characteristics; the issue of cultural evolution is not an issue here. But this shared assumption about the unit of replication simply cuts no ice. That genes are passed along leaves open the question of what causes their differential transmission (*pace* Dawkins, 1976). This is not to say that facts about heritability are irrelevant to the question of how selection at different levels may produce cumulative evolution (see Lewontin,

1970, for this kind of argument). But any such argument must do more than merely point out that genes are the devices by which characteristics are inherited.

## 7. BETWEEN SCYLLA AND CHARYBDIS

The Stock market crash, which was a social fact, was caused by market conditions, which constitute other social facts. These market conditions, in turn, were caused by individual interactions. By transitivity of causality, the individual interactions caused the stock market crash. *If social facts cause something, so do individual facts.* Once we decide to avoid atomism on the one hand and hypostasis on the other, the sensible middle course appears to provide no asymmetry between the social and the individual; *both* are causally efficacious.

Yet, it is emphatically *not* the case that if group selection causes something, so does individual selection. Group selection and individual selection are objectively distinct forces. Individual selection does not require an atomistic view of the organism; it does not require one to ignore the fact that organismic fitness is context sensitive. Individual selection is a process that a sensible individualist can embrace. Similarly, group selection does not require a reification of the group; it does not force one to suppose that groups are something above and beyond the interactions of their member individuals and the environment. Group selection is a process that a sensible holist can embrace. And, best of all, it is a substantive empirical question what the role and importance of these two forms of selection has been in the history of evolution.

Holism and individualism in the social sciences should have such luck. To move beyond truisms to nontrivial empirical issues, holists and individualists need to formulate their dispute with reference to specific social forces. Although there is no real question involved in asking whether a certain evolutionary outcome was caused by properties of groups or by properties of individuals, there *is* a substantive question involved in asking whether that outcome resulted from individual selection or from group selection. In the same way, societies change because of the way their constituent groups interact, and these groups, ultimately, are caused to be the way they are by the individuals they contain. The debate between holism and individualism might become fruitful if specific mechanisms were considered and the question were then posed with respect to them: does their impact on individuals correspond to the causal structure we have identified in group selection processes, or does their activity represent a form of individual selection? Although it is probably a mistake to try to mimic the units of selection debate too closely, and there is no reason why the holism/individualism controversy *must* be recast in its terms, let's explore, in conclusion, what individual and group selection would look like in the case of social processes mediated by cultural, rather than genetic, evolution.

Social institutions can be viewed as selection mechanisms. They discriminate among individuals by virtue of their having certain properties and differentially distribute effects on that basis. This description encompasses a great many, diverse, social processes and appears to be nontendentious, in that it is consistent with the outlook of neoclassical and Marxist social thought alike.

As an example, consider a college admissions test. The test discriminates among individuals, and, on the basis of that discrimination, the individual is either admitted or not to a particular college. Is this a case of individual or of group selection? The test tests *individuals*, of course, but that doesn't show that it embodies a kind of individual selection. And individuals are influenced in their ability to do well on the test by the groups to which they belong. But that doesn't show that the test is a form of group selection. As we have seen earlier, neither the fact that individuals are the material basis of groups nor the fact that individual properties are context sensitive suffices to decide the issue between individual and group selection.

The overall ability to do well on the test can presumably be broken down into a number of component abilities. Are there component abilities which an individual has, simply by virtue of belonging to a group of some particular kind? Are the properties of the group which have this effect on individual ability the result of interaction among individuals? Could two individuals who are otherwise similar differ in ability simply by virtue of their belonging to different groups? Could two individuals who are otherwise quite different possess the same component ability simply by virtue of their belonging to the same group? If the answers to these questions are yes, then the admissions test would appear to have the earmarks of a mechanism of group selection.

Although each of us probably thinks

that he or she can readily answer the above questions, a note of caution is in order. The mechanisms of selection processes are often difficult to discern, and it is a mistake to think that one can conclusively identify the character of a mechanism from the kinds of results it produces. Perhaps the admissions test gives greater than proportional representation to some particular group; it doesn't follow that the test involves group selection for membership in that group. In this case, a serious assessment of what the test is doing must be based on a serious understanding of the various abilities that affect the ability to do well on the test. Characterizing these components is a highly nontrivial task, one which we have barely begun to discharge. Although it is transparent that the admissions test is a form of selection, the character of this selection process is in many ways extremely opaque.

Another application to social progress that can be made of our distinction between group and individual selection involves the idea of the selection of selection processes. Besides wanting to answer the question of how the admissions test *works*, one would also like to know where it came from—how it came to be used as the admissions test. Even if it were true that the admissions test embodied a form of individual selection, the possibility remains open that it evolved by a process of group selection. Perhaps part of the cause of its being used is that it has certain group-level results; this may be true even if the test does not make its discriminations on the basis of group membership.

Marxist critiques of "bourgeois" social science often have two components (Keat and Urry, 1977). First, bourgeois social science is allegedly too individualistic in its orientation, seeing the individual rather than the group as the correct unit of analysis. Secondly, it is claimed to be superficial in the kinds of questions it asks about society, typically focusing on issues concerning the regularities that social institutions obey, rather than on more structural questions having to do with why those institutions are as they are. These two lines of criticism are not unrelated, of course, since for Marxists an explanation of why particular social institutions have the form they do must crucially involve considerations of class conflict. From this point of view, the results of bourgeois social theory need not be false, but they must be incomplete. This means that if they are, mistakenly, taken to be complete, they will offer a distorted view of social life.

One might interpret this point of view as holding that social institutions, at least in bourgeois society, embody a form of individual selection, but that they evolved by a process of group selection. One of the differences between bourgeois and feudal society may consist in which properties of individuals determine how social institutions treat them. Whereas membership in particular social groups was used to decide all manner of social sortings out, these

criteria are much less often the ones which are directly invoked in bourgeois society. Rather, the mechanisms have shifted toward the structure of individual selection. But this by no means implies that those social institutions do not themselves constitute a form of class interest, since they may have evolved by a process of group selection. From this point of view, holism and individualism may each be correct in a limited domain, if each is understood as claiming that certain sorts of selection processes are at work in a given society.

Although this articulation of the holism versus individualism debate is not the only one possible, it does have one virtue. It yields a conception of individualism which is untainted by atomism and a conception of holism which is unspoiled by hypostasis. In so doing, it turns the social science dispute into what it ought to be—a question about the character of social causation which is not decidable by a priori argument but can only be addressed by the assessment of evidence and the development of theories which are specifically sociological. This reformulation makes the dispute harder than it was before; the road away from truisms and toward contentful hypotheses about causal mechanisms is never an easy one. But this presumably is a price that an explanatory science willingly pays.

REFERENCES

Brodbeck, M., ed., 1968, *Readings in Philosophy of the Social Sciences*, New York, MacMillan.

Crow, J., 1979, Genes that violate Mendel's rules. *Scientific American*, 240 (2): 134–146.

—— and Kimura, M., 1970, *An Introduction to Population Genetics Theory*. New York, Harper and Row.

Dawkins, R., 1976, *The Selfish Gene*, Oxford: Oxford University Press.

Dretske, F., and Snyder, A., 1972, Causal irregularity. *Philosophy of Science*, 39: 69–71.

Fisher, R., 1930, *The Genetical Theory of Natural Selection*, New York, Dover.

Fodor, J., 1976, *The Language of Thought*, New York, Thomas Crowell.

Gould, S., 1980, Is a new and general theory of evolution emerging? *Paleobiology*, 6: 119–130.

Hull, D., 1980, Individuality and selection. *Annual Review of Ecology and Systematics*, 11: 311–332.

———, 1981, The herd as a Means. In *PSA 1980*, Volume 2. Eds., P. D. Asquith and R. N. Giere. East Lansing, Mich.: Philosophy of Science Association, Pp. 73–92.

Keat, R., and Urry, J., 1977, *Social Theory as Science*, London: Routledge and Kegan Paul.

Levene, H., Pavlovsky, O., and Dobzhansky, T., 1954, Interaction of the adaptive values in polymorphic experimental populations of *Drosophila pseudoobscura*. *Evolution*, 8: 335–349.

Levins, R., 1970, Extinction. In *Some Mathematical Questions in Biology*, Volume 2, Ed. M. Gerstenhaber, Providence, R.I., American Mathematical Society, Pp. 75–108.

———, 1975, Evolution in communities near equilibrium. In *Ecology and Evolution of Communities*. Eds., M. Cody and J. Diamond, Cambridge, Mass., Harvard University Press, 16–50.

Lewontin, R., 1970, The units of selection. *Annual Review of Ecology and Systematics*, 1: 1–18.

———, 1978, Adaptation, *Scientific American*, 229 (3): 156–169.

——— and Dunn, R., 1960, The evolutionary dynamics of a polymorphism in the House Mouse. *Genetics*, 45: 705–722.

Mackie, J., 1974, *The Cement of the Universe*, Oxford, Clarendon Press.

Mills, S., and Beatty, J., 1979, The propensity interpretation of fitness. *Philosophy of Science*, 46: 263–286.

Putnam, H., 1967, Psychological Predicates, In *Art, Mind and Religion*, ed., W. J. Capitan and D. D. Merrill, Pittsburgh, University of Pittsburgh Press, 37–48. (Reprinted as "The nature of mental states." In *Mind, Language, and Reality* (*Philosophical Papers*, Volume 2), Cambridge, Cambridge University Press, 1975, pp. 429–440.)

Rosenberg, A., 1978, The supervenience of biological concepts, *Philosophy of Science*, 45: 368–386.

Sober, E., 1980, Evolution, population thinking, and essentialism. *Philosophy of Science*, 47: 350–383.

———, 1981a, Evolutionary theory and the ontological status of properties. *Philosophical Studies*, 40: 147–176.

———, 1981b, The principle of parsimony. *British Journal for the Philosophy of Science*, 32: 145–156.

Stanley, S., 1975, A theory of evolution above the species' level. *Proceedings of the National Academy of Science, USA*, 72: 646–650.

Wade, M., 1976, Group selection among laboratory populations of *Tribolium*. *Proceedings of the National Academy of Sciences, USA*, 73: 4604–4607.

1978, A critical review of the models of group selection. *Quarterly Review of Biology*, 53: 101–114.

———, 1979, The evolution of social interactions by family selection. *American Naturalist*, 113: 399–417.

Williams, G., 1966, *Adaptation and Natural Selection*, Princeton, New Jersey, Princeton University Press.

Wilson, D., 1980, *The Natural Selection of Populations and Communities*, Menlo Park, Calif., Benjamin/Cummings Publishing Co.

Wimsatt, W., 1980, Reductionistic research strategies and their biases in the units of selection controversy. In *Scientific Discovery: Case Studies*, ed., T. Nickels, Dordrecht, North Holland, Reidel, 213–259.

Wright, S., 1931, Evolution in Mendelian populations, *Genetics*, 16: 97–159.

———, 1978, *Evolution and the Genetics of Populations.* (*Variability Within and Among Natural Populations.* Volume IV.) Chicago: University of Chicago Press.

Wynne-Edwards, V., 1962, *Animal Dispersion in Relation to Social Behavior*, Edinburgh: Oliver and Boyd.

# 13

# Artifact, Cause, and Genic Selection

## ELLIOTT SOBER AND RICHARD C. LEWONTIN

*Several evolutionary biologists have used a parsimony argument to argue that the single gene is the unit of selection. Since all evolution by natural selection can be represented in terms of selection coefficients attaching to single genes, it is, they say, "more parsimonious" to think that all selection is selection for or against single genes. We examine the limitations of this genic point of view, and then relate our criticisms to a broader view of the role of causal concepts and the dangers of reification in science.*

### INTRODUCTION

Although predicting an event and saying what brought it about are different, a science may yet hope that its theories will do double duty. Ideally, the laws will provide a set of parameters which facilitate computation and pinpoint causes; later states of a system can be predicted from its earlier parameter values, where these earlier parameter values are the ones which cause the system to enter its subsequent state.

In this paper we argue that these twin goals are not jointly attainable by some standard ideas used in evolu-tionary theory. The idea that natural selection is always, or for the most part, selection for and against single genes has been vigorously defended by George C. Williams (*Adaptation and Natural Selection*) and Richard Dawkins (*The Selfish Gene*). Although models of evolutionary processes conforming to this view of genic selection may permit computation, they often misrepresent the causes of evolution. The reason is that genic selection coefficients are *artifacts*, not causes, of population dynamics. Since the gene's-eye point of view exerts such a powerful influence both within biology and in popular discussions of

This paper was written while the authors held grants, respectively, from the University of Wisconsin Graduate School and the John Simon Guggenheim Foundation and from the Department of Energy (DE-AS02-76EV02472). We thank John Beatty, James Crow, and Steven Orzack for helpful suggestions.

sociobiology, it is important to show how limited it is. Our discussion will not focus on cultural evolution or on group selection, but rather will be restricted to genetic cases of selection in a single population. The selfish gene fails to do justice to standard textbook examples of Darwinian selection.

The philosophical implications and presuppositions of our critique are various. First, it will be clear that we reject a narrowly instrumentalist interpretation of scientific theories; models of evolutionary processes must do more than correctly predict changes in gene frequencies. In addition, our arguments go contrary to certain regularity and counterfactual interpretations of the concepts of causality and force. To say that *a* caused *b* is to say more than that any event that is relevantly similar to *a* would be followed by an event that is relevantly similar to *b* (we ignore issues concerning indeterministic causation); and to say that a system of objects is subject to certain forces is to say more than that they will change in various ways, as long as nothing interferes. And lastly, our account of what is wrong with genic selection coefficients points to a characterization of the conditions under which a predicate will pick out a real property. Selfish genes and "grue" emeralds bear a remarkable similarity.

## 1. THE "CANONICAL" OBJECTS OF EVOLUTIONARY THEORY

The Modern Synthesis received from Mendel a workable conception of the mechanism of heredity. But as important as this contribution was, the role of Mendelian "factors" was more

profound. Not only did Mendelism succeed in filling in a missing link in the three-part structure of variation, selection, and transmission; it also provided a canonical form in which *all* evolutionary processes could be characterized. Evolutionary models must describe the interactions of diverse forces and phenomena. To characterize selection, inbreeding, mutation, and sampling error in a single predictive theoretical structure, it is necessary to describe their respective effects in a common currency. Change in gene frequencies is the "normal form" in which all these aspects are to be represented; therefore, genes might be termed the canonical objects of evolutionary theory.

Evolutionary phenomena can be distilled into a tractable mathematical form by treating them as preeminently genetic. It by no means follows from this that the normal form characterization captures everything that is biologically significant. In particular, the computational adequacy of genetic models leaves open the question whether they also correctly identify the causes of evolution. The canonical form of the models has encouraged many biologists to think of all natural selection as genic selection, but there has always been a tradition within the Modern Synthesis which thinks of natural selection differently and holds this gene's-eye view to be fundamentally distorted.

Ernst Mayr perhaps typifies this perspective. Although it is clear that selection has an *effect* on gene frequencies, it is not so clear that natural selection is always selection for or against particular genes. Mayr has given two reasons for thinking that the idea of genic selection is wrong.

One of the interesting things about his criticisms is their simplicity; they do not report any recondite facts about evolutionary processes but merely remind evolutionary theorists of what they already know (although perhaps lose sight of at times). As we will see, genic selectionists have ready replies for these criticisms.

The first elementary observation is that "natural selection favors (or discriminates against) phenotypes, not genes or genotypes" (1963, p. 184). Protective coloration and immunity from DDT are phenotypic traits. Organisms differ in their reproductive success under natural selection because of their phenotypes. If those phenotypes are heritable, then natural selection will produce evolutionary change (*ceteris paribus,* of course). But genes are affected by natural selection only indirectly. So the gene's-eye view, says Mayr, may have its uses, but it does not correctly represent how natural selection works.

Mayr calls his second point *the genetic theory of relativity* (1963, p. 296). This principle says that "no gene has a fixed selective value, the same gene may confer high fitness on one genetic background and be virtually lethal on another." Should we conclude from this remark that there is never selection for single genes or that a single gene simultaneously experiences different selection pressures in different genetic backgrounds? In either case, the lesson here seems to be quite different from that provided by Mayr's first

point—which was that phenotypes, not genotypes, are selected for. In this case, however, it seems to be gene complexes, rather than single genes, which are the objects of selection.

Mayr's first point about phenotypes and genotypes raises the following question: if we grant that selection acts "directly" on phenotypes and only "indirectly" on genotypes, why should it follow that natural selection is not selection for genetic attributes? Natural selection is a causal process; to say that there is selection for some (genotypic or phenotypic) trait $X$ is to say that having $X$ causes differential reproductive success (*ceteris paribus*).[1] So, if there is selection for protective coloration, this merely means that protective coloration generates a reproductive advantage. But suppose that this phenotype is itself caused by one or more genes. Then having those genes causes a reproductive advantage as well. Thus, if selection is a causal process, in acting on phenotypes it also acts on the underlying genotypes. Whether this is "direct" may be important, but it doesn't bear on the question what is and what is not selected for. Selection, in virtue of its causal character and on the assumption that causality is transitive, seems to block the sort of asymmetry that Mayr demands. Asking whether phenotypes or genotypes are selected for seems to resemble asking whether a person's death was caused by the entry of the bullet or by the pulling of the trigger.

Mayr's second point—his genetic

1. The "*ceteris paribus*" is intended to convey the fact that selection for $X$ can fail to bring about greater reproductive success for objects that have $X$, if countervailing forces act. Selection for $X$, against $Y$, and so on, are component forces that combine vectorially to determine the dynamics of the population.

principle of relativity—is independent of the alleged asymmetry between phenotype and genotype. It is, of course, not in dispute that a gene's fitness depends on its genetic (as well as its extrasomatic) environment. But does this fact show that there is selection for gene complexes and not for single genes? Advocates of genic selection tend to acknowledge the relativity but to deny the conclusion that Mayr draws. Williams (1966, pp. 56–57) gives clear expression to this common reaction when he writes:

> Obviously it is unrealistic to believe that a gene actually exists in its own world with no complications other than abstract selection coefficients and mutation rates. The unity of the genotype and the functional subordination of the individual genes to each other and to their surroundings would seem at first sight, to invalidate the one-locus model of natural selection. Actually these considerations do not bear on the basic postulates of the theory. No matter how functionally dependent a gene may be, and no matter how complicated its interactions with other genes and environmental factors, it must always be true that a given gene substitution will have an arithmetic mean effect on fitness in any population. One allele can always be regarded as having a certain selection coefficient relative to another at the same locus at any given point in time. Such coefficients are numbers that can be treated algebraically, and conclusions inferred for one locus can be iterated over all loci. Adaptation can thus be attributed to the effect

of selection acting independently at each locus.

Dawkins (1976, p. 40) considers the same problem: how can single genes be selected for, if genes build organisms only in elaborate collaboration with each other and with the environment? He answers by way of an analogy:

> One oarsman on his own cannot win the Oxford and Cambridge boat race. He needs eight colleagues. Each one is a specialist who always sits in a particular part of the boat—bow or stroke or cox, etc. Rowing the boat is a cooperative venture, but some men are nevertheless better at it than others. Suppose a coach has to choose his ideal crew from a pool of candidates, some specializing in the bow position, others specializing as cox, and so on. Suppose that he makes his selection as follows. Every day he puts together three new trial crews, by random shuffling of the candidates, for each position, and he makes the three crews race against each other. After some weeks of this it will start to emerge that the winning boat often tends to contain the same individual men. These are marked up as good oarsmen. Other individuals seem consistently to be found in slower crews, and these are eventually rejected. But even an outstandingly good oarsman might sometimes be a member of a slow crew, either because of the inferiority of the other members, or because of bad luck— say a strong adverse wind. It is only *on average* that the best men tend to be in the winning boat.

The oarsmen are genes. The rivals for each seat in the boat are alleles potentially capable of occupying the same slot along the length of a chromosome. Rowing fast corresponds to building a body which is successful at surviving. The wind is the external environment. The pool of alternative candidates is the gene pool. As far as the survival of any one body is concerned, all its genes are in the same boat. Many a good gene gets into bad company, and finds itself sharing a body with a lethal gene, which kills the body off in childhood. Then the good gene is destroyed along with the rest. But this is only one body, and replicas of the same good gene live on in other bodies which lack the lethal gene. Many copies of good genes are dragged under because they happen to share a body with bad genes, and many perish through other forms of ill luck, say when their body is struck by lightning. But by definition luck, good and bad, strikes at random, and a gene which is consistently on the losing side is not unlucky; it is a bad gene.

Notice that this passage imagines that oarsmen (genes) are good and bad pretty much *in*dependently of their context. But even when fitness is heavily influenced by context, Dawkins still feels that selection functions at the level of the single gene. Later in the book (pp. 91–92), he considers what would happen if a team's performance were improved by having the members communicate with each other. Suppose that half of the oarsmen spoke only English and the other half spoke only German:

What will emerge as the overall best crew will be one of the two stable states—pure English or pure German, but not mixed. Superficially it looks as though the coach is selecting whole language groups *as units*. This is not what he is doing. He is selecting individual oarsmen for their apparent ability to win races. It so happens that the tendency for an individual to win races depends on which other individuals are present in the pool of candidates.

Thus Dawkins follows Williams in thinking that genic selectionism is compatible with the fact that a gene's fitness depends on context.

Right after the passage just quoted, Dawkins says that he favors the perspective of genic selectionism because it is more "parsimonious." Here, too, he is at one with Williams (1966), who uses parsimony as one of two main lines of attack against hypotheses of group selection. The appeal to simplicity may confirm a suspicion that already arises in this context: perhaps it is a matter of taste whether one prefers the single gene perspective or the view of selection processes as functioning at a higher level of organization. As long as we agree that genic fitnesses depend on context, what difference does it make how we tell the story? As natural as this suspicion is in the light of Dawkins' rowing analogy, it is mistaken. Hypotheses of group selection can be genuinely incompatible with hypotheses of organismic selection (Sober, 1980), and, as we will see in what follows, claims of single gene selection are at times incompatible with claims that gene complexes are selected for and against. Regardless of one's aesthetic inclinations and regardless of whether one thinks of parsimony as a "real" reason for hypothesis choice, the general perspective of genic selectionism is mistaken for biological reasons.[2]

Before stating our objections to genic selectionism, we want to make clear one defect that this perspective does *not* embody. A quantitative genetic model that is given at any level can be recast in terms of parameters that attach to genes. This genic representation will correctly trace the trajectory of the population as its gene frequencies change. In a minimal sense (to be made clear in what follows), it will be "descriptively adequate". Since the parameters encapsulate information about the environment, both somatic and extrasomatic, genic selectionism cannot be accused of ignoring the complications of linkage or of thinking that genes exist in a vacuum. The defects of genic selectionism concern its distortion of causal processes, not whether its models allow one to predict future states of the population.[3]

The causal considerations which will play a preeminent role in what follows are not being imposed from without but already figure centrally in evolutionary theory. We have already mentioned how we understand the idea of *selection for X*. Our causal construal is natural in view of how the phenomena of linkage and pleiotropy are understood (see Sober, 1981a). Two genes may be linked together on the same chromosome, and so selection for one may cause them both to increase in frequency. Yet the linked gene—the "free rider"—may be neutral or even deleterious; there was no selection *for it*. In describing pleiotropy, the same distinction is made. Two phenotypic traits may be caused by the same underlying gene complex, so that selection for one leads to a proliferation of both. But, again, there was no selection for the free rider. So it is a familiar idea that two traits can attach to exactly the same organisms and yet differ in their causal roles in a selection process. What is perhaps less familiar is that two sets of selection coefficients may both attach to the same population and yet differ in their causal roles— the one causing change in frequencies, the other merely reflecting the changes that ensue.

## 2. AVERAGING AND REIFICATION

Perhaps the simplest model exhibiting the strategy of averaging recommended by Williams and Dawkins is used in describing heterozygote superiority. In organisms whose chromosomes

2. In the passages quoted, Williams and Dawkins adopt a very bold position: any selection process which *can* be represented as genic selection *is* genic selection. Dawkins never draws back from this monolithic view, although Williams' more detailed argumentation leads him to hedge. Williams allows that group selection (clearly understood to be an alternative to genic selection) is possible and has actually been documented once (see his discussion of the t-allele). But *all* selection processes—including group selection—can be "represented" in terms of selection coefficients attaching to single genes. This means that the representation argument proves far too much.

3. Wimsatt (1980) criticizes genic selectionist models for being computationally inade-

|                               | $AA$            | $Aa$            | $aa$            |
| ----------------------------- | --------------- | --------------- | --------------- |
| Proportion before selection   | $p^2$           | $2pq$           | $q^2$           |
| Fitness                       | $w_1$           | $w_2$           | $w_3$           |
| Proportion after selection    | $\dfrac{p^2 w_1}{\overline{W}}$ | $\dfrac{2pq w_2}{\overline{W}}$ | $\dfrac{q^2 w_3}{\overline{W}}$ |

come in pairs, individuals with different genes (or alleles) at the same location on two homologous chromosomes are called heterozygotes. When a population has only two alleles at a locus, there will be one heterozygote form ($Aa$) and two homozygotes ($AA$ and $aa$). If the heterozygote is superior in fitness to both homozygotes, then natural selection may modify the frequencies of the two alleles $A$ and $a$ but will not drive either to fixation (i.e., 100 percent), since reproduction by heterozygotes will inevitably replenish the supply of homozygotes, even when homozygotes are severely selected against. A textbook example of this phenomenon is the sickle trait in human beings. Homozygotes for the allele controlling the trait develop severe anemia that is often fatal in childhood. Heterozygotes, however, suffer no deleterious effects, but enjoy a greater than average resistance to malaria. Homozygotes for the other allele have neither the anemia nor the immunity, and so are intermediate in fitness. Human populations with both alleles that live in malarial areas have remained polymorphic, but with the eradication of malaria, the sickle cell allele has been eliminated.

Population genetics provides a simple model of the selection process

that results from heterozygotes having greater viability than either of the homozygotes (Li, 1955). Let $p$ be the frequency of $A$ and $q$ be the frequency of $a$ (where $p + q = 1$). Usually, the maximal fitness of $Aa$ is normalized and set equal to 1. But for clarity of exposition we will let $w_1$ be the fitness of $AA$, $w_2$ be the fitness of $Aa$, and $w_3$ be the fitness of $aa$. These genotypic fitness values play the mathematical role of transforming genotype frequencies before selection into genotype frequencies after selection (see above). Here, $\overline{W}$, the average fitness of the population is $p^2 w_1 + 2pq w_2 + q^2 w_3$. Assuming random mating, the population will move towards a stable equilibrium frequency $\hat{p}$ where

$$\hat{p} = \frac{w_3 - w_2}{(w_1 - w_2) + (w_3 - w_2)}.$$

It is important to see that this model attributes fitness values and selection coefficients to diploid genotypes and not to the single genes $A$ and $a$. But, as genic selectionists are quick to emphasize, one can always define the required parameters. Let us do so.

We want to define $W_A$, which is the fitness of $A$. If we mimic the mathematical role of genotype fitness values in the previous model, we will require that $W_A$ obey the following condition:

---

quate and for at best providing a kind of "genetic bookkeeping" rather than a "theory of evolutionary change." Although we dissent from the first criticism, our discussion in what follows supports Wimsatt's second point.

$W_A$ × frequency of $A$ before selection = frequency of $A$ after selection × $\overline{W}$

Since the frequency of $A$ before selection is $p$ and the frequency of $A$ after selection is

$$\frac{w_1 p^2 + w_2 pq}{\overline{W}},$$

it follows that

$$W_A = w_1 p + w_2 q.$$

By parity of reasoning,

$$W_a = w_3 q + w_2 p.$$

Notice that the fitness values of single genes are just weighted averages of the fitness values of the diploid genotypes in which they appear. The weighting is provided by their frequency of occurrence in the genotypes in question. The genotypic fitnesses specified in the first model are *constants;* as a population moves toward its equilibrium frequency, the selection coefficients attaching to the three diploid genotypes do not change. In contrast, the expression we have derived for allelic fitnesses says that allelic fitnesses change as a function of their own frequencies; as the population moves toward equilibrium, the fitnesses of the alleles must constantly be recomputed.

Heterozygote superiority illustrates the principle of genetic relativity. The gene $a$ is maximally fit in one context (namely, when accompanied by $A$) but is inferior when it occurs in another (namely, when it is accompanied by another copy of itself). In spite of this, we can average over the two different contexts and provide the required representation in terms of genic fitness and genic selection.

In the diploid model discussed first, we represented the fitness of the three genotypes in terms of their *viability* —that is, in terms of the proportion of individuals surviving from egg to adult. It is assumed that the actual survivorship of a class of organisms sharing the same genotype precisely represents the fitness of that shared genotype. This assumes that random drift is playing no role. Ordinarily, fitness *cannot* be identified with actual reproductive success (Brandon, 1978: Mills and Beatty, 1979; Sober, 1981a). The same point holds true, of course, for the fitness coefficients we defined for the single genes.[4]

Of the two descriptions we have constructed of heterozygote superiority, the first model is the standard one; in it, *pairs* of genes are the bearers of fitness values and selection coefficients. In contrast to this diploid model, our second formulation adheres strictly to the dictates of genic selectionism, according to which it is *single genes* which are the bearers of

---

4. We see from this that Dawkins' remark that a gene that is "consistently on the losing side is not unlucky; it's a bad gene" is not quite right. Just as a single genotoken (and the organism in which it is housed) may enjoy a degree of reproductive success that is not an accurate representation of its fitness, so a set of genotokens (which are tokens of the same genotype) may encounter the same fate. Fitness and actual reproductive success are guaranteed to be identical only in models which ignore random drift and thereby presuppose an infinite population.

the relevant evolutionary properties. We now want to describe what each of these models will say about a population that is at its equlibrium frequency.

Let's discuss this situation by way of an example. Suppose that both homozygotes are lethal. In that case, the equilibrium frequency is .5 for each of the alleles. Before selection, the three genotypes will be represented in proportions 1/4, 1/2, 1/4, but after selection the frequencies will shift to 0, 1, 0. When the surviving heterozygotes reproduce, Mendelism will return the population to its initial 1/4, 1/2, 1/4 configuration, and the population will continue to zig-zag between these two genotype configurations, all the while maintaining each allele at .5. According to the second, single gene, model, at equilibrium the fitnesses of the two genes are both equal to 1 and the selection coefficients are therefore equal to zero. At equilibrium, no selection occurs, on this view. Why the populations's *genotypic configuration* persists in zig-zagging, the gene's eye point of view is blind to see; it must be equally puzzling why $\overline{W}$, the average fitness of the population, also zig-zags. However, the standard diploid model yields the result that selection occurs when the population is at equilibrium, just as it does at other frequencies, favoring the heterozygote at the expense of the homozygotes. Mendelism *and selection* are the causes of the zig-zag. Although the models are computationally equivalent in their prediction of gene frequencies, they are not equivalent when it comes to saying whether or not selection is occurring.

It is hard to see how the adequacy of the single gene model can be defended in this case. The biological term for the phenomenon being described is apt. We are talking here about *heterozygote superiority,* and both terms of this label deserve emphasis. The heterozygote—i.e., the diploid genotype (not a single gene) —is superior *in fitness* and, therefore, enjoys a selective advantage. To insist that the single gene is always the level at which selection occurs obscures this and, in fact, generates precisely the wrong answer to the question of what is happening at equilibrium. Although the mathematical calculations can be carried out in the single gene model just as they can in the diploid genotypic model, the phenomenon of heterozygote superiority cannot be adequately "represented" in terms of single genes. This model does not tell us what is patently obvious about this case: even at equilibrium, what happens to gene frequencies is an artifact of selection acting on diploid genotypes.

One might be tempted to argue that in the heterozygote superiority case, the kind of averaging we have criticized is just an example of frequency dependent selection and that theories of frequency dependent selection are biologically plausible and also compatible with the dictates of genic selectionism. To see where this objection goes wrong, one must distinguish genuine from spurious cases of frequency dependent selection. The former occurs when the frequency of an allele has some *biological impact* on its fitness; an example would be the phenomenon of mimicry in which the rarity of a mimic enhances its fitness. Here one can tell a biological story explaining why the fitness values have the mathematical form they do. The case of heterozygote superiority is altogether different; here frequencies are taken into account simply as a mathematical contrivance, the only point being to get the parameters to multiply out in the right way.

The diploid model is, in a sense, more contentful and informative than the single gene model. We noted before that from the *constant* fitness values of the three genotypes we could obtain a formula for calculating the fitnesses of the two alleles. Allelic fitnesses are implied by genotype fitness values and allelic frequencies; since allelic frequencies change as the population moves toward equilibrium, allelic fitnesses must constantly be recomputed. However, the derivation in the opposite direction cannot be made.[5] One cannot deduce the fitnesses of the genotypes from allelic fitnesses and frequencies. This is especially evident when the population is at equilibrium. At equilibrium, the allelic fitnesses are identical. From this information alone, we cannot tell whether there is no selection at all or whether some higher level selection process is taking place. Allelic frequencies plus genotypic fitness imply allelic fitness values, but allelic frequencies plus allelic fitness values do not imply genotypic fitness values. This derivational asymmetry suggests that the genotypic description is more informative.

Discussions of reductionism often suggest that theories at lower levels of organization will be more detailed and informative than ones at higher levels. However, here, the more contentful, constraining model is provided at the higher level. The idea that genic selection models are "deeper" and describe the fundamental level at which selection "really" occurs is simply not universally correct.

The strategy of averaging fosters the illusion that selection is acting at a lower level of organization than it in fact does. Far from being an idiosyncratic property of the genic model of heterozygote superiority just discussed, averaging is a standard technique in modeling a variety of selection processes. We will now describe another example in which this technique of representation is used. The example of heterozygote superiority focused on differences in genotypic *viabilities.* Let us now consider the way differential fertilities can be modeled for one locus with two alleles. In the fully general case, fertility is a property of a mating pair, not of an individual. It may be true that a cross between an *AA* male and *aa* female has an expected number of offspring different from a cross between an *AA* female and an *aa* male. If fitnesses are a unique function of the pair, the model must represent nine possible fitnesses, one for each mating pair. Several special cases permit a reduction in dimensionality. If the sex of a genotype does not affect of its fertility, then only six fitnesses need be given;

5. If the heterozygote fitness is set equal to 1, the derivation is possible for the one locus two allele case considered. But if more than two alleles are considered, the asymmetry exists even in the face of normalization.

| Chromosome EF | Chromosome CD | | |
|---|---|---|---|
| | ST/ST | ST/BL | BL/BL |
| ST/ST | 0.791 | 1.000 | 0.834 |
| ST/TD | 0.670 | 1.006 | 0.901 |
| TD/TD | 0.657 | 0.657 | 1.067 |

and if fertility depends only on one of the sexes, say the females, the three female genotypes may be assigned values which fix the fertilities of all mating pairs.

But even when these special cases fail to obtain, the technique of averaging over contexts can nevertheless provide us with a fitness value for each genotype. Perhaps an *aa* female is highly fertile when mated with an *Aa* male but is much less so when mated with an *AA* male; perhaps *aa* females are quite fertile on average, but *aa* males are uniformly sterile. No matter—we can merely average over all contexts and find the average effect of the *aa* genotype. This number will fluctuate with the frequency distributions of the different mating pairs. Again, the model appears to locate selection at a level lower than what might first appear to be the case. Rather than assigning fertilities to mating pairs, we now seem to be assigning them to genotypes. This mathematical contrivance is harmless as long as it does not lead us to think that selection really acts at this lower level of organization.[6]

Our criticism of genic selectionism has so far focused on two forms of selection at a single locus. We now need to take account of how a multilocus theory can imply that selection

is not at the level of the selfish gene. The pattern of argument is the same. Even though the fitness of a pair of genes at one locus may depend on what genes are found at other loci, the technique of averaging may still be pressed into service. But the selection values thereby assigned to the three genotypes at a single locus will be artifacts of the fitnesses of the nine genotype complexes that exist at the two loci. As in the examples we already described, the lower-level selection coefficients will change as a function of genotype frequencies, whereas the higher-level selection coefficients will remain constant. An example of this is provided by the work of Lewontin and White (reported in Lewontin, 1974) on the interaction of two chromsome inversions found in the grasshopper *Moraba scura*. On each of the chromosomes of the EF pair, Standard (ST) and Tidbinbilla (TD) may be found. On the CD chromsome pair, Standard (ST) and Blundell (BL) are the two alternatives. The fitness values of the nine possible genotypes were estimated from nature (as above). Notice that there is heterozygote superiority on the CD chromosome if the EF chromosome is either ST/ST or ST/TD, but that BL/BL dominance ensues when the EF chromosome is

---

6. The averaging of effects can also be used to foster the illusion that a group selection process is really just a case of individual selection. But since this seems to be a relatively infrequent source of abuse, we will not take the space to spell out an example.

homozygous for TD. Moreover, TD/TD is superior when in the context BL/BL but is inferior in the other contexts provided by the CD pair. These fitness values represent differences in viability, and again the inference seems clear that selection acts on multilocus genotypic configurations and not on the genotype at a single locus, let alone on the separate genes at that locus.

## 3. INDIVIDUATING SELECTION PROCESSES

The examples in the previous section have a common structure. We noted that the fitness of an object (a gene, a genotype) varied significantly from context to context. We concluded that selection was operating at a level higher than the one posited by the model—at the level of genotypes in the case of heterozygote superiority, at the level of the mating pair in the fertility model, and at the level of pairs of chromosome inversions in the *Moraba scura* example. These analyses suggest the following principle: *if the fitness of X is context sensitive, then there is not selection for X; rather, there is selection at a level of organization higher than X.*

We believe that this principle requires qualification. To see why context sensitivity is not a *sufficient* condition for higher level selection, consider the following example. Imagine a dominant lethal gene: it kills any organisms in which it is found unless the organism also has a suppressor gene at another locus. Let's consider two populations, In the first population, each organism is homozygous for a suppressor gene which prevents copies of the lethal gene from having any effect. In the second population, no organism has a suppressor, so, whenever the lethal gene occurs, it is selected against. A natural way of describing this situation is that there is selection against the lethal gene in one population, but, in the other, there is no selection going on at all. It would be a mistake (of the kind we have already examined) to think that there is a single selection process at work here against the lethal gene, whose magnitude we calculate by averaging over the two populations. However, we do not conclude from this that there is a selection process at work at some higher level of organization than the single gene. Rather, we conclude that there are *two* populations; in one *genic* selection occurs, and in the other *nothing* occurs. So the context sensitivity of fitness is an ambiguous clue. If the fitness of *X* depends on genetic context, this may mean that there is a single selection process at some higher level, *or* it may mean that there are several different selection processes at the level of *X*. Context sensitivity does not suffice for there to be selection at a higher level.[7]

Thus, the fitness of an object can be sensitive to genetic context for at least two reasons. How are they to be

---

7. The argument given here has the same form as one presented in Sober (1980), which showed that the following is not a sufficient condition for group selection: there is heritable variation in the fitness of groups in which the fitness of an organism depends on the character of the group it is in.

distinguished? This question leads to an issue at the foundation of *all* evolutionary models. What unites a set of objects as all being subject to a single selection process? Biological modeling of evolution by natural selection is based on three necessary and sufficient conditions (Lewontin, 1970): a given set of objects must exhibit variation; some individuals must be fitter than others; and there must be correlation between the fitness of parents and the fitness of offspring. Here, as before, we will identify fitness with actual reproductive success, subject to the proviso that these will coincide only in special cases. Hence, evolution by natural selection exists when and only when there is heritable variation in fitness.

Using these conditions presupposes that some antecedent decision has been made about which objects can appropriately be lumped together as participating in a single selection process (or, put differently, the conditions are not sufficient after all.) Biologists do not talk about a *single* selection process subsuming widely scattered organisms of different species which are each subject to quite different local conditions. Yet, such a gerrymandered assemblage of objects may well exhibit heritable variation in fitness. And even within the same species, it would be artificial to think of two local populations as participating in the same selection process because one encounters a disease and the other experiences a food shortage as its principal selection pressure. Admittedly, the gene frequencies can be tabulated and pooled, but in some sense the relation of organisms to environments is too heterogeneous for this kind of averaging to be more than a mathematical contrivance.

It is very difficult to spell out necessary and sufficient conditions for when a set of organisms experience "the same" selection pressure. They need not compete with each other. To paraphrase Darwin, two plants may struggle for life at the edge of a desert, and selection may favor the one more suited to the stressful conditions. But it needn't be the case that some resource is in short supply, so that the amount expropriated by one reduces the amount available to the other. Nor need it be true that the two organisms be present in the same geographical locale; organisms in the semi-isolated local populations of a species may experience the same selection pressures. What seems to be required, roughly, is that some common causal influence impinge on the organisms. This sameness of causal influence is as much determined by the biology of the organisms as it is by the physical characteristics of the environment. Although two organisms may experience the same temperature fluctuations, there may be no selective force acting on both. Similarly, two organisms may experience the same selection pressure (for greater temperature tolerance, say) even though the one is in a cold environment and the other is in a hot one. Sameness of causal influence needs to be understood biologically.

For all the vagueness of this requirement, let us assume that we have managed to single out the class of objects which may properly be viewed as participating in a single selection process. To simplify matters, let us suppose that they are all organisms within the same breeding population. What, then, will tell us whether selection is at the level of the single gene or at the level of gene complexes? To talk about either of these forms of selection is, in a certain important but nonstandard sense, to talk about "group selection." Models of selection do not concern single organisms or the individual physical copies of genes (i.e., geno*tokens*) that they contain. Rather, such theories are about groups of organisms which have in common certain geno*types*. To talk about selection for *X*, where *X* is some single gene or gene cluster, is to say something about the effect of having *X* and of lacking *X* on the relevant subgroups of the breeding population. If there is selection for *X*, every object which has *X* has its reproductive chances augmented by its possessing *X*. This does not mean that every organism which has *X* has precisely the same overall fitness, nor does it mean that every organism must be affected in precisely the same way (down to the minutest details of developmental pathways). Rather, what is required is that the effect of *X* on each organism be in the same direction as far as its overall fitness is concerned. Perhaps this characterization is best viewed as a limiting ideal. To the degree that the population conforms to this requirement,

it will be appropriate to talk about genic selection. To the degree that the population falls short of this, it will be a contrivance to represent matters in terms of genic selection.[8]

It is important to be clear on why the context sensitivity of a gene's effect on organismic fitness is crucial to the question of genic selection. Selection theories deal with groups of single organisms and not with organisms taken one at a time. It is no news that the way a gene inside of a single organism will affect that organism's phenotype and its fitness depends on the way it is situated in a context of background conditions. But to grant this fact of context sensitivity does not impugn the claim of causation; striking the match caused it to light, even though the match had to be dry and in the presence of oxygen for the cause to produce the effect.

Selection theory is about geno*types* not geno*tokens*. We are concerned with what properties are selected for and against in a population. We do not describe single organisms and their physical constituents one by one. It is for this reason that the question of context sensitivity becomes crucial. If we wish to talk about selection for a single gene, then there must be such a thing as *the* causal upshot of possessing that gene. A gene which is beneficial in some contexts and deleterious in others will have many *organismic*

8. The definition of genic selection just offered is structurally similar to the definition of group selection offered in Sober (1980). There, the requirement was that for there to be selection for groups which are *X*, it must be the case that every organism in a group that is *X* has one component of its fitness determined by the fact that it is in a group which is *X*. In group selection, organisms within the same group are bound together by a common group characteristic just as in genic selection organisms with the same gene are influenced in the same way by their shared characteristic.

effects. But at *the population level,* there will be no selection for or against that gene.

It is not simply the averaging over contexts which reveals the fact that genic selection coefficients are pseudoparameters; the fact that such parameters *change* in value as the population evolves while the biological relations stay fixed also points to their being artifacts. In the case of heterozygote superiority, genotypic fitnesses remain constant, mirroring the fact that the three genotypes have a uniform effect on the viability of the organisms in which they occur. The population is thereby driven to its equilibrium value while genic fitness values are constantly modified. A fixed set of biological relationships fuels both of these changes; the evolution of genic fitness values is effect, not cause.[9]

Are there real cases of genic selection? A dominant lethal—a gene which causes the individual to die regardless of the context in which it occurs—would be selected against. And selection for or against a phenotypic trait controlled by a single locus having two alleles might also be describable in terms of genic selection, provided that the heterozygote is intermediate in fitness between the two homozygotes. In addition, meiotic drive, such as is found in the house mouse *Mus musculus,* similarly seems to involve genic selection (Lewontin and Dunn, 1960). Among heterozygote males, the proportion of t-alleles in the sperm pool is greater than 1/2. Chromosomes with the t-allele have enhanced chances of representation in the gamete pool, and this directional effect seems to hold true regardless of what other genes are present at other loci.[10] At this level, but not at the others at which the t-allele affects the population, it is appropriate to talk about genic selection.

We so far have construed genic selection in terms of the way that having or lacking a gene can affect the reproductive chances of organisms. But there is another possibility—

9. In our earlier discussion of Mayr's ideas, we granted that selection usually acts "directly" on phenotypes and only "indirectly" on genotypes. But given the transitivity of causality, we argued that this fact is perfectly compatible with the existence of genotypic selection. However, our present discussion provides a characterization of when phenotypic selection can exist without there being any selection at the genotypic level. Suppose that individuals with the same genotype in a population end up with different phenotypes, because of the different microenvironments in which they develop. Selection for a given phenotype may then cross-classify the genotypes, and by our argument above, there will be no such thing as *the* causal upshot of a genotype. Averaging over effects will be possible, as always, but this will not imply genotypic selection. It is important to notice that this situation can allow evolution by natural selection to occur; gene frequencies can change in the face of phenotypic selection that is not accompanied by any sort of genotypic selection. Without this possibility, the idea of phenotypic selection is deprived of its main interest. There is no reason to deny that there can be selection for phenotypic differences that have no underlying genetic differences, but this process will not produce any change in the population (ignoring cultural evolution and the like).

10. Genes at other loci which modify the intensity of segregator distortion are known to exist in *Drosophila;* the situation in the house mouse is not well understood. Note that the

namely, that genes differentially proliferate even though they have *no* effect on the phenotypes of organisms. A considerable quantity of DNA has no known function; Orgel and Crick (1980) and Doolittle and Sapienza (1980) have suggested that this DNA may in fact be "junk." Such "selfish DNA," as they call it, could nonetheless undergo a selection process, provided that some segments are better replicators than others. Although these authors associate their ideas with Dawkins' selfish gene, their conception is far more restrictive. For Dawkins, *all* selection is genic selection, whereas for these authors, selfish DNA is possible only when the differential replication of genes is not exhaustively accounted for by the differential reproductive success of organisms.

Standard ways of understanding natural selection rule out rather than substantiate the operation of genic selection. It is often supposed that much of natural selection is *stabilizing selection,* in which an intermediate phenotype is optimal (e.g., birth weight in human beings). Although the exact genetic bases of such phenotypes are frequently unknown, biologists often model this selection process as follows. It is hypothesized that the phenotypic value is a monotone increasing function of the number of "plus alleles" found at a number of loci. Whether selection favors the presence of plus genes at one locus depends on how many such genes exist at other loci. Although this model does not view heterozygote superiority as the most common fitness relation *at a locus,* it nevertheless implies that a *heterogeneous genome* is superior in fitness. Exceptions to this intermediate optimum model exist, and the exact extent of its applicability is still an open question. Still, it appears to be widely applicable. If it is generally correct, we must conclude that the conditions in which genic selection exists are extremely narrow. Genic selection is not impossible, but the biological constraints on its operation are extremely demanding.

Although it is just barely conceivable that a critique of a scientific habit of thought might be devoid of philosophical presuppositions, our strictures against genic selectionism are not a case in point. We have described selection processes in which genic selection coefficients are *reifications;* they are artifacts, not causes, of evolution. For this to count as a criticism, one must abandon a narrowly instrumentalist view of scientific theories; this we gladly do, in that we assume that selection theory ought to pinpoint causes as well as facilitate predictions.

But even assuming this broadly noninstrumentalist outlook, our criticisms are philosophically partisan in additional ways. In that we have argued that genic selection coefficients are often "pseudoproperties" of genes, our criticisms of the gene's eye point of view are connected with more general metaphysical questions about the ontological status of properties. Some of these we take up in the following section. And in that we have understood "selection for" as a

---

existence of such modifiers is consistent with genic selection, as long as they do not affect the *direction* of selection.

causal locution, it turns out that our account goes contrary to certain regularity analyses of causation. In populations in which selection generated by heterozygote superiority is the only evolutionary force, it is true that gene frequencies will move to a stable equilibrium. But this law-like regularity does not imply that there is selection for or against any individual gene. To say that "the gene's fitness value caused it to increase in frequency" is not simply to say that "any gene with that fitness value (in a relevantly similar population) would increase in frequency," since the former is false and the latter is true. Because we take natural selection to be a force of evolution, these remarks about causation have implications (explored below in section 5) for how the concept of force is to be understood.

## 4. PROPERTIES

The properties, theoretical magnitudes, and natural kinds investigated by science ought not to be identified with the meanings that terms in scientific language possess. Nonsynonymous predicates (like "temperature" and "mean kinetic energy" and like "water" and "$H_2O$") may pick out the same property, and predicates which are quite meaningful (like "phlogiston" and "classical mass") may fail to pick out a property at all. Several recent writers have explored the idea that properties are to be individuated by their potential causal efficacy (Achinstein, 1974; Armstrong, 1978; Shoemaker, 1980; and Sober, 1982b). Besides capturing much of the intuitive content of our informal talk of properties, this view

also helps explicate the role of property-talk in science (Sober, 1981a). In this section, we will connect our discussion of genic selectionism with this metaphysical problem.

The definitional power of ordinary and scientific language allows us to take predicates each one of which picks out properties and to construct logically from these components a predicate which evidently does not pick out a property at all. An example of this is that old philosophical chestnut, the predicate "grue." We will say that an object is grue at a given time if it is green and the time is before the year 2000, or it is blue and the time is not before the year 2000. The predicate "grue" is defined from the predicates "green," "blue," and "time," each of which, we may assume for the purposes of the example, picks out a "real" property. Yet "grue" does not. A theory of properties should explain the basis of this distinction.

The difference between real and pseudo property is not captured by the ideas that animate the metaphysical issues usually associated with doctrines of realism, idealism, and conventionalism. Suppose that one adopts a "realist" position toward color and time, holding that things have the colors and temporal properties they do independently of human thought and language. This typical realist declaration of independence (Sober, 1982a) will then imply that objects which are grue are so independently of human thought and language as well. In this sense, the "reality" of grulers is ensured by the "reality" of colors and time. The distinction between real properties and pseudo-properties must be sought elsewhere.

Another suggestion is that properties can be distinguished from non-properties by appeal to the idea of *similarity* or of *predictive power.* One might guess that green things are more similar to each other than grue things are to each other, or that the fact that a thing is green is a better predictor of its further characteristics than the fact that it is grue. The standard criticism of these suggestions is that they are circular. We understand the idea of similarity in terms of shared *properties,* and the idea of predictive power in terms of the capacity to facilitate inference of further *properties.* However, a more fundamental difficulty with these suggestions presents itself: even if grue things happened to be very similar to each other, this would not make grue a real property. If there were no blue things after the year 2000, then the class of grue things would simply be the class of green things before the year 2000. The idea of similarity and the idea of predictive power fail to pinpoint the *intrinsic* defects of nonproperties like grue. Instead, they focus on somewhat accidental facts about the objects which happen to exist.

Grue is not a property for the same reason that genic selection coefficients are pseudoparameters in models of heterozygote superiority. The key idea is not that nonproperties are mind-dependent or are impoverished predictors; rather, they cannot be causally efficacious. To develop this idea, let's note a certain similarity between grue and genic selection coefficients. We pointed out before that genotype fitnesses plus initial genotype frequencies in the population causally determine the gene frequencies after selection. These same parameters also permit the mathematical derivation of genic fitness values, but, we asserted, these genic fitness values are artifacts; they do not cause the subsequent alteration in gene frequencies. The structure of these relationships is as follows.

genotype fitness values and
frequencies at time $t$

genic frequencies at $t + 1$

genic fitness values at time $t$

Note that there are two different kinds of determination at work here. Genic fitness values at a given time are not *caused* by the genotypic fitness values at the same time. We assume that causal relations do not obtain between simultaneous events; rather, the relationship is one of logical or mathematical deducibility (symbolized by a broken line). On the other hand, the relation of initial genotype fitnesses and frequencies and subsequent gene frequencies is one of causal determination (represented by a solid line).

Now let's sketch the causal relations involved in a situation in which an object's being green produces some effect. Let the object be a grasshopper. Suppose that it matches its grassy background and that this protective coloration hides it from a hungry predator nearby. The relationships involved might be represented as follows.

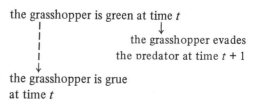

the grasshopper is green at time $t$

the grasshopper evades
the predator at time $t + 1$

the grasshopper is grue
at time $t$

Just as in the above case, the object's color at the time *logically implies* that it is grue at that time but is the *cause* of its evading the predator at a subsequent time. And just as genic fitness values do not cause changes in gene frequencies, so the grasshopper's being grue does not cause it to have evaded its predator.

Our assessment of genic selectionism was not that genic fitness values are *always* artifactual. In cases other than that of heterozygote superiority—say, in the analysis of the *t*-allele—it may be perfectly correct to attribute causal efficacy to genic selection coefficients. So a predicate can pick out a real (causally efficacious) property in one context and fail to do so in another. This does not rule out the possibility, of course, that a predicate like "grue" is *globally artifactual*. But this consequence should not be thought to follow from a demonstration that grue is artifactual in a single kind of causal process.

The comparison of grue with genic selection is not meant to solve the epistemological problems of induction that led Goodman (1965) to formulate the example. Nor does the discussion provide any a priori grounds for distinguishing properties from nonproperties. Nor is it even a straightforward and automatic consequence of the truth of any scientific model that grue is artifactual, or that the idea of causal efficacy captures the metaphysical distinction at issue. Instead, the point is that a certain natural interpretation of a biological phenomenon helps to indicate how we ought to understand a rather abstract metaphysical issue.[11]

## 5. FORCES

Our arguments against genic selectionism contradict a standard positivist view of the concept of force. Positivists have often alleged that Newtonian mechanics tells us that forces are not "things," but that claims about forces are simply to be understood as claims about how objects actually behave, or would behave, if nothing else gets in the way. An exhaustive catalog of the forces acting on a system is to be understood as simply specifying a set of counterfactuals that describe objects.[12]

A Newtonian theory of forces will

11. Another consequence of this analogy is that one standard diagnosis of what is wrong with "grue" fails to get to the heart of the matter. Carnap (1947) alleged that "green," unlike "grue," is purely qualitative, in that it makes no essential reference to particular places, individuals, or times. Goodman (1965) responded by pointing out that *both* predicates can be defined with reference to the year 2000. But a more fundamental problem arises: even if "grue" were, in some sense, not purely qualitative, this would not provide a fully general characterization of when a predicate fails to pick out a real property. Genic selection coefficients are "purely qualitative" if genotypic coefficients are, yet their logical relationship to each other exactly parallels that of "grue" to "green." Predicates picking out real properties can be "gruified" in a purely qualitative way: Let $F$ and $G$ be purely qualitative and be true of all the objects sampled (the emeralds, say). The predicate "($F$ and $G$) or ($-F$ and $-G$)" is a gruification of $F$ and poses the same set of problems as Goodman's "grue."

12. Joseph (1980) has argued that this position, in treating the distribution of objects as

characterize each force in its domain in terms of the changes it would produce, were it the only force at work. The theory will take pair-wise combinations of forces and describe the joint effects that the two forces would have were they the only ones acting on a system. Then the forces would be taken three at a time, and so on, until a fully realistic model is constructed, one which tells us how real objects, which after all are subject to many forces, can be expected to behave. Each step in this program may face major theoretical difficulties, as the recent history of physics reveals (Cartwright, 1980b; Joseph, 1980).

This Newtonian paradigm is a hospitable home for the modeling of evolutionary forces provided in population genetics. The Hardy-Weinberg Law says what happens to gene frequencies when no evolutionary forces are at work. Mutation, migration, selection, and random drift are taken up one at a time, and models are provided for their effects on gene frequencies when no other forces are at work. Then these (and other) factors are taken up in combination. Each of these steps increases the model's realism. The culmination of this project would be a model that

simultaneously represents the interactions of all evolutionary forces.

Both in physics and in population genetics, it is useful to conceive of forces in terms of their *ceteris paribus* effects. But there is more to a force than the truth of counterfactuals concerning change in velocity, or change in gene frequencies. The laws of motion describe the *effects* of forces, but they are supplemented by source laws which describe their *causes*. The standard genotypic model of heterozygote superiority not only says what will happen to a population, but also tells us what makes the population change.

It is quite true that when a population moves to an equilibrium value due to the selection pressures generated by heterozygote superiority, the alleles are "disposed" to change in frequency in certain ways.[13] That is, the frequencies *will* change in certain ways, as long as no other evolutionary forces impinge. Yet there is no force of genic selection at work here. If this is right, then the claim that genic selection is occurring must involve more than the unproblematic observations that gene frequencies are disposed to change in certain ways.

There is something more to the concept of force because it involves the idea of *causality*, and there is

given and then raising epistemological problems about the existence of forces, is committed to the existence of an asymmetry between attributions of quantities of *mass* to points in space-time and attributions of quantities of *energy* thereto. He argues that this idea, implicit in Reichenbach's (1958) classic argument for the conventionality of geometry, contradicts the relativistic equivalence of mass and energy. If this is right, then the positivistic view of force just described, far from falling out of received physical theory, in fact contradicts it.

13. For the purpose of this discussion, we will assume that attributions of dispositions and subjunctive conditionals of certain kinds are equivalent. That is, we will assume that to say that $x$ is disposed to $F$ is merely to say that if conditions were such-and-such, $x$ would $F$.

more to the idea of causality than is spelled out by such counterfactuals as the ones cited above. Suppose that something pushes (i.e., causally interacts with in a certain way) a billiard ball due north, and something else pushes it due west. Assuming that nothing else gets in the way, the ball will move northwest. There are two "component" forces at work here, and, as we like to say, one "net" force. However, there is a difference between the components and the resultant. Although something pushes the ball due north and something else pushes it due west, nothing pushes it northwest. In a sense, the resultant force is not a force at all, if by force we mean a causal agency. The resultant force is an artifact of the forces at work in the system. For mathematical purposes this distinction may make no difference. But if we want to understand why the ball moves the way it does, there is all the difference in the world between component and net.[14]

The "force" of genic selection in the evolutionary process propelled by heterozygote superiority is no more acceptable than the resultant "force" which is in the northwesterly direction. In fact, it is much worse. The resultant force, at least, is defined from the same conceptual building blocks as the component forces are. Genic selection coefficients, however, are gerrymandered hodgepodges, conceptually and dynamically quite unlike the genotypic selection coefficients that go into their construction. For genic selection coefficients are defined in terms of genotypic selection coefficients *and* gene frequencies. As noted before, they vary as the population changes in gene frequency, whereas the genotypic coefficients remain constant. And if their uniform zero value at equilibrium is interpreted as meaning that no selection is going on, one obtains a series of false assertions about the character of the population.

The concept of force is richer than that of disposition. The array of forces that act on a system uniquely determine the disposition of that system to change, but not conversely. If natural selection is a force and fitness is a disposition (to be reproductively successful), then the concept of selection is richer than that of fitness. To say that objects differ in fitness is not yet to say *why* they do so. The possible causes of such differences may be various, in that many different combinations of selection pressures acting at different levels of organization can have the same instantaneous effect on gene frequencies. Although selection coefficients and fitness values are interdefinable mathematically (so that, typically, $s = 1 - w$), they play different conceptual roles in evolutionary theory (Sober, 1980).

14. This position is precisely the opposite of that taken by Cartwright (1980a), who argues that net forces, rather than component forces, are the items which really exist. Cartwright argues this by pointing out that the billiard ball moves northwesterly and not due north or due west. However, this appears to conflate the *effect* of a force with the force or forces actually at work.

## REFERENCES

Achinstein, P., 1974, The identity of properties. *American Philosophical Quarterly*, 11, 4: 257–275.

Armstrong, D., 1978, *Universals and Scientific Realism*, Cambridge, Cambridge University Press.

Carnap, R., 1947, On the application of inductive logic. *Philosophy and Phenomenological Research*, 8: 133–147.

Cartwright, N., 1980a, Do the laws of nature state the facts? *Pacific Philosophical Quarterly*, 61, 1: 75–84.

——, 1980b, The truth doesn't explain much. *American Philosophical Quarterly*, 17, 2: 159–163.

Dawkins, R., 1976, *The Selfish Gene*, Oxford, Oxford University Press.

Doolittle, W., and Sapienza, C., 1980, Selfish genes, the phenotype paradigm, and genome evolution. *Nature*, 284: 601–603.

Fisher, R., 1930, *The Genetical Theory of Natural Selection*, New York, Dover.

Goodman, N., 1965, *Fact, Fiction, and Forecast*, Indianapolis, Bobbs Merrill.

Joseph, G., 1979, Riemannian geometry and philosophical conventionalism. *Australasian Journal of Philosophy*, 57, 3: 225–236.

——, 1980, The many sciences and the one world. *Journal of Philosophy*, LXXVII, 12: 773–790.

Lewontin, R., 1970, The units of selection. *Annual Review of Ecology and Systematics*, 1, 1: 1–14.

——, 1974, *The Genetic Basis of Evolutionary Change*, New York, Columbia University Press.

——, and Dunn, L., 1960, The evolutionary dynamics of a polymorphism in the House Mouse. *Genetics*, 45: 705–722.

Li, C., 1955, *Population Genetics*, Chicago, University of Chicago Press.

Mayr, E., 1963, *Animal Species and Evolution*, Cambridge, Harvard University Press.

Mills, S., and Beatty, J., 1979, The propensity interpretation of fitness. *Philosophy of Science*, 46: 263–286.

Orgel, L., and Crick, F., 1980, Selfish DNA: The ultimate parasite. *Nature*, 284: 604–607.

Reichenbach, H., 1958, *The Philosophy of Space and Time*, New York, Dover.

Shoemaker, S., 1980, Causality and properties, in P. van Inwagen, ed., *Essays in Honor of Richard Taylor*, Dordrecht, Reidel.

Sober, E., 1980, Holism, individualism and the units of selection, in P. Asquith and R. Giere, eds., *PSA*, 1980, vol. 2, Proceedings of the 1980 Biennial Meeting of the Philosophy of Science Association; East Lansing, Michigan.

——, 1981a, Evolutionary theory and the ontological status of properties, *Philosophical Studies*, 40: 147–176.

——, 1982a, Realism and independence. *Noûs*, 16, 3: 369–386.

——, 1982b, Why logically equivalent predicates may pick out different properties. *American Philosophical Quarterly*, 19, 2: 183–189.

Williams, G., 1966, *Adaptation and Natural Selection*, Princeton, Princeton University Press.

Wimsatt, W., 1980, Reductionistic research strategies and their biases in the units of selection controversy. In T. Nickles, ed., *Scientific Discovery*, vol. 2, *Case Studies*, Dordrecht, Reidel.

# IV
# ADAPTATION

# 14

# Adaptation

∽∿∽∿∽∿∽∿∽∿∽∿∽∿∽∿∽∿∽∿∽∿∽∿∽∿∽∿∽

## RICHARD C. LEWONTIN

EVERY THEORY OF THE WORLD that is at all powerful and covers a large domain of phenomena carries immanent within itself its own caricature. If we are to give a satisfactory explanation of a wide range of events in the world in a wide variety of circumstances, then necessarily our theory must contain some logically very powerful element that is flexible enough to be applicable in so many situations. Yet the very logical power of such a system is also its greatest weakness, for a theory that can explain everything explains nothing. It ceases to be a theory of the contingent world and becomes instead a vacuous metaphysic that generates not only all possible worlds, but all conceivable ones. The narrow line that separates a genuinely fruitful and powerful theory from its sterile caricature is crossed over and over again by vulgarizers who seize upon the powerful explanatory element and, by its indiscriminate use, destroy its usefulness. In doing so, however, they reveal underlying weaknesses in the theories themselves, leading to their reformulation.

This element of immanent caricature is certainly present in three theoretical structures that have had immense effects on twentieth-century bourgeois thought: Marxism, Freudianism, and Darwinism. Marx's historical materialism has been caricatured by the vulgar economism that attempts to explain the smallest detail of human history as a direct consequence of economic forces. Freud's ideas of sublimation, transference, reversal, and repression allow any form of overt behavior to be explained as a direct or transformed manifestation of any arbitrary psychological cause. In Darwinism the element that is both central to the evolutionary world view and yet so powerful that it can destroy Darwinism as a testable theory is that of *adaptation.* Adaptation is a concept that not only characterizes explanations of the evolution of life forms but also reappears in cultural theory as functionalism. It is the concept that there exist certain "problems" to be "solved" by organisms and by societies and that the actual forms of biological and social organizations that we see in the world are "solutions" to these "problems." Describing adaptation in the modern terms of solutions to problems should not mask the fact that adaptation is an inheritance from a much older world view that was characteristic of the aristocratic and fixed world before the European bourgeois revolution. This was the view that the entire universe, including living organisms and especially the human species and its social organization, were perfectly fit to serve a higher purpose. "The heavens declare the glory of God and the firmament showeth his handiwork" are the words of the Psalmist. The universe was the work of a Divine Creator and its parts were made by Him to fit together in a harmonious way, each part subserving the higher function. In the view of some Christians the primary object of this creation was man, whose nature was carefully fashioned to allow a new and more trustworthy race of angels to develop, while the rest of the living world was designed to serve him. Cows were ideally designed to provide him with milk and trees to give shade and shelter. The most important political consequence of this world view was the legitimation that it provided for social organization. Lords and serfs, masters and slaves, represented a division of power and labor that was necessary for the proper functioning of society and the working out of the divine plan.

The belief that organisms were marvelously fit to their environments

and that each part of an organism was exquisitely adjusted to serve a special function in the body, just as parts of the body politic were perfectly fit to serve the needs of "society," was carried over into modern biological and anthropological thought. All that was changed was the explanation. Having rejected the Supreme Designer as being responsible for the world's perfection, Darwin needed to show that evolution by natural selection could lead to the same end:

> In considering the origin of species, it is quite conceivable that a naturalist ... might come to the conclusion that each species ... had descended, like varieties, from other species. Nevertheless, such a conclusion, even if well founded, would be unsatisfactory until it could be shown how the innumerable species inhabiting this world have been modified, so as to acquire that perfection of structure and coadaptation which most justly excites our admiration.
>
> [*On the Origin of Species,* 1859, p. 3]

Indeed, in Chapter VI where he dealt with "Difficulties of the Theory," Darwin realized that "organs of extreme perfection and complication" were a critical test case for his theory: "To suppose that the eye, with all its inimitable contrivances for adjusting the focus to different distances, for admitting different amounts of light, and for the correction of spherical and chromatic aberration, could have been formed by natural selection seems, I freely confess, absurd in the highest degree" (*On the Origin of Species*, 1859, p. 186).

But such "organs of perfection" are only the extreme and obvious results of the process of natural selection which lies at the center of the Darwinian evolutionary theory. For Darwin the origin of species was the result of a continuing process of adaptation which, at the same time that it produced new species, produced organisms whose parts were in harmony with each other so that the organism as a whole was in harmony with its environment.

### BEING ADAPTED
### AND BECOMING ADAPTED

The concept of adaptation implies that there is a preexistent form, problem, or ideal to which things are fitted by a dynamical process. The process is *adaptation* and the end result is the state of being *adapted*. Thus a key may be adapted to fit a lock by cutting and filing it, or a part made for one model of a machine may be used in a different model by an adaptor that alters its shape. There cannot be adaptation without the ideal model to which the adaptation is taking place. Thus the very use of the notion of adaptation inevitably carries over into modern biology the theological view of a preformed physical world to which organisms were fitted. When the world was explained as the product of a Divine Will, there was no difficulty with such a concept, since the creation myth arranged for the physical world to be produced first and the organisms were then made to fit into that world. It was the mind of the Divine Artificer that created both the physical world and the organisms that populated it so the "problems" to be "solved" and the "solutions" were products of the same schema. God posed the problems and He gave the answers. He made oceans and gave fish fins to swim in them. He made the air and put wings on birds to fly in it. Having created the locks, *il Alto Fattore,* made the keys to fit them. With the advent of evolutionary explanations, however, serious problems arise for the concept of adaptation. While the physical universe certainly predated living organisms, what are the physical schemata to which organisms are adapting and adapted? Are there really preexistent "problems" to which the evolution of organisms provides "solutions"? This is the problem of the *ecological niche.* The niche is a multidimensional description of all the relations entered into by an organism with the surrounding world. What kind of food, and in what quantities, does the organism eat; what is its pattern of spatial movement; where does it reproduce; at what times of day and during what seasons is it active? To maintain that organisms adapt to the environment is to maintain that such ecological niches exist in the absence of organisms and that evolution consists in filling these empty and preexistent niches. But the external world can be divided up in a noncountable infinity of ways so that there is a noncountable infinity of conceivable ecological niches. Unless there is a preferred and correct way in which to partition the world, the idea of an ecological niche without an organism filling it loses all meaning.

The alternative is that ecological niches are defined only by the organisms living in them, but this raises serious difficulties for the concept of adaptation. Adaptation cannot be a process of the gradual fitting of an organism to the environment if the specific environmental configuration, the ecological niche, to which it is being adapted, does not already exist. If organisms define their own niches, then all species are already adapted and evolution cannot be seen as the process of *becoming* adapted.

Indeed, even if we put aside the problem of the existence of ecological niches, there are difficulties in seeing evolution as a process of adaptation. All extant species, for a very large part of their evolutionary histories, neither increase nor decrease in numbers and range. If a species were to increase on the average by even a small fraction of a percent per generation, it would soon fill the world and crowd out all other organisms. Conversely, if it were decreasing on the average, it would soon become extinct. Thus for long periods of their evolutionary lifetimes species are adapted in the sense that they make a living and are replacing themselves. At the same time, the species are evolving—changing their morphologies, physiologies, behaviors. The problem is how species can be at all times both adapting and adapted.

A solution to the paradox has been that the environment is constantly decaying with respect to the existing organisms so that the organisms must evolve to maintain their state of adaptation. Evolutionary adaptation is then an infinitesimal process in which the organism tracks the ever-changing environment, always lagging slightly behind, always adapting to the most recent environment, but always at the

mercy of further historical change. Both the occasional sudden increases in abundance and range of a species and the inevitable extinction of all species can be explained in this way. If the environment should change in such a way that the present physiology and behavior of a species by chance makes it reproductively very successful, it may spread very rapidly. This is the situation of species that have colonized a new continent, as, for example, the rabbit in Australia, finding there by sheer chance environmental conditions (including the lack of competitors) to which it is better adapted than it had been to its native habitat. Eventually, of course, such a species either uses up some resource that had been in great excess of its needs, or otherwise alters the environment by its own activity so that it is no longer able to increase in numbers. The alternative, that the environment has remained unchanged, but that the species by chance acquired a character that enabled it to utilize a previously untapped resource, is very much less likely. Such favorable mutations, or "hopeful monsters," may nevertheless have occurred—as, for example, in the evolution of fungus gardening by ants.

The simple view that the external environmental changes by some dynamic of its own and is tracked by the organisms takes no account of the effect that organisms have on the environment. The activity of all living forms transforms the external world in ways that both promote and inhibit the organism's life. Nest building, trail and boundary marking, the creation of entire habitats as in the dam-building activity of beavers—all increase the possibilities of life for their creators. On the other hand, the universal character of organisms is that they are self-limited in their increase because they quickly use up food and space resources. In this way the environment is a product of the organism, just as the organism is a product of the environment. The organism adapts the environment in the short term to its own needs, as, for example, by nest building, but in the long term the organism must adapt to an environment that is changing, partly by the organisms' own activity, in ways that are distinctive to the species.

In human evolution the relation between organism and environment has become virtually reversed in adaptation. Cultural invention replaces genetic change as the effective source of variation. Consciousness makes analysis and deliberate alterations in practice possible. As a result, the adaptation of environment to organism has become the dominant mode. Beginning with the usual relation, in which slow genetic adaptation to an almost independently changing environment was dominant, the line leading to *Homo sapiens* passed to a stage where conscious activity made adaptation of the environment to the organism's needs an integral part of the biological evolution of the species. As Engels observed in *The Part Played by Labor in the Transition from Ape to Man,* the human hand is as much a product of human labor as it is an instrument of that labor. Finally, the human species has passed to the stage where adaptation of the environment to the organism has come to be completely dominant, marking off *Homo sapiens* from all other life. It is this phenomenon, rather than any lucky change in the external world, that is responsible

for the rapid expansion of the human species in historical time.

Extinction may be seen as the eventual failure of adaptation in an already adapted species because the environment has changed in such a way that genetic or plastic changes in the species are unable to keep up. The response of a species to environmental alteration is limited by the morphological, physiological, and behavioral plasticity given by its present biology and by genetic changes that may occur by mutations and natural selection. This phenotypic and genetic plasticity is not infinite in kind but, more important, is limited in rate of response, so that the environment is sure eventually to alter in a way and at a rate that outdistances the species' adaptive response. More than 99.9 percent of all species that ever existed are extinct, and all are sure to be extinguished eventually.

The theory of environmental tracking does not, however, solve the problem of evolution. It cannot explain, for example, the immense diversification of organisms that has occurred. If evolution were only the successive modification of species to keep up with a constantly changing environment, then it is difficult to see how the land came to be populated from the water and the air from the land, or why homoiotherms evolved at the same time that poikilotherms were abundant. There is no consistent way in which this evolutionary diversification can be described as a process of adaptation unless we can describe preferred ways of dividing up the multidimensional niche space toward which species were evolving and, therefore, adapting, That is, the concept of adaptation is informative only if it has some predictive power. It must be possible to construct a priori ecological niches before organisms are known to occupy them, and then to describe the evolution of organisms toward these niches as adaptation. The exploration of other planets does provide the possibility of making such a priori predictions; yet it also illustrates the epistemological difficulties involved. If there really are preexistent niches to which organisms adapt, then it ought to be possible to predict the kind of organisms that will be discovered (if any) on Mars or Venus from an examination of the physical environments of those planets. Predictions are, in fact, being made in the building of devices that are meant to detect life on these planets, since the detection depends upon the growth of hypothetical organisms in defined nutrient solutions. These solutions, however, are based on the physiology of *terrestrial* microorganisms, so that the devices will detect only those extraterrestrial life forms that conform to the ecological niches already defined on earth. If life on other planets has partitioned the environment in ways that are radically different from those on earth, those living forms will remain unrecorded. There is no way to use adaptation as the central principle of evolution without recourse to a predetermination of the states of nature to which this adaptation occurs, yet there seems no way to choose these states of nature, except by reference to already existing organisms.

### SPECIFIC ADAPTATIONS

Evolutionists, having accepted that evolution is a process of adaptation, regard each aspect of the morphology, physiology, and behavior of organisms as specific adaptations, subserving the state of total adaptation of the entire organism. Thus fins are an adaptation for swimming, wings for flying, and legs for walking. Just as the notion of adaptation as a state of being of an organism requires a predetermination of the ecological niche, so, even more clearly, the assignment of the adaptive significance of an organ or behavior pattern presumes that a "problem" exists to which the character is a "solution." Fins, wings, and legs are the organism's solutions to the problem of locomotion in three different media. Such a view amounts to the construction of a description of the external environment and a description of the organism such that they can be mapped into each other by statements about function. In practice the construction may begin with either environment or organism, and the functional statement is then used to construct the corresponding structure in the other domain. That is, "problems" may be enumerated first and then the organism partitioned into "solutions," or a particular trait of an organism may be assumed to be a "solution" and the "problem" reconstructed from it. For example, the correct mutual recognition of males and females of the same species is regarded as a problem, since the failure to make this identification would result in the wastage of gametes and energy in a fruitless attempt to produce viable offspring from an interspecific mating. A variety of characters of organisms, such as color markings, temporal pattern of activity, vocalizations as in the "mating call" of frogs, courtship rituals, odor, etc., can then be explained as specific adaptations for solving this universal problem. Conversely, the large erect bony plates along the mid-dorsal line of the dinosaur *Stegosaurus* constitute a character that demands adaptive explanation, and they have been variously proposed as a solution to the problems of defense either by actually interfering with a predator's attack or by making the animal appear larger in profile, the problem of recognition in courtship, and the problem of temperature regulation by acting as cooling fins.

Hidden in adaptive analyses are a number of assumptions that go back to theistic views of nature and to a naive Cartesianism. First, it must be assumed that the partitioning of organisms into traits and the partitioning of environment into problems has a real basis and is not simply the reification of intuitive human categories. In what natural sense is a fin, leg, or wing an individual trait whose evolution can be understood in terms of the particular problem it solves? If the leg is a trait, is each part of the leg also a trait? At what level of subdivision do the boundaries no longer correspond to "natural" divisions? Perhaps the topology as a whole is incorrect. For example, the ordinary physical divisions of the brain correspond in a very rough way to the localization of some central nervous functions, but the memory of events appears to be diffusely stored so that particular memories are not to be found in particular microscopic regions. As we move from anatomical features to descriptions of behavior, the danger of reification becomes greater. Animal behavior is described by such categories as aggression, altruism, parental investment, warfare, slave-making, and cooperation, and each of these "organs of behavior" is provided with an adaptive explanation by finding the problem to which it is a solution (Wilson, 1975). Alternatively, the "problems" to be solved in adaptation also may be arbitrary reifications. For example, by extension from human behavior in some societies, other animals are said to have to cope with "parent-offspring conflict," a conflict that arises because parents and offspring are not genetically identical but both are motivated by natural selection to spread their genes (Trivers, 1974). A whole variety of manifest behaviors, such as the pattern of feeding of offspring by parents, are explained in this way. Thus the noise-making of immature birds or humans is a device to coerce parents into feeding their offspring who otherwise would go untended by the selfish parents.

A second hidden assumption is that characters can be isolated in an adaptive analysis and that although there may be interaction among characters, these interactions are secondary and represent constraints on the adaptation of each character separately. Similarly, each environmental problem to be solved is isolated and its solution regarded as independent of other interactions with the environment which can be, at most, constraints on the solution. Obviously, a *ceteris paribus* argument is necessary for adaptive reconstructions, otherwise all traits would need to be considered in the solution to all problems, and vice versa, leading to a kind of complex systems analysis of the whole organism in its total environment. The entire trend of adaptive evolutionary arguments is toward a Cartesian analysis into separate parts, each with its separate function.

The third hidden assumption is that all aspects of an organism are adaptive. The methodological program of adaptive explanation demands an a priori commitment to such explanations for all traits that can be described. This commitment establishes the problematic of the science, which is to *find* the adaptation, not to ask whether it exists at all. The problematic is an inheritance from the concept of the world as having been designed by a rational creator so that all aspects of it have a function and can be rationalized. The problem of explanation is to reveal the workings of this rational system.

It is in the assumption that all traits, arbitrarily described, are adaptive that the weakness of evolutionary theory is manifest. If the assumption is allowed to stand, then adaptive explanations simply become a test of the ingenuity of theorists and the tolerance of intellectuals for tortured and absurd stories. Again, it is in behavioral traits that the greatest scope for rationalization appears, as, for example, explanations of the supposed mass suicide of the lemmings by drowning, as being a population regulation device that is adaptive for the species as a whole. If, on the other hand, the assumption is dropped, then traits that are difficult to rationalize can be declared nonadaptive, allowing evolutionists to explain just those traits that seem most obviously to fit their mode of explanation, relegating the others to the category of "non-Darwinian" (King and Jukes, 1969). A large part

of the variation in protein structure between species is now regarded as random, irrational, and non-Darwinian by some evolutionists (Kimura and Ohta, 1971), but this is bitterly contested by conventional Darwinians who accept that adaptationist methodological program without reserve (Ford, 1975).

Given the assumptions of the adaptationist program, there are great difficulties and ambiguities in determining the adaptation of a given organ. Every trait is involved in a variety of functions; yet it cannot be claimed that it is an adaptation for all. Thus a whale's flipper can destroy a small whaling boat, but no one would argue that the flipper is an adaptation for destroying surface predators rather than for swimming. Nor does the habitual and "natural" use of an organ necessarily imply that it is an adaptation for that purpose. The Green Turtle, *Chelonia mylas,* uses its front flippers to propel itself over dry sand to an egg-laying site above high-water mark, and then digs a deep hole for the eggs in a slow and clumsy way using its hind flippers as a trowel. But this use of these swimming paddles is *faute de mieux,* and they cannot be regarded as adaptations either to land locomotion or hole digging. If sufficiency of an organ is not a sufficient condition of its being an adaptation, neither is necessity of an organ a necessary condition. Every terrestrial animal above the size of an insect must have lungs because the passive transpiration of gases across the skin or by a tracheal system would not suffice for respiration in a large volume. Lungs can properly be said to be an adaptation for breathing because without them, the animal would suffocate. But most adaptations are not so essential. The striping of zebras may be an adaptation to protective camouflage in tall grass, but is by no means certain that a species of un-striped zebras would become extinct from predation, or even that they would be less numerous.

The problem of judging the adaptive importance of a trait from its use becomes more difficult when the use must itself be reconstructed. The bony plates of *Stegosaurus* may have been used for temperature regulation, predator protection, and species recognition—all simultaneously. Nor is this doubt restricted to extinct forms. Modern lizards have erectile "sails" along their dorsal lines and inflatable gular pouches that are brightly colored. These may serve as both aggressive display and sexual recognition signals, and the dorsal spines may also be heat regulators. In principle, experiments can be done on living lizards to determine the effect of removal or alteration of these characters, but in practice the interpretation of such alterations is dangerous, since it is not clear whether the alteration has introduced an extraneous variable. Even if an organ could be shown to function in a variety of ways, the question of its adaptation is not settled because of the implied historical causation in the theory of adaptation. The judgment whether the gular pouch of a lizard is an adaptation for species recognition depends upon whether natural selection is supposed to have operated on the pouch as a recognition signal through the more frequent correct matings of individuals with the pouch. If, when the pouch reached a certain size, it also,

incidentally, frightened predators, the pouch would be a preadaptation for this latter purpose. The distinction between those uses for which an organ or trait is an adaptation and those for which it is a preadaptation could be made only on historical grounds by a reconstruction of the actual forces of natural selection which, even for extant organisms, is impossible.

In the absence of actual historical data on natural selection, the argument that a trait is an adaptation rests on an analysis of the organism as a machine for solving postulated problems. By using principles of engineering, a design analysis is performed, and the characteristics of the postulated design are compared with the organ in question. Thus the postulate that the dorsal plates of *Stegosaurus* are indeed adaptations for heat exchange rests on the porous nature of the bone, suggesting a large amount of blood circulation; the larger size of the plates over the most massive part of the body where heat production is greatest; the alternating unpaired arrangement of the plates to the left and right side of the midline, suggesting the proper placement of cooling fins; and the constriction of the plates at their base, where they are nearest the heat source and thus are inefficient radiators. A more quantitative engineering analysis is sometimes made, proposing that the organ or character is actually optimal for its postulated purpose. Thus Leigh (1971), on the basis of hydrodynamic principles, showed that sponges were the optimal shape on the supposition that the problem for the sponge was to process the maximum amount of food-containing water per unit time. The fit is not always perfect, however. Orians (1976) has calculated the optimal distribution of food-item sizes for a bird that must search and catch prey and then return with it to a nest ("central place foraging"). A comparison of the prey caught with the distribution of available prey sizes did, indeed, show that birds do not take food items at random but are biased toward larger items, although they did not behave according to the calculated optimum. The explanation offered for the failure of a close fit is that the birds spend less time searching for optimal prey than they should if the behavior were a pure adaptation to feeding efficiency because of the competing demand to visit the nest often enough to discourage predators. This is a paradigm for adaptive reconstruction. The problem is orginally posed as efficiency of food gathering. A deviation of behavior from random, in the direction predicted, is regarded as strong support for the adaptive explanation of the behavior, and the discrepancy from the predicted optimum is accounted for by an ad hoc secondary problem which acts as a constraint on the solution to the first. There is no methodological rule that instructs the theorist in how far the deviation of observation from prediction must be before the original adaptive explanation is abandoned altogether. By allowing the theorist to postulate various combinations of "problems" to which manifest traits are optimal "solutions," the adaptationist program makes of adaptation a metaphysical postulate, not only incapable of refutation but necessarily confirmed by every observation. This is the caricature that was immanent in

Darwin's insight that evolution is the product of natural selection.

## NATURAL SELECTION AND ADAPTATION

A sufficient mechanism for evolution by natural selection is contained in three propositions:

a. There is variation in morphological, physiological, and behavioral traits among members of a species (the principle of variation);
b. the variation is in part heritable, so that individuals resemble their relations more than they resemble unrelated individuals, and, in particular, offspring resemble their parents (the principle of heredity);
c. different variants leave different numbers of offspring either in immediate or in remote generations (the principle of differential fitness).

Any trait for which these three principles apply may be expected to evolve. That is, the frequency of different variant forms in the species will change, although it does not follow in all cases that one form of the trait will displace all others. There may be stable intermediate equilibria at which two or more variant forms coexist at a characteristic stationary frequency. It is important to note that all three conditions are necessary as well as sufficient conditions for evolution by natural selection. If there is no differential reproductive success of different variants, then, of course, there is no natural selection, but especially essential is the existence of *heritable variation.* If variation exists but is not passed from parent to offspring, then the differential reproductive successes of different forms is irrelevant, since all forms will produce the same distribution of types in the next generation.

These necessary and sufficient principles for evolution by natural selection contain no reference to adaptation. The postulate of adaptation was added as a fourth proposition by Darwin as an explanation of the mechanical cause of the phenomenon of differential reproduction and survival. The "struggle for existence," according to Darwin, was the result of the tendency of species to reproduce in excess of the resources available to them, an idea he got from reading Malthus' *Essay on Population*. The struggle would result in the victory of those individuals whose morphologies, physiologies, and behaviors allowed them to appropriate a greater share of the resources in short supply, or who could survive and reproduce on a lower resource level, or who could utilize a resource that was unsuitable for their competitors. These last two forms of the struggle for existence freed the principle from dependence upon actual struggle between individuals:

> I should premise that I use the term Struggle for Existence in a large and metaphorical sense . . . Two canine animals in a time of dearth may be truly said to struggle with each other which shall get food and live. But a plant at the edge of the desert is said to struggle for life against the drought.
>
> [*On the Origin of Species*, 1859. p. 62]

Given this struggle in its "large and metaphorical sense," it should be possible to predict which of two individuals will better survive and reproduce by an engineering analysis. So by a study of the bones and muscles of the legs of a zebra and by the application of simple mechanical principles, it should be possible to predict which of two zebras could run faster and therefore better escape predators. Furthermore it is in principle possible to predict the direction of evolution of leg muscles and bones by a local differential analysis, since for any two slightly different shapes the superior one can be discerned.

The struggle for existence also redirects the idea of adaptation from an absolute to a relative criterion. So long as organisms are considered only in relation to their ecological niche, they are either adapted, in which case they will persist, or they are unadapted and on their way to extinction. But if individuals of the same species are considered in relation to each other, they are competing for the same set of resources or struggling to reproduce in the same unfavorable environment (the plants at the edge of the desert), and their relative adaptation becomes relevant. Two forms of the same species might both be absolutely adapted in the sense that the species might persist if it were made up entirely of either form, yet when placed in competition the greater adaptation of one would lead to the extinction of the other. By the same consideration, the relative adaptation of two distinct species cannot, in general, be considered because species are never competing with each other in the same exclusionary way as forms of the same species. If two species did indeed overlap so much in their ecological niches that their abundances were critically determined by the same limiting resource, one would become extinct in the competition. Occasionally, of course, one species does extinguish another following a new introduction, as, for example, the Mediterranean fruit fly which was extinguished in eastern Australia by the sudden southward spread of the Queensland fruit fly, a very close relative laying its eggs in the same cultivated fruit.

The engineering approach to differential fitness, at first sight, removes the apparent tautology in the theory of natural selection. Without this design analysis, Darwinian theory would simply state that the more fit individuals leave more offspring in future generations and then make the determination of relative fitness from the number of offspring left by different individuals. Since in a finite world of contingent events, there will always be some individuals who will, even by chance, leave more offspring than others, then a posteriori there will be tautological fitness differences between individuals. All that can be said is that evolution occurs because evolution occurs. The design analysis, however, makes a priori fitness determination possible and therefore the judgment of relative adaptation of two forms can be made in the absence of any prior knowledge of their reproductive performances. Or can it?

The conditions for evolutionary prediction from relative adaptation analysis are the same as for judging absolute adaptation. A change in length of the long bones of zebras' legs, allowing them to run faster, will be favored in evolution provided (a) that running speed is really the "problem" to be solved by the zebra, (b) that the change in speed does not have countervailing adverse effects on the animal's adaptation to "solving" other "problems" set by the environment, and (c) that lengthening the bone does not produce countervailing direct developmental or physiological effects on other organs or on its own function. Even though lions prey on zebra, it is not necessarily true that faster zebra will escape more easily, since it is by no means certain that lions are speed limited in their ability to catch prey. Moreover, greater speed may be at the expense of metabolic efficiency, so that if zebra are food limited, the "problem" of feeding may be made worse for the zebra by "solving" the "problem" of escaping from predators. Finally, longer shank bones may be more easily broken, cost more developmental energy to produce, and create a whole series of problems of integrated morphology. Relative adaptation, like the judgment of absolute adaptation, must be a *ceteris paribus* argument, and since all other things are never equal, the final judgment as to whether a particular change in a trait produces relatively greater adaptation will depend upon the net effect on the entire organism. The alternative would be to maintain that the engineering analysis in relation to a predetermined "problem" will be taken as *defining* adaptation, irrespective of whether it is of net benefit to the organism. Such a solution would decouple adaptation from evolution and make it into a pure intellectual game.

The serious methodological difficulties in the use of adaptive arguments should not blind us to the fact that many features of organisms clearly are adaptations to obvious environmental "problems." It is no accident that fish have fins, aquatic mammals have altered their appendages to form finlike flippers, ducks, geese, and seabirds have webbed feet, penguins have paddle-like wings, and even seasnakes, lacking fins, are flattened in cross-section. It is obvious that all these traits are adaptations for aquatic locomotion, so that the reproductive fitness of the ancestor of these forms must have been increased by the gradual modification of their appendages in an adaptive way. It follows that the *ceteris paribus* argument must be true reasonably often or else no progressive alteration to form such structures could occur. Therefore, the mapping of character states into net reproductive fitness must have two characteristics: *continuity* and *quasi-independence*. By continuity we mean that very small changes in a character result in very small changes in the ecological relations of the organism and therefore very small changes in reproductive fitness. Neighborhoods in character-space map into neighorhoods in fitness-space. So a very slight change in the shape of a fin or mammalian appendage to make it finlike cannot cause a dramatic change in the sexual recognition pattern, or make the organism attractive to a completely new set of predators. By quasi-independence we mean that there exist a large variety of developmental paths by which a given character may

change, and although some of these may give rise to countervailing changes in other organs and in other aspects of the ecological relations of the organism, a non-negligible proportion of these paths will not result in countervailing effects of sufficient magnitude to overcome the increase in fitness from the adaptation. In genetic terms, quasi-independence means that a variety of mutations may occur all with the same effect on the primary character but with different effects on other characters, some set of which will not be at a net disadvantage.

## NON-ADAPTIVE CHARACTERS AND THE FAILURE OF ADAPTATION

While the principles of continuity and quasi-independence can be used to explain adaptive trends in characters that have actually occurred, they cannot be used indiscriminately to assert that all characters are adaptive, or that a character that ought to evolve because it would be adaptive will necessarily make its appearance. The lack of continuity and quasi-independence may, in fact, be powerful deterrents to adaptive trends. That adaptation has occurred seems obvious. That it does so most of the time or even very often is completely unclear. The adaptationist program is so much a part of the vulgarization of Darwinism that an increasing amount of evolutionary theory consists in the uncritical application of the program to both manifest and postulated traits of organisms. A paradigm is the argument by Wilson (1975) that indoctrinability ("Human beings are absurdly easy to indoctrinate

... they seek it"—p. 562) and blind faith ("Men would rather believe than know"—p. 561) are adaptive consequences of human evolution, since conformist individuals would more often submit to the common goals of the group, guaranteeing support rather than hostility and thus increasing their reproductive fitness. Two socially determined behaviors are universalized, made part of "human nature," and then an argument for their adaptive evolution is constructed. Putting aside the question of the universality of indoctrinability and blind faith, the claim that they are the product of adaptive evolution requires that there has been heritable variation for these traits in human evolutionary biology, that conformists really would leave more offspring, all other things being equal, and, finally, that all other things *are* equal. None of these propositions is susceptible of test. There is no evidence of any present genetic variation for conformism, but that is not compelling, since it is genetic variation in the evolutionary past that is required. Nor is there any reason to suppose that conformism is a separate trait and not simply a culturally defined concept that has been reified by the biologist. The alternative is to recognize that "conformism" is a "trait" only by abstract construction, that it is one of the possible ways of describing some aspect of the behavior of some individuals at some times, and that it is a consequence of the evolution of a complex central nervous system. That is, the adaptive trait is the extremely highly developed central nervous organization, and the appearance of conformity as a manifestation of that complexity is en-

tirely epiphenomenal. A parallel situation for morphological characters has long been recognized in the phenomenon of *allometry:* the rate of growth of different organs is different, so larger organisms do not have all their parts proportionately larger. For example, in primates tooth size grows larger more slowly between species than does body size, so that large primates have proportionately smaller teeth than small primates. This relationship is constant across all primates, and it would be erroneous to argue that for some special adaptive reason gorillas have been selected for relatively small teeth. Developmental correlations tend to be quite conservative in evolution, and many so-called "adaptive trends" turn out on closer examination to be purely allometric.

Reciprocally, the increase of certain traits in a population by natural selection is not in itself a guide to adaptation. A mutation that doubled the egg-laying rate in an insect, limited by the amount of food available to the immature stages, would very rapidly spread through the population. Yet the end result would be a population with the same adult density as before but twice the density of early immatures and much greater competition between larval stages. If there are periodic severe shortages of food, the probability of extinction of the population will be greater than it was when larval competition was less. Moreover, predators may switch their search images to the larvae of this species now that they are more abundant, and epidemic diseases may more easily spread. It would be difficult to say precisely to what environmental "problem" the increase in fecundity was a "solution."

## ADAPTATION AS IDEOLOGY

The caricature of Darwinian adaptationism that sees all characteristics, real or constructed, as optimal solutions to problems has more in common with the ideology of the sixteenth century than with that of the nineteenth. Before the rising power and eventual victory of the bourgeoisie, the state and unchanging world were seen and justified as a manifestation of divine will. The relations among men, and between men and nature, were unchangeably just and rational because the Author of All Things was unchanging and supremely just and rational. There was, moreover, an organic unity of relationships, for example of lord and serf and of both to the land, which could not be broken, since they were all part of an articulated plan. This ideology, which was at the same time a conscious legitimation of the social order and its unconscious product, necessarily came under attack by the ideologues of the increasingly powerful commercial bourgeoisie. The success of commercial and manufacturing interests required that men could rise as high in status and power as their entrepreneurial activities could take them, and that money, land, and labor power be freed of their traditional rigid relationships. It had to be possible to alienate land for primary production and, by the same process, to give over to the laborer possession of his own labor and the possibility to carry it to the centers of manufacturing where it could be sold in the labor market. Thus, the ideology of the Enlightment emphasized progress rather than stasis, becoming rather than being, and the freedom and disarticulation of parts of the world, rather than their indissoluble unity. Dr. Pangloss, who believed that even the death of thousands in the Lisbon earthquake was a proof that this was the "meilleur des mondes possibles," symbolized the foolishness of the old ideology. Descartes' *bête machine* and La Mettrie's *homme machine* provided the program for the analysis of nature by its dissection and disarticulation into separate causes and effects.

Darwin's work came at the end of the successful struggle of the bourgeoisie to make a world appropriate to its own activities. The middle of the nineteenth century was a time of immense expansion of production and of wealth. Darwin's maternal grandfather, Josiah Wedgwood, had started as a potter's apprentice and had become one of the great Midland industrialists who epitomized the flowering of an exuberant capitalism. Mechanical invention and a free labor market underlay the required growth of capital and the social and physical transformation of Europe. Herbert Spencer's *Progress: Its Law and Causes,* expressed the midnineteenth century belief in the inevitability of change and progress. The theory of the evolution of organic life that Darwin developed was an expression of these same ideological elements. It emphasized that change and instability were characteristic of the living world (and of the inorganic world as well, since the earth itself was being built up and broken down by geological processes).

Adaptation, for Darwin, was a process of becoming rather than a state of final optimality. Progress through successive improvement of mechanical relations was the characteristic of evolution in this scheme. It must be remembered that for Darwin, the existence of "organs of extreme perfection and complication" was a difficulty of his theory, not a proof of it. He called attention to the numerous rudimentary and imperfect forms of these organs that were present in living species. The idea that the analysis of living forms would show them, in general, to have optimal characters, would have been quite foreign to Darwin. A demonstration of universal optimality could only have been a blow against his progressivist theory and a return to ideas of special creation. At the end of *The Origin of Species,* he wrote:

> When I view all beings not as special creations, but as lineal descendants of some few beings which lived long before the first bed of the Cambrian system was deposited, they seem to me to become ennobled. Judging from the past, we may safely infer that not one living species will transmit its unaltered likeness to a distant futurity ... And as natural selection works solely by and for the good of each being, all corporeal and mental endowments *will tend to progress toward perfection.*
>
> [p. 489]

Even as Darwin wrote, however, a "spectre was haunting Europe." The successful revolutions of the eighteenth century were in danger of being overturned by yet new revolu-

tions. The resistance by the now dominant bouregoisie to yet further social progress required a change in the legitimating ideology. It was indeed true that the rise of the middle classes had been a progressive one, but it was also the last progressive change. Liberal, democratic, entrepreneurial man was the highest form of civilization and, for them, reexamination of history would show that all along the development of society had been tending toward the present state of perfect adaptation. Pangloss was right, after all, only a bit premature. Liberal social theory of the last part of the nineteenth century and of the twentieth has emphasized dynamic equilibrium and optimality. Individuals may rise and fall in the social system, but the system itself is seen as stable, and as close to perfect as any system can be. It is efficient, just, and productive of the greatest good for the greatest number. At the same time, the Cartesian mechanical analysis by disarticulation of parts and separation of causes has been maintained from the earlier world view. The ideology of equilibrium and dynamic stability characterizes modern evolutionary theory as much as it does bourgeois economics and political theory. Whig history is mimicked by Whig biology. The modern adaptationist program with its attempt to demonstrate that organisms are at or near their expected optima, leads to the consequence that although species come into existence and become extinct, nothing really new is happening in evolution. In contrast to Darwin, modern adaptationists regard the existence of optimal structures, perfect adaptation, as the evidence of evolution by natural selection. There is no progress, because there is nothing to improve. Natural selection simply keeps the species from falling too far behind the constantly but slowly changing environment. There is a striking similarity between this view of evolution and the claim that modern market society is the most rational organization possible, that although individuals may rise or fall in the social hierarchy on their individual merits, there is a dynamic equilibrium of social classes and that technological and social change occur only insofar as they are needed to keep up with a decaying environment.

## REFERENCES

Darwin, Ch. S., 1859, *On the Origin of Species by Means of Natural Selection,* London, Murray.

Engels, F., 1876, *Der Anteil der Arbeit an der Menschwerdung des Affen,* in *Die Neue Zeit,* XIV, 1896, 2, 545–54.

Ford, E. B., 1975, *Ecological Genetics,* London, Chapman and Hall.

Kimura, M. and Ohta, T., 1971, *Theoretical Aspects of Population Genetics,* Princeton, N.J., Princeton University Press.

King, J. L. and Jukes, T. H., 1965, Non-Darwinian evolution: random fixation of selectively neutral mechanisms, in *Science,* 164: 788–98.

Leigh, E., 1971, *Adaptation and Diversity,* San Francisco, Freeman.

Orians, G., 1976, The Strategy of central-place foraging, in *Analysis of Ecological Systems,* Columbus, Ohio, Ohio State University Press.

Spencer, H., 1857, Progress: Its laws and cause, in *Westminster Review,* 17: 445–85.

Trivers, R., 1974, Parent offspring conflict, in *American Zoologist,* 14: 249–64.

Wilson, E. O., 1975, *Sociobiology: The New Synthesis,* Cambridge, Mass., Harvard University Press.

# 15

# The Spandrels of San Marco and the Panglossian Paradigm: A Critique of the Adaptationist Programme

STEPHEN JAY GOULD AND RICHARD C. LEWONTIN

*An adaptationist programme has dominated evolutionary thought in England and the United States during the past forty years. It is based on faith in the power of natural selection as an optimizing agent. It proceeds by breaking an organism into unitary "traits" and proposing an adaptive story for each considered separately. Trade-offs among competing selective demands exert the only brake upon perfection; nonoptimality is thereby rendered as a result of adaptation as well. We criticize this approach and attempt to reassert a competing notion (long popular in continental Europe) that organisms must be analyzed as integrated wholes, with Baupläne so constrained by phyletic heritage, pathways of deveopment, and general architecture that the constraints themselves become more interesting and more important in delimiting pathways of change than the selective force that may mediate change when it occurs. We fault the adaptationist programme for its failure to distinguish current utility from reasons for origin (male tyrannosaurs may have used their diminutive front legs to titillate female partners, but this will not explain why they got so small); for its unwillingness to consider alternatives to adaptive stories; for its reliance upon plausibility alone as a criterion for accepting speculative tales; and for its failure to consider adequately such competing themes as random fixation of alleles, production of nonadaptive structures by developmental correlation with selected features (allometry, pleiotropy, material compensation, mechanically forced correlation), the separability of adaptation and selection, multiple adaptive peaks, and current utility as an epiphenomenon of nonadaptive structures. We support Darwin's own pluralistic approach to identifying the agents of evolutionary change.*

## 1. INTRODUCTION

The great central dome of St. Mark's Cathedral in Venice presents in its mosaic design a detailed iconography expressing the mainstays of Christian faith. Three circles of figures radiate out from a central image of Christ: angels, disciples, and virtues. Each circle is divided into quadrants, even though the dome itself is radially symmetrical in structure. Each quad-

Figure 1. One of the four spandrels of St Mark's; seated evangelist above, personification of river below.

rant meets one of the four spandrels in the arches below the dome. Span-drels—the tapering triangular spaces formed by the intersection of two rounded arches at right angles (figure 1)—are necessary architectural by-products of mounting a dome on rounded arches. Each spandrel contains a design admirably fitted into its tapering space. An evangelist sits in the upper part flanked by the heavenly cities. Below, a man representing one of the four biblical rivers (Tigris, Euphrates, Indus, and Nile) pours water from a pitcher in the narrowing space below his feet.

The design is so elaborate, harmonious, and purposeful that we are tempted to view it as the starting point of any analysis, as the cause in some sense of the surrounding architecture. But this would invert the proper path of analysis. The system begins with an architectural constraint: the necessary four spandrels and their tapering triangular form. They provide a space in which the mosaicists worked; they set the quadripartite symmetry of the dome above.

Figure 2. The ceiling of King's College Chapel.

Such architectural constraints a-bound, and we find them easy to understand because we do not impose our biological biases upon them. Every fan-vaulted ceiling must have a series of open spaces along the midline of the vault, where the sides of the fans intersect between the pillars (figure 2). Since the spaces must exist, they are often used for ingenious ornamental effect. In King's College Chapel in Cambridge, for example, the spaces contain bosses alternately embellished with the Tudor rose and portcullis. In a sense, this design represents an "adaptation," but the architectural constraint is clearly primary. The spaces arise as a necessary by-product of fan vaulting; their appropriate use is a secondary effect. Anyone who tried to argue that the structure exists because the alternation of rose and portcullis makes so much sense· in a Tudor chapel would be inviting the same ridicule that Voltaire heaped on Dr Pangloss: "Things cannot be other than they are . . . Everything is made for the best purpose. Our noses were made to carry spectacles, so we have spectacles. Legs were clearly intended for breeches, and we wear them." Yet evolutionary biologists, in their tendency to focus exclusively on immediate adaptation to local conditions, do tend to ignore architectural constraints and perform just such an inversion of explanation.

As a closer example, recently featured in some important biological literature on adaptation, anthropologist Michael Harner has proposed (1977) that Aztec human sacrifice arose as a solution to chronic shortage of meat (limbs of victims were often consumed, but only by people of high status). E. O. Wilson (1978) has used this explanation as a primary illustration of an adaptive, genetic predisposition for carnivory in humans. Harner and Wilson ask us to view an elaborate social system and a complex set of explicit justifications involving myth, symbol, and tradition as mere epiphenomena generated by the Aztecs as an unconscious rationalization masking the "real" reason for it all: need for protein. But Sahlins (1978) has argued that human sacrifice represented just one part of an elaborate cultural fabric that, in its entirety, not only represented the material expression of Aztec cosmology, but also performed such utilitarian functions as the maintenance of social ranks and systems of tribute among cities.

We strongly suspect that Aztec cannibalism was an "adaptation" much like evangelists and rivers in spandrels, or ornamented bosses in ceiling spaces: a secondary epiphnomenon representing a fruitful use of available parts, not a cause of the entire system. To put it crudely: a system developed for other reasons generated an increasing number of fresh bodies; use might as well be made of them. Why invert the whole system in such a curious fashion and view an entire culture as the epiphe-nomenon of an unusual way to beef up the meat supply. Spandrels do not exist to house the evangelists. Moreover, as Sahlins argues, it is not even clear that human sacrifice was an adaptation at all. Human cultural practices can be orthogenetic and drive toward extinction in ways that Darwinian processes, based on genetic selection, cannot. Since each new monarch had to outdo his predecessor in even more elaborate and copious sacrifice, the practice was beginning to stretch resources to the breaking point. It would not have been the first time that a human culture did itself in. And, finally, many experts doubt Harner's premise in the first place (Ortiz de Montellano, 1978). They argue that other sources of protein were not in short supply, and that a practice awarding meat only to privileged people who had enough anyway, and who used bodies so inefficiently (only the limbs were consumed, and partially at that) represents a mighty poor way to run a butchery.

We deliberately chose nonbiological examples in a sequence running from remote to more familiar: architecture to anthropology. We did this because the primacy of architectural constraint and the epiphenomenal nature of adaptation are not obscured by our biological prejudices in these examples. But we trust that the message for biologists will not go unheeded: if these had been biological systems, would we not, by force of habit, have regarded the epiphenomenal adaptation as primary and tried to build the whole structural system from it?

## 2. THE ADAPTATIONIST PROGRAMME

We wish to question a deeply engrained habit of thinking among students of evolution. We call it the adaptationist programme, or the Panglossian paradigm. It is rooted in a notion popularized by A. R. Wallace and A. Weismann, (but not, as we shall see, by Darwin) toward the end of the nineteenth century: the near omnipotence of natural selection in forging organic design and fashioning the best among possible worlds. This programme regards natural selection as so powerful and the constraints upon it so few that direct production of adaptation through its operation becomes the primary cause of nearly all organic form, function, and behavior. Constraints upon the pervasive power of natural selection are recognized of course (phyletic inertia primarily among them, although immediate architectural constraints, as discussed in the last section, are rarely acknowledged). But they are usually dismissed as unimportant or else, and more frustratingly, simply acknowledged and then not taken to heart and invoked.

Studies under the adaptationist programme generally proceed in two steps:

(1) An organism is atomized into "traits" and these traits are explained as structures optimally designed by natural selection for their functions. For lack of space, we must omit an extended discussion of the vital issue "What is a trait?" Some evolutionists may regard this as a trivial, or merely a semantic problem. It is not. Organisms are integrated entities, not collections of discrete objects. Evolutionists have often been led astray by inappropriate atomization, as D'Arcy Thompson (1942) loved to point out. Our favorite example involves the human chin (Gould, 1977, pp. 381–382; Lewontin, 1978). If we regard the chin as a "thing," rather than as a product of interaction between two growth fields (alveolar and mandibular), then we are led to an interpretation of its origin (recapitulatory) exactly opposite to the one now generally favored (neotenic).

(2) After the failure of part-by-part optimization, interaction is acknowledged via the dictum that an organism cannot optimize each part without imposing expenses on others. The notion of "trade-off" is introduced, and organisms are interpreted as best compromises among competing demands. Thus interaction among parts is retained completely within the adaptationist programme. Any suboptimality of a part is explained as its contribution to the best possible design for the whole. The notion that suboptimality might represent anything other than the immediate work of natural selection is usually not entertained. As Dr Pangloss said in explaining to Candide why he suffered from venereal disease: "It is indispensable in this best of worlds. For if Columbus, when visiting the West Indies, had not caught this disease, which poisons the source of generation, which frequently even hinders generation, and is clearly opposed to the great end of Nature, we should have neither chocolate nor cochineal." The adaptationist programme is truly Panglossian. Our world may not be good in an abstract sense, but it is the very best we could have. Each trait plays its part and must be as it is.

At this point, some evolutionists will protest that we are caricaturing their view of adaptation. After all, do they not admit genetic drift, allometry, and a variety of reasons for nonadaptive evolution? They do, to be sure, but we make a different point. In natural history, all possible things happen sometimes; you generally do not support your favored phenomenon by declaring rivals impossible in theory. Rather, you acknowledge the rival but circumscribe its domain of action so narrowly that it cannot have any importance in the affairs of nature. Then, you often congratulate yourself for being such an undogmatic and ecumenical chap. We maintain that alternatives to selection for best overall design have generally been relegated to unimportance by this mode of argument. Have we not all heard the catechism about genetic drift: it can only be important in populations so small that they are likely to become extinct before playing any sustained evolutionary role (but see Lande, 1976).

The admission of alternatives in principle does not imply their serious consideration in daily practice. We all say that not everything is adaptive; yet, faced with an organism, we tend to break it into parts and tell adaptive stories as if trade-offs among competing, well designed parts were the only constraint upon perfection for each trait. It is an old habit. As Romanes complained about A. R. Wallace in 1900: "Mr. Wallace does not expressly maintain the abstract impossibility of laws and causes other than those of utility and natural selection . . . Nevertheless, as he nowhere recognizes any other law or cause . . . he prac-

tically concludes that, on inductive or empirical grounds, there is *no* such other law or cause to be entertained."

The adaptationist programme can be traced through common styles of argument. We illustrate just a few; we trust they will be recognized by all:

(1) If one adaptive argument fails, try another. Zig-zag commissures of clams and brachiopods, once widely regarded as devices for strengthening the shell, become sieves for restricting particles above a given size (Rudwick, 1964). A suite of external structures (horns, antlers, tusks) once viewed as weapons against predators, become symbols of intraspecific competition among males (Davitashvili, 1961). The eskimo face, once depicted as "cold engineered" (Coon, et al., 1950), becomes an adaptation to generate and withstand large masticatory forces (Shea, 1977). We do not attack these newer interpretations; they may all be right. We do wonder, though, whether the failure of one adaptive explanation should always simply inspire a search for another of the same general form, rather than a consideration of alternatives to the proposition that each part is "for" some specific purpose.

(2) If one adaptive argument fails, assume that another must exist; a weaker version of the first argument. Costa & Bisol (1978), for example, hoped to find a correlation between genetic polymorphism and stability of environment in the deep sea, but they failed. They conclude (1978, pp. 132, 133): "The degree of genetic polymorphism found would seem to indicate absence of correlation with the particular environmental

factors which characterize the sampled area. The results suggest that the adaptive strategies of organisms belonging to different phyla are different."

(3) In the absence of a good adaptive argument in the first place, attribute failure to imperfect understanding of where an organism lives and what it does. This is again an old argument. Consider Wallace on why all details of color and form in land snails must be adaptive, even if different animals seem to inhabit the same environment (1899, p. 148): "The exact proportions of the various species of plants, the numbers of each kind of insect or of bird, the peculiarities of more or less exposure to sunshine or to wind at certain critical epochs, and other slight differences which to us are absolutely immaterial and unrecognizable, may be of the highest significance to these humble creatures, and be quite sufficient to require some slight adjustments of size, form, or color, which natural selection will bring about."

(4) Emphasize immediate utility and exclude other attributes of form. Fully half the explanatory information accompanying the full-scale Fibreglass Tyrannosaurus at Boston's Museum of Science reads: "Front legs a puzzle: how Tyrannosaurus used its tiny front legs is a scientific puzzle; they were too short even to reach the mouth. They may have been used to help the animal rise from a lying position." (We purposely choose an example based on public impact of science to show how widely habits of the adaptationist programme extend. We are not using glass beasts as straw men; similar arguments and relative emphases, framed in different words, appear regularly in the professional literature.) We don't doubt that Tyrannosaurus used its diminutive front legs for something. If they had arisen de novo, we would encourage the search for some immediate adaptive reason. But they are, after all, the reduced product of conventionally functional homologues in ancestors (longer limbs of allosaurs, for example). As such, we do not need an explicitly adaptive explanation for the reduction itself. It is likely to be a developmental correlate of allometric fields for relative increase in head and hindlimb size. This nonadaptive hypothesis can be tested by conventional allometric methods (Gould, 1974, in general; Lande, 1978, on limb reduction) and seems to us both more interesting and fruitful than untestable speculations based on secondary utility in the best of possible worlds. One must not confuse the fact that a structure is used in some way (consider again the spandrels, ceiling spaces, and Aztec bodies) with the primary evolutionary reason for its existence and conformation.

## 3. TELLING STORIES

All this is a manifestation of the rightness of things, since if there is a volcano at Lisbon it could not be anywhere else. For it is impossible for things not to be where they are, because everything is for the best.
———[Dr Pangloss on the great Lisbon earthquake of 1755, in which up to 50,000 people lost their lives]

We would not object so strenuously to the adaptationist programme if its invocation, in any particular case, could lead in principle to its rejection for want of evidence. We might still view it as restrictive and object to its status as an argument of first choice. But if it could be dismissed after failing some explicit test, then alternatives would get their chance. Unfortunately, a common procedure among evolutionists does not allow such definable rejection for two reasons. First, the rejection of one adaptive story usually leads to its replacement by another, rather than to a suspicion that a different kind of explanation might be required. Since the range of adaptive stories is as wide as our minds are fertile, new stories can always be postulated. And if a story is not immediately available, one can always plead temporary ignorance and trust that it will be forthcoming, as did Costa & Bisol (1978), cited above. Secondly, the criteria for acceptance of a story are so loose that many pass without proper confirmation. Often, evolutionists use *consistency* with natural selection as the sole criterion and consider their work done when they concoct a plausible story. But plausible stories can always be told. The key to historical research lies in devising criteria to identify proper explanations among the substantial set of plausible pathways to any modern result.

We have, for example (Gould, 1978) criticized Barash's (1976) work on aggression in mountain bluebirds for this reason. Barash mounted a stuffed male near the nests of two pairs of bluebirds while the male was out foraging. He did this at the same nests on three occasions at ten-day intervals: the first before eggs were laid, the last two afterwards. He then counted aggressive approaches of the returning male toward both the model and the female. At time one, aggression was high toward the model and lower toward females but substantial in both nests. Aggression toward the model declined steadily for times two and three and plummeted to near zero toward females. Barash reasoned that this made evolutionary sense, since males would be more sensitive to intruders before eggs were laid than afterward (when they can have some confidence that their genes are inside). Having devised this plausible story, he considered his work as completed (1976, pp. 1099, 1100):

The results are consistent with the expectations of evolutionary theory. Thus aggression toward an intruding male (the model) would clearly be especially advantageous early in the breeding season, when territories and nests are normally defended . . . The initial aggressive response to the mated female is also adaptive in that, given a situation suggesting a high probability of adultery (i.e., the presence of the model near the female) and assuming that replacement females are available, obtaining a new mate would enhance the fitness of males . . . The decline in male-female aggressiveness during incubation and fledgling stages could be attributed to the impossibility of being cuckolded after the eggs have been laid . . . The results are consistent with an evolutionary interpretation.

They are indeed consistent, but what about an obvious alternative, dismissed without test by Barash? Male returns at times two and three, approaches the model, tests it a bit, recognizes it as the same phoney he saw before, and doesn't bother his female. Why not at least perform the obvious test for this alternative to a conventional adaptive story: expose a male to the model for the *first* time after the eggs are laid?

After we criticized Barash's work, Morton et al. (1978) repeated it, with some variations (including the introduction of a female model), in the closely related eastern bluebird *Sialia sialis*. "We hoped to confirm", they wrote, that Barash's conclusions represent "a widespread evolutionary reality, at least within the genus *Sialia*. Unfortunately, we were unable to do so." They found no "anticuckoldry" behavior at all: males never approached their females aggressively after testing the model at any nesting stage. Instead, females often approached the male model and, in any case, attacked female models more than males attacked male models. "This violent response resulted in the near destruction of the female model after presentations and its complete demise on the third, as a female flew off with the model's head early in the experiment to lose it for us in the brush" (1978, p. 969). Yet, instead of calling Barash's selected story into question, they merely devise one of their own to render both results in the adaptationist mode. Perhaps, they conjecture, replacement females are scarce in their species and abundant in Barash's. Since Barash's males can replace a potentially "unfaithful" female, they can afford to be choosy and possessive. Eastern bluebird males

are stuck with uncommon mates and had best be respectful. They conclude: "If we did not support Barash's suggestion that male bluebirds show anticuckoldry adaptations, we suggest that both studies still had 'results that are consistent with the expectations of evolutionary theory' (Barash 1976, p. 1099), as we presume any careful study would." But what good is a theory that cannot fail in careful study (since by 'evolutionary theory', they clearly mean the action of natural selection applied to particular cases, rather than the fact of transmutation itself)?

## 4. THE MASTER'S VOICE RE-EXAMINED

Since Darwin has attained sainthood (if not divinity) among evolutionary biologists, and since all sides invoke God's allegiance, Darwin has often been depicted as a radical selectionist at heart who invoked other mechanisms only in retreat, and only as a result of his age's own lamented ignorance about the mechanisms of heredity. This view is false. Although Darwin regarded selection as the most important of evolutionary mechanisms (as do we), no argument from opponents angered him more than the common attempt to caricature and trivialize his theory by stating that it relied exclusively upon natural selection. In the last edition of the *Origin*, he wrote (1872, p. 395):

As my conclusions have lately been much misrepresented, and it has been stated that I attribute the modification of species exclusively to natural selection, I may be permitted to remark that in the

*Darwin*

first edition of this work, and subsequently, I placed in a most conspicuous position—namely at the close of the Introduction—the following words: "I am convinced that natural selection has been the main, but not the exclusive means of modification." This has been of no avail. Great is the power of steady misinterpretation.

Romanes, whose once famous essay (1900) on Darwin's pluralism versus the panselectionism of Wallace and Weismann deserves a resurrection, noted of this passage (1900, p. 5): "In the whole range of Darwin's writings there cannot be found a passage so strongly worded as this: it presents the only note of bitterness in all the thousands of pages which he has published." Apparently, Romanes did not know the letter Darwin wrote to *Nature* in 1880, in which he castigated Sir Wyville Thomson for caricaturing his theory as panselectionist (1880, p. 32):

I am sorry to find that Sir Wyville Thomson does not understand the principle of natural selection . . . If he had done so, he could not have written the following sentence in the Introduction to the Voyage of the Challenger: "The character of the abyssal fauna refuses to give the least support to the theory which refers the evolution of species to extreme variation guided only by natural selection." This is a standard of criticism not uncommonly reached by theologians and metaphysicians when they write on scientific subjects, but is something new as coming from a naturalist . . . Can Sir Wyville Thomson

name any one who has said that the evolution of species depends only on natural selection? As far as concerns myself, I believe that no one has brought forward so many observations on the effects of the use and disuse of parts, as I have done in my "Variation of Animals and Plants under Domestication"; and these observations were made for this special object. I have likewise there adduced a considerable body of facts, showing the direct action of external conditions on organisms.

We do not now regard all of Darwin's subsidiary mechanisms as significant or even valid, though many, including direct modification and correlation of growth, are very important. But we should cherish his consistent attitude of pluralism in attempting to explain Nature's complexity.

### 5. A PARTIAL TYPOLOGY OF ALTERNATIVES TO THE ADAPTATIONIST PROGRAMME

In Darwin's pluralistic spirit, we present an incomplete hierarchy of alternatives to immediate adaptation *alternatives* for the explanation of form, function, *b* and behavior.

(1) No adaptation and no selection *1.* at all. At present, population geneticists are sharply divided on the question of how much genetic polymorphism within populations and how much of the genetic differences between species is, in fact, the result of natural selection as opposed to purely random factors. Populations are finite in size, and the isolated populations that form the first step in the speciation process are often

founded by a very small number of individuals. As a result of this restriction in population size, frequencies of alleles change by *genetic drift,* a kind of random genetic sampling error. The stochastic process of change in gene frequency by random genetic drift, including the very strong sampling process that goes on when a new isolated population is formed from a few immigrants, has several important consequences. First, populations and species will become genetically differentiated, and even fixed for different alleles at a locus in the complete absence of any selective force at all.

Secondly, alleles can become fixed in a population *in spite of natural selection.* Even if an allele is favored by natural selection, some proportion of populations, depending upon the product of population size $N$ and selection intensity $s$, will become homozygous for the less fit allele because of genetic drift. If $Ns$ is large, this random fixation for unfavorable alleles is a rare phenomenon, but if selection coefficients are on the order of the reciprocal of population size ($Ns = 1$) or smaller, fixation for deleterious alleles is common. If many genes are involved in influencing a metric character like shape, metabolism, or behavior, then the intensity of selection on each locus will be small and $Ns$ per locus may be small. As a result, many of the loci may be fixed for nonoptimal alleles.

Thirdly, new mutations have a small chance of being incorporated into a population, even when selectively favored. Genetic drift causes the immediate loss of most new mutations after their introduction. With a selection intensity $s$, a new favorable mutation has a probability of only $2s$ of ever being incorporated. Thus one cannot claim that, eventually, a new mutation of just the right sort for some adaptive argument will occur and spread. "Eventually" becomes a very long time if only one in 1,000 or one in 10,000 of the "right" mutations that do occur ever get incorporated in a population.

(2) No adaptation and no selection on the part at issue; form of the part is a correlated consequence of selection directed elsewhere. Under this important category, Darwin ranked his "mysterious" laws of the "correlation of growth." Today, we speak of pleiotropy, allometry, "material compensation" (Rensch, 1959, pp. 179–187) and mechanically forced correlations in D'Arcy Thompson's sense (1942; Gould 1971). Here we come face to face with organisms, as integrated wholes, fundamentally not decomposable into independent and separately optimized parts.

Although allometric patterns are as subject to selection as static morphology itself (Gould, 1966), some regularities in relative growth are probably not under immediate adaptive control. For example, we do not doubt that the famous 0.66 interspecific allometry of brain size in all major vertebrate groups represents a selected "design criterion," though its significance remains elusive (Jerison, 1973). It is too repeatable across too wide a taxonomic range to represent much else than a series of creatures similarly well designed for their different sizes. But another common allometry, the 0.2 to 0.4 intraspecific scaling among homeothermic adults differing in body size, or among races within a species, probably does not require a selectionist story, though many, including one of us, have tried to provide one (Gould, 1974). R. Lande (personal communication) has used the experiments of Falconer (1973) to show that selection upon *body size alone* yields a brain-body slope across generations of 0.35 in mice.

More compelling examples abound in the literature on selection for altering the timing of maturation (Gould, 1977). At least three times in the evolution of arthropods (mites, flies, and beetles), the same complex adaptation has evolved, apparently for rapid turnover of generations in strongly *r*-selected feeders on super-abundant but ephemeral fungal resources: females reproduce as larvae and grow the next generation within their bodies. Offspring eat their mother from inside and emerge from her hollow shell, only to be devoured a few days later by their own progeny. It would be foolish to seek adaptive significance in paedomorphic morphology per se; it is primarily a by-product of selection for rapid cycling of generations. In more interesting cases, selection for small size (as in animals of the interstitial fauna) or rapid maturation (dwarf males of many crustaceans) has occurred by progenesis (Gould, 1977, pp. 324–336), and descendant adults contain a mixture of ancestral juvenile and adult features. Many biologists have been tempted to find primary adaptive meaning for the mixture, but it probably arises as a by-product of truncated maturation, leaving some features "behind" in the larval state, while allowing others, more strongly

correlated with sexual maturation, to retain the adult configuration of ancestors.

(3) The decoupling of selection and adaptation.

(i) Selection without adaptation. Lewontin (1979) has presented the following hypothetical example: "A mutation which doubles the fecundity of individuals will sweep through a population rapidly. If there has been no change in efficiency of resource utilization, the individuals will leave no more offspring than before, but simply lay twice as many eggs, the excess dying because of resource limitation. In what sense are the individuals or the population as a whole better adapted than before? Indeed, if a predator on immature stages is led to switch to the species now that immatures are more plentiful, the population size may actually decrease as a consequence, yet natural selection at all times will favour individuals with higher fecundity."

(ii) Adaptation without selection. Many sedentary marine organisms, sponges and corals in particular, are well adapted to the flow régimes in which they live. A wide spectrum of "good design" may be purely phenotypic in origin, largely induced by the current itself. (We may be sure of this in numerous cases, when genetically identical individuals of a colony assume different shapes in different microhabitats.) Larger patterns of geographic variation are often adaptive and purely phenotypic as well. Sweeney and Vannote (1978), for example, showed that many hemimetabolous aquatic insects reach smaller adult size with reduced fecundity when they grow at temperatures above and below their optima.

Coherent, climatically correlated patterns in geographic distribution for these insects—so often taken as a priori signs of genetic adaptation—may simply reflect this phenotypic plasticity.

"Adaptation"—the good fit of organisms to their environment—can occur at three hierarchical levels with different causes. It is unfortunate that our language has focused on the common result and called all three phenomena "adaptation": the differences in process have been obscured, and evolutionists have often been misled to extend the Darwinian mode to the other two levels as well. First, we have what physiologists call "adaptation": the phenotypic plasticity that permits organisms to mold their form to prevailing circumstances during ontogeny. Human "adaptations" to high altitude fall into this category (while others, like resistance of sickling heterozygotes to malaria, are genetic, and Darwinian). Physiological adaptations are not heritable, though the capacity to develop them presumably is. Secondly, we have a "heritable" form of non-Darwinian adaptation in humans (and, in rudimentary ways, in a few other advanced social species): cultural adaptation (with heritability imposed by learning). Much confused thinking in human sociobiology arises from a failure to distinguish this mode from Darwinian adaptation based on genetic variation. Finally, we have adaptation arising from the conventional Darwinian mechanism of selection upon genetic variation. The mere existence of a good fit between organism and environment is insufficient for inferring the action of natural selection.

(4) Adaptation and selection but no selective basis for differences among adaptations. Species of related organisms, or subpopulations within a species, often develop different adaptations as solutions to the same problem. When "multiple adaptive peaks" are occupied, we usually have no basis for asserting that one solution is better than another. The solution followed in any spot is a result of history; the first steps went in one direction, though others would have led to adequate prosperity as well. Every naturalist has his favorite illustration. In the West Indian land snail Cerion, for example, populations living on rocky and windy coasts almost always develop white, thick, and relatively squat shells for conventional adaptive reasons. We can identify at least two different developmental pathways to whiteness from the mottling of early whorls in all Cerion, two paths of thickened shells and three styles of allometry leading to squat shells. All 12 combinations can be identified in Bahamian populations, but would it be fruitful to ask why—in the sense of optimal design rather than historical contingency—Cerion from eastern Long Island evolved one solution, and Cerion from Acklins Island another?

(5) Adaptation and selection, but the adaptation is a secondary utilization of parts present for reasons of architecture, development or history. We have already discussed this neglected subject in the first section on spandrels, spaces, and cannibalism. If blushing turns out to be an adaptation affected by sexual selection in humans, it will not help us to understand why blood is red. The immediate utility of an organic

structure often says nothing at all about the reason for its being.

## 6. ANOTHER, AND UNFAIRLY MALIGNED, APPROACH TO EVOLUTION

In continental Europe, evolutionists have never been much attracted to the Anglo-American penchant for atomizing organisms into parts and trying to explain each as a direct adaptation. Their general alternative exists in both a strong and a weak form. In the strong form, as advocated by such major theorists as Schindewolf (1950), Remane (1971), and Grassé (1977), natural selection under the adaptationist programme can explain superficial modifications of the *Bauplan* that fit structure to environment: why moles are blind, giraffes have long necks, and ducks webbed feet, for example. But the important steps of evolution, the construction of the *Bauplan* itself and the transition between *Baupläne,* must involve some other unknown, and perhaps "internal," mechanism. We believe that English biologists have been right in rejecting this strong form as close to an appeal to mysticism.

But the argument has a weaker—and paradoxically powerful—form that has not been appreciated, but deserves to be. It also acknowledges conventional selection for superficial modifications of the *Bauplan.* It also denies that the adaptationist programme (atomization plus optimizing selection on parts) can do much to explain *Baupläne* and the transitions between them. But it does not therefore resort to a fundamentally unknown process. It holds instead that the basic body plans of organisms are so integrated and so replete with constraints upon adaptation (categories 2 and 5 of our typology) that conventional styles of selective arguments can explain little of interest about them. It does not deny that change, when it occurs, may be mediated by natural selection, but it holds that constraints restrict possible paths and modes of change so strongly that the constraints themselves become much the most interesting aspect of evolution.

Rupert Riedl, the Austrian zoologist who has tried to develop this thesis for English audiences (1977 and 1975, translated into English by R. Jeffries in 1978) writes:

The living world happens to be crowded by universal patterns of organization which, most obviously, find no direct explanation through environmental conditions or adaptive radiation, but exist primarily through universal requirements which can only be expected under the systems conditions of complex organization itself ... This is not self-evident, for the whole of the huge and profound thought collected in the field of morphology, from Goethe to Remane, has virtually been cut off from modern biology. It is not taught in most American universities. Even the teachers who could teach it have disappeared.

Constraints upon evolutionary change may be ordered into at least two categories. All evolutionists are familiar with *phyletic constraints,* as embodied in Gregory's classic distinction (1936) between habitus and heritage. We acknowledge a kind of phyletic inertia in recognizing, for example, that humans are not optimally designed for upright posture because so much of our *Bauplan* evolved for quadrupedal life. We also invoke phyletic constraint in explaining why no molluscs fly in air and no insects are as large as elephants.

(2) *Developmental* constraints, a subcategory of phyletic restrictions, may hold the most powerful rein of all over possible evolutionary pathways. In complex organisms, early stages of ontogeny are remarkably refractory to evolutionary change, presumably because the differentiation of organ systems and their integration into a functioning body is such a delicate process so easily derailed by early errors with accumulating effects. Von Baer's fundamental embryological laws (1828) represent little more than a recognition that early stages are both highly conservative and strongly restrictive of later development. Haeckel's biogenetic law, the primary subject of late nineteenth century evolutionary biology, rested upon a misreading of the same data (Gould, 1977). If development occurs in integrated packages and cannot be pulled apart piece by piece in evolution, then the adaptationist programme cannot explain the alteration of developmental programmes underlying nearly all changes of *Bauplan.*

The German palaeontologist A. Seilacher, whose work deserves far more attention than it has received, has emphasized what he calls "*bau-technischer,* or *architectural,* constraints (Seilacher, 1970). These arise not from former adaptations retained

(3)

in a new ecological setting (phyletic constraints as usually understood), but as architectural restrictions that never were adaptations but rather were the necessary consequences of materials and designs selected to build basic *Baupläne.* We devoted the first section of this paper to nonbiological examples in this category. Spandrels must exist once a blueprint specifies that a dome shall rest on rounded arches. Architectural constraints can exert a far-ranging influence upon organisms as well. The subject is full of potential insight because it has rarely been acknowledged at all.

In a fascinating example, Seilacher (1972) has shown that the divaricate form of architecture (figure 3) occurs again and again in all groups of molluscs, and in brachiopods as well. This basic form expresses itself in a wide variety of structures: raised ornamental lines (not growth lines because they do not conform to the mantle margin at any time), patterns of coloration, internal structures in the mineralization of calcite and incised grooves. He does not know what generates this pattern and feels that traditional and nearly exclusive focus on the adaptive value of each manifestation has diverted attention from questions of its genesis in growth and also prevented its recognition as a general phenomenon. It must arise from some characteristic pattern of inhomogeneity in the growing mantle, probably from the generation of interference patterns around regularly spaced centers; simple computer simulations can generate the form in this manner (Waddington

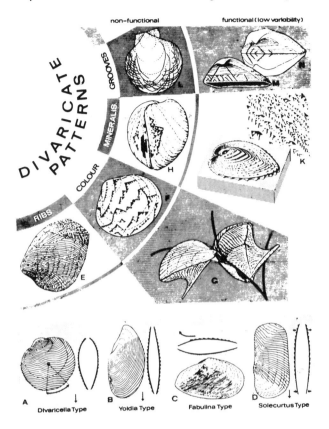

Figure 3. The range of divaricate patterns in molluscs. E, F, H, and L are nonfunctional in Seilacher's judgement. A–D are functional ribs (but these are far less common that nonfunctional ribs of the form E). G is the mimetic *Arca zebra*. K is *Corculum*. See text for details.

and Cowe, 1969). The general pattern may not be a direct adaptation at all.

Seilacher then argues that most manifestations of the pattern are probably nonadaptive. His reasons vary but seem generally sound to us. Some are based on field observations: color patterns that remain invisible because clams possessing them either live buried in sediments or remain covered with a periostracum so thick that the colors cannot be seen. Others rely on more general prin-

ciples: presence only in odd and pathological individuals, rarity as a developmental anomaly, excessive variability compared with much reduced variability when the same general structure assumes a form judged functional on engineering grounds.

In a distinct minority of cases, the divaricate pattern becomes functional in each of the four categories (figure 3). Divaricate ribs may act as scoops and anchors in burrowing (Stanley, 1970), but they are not properly arranged for such function in most clams. The color chevrons are mimetic in one species (*Pteria zebra*) that lives on hydrozoan branches; here the variability is strongly reduced. The mineralization chevrons are probably adaptive in only one remarkable creature, the peculiar bivalve *Corculum cardissa* (in other species they either appear in odd specimens or only as post-mortem products of shell erosion). This clam is uniquely flattened in an anterio-posterior direction. It lies on the substrate, posterior up. Distributed over its rear end are divaricate triangles of mineralization. They are translucent, while the rest of the shell is opaque. Under these windows dwell endosymbiotic algae!

All previous literature on divaricate structure has focused on its adaptive significance (and failed to find any in most cases). But Seilacher is probably right in representing this case as the spandrels, ceiling holes, and sacrificed bodies of our first section. The divaricate pattern is a fundamental architectural constraint. Occasionally, since it is there, it is used to beneficial effect. But we cannot understand the pattern or its evolutionary meaning by viewing these infrequent and secondary adaptations as a reason for the pattern itself.

Galton (1909, p. 257) contrasted the adaptationist programme with a focus on constraints and modes of development by citing a telling anecdote about Herbert Spencer's fingerprints:

Much has been written, but the last word has not been said, on the rationale of these curious papil-lary ridges; why in one man and in one finger they form whorls and in another loops. I may mention a characteristic anecdote of Herbert Spencer in connection with this. He asked me to show him my Laboratory and to take his prints, which I did. Then I spoke of the failure to discover the origin of these patterns, and how the fingers of unborn children had been dissected to ascertain their earliest stages, and so forth. Spencer remarked that this was beginning in the wrong way; that I ought to consider the purpose the ridges had to fulfil, and to work backwards. Here, he said, it was obvious that the delicate mouths of the sudorific glands required the protection given to them by the ridges on either side of them, and therefrom he elaborated a consistent and ingenious hypothesis at great length. I replied that his arguments were beautiful and deserved to be true, but it happened that the mouths of the ducts did not run in the valleys between the crests, but along the crests of the ridges themselves.

We feel that the potential rewards of abandoning exclusive focus on the adaptationist programme are very great indeed. We do not offer a council of despair, as adaptationists have charged; for nonadaptive does not mean nonintelligble. We welcome the richness that a pluralistic approach, so akin to Darwin's spirit, can provide. Under the adaptationist programme, the great historic themes of developmental morphology and *Bauplan* were largely abandoned; for if selection can break any correlation and optimize parts separately, then an organism's integration counts for little. Too often, the adaptationist programme gave us an evolutionary biology of parts and genes, but not of organisms. It assumed that all transitions could occur step by step and underrated the importance of integrated developmental blocks and pervasive constraints of history and architecture. A pluralistic view could put organisms, with all their recalcitrant yet intelligible complexity, back into evolutionary theory.

## REFERENCES

Baer, K. E. von, 1828, *Entwicklungsgeschichte der Tiere,* Königsberg: Bornträger.

Barash, D. P., 1976, Male response to apparent female adultery in the mountain-bluebird: an evolutionary interpretation, *Am. Nat.,* 110: 1097–1101.

Coon, C. S., Garn, S. M., and Birdsell, J. B., 1950, *Races,* Springfield Oh., C. Thomas.

Costa, R., and Bisol, P. M., 1978, Genetic variability in deep-sea organisms, *Biol. Bull.,* 155: 125–133.

Darwin, C., 1872, *The origin of species,* London, John Murray.

——, 1880, Sir Wyville Thomson and natural selection, *Nature,* London, 23: 32.

Davitashvili, L. S., *Teoriya polovogo otbora* [Theory of sexual selection], Moscow, Akademii Nauk.

Falconer, D. S., 1973, Replicated selection for body weight in mice, *Genet. Res.,* 22: 291–321.

Galton, F., 1909, *Memories of my life,* London, Methuen.

Gould, S. J., 1966, Allometry and size in ontogeny and phylogeny, *Biol. Rev.,* 41: 587–640.

——, 1971, D'Arcy Thompson and the science of form, *New Literary Hist.,* 2, no. 2, 229–258.

——, 1974, Allometry in primates, with emphasis on scaling and the evolution of the brain. In *Approaches to primate paleobiology, Contrib. Primatol.,* 5: 244–292.

——, 1977, *Ontogeny and phylogeny,* Cambridge, Ma., Belknap Press.

——, 1978, Sociobiology: the art of story-telling, *New Scient.,* 80: 530–533.

Grassé, P. P., 1977, *Evolution of living organisms,* New York, Academic Press.

Cregory, W. K., 1936, Habitus factors in the skeleton fossil and recent mammals, *Proc. Am. phil. Soc.,* 76: 429–444.

Harner, M., 1977, The ecological basis for Aztec sacrifice. *Am. Ethnologist,* 4: 117–135.

Jerison, H. J., 1973, *Evolution of the brain and intelligence,* New York, Academic Press.

Lande, R., 1976, Natural Selection and random genetic drift in phenotypic evolution, *Evolution,* 30: 314–334.

——, 1978, Evolutionary mechanisms of limb loss in tetrapods, *Evolution,* 32: 73–92.

Lewontin, R. C., 1978, Adaptation, *Scient. Am.,* 239 (3): 156–169.

——, 1979, Sociobiology as an adaptationist program, *Behav. Sci.,* 24: 5–14.

Morton, E. S., Geitgey, M. S., and McGrath, S., 1978, On Bluebird 'responses to apparent female adultery'. *Am. Nat.,* 112: 968–971.

Ortiz de Montellano, B. R., 1978, Aztec cannibalism: an ecological necessity? *Science,* 200: 611–617.

Remane, A., 1971, *Die Grundlagen des natürlichen Systems der vergleichenden Anatomie und der Phylogenetik.* Königstein-Taunus: Koeltz.

Rensch, B., 1959, *Evolution above the species level,* New York, Columbia University Press.

Riedl, R., 1975, *Die Ordnung des Lebendigen,* Hamburg, Paul Parey, tr. R. P. S. Jefferies, *Order in Living Systems: A*

*Systems Analysis of Evolution,* New York, Wiley, 1978.

——, 1977, A systems-analytical approach to macro-evolutionary phenomena, *Q. Rev. Biol.,* 52: 351–370.

Romanes, G. J., 1900, The Darwinism of Darwin and of the post-Darwinian schools. In *Darwin, and after Darwin,* vol. 2, new edn., London, Longmans, Green and Co.

Rudwick, M. J. S., 1964, The function of zig-zag deflections in the commissures of fossil brachiopods, *Palaeontology,* 7: 135–171.

Sahlins, M., 1978, Culture as protein and profit, *New York review of books,* 23: Nov., pp. 45–53.

Schindewolf, O. H., 1950, *Grundfragen der Paläontologie,* Stuttgart, Schweizerbart.

Seilacher, A., 1970, Arbeitskonzept zur Konstruktionsmorphologie, *Lethaia,* 3: 393–396.

——, 1972, Divaricate patterns in pelecypod shells, *Lethaia,* 5: 325–343.

Shea, B. T., 1977, Eskimo craniofacial morphology, cold stress and the maxillary sinus, *Am. J. phys. Anthrop.,* 47: 289–300.

Stanley, S. M., 1970, Relation of shell form to life habits in the Bivalvia (Mollusca). *Mem. geol. Soc. Am.,* no. 125, 296 pp.

Sweeney, B. W., and Vannote, R. L., 1978, Size variation and the distribution of hemimetabolous aquatic insects: two thermal equilibrium hypotheses. *Science,* 200: 444–446.

Thompson, D. W., 1942, *Growth and Form,* New York, Macmillan.

Waddington, C. H., and Cowe, J. R., 1969, Computer simulation of a molluscan pigmentation pattern, *J. theor. Biol.,* 25: 219–225.

Wallace, A. R., 1899, *Darwinism,* London, Macmillan.

Wilson, E. O., 1978, *On human nature,* Cambridge, Ma., Harvard University Press.

# 16

# A Critique of Optimization
# Theory in Evolutionary Biology

*ᴏᴠᴏᴠᴏᴠᴏᴠᴏᴠᴏᴠᴏᴠᴏᴠᴏᴠᴏᴠᴏᴠᴏᴠᴏᴠᴏᴠᴏᴠᴏᴠ*

## GEORGE F. OSTER AND EDWARD O. WILSON

OPTIMIZATION ARGUMENTS HAVE been the central theoretical tool in our investigations of caste evolution.[1] We now feel obliged to present a critical review of this method, extending the discussion to the broader field of evolutionary biology, in order to evaluate its strengths and weaknesses objectively. Such a review needs to be conducted carefully and repeatedly, because optimization arguments are the foundation upon which a great deal of theoretical biology now rests. Indeed, biologists view natural selection as an optimizing process virtually by definition. This use of the concept of optimality does not, however, require that natural selection create phenotypes "better" than their predecessors in any absolute sense. Rather, we suppose only that physiological design features providing a mortality or fecundity advantage in the local environment will be amplified.

The use of optimization models in biology has its intellectual roots in the development of classical physics during the nineteenth century. The laws of motion in classical mechanics and electromagnetic theory were at first formulated "locally." That is, they were written as differential equations that projected the future trajectory of the system from its present state "one step at a time." Later it was observed that the laws of motion could also be expressed in a global form as extremum principles. This seemed to lend an aura of predesign to the laws of physics, since the solutions to the equations of motion produced trajectories that minimized some quantity. It was easy for some to read divine intent into the evolution of physical laws like Fermat's Principle and the Second Law of Thermodynamics, which describe nature apparently striving to behave in so economical a fashion.

As the mathematical structure of physical laws was elucidated, the fragility of extremum principles became apparent. Although they lost their metaphysical significance, they retained their esthetic appeal and in many cases still possess the practical

---

1. In Oster and Wilson (1978). This essay is chapter 8 of that book. Chapter references in this essay refer to other chapters of that book.

advantage of superior computational efficiency.

Joel Cohen has noted ruefully that "physics-envy is the curse of biology." This has been nowhere more true than in evolutionary theory. Biologists have never tired of seeking global designs in nature, but it was not until the advent of modern population genetics that such teleological views were given a concrete form. The crucial step was taken by R. A. Fisher, who showed that the equation governing the change of gene frequencies at a single locus with two alleles under natural selection could be expressed as an extremum principle: the Fundamental Theorem of Natural Selection. The solution to the equations maximized a quantity that Fisher sagaciously called "Fitness." Never mind the delicacy of such a result—constant environment, limited types of density or frequency dependence, no linkage or epistasis, and so forth—teleology had been given mathematical respectability! Efforts to generalize Fisher's result have been ceaseless. They have also been largely unsuccessful, and the validity of the whole enterprise might have been questioned had not a new infusion of optimism come from a completely different quarter.

Analogies between biological evolution and economics had been noted by Marshall and Keynes in the 1920's and by Huxley in the 1930's (Rapport and Turner, 1977). This viewpoint was consistent with the attention then being given competition theory by Lotka, Volterra, Gause, and others, as well as the concurrent development of mathematical population genetics by Fisher, Haldane, and Wright. How much biologists' intuitions about the workings of nature were really influenced by the economic and political currents of the day is difficult to say (but see Lewontin, 1977). However, to the extent that economic notions penetrated biological thought, so too did optimization arguments, because the entire science of economics deals with the optimal allocation of scarce resources.

The conceptual exchange between economics and biology was reciprocal. Milton Friedman argued in his influential book *Essays in Positive Economics* (1953) that the evolution of business firms proceeds by a process of natural selection for the most efficient. To some economists this conception implies that the present distribution of business economic power is the outcome of inevitable forces. The implications of such a view have caused some politically radical biologists to reject economics-type thinking and, hence, optimization arguments in order to avoid sullying their scientific investigations with what they consider unsavory assumptions (see, e.g., Allen et al., 1976).

The most recent impetus to optimization modeling in ecology has come from engineering, and especially operations research. Indeed, most of the models we have developed in this book employ mathematical techniques developed to solve problems in engineering and industrial design. The crucial difference between engineering and evolutionary theory is that the former seeks to design a machine or an operation in the most efficient form, while the latter seeks to infer "nature's design" already created by natural selection.

In order to employ engineering optimization models, the biologist tries to interpret living forms as in some

sense the "best." But just what constitutes superiority is rarely made explicit and is seldom very clearly
understood. In effect the biologist
"plays God": he redesigns the biological system, including as many of
the relevant quantities as possible, and
then checks to see if his own optimal
design is close to that observed in nature. If the two correspond, then nature can be regarded as reasonably well
understood. If they fail to correspond
to any degree (a frequent result), the
biologist revises the model and tries
again. Thus optimization models are a
method for organizing empirical evidence, making educated guesses as to
how evolution might have proceeded,
and suggesting avenues for further
empirical research. At the same time,
the concept of optimization can be
given a precise definition only in
mathematical language. Therefore, it
might be a good idea to apply the term
only to mathematical models and not
to real world phenomena. Otherwise,
like the notions of "instinct" and
"drive" in ethology, the concept
can create more problems than it
solves.

In order to justify this pragmatic
view of optimization, we feel it would
be desirable to take a critical look at
such models as they are most frequently used in ecology and evolutionary biology. There is little chance that
contemporary biologists will repeat
the mistake of the early physicists and
see in their abstractions the hand of
God. But they are still prone to see
His surrogate in the automatic process leading populations to ever higher
levels of adaptation. We will show that
not even this relatively modest theme
can be translated into mathematically
sound arguments.

## 1. THE STRUCTURE OF OPTIMIZATION MODELS

The goal of optimization models is to
determine the "best way" to allocate
a scarce resource among various alternatives. In order to employ quantitative language—that is, mathematics—
we must give precise definitions to
the innocuous-sounding terms "best
way" and "scarce resource." Unfortunately, ambiguity can be eliminated
only by introducing a plethora of supporting definitions and concepts. We
do not propose to lead the reader
through this mathematical labyrinth
now; instead, we urge anyone interested in the deeper issues to consult
one of the many texts on mathematical programming (e.g., Dorny, 1975;
Varaiya, 1973; Intriligator, 1971;
Bryson and Ho, 1975; Leitmann,
1966). The discussion to follow will
be brief and largely informal.

The simplest optimization scheme—
one employed several times in the
present book—is shown in Figure 1
and can be verbalized as follows:

Maximize $\mathfrak{F}(x_1, x_2)$, subject to the
constraints:

$$f(x_1, x_2) \leqslant \text{constant}; \quad x_1, x_2 \geqslant 0.$$

This very elementary picture is probably what most biologists have in
mind when thinking about optimization models. However, when applied
to evolutionary problems it can be
seriously misleading, to the point of
subverting our intuitions. Let us therefore dissect the picture and discuss
some of the complications that intrude
during the modeling of evolutionary
processes.

Optimization models consist formally of 4 components: (1) a state

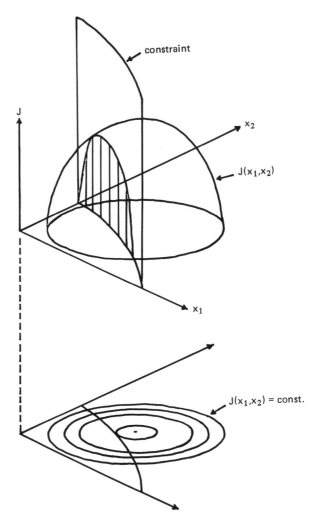

Figure 1. The simplest optimization scheme.

space; (2) a set of strategies; (3) one or more optimization criteria, or fitness functions; and (4) a set of constraints. Each will now be discussed in turn.

### (1) How do We Define the State of a System?

A basic goal of virtually all mathematical models is to make as complete a description as possible of the system with reference to the questions being asked. By this we mean precisely the following (Desoer, 1970). Denote by $\mathbf{x}(t) = (x_1, x_2, \ldots, x_n)$ a collection of measurements performed on a system at some time $t$. If these are sufficient to compute all the future values of the measurements, $x_i(t)$, then the collection $(x_1(t) \ldots, x_n(t)$ are said to constitute a "state description."[2] For example, most population models employ simply the population number $\mathfrak{N}_i(t)$ $(i = 1, 2, \ldots n)$ of each of $n$ species as a state description. It is clear, however, that a census alone is generally not sufficient to predict the future growth of the population. Age structure, nutritional state, sex ratios, and a host of other variables are also crucial for precise population projections.

Among physical scientists there is general agreement, based on generations of empirical evidence, on what constitutes a state description. For biological populations the situation is not so clear, both because the experimental data are too sketchy and because the complexity of the system is much greater. Therefore, biologists are forced to rely more on intuition and personal experience in selecting the right population descriptors. We have defined the state of a social insect colony by the caste distribution function, $n(t,a,s,x)$, because age, size, and activity distributions seem to us to be the most distinctive physical features of a social insect colony. Yet the CDF cannot be a complete state description, because other phenotypic variables must also be important, including nutritional state, allometric measurements other than size, and a plethora of behavioral traits and chemical substances about which it is only possible to speculate at the present time. Furthermore, our ergonomic analysis has neglected the genetic structure almost entirely. The models are therefore endemically provisional. They are at

best imprecise and will certainly have to be revised with each accretion of empirical evidence. But the situation is merely that which prevails throughout the remainder of evolutionary biology; our state description may be necessary, but it is surely not sufficient.

## (2) What Are the Strategies?

The formulation of a mathematical optimization model requires an exhaustive specification of the allowable strategies. That is, the set of strategies must be just as complete as the set of states. For example, in the analysis of reproductive strategies, introduced in chapter 5, the strategic variable is the parameter $0 \leqslant u(t) \leqslant 1$, which determines the fractional allocation of resources to workers as opposed to queens. This requirement makes it very difficult to apply optimization models to evolutionary processes. The reason is clear. The fundamental source of new adaptive strategies is mutation and recombination; natural selection acts only to delete the least "fit" individuals. There is no way to anticipate what new strategies can be generated by these genetic processes; they might include a new enzyme, a modified appendage suited for a new function, a novel behavioral response, and so forth. The very combinatoric richness of the possible molecular permutations in the genetic code makes it impossible to enumerate even a small fraction of the allowable strategies. The essential innovative nature of the evolutionary process precludes an exhaustive list of allowable strategies. The strategy set

---

2. A more general definition of "state" is required for systems where stochastic effects are important.

is always changing, new ones being added and old ones deleted. On the face of it, this property appears to be incompatible with the requirements of optimization models, since the strategy set cannot be specified a priori.

Most of the models we have developed to describe caste structure have been economic in conception: they aimed at specifying the optimal allocation of scarce resources among a predetermined set of alternatives. These alternatives consisted mostly of guesses based on our knowledge of natural history. In some sense we had to anticipate the allowable strategic alternatives by reconstructing evolution in our imagination. Our treatment of foraging strategies and caste structure exemplifies the difficulties. By assuming that allometry alone determines foraging effectiveness, we could construct efficiency curves based on a single-size parameter and thus establish a relationship between caste polymorphism and ergonomic efficiency. However, many species can do equally well with monomorphism simply by employing increased behavioral flexibility and recruitment. Such a strategic alternative would be hard to anticipate with reference to any particular species.

This limitation on mathematical models has been appreciated by workers in the field of prebiotic molecular evolution. The combinatoric possibilities of carbon chemistry make it impossible to enumerate all of the alternative compounds generated by even very low molecular weight substances. Thus efforts have been initiated to develop theories that can anticipate the most probable reaction pathways from specified initial conditions. At the levels of organismic and colonial evolution such a program is clearly not feasible. However, comparative studies can, to some extent, play the same role. For example, by examining the foraging strategies of a large number of ant species, we can get a reasonable idea of the range of strategic possibilities. This has obvious limitations, of course, but in the case of ants at least, the following two conditions are met: (1) there are a large number of comparable species, in fact over 10,000; and (2) they appear to be evolutionarily static, since a large percentage of living genera and even species groups date back to Oligocene times. Thus we can guess that adaptive radiation has long since exhausted most of the genetically feasible solutions, and we can hope to enumerate a fairly complete strategy set. In other taxonomic groups for which comparative data are scantier, such an assumption is more risky.

It should now be apparent that the central process of evolution—speciation—is relatively impervious to mathematical treatment, since optimization models can only *compare* strategies. The evaluation of qualitatively new adaptive innovations requires most of the answer to be known ahead of time.

Finally, optimization models frequently beg the question of greatest biological interest: how the optimal strategy is implemented. In the reproductive strategy model we ascertained that the "bang-bang" strategy was the best: the colony should produce all of the queens and drones at the very end of each season. Nowhere, however, was the question addressed of how the queen recognizes the optimal switching time. Some sort of biological clock is probably involved, but it was not necessary for our purposes to postulate any particular mechanism. We had only to assume that the genetic potential of the species was sufficient to implement the optimal strategy, whatever the strategy turned out to be. This robust quality is both a strength and a weakness of economic optimization models. On the one hand, they circumvent the details of genetic or physiological mechanisms, yet on the other, they force us to assume that at the lower genetic and physiological levels of organization sufficient flexibility exists to realize all of the strategies.

### (3) What Should Be Optimized?

It is tautological nonsense to say that fitness is maximized: what is fitness? The only unequivocal definition involves maximizing the number of genes projected into future generations. Unfortunately, it is almost always impossible to compute this explicitly in any but the most trivial cases. Hence one is forced to deal with more macroscopic quantities which, presumably, affect genetic fitness in a direct way. In the reproductive strategy model discussed in chapter 5, we used as a fitness criterion the number of alates produced by season's end. There can be little quarrel with this as a fitness measure, all other things being equal. Unfortunately, the life cycle is seldom so clear as in this case, and all other things are seldom equal. The central goal of most of our models has been to see how caste structure could be used as an instrument for increasing the ergonomic efficiency of a colony. Presumably, the greater the efficiency, the greater the reproductive crop and, hence, the higher the genetic fitness.

However, the connection is not so direct as it might seem at first. A colony has other contingencies to meet in addition to energy procurement if it is to maximize its reproductive fitness. One of the most crucial of these contingencies is colony defense; effort expended in that direction must be at the expense of ergonomic efficiency. The situation is typical of optimization models in evolution; there are generally many fitness "components" to be simultaneously maximized if the "overall" genetic fitness is to be maximized. For the purposes of our discussion we can classify the situation according to the number of fitness criteria and the number of "players," or decision makers, as shown in Figure 2. Generally, evolutionary optimization models fall into the category of multicriteria optimization or game theory, depending on whether the situation is one of cooperation or conflict between the participants. Team theory may have some relevance to models of symbiosis, such as the relation between ants and the aphids they attend.

In our analyses we have learned repeatedly that the conflicting nature of these various criteria prohibits their simultaneous optimization. Risk counters return, specialization on one class of food items counters exploitation of other classes, and so forth. Therefore, maximum genetic fitness can be achieved only by the best compromise solution to these conflicting interests.

When dealing with multiple-criteria decision problems in economics or physics, one can frequently use a utility function that establishes a common currency for each criterion, such as dollars or energy. This is seldom

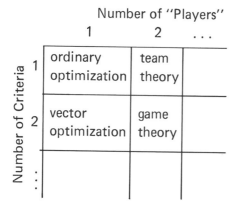

Figure 2. Classification of optimization models by the number of players and the number of criteria.

possible in evolutionary models, since the "payoff" is in the genes of distant future generations. How can one calibrate present actions by such a standard? We are forced to employ ad hoc methods of combining fitness components into an overall adaptive function. One common method is to combine the separate optimization criteria linearly, or log-linearly by weighting functions that measure the relative importance of each component (Levins, 1968). This procedure is nevertheless usually quite arbitrary, and the assignment of the weighting factors is a difficult problem in itself. Such a procedure has enabled us at the very least to classify the life styles of various species according to the relative degree of risk they could accept. Yet it has to be admitted that the choice of a set of macroscopic fitness functions and of a method for combining them into an overall utility function is another exercise in concealed teleology, and as such it can be no better than our taste and personal experience.

The problem is complicated further by conditions of conflict between species or between individuals of the same species, the game classification in Figure 2. There is simply no way of combining the fitness criteria of the various parties, because their genetic interests are basically incommensurable. Therefore, one must examine carefully just what optimization could mean in such circumstances. Optimization is perhaps not even the right word, since nothing is really maximized. We might instead ask how the conflicting interests between the parties can be resolved in a stable fashion. An equitable solution, while desirable in human affairs, is certainly irrelevant from an evolutionary standpoint.

In chapter 3 we diagnosed the worker-queen conflict as a competitive or Nash equilibrium. The idea was to find a set of strategies wherein neither party could gain by changing strategies —provided the other party also maintained the same strategy. However, such a condition is usually only local. That is, the Nash equilibrium is resistant to *small* deviations by either party. When the "players" in the game are populations of individuals, then the Nash equilibrium will usually be evolutionarily stable as well. This is simply because any mutation giving rise to a strategic deviation will always be rare, i.e., be manifest in only a small segment of the population "strategy." When the players are individuals, however, the Nash point may not be evolutionarily stable. This is because derivations in strategy are produced by variations in phenotypic traits in the players, and these in turn are brought about in part by changes in the underlying genetic properties. For the most part such changes will indeed be gradual and small, and thus the Nash equilibrium may well be stable in the evolutionary sense—an Evolutionary Stable Strategy in the sense of Maynard Smith (1974, 1976). However, we cannot overlook the possibility of a larger-scale strategic change on the part of one player (population) due to a saltational migration or genetic event. Recent work suggests that major phenotypic revisions can occur as a result of regulator gene changes which alter the sequence and timing of developmental events. To the extent that such phenomena play a major role in evolution, global stability criteria will have to be devised to handle large strategic perturbations.

Even if an unequivocal definition of evolutionary stable strategies is devised, there is no guarantee that an equilibrium strategy can be achieved. That is, achieving and maintaining the Nash point may not be consistent with the laws of Mendelian inheritance. We will return to this question shortly when the role of constraints is examined.

Finally, in addition to the difficulties of the modeling procedure itself, we encounter formidable mathematical obstacles in computing an optimal solution whenever the optimization criterion is made complex to any degree. For example, if the fitness functions are not convex, local maxima can fit together to yield cyclic optimal global strategies (Ekeland, 1977).

### (4) What Are the Constraints?

No biological optimization can be unconstrained. Resources are bounded, and such key quantities as population size and gene frequency must be nonnegative. In other words, the optimal feasible solution is always less advantageous than the optimal conceivable solution. Thus completeness of the constraint set is as crucial as completeness of the state and strategy sets. Neglecting or overlooking a constraint can change the qualitative predictions of the model. Yet to enumerate all of the constraints, much less to write mathematical expressions for them, poses formidable technical problems.

From a mathematical viewpoint, the problem of enumerating all of the constraints is not distinct from that of enumerating the state variables. A constraint means that some of the state variables may be redundant and the problem could have been formulated with fewer state variables. For example, a ball rolling in a bowl only requires two coordinates to specify its position, not three. However, in the model procedures of evolutionary theory it is usually impossible to know a priori the theoretical minimum number of state variables required.

For the purposes of the present discussion the constraints can be meaningfully classified as shown in Figure 3. The simplest case (the first entry in the classification matrix) has already been adequately captured by Figure 1 and needs no further comment. When constraint equations vary in time, the picture becomes much more complicated. For example, consider the model (chapter 2) in which we estimated the optimal reproductive strategy for an annual social insect colony. The mathematical structure of the model is summarized in Figure 4. In order to maximize the number of reproductives, $Q(T)$, by the end of the season, it is necessary to find the optimal time course of

|  | Deterministic | Stochastic |
|---|---|---|
| Static | linear or nonlinear programming | stochastic programming |
| Dynamic | optimal control theory, dynamic programming, calculus of variations | stochastic-dynamic optimization |

Figure 3. A classification of constraints in optimization models. The mathematical techniques used to study them are given in the entries.

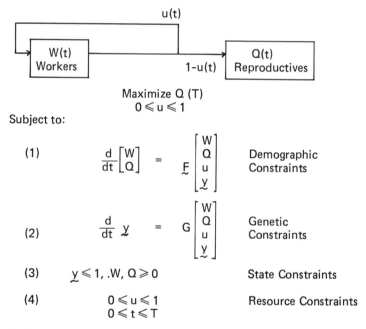

Maximize Q (T)
$$0 \leqslant u \leqslant 1$$

Subject to:

(1) $\quad \dfrac{d}{dt}\begin{bmatrix} W \\ Q \end{bmatrix} = \underset{\sim}{F}\begin{bmatrix} W \\ Q \\ u \\ y \end{bmatrix}$  Demographic Constraints

(2) $\quad \dfrac{d}{dt}\, \underset{\sim}{y} = G\begin{bmatrix} W \\ Q \\ u \\ y \end{bmatrix}$  Genetic Constraints

(3) $\quad \underset{\sim}{y} \leqslant 1, \, .W, \, Q \geqslant 0$  State Constraints

(4) $\quad 0 \leqslant u \leqslant 1$  Resource Constraints
$\quad\quad\ 0 \leqslant t \leqslant T$

Figure 4. The structure of the optimal reproductive strategy model for an annual social insect colony. The vectory $y = (y_1 y_2, \ldots)$ refers to the genetic state variables, and $u(t) = (u_1 u_2, \ldots)$ are the strategic (control) parameters that replace them in the optimization model.

resource allocation, $u(t)$. This allocation must be consistent with a number of static and dynamic constraints. The only dynamic constraints explicitly accounted for in the original treatment were the demographic equations for colony growth. Thus in realizing the optimal strategy we consciously begged the question of physiological and genetic considerations. In principle, a set of dynamic equations describing the gene frequencies that affect the production of alates should have been appended to the demographic equations. These genetic variables are denoted in Figure 4 by the vector $\mathbf{y}(t) = y_1, y_2, \ldots$

Yet a moment's reflection reveals that if we could indeed write the explicit equations governing all the genetic quantities, then there would be no optimization problem to solve. We would then need only to watch the state variables $(\mathbf{x}(t), \mathbf{y}(t))$ evolve in time, and see where natural selection takes the system. The whole purpose of an optimization model is to circumvent the virtually impossible task of writing explicit expressions for "microscopic" details such as gene frequencies. That is, we replace the equations for the microscopic constraints, given by Equations (2) in Figure 4, by a "macroscopic" fitness criterion, in this particular case $Q(T)$. However, we must then supply the model with a complete set of control parameters, $u(t)$, that take the place of the genetic equations. This procedure carries with it some strong assumptions. *Most importantly, it requires that the optimal solution be consistent with the genetic constraint equations.* Although this basic assumption underlies virtually all ecological optimization models, there is no guarantee that it is valid.[3]

Indeed, it is not difficult to produce models in which the correspondence assumption is demonstrably false (Slatkin, 1977; Auslander, Guckenheimer, and Oster, 1978). Quite strong mathematical restrictions must be placed on the constraint Equations (1), (2), and (3) in order to characterize their solutions by an extremum principle. The assumption, frequently made, that "sufficient genetic flexibility exists to realize the optimal ecological strategy" is an assertion of pure faith. Rather like the concept of the Holy Trinity, it must be believed to be understood. Let us be clear about this difficulty, since it is central to all optimization models presumed to operate in ecological time. The fitness functions and control parameters must summarize all of the relevant microscopic details not explicitly accounted for in the model. The procedure is analogous to the practice in chemistry of summarizing the aggregate effects of microscopic molecular activity by a macroscopic quantity such as free energy or entropy, which is extremized at the thermodynamic equilibrium. It is easily possible that the microscopic equations are mathematically incompatible with the existence of a macroscopic fitness function. For example, even in the simplest case of 1 locus and 2 alleles, it is known that the genetic equations of motion do not support an extremum formulation except in very restricted situations (Shashahani, 1978). For the most part, ecologically relevant traits will surely be polygenic.

The mathematical treatment of such gene ensembles appears prohibitive at this time. If we regard the system as autonomous and simply follow the evolution of gene frequencies and population, then it is only under very special assumptions concerning the form of the population and genetic

---

3. Hidden in most optimization models is an assumption that the underlying genetic equations are close to a "gradient flow" i.e., that the system is asymptotically stable in a global sense (Oster and Rocklin, 1978).

dynamics that the system evolves so as to maximize anything. For example, if one assumes the simplest sort of density and frequency dependent selection then one can show that the equilibrium gene frequency configuration corresponds to a (local) maximum of the total population (cf., Roughgarden, 1978). This result, however, cannot be regarded as a general feature of population models (Oster and Rocklin, 1978). Thus it is unlikely that a mathematically convincing link will soon be established between the microscopic dynamics of genotypes and the macroscopic processes of phenotype selection, although attempts in this direction are being made (see, for example, Rocklin and Oster, 1976; Levin and Udovic, 1977).

Another justification for neglecting genetic dynamics frequently invoked during ecological modeling is the supposed disparity of time scales. If one could assert that evolutionary processes involving gene changes occur at rates much slower than demographic changes, then he would be justified in transforming the dynamic equations for gene frequencies into static constraints. In Figure 4 this corresponds to setting $y = 0$. In formal terms, it appears that we might then eliminate the genetic values from consideration by simply solving $0 = g(x,y,t)$ for $y = y(x,t)$, where x are the demographic state variables $W$ and $Q$, and substituting for y in constraint Equation (1). But several things can go wrong with this program. First, the model approximation may not be valid; genetic and demographic processes can be tightly coupled. For example, density-dependent natural selection can alter gene frequencies significantly within a few genera-

tions—well within ecological time—and contribute to demographic changes which in turn shift the program of natural selection. This has been demonstrated in theory and actually occurs in populations of insects, rodents, and other prolifically breeding organisms (see Pimentel, 1968; Nolte et al., 1969; Krebs et al., 1973; Auslander et al., 1978). Second, the "static" genetic constraints might not be well behaved. Saltational changes due to gene migration, mutation, or recombination can produce a constraint surface with folds (or "catastrophes" in current mathematical jargon). This phenomenon enormously complicates the demographic dynamics. The likelihood of formulating an optimization scheme that captures such constraints is remote.

Still another pitfall is the possibility that the dynamic equations will exhibit chaotic behavior. Even though the equations may be perfectly deterministic, they can still behave in a wholly unpredictable way, in effect resembling a random process. Such bizarre behavior could prove quite common in ecological models when nonlinear density or frequency dependent effects are important (see May, 1976; May and Oster, 1976; Guckenheimer et al., 1977; Oster and Guckenheimer, 1977). No methods exist at the present time for the solution of optimization models when the constraints exhibit chaotic or discontinuous behavior. It is not even clear how extrema could be mathematically defined in an unambiguous way.

Short of chaotic behavior, the dynamic constraints can exhibit periodic behavior of various kinds. Population cycles are in fact a commonplace in ecology. Such responses generally preclude characterization by an optimization formulation, although cycle averages alone can be meaningful in some cases (see Holmes and Rand, 1977).

Finally, we must confront the problem of stochastic influences in the model equations. No realistic model can afford to neglect them, since all organisms evolve in a more or less uncertain environment. It was shown during our earlier analysis of caste ratios that reasoning with deterministic averages can yield totally erroneous conclusions. When an attempt is made to include stochastic effects, the mathematical arsenal is quickly exhausted, because stochastic optimization theory is still at a primitve state. Moreover, there is an important distinction between the definitions of the system "state" in deterministic and stochastic models. The state of a stochastic model is described not by a vector of precise measurements, $x(t)$, but by a probability distribution. Furthermore, maximizing ergonomic efficiency is generally *not* the same as maximizing reproductive fitness when stochastic influences are important. This is because the conversion of energy and labor into reproductive alates is expected to be nonlinear. Finally, the fitness maximum may be rather broad and flat. Thus selection might act—but weakly—in the vicinity of such optima, and one might expect the system to be dominated by rather large random deviations from the equilibrium. Computing the evolution of the system as well as validating the model empirically poses serious mathematical and experimental difficulties (cf. Astrom, 1970; Melsa and Sage, 1973). Some progress has been made in the case of static optimization by employing only mean-variance analysis to construct deterministic equivalent models. Even there, however, linear models became nonlinear. In the dynamic case the difficulties are enormously greater, and at the present stage we can hope for little more than exhaustive computer studies. These, in turn, will require large quantities of real data.

The evolutionary success of a species is determined by how successfully it tracks its physical and biological environment. Probabilistic uncertainty enters into the evolutionary process at every level, from randomness generated by allelic segregation, recombination, and mutation, through chance processes in mating patterns and such ecological forces as weather, climate, predation, and resource abundance. Cohen (1976) has presented some disturbing examples of how these probabilistic processes can generate a fundamentally unpredictable evolutionary process. Such models hint that historical accident plays a more decisive role in evolution than we have been willing to accept. Given our present state of knowledge, a scientist attempting to be scrupulously objective must remain skeptical concerning the potential long-range predictive power of evolutionary and ecological theory.

Yet for all the pessimism justified by these mathematical arguments, evolution cannot be demoted to the status of a random meander. Order and progression in evolutionary processes are consistently perceived. Nevertheless, we must always bear in mind the crucial fact that evolution is a history-dependent process. Adaptations are not "designed" *de novo* by nature. Rather, they are jury-rigged, using the material available at the time. Evolution, in the words of Jacob (1977), is a "tinkerer," not an engineer! As systems become more complex, the historical accidents play a more and more central role in determining the evolutionary path they will follow. This is simply because the combinatorial number of possible alternatives multiplies enormously as the complexity of the overall system increases through the addition and increase in size of the subsystems. When we compare the complexity of the eucaryotic genome with that of the hydrogen atom, it is little wonder that the physical scientist can perceive the "economy of nature" more clearly than the biologist!

## 2. WHAT IS THE PROPER ROLE FOR OPTIMIZATION MODELS IN EVOLUTIONARY BIOLOGY?

When all of the canonical limitations are taken into account (see Table 1), we are left with a more modest view of the role of optimization models in biology. Rather than a grand scheme for predicting the course of natural selection, optimization theory constitutes no more than a tactical tool for making educated guesses about evolutionary trends. If we wish to view evolution as an "optimizing" process and to retain mathematical modeling as an analytic tool, we are at least forced to admit that an element of teleology has entered the theorizing. This appears to be unavoidable and should be frankly admitted at the outset. The prudent course is to regard optimization models as provisional guides to further empirical research, and not necessarily as the key to deeper laws of nature. The proper role of optimization

models, in our view, is to provide the means for recreating short-term evolution in the imagination. To this end a three-step procedure is suggested:

(1) Construct a model on the basis of natural history experience and intuition. A complete model will contain four components: a state description, a collection of strategic alternatives, a set of fitness criteria, and a set of constraints.

(2) Work out the logical conclusions of the model.

(3) Compare the conclusions with empirical observations. Thus the model-building process can be analogous to a blind experiment in the laboratory. Models sufficiently complex to produce nonobvious results are also able to compete with one another, just as one experiment discloses the operation of the crucial process and another does not. If the surviving model then sufficiently matches the available empirical data, the biologist is provisionally justified in feeling that he has captured the essential features of the phenomenon. At the very least his intuition has been tested and enhanced. If the predictions of a particular model are erroneous, he knows precisely how to revise it, since the assumptions were made explicit and clear at the outset. Thus, step by step the investigator proceeds by the "strong inference" method that has proven so effective in the physical sciences and molecular biology (Platt, 1964).

Optimization models, while useful for surmising "strategies", can hardly ever address the most basic question of "tactics." That is, the particular chemical or physiological mechanism a species employs to implement a strategy is contingent on its history.

TABLE 1

The four essential components of optimization models and the difficulties encountered in specifying them in evolutionary biology.

| Model Component | Difficulties |
| --- | --- |
| 1. State space | a. Generally impossible to give a complete objective description. |
| 2. Strategy set | a. More genetic combinations are possible than can be specified. |
| | b. Usually not possible to specify how the strategy could be implemented. |
| 3. Optimization criteria | a. All tractable definitions are indirect. |
| | b. Difficult to define a common currency for multiple criteria and a proper global notion of equilibrium in conflict situations. |
| | c. Severe mathematical difficulties if the optimization criteria are not simple. |
| 4. Constraints | a. Impossible to enumerate all of the constraints. |
| | b. Constraints may be incompatible with stable optimum. |
| | c. Dynamic constraints may behave chaotically; static constraints may have "catastrophes." |
| | d. Mathematical machinery for handling stochastic effects is not well developed. |

Moreover, an optimization model cannot even find the best of all possible strategies; it can only eliminate many inferior ones from among a preselected set. The only practicable goal is to identify local optima, conditioned by the organism's ecological niche. If it is sound, the resulting theory is at least as explicit as the inductive formulations of natural history, and more vulnerable to falsification. Above all, it is linked to other generalizations by a shorter chain of logical steps. Good theory, being the product of human imagination, is also that which provides the greatest esthetic satisfaction, perhaps because esthetics is to some extent the judgment of formulations that work in the realm of the poorly understood. P. A. M. Dirac has said that physical theories with some mathematical beauty are also the ones most likely to be correct. However, in view of the dominant role of history in determining the awesome complexity of the biological world, we cannot as a rule expect mathematical theories in biology to have the same elegant simplicity we find in physical theories.

SUMMARY

In examining a behavioral or physiological trait, we can adopt one or the other of three philosophies: (1) the trait is a priori "adaptive," that is, has been molded by particular agents of natural selection; (2) it is a random event; (3) it is partially adaptive and partially an epiphenomenon created by stronger selection occuring on other traits. Much of the present discussion has been devoted to the demonstration that only (3) is supportable. Given the complexity of the eucaryotic genome, the finiteness of population sizes, and the limited number of generations in which evolution can occur, it is simply impossible for selection to optimize independently every aspect of an organism's genotype. But the nihilistic opposite of this conception is equally untenable, unless we are willing to deny the underpinnings of a large part of modern biology.

Some traits are certainly adaptive and have been "optimized" in the literal terms of elementary Neo-Darwinian theory, while others are random manifestations in the sense of being selectively neutral or secondary, canalized consequences of the existing genetic order. In our opinion, the way forward in evolutionary theory is not through the formulation of global statements about the evolutionary process, but through the prudent choice of paradigmatic examples that permit the role of natural selection to be analyzed with unusual clarity. Social insects appear to offer many such opportunities; they are the "squid axon" of evolutionary sociobiology.

In an effort to temper our own strongly selectionist view of the evolution of caste systems, we have in this discussion presented a critique of the use of optimization theory in evolutionary biology. We conclude that the mathematical techniques of the theory cannot be used to make long-range predictions of evolutionary processes. Indeed, the concept of lone optima toward which many species can

be said to be moving along certain trajectories appears to be an unsupportable metaphysical notion. Nevertheless, in many cases the course of evolution might be predicted over *short* distances. And if carefully constructed, optimization models can be used to identify local optima and to test adaptation hypotheses in a more rigorous manner than is possible by inductive natural history. The construction and testing of models is a potentially powerful technique analogous to blind experimentation conducted in the laboratory.

Good optimization theory requires a systematic accounting of four components of its models: a state space, incorporating all of the relevant variables; a set of conceivable strategies; one or more optimization criteria, or fitness functions, precisely defined; and a set of constraints that limit the approach to the idealized local optima. Biological systems create special difficulties for the characterization of each of these four components.

In spite of the difficulties, we feel justified in making optimization theory the cornerstone of caste theory. The rigidity of insect caste systems, their stability through evolutionary time, and the existence of literally thousands of species that can be examined as independent evolutionary experiments make the theoretical enterprise feasible. If the effort is successful, a deeper understanding of social evolution in the insects will be gained and general evolutionary theory can be additionally tested and refined.

## REFERENCES

Allen, E. et al. (35 authors, comprising the Sociobiology Study Group of Science for the People). 1976. Sociobiology—another biological determinism. *Bioscience,* 26 (3): 182, 184–186.

Astrom, K. 1970. *Introduction to Stochastic Control Theory,* New York, Academic Press.

Auslander, D., J. Guckenheimer, and G. Oster. 1978. Random evolutionarily stable strategies. *Theoretical Population Biology,* 13 (2): 276–293.

Bryson, A. and Y. Ho. 1975. *Applied Optimal Control,* New York, Wiley.

Desoer, C. 1970. *Notes for a Second Course in Systems Theory,* New York, Van Nostrand.

Dorny, C. 1975. *A Vector Space Approach to Models and Optimization,* New York, Wiley.

Ekeland, I. 1977. Discontinuités des champs Hamiltoniens et existance de solutions optimales en calcul de variations. *Publications Mathématiques,* 47: 5–32.

Friedman, M. 1953. *Essays in Positive Economics,* Chicago, University of Chicago Press.

Guckenheimer, J., G. G. Oster, and A. Ipaktchi. 1977. The dynamics of density dependent population models. *Journal of Mathematical Biology,* 4: 101–147.

Holmes, P. and D. Rand. 1977. "Bifurcation of the forced van der Pol oscillator." Unpublished manuscript.

Intriligator, M. 1971. *Mathematical Optimization and Economic Theory,* Englewood Cliffs, N.J., Prentice-Hall.

Jacob, F. 1977. Evolution and tinkering. *Science,* 196: 1161–1166.

Krebs, C. J., M. S. Gaines, B. L. Keller, J. H. Myers, and R. H. Tamarin. 1973. Population cycles in small rodents. *Science,* 179: 35–44.

Leitmann, G. 1966. *An Introduction to Optimal Control,* New York, McGraw-Hill.

Levin, S. and J. D. Udovic. 1977. A mathematical model of coevolving populations. *American Naturalist,* 111: 657–675.

Levins, R. 1968. "The limits of optimization," unpublished.

Lewontin, R. 1977. Adaptation. In *The Italian Encyclopedia,* Turin, Einaudi.

May, R. 1976. Simple mathematical models with very complicated dynamics. *Nature,* 261: 459–467.

May, R. and G. Oster. 1976. Bifurcations and dynamic complexity in simple ecological models. *American Naturalist,* 110: 573–599.

Maynard Smith, J. 1974. The theory of games and the evolution of animal conflicts. *Journal of Theoretical Biology,* 47: 209–221.

Maynard Smith, J. 1976. Evolution and the theory of games. *American Scientist,* 64 (1): 41–45.

Melsa, J. and A. Sage. 1973. *An Introduction to Probability and Stochastic Process,* Englewood Cliffs, N.J., Prentice-Hall.

Nolte, D. J., I. Desi, and B. Meyers. 1969. Genetic and environmental factors affecting chiasma formation in locusts. *Chromosoma,* Berlin, 27 (2): 145–55.

Oster, G. F. and J. Guckenheimer. 1977. Bifurcation behavior of population models. In J. Marsden and M. McCracken, eds., *The Hopf Bifurcation, Lecture Notes in Mathematics,* Vol. 19, pp. 327–53. New York, Springer-Verlag.

Oster, G. F. and S. Rocklin. 1978. Optimization models in evolutionary biology. In S. Levin, ed., *Some Mathematical Questions in Biology.* American Mathematical Society, Providence, Rhode Island.

Oster, G. F., and E. O. Wilson. 1978. *Caste and Ecology in the Social Insects.* Princeton, N.J., Princeton University Press.

Pimentel, D. 1968. Population regulation and genetic feedback. *Science,* 159: 1432–1437.

Platt, J. R. 1964. Strong inference. *Science,* 146: 347–353.

Rapport, D. and J. Turner. 1977. Economic models in ecology. *Science,* 195: 367–373.

Rocklin, S. and G. F. Oster. 1976. Competition between phenotypes. *Journal of Mathematical Biology,* 3: 225–261.

Roughgarden, J. 1978. *Theory of Population Genetics and Evolutionary Ecology, an Introduction.* New York, Macmillan.

Shashahani, S. 1978. A new mathematical framework for the study of linkage and selection. *Bulletin of the American Mathematical Society.*

Slatkin, M. 1977. On the equilibrium of fitness by natural selection. Unpublished manuscript.

Varaiya, P. 1973. *Notes on Optimization,* New York, Van Nostrand Reinhold.

# 17

# Optimization Theory in Evolution

JOHN MAYNARD SMITH

## INTRODUCTION

In recent years there has been a growing attempt to use mathematical methods borrowed from engineering and economics in interpreting the diversity of life. It is assumed that evolution has occurred by natural selection, hence that complex structures and behaviors are to be interpreted in terms of the contribution they make to the survival and reproduction of their possessors—that is, to Darwinian fitness. There is nothing particularly new in this logic, which is also the basis of functional anatomy, and indeed of much physiology and molecular biology. It was followed by Darwin himself in his studies of climbing and insectivorous plants, of fertilization mechanisms and devices to ensure cross-pollination.

What is new is the use of such mathematical techniques as control theory, dynamic programming, and the theory of games to generate a priori hypotheses, and the application of the method to behaviors and life history strategies. This change in method has led to the criticism (e.g., Lewontin, 54, 55) that the basic hypothesis of adaptation is untestable and therefore unscientific, and that the whole program of functional explanation through optimization has become a test of ingenuity rather than an inquiry into truth. Related to this is the criticism that there is no theoretical justification for any maximization principles in biology, and therefore that optimization is no substitute for an adequate genetic model.

My aim in this review is not to summarize the most important conclusions reached by optimization methods, but to discuss the methodology of the program and the criticisms that have been made of it. In doing so, I have taken as my starting point two articles by Lewontin (54, 55). I disagree with some of the views he expresses, but I believe that the development of evolution theory could benefit if workers in optimization paid serious attention to his criticisms.

I first outline the basic structure of optimization arguments, illustrating this with three examples, namely the sex ratio, the locomotion of mammals, and foraging behavior. I then discuss the possibility that some variation may be selectively neutral, and some structures maladaptive. I summarize and comment on criticisms made by Lewontin. The most damaging, undoubtedly, is the difficulty of testing the hypotheses that are generated. The next section therefore discusses the methodology of testing; in this section I have relied heavily on the arguments of Curio (23). Finally I discuss mathematical methods. The intention here is not to give the details of the mathematics, but to identify the kinds of problems that have been attacked and the assumptions that have been made in doing so.

## THE STRUCTURE OF
## OPTIMIZATION MODELS

In this section I illustrate the argument with three examples: (*a*) the sex ratio, based on Fisher's (28) treatment and later developments by Hamilton (34), Rosado and Robertson (85), Trivers and Willard (96), and Trivers and Hare (95); (*b*) the gaits of mammals—given a preliminary treatment by Maynard Smith and Savage (66), and further analyzed in several papers in Pedley (78); (*c*) foraging strategies. Theoretical work on them originated with the papers of Emlen (27) and MacArthur and Pianka (57). I have relied heavily on a recent review by Pyke et al. (81). These authors suggest that models have in the main been concerned with four problems: choice by the animal of which types of food to eat (optimal diet); choice of which patch type to

feed in; allocation of time to different patches; pattern and speed of movement. In what follows I shall refer only to two of those—optimal diet and allocation of time to different patches.

All optimization models contain, implicitly or explicitly, an assumption about the "constraints" that are operating, an optimization criterion, and an assumption about heredity. I consider these in turn.

### The Constraints: Phenotype Set and State Equations

The constraints are essentially of two kinds. In engineering applications, they concern the "strategy set," which specifies the range of control actions available, and the "state equations," which specify how the state of the system being controlled changes in time. In biological applications, the strategy set is replaced by an assumption about the set of possible phenotypes on which selection can operate.

It is clearly impossible to say what is the "best" phenotype unless one knows the range of possibilities. If there were no constraints on what is possible, the best phenotype would live forever, would be impregnable to predators, would lay eggs at an infinite rate, and so on. It is therefore necessary to specify the set of possible phenotypes, or in some other way describe the limits on what can evolve. The "phenotype set" is an assumption about what can evolve and to what extent; the "state equations" describe features of the situation that are assumed not to change. This distinction will become clearer when particular examples are discussed. Let us consider the three problems in turn.

*Sex Ratio.* For the sex ratio, the simplest assumption is that a parent can produce a fixed number $N$ of offspring, and that the probability $S$ that each birth will be a male can vary from parent to parent over the complete range from 0 to 1; the phenotype set is then the set of values of $S$ over this range. Fisher (28) extended this by supposing that males and females "cost" different amounts; i.e. he supposed that a parent could produce $\alpha$ males and $\beta$ females, where $\alpha$ and $\beta$ are constrained to lie on or below the line $\alpha + \beta k = N$, and $k$ is the cost of a female relative to that of a male. He then concluded that the parent should equalize expenditure on males and females. MacArthur (56) further broadened the phenotype set by insisting only that $\alpha$ and $\beta$ lie on or below a line of arbitrary shape, and concluded that a parent should maximize $\alpha\beta$. A similar assumption was used by Charnov et al. (11) to analyze the evolution of hermaphroditism as opposed to dioecy. Finally, it is possible to ask (97) what is the optimal strategy if a parent can choose not merely a value of $S$, hence of the expected sex ratio, but also the variance of the sex ratio.

The important point in the present context is that the optimal solution depends on the assumption made. For example, Crow and Kimura (21) conclude that the sex ratio should be unity, but they do so for a model that assumes that $N = \alpha + \beta$ is a constant.

*Gaits.* In the analysis of gaits, it is assumed that the shapes of bones can vary but the mechanical properties of bone, muscle, and tendon cannot. It is also assumed that changes must be gradual; thus the gaits of ostrich, antelope, and kangaroo are seen as different solutions to the same problem, not as solutions to different problems—i.e., they are different "adaptive peaks" (101).

*Foraging Strategy.* In models of foraging behavior, a common assumption is that the way in which an animal allocates its time among various activities (e.g., consuming one prey item rather than another, searching in one kind of patch rather than another, moving between patches rather than continuing to search in the same one) can vary, but the efficiency with which it performs each act cannot. Thus, for example, the length of time it takes to "handle" (capture and consume) a given item, the time and energy spent in moving from place to place, and the time taken to find a given prey item at a given prey density are taken as invariant. Thus the models of foraging so far developed treat the phenotype set as the set of possible behavioral strategies, and treat structure and locomotory or perceptual skills as constants contributing to the state equations (which determine how rapidly an animal adopting some strategy acquires food). In principle there is no reason why optimization models should not be applied to the evolution of structure or skill also; it is simply a question of how the phenotype set is defined.

### The Optimization Criterion

Some assumption must then be made concerning what quantity is being maximized. The most satisfactory is the inclusive fitness (see the section on Games Between Relatives, below, page 310); in many contexts the individual fitness (expected number of offspring) is equally good. Often, as in the second and third of my examples, neither criterion is possible, and some other assumption is needed. Two points must be made. First, the assumption about what is maximized is an assumption about what selective forces have been responsible for the trait; second, this assumption is part of the hypothesis being tested.

In most theories of sex ratio the basic assumption is that the ratio is determined by a gene acting in a parent, and what is maximized is the number of copies of that gene in future generations. The maximization has therefore a sound basis. Other maximization criteria have been used. For example, Kalmus and Smith (41) propose that the sex ratio maximizes the probability that two individuals meeting will be of different sexes; it is hard to understand such an eccentric choice when the natural one is available.

An equally natural choice—the maximization of the expected number of offspring produced in a lifetime—is available in theories of the evolution of life history strategies. But often no such easy choice is available.

In the analysis of gaits, Maynard Smith and Savage (66) assumed that the energy expenditure at a given speed would be minimized (or, equivalently, that the speed for a given energy expenditure was maximized). This led to the prediction that the proportion of time spent with all four legs off the ground should increase with speed and decrease with size.

In foraging theory, the common assumption is that the animal is maximizing its energy intake per unit time spent foraging. Schoener (87) points out that this is an appropriate choice, whether the animal has a fixed energy requirement and aims to minimize the time spent feeding so as to leave more time for other activities ("time minimizers"), or has a fixed time in which to feed during which it aims to maximize its energy gain ("energy maximizers"). There will, however, be situations in which this is not an appropriate choice. For example, there may be a higher risk of predation for some types of foraging than for others. For some animals the problem may be not to maximize energy intake per unit time, but to take in a required amount of energy, protein, etc. without taking an excess of any one of a number of toxins (S. A. Altmann, personal communication).

Pyke et al. (81) point out that the optimal strategy depends on the time scale over which optimization is carried out, for two reasons. First, an animal that has sole access to some resource (e.g., a territory-holder) can afford to manage that resource so as to maximize its yield over a whole season. Second, and more general, optimal behavior depends on a knowledge of the environment, which can be acquired only by experience; this means that in order to acquire information of value in the long run, an animal may have to behave in a way that is inefficient in the short run.

Having considered the phenotype set and the optimization criterion, a word must be said about their relationship to Levins' (51) concept of a fitness set. Levins was explicitly concerned with defining fitness "in such a way that interpopulation selection would be expected to change a species towards the optimum (maximum fitness) structure." This essentially group-selectionist approach led him to conclusions (e.g. for the conditions for a stable polymorphism) different from those reached from the classic analysis of gene frequencies (93). Nevertheless, Levins' attempt to unite ecological and genetic approaches did lead him to recognize the need for the concept of a fitness set—i.e., the set of all possible phenotypes, each phenotype being characterized by its (individual) fitness in each of the environments in which it might find itself.

Levins' fitness set is thus a combination of what I have called the phenotype set and of a measure of the fitness of each phenotype in every possible environment. It did not allow for the fact that fitnesses may be frequency-dependent (see the section on Games, below, page 307). The valuable insight in Levins' approach is that it is possible to discuss what course phenotypic evolution may take only if one makes explicit assumptions about the constraints on what phenotypes are possible. It may be better to use the term "phenotype set" to define these constraints, both because a description of possible phenotypes is a process prior to and separable from an estimation of their fitnesses, and because of the group-selectionist associations of the term "fitness set."

### An Assumption about Heredity

Because natural selection cannot produce adaptation unless there is heredity, some assumption, explicit or otherwise, is always present. The nature of this assumption can be important. Fisher (28) assumed that the sex ratio was determined by autosomal genes expressed in the parent, and that mating was random. Hamilton (34) showed that the predicted optima are greatly changed if these assumptions are altered. In particular, he considered the effects of inbreeding, and

of genes for meiotic drive. Rosado and Robertson (85), Trivers and Willard (96), and Trivers and Hare (95) have analyzed the effects of genes acting in the children and (in Hymenoptera) in the sterile castes.

It is unusual for the way in which a trait is inherited to have such a crucial effect. Thus in models of mammalian gaits no explicit assumption is made; the implicit assumption is merely that like begets like. The same is true of models of foraging, although in this case "heredity" can be cultural as well as genetic—e.g. (72), for the feeding behavior of oyster-catchers.

The question of how optimization models can be tested is the main topic of the next three sections. A few preliminary remarks are needed. Clearly, the first requirement of a model is that the conclusions should follow from the assumptions. This seems not to be the case, for example, for Zahavi's (102) theory of sexual selection (61). A more usual difficulty is that the conclusions depend on unstated assumptions. For example, Fisher does not state that his sex ratio argument assumes random mating, and this was not noticed until Hamilton's 1967 paper (34). Maynard Smith and Price (65) do not state that the idea of an ESS (evolutionary stable strategy) assumes asexual inheritance. It is probably true that no model ever states all its assumptions explicitly. One reason for writing this review is to encourage authors to become more aware of their assumptions.

A particular model can be tested either by a direct test of its assumptions or by comparing its predictions with observation. The essential point is that in testing a model we are testing *not* the general proposition that nature optimizes, but the specific hypotheses about constraints, optimization criteria, and heredity. Usually we test whether we have correctly identified the selective forces responsible for the trait in question. But we should not forget hypotheses about constraints or heredity. For example, the weakest feature of theories concerning the sex ratio is that there is little evidence for the existence of genetic variance of the kind assumed by Fisher—for references, see (63). It may be for this reason that the greatest successes of sex ratio theory (34, 95) have concerned Hymenoptera, in which it is easy to see how genes in the female parent can affect the sex of her children.

### NEUTRALITY AND MALADAPTATION

I have said that when testing optimization models, one is not testing the hypothesis that nature optimizes. But if it is not the case that the structure and behavior of organisms are nicely adapted to ensure their survival and reproduction, optimization models cannot be useful. What justification have we for assuming this?

The idea of adaptation is older than Darwinism. In the form of the argument from design, it was a buttress of religious belief. For Darwin the problem was not to prove that organisms were adapted but to explain how adaptation could arise without a creator. He was quite willing to accept that some characteristics are "selectively neutral." For example, he says (26) of the sterile dark red flower at the center of the umbel of the wild carrot: "That the modified central flower is of no functional importance to the plant is almost certain." Indeed,

Darwin has been chided by Cain (8) for too readily accepting Owen's argument that the homology between bones of limbs of different vertebrates is nonadaptive. For Darwin the argument was welcome, because the resemblance could then be taken as evidence for genetic relationship (or, presumably, for a paucity of imagination on the part of the creator). But Cain points out that the homology would not have been preserved if it were not adaptive.

Biologists differ greatly in the extent to which they expect to find a detailed fit between structure and function. It may be symptomatic of the times that when, in conversation, I raised Darwin's example of the carrot, two different functional explanations were at once suggested. I suspect that these explanations were fanciful. But however much one may be in doubt about the function of the antlers of the Irish Elk or the tail of the peacock, one can hardly suppose them to be selectively neutral. In general, the structural and behavioral traits chosen for functional analysis are of a kind that rules out neutrality as a plausible explanation. Curio (23) makes the valid point that the ampullae of Lorenzini in elasmobranchs were studied for many years before their role in enabling a fish to locate prey buried in the mud was demonstrated (40), yet the one hypothesis that was never entertained was that the organ was functionless. The same could be said of Curio's own work (24) on the function of mobbing in birds; behavior so widespread, so constant, and so apparently dangerous calls for a functional explanation.

There are, however, exceptions to the rule that functional investigations are carried out with the aim of identifying particular selective forces, and not of demonstrating that traits are adaptive. The work initiated by Cain and Sheppard (9) on shell color and banding in *Cepaea* was in part aimed at refuting the claim that the variation was selectively neutral and explicable by genetic drift. To that extent the work was aimed at demonstrating adaptation as such; it is significant, however, that the work has been most successful when it has been possible to identify a particular selection pressure (e.g., predation by thrushes).

At present, of course, the major argument between neutral and selective theories concerns enzyme polymorphism. I cannot summarize the argument here, but a few points on methodology are relevant. The argument arose because of the formulation by Kimura (43) and King and Jukes (44) of the "neutral" hypothesis; one reason for proposing it was the difficulty of accounting for the extensive variation by selection. Hence the stimulus was quite different from that prompting most functional investigations; it was the existence of widespread variation in a trait of no obvious selective significance.

The neutral hypothesis is a good "Popperian" one; if it is false, it should be possible to show it. In contrast, the hypothesis of adaptation is virtually irrefutable. In practice, however, the statistical predictions of the neutral theory depend on so many unknowns (mutation rates, the past history of population number and structure, hitch-hiking from other loci) that it has proved hard to test (53). The difficulties have led some geneticists (e.g., 14) to propose that the only way in which the matter can be settled is by the classical methods of ecological genetics—i.e., by identifying the specific selection pressures associated with particular enzyme loci. The approach has had some success but is always open to the objection that the loci for which the neutral hypothesis has been falsified are a small and biased sample.

In general, then, the problems raised by the neutral mutation theory and by optimization theory are wholly different. The latter is concerned with traits that differ between species and that can hardly be selectively neutral but whose selective significance is not fully understood.

A more serious difficulty for optimization theory is the occurrence of maladaptive traits. Optimization is based on the assumption that the population is adapted to the contemporary environment, whereas evolution is a process of continuous change. Species lag behind a changing environment. This is particularly serious when studying species in an environment that has recently been drastically changed by man. For example, Lack (48) argued that the number of eggs laid by a bird maximizes the number of surviving young. Although there is much supporting evidence, there are some apparent exceptions. For example, the gannet *Sula bassana* lays a single egg. Studying gannets on the Bass Rock, Nelson (71) found that if a second egg is added, the pair can successfully raise two young. The explanation can hardly be a lack of genetic variability, because species nesting in the Humboldt current off Peru lay lay two or even three eggs and successfully raise the young.

Lack (48) suggests that the environment for gannets may recently have improved, as evidenced by the recent increase in the population on the Bass Rock. Support for this interpretation comes from the work of Jarvis (39) on the closely related *S. capensis* in South Africa. This species typically lays one egg, but one percent of nests contain two. Using methods similar to Nelson's, Jarvis found that a pair can raise two chicks to fledgings, but that the average weight of twins was lower than singles, and in each nest one twin was always considerably lighter than its fellow. There is good evidence that birds fledging below the average weight are more likely to die soon after. Difficulties of a similar kind arise for the Glaucous Gull (see 45).

The undoubted existence of maladaptive traits, arising because evolutionary change is not instantaneous, is the most serious obstacle to the testing of optimization theories. The difficulty must arise; if species were perfectly adapted, evolution would cease. There is no easy way out. Clearly a wholesale reliance on evolutionary lag to save hypotheses that would otherwise be falsified would be fatal to the whole research program. The best we can do is to invoke evolutionary lag sparingly, and only when there are independent grounds for believing that the environment has changed recently in a relevant way.

What then is the status of the concept of adaptation? In the strong form—that all organs are perfectly adapted—it is clearly false; the vermiform appendix is sufficient to refute it. For Darwin, adaptation was an obvious fact that required an explanation; this still seems a sensible point of view. Adaptation can also be seen as a necessary consequence of natural selection. The latter I regard as a refutable

scientific theory (60); but it must be refuted, if at all, by genetic experiment and not by the observation of complex behavior.

### CRITIQUES OF OPTIMIZATION THEORY

Lewontin (55) raises a number of criticisms, which I discuss in turn.

#### Do Organs Solve Problems?

Most organs have many functions. Therefore, if a hypothesis concerning function fails correctly to predict behavior, it can always be saved by proposing an additional function. Thus hypotheses become irrefutable and metaphysical, and the whole program merely a test of ingenuity in conceiving possible functions. Three examples follow: the first is one used by Lewontin.

Orians and Pearson (73) calculated the optimal food item size for a bird, on the assumption that food intake is to be maximized. They found that the items diverged from random in the expected direction, but did not fit the prediction quantitatively. They explained the discrepancy by saying that a bird must visit its nest frequently to discourage predators. Lewontin (54) comments:

> This is a paradigm for adaptive reconstruction. The problem is originally posed as efficiency for food-gathering. A deviation of behavior from random, in the direction predicted, is regarded as strong support for the adaptive explanation of the behavior and the discrepancy from the predicted optimum is accounted for by an ad hoc secondary

problem which acts as a constraint on the solution to the first. . . . By allowing the theorist to postulate various combinations of "problems" to which manifest traits are optimal "solutions", the adaptationist programme makes of adaptation a metaphysical postulate, not only incapable of refutation, but necessarily confirmed by every observation. This is the caricature that was immanent in Darwin's insight that evolution is the product of natural selection.

It would be unfair to subject Orians alone to such criticism, so I offer two further examples from my own work.

First, as explained earlier, Maynard Smith and Savage (66) predicted qualitative features of mammalian gaits. However, their model failed to give a correct quantitative prediction. I suspect that if the model were modified to allow for wind resistance and the visco-elastic properties of muscle, the quantitative fit would be improved; at present, however, this is pure speculation. In fact, it looks as if a model that gives quantitatively precise predictions will be hard to devise (1).

Second, Maynard Smith and Parker (64) predicted that populations will vary in persistence or aggressiveness in contest situations, but that individuals will not indicate their future behavior by varying levels of intensity of display. Rohwer (84) describes the expected variability in aggressivity in the Harris sparrow in winter flocks, but also finds a close correlation between aggressivity and a signal (amount of black in the plumage). I could point to the first observation as a confirmation of our theory, and explain how, by altering the model (by changing

the phenotype set to permit the detection of cheating), one can explain the second.

What these examples, and many others, have in common is that a model gives predictions that are in part confirmed by observation but that are contradicted in some important respect. I agree with Lewontin that such discrepancies are inevitable if a simple model is used, particularly a model that assumes each organ or behavior to serve only one function. I also agree that if the investigator adds assumptions to his model to meet each discrepancy, there is no way in which the hypothesis of adaptation can be refuted. But the hypothesis of adaptation is not under test.

What is under test is the specific set of hypotheses in the particular model. Each of the three example models above has been falsified, at least as a complete explanation of these particular data. But since all have had qualitative success, it seems quite appropriate to modify them (e.g. by allowing for predation, for wind resistance, for detection of cheating). What is not justified is to modify the model and at the same time to claim that the model is confirmed by observation. For example, Orians would have to show that his original model fits more closely in species less exposed to predation. I would have to show that Rohwer's data fit the "mixed ESS" model in other ways—in particular, that the fitness of the different morphs are approximately equal. If, as may well be the case, the latter prediction of the ESS model does not hold, it is hard to see how it could be saved.

If the ESS model proves irrelevant to the Harris sparrow, it does not follow, however, that it is never relevant. By analogy, the assertion is logically correct that there will be a stable polymorphism if the heterozygote at a locus with two alleles is fitter than either homozygote. The fact that there are polymorphisms not maintained by heterosis does not invalidate the logic. The (difficult) empirical question is whether polymorphisms are often maintained by heterosis. I claim a similar logical status for the prediction of a mixed ESS.

In population biology we need simple models that make predictions that hold qualitatively in a number of cases, even if they are contradicted in detail in all of them. One can say with some confidence, for example, that no model in May's *Stability and Complexity in Model Ecosystems* describes exactly any actual case, because no model could ever include all relevant features. Yet the models do make qualitative predictions that help to explain real ecosystems. In the analysis of complex systems, the best we can hope for are models that capture some essential feature.

To summarize my comments on this point, Lewontin is undoubtedly right to complain if an optimizer first explains the discrepancy between theory and observation by introducing a new hypothesis, and then claims that his modified theory has been confirmed. I think he is mistaken in supposing that the aim of optimization theories is to confirm a general concept of adaptation.

### Is There Genetic Variance?

Natural selection can optimize only if there is appropriate genetic variance. What justification is there for assuming

the existence of such variance? The main justification is that, with rare exceptions, artificial selection has always proved effective, whatever the organism or the selected character (53).

A particular difficulty arises because genes have pleiotropic effects, so that selection for trait A may alter trait B; in such cases, any attempt to explain the changes in B in functional terms is doomed to failure. There are good empirical grounds for doubting whether the difficulty is as serious as might be expected from the widespread nature of pleiotropy. The point can best be illustrated by a particular example. Lewontin (54) noted that in primates there is a constant allometric relationship between tooth size and body size. It would be a waste of time, therefore, to seek a functional explanation of the difference between the tooth size of the gorilla and of the rhesus monkey, since the difference is probably a simple consequence of the difference in body size.

It is quite true that for most teeth there is a constant allometric relationship between tooth and body size, but there is more to it than that (36). The canine teeth (and the teeth occluding with them) of male primates are often larger than those of females, even when allowance has been made for the difference in body size. This sex difference is greater in species in which males compete for females than in monogamous species, and greater in ground-living species (which are more exposed to predation) than in arboreal ones. Hence, there is sex-limited genetic variance for canine tooth size, independent of body size, and the behavioral and ecological correlations suggest that this variance has been the basis of adaptation. It would be odd if there were tooth-specific, sex-limited variance, but no variance for the relative size of the teeth as a whole. However, there is some evidence for the latter. The size of the cheek teeth in females (relative to the size predicted from their body size) is significantly greater in those species with a higher proportion of leaves (as opposed to fruit, flowers, or animal matter) in their diets.

Thus, although at first sight the data on primate teeth suggest that there may be nothing to explain in functional terms, a more detailed analysis presents quite a different picture. More generally, changes in allometric relationships can and do occur during evolution (30).

I have quoted Lewontin as a critic of adaptive explanation, but it would misinterpret him to imply that he rejects all such explanations. He remarks (54) that "the serious methodological difficulties in the use of adaptive arguments should not blind us to the fact that many features of organisms are adaptations to obvious environmental 'problems.'" He goes on to argue that if natural selection is to produce adaptation, the mapping of character states into fitnesses must have two characteristics: "continuity" and "quasi-independence." By continuity is meant that small changes in a character result in small changes in the ecological relations of the organism; if this were not so, it would be hard to improve a character for one role without ruining it for another. By quasi-independence is meant that the developmental paths are such that a variety of mutations may occur, all with the same effect on the primary character, but with different effects on other characters.

It is hard to think of better evidence for quasi-independence than the evolution of primate canines.

To sum up this point, I accept the logic of Lewontin's argument. If I differ from him (and on this point he is his own strongest critic), it is in thinking that genetic variance of an appropriate kind will usually exist. But it may not always do so.

It has been an implicit assumption of optimization models that the optimal phenotype can breed true. There are two kinds of reasons why this might not be true. The first is that the optimal phenotype may be produced by a heterozygote. This would be a serious difficulty if one attempted to use optimization methods to analyze the genetic structure of populations, but I think that would be an inappropriate use of the method. Optimization models are useful for analyzing phenotypic evolution, but not the genetic structuring of populations. A second reason why the optimal phenotype may not breed true is more serious: the evolutionarily stable population may be phenotypically variable. (This point is discussed further in the section on Games, below, page 307.)

The assumption concerning the phenotype set is based on the range of variation observable within species, the phenotypes of related species, and on plausible guesses at what phenotypes might arise under selection. It is rare to have any information on the genetic basis of the phenotypic variability. Hence, although it is possible to introduce specific genetic assumptions into optimization models (e.g. 2, 89), this greatly complicates the analysis. In general, the assumption of "breeding true" is reasonable in particular applications; models in which genes appear explicitly need to be analyzed to decide in what situations the assumption may mislead us.

### The Effects of History

If, as Wright (101) suggested, there are different "adaptive peaks" in the genetic landscape, then depending on initial conditions, different populations faced with identical "problems" may finish up in different stable states. Such divergence may be exaggerated if evolution takes the form of a "game" in which the optimal phenotype for one individual depends on what others are doing (see the section on Games, below). An example is Fisher's (28) theory of sexual selection, which can lead to an "auto-catalytic" exaggeration of initially small differences. Jacob (38) has recently emphasized the importance of such historical accidents in evolution.

As an example of the difficulties that historical factors can raise for functional explanations, consider the evolution of parental care. A simple game-theory model (62) predicts that for a range of ecological parameters either of two patterns would be stable: male parental care only, or female care only. Many fish and amphibia show one or the other of these patterns. At first sight, the explanation of why some species show one pattern and others the other seems historical; the reasons seem lost in an unknown past. However, things may not be quite so bad. At a recent discussion of fish behavior at See-Wiesen the suggestion emerged that if uniparental care evolved from no parental care, it would be male care, whereas if it evolved from biparental care it would

be female care. This prediction is plausible in the light of the original game-theory model, although not a necessary consequence of it. It is, however, testable by use of the comparative data; if it is true, male care should occur in families that also include species showing no care, and female care in families that include species showing biparental care. This may not prove to be the case; the example is given to show that even if there are alternative adaptive peaks, and in the absence of a relevant fossil record, it may still be possible to formulate testable hypotheses.

### What Optimization Criterion Should One Use?

Suppose that, despite all difficulties, one has correctly identified the "problem." Suppose, for example, that in foraging it is indeed true that an animal should maximize $E$, its rate of energy intake. We must still decide in what circumstances to maximize $E$. If the animal is alone in a uniform environment, no difficulty arises. But if we allow for competition and for a changing environment, several choices of optimization procedure are possible. For example, three possibilities arise if we allow just for competition:

(1) The "maximum" solution: Each animal maximizes $E$ on the assumption that other individuals behave in the least favorable way for it.
(2) The "Pareto" point: The members of the population behave so that no individual can improve its intake without harming others.
(3) The ESS: The members of the

population adopt feeding strategy *I* such that no mutant individual adopting a strategy other than *I* could do better than typical members.

These alternatives are discussed further in the section on Games, below. For the moment, it is sufficient to say that the choice among them is not arbitrary, but follows from assumptions about the mode of inheritance and the population structure. For individual selection and parthenogenetic inheritance, the ESS is the appropriate choice.

Lewontin's criticism would be valid if optimizers were in the habit of assuming the truth of what Haldane once called "Pangloss's theorem," which asserts that animals do those things that maximize the chance of survival of their species. If optimization rested on Pangloss's theorem it would be right to reject it. My reason for thinking that Lewontin regards optimization and Pangloss's theorem as equivalent is that he devotes the last section of his paper to showing that in *Drosophila* a characteristic may be established by individual selection and yet may reduce the competitive ability of the population relative to others. The point is correct and important, but in my view does not invalidate most recent applications of optimization.

## THE METHODOLOGY OF TESTING

The crucial hypothesis under test is usually that the model correctly incorporates the selective forces responsible for the evolution of a trait. Optimization models sometimes make fairly precise quantitative predictions that can be tested. However, I shall discuss the question how functional explanations can be tested more generally, including cases in which the predictions are only qualitative. It is convenient to distinguish comparative, quantitative, and individual-variation methods.

### Comparative Tests

Given a functional hypothesis, there are usually testable predictions about the development of the trait in different species. For example, two main hypothesis have been proposed to account for the greater size of males in many mammalian species: It is a consequence of competition among males for females; or it arises because the two sexes use different resources. If the former hypothesis is true, dimorphism should be greater in harem-holding and group-living species, whereas if the latter is true it should be greater in monogamous ones, and in those with a relatively equal adult sex ratio.

Clutton-Brock et al. (16) have tested these hypotheses by analyzing 42 species of primates (out of some 200 extant species) for which adequate breeding data are available. The data are consistent with the sexual selection hypothesis, and show no sign of the trend predicted by the resource differentiation hypothesis. The latter can therefore be rejected, at least as a major cause of sexual dimorphism in primates. It does not follow that inter-male competition is the only relevant selective factor (82). Nor do their observations say anything about the causes of sexual dimorphism in other groups. It is interesting (though not strictly relevant at this point) that the analysis also showed a strong correlation between female body size and degree of dimorphism. This trend, as was first noted by Rensch (83), occurs in a number of taxa, but has never received an entirely satisfactory explanation.

The comparative method requires some criterion for inclusion of species. This may be purely taxonomic (e.g. all primates, all passerine birds), or jointly taxonomic and geographic (e.g., all African ungulates, all passerines in a particular forest). Usually, some species must be omitted because data are not available. Studies on primates can include a substantial proportion of extant species (16, 68); in contrast, Schoener (86), in one of the earliest studies of this type, included all birds for which data were available and which also met certain criteria of territoriality, but he had to be content with a small fraction of extant species. It is therefore important to ask whether the sample of species is biased in ways likely to affect the hypothesis under test. Most important is that there be some criterion of inclusion, since otherwise species may be included simply because they confirm (or contradict) the hypothesis under test.

Most often, limitations of data will make it necessary to impose both taxonomic and geographic criteria. This need not prevent such data from being valuable, either in generating or in testing hypotheses; examples are analyses of flocking in birds (7, 31) and of breeding systems in forest plants (3, 4).

A second kind of difficulty concerns the design of significance tests. Different species cannot always be treated as statistically independent. For example, all gibbons are monogamous, and all are arboreal and frugivorous, but since all may be descended from a single ancestor with these properties, they should be treated as a single case in any test of association (not that any is suspected). To take an actual example of this difficulty, Lack (49) criticized Verner and Willson's (98) conclusion that polygamy in passerines is associated with marsh and prairie habitats on the grounds that many of the species concerned belong to a single family, the Icteridae.

Statistical independence and other methodological problems in analyzing comparative data are discussed by Clutton-Brock and Harvey (17). In analyzing the primate data, they group together as a single observation all congeneric species belonging to the same ecological category. This is a conservative procedure, in that it is unlikely to find spurious cases of statistical significance. Their justification for treating genera, but not families, as units is that for their data there are significant differences between genera within families for seven of the eight ecological and behavioral variables, but significant additional variation between families for only two of them. It may be, however, that a more useful application of statistical methods is their use (17) of partial regression, which enables them to examine the effects of a particular variable when the effects of other variables have been removed, and to ask how much of the total variation in some trait is accounted for by particular variables.

### Quantitative Tests

Quantitative tests can be illustrated by reference to some of the predictions of foraging theory. Consider first the problem of optimal diet. The following model situation has been widely assumed. There are a number of different kinds of food items. An animal can search simultaneously for all of them. Each item has a characteristic food value and "handling time" (the time taken to capture and consume it). For any given set of densities and hence frequencies of encounter, the animal must only decide which items it should consume and which ignore.

Pyke et al. (81) remark that no fewer than eight authors have independently derived the following basic result. The animal should rank the items in order of $V =$ food value/handling time. Items should be added to the diet in rank order, provided that for each new item the value of $V$ is greater than the rate of food intake for the diet without the addition. This basic result leads to three predictions:

(1) Greater food abundance should lead to greater specialization. This qualitative prediction was first demonstrated by Ivlev (37) for various fish species in the laboratory, and data supporting it have been reviewed by Schoener (87). Curio (25) quotes a number of cases that do not fit.

(2) For fixed densities, a food type should either be always taken, or never taken.

(3) Whether a food item should be taken is independent of its density, and depends on the densities of food items of higher rank.

Werner and Hall (100) allowed bluegill sunfish to feed on *Daphnia* of three different size classes; the diets observed agreed well with the predictions of the model. Krebs et al. (47) studied Great Tits foraging for parts of mealworms on a moving conveyor belt. They confirmed prediction 3 but not 2; that is, they found that whether small pieces were taken was independent of the density of small pieces, but, as food abundance rose, small pieces were dropped only gradually from the diet. Goss-Custard (29) has provided field evidence confirming the model from a study of redshank feeding on marine worms of different sizes, and Pulliam (80) has confirmed it for Chipping Sparrows feeding on seeds.

Turning to the problem of how long an animal should stay in a patch before moving to another, there is again a simple prediction, which Charnov (10) has called the "Marginal Value Theorem" (the same theorem was derived independently by Parker and Stuart (77) in a different context). It asserts that an animal should leave a patch when its rate of intake in the patch (its "marginal" rate) drops to the average rate of intake for the habitat as a whole. It is a corollary that the marginal rate should be the same for all patches in the habitat. Two laboratory experiments on tits (20, 46) agree well with the prediction.

A more general problem raised by these experiments is discussed by Pyke et al. (81). How does an animal estimate the parameters it needs to know before it can perform the required optimization? How much time should it spend acquiring information? Sometimes these questions may receive a simple answer. Thus the results of Krebs et al. (46) suggest that a bird leaves a patch if it has not found an item of food for some fixed period $\tau$ (which varied with the overall abundance of food). The bird seems to be using $\tau$, or rather $1/\tau$, as an estimate of its marginal capture rate. But not all cases are so simple.

### Individual Variation

The most direct way of testing a hypothesis about adaptation is to compare individuals with different phenotypes, to see whether their fitnesses vary in the way predicted by the hypothesis. This was the basis of Kettlewell's (42) classic demonstration of selection on industrial melanism in moths. In principle, the individual differences may be produced by experimental interference (Curio's (23) "method of altering a character") or they may be genetic or of unknown origin (Curio's "method of variants"). Genetic differences are open to the objection that genes have pleiotropic effects, and occasionally are components of supergenes in which several closely linked loci affecting the same function are held in linkage disequilibrium, so that the phenotypic difference responsible for the change in fitness may not be the one on which attention is concentrated. This difficulty, however, is trivial compared to that which arises when two species are compared.

The real difficulty in applying this method to behavioral differences is that suitable individual differences are often absent and experimental interference is impractical. Although it is hard to alter behavior experimentally, it may be possible to alter its consequences. Tinbergen et al. (94) tested the idea that gulls remove egg shells from the nest because the shells attract predators to their eggs and young; they placed egg shells close to eggs and recorded a higher predation rate.

However, the most obvious field of application of this method arises when a population is naturally variable. Natural variation in a phenotype may be maintained by frequency-dependent selection; in game-theoretical terms, the stable state may be a mixed strategy. If a particular case of phenotypic variability (genetic or not) is thought to be maintained in this way, it is important to measure the fitnesses of individuals with different phenotypes. At a mixed ESS (which assumes parthenogenetic inheritance) these fitnesses are equal; with sexual reproduction, exact equality is not guaranteed, but approximate equality is a reasonable expectation (91). If the differences are not genetic, we still expect a genotype to evolve that adopts the different strategies with frequencies that equalize their payoffs.

The only test of this kind known to me is Parker's (76) measurement of the mating success of male dungflies adopting different strategies. His results are consistent with a "mixed ESS" interpretation; it is not known whether the differences are genetic. The importance of tests of this kind lies in the fact that phenotypic variability can have other explanations; for example, it may arise from random environmental effects, or from genes with heterotic effects. In such cases, equality of fitness between phenotypes is not expected.

During the past twenty years there has been a rapid development of mathematical techniques aimed at solving problems of optimization and control arising in economics and engineering. These stem from the concepts of "dynamic programming" (5) and of the "maximum principle" (79). The former is essentially a computer procedure to seek the best control policy in particular cases without the hopelessly time-consuming task of looking at every possibility. The latter is an extension of the classic methods of the calculus of variations that permits one to allow for "inequality" constraints on the state and control variables (e.g., in the resource allocation model discussed below, the proportion $u$ of the available resources allocated to seeds must obey the constraint $u < 1$).

This is not the place to describe these methods, even if I were competent to do so. Instead, I shall describe the kinds of problems that can be attacked. If a biologist has a problem of one of these kinds, he would do best to consult a mathematician. For anyone wishing to learn more of the mathematical background, Clark (12) provides an excellent introduction.

I discuss in turn "optimization," in which the problem is to choose an optimal policy in an environment without competitors; "games," in which the environment includes other "players" who are also attempting to optimize something; and "games of inclusive fitness," in which the "players" have genes in common. I shall use as an illustration the allocation of resources between growth and reproduction.

## Optimization

*Choice of a Single Value.* The simplest type of problem, which requires for its solution only the technique of differentiation, is the choice of a value for a single parameter. For example, in discussing the evolution of gaits, Maynard Smith and Savage (66) found an expression for $P$, the power output, as a function of the speed $V$, of size $S$, and of $J$, the fraction of time for which all four legs are off the ground. By solving the equation $dP/dJ = 0$, an equation $J = f(V, S)$ was obtained, describing the optimum gait as a function of speed and size.

Few problems are as simple as this, but some more complex cases can be reduced to problems of this kind, as will appear below.

*A Simple Problem in Sequential Control.* Most optimization theory is concerned with how a series of sequential decisions should be taken. For example, consider the growth of an annual plant (19, 69). The rate at which the plant can accumulate resources depends on its size. The resources can be allocated either for further growth, or to seeds, or divided between them. For a fixed starting size and length of season, how should the plant allocate its resources so as to maximize the total number of seeds produced?

In this problem the "state" of the system at any time is given simply by the plant's size, $x$; the "control variable" $u(t)$ is the fraction of the incoming resource allocated to seeds at time $t$; the "constraints" are the

the initial size, the length of the season, the fact that $u(t)$ must lie between 0 and 1, and the "state equation,"

1.        $dx/dt = F[x(t), u(t)]$,

which describes how the system changes as a function of its state and of the control variable.

If equation 1 is linear in $u$, it can be shown that the optimal control is "bang-bang"—that is, $u(t) = 0$ up to some critical time $t^*$, and subsequently $u(t) = 1$. The problem is thus reduced to finding the single value, $t^*$. But if equation 1 is nonlinear, or has stochastic elements, the optimal control may be graded.

*More Complex Control Problems.* Consider first the "state" of the system. This may require description by a vector rather than by a single variable. Thus suppose the plant could also allocate resources to the production of toxins that increased its chance of survival. Then its state would require measures of both size and toxicity. The state description must be sufficient for the production of a state equation analogous to equation 1. The state must also include any information used in determining the control function $u(t)$. This is particularly important when analyzing the behavior of an animal that can learn. Thus suppose that an animal is foraging and that its decisions on whether to stay in a given patch or to move depend on information it has acquired about the distribution of food in patches; then this information is part of the state of the animal. For a discussion, see (20).

Just as the state description may be multidimensional, so may the control function; for example, for the toxic plant the control function must specify the allocation both to seeds and to toxins.

The state equation may be stochastic. Thus the growth of a plant depends on whether it rains. A plant may be supposed to "know" the probability of rain (i.e., its genotype may be adapted to the frequency of rain in previous generations) but not whether it will actually rain. In this case, a stochastic state equation may require a graded control. This connection between stochasticity and a "compromise" response as opposed to an all-or-none one is a common feature of optimal control. A second example is the analysis by Oster and Wilson (75) of the optimal division into castes in social insects: A predictable environment is likely to call for a single type of worker, while an uncertain one probably calls for a division into several castes.

*Reverse Optimality.* McFarland (67) has suggested an alternative approach. The typical one is to ask how an organism should behave in order to maximize its fitness. Mathematically, this requires that one define an "objective function" that must be maximized" ("objective" here means "aim" or "goal"); in the plant example, the objective function is the number of seeds produced, expressed as a function of $x$ and $u(t)$. But a biologist may be faced with a different problem. Suppose that he knew, by experiment, how the plant actually allocates its resources. He could then ask what the plant is actually maximizing. If the plant is perfectly adapted, the objective function so obtained should correspond to what Sibly and McFarland (88) call the "cost function"—i.e., the function that should be maximized if the organism is maximizing its fitness. A discrepancy would indicate maladaptation.

There are difficulties in seeing how this process of reverse optimality can be used. Given that the organism's behavior is "consistent" (i.e., if it prefers $A$ to $B$ and $B$ to $C$, it prefers $A$ to $C$), it is certain that its behavior maximizes *some* objective function; in general there will be a set of functions maximized. Perfect adaptation then requires only that the cost function correspond to one member of this set. A more serious difficulty is that it is not clear what question is being asked. If a discrepancy is found, it would be hard to say whether this was because costs had been wrongly measured or because the organism was maladapted. This is a particular example of my general point that it is not sensible to test the hypothesis that animals optimize. But it may be that the reverse optimality approach will help to analyze how animals in fact make decisions.

### Games

Optimization of the kind just discussed treats the environment as fixed, or as having fixed stochastic properties. It corresponds to that part of population genetics that assumes fitnesses to be independent of genotype frequencies. A number of selective processes have been proposed as frequency-dependent, including predation (13, 70) and disease (15, 32). The maintenance of polymorphism in a varied environment (50) is also best seen as a case of frequency-dependence (59). The concept can be applied directly to phenotypes.

The problem is best formulated in terms of the theory of games, first developed (99) to analyze human conflicts. The essence of a game is that the best strategy to adopt depends on what one's opponent will do; in the context of evolution, this means that the fitness of a phenotype depends on what others are present; i.e., fitnesses are frequency-dependent.

The essential concepts are those of a "strategy" and a "payoff matrix." A strategy is a specification of what a "player" will do in every situation in which it may find itself; in the plant example, a typical strategy would be to allocate all resources to growth for 20 days, and then divide resources equally between growth and seeds. A strategy may be "pure" (i.e., without chance elements) or "mixed" (i.e. of the form "do $A$ with probability $p$ and $B$ with probability $1-p$," where $A$ and $B$ are pure strategies).

The "payoff" to an individual adopting strategy $A$ in competition to one adopting $B$ is written $E(A, B)$, which expresses the expected *change* in the fitness of the player adopting $A$ if his opponent adopts $B$. The evolutionary model is then of a population of individuals adopting different strategies. They pair off at random, and their fitnesses change according to the payoff matrix. Each individual then produces offspring identical to itself, in numbers proportional to the payoff it has accumulated. Inheritance is thus parthenogenetic, and selection acts on the individual. It is also assumed that the population is infinite, so that the chance of meeting an opponent adopting a particular strategy is independent of one's own strategy.

The population will evolve to an evolutionarily stable strategy, or ESS, if one exists (64). An ESS is a strategy that, if almost all individuals adopt it, no rare mutant can invade. Thus let $I$ be an ESS, and $J$ a rare mutant strategy of frequency $p \ll 1$. Writing the fitnesses of $I$ and $J$ as $W(I)$ and $W(J)$,

$$W(I) = C + (1-p) E (I, I) + p E (I, J);$$
$$W(J) = C + (1-p) E (J, I) + p E (J, J).$$

In these equations $C$ is the fitness of an individual before engaging in a contest. Since $I$ is an ESS, $W(I) > W(J)$ for all $J \neq I$; that is, remembering that $p$ is small, *either*

2.   $E (I, I) > E (J, I)$, *or*
     $E (I, I) = E (J, I)$ and $E (I, J) > E (J, J)$.

These conditions (expressions 2) are the definition of an ESS.

Consider the matrix in Table 1. For readers who prefer a biological interpretation, $A$ is "Hawk" and $B$ is "Dove"; thus $A$ is a bad strategy to adopt against $A$, because of the risk of serious injury, but a good strategy to adopt against $B$, and so on.

The game has no pure ESS, because $E (A, A) < E (B, A)$ and $E (B, B) < E (A, B)$. It is easy to show that the mixed strategy—playing $A$ and $B$ with equal probability—is an ESS. It is useful to compare this with other "solutions," each of which has a possible biological interpretation:

*The Maximin Solution.* This is the pessimist's solution, playing the strategy that minimizes your losses if your opponent does what is worst for you. For our matrix, the maximin strategy is always to play $B$. Lewontin (52)

Table 1. Payoff matrix for a game; the values in the matrix give the payoff to Player 1

|          |   | Player 2 |   |
|----------|---|----------|---|
|          |   | A        | B |
| Player 1 |   |          |   |
| A        |   | 1        | 5 |
| B        |   | 2        | 4 |

suggested that this strategy is appropriate if the "player" is a species and its opponent nature: The species should minimize its chance of extinction when nature does its worst. This is the "existential game" of Slobodkin and Rapoport (92). It is hard to see how a species could evolve this strategy, except by group selection. (Note that individual selection will not necessarily minimize the chance of death: A mutant that doubled the chance that an individual would die before maturity, but that quadrupled its fecundity if it did survive, would increase in frequency.)

*The Nash Equilibrium.* This is a pair of strategies, one for each player, such that neither would be tempted to change his strategy as long as the other continues with his. If in our matrix, player 1 plays $A$ and 2 plays $B$, we have a Nash equilibrium; this is also the case if 1 plays $B$ and 2 plays $A$. A population can evolve to the Nash point if it is divided into two classes, and if members of one class compete only with members of the other. Hence, it is the appropriate equilibrium in the "parental investment" game (62), in which all contests are between a male and a female. The ESS is subject to the added constraint that both players must adopt the same strategy.

*The Group Selection Equilibrium.*
If the two players have the same genotype, genes in either will be favored
that maximize the sum of their payoffs. For our matrix both must play
strategy *B*. The problem of the stable
strategy when the players are related
but not identical is discussed in the
section on Games Between Relatives,
below.

It is possible to combine the
game-theoretical and optimization approaches. Mirmirani and Oster (69)
make this extension in their model of
resource allocation in plants. They ask
two questions. What is the ESS for a
plant growing in competition with
members of its own species? What is
the ESS when two species compete
with one another?

Thus consider two competing plants
whose sizes at time $t$ are $P_1$ and $P_2$.
The effects of competition are allowed
for by writing

3. $$dP_1/dt = (r_1 - e_1 P_2)(1 - u_1) P_1,$$
$$dP_2/dt = (r_2 - e_2 P_1)(1 - u_2) P_2,$$

where $u_1$ and $u_2$ are the fractions of
the available resources allocated to
seeds. Let $J_1[u_1(t), u_2(t)]$ be the total
seed production of plant 1 if it adopts
the allocation strategy $u_1(t)$ and its
competitor adopts $u_2(t)$. Mirmirani
and Oster seek a stable pair of strategies $u_1^*(t)$, $u_2^*(t)$, such that

4. $$J_1[u_1(t), u_2^*(t)] \leqslant J_1[u_1^*(t),$$
$$u_2^*(t)], \text{ and}$$
$$J_2[u_1^*(t), u_2(t)] \leqslant J_2[u_1^*(t),$$
$$u_2^*(t)].$$

That is, they seek a Nash equilibrium, such that neither competitor
could benefit by unilaterally altering
its strategy. They find that the optimal
strategies are again "bang-bang," but

with earlier switching times than in
the absence of competition. Strictly,
the conditions indicated by expressions
4 are correct only when there is
competition between species, and
when individuals of one species compete only with individuals of the
other; formally this would be so if the
plants grew alternately in a linear
array. The conditions indicated by expressions 4 are not appropriate for
intra-specific competition, since they
permit $u_1^*(t)$ and $u_2^*(t)$ to be different, which could not be the case unless individuals of one genotype
competed only with individuals of
the other. For intra-specific competition $(r_1 = r_2, e_1 = e_2)$, the ESS is given
by

5. $$J_1[u_1(t), u_1^*(t)] \leqslant J_1[u_1^*(t),$$
$$u_1^*(t)].$$

As it happens, for the plant growth
example equations 4 and 5 give the
same control function, but in general
this need not be so.

The ESS model assumes parthenogenetic inheritance, whereas most
interesting populations are sexual. If
the ESS is a pure strategy, no difficulty arises; a genetically homogeneous
sexual population adopting the strategy will also be stable. If the ESS is a
mixed strategy that can be achieved
by a single individual with a variable
behavior, there is again no difficulty.
If the ESS is a mixed one that can be
achieved only by a population of pure
strategists in the appropriate frequencies, two difficulties arise:

(1) Even with the parthenogenetic
model, the conditions expressed
in expressions 2 do not guarantee stability. (This was first
pointed out to me by Dr. C.

Strobeck.) In such cases, therefore, it is best to check the stability of the equilibrium, if necessary by simulation; so far, experience suggests that stability, although not guaranteed, will usually be found.

(2) The frequency distribution may be one that is incompatible with the genetic mechanism. This difficulty, first pointed out by Lewontin (52), has recently been investigated by Slatkin (89–91) and by Auslander et al. (2). It is hard to say at present how serious it will prove to be; my hope is that a sexual population will usually evolve a frequency distribution as close to the ESS as its genetic mechanism will allow.

*Games Between Relatives*

The central concept is that of "inclusive fitness" (33). In classical population genetics we ascribe to a genotype $I$ a "fitness" $W$, corresponding to the expected number of offspring produced by $I$. If, averaged over environments and genetic backgrounds, the effect of substituting allele $A$ for $a$ is to increase $W$, allele $A$ will increase in frequency. Following Oster et al. (74) but ignoring unequal sex ratios, Hamilton's proposal is that we should replace $W_i$ by the inclusive fitness, $Z_i$, where

6.     $$Z_i = \sum_{j=1}^{R} r_{ij} W_j,$$

where the summation is over all $R$ relatives of $I$; $r_{ij}$ is the fraction of $J$'s genome that is identical by descent to alleles in $I$; and $W_j$ is the expected number of offspring of the $j$th relative

of $I$. (If $J = I$, then equation 6 refers to the component of inclusive fitness from an individual's own offspring.)

An allele $A$ will increase in frequency if it increases $Z$ rather than just $W$. Three warnings are needed:

(1) It is usual to calculate $r_{ij}$ from the pedigree connecting $I$ and $J$ (as carried out, for example, by Malécot (58)). However, if selection is occurring, $r_{ij}$ so estimated is only approximate, as are predictions based on equation 6 (35).

(2) Some difficulties arose in calculating appropriate values of $r_{ij}$ for haplo-diploids; these were resolved by Crozier (22).

(3) If the sex ratio is not unity, additional difficulties arise (74).

Mirmirani and Oster (69) have extended their plant-growth model along these lines to cover the case when the two competitors are genetically related. They show that as $r$ increases, the switching time becomes earlier and the total yield higher.

CONCLUSION

The role of optimization theories in biology is not to demonstrate that organisms optimize. Rather, they are an attempt to understand the diversity of life.

Three sets of assumptions underlie an optimization model. First, there is an assumption about the kinds of phenotypes or strategies possible (i.e., a "phenotype set"). Second, there is an assumption about what is being maximized; ideally this should be the inclusive fitness of the individual, but often one must be satisfied with some component of fitness (e.g., rate of energy intake while foraging). Finally, there is an assumption, often tacit, about the mode of inheritance and the population structure; this will determine the type of equilibrium to which the population will move.

In testing an optimization model, one is testing the adequacy of these hypotheses to account for the evolution of the particular structures or patterns of behavior under study. In most cases the hypothesis that variation in the relevant phenotypes is selectively neutral is not a plausible alternative, because of the nature of the phenotypes chosen for study. However, it is often a plausible alternative that the phenotypes are not well adapted to current circumstances because the population is lagging behind a changing environment; this is a serious difficulty in testing optimization theories.

The most damaging criticism of optimization theories is that they are untestable. There is a real danger that the search for functional explanations in biology will degenerate into a test of ingenuity. An important task, therefore, is the development of an adequate methodology of testing. In many cases the comparative method is the most powerful; it is essential, however, to have clear criteria for inclusion or exclusion of species in comparative tests, and to use statistical methods with the same care as in the analysis of experimental results.

Tests of the quantitative predictions of optimization models in particular populations are beginning to be made. It is commonly found that a model correctly predicts qualitative features of the observations, but is contradicted in detail. In such cases the Popperian view would be that the original model has been falsified. This is correct, but it does not follow that the model should be abandoned. In the analysis of complex systems it is most unlikely that any simple model, taking into account only a few factors, can give quantitatively exact predictions. Given that a simple model has been falsified by observations, the choice lies between abandoning it and modifying it, usually by adding hypotheses. There can be no simple rule by which to make this choice; it will depend on how persuasive the qualitative predictions are, and on the availability of alternative models.

Mathematical methods of optimization have been developed with engineering and economic applications in mind. Two theoretical questions arise in applying these methods in biology. First, in those cases in which the fitnesses of phenotypes are frequency-dependent, the problem must be formulated in game-theoretical terms; some difficulties then arise in deciding to what type of equilibrium a population will tend. A second and related set of questions arise when specific genetic assumptions are incorporated into the model, because it may be that a population with the optimal phenotype cannot breed true. These questions need further study, but at present there is no reason to doubt the adequacy of the concepts of optimization and of evolutionary stability for studying phenotypic evolution.

ACKNOWLEDGMENTS

My thanks are due Dr. R. C. Lewontin for sending me two manuscripts that formed the starting point of this review, and Drs. G. Oster and R. Pulliam for their comments on an earlier draft. I was also greatly helped by preliminary discussions with Dr. E. Curio.

## REFERENCES

1. Alexander, R. M., 1977, Mechanics and scaling of terrestrial locomotion. In *Scale Effects in Animal Locomotion*, ed. T. J. Pedley, London, Academic Press, pp. 93–110.

2. Auslander, D., J. Guckenheimer, and G. Oster, 1978, Random evolutionarily stable strategies. *Theor. Pop. Biol.*, 13 (2): 276–293.

3. Baker, H. G., 1959, Reproductive methods as factors in speciation in flowering plants. *Cold Spring Harbor Symp. Quant. Biol.*, 24: 177–191.

4. Bawa, K. S. and P. A. Opler, 1975, Dioecism in tropical forest trees. *Evolution*, 29: 167–179.

5. Bellman, R., 1957, *Dynamic Programming*, Princeton, N.J., Princeton University Press.

6. Bishop, D. T. and C. Cannings, 1978, A generalized war of attrition. *J. Theor. Biol.*, 70: 85–124.

7. Buskirk, W. H., 1976, Social systems in tropical forest avifauna. *Am. Nat.*, 110: 293–310.

8. Cain, A. J., 1964, The perfection of animals. In *Viewpoints in Biology*, ed. J. D. Carthy and C. L. Duddington, 3: 36–63.

9. ——, and P. H. Sheppard, 1954, Natural selection in *Cepaea*. *Genetics*, 39: 89–116.

10. Charnov, E. L., 1976, Optimal foraging, the marginal value theorem. *Theor. Pop. Biol.* 9: 129–136.

11. ——, J. Maynard Smith, and J. J. Bull, 1976, Why be an hermaphrodite? *Nature*, 263: 125–126.

12. Clark, C. W., 1976, *Mathematical Bioeconomics*, N.Y., Wiley.

13. Clarke, B., 1962, Balanced polymorphism and the diversity of sympatric species. In *Taxonomy and Geography*, ed. D. Nichols, London, Syst. Assoc. Publ., 4: 47–70.

14. ——, 1975, The contribution of ecological genetics to evolutionary theory: detecting the direct effects of natural selection on particular polymorphic loci. *Genetics*, 79: 101–113.

15. ——, 1976, The ecological genetics of host-parasite relationships. In *Genetic Aspects of Host-Parasite Relationships*, ed. A. E. R. Taylor and R. Muller, Oxford, Blackwell, pp. 87–103.

16. Clutton-Brock, T. H., and P. H. Harvey, 1977, Primate ecology and social organisation. *J. Zool.*, London, 183: 1–39.

17. ——, T. H. and P. H. Harvey, 1977, Species differences in feeding and ranging behaviour in primates. In *Primate Ecology*, ed. T. H. Clutton-Brock, London, Academic, pp. 557–584.

18. ——, T. H., P. H. Harvey, and B. Rudder, 1977, Sexual dimorphism, socioeconomic sex ratio and body weight in primates. *Nature*, 269: 797–800.

19. Cohen, D., 1971, Maximising final yield when growth is limited by time or by limiting resources. *J. Theor. Biol.*, 33: 299–307.

20. Cowie, R. J., 1977, Optimal foraging in great tits (*Parus major*). *Nature*, 268: 137–139.

21. Crow, J. F. and M. Kimura, 1970, *An Introduction to Population Genetics Theory*, N.Y., Harper & Row.

22. Crozier, R. H., 1970, Coefficients of relationship and the identity of genes by descent in the Hymenoptera. *Am. Nat.*, 104: 216–217.

23. Curio, E., 1973, Towards a methodology of teleonomy. *Experientia*, 29: 1045–1058.

24. ——, 1975, The functional organization of anti-predator behaviour in the pied flycatcher: a study of avian visual perception. *Anim. Behav.*, 23: 1–115.

25. ——, 1976, *The Ethology of Predation*, Berlin, Springer-Verlag.

26. Darwin, C., 1877, *The Different Forms of Flowers on Plants of the Same Species*, London, John Murray.

27. Emlen, J. M., 1966, The role of time and energy in food preference. *Am. Nat.*, 100: 611–617.

28. Fisher, R. A., 1930, *The Genetical Theory of Natural Selection*, London: Oxford Univ. Press, 291 pp.

29. Goss-Custard, J. D., 1977, Optimal foraging and the size selection of worms by redshank, *Tringa totanus*, in the field. *Anim. Behav.* 25: 10–29.

30. Gould, S. J., 1971, Geometric scaling in allometric growth: a contribution to the problem of scaling in the evolution of size. *Am. Nat.*, 105: 113–116.

31. Grieg-Smith, P. W., 1978, The formation, structure and feeding of insectivorous bird flocks in West African savanna woodland. *Ibis*, 121 (3): 284–297.

32. Haldane, J. G. S., 1949, Disease and evolution. *Ric. Sci.*, Suppl., 19: 68–76.

33. Hamilton, W. D., 1964, The genetical theory of social behavior. I and II. *J. Theor. Biol.*, 7: 1–16; 17–32.

34. ——, 1967, Extraordinary sex ratios. *Science*, 156: 477–488.

35. ——, 1972, Altruism and related phenomena, mainly in social insects. *Ann. Rev. Ecol. Syst.*, 3: 193–232.

36. Harvey, P. H., M. Kavanagh, and T. H. Clutton-Brock, 1978, Sexual dimorphism in primate teeth. *J. Zool.*, 186: 475–485.

37. Ivlev, V. S., 1961, *Experimental Ecology of the Feeding of Fishes*, New Haven, Yale Univ. Press.

38. Jacob, F., 1977, Evolution and tinkering, *Science*, 196: 1161–1166.

39. Jarvis, M. J. F., 1974, The ecological significance of clutch size in the South African gannet [*Sula capensis*. (Lichtenstein)]. *J. Anim. Ecol.*, 43: 1–17.

40. Kalmijn, A. J., 1971, The electric sense of sharks and rays. *J. Exp. Biol.*, 55: 371–383.

41. Kalmus, H. and C. A. B. Smith, 1960, Evolutionary origin of sexual differentiation and the sex-ratio. *Nature*, 186: 1004–1006.

42. Kettlewell, H. B. D., 1956, Further selection experiments on industrial melanism in the Lepidoptera. *Heredity*, 10: 287–301.

43. Kimura, M., 1968, Evolutionary rate at the molecular level. *Nature*, 217: 624–626.

44. King, J. L. and T. H. Jukes, 1969, Non-Darwinian Evolution: random fixation of selectively neutral mutations. *Science*, 164: 788–798.

45. Krebs, C. J., 1972, *Ecology*, N.Y., Harper & Row.

46. Krebs, J. R., J. C. Ryan, and E. L. Charnov, 1974, Hunting by expectation or optimal foraging? A study of patch use by chickadees. *Anim. Behav.*, 22: 953–964.

47. ——, J. T. Ericksen, M. I. Webber, and E. L. Charnov, 1977, Optimal prey selection in the Great Tit (*Parus major*). *Anim. Behav.*, 25: 30–38.

48. Lack, D., 1966, *Population Studies of Birds*, Oxford, Clarendon Press.

49. ——, 1968, *Ecological Adaptations for Breeding in Birds*, London, Methuen.

50. Levene, H., 1953, Genetic equilibrium when more than one ecological niche is available. *Am. Nat.*, 87: 131–133.

51. Levins, R., 1962, Theory of fitness in a heterogeneous environment. I. The fitness set and adaptive function. *Am. Nat.*, 96: 361–373.

52. Lewontin, R. C., 1961, Evolution and the theory of games. *J. Theor. Biol.*, 1: 382–403.

53. ——, 1974, *The Genetic Basis of Evolutionary Change*, N.Y., Columbia Univ. Press.

54. ——, 1977, Adaptation. In *The Enciclopedia Einaudi*, Torino, Giulio Einaudi Edition.

55. ——, 1978, Fitness, survival and optimality. In *Analysis of Ecological Systems*, ed. D. H. Horn, R. Mitchell, G. R. Stairs, Columbus, Oh., Ohio State Univ. Press.

56. MacArthur, R. H., 1965, Ecological consequences of natural selection. In *Theoretical and Mathematical Biology*, ed. T. Waterman, H. Morowritz, N.Y., Blaisdell.

57. ——, R. H. and E. R. Pianka, 1966, On optimal use of a patch environment. *Am. Nat.*, 100: 603–609.

58. Malécot, G., 1969, *The Mathematics of Heredity*, transl. D. M. Yer-

manos. San Francisco, W. H. Freeman, 88 pp.

59. Maynard Smith, J., 1962, Disruptive selection, polymorphism and sympatric speciation. *Nature,* 195: 60–62.

60. ———, 1969, The status of neo-Darwinism. In *Towards a Theoretical Biology. 2: Sketches,* ed. C. H. Waddington, Edinburgh, Edinburgh Univ. Press, pp. 82–89.

61. ———, 1976, Sexual selection and the handicap principle. *J. Theor. Biol.,* 57: 239–242.

62. ———, 1977, Parental investment—a prospective analysis. *Anim. Behav.,* 25: 1–9.

63. ———, 1978, *The Evolution of Sex,* London, Cambridge Univ. Press. In press.

64. ———, and G. A. Parker, 1976, The logic of asymmetric contests. *Anim. Behav.,* 24: 159–175.

65. ———, and G. R. Price, 1973, The logic of animal conflict. *Nature,* 246: 15–18.

66. ———, and R. J. G. Savage, 1956, Some locomotory adaptations in mammals. *Zool. J. Linn. Soc.,* 42: 603–622.

67. McFarland, D. J., 1977, Decision making in animals. *Nature,* 269: 15–21.

68. Milton, K. and M. L. May, 1976, Bodyweight, diet and home range area in primates. *Nature,* 259: 459–462.

69. Mirmirani, M. and G. Oster, 1978, Competition, kin selection and evolutionarily stable strategies. *Theor. Pop. Biol.,* 13 (3): 304–339.

70. Moment, G., 1962, Reflexive selection: a possible answer to an old puzzle. *Science,* 136: 262–263.

71. Nelson, J. B., 1964, Factors influencing clutch size and chick growth in the North Atlantic Gannet, *Sula bassana. Ibis,* 106: 63–77.

72. Norton-Griffiths, M., 1969, The organization, control and development of parental feeding in the oystercatcher (*Haematopus ostralegus*). *Behavior,* 34: 55–114.

73. Orians, G. H. and N. E. Pearson, 1978, On the theory of central place foraging. In *Analysis of Ecological Systems,* ed. D. H. Horn, R. Mitchell, G. R. Stairs, Columbus, Ohio State Univ. Press.

74. Oster, G., I. Eshel, and D. Cohen, 1977, Worker-queen conflicts and the evolution of social insects. *Theor. Pop. Biol.,* 12: 49–85.

75. ———, and E. O. Wilson, 1978, *Caste and Ecology in the Social Insects,* Princeton, N.J., Princeton Univ. Press. In press.

76. Parker, G. A., 1974, The reproductive behaviour and the nature of sexual selection in *Scatophaga stercoraria* L. IX. Spatial distribution of fertilization rates and evolution of male search strategy within the reproductive area. *Evolution,* 28: 93–108.

77. ———, and R. A. Stuart, 1976, Animal behaviour as a strategy optimizer: evolution of resource assessment strategies and optimal emigration thresholds. *Am. Nat.,* 110: 1055–1076.

78. Pedley, T. J., 1977, *Scale Effects in Animal Locomotion,* London, Academic Press.

79. Pontryagin, L. S., V. S. Boltyanskii, R. V. Gamkrelidze, and E. F. Mishchenko, 1962, *The Mathematical Theory of Optimal Processes,* N.Y., Wiley.

80. Pulliam, H. R., 1978, Do chipping sparrows forage optimally? A test of optimal foraging theory in nature. *Am. Nat.* In press.

81. Pyke, G. H., H. R. Pulliam, and E. L. Charnov, 1977, Optimal foraging: a selective review of theory and tests. *Q. Rev. Biol.,* 52: 137–154.

82. Ralls, K., 1976, Mammals in which females are larger than males. *Q. Rev. Biol.,* 51: 245–276.

83. Rensch, B., 1959, *Evolution above the Species Level,* New York, Columbia Univ. Press.

84. Rohwer, S., 1977, Status signaling in Harris sparrows: some experiments in deception. *Behaviour,* 61: 107–129.

85. Rosado, J. M. C. and A. Robertson, 1966, The genetic control of sex ratio. *J. Theor. Biol.,* 13: 324–329.

86. Schoener, T. W., 1968, Sizes of feeding territories among birds. *Ecology,* 49: 123–141.

87. ———, 1971, Theory of feeding strategies. *Ann. Rev. Ecol. Syst.* 2: 369–404.

88. Sibly, R. and D. McFarland, 1976, On the fitness of behaviour sequences. *Am. Nat.*, 110: 601–617.

89. Slatkin, M., 1978, On the equilibration of fitnesses by natural selection. *Am. Nat.*, 112: 845–859.

90. ——, 1979, The evolutionary response to frequency- and density-dependent interactions, *Am. Nat.*, 114: 384–398.

91. ——, 1979, Frequency- and density-dependent selection on a quantitative character, *Genetics,* 93: 755–771.

92. Slobodkin, L. B. and A. Rapoport, 1974, An optimal strategy of evolution. *Q. Rev. Biol.*, 49: 181–200.

93. Strobeck, C., 1975, Selection in a fine-grained environment. *Am. Nat.*, 109: 419–425.

94. Tinbergen, N., G. J. Broekhuysen, F. Feekes, J. C. W. Houghton, H, Kruuk, and E. Szule, 1963, Egg shell removal by the Black-headed Gull, *Larus ribidundus* L.: a behaviour component of camouflage. *Behaviour,* 19: 74–117.

95. Trivers, R. L. and H. Hare, 1976, Haplodiploidy and the evolution of social insects. *Science,* 191: 249–263.

96. ——, and D. E. Willard, 1973, Natural selection of parental ability to vary the sex ratio of offspring. *Science,* 179: 90–92.

97. Verner, J., 1965, Selection for sex ratio. *Am. Nat.*, 19: 419–421.

98. ——, and M. F. Willson, 1966, The influence of habitats on mating systems of North American passerine birds. *Ecology,* 47: 143–147.

99. Von Neumann, J. and O. Morgenstern, 1953, *Theory of Games and Economic Behavior,* Princeton, N.J., Princeton Univ. Press.

100. Werner, E. E. and D. J. Hall, 1974, Optimal foraging and size selection of prey by the bluegill sunfish (*Lepomis mochrochirus*). *Ecology,* 55: 1042–1052.

101. Wright, S., 1932, The roles of mutation, inbreeding, crossbreeding and selection in evolution. *Proc. Sixth Int. Congr. Genet.,* 1: 356–366.

102. Zahavi, A., 1975, Mate selection—a selection for a handicap. *J. Theor. Biol.,* 53: 205–214.

# V
# FUNCTION AND TELEOLOGY

# 18

# The Structure
# of Teleological Explanations

ERNEST NAGEL

THE ANALYTIC METHODS OF THE modern natural sciences are universally admitted to be competent for the study of all nonliving phenomena, even those which, like cosmic rays and the weather, are still not completely understood. Moreover, attempts at unifying special branches of physical science, by reducing their several systems of explanation to an inclusive theory, are generally encouraged and welcomed. During the past four centuries these methods have also been fruitfully employed in the study of living organisms; and many features of vital processes have been successfully explained in physicochemical terms. Outstanding biologists as well as physical scientists have therefore concluded that the methods of the physical sciences are fully adequate to the materials of biology, and many of these scientists have been confident that eventually the whole of biology would become simply a chapter of physics and chemistry.

But, despite the undeniable successes of physicochemical explanations in the study of living things, biologists of unquestioned competence continue to regard such explanations as not entirely adequate for the subject matter of biology. Most biologists are in general agreement that vital processes, like nonliving ones, occur only under determinate physicochemical conditions and form no exceptions to physicochemical laws. Some of them nevertheless maintain that the mode of analysis required for understanding living phenomena is fundamentally different from that which obtains in the physical sciences. Opposition to the systematic absorption of biology into physics and chemistry is sometimes based on the practical ground that it does not conform to the correct strategy of biological research. However, such opposition is often also supported by theoretical arguments which aim to show that the reduction of biology to physicochemistry is inherently impossible. Biology has long been an area in which crucial issues in the logic of explanation have been the subject of vigorous debate.

In any event, it is instructive to examine some of the reasons biologists commonly advance for the claim that the logic of explanatory concepts in biology is distinctive of the science and that biology is an inherently autonomous discipline.

What are the chief supports for this claim?

1. Let us first dispose of two less weighty ones. Although it is difficult to formulate in precise terms the generic differences between the living and the nonliving, no one seriously doubts the obvious fact that there are such differences. Accordingly, the various "life sciences" are concerned with special questions that are patently unlike those with which physics and chemistry deal. In particular, biology studies the anatomy and physiology of living things, and investigates the modes and conditions of their reproduction, development, and decay. It classifies vital organisms into types or species; and it inquires into their geographic distribution, their lines of descent, and the modes and conditions of their evolutionary changes. Biology also analyzes organisms as structures of interrelated parts and seeks to discover what each part contributes to the maintenance of the organism as a whole. Physics and chemistry, on the other hand, are not specifically concerned with such problems, although the subject matter of biology also falls within the province of these sciences. Thus a stone and a cat when dropped from a height exhibit behaviors which receive a common formulation in the laws of mechanics; and cats as well as stones therefore belong to the subject matter of physics. Nevertheless, cats possess structural features and engage

in processes in which physics and chemistry, at any rate in their current form, are not interested. Stated more formally, biology employs expressions referring to identifiable characteristics of living phenomena (such as 'sex,' 'cellular division,' 'heredity,' or 'adaptation') and asserts laws containing them (such as 'Hemophilia among humans is a sex-linked hereditary trait') that do not occur in the physical sciences and are not at present definable or derivable within these sciences. Accordingly, while the subject matter of biology and the physical sciences is not disparate, and though biology makes use of distinctions and laws borrowed from the physical sciences, the two sciences do not at present coincide.

It is no less evident that the techniques of observation and experimentation in biology are in general different from those current in the physical sciences. To be sure, some tools and techniques of observations, measurement, and calculation (such as lenses, balances, and algebra) are used in both groups of disciplines. But biology also requires special skills (such as those involved in the dissection of organic tissues) which serve no purpose in physics; and physics employs techniques (such as those needed for handling high-voltage currents) that are irrelevant in present-day biology. A physical scientist untrained in the special techniques of biological research is no more likely to perform a biological experiment successfully than is a pianist untutored in playing wind instruments likely to perform well on an oboe.

These differences between the special problems and techniques of the physical and biological sciences are sometimes cited as evidence for the inherent autonomy of biology, and for the claim that the analytical methods of physics are not fully adequate to the objectives of biological inquiry. However, though the differences are genuine, they certainly do not warrant such conclusions. Mechanics, electromagnetism, and chemistry, for example, are prima facie distinct branches of physical science, in each of which different special problems are pursued and different techniques are employed. But as we have seen, these are not sufficient reasons for maintaining that each of those divisions of physical science is an autonomous discipline. If there is a sound basis for the alleged absolute autonomy of biology, it must be sought elsewhere than in the differences between biology and the physical sciences which have been noted thus far.

2. What, then, are the weightier reasons which support that allegation? The main ones appear to be as follows. Vital processes have a prima facie purposive character; organisms are capable of self-regulation, self-maintenance, and self-reproduction, and their activities seem to be directed toward the attainment of goals that lie in the future. It is usually admitted that one can study and formulate the morphological characteristics of plants and animals in a manner comparable with the way physical sciences investigate the structural traits of nonliving things. Thus the categories of analysis and explanation in physics are generally held to be adequate for studying the gross and minute anatomy of the human kidney, or the serial order of its development. But morphological studies are only one part of the biologist's task, since it also includes inquiry into the *functions* of structures in sustaining the activities of the organism as a whole. Thus biology studies the role played by the kidney and its microscopic structure in preserving the chemical composition of the blood, and thereby in maintaining the whole body and its other parts in their characteristic activities. It is such manifestly "goal-directed" behavior of living things that is often regarded as requiring a distinctive category of explanation in biology.

Moreover, living things are organic wholes, not "additive systems" of independent parts, and the behavior of these parts cannot be properly understood if they are regarded as so many isolable mechanisms. The parts of an organism must be viewed as internally related members of an integrated whole. They mutually influence one another, and their behavior regulates and is regulated by the activities of the organism as a whole. Some biologists have argued that the coordinated, adaptive behavior of living organisms can be explained only by assuming a special vitalistic agent; others believe that an explanation is possible in terms of the hierarchical organization of internally related parts of the organism. But in either case, so it is frequently claimed, biology cannot dispense with the notion of organic unity; and in consequence it must use modes of analysis and formulation that are unmistakably *sui generis*.

Accordingly, two main features are commonly alleged to differentiate biology from the physical sciences in an essential way. One is the dominant place occupied by *teleological* explanations in biological inquiry.

The other is the use of conceptual tools uniquely appropriate to the study of systems whose total behavior is not the resultant of the activities of independent components. We must now examine these claims in some detail.*

Almost any biological treatise or monograph yields conclusive evidence that biologists are concerned with the functions of vital processes and organs in maintaining characteristic activities of living things. In consequence, if "teleological analysis" is understood to be an inquiry into such functions, and into processes directed toward attaining certain end-products, then undoubtedly teleological explanations are pervasive in biology. In this respect, certainly, there appears to be a marked difference between biology and the physical sciences. It would surely be an oddity on the part of a modern physicist were he to declare, for example, that atoms have outer shells of electrons in order to make chemical unions between themselves and other atoms possible. In ancient Aristotelian science, categories of explanations suggested by the study of living things and their activities (and in particular by human art) were made canonical for all inquiry. Since nonliving as well as living phenomena were thus analyzed in teleological terms—an analysis which made the notion of final cause focal—Greek science did not assume a fundamental cleavage between biology and other natural science. Modern science, on the other hand, regards final causes to be vestal virgins which bear no fruit in the study of physical and chemical phenomena; and, because of the asso-

ciation of teleological explanations with the doctrine that goals or ends of activity are dynamic agents in their own realizations, it tends to view such explanations as a species of obscurantism. But does the presence of teleological explanations in biology and their apparent absence from the physical sciences entail the absolute autonomy of the former? We shall try to show that it does not.

1. Quite apart from their association with the doctrine of final causes, teleological explanations are sometimes suspect in modern natural science because they are assumed to invoke purposes or ends-in-view as causal factors in natural processes. Purposes and deliberate goals admittedly play important roles in human activities, but there is no basis whatever for assuming them in the study of physicochemical and most biological phenomena. However, as has already been noted, a great many explanations counted as teleological do not postulate any purposes or ends-in-view; for explanations are often said to be "teleological" only in the sense that they specify the *functions* which things or processes possess. Most contemporary biologists certainly do not impute purposes to the organic parts of living things whose functions are investigated; most of them would probably also deny that the means-ends relationships discovered in the organization of living creatures are the products of some deliberate plan on the part of a purposeful agent, whether divine or in some other manner supranatural. To be sure, there are biologists who

*The second claim is taken up by Nagel in Essay 22, below. [*Editor.*]

postulate psychic states as concomitants and even as directive forces of all organic behavior. But such biologists are in a minority; and they usually support their views by special considerations that can be distinguished from the facts of functional or teleological dependencies which most biologists do not hesitate to accept. Since the word 'teleology' is ambiguous, confusion and misunderstandings would doubtless be prevented if the word were eliminated from the vocabulary of biology. But biologists do use it, and say they are giving a teleological explanation when, for example, they explain that the function of the alimentary canal in vertebrates is to prepare ingested materials for absorption into the bloodstream. The crucial point is that when biologists do employ teleological language, they are not necessarily committing the pathetic fallacy or lapsing into anthropomorphism.

We shall therefore assume that teleological (or functional) statements in biology normally neither assert nor presuppose in the materials under discussion either manifest or latent purposes, aims, objectives, or goals. Indeed, it seems safe to suppose that biologists would generally deny they are postulating any conscious or implicit ends-in-view even when they employ such words as 'purpose' in their functional analyses—as when the 'purpose' (i.e., the function) of kidneys in the pig is said to be that of eliminating various waste products from the bloodstream of the organism. On the other hand, we shall adopt as the mark of a teleological statement in biology, and as the feature that distinguishes such statements from nonteleological ones, the

occurrence in the former but not in the latter of such typical locutions as 'the function of,' 'the purpose of,' 'for the sake of,' 'in order that,' and the like—more generally, the occurrence of expressions signifying a means-ends nexus.

Nevertheless, despite the prima facie distinctive character of teleological (or functional) explanations, we shall first argue that they can be reformulated, without loss of asserted content, to take the form of nonteleological ones, so that in an important sense teleological and nonteleological explanations are equivalent. To this end, let us consider a typical teleological statement in biology, for example, 'The function of chlorophyll in plants is to enable plants to perform photosynthesis (i.e., to form starch from carbon dioxide and water in the presence of sunlight).' This statement accounts for the presence of chlorophyll (a certain substance $A$) in plants (in every member $S$ of a class of systems, each of which has a certain organization $C$ of component parts and processes). It does so by declaring that, when a plant is provided with water, carbon dioxide, and sunlight (when $S$ is placed in a certain "internal" and "external" environment $E$), it manufactures starch (a certain process $P$ takes place yielding a definite product or outcome) only if the plant contains chlorophyll. The statement usually carries with it the additional tacit assumption that without starch the plant cannot continue its characteristic activities, such as growth and reproduction (it cannot maintain itself in a certain state $G$); but for the present we shall ignore this further claim.

Accordingly, the teleological statement is a telescoped argument, so that when the content is unpacked it can be rendered approximately as follows: When supplied with water, carbon dioxide, and sunlight, plants produce starch; if plants have no chlorophyll, even though they have water, carbon dioxide, and sunlight, they do not manufacture starch; hence plants contain chlorophyll. More generally, a teleological statement of the form 'The function of $A$ in a system $S$ with organization $C$ is to enable $S$ in environment $E$ to engage in process $P$' can be formulated more explicitly by: Every system with organization $C$ and in environment $E$ engages in process $P$; if $S$ with organization $C$ and in environment $E$ does not have $A$, then $S$ does not engage in $P$; hence, $S$ with organization $C$ must have $A$.

It is clearly not relevant in the present context to inquire whether the premises in this argument are adequately supported by competent evidence. However, because the issue is sometimes raised in discussions of teleological explanations, at least passing notice deserves to be given to the question whether chlorophyll is really necessary to plants and whether they could not manufacture starch (or other substances essential for their maintenance) by some alternative process not requiring chlorophyll. For if the presence of chlorophyll is not actually necessary for the production of starch (or if plants can maintain themselves without the mechanisms of photosynthesis), so it has been urged, the second premise in the above argument is untenable. The premise would then have to be modified; and in its emended form it would

assert that chlorophyll is an element in a set of conditions that is *sufficient* (but not necessary) for the production of starch. In that case, however, the new argument with the emended premise would be invalid, so that the proposed teleological explanation of the presence of chlorophyll in plants would apparently be unsatisfactory.

This objection is in part well-taken. It is certainly *logically* possible that plants might maintain themselves without manufacturing starch, or that processes in living organisms might produce starch without requiring chlorophyll. Indeed, there are plants (the funguses) that can flourish without chlorophyll; and in general, there is more than one way of skinning a cat. On the other hand, the above teleological explanation of the occurrence of chlorophyll in plants is presumably concerned with living organisms having certain determinate forms of organization and definite modes of behavior—in short, with the so-called "green plants." Accordingly, although living organisms (plants as well as animals) capable of maintaining themselves without processes involving the operation of chlorophyll are both abstractly and physically possible, there appears to be no evidence whatever that in view of the limited capacities green plants possess as a consequence of their *actual* mode of organization, these organisms can live without chlorophyll.

Two important complementary points thus emerge from these considerations. In the first place, teleological analyses in biology (or in other sciences in which such analyses are pursued) are not explorations of merely logical possibilities, but deal with the actual functions of definite components in concretely given living systems. In the second place, on pain of failure to recognize the possibility of alternative mechanisms for achieving some end-product, and of unwittingly (and perhaps mistakenly) assuming that a process known to be indispensable in a given class of systems is also indispensable in a more inclusive class, a teleological explanation must articulate with exactitude both the character of the end-product and the defining traits of the systems manifesting them, relative to which the indicated processes are supposedly indispensable.

In any event, however, the above teleological account of chlorophyll, in its expanded form, is simply an illustration of an explanation that conforms to the deductive model, and contains no locution distinctive of teleological statements. Accordingly, the initial, unexpanded statement about chlorophyll appears to assert nothing that is not asserted by 'Plants perform photosynthesis only if they contain chlorophyll,' or alternatively by 'A necessary condition for the occurrence of photosynthesis in plants is the presence of chlorophyll.' These latter statements do not explicitly ascribe a function to chlorophyll, and in that sense are therefore not teleological formulations. If this example is taken as a paradigm, it seems that when a function is ascribed to a constituent element in an organism, the content of the teleological statement is fully conveyed by another statement that is not explicitly teleological and that simply asserts a necessary (or possibly a necessary and sufficient) condition for the occurrence of a certain trait or activity of the organism. In the

light of this analysis, therefore, a teleological explanation in biology indicates the *consequences* for a given biological system of a constituent part or process; the equivalent nonteleological formulation of this explanation, on the other hand, states some of the *conditions* (sometimes, but not invariably, in physicochemical terms) under which the system persists in its characteristic organization and activities. The difference between a teleological explanation and its equivalent nonteleological formulation is thus comparable to the difference between saying that $Y$ is an effect of $X$, and saying that $X$ is a cause or condition of $Y$. In brief, the difference is one of selective attention, rather than of asserted content.

This point can be reinforced by another consideration. If a teleological explanation had an asserted content different from the content of every conceivable nonteleological statement, it would be possible to cite procedures and evidence employed for establishing the former that differ from the procedures and evidence required for warranting the latter. But in point of fact there appear to be no such procedures and evidence. Consider, for example, the teleological statement 'The function of the leucocytes in human blood is to defend the body against foreign microorganisms.' Now whatever may be the evidence that warrants this statement, that evidence also confirms the nonteleological statement 'Unless human blood contains a sufficient number of leucocytes, certain normal activities of the body are impaired,' and conversely. If this is so, however, there is a strong presumption that the two statements do not differ in factual content. More generally, if, as seems to be the case, the conceivable evidence for any given teleological explanation is identical with the conceivable evidence for a certain nonteleological one, the conclusion appears inescapable that those statements cannot be distinguished with respect to what they *assert,* even though they are distinguishable in other ways.

2. This proposed equation of teleological and nonteleological explanations must nevertheless face a fundamental objection. Many biologists would perhaps admit that a teleological statement *implies* a certain nonteleological one; but some of them, at any rate, are prepared to maintain that the latter statement generally does not in turn imply the former one, and that in consequence the alleged equivalence between the statements does not in fact hold.

The claim that there is indeed no such equivalence can be forcefully presented as follows. If there were such an equivalence, not only could a teleological explanation be replaced by a nonteleological one, but conversely a nonteleological explanation could also be replaced by a teleological one. In consequence, the customary statements of laws and theories in the physical sciences would be translatable without change in asserted content into teleological formulations. In point of fact, however, modern physical science does not appear to sanction such reformulations. Indeed, most physical scientists would doubtless resist the introduction of teleological statements into their disciplines as a misguided attempt to reinstate the point of view of Greek and medieval science. For example, the statement 'The volume of a gas at constant temperature varies inversely with its pressure' is a typical physical law, which is entirely free of teleological connotations. If it were equivalent to a teleological statement, its equivalent (constructed on the model of the example adopted above as paradigmatic) would presumably be 'The function of a varying pressure in a gas at constant temperature is to produce an inversely varying volume of the gas,' or perhaps 'Every gas at constant temperature under a variable pressure alters its volume in order to keep the product of the pressure and the volume constant.' But most physicists would undoubtedly regard these formulations as preposterous, and at best as misleading. Accordingly, if no teleological statement can correctly translate a law of physics, the contention that for every teleological statement a logically equivalent nonteleological one can be constructed seems hardly tenable. There must therefore be some important difference between teleological and nonteleological statements, so the objection concludes, which the discussion has thus far failed to make explicit.

The difficulty just expounded cannot be disposed of easily. To assess it adequately, we must consider the type of subject matter in which teleological analyses are currently undertaken, and in which teleological explanations are not rejected ostensibly as a matter of general principle.

a. The attitude of physical scientists toward teleological formulations in their disciplines is doubtless as alleged in this objection. Nevertheless, this fact is not completely decisive on the point at issue. Two comments are in order which tend to weaken its critical force.

In the first place, it is not entirely accurate to maintain that the physical sciences never employ formulations that have at least the appearance of teleological statements. As is well known, some physical laws and theories are often expressed in so-called "isoperimetric" or "variational" form, rather than in the more familiar manner of numerical or differential equations. When laws and theories are expressed in this fashion, they strongly resemble teleological formulations, and have in fact been frequently assumed to express a teleological ordering of events and processes. For example, an elementary law of optics states that the angle of incidence of a light ray reflected by a surface is equal to the angle of reflection. However, this law can also be rendered by the statement that a light ray travels in such a manner that the length of its actual path (from its source to reflecting surface to its terminus) is the minimum of all possible paths. More generally, a considerable part of classical as well as contemporary physical theory can be stated in the form of "extremal" principles. These principles assert that the actual development of a system proceeds in such a manner as to minimize or maximize some magnitude which represents the possible configurations of the system.[1]

The discovery that the principles of mechanics can be given such extremal formulations was once considered as evidence for the operation of a divine plan throughout nature. This view was made prominent by Maupertuis, an eighteenth-century thinker who was perhaps the first to state mechanics in variational form; and it was widely accepted in the eighteenth and nineteenth centuries. Such theological interpretations of extremal principles are now almost universally recognized to be entirely gratuitous; and, with rare exceptions, physicists today do not accept the earlier claim that extremal principles entail the assumption of a plan or purpose animating physical processes. The use of such principles in physical science nevertheless does show that the dynamical structure of physical systems can be formulated so as to make focal the effect of constituent elements and subsidiary processes upon certain global properties of the system taken as a whole. If physical scientists dislike teleological language in their own disciplines, it is not because they regard teleological notions in this sense as foreign to their task. Their dislike stems in some measure from the fear that, except when such teleological language is made rigorously precise through the use of quantitative formulations, it is apt to be misunderstood as connoting the operation of purposes.

In the second place, the physical sciences, unlike biology, are in

1. Cf. A. D'Abro, *The Decline of Mechanism in Modern Physics,* New York (1939), chap. 18; Adolf Kneser, *Das Prinzip der kleinsten Wirkung,* Leipzig, 1928; Wolfgang Yourgrau and Stanley Mandelstam, *Variational Principles in Dynamics and Quantum Theory,* London, 1955. It can, in fact, be shown that when certain very general conditions are satisfied, all quantitative laws can be given an "extremal" formulation.

general not concerned with a relatively special class of organized bodies, and they do not investigate the conditions making for the persistence of some selected physical system rather than of others. When a biologist ascribes a function to the kidney, he tacitly assumes that it is the kidney's contribution to the maintenance of the living animal which is under discussion; and he ignores as irrelevant to his primary interest the kidney's contribution to the maintenance of any other system of which it may also be a constituent. On the other hand, a physicist generally attempts to discuss the effects of solar radiation upon a wide variety of things; and he is reluctant to ascribe a "function" to the sun's radiation, because no one physical system of which the sun is a part is of greater interest to him than is any other such system. And similarly for the law relating the pressure and volume of a gas: if a physicist views with suspicion the formulation of this law in functional or teleological language, it is because (in addition to the reasons which have been or will be discussed) he does not regard it as his business to assign special importance, even if only by vague suggestion, to one rather than another consequence of varying pressures in a gas.

b. However, the discussion thus far can be accused, with some justice, of naiveté if not of irrelevance, on the ground that it has ignored completely the fundamental point, namely, the "goal-directed" character of organic systems. It is because living things exhibit in varying degrees adaptive and regulative structures and activities, while the systems studied in the physical sciences do not—so it is frequently claimed—that teleological explanations are peculiarly appropriate for biological systems but not for physical systems. Thus, because the solar system, or any other system of which the sun is a part, does not tend to persist in some integrated pattern of activities in the face of environmental changes, and because the constituents of the system do not undergo mutual adjustments so as to maintain this pattern in relative independence from the environment, it is preposterous to ascribe any function to the sun or to solar radiation. Nor does the fact that physics can state some of its theories in the form of external principles, so the objection continues, minimize the difference between biological and purely physical systems. It is true that a physical system develops in such a way as to minimize or maximize a certain magnitude which represents a property of the system as a whole. But physical systems are not organized to maintain, in the face of considerable alterations in their environment, some *particular* extremal values of such magnitudes, or to develop under widely varying conditions in the direction of realizing some particular values of such magnitudes.

Biological systems, on the other hand, do possess such organization, as a single example (which could be matched by an indefinite number of others) makes quite clear. The human body maintains many of its characteristics in a relatively steady state (or homeostasis) by complicated but coordinated physiological processes. Thus the internal temperature of the body must remain fairly constant if it is not to be fatally injured. In point of fact, the temperature of the normal human being varies during a day only from about 97.3°F to 99.1°F, and cannot fall much below 75°F or rise much above 110°F without permanent injury to the body. However, the temperature of the external environment can fluctuate much more widely than this; and it is clear from elementary physical considerations that the body's characteristic activities would be profoundly impaired or curtailed unless it were capable of compensating for such environmental changes. But the body is indeed capable of doing just this; and in consequence its normal activities can continue in relative independence of the temperature of the environment—provided, of course, that the environmental temperature does not fall outside a certain interval of magnitudes. The body achieves this homeostasis by means of a number of mechanisms, which serve as a series of defenses against shifts in the internal temperature. Thus the thyroid gland is one of several that control the body's basal metabolic rate (which is the measure of the heat produced by combustion in various cells and organs); the heat radiated or conducted through the skin depends on the quantity of blood flowing through peripheral vessels, a quantity which is regulated by dilation or contraction of these vessels; sweating and the respiration rate determine the quantity of moisture that is evaporated, and so affect the internal temperature; adrenaline in the blood also stimulates internal combustion, and its

secretion is affected by changes in the external temperature; and automatic muscular contractions involved in shivering are an additional source of internal heat. There are thus physiological mechanisms in the body that automatically preserve its internal temperature, despite disturbing conditions in the body's internal and external environment.[2]

Three separate questions, frequently confounded, are raised by such facts of biological organization. (1) Is it possible to formulate in general but fairly precise terms the distinguishing structure of "goal-directed" systems, but in such a way that the analysis is neutral with respect to assumptions concerning the existence of purposes or the dynamic operation of goals as instruments in their own realization? (2) Does the fact, if it is a fact, that teleological explanations are customarily employed only in connection with "goal-directed" systems constitute relevant evidence for deciding the issue of whether a teleological explanation is equivalent to some nonteleological one? (3) Is it possible to explain in purely physicochemical terms—that is, exclusively in terms of the laws and theories of current physics and chemistry—the operations of biological systems? This third question will not concern us for the present, though we shall return to it later; but the other two require our immediate attention.

i. Since antiquity there have been many attempts at constructing machines and physical systems simulating the behavior of living organisms

2. Cf. Walter B. Cannon, *The Wisdom of the Body*, New York, 1932, chap. 12.

in one respect or another. None of these attempts has been entirely successful, for it has not been possible thus far to manufacture in the workshop and out of inorganic materials any device that acts fully like a living body. Nevertheless, it has been possible to construct physical systems that are self-maintaining and self-regulating in respect to certain of their features, and which therefore resemble living organisms in at least this one important characteristic. In an age in which servomechanics (governors on engines, thermostats, automatic airplane pilots, electronic calculators, radar-controlled anti-aircraft firing devices, and the like) no longer excite wonder, and in which the language of cybernetics and "negative feedbacks" has become widely fashionable, the imputation of "goal-directed" behavior to purely physical systems certainly cannot be rejected as an absurdity. Whether "purposes" can also be imputed to such physical systems, as some expounders of cybernetics claim,[3] is perhaps doubtful, though the question is in large measure a semantic one; and in any event, this further issue is not relevant to the present context of discussion. Moreover, it is worth noting that the possibility of constructing self-regulating physical systems does not, by itself, constitute a proof that the activities of living organisms can be explained in exclusively physicochemical terms. Nevertheless, the fact that such systems have been constructed does suggest that there is no sharp demarcation setting off the teleological organizations, often assumed to be distinctive of living things, from the goal-directed organizations of many physical systems. At the minimum, that fact offers strong support for the presumption that the teleologically organized activities of living organisms and of their parts can be analyzed without requiring the postulation of purposes or goals as dynamic agents.

With the homeostasis of the temperature of the human body as an exemplar, let us now state in general terms the formal structure of systems possessing a goal-directed organization.[4] The characteristic feature of such systems is that they continue to manifest a certain state or property $G$ (or that they exhibit a persistence of development "in the direction" of attaining $G$) in the face of a relatively extensive class of changes in their external environments or in some of their internal parts—changes which, if not compensated for by internal modification in the system, would result in the disappearance of $G$ (or

3. Cf. Arturo Rosenblueth, Norbert Wiener, Julian Bigelow, "Behavior, Purpose and Teleology," *Philosophy of Science,* 10 (1943); Norbert Wiener, *Cybernetics,* New York, 1948; A. M. Turing, "Computing Machines and Intelligence," *Mind,* 59 (1950); Richard Taylor, "Comments on a Mechanistic Conception of Purposefulness," *Philosophy of Science,* 17 (1950), and the reply by Rosenblueth and Wiener with a rejoinder by Taylor in the same volume.

4. The following discussion is heavily indebted to G. Sommerhoff, *Analytical Biology,* London, 1950. See also Alfred J. Lotka, *Elements of Physical Biology,* New York, 1926, chap. 25; W. Ross Ashby, *Design for a Brain,* London, 1953, and *An Introduction to Cybernetics,* London, 1956; and R. B. Braithwaite, *Scientific Explanation,* Cambridge, 1954, chap. 10.

in an altered direction of development of the systems). The abstract pattern of organization of such systems can be formulated with considerable precision, although only a schematic statement of that pattern can be presented in what follows.

Let $S$ be some system, $E$ its external environment, and $G$ some state, property, or mode of behavior that $S$ possesses or is capable of possessing under suitable conditions. Assume for the moment (this assumption will eventually be relaxed) that $E$ remains constant in all relevant respects, so that its influence upon the occurrence of $G$ in $S$ may be ignored. Suppose also that $S$ is analyzable into a structure of parts or processes, such that the activities of a certain number (possibly all) of them are causally relevant for the occurrence of $G$. For the sake of simplicity, assume that there are just three such parts, each capable of being in one of several distinct conditions or states. The state of each part at any given time will be represented by the predicates '$A_x$,' $B_y$,' and '$C_z$' respectively, with numerical values of the subscripts to indicate the different particular states of the corresponding parts. Accordingly, '$A_x$,' '$B_y$,' and '$C_z$' are state variables, though they are not necessarily numerical variables, since numerical measures may not be available for representing the states of the parts; and the state of $S$ that is causally relevant to $G$ at any given time will thus be expressed by a specialization of the matrix '$(A_x B_y C_z)$.' The state variables may, however, be quite complex in form—for example, '$A_x$' may represent the state of the peripheral blood vessels in a human body at a given time—

and they may be either individual or statistical coordinates. But in order to avoid inessential complications in exposition, we shall suppose that whatever the nature of the state variables, in respect to the states they represent $S$ is a deterministic system: the states of $S$ change in such a way that, if $S$ is in the same state at any two different moments, the corresponding states of $S$ after equal lapses of time from those moments will also be the same.

One further important general assumption must also be made explicit. Each of the state variables can be assigned any particular "value" to characterize a state, provided the value is compatible with the known character of the part of $S$ whose state the variable represents. In effect, therefore, the values of '$A_x$' must fall into a certain restricted class $K_A$; and there are similar classes $K_B$ and $K_C$ for the permissible values of the other two state variables. The reason for these restrictions will be clear from an example. If $S$ is the human body, and '$A_x$' states the degree of dilation of peripheral blood vessels, it is obvious that this degree cannot exceed some maximum value; for it would be absurd to suppose that the blood vessels might have a mean diameter of, say, five feet. On the other hand, the possible values of one state variable *at a given time* will be assumed to be independent of the possible values of the other state variables *at that time*. This assumption must not be misunderstood. It does not assert that the value of a variable at one time is independent of the values of the other variables at some *other* time; it merely stipulates that the value of a variable at some specified instant is not a function of the values of the other variables *at that very same instant*. The assumption is the one normally made for state variables, and is introduced in part to avoid redundant coordinates of state. For example, the state variables in classical mechanics are the position and the momentum coordinates of a particle at an instant. Although the position of a particle at one instant will in general depend on its momentum (and position) at some *previous* time, the position at a given instant is not a function of the momentum *at that given instant*. If the position were such a function of the momentum, it is clear that the state of a particle in classical mechanics could be specified by just one state variable (the momentum), so that mention of the position would be redundant. In our present discussion we are similarly assuming that none of the state variables is dispensable, so that any combination of simultaneous values of the state variables yields a permissible specialization of the matrix '$(A_x B_y C_z)$,' provided that the values of the variables belong to the classes $K_A$, $K_B$, and $K_C$, respectively. This is tantamount to saying that, apart from the proviso, the state of $S$ stipulated to be causally relevant to $G$ must be so analyzed that the state variables employed for describing the state at a given time are mutually independent of one another.

Suppose now that if $S$ is in the state $(A_0 B_0 C_0)$ at some initial time, then either $S$ has the property $G$, or else a sequence of changes occurs in $S$ as a consequence of which $S$ will possess $G$ at some subsequent time. Let us call such an initial state of $S$ a "causally effective state with respect to $G$," or a "$G$-state" for short. Not every possible state of $S$ need be a $G$-state, for one of the causally relevant parts of $S$ may be in a certain state at a given time, such that *no* combination of possible states of the other parts will yield a $G$-state for $S$. Thus, suppose that $S$ is the human body, $G$ the property of having an internal temperature lying in the range 97° to 99° F, $A_x$ again the state of the peripheral blood vessels, $B_y$ the state of the thyroid glands, and $C_z$ the state of the adrenal glands. It may happen that $B_y$ assumes a value (e.g., corresponding to acute hyperactivity) such that for no possible values of $A_x$ and $C_z$, respectively, will $G$ be realized. It is of course also conceivable that no possible state of $S$ is a $G$-state, so that in fact $G$ is never realized in $S$. For example, if $S$ is the human body and $G$ the property of having an internal temperature lying in the range from 150° to 160°, then there is no $G$-state for $S$. On the other hand, more than one possible state of $S$ may be a $G$-state. But if there is more than one possible $G$-state, then (since $S$ has been assumed to be a deterministic system) the one that is realized at a given time is uniquely determined by the actual state of $S$ at some previous time. The case in which there is more than one such possible $G$-state for $S$ is of particular relevance to the present discussion, and we must now consider it more closely.

Assume once more that at some initial time $t_0$, the system $S$ is in the $G$-state $(A_0 B_0 C_0)$. Suppose, however, that a change occurs in $S$ so that in consequence $A_0$ is caused to to vary, with the result that at time $t_1$ subsequent to $t_0$ the state variable '$A_x$' has some other value. What value it will have at $t_1$ will in general depend on the particular changes that have taken place in $S$. We shall assume, however, that $S$ will continue to be in a $G$-state at time $t_1$, provided that the values of '$A_x$' at $t_1$ fall into a certain class $K_A'$ (a subclass of $K_A$) containing more than one member, and provided also that certain further changes take place in the other state variables. To fix our ideas, suppose that $A_1$ and $A_2$ are the only possible members of $K_A'$; and assume also that neither $(A_1 B_0 C_0)$ nor $(A_2 B_0 C_0)$ is a $G$-state. In other words, if $A_0$ were changed into $A_3$ (a member of $K_A$ but not of $K_A'$), $S$ would no longer be in a $G$-state; but even though the new value of '$A_x$' falls into $K_A'$, if this were the only change in $S$ the system would also no longer be in a $G$-state at time $t_1$. Let us assume, however, $S$ to be so contsituted that if $A_0$ is caused to vary so that the value of '$A_x$' at time $t_1$ falls into $K_A'$, there will be further compensatory changes in the values of some or all of the other state variables such that $S$ continues to be in a $G$-state.

These further changes are stipulated to be of the following kind. If, as a concomitant of the change in $A_0$, the values of '$B_y$' and '$C_z$' at time $t_1$ fall into certain classes $K_B'$ and $K_C'$, respectively (where of course $K_B'$ is a subclass, though not necessarily a proper subclass, of $K_B$, and $K_C'$ is a subclass of $K_C$), then for each value in $K_A'$ there is a unique pair of values, one member of the pair belonging to $K_B'$ and the other to $K_C'$, such that for those values $S$ continues to be in a $G$-state at

time $t_1$. These pairs of values can be taken to be elements in a certain class $K_{BC}'$. On the other hand, were the altered values of '$B_y$' and '$C_z$' not accompanied by the indicated changes in the value of '$A_x$,' the system $S$ would no longer be in a $G$-state at time $t_1$. In terms of the notation just introduced, accordingly, if at time $t_1$ the state variables of $S$ have values such that two of them are members of a pair belonging to the class $K_{BC}'$ while the value of the third variable is not the corresponding element in $K_A'$, then $S$ is not in a $G$-state. For example, suppose that, when $A_0$ changes into $A_1$, the initial $G$-state $(A_0 B_0 C_0)$ is changed into the $G$-state $(A_1 B_1 C_1)$, but that $(A_0 B_1 C_1)$ is not a $G$-state; and suppose also that when $A_0$ changes into $A_2$, the initial $G$-state is changed into the $G$-state $(A_2 B_1 C_2)$, with $(A_0 B_1 C_2)$ not a $G$-state. In this example, $K_A'$ is the class $(A_1, A_2)$; $K_B'$ is the class $(B_1)$; $K_C'$, is the class $(C_1, C_2)$; and $K_{BC}'$ is the class of pairs $[(B_1, C_1), (B_1, C_2)]$, with $A_1$ corresponding to the pair $(B_1, C_1)$ and $A_2$ to the pair $(B_1, C_2)$.

Let us now bring together these various points, and introduce some definitions. Assume $S$ to be a system satisfying the following conditions: (1) $S$ can be analyzed into a set of related parts or processes, a certain number of which (say three, namely $A$, $B$, and $C$) are causally relevant to the occurrence in $S$ of some property or mode of behavior $G$. At any time the state of $S$ causally relevant to $G$ can be specified by assigning values to a set of state variables '$A_x$,' '$B_y$,' and '$C_z$.' The values of the state variables for any given time can be assigned independently of one another; but the possible values of each variable are restricted, in virtue of the nature of $S$, to certain classes of values $K_A$, $K_B$, and $K_C$, respectively. (2) If $S$ is in a $G$-state at a given initial instant $t_0$ falling into some interval of time $T$, a change in any of the state variables will in general take $S$ out of the $G$-state. Assume that a change is initiated in one of the state variables (say the parameter '$A$'); and suppose that in fact the possible values of the parameter at time $t_1$ within the interval $T$ but later than $t_0$ fall into a certain class $K_A'$, with the proviso that if this were the sole change in the state of $S$ the system would be taken out of its $G$-state. Let us call this initiating change a "primary variation" in $S$. (3) However, the parts $A$, $B$, and $C$ of $S$ are so related that, when the primary variation in $S$ occurs, the remaining parameters also vary, and in point of fact their values at time $t_1$ fall into certain classes $K_B'$ and $K_C'$, respectively. These changes induced in $B$ and $C$ thus yield unique pairs of values for their parameters at time $t_1$, the pairs being elements of a class $K_{BC}'$. Were these latter changes the only ones in the initial $G$-state of $S$, unaccompanied by the indicated primary variation in $S$, the system would not be in a $G$-state at time $t_1$. (4) As a matter of fact, however, the elements of $K_A'$ and $K_{BC}'$ correspond to each other in a uniquely

reciprocal manner, such that, when $S$ is in a state specified by these corresponding values of the state variables, the system is in a $G$-state at time $t_1$. Let us call the changes in the state of $S$ induced by the primary variation and represented by the pairs of values in $K_{BC}'$ the "adaptive variations" of $S$ with respect to the primary variation of $S$ (i.e., with respect to possible values of the parameter '$A$' in $K_A'$). Finally, when a system $S$ satisfies all these assumptions for every pair of initial and subsequent instants in the interval $T$, the parts of $S$ causally relevant to $G$ will be said to be "directively organized during the interval of time $T$ with respect to $G$"—or, more shortly, to be "directively organized," if the reference to $G$ and $T$ can be taken for granted.

This discussion of directively organized systems has been based on several simplifying assumptions. However, the analysis can be readily generalized for a system requiring the use of any number of state variables (including numerical ones), for changes in the state of a system that are initiated in more than one of the causally relevant parts of the system, and for continuous as well as discrete series of transitions from one $G$-state of a system to another.[5] Indeed, it is not difficult to develop within this framework of analysis the notion of a system exhibiting self-regulatory behaviors with respect

5. When the state coordinates are assumed to be numerical, it is possible to formulate the conditions for a directively organized system as follows:

Let $S$ be a system, $G$ a trait of $S$, and '$x_1$,' '$x_2$,' . . . , '$x_n$' the state variables for $G$. The variables are stipulated to be independent and continuous functions of the time; and superscripts will indicate their values at any given time $t$.

(a) If $S$ is a deterministic system with respect to $G$, the state of $S$ at time $t$ is uniquely determined by the state of $S$ at some preceding time $t_0$. Hence

to several $G$'s at the same time, alternative (and even incompatible) $G$'s at different times, a set of $G$'s constituting a hierarchy on the basis of some postulated scale of "relative importance," or more generally a set of $G$'s whose membership changes with time and circumstance. But apart

$$x_1{}^t = f_1(x_1{}^{t_0}, \ldots, x_n{}^{t_0}, t - t_0)$$
$$\cdots\cdots\cdots\cdots\cdots\cdots\cdots\cdots$$
$$x_i{}^t = f_i(x_1{}^{t_0}, \ldots, x_n{}^{t_0}, t - t_0)$$
$$\cdots\cdots\cdots\cdots\cdots\cdots\cdots\cdots$$
$$x_n{}^t = f_n(x_1{}^{t_0}, \ldots, x_n{}^{t_0}, t - t_0)$$

where the $f_i$'s are single-valued functions of their arguments. Their first derivates with respect to the time are also single-valued functions of their arguments and of no other functions of the time.

(b) Since the special character of $S$ imposes restrictions on the values of the state variable, the values of each variable '$x_i$' will fall within an interval determined by a pair of numbers $a_i$ and $b_i$. That is,

$$a_i \leq x_i \leq b_i$$

with $i \leq 1, 2, \ldots, n$, or alternately

$$x_i \in \Delta x_i$$

where $\Delta x_i$ is some definite interval and '$\epsilon$' is the usual sign for class membership.

(c) If $S$ is in a $G$-state at a given time $t$ falling into a given period of time $T$, the state variable must satisfy a set of conditions or equations. That $S$ is in a $G$-state at time $t$ can be expressed by requirement:

$$g_1(x_1{}^t, \ldots, x_n{}^t) = 0$$
$$\cdots\cdots\cdots\cdots\cdots\cdots$$
$$g_r(x_1{}^t, \ldots, x_n{}^t) = 0$$

where each $g_j$ ($j = 1, 2, \ldots, r$) is a function differentiable with respect to each of the state variables, and $r < n$.

(d) The values of each state variable '$x_i{}^t$' satisfying these equations defining a $G$-state of $S$ fall into certain restricted intervals:

$$a_i \leq a_i{}^G \leq x_i{}^t \leq b_i{}^G \leq b_i$$

or alternately:

$$x_i{}^t \in \Delta x_i{}^G$$

where $\Delta x_i{}^G$ falls into the interval $\Delta x_i$.

(e) Assume that $S$ is in a $G$-state at the initial time $t_0$ during the period $T$, and that a change takes place in the value of some state variable '$x_k$' so that at time $t$ later than $t_0$ in $T$ its value is $x_k{}^t$. The condition that this change is a $G$-preserving change (so that $x_k{}^t \in \Delta x_k{}^G$), is that for each function $g_j$:

$$\frac{\partial g_j}{\partial x_k{}^{t_0}} = \frac{\partial g_j}{\partial x_1{}^t}\frac{\partial x_1{}^t}{\partial x_k{}^{t_0}} + \frac{\partial g_j}{\partial x_2{}^t}\frac{\partial x_2{}^t}{\partial x_k{}^{t_0}} + \cdots + \frac{\partial g_j}{\partial x_n{}^t}\frac{\partial x_n{}^t}{\partial x_k{}^{t_0}} = 0$$

(f) The system $S$ is directively organized with respect to $G$ during $T$ if, when such $G$-preserving changes occur in any given state variable '$x_k$,' there are compensating variations in one or more of the other state variables. Accordingly, there must be at least one function $g_j$ such that in the partial differential equations just mentioned there are at least two non-vanishing summands. That is, there are at least two summands in one or more of these equations such that

$$\frac{\partial g_j}{\partial x_i{}^t}\frac{\partial x_i{}^t}{\partial x_k{}^{t_0}} \neq 0$$

from complexity, nothing immediately relevant would be gained by such extensions of the analysis; and the schematic and incompletely general definitions that have been presented will suffice for our purposes.

It will in any case be clear from the above account that if $S$ is directively organized, the persistence of $G$ is in an important sense independent of the variations in any one of the causally relevant parts of $S$, provided that these variations do not exceed certain limits. For although by hypothesis the occurrence of $G$ in $S$ depends upon $S$ being in a $G$-state, and therefore upon the state of the causally relevant parts of $S$, an alteration in the state of one of those parts may be compensated by induced changes in one or more of the other causally relevant parts, so as to preserve $S$ in its assumed $G$-state. The prima facie distinctive character of so-called "goal-directed" or teleological systems is thus formulated by the stated conditions for a directively organized system. The above analysis has therefore shown that the notion of a teleological system can be explicated in a manner not requiring the adoption of teleology as a fundamental or unanalyzable category. What may be called the "degree of directive organization" of a system, or perhaps the "degree of persistence" of some trait of a system, can also be made explicit in terms of the above analysis. For the property $G$ is maintained in $S$ (or $S$ persists in its development, which eventuates in $G$) to the extent that the range $K_A{}'$ of the possible primary variations is associated with the range of induced compensatory changes $K_{BC}{}'$ (i.e., the adaptive

variations) such that $S$ is preserved in its $G$-state. The more inclusive the range $K_A{}'$ that is associated with such compensatory changes, the more is the persistence of $G$ independent of variations in the state of $S$. Accordingly, on the assumption that it is possible to specify a measure for the range $K_A{}'$, the "degree of directive organization" of $S$ with respect to variations in the state parameter '$A$' could be defined as the measure of this range.

We may now relax the assumption that the external environment $E$ has no influence upon $S$. But in dropping this assumption, we merely complicate the analysis, without introducing anything novel into it. For suppose that there is some factor in $E$ which is causally relevant to the occurrence of $G$ in $S$, and whose state at any time can be specified by some determinate form of the state variable '$F_w$.' Then the state of the enlarged system $S$ (consisting of $S$ together with $E$) which is causally relevant to the occurrence of $G$ in $S$ is specified by some determinate form of the matrix '$(A_x B_y C_z F_w)$,' and the discussion proceeds as before. However, it is generally not the case that a variation in any of the internal parts of $S$ produces any significant variation in the environmental factors. What usually is the case is that the environmental factors vary quite independently of the internal parts; they do not undergo changes which compensate for changes in the state of $S$; and, while a limited range of changes in them may be compensated by changes in $S$ so as to preserve $S$ in some $G$-state, most of the states which environmental factors are capable of assuming cannot be so compensated by changes in $S$. It is customary, therefore, to talk of the "degree of plasticity" or the "degree of adaptability" of organic systems in relation to their environments, and not conversely. However, it is possible to define these notions without special reference to organic systems, in a manner analogous to the definition of the "degree of directive organization" of a system already suggested. Thus, suppose that the variations in the environmental state variable '$F$', assumed to be compensated by further changes in $S$ so as to preserve $S$ in some $G$-state, all fall into the class $K_F'$. If an appropriate measure for the magnitude of this class could be devised, the "degree of plasticity" of $S$ with respect to the maintenance of some $G$ in relation to $F$ could then be defined as equal to the measure of $K_F'$.

This must suffice as an outline of the abstract structure of goal-directed or teleological systems. The account given deliberately leaves undiscussed the detailed mechanisms involved in the operation of particular teleological systems; and it simply assumes that all such systems can in principle be analyzed into parts which are causally relevant to the maintenance of some feature in those systems, and which stand to each other and to environmental factors in determinate relations capable of being formulated as general laws. The discovery and analysis of such detailed mechanisms is the task of specialized scientific inquiry. Accordingly, since the above account deals only with what is assumed to be the common distinctive structure of teleological sytems, it is also entirely neutral on such substantive issues as to whether the oper-

ations of all teleological systems can be explained in exclusively physicochemical terms. On the other hand, if the account is at least approximately adequate, it requires a positive answer to the question whether the distinguishing features of goal-directed systems can be formulated without invoking purposes and goals as dynamic agents.

There is, however, one further matter that must be briefly discussed. The definitions of directively organized systems has been so stated that it can be used to characterize both biological and nonvital systems. It is in fact easy to cite illustrations for the definition from either domain. The human body with respect to homeostasis of its internal temperature is an example from biology; a building equipped with a furnace and thermostat is an example from physicochemistry. Nevertheless, although the definition is not intended to distinguish between vital and nonvital teleological systems—for the differences between such systems must be stated in terms of the specific material composition, characteristics, and activities they manifest—it *is* intended to set off systems having a prima facie "goal-directed" character from systems usually not so characterized. The question therefore remains whether the definition does achieve this aim, or whether on the contrary it is so inclusive that almost *any* system (whether it is ordinarily judged to be goal-directed or not) satisfies it.

Now there are certainly many physicochemical systems that are not ordinarily regarded as being "goal-directed" but that nevertheless appear to conform to the definition of directively organized systems proposed above. Thus, a pendulum at rest, an elastic solid, a steady electric current flowing through a conductor, a chemical system in thermodynamic equilibrium, are obvious examples of such systems. It seems therefore that the definition of directive organization—and in consequence the proposed analysis of "goal-directed" or "teleological" systems—fails to attain its intended objective. However, two comments are in order on the point at issue. In the first place, though we admittedly do distinguish between systems that are goal-directed and those which are not, the distinction is highly vague, and there are many systems which cannot be classified definitely as one kind rather than another. Thus, is the child's toy sometimes known as the "walking beetle"—which turns aside when it reaches the edge of a table and fails to fall off, because an idle wheel is then brought into play through the action of an "antenna"—a goal-directed system or not? Is a virus such a system? Is the system consisting of members of some biological species that has undergone evolutionary development in a steady direction (e.g., the development of gigantic antlers in the male Irish elk), a goal-directed one? Moreover, some systems have been classified as "teleological" at one time and in relation to one body of knowledge, only to be reclassified as "nonteleological" at a later time, as knowledge concerning the physics of mechanisms improved. "Nature does nothing in vain" was a maxim commonly accepted in pre-Newtonian physics, and on the basis of the doctrine of "natural places"

even the descent of bodies and the ascent of smoke were regarded as goal-directed. Accordingly, it is at least an open question whether the current distinction between systems that are goal-directed and those that are not invariably has an identifiable objective basis (i.e., in terms of differences between the actual organizations of such systems), and whether the *same* system may not often be classified in alternative ways depending on the perspective from which it is viewed and on the antecedent assumptions adopted for analyzing its structure.

In the second place, it is by no means certain that physical systems such as the pendulum at rest, which is not usually regarded as goal-directed, really do conform to the definition of "directively organized" systems proposed above. Consider a simple pendulum that is initially at rest and is then given a small impulse (say by a sudden gust of wind); and assume that apart from the constraints of the system and the force of gravitation the only force that acts on the bob is the friction of the air. Then on the usual physical assumptions, the pendulum will perform harmonic oscillations with decreasing amplitudes, and finally assume its initial position of rest. The system here consists of the pendulum and the various forces acting on it, while the property $G$ is the state of the pendulum when it is at rest at the lowest point of its path of oscillation. By hypothesis, its length and the mass of the bob are fixed, and so is the force of gravitation acting on it, as well as the coefficient of damping; the variables are the impulsive force of the gust of wind, and the restoring force which operates on the bob as a consequence of the constraints of the system and of the presence of the gravitational field. However—and this is the crucial point—these two forces are *not* independent of one another. Thus, if the effective component of the former has a certain magnitude, the restoring force will have an equal magnitude with an opposite direction. Accordingly, if the state of the system at a given time were specified in terms of state variables which take these forces as values, these state variables would not satisfy one of the stipulated conditions for state variables of directively organized systems; for the value of one of them at a given time is uniquely determined by the value of the other at that same time. In short, the values of these proposed state variables at any given instant are not independent.[6] It therefore follows that the simple pendulum is not a directively organized system in the sense of the definition presented. Moreover, it is also possible to show in a similar manner that a number of other systems, generally regarded as nonteleological ones, fail to satisfy that definition. Whether one could show this for all systems currently so regarded is admittedly an open question. However, since there are at least some systems not usually characterized as teleological which must also be so characterized on the bases of the definition, the label of 'directively organized system' whose meaning the definition explicates does not apply to everything whatsoever, and it does not baptize a distinction without a difference. There are therefore some grounds for claiming that the definition achieves what it was designed to achieve, and that it formulates the abstract structure commonly held to be distinctive of "goal-directed" systems.

ii. We can now settle quite briefly the second question, on page 329, we

---

6. This can be shown in greater detail by considering the usual mathematical discussion of the simple pendulum. If $l$ is the length of the pendulum, $m$ the mass of its bob, $g$ the constant force of gravity, $k$ the coefficient of damping due to air resistance, $t$ the time as measured from some fixed instant, and $s$ the distance of the bob along its path of oscillation from the point of initial rest, the differential equation of motion of the pendulum (on the assumption that the amplitude of vibration is small) is

$$m\frac{d^2 s}{dt^2} + k\frac{ds}{dt} + \frac{mg}{l}s = 0$$

If at time $t_0$ the pendulum is at rest, both $s_0$ and $v_0 \left[= \left(\frac{ds}{dt}\right)_0\right]$ are zero, so that

$$\left(m\frac{d^2 s}{dt^2}\right)_0 = 0;$$

i.e., no unbalanced forces are acting on the bob. Suppose now that at time $t_1$ the bob is at $s_1$ with a velocity $v_1$; the restoring force will then be

$$\left(m\frac{d^2 t}{dt^2}\right)_1 = -kv_1 - \frac{mg}{l}s_1$$

But an impulsive force $F_1$ communicated to the bob at time $t_1$ uniquely determines the velocity $v_1$ and the position $s_1$ of the bob at that time. Hence the restoring force can be calculated, so that it is uniquely determined by the impulsive force.

undertook to examine, namely, whether the fact that teleological explanations are usually advanced only in connection with "goal-directed" systems affects the claim that, in respect to its asserted content, every teleological explanation is translatable into an equivalent nonteleological one. The answer is clearly in the negative, if such systems are analyzable as directively organized ones in the sense of the above definition. For on the supposition that the notion of a goal-directed system can be explicated in the proposed manner, the characteristics that ostensibly distinguish such systems from those not goal-directed can be formulated entirely in nonteleological language. In consequence, every statement about the subject matter of a teleological explanation can in principle be rendered in nonteleological language, so that such explanations together with all assertions about the contexts of their use are translatable into logically equivalent nonteleological formulations.

Why, then, does it seem odd to render physical statements such as Boyle's law in teleological form? The answer is plain, if indeed teleological statements (and in particular, teleological explanations) are normally advanced only in connection with subject matters that are assumed to be directively organized. The oddity does not stem from any difference between the explicitly asserted content of a physical law and of its purported teleologically formulated equivalent. A teleological version of Boyle's law appears strange and unacceptable because such a formulation would usually be construed as resting on the assumption that a gas enclosed in a volume is a directively organized system, in contradiction to the normally accepted assumption that a volume of gas is not such a system. In a sense, therefore, a teleological explanation does connote more than does its prima facie equivalent nonteleological translation. For the former presupposes, while the latter normally does not, that the system under consideration in the explanation is directively organized. Nevertheless, if the above analysis is generally sound, this "surplus meaning" of teleological statements can always be expressed in nonteleological language.

3. On the hypothesis that a teleological explanation can always be translated, with respect to what it explicitly asserts, into an equivalent nonteleological one, let us now make more explicit in what way two such explanations nevertheless do differ. The difference appears to be as follows: Teleological explanations focus attention on the culminations and products of specific processes, and in particular upon the contributions of various parts of a system to the maintenance of its global properties or modes of behavior. They view the operations of things from the perspective of certain selected "wholes" or integrated systems to which the things belong; and they are therefore concerned with characteristics of the parts of such wholes, only insofar as those traits of the parts are relevant to the various complex features or activities assumed to be distinctive of those wholes. Nonteleological explanations, on the other hand, direct attention primarily to the conditions under which specified processes are initiated or persist,

and to the factors upon which the continued manifestations of certain inclusive traits of a system are contingent. They seek to exhibit the integrated behaviors of complex systems as the resultants of more elementary factors, frequently identified as constituent parts of those systems; and they are therefore concerned with traits of complex wholes almost exclusively to the extent that these traits are dependent on assumed characteristics of the elementary factors. In brief, the difference between teleological and nonteleological explanations, as has already been suggested, is one of emphasis and perspective in formulation.

If this account is sound, the use of teleological explanations in the study of directively organized systems is as congruent with the spirit of modern science as is the use of nonteleological ones. This conclusion is confirmed by an examination of two currently held assessments of teleological explanations, one suggesting a limit to the value of such explanations, the other objecting in principle to their use.

a. The claim has been advanced that, although teleological explanations are in general legitimate, they are useful only when the knowledge we happen to possess of directively organized systems is of a certain kind.[7] Our available information about the range of environmental changes to which such a system can make adaptive responses (i.e., about what we have called the "plasticity" of goal-directed systems) may have two sources. It may have the status simply of an extrapolation to a given system

of inductive generalizations obtained from a direct experimental study of quite similar systems. For example, the knowledge we have at present concerning the plasticity of a particular human organism in maintaining its internal temperature in the face of changes in temperature of the environment is based on our familiarity with the adaptive responses of other human bodies. On the view under consideration, teleological explanations in such cases are valuable, since they enable us to predict certain future behaviors of a given system from our knowledge concerning the past behaviors of similar systems— future behaviors that would otherwise not be predictable in the assumed state of our knowledge. On the other hand, our information about the plasticity of a given system may have the status of a body of deductions from previously established causal laws concerning the mechanisms embodied in the system. In such cases, adaptive responses of a given system to environmental changes can be calculated with the help of general assumptions, and can be predicted without any familiarity with the past behaviors of similar systems. In consequence, teleological explanations in such cases are said to have little if any value.

Although the distinction between these two types of sources of available knowledge concerning the plasticity of directively organized systems is patently sound, it is nevertheless not evident why the line between valuable and useless teleological explanations should be drawn in the indicated manner. Questions about

7. R. B. Braithwaite, *Scientific Explanation*, pp. 333 ff.

the value of an explanation are not decided by reference to the logical source of the explanatory premises, and can be answered only by examining the effective role an explanation plays in inquiry and in the communication of ideas. It is in any event far from certain that teleological explanations for goal-directed systems concerning which we possess theoretically based knowledge are invariably or normally regarded as otiose. For there are in fact many artificial self-regulating systems (such as engines with governors regulating their speed) whose plasticity can be deduced from general theoretical assumptions. Teleological explanations for various features of such systems nevertheless continue to fill many pages in technical treatises about those systems, and there is no good reason to suppose that the explanations are commonly regarded as so much worthless lumber.

b. It is sometimes objected, however, that teleological explanations are inexcusably parochial. They are based, so it is argued, on a tacit assumption that a special set of complex systems have a privileged status; and in consequence such explanations make focal the role of things and processes in maintaining just those systems and no others. Processes have no inherent termini, the objection continues, and cannot rightly be supposed to contribute exclusively to the maintenance of some unique set of wholes. It is therefore misleading to say, for example, that *the* function of the white cells in the human blood is to defend the human body against foreign microorganisms. This is undoubtedly *a* function of the leucocytes; and this particular activity may even be said

to be *the* function of these cells from the perspective of the human body. But leucocytes are elements in other systems as well; for example, they are parts of the blood stream considered in isolation from the rest of the body, of the system composed of some virus colony together with these white cells, or of the more inclusive and complex solar system. These other systems are also capable of persisting in their "normal" organization and activities only under definite conditions; and, from the perspective of the maintenance of these numerous other systems, the leucocytes possess other functions.

One obvious reply to this objection is in the form of a *tu quoque.* It is as legitimate to focus attention on consequences, culminations, and uses as it is on antecedents, starting points, and conditions. Processes do not have inherent termini, but neither do they have absolute beginnings. Things and processes are in general not elements engaged in maintaining some exclusively unique whole, but neither are wholes analyzable into an exclusively unique set of constituents. It is nevertheless intellectually profitable in causal inquiries to focus attention on certain earlier stages in the development of a process rather than on later ones, and on one set of constituents of a system rather than on another set. Similarly, it is illuminating to select as the point of departure for the investigation of some problems certain complex wholes rather than others. Moreover, as we have seen, some things are parts of directively organized systems, but do not appear to be parts of more than one such system. The study of the unique functions of parts in such unique directively organized systems is therefore not a preoccupation that assigns without warrant a special importance to certain particular systems. On the contrary, it is an inquiry that is sensitive to fundamental and objectively identifiable differences in subject matter.

There is nevertheless a point to the objection. For the refractive influence of provincial human interests on the construction of teleological explanations is perhaps more often overlooked than it is in the case of nonteleological analyses. In consequence, certain end-products of processes and certain directions of change are frequently assumed to be inherently "natural," "essential," or "proper," while all others are then labeled as "unnatural," "accidental," or even "monstrous." Thus, the development of corn seeds into corn plants is sometimes said to be natural, while their transformation into the flesh of birds or men is asserted to be merely accidental. In a given context of inquiry, and in the light of the problem which initiates it, there may be ample justification for ignoring all but one direction of possible changes and all but one system of activities to whose maintenance things and processes contribute. But such disregard of other functions that things may have, and of other wholes of which things may be parts, does not warrant the conclusion that what is ignored is less genuine or natural than what receives selective attention.

4. One final point in connection with teleological explanation in biology must be briefly noted. As has already been mentioned, some biologists maintain that the distinctive character of biological explanations appears in physiological inquiries, in which the functions of organs and vital processes are under investigation, even though most biologists are quite prepared to admit that no special categories of explanation are required in morphology or the study of structural traits. Accordingly, great stress has been placed by some writers on the contrast between structure and function, and on the difficulties in assessing the relative importance of each as a determinant of living phenomena. It is generally conceded that "the development of functions goes hand in hand with the development structure," that living activity does not occur apart from a material structure, nor does vital structure exist save as a product of protoplasmic activity. In this sense, structure and function are commonly regarded as "inseparable aspects" of biological organization. Nevertheless, eminent biologists believe it is still an unresolved and perhaps insoluble problem "to what extent structures may modify functions or functions structures"; they regard the contrast between structure and function as presenting a "dilemma."[8]

But what is this contrast, why do its terms raise an apparently

8. Cf. Edwin G. Conklin, *Heredity and Environment* (Princeton, 1922), p. 32, and Edmund B. Wilson, *The Cell* (New York, 1925), p. 670. In a later volume Conklin declared that "the relation of mechanism to finalism is not unlike that of structure to function—they are two aspects of organization. The mechanistic conception of life is in the main a structural aspect, the teleological view looks chiefly to ultimate function. These two aspects of life are not antagonistic, but complementary." *Man: Real and Ideal* (New York, 1943), p. 117.

irresolvable issue, and what does one of its terms cover which allegedly requires a mode of analysis and explanation specific to biology? Let us first remind ourselves in what way a morphological study of some biological organ, say the human eye, differs from the corresponding physiological investigation. A structural account of the eye usually consists in a description of its gross and minute anatomy. Such an account therefore specifies the various parts of the organ, their shapes and relative spatial arrangements with respect to each other and other parts of the body, and their cellular and physicochemical compositions. The phrase "structure of the eye" therefore ordinarily signifies the spatial organization of its parts, together with the physicochemical properties of each part. On the other hand, a physiological account of the organ specifies the activities in which its various parts can or do participate, and the role these parts play in vision. For example, the ciliary muscles are shown to be capable of contracting and slackening, so that because of their connection with the suspensory ligament the curvature of the lens can be accommodated to near and far vision; or the lachrymal glands are identified as the sources of fluids which lubricate and cleanse the conjunctival membranes. In general, therefore, physiology is concerned with the character, the order, and the consequences of the activities in which the parts of the eye may be engaged.

If this example is typical of the way biologists employ the terms, the contrast between structure and function is evidently a contrast between the *spatial* organization of anatomically distinguishable parts of an organ and the *temporal* (or spatiotemporal) organization of changes in those parts. What is investigated under each term of the contrasting pair is a mode of organization or a type of order. In the one case the organization is primarily if not exclusively a spatial one, and the object of the investigation is to ascertain the spatial distribution of organic parts and the modes of their linkage. In the other case the organization has a temporal dimension, and the aim of the inquiry is to discover sequential and simultaneous orders of change in the spatially ordered and linked parts or organic bodies. It is evident, therefore, that structure and function (in the sense in which biologists appear to use these words) are indeed "inseparable." For it is difficult to make sense of any supposition that a system of activities having a temporal organization is not also a system of spatially structured parts manifesting these activities. In any event, there is obviously no antithesis between an inquiry directed to the discovery of the spatial organization of organic parts and an inquiry addressed to ascertaining the spatiotemporal structures that characterize the activities of those parts.

A comparable distinction between types of inquiries can also be introduced in the physical sciences. Descriptive physical geography, for example, is concerned primarily with the spatial distribution and spatial relations of mountains, plains, rivers, and oceans; historical geology and geophysics, on the other hand, investigate the temporal and dynamic orders of change in which such geographic features are involved. Accordingly, if inquiries into structure and function are antithetical in biology, a comparable antithesis would also occur within the nonbiological sciences. Every inquiry involves discriminating selection from the great variety of patterns of relations embodied in a subject matter; and it is both convenient and unavoidable to direct some inquiries to one kind of pattern and other inquiries to different kinds. There seems to be no reason for generating a fundamental puzzle from the fact that living organisms exhibit both a spatial and a spatiotemporal structure of their parts.

What, then, is the unsolved or irresolvable issue raised by the biological distinction between structure and function? Two questions can be distinguished in this connection. It may be asked, in the first place, what spatial structures are required for the exercise of specified functions, and whether a change in the pattern of activities of an organism or of its parts is associated with any change in the distribution and spatial organization of the constituents of that system. This is obviously a matter to be settled by detailed empirical inquiry, and, though there are innumerable unsettled problems in this connection, they do not raise issues of fundamental principle. One school of philosophers and biological theorists, for example, maintains that the development of certain comparable organs in markedly different species can be explained only on the assumption of a "vital impulse" that directs evolution toward the attainment of some future function. Thus the fact that the eyes of the octopus and of man are anatomically similar, though the evolution of each species

from eyeless ancestors has followed different lines of development, has been offered as evidence for the claim that no explanation of this convergence is possible in terms of the mechanisms of chance variation and adaptations. That fact has in consequence been used to support the view that there is an "undivided original vital impulse" which so acts on inert matter as to create appropriate organs for the function of vision.[9] But even this hypothesis, however vague and otherwise unsatisfactory it may be, involves in part factual issues; and if most biologists reject it, they do so largely because the available factual evidence supports more adequately a different theory of evolutionary development.

In the second place, one may ask just why it is that a given structure is associated with a certain set of functions, or conversely. Now this question may be understood as a demand for an explanation, perhaps in physicochemical terms, for the fact that when a living body has a given spatial organization of its parts it exhibits certain patterns of activities. When the question is so understood, it is far from being a preposterous one. Although we may not possess answers to it in most cases, we do have reasonably adequate answers in at least a few others, so that we have some ground for the presumption that our ignorance is not necessarily permanent. However, such explanations must contain as premises not only statements about the physicochemical constitution of the parts of a living

9. Cf. H. Bergson, *Creative Evolution,* New York, 1911, chap. 1, and the brief but incisive critique of views similar to those of Bergson in George G. Simpson, *The Meaning of Evolution,* New Haven, 1949, chap. 12. See also Theodosius Dobzhansky, *Evolution, Genetics and Man,* New York and London, 1955, chap. 14.

thing and about the spatial organization of these parts, but also statements of physicochemical laws or theories. Moreover, at least some of these latter premises must assert connections between the spatial organization of physicochemical systems and the temporal patterns of activities. But if the question continues to be pressed, and an explanation is demanded for these latter connections as well, an impasse is finally reached. For the demand then in effect assumes that the temporal or causal structure of physical processes is deducible simply from the spatial organization of physical systems, or conversely; and neither assumption is in fact tenable.

It is possible, analogously, to give a quite accurate account of the spatial relations in which the various parts of a clock stand to one another. We can specify the sizes of its cogwheels, the location of the mainspring and the escapement wheel, and so on. But although such knowledge of the clock's spatial structure is indispensable, it is not sufficient for understanding how the clock will operate. We must also know the laws of mechanics, which formulate the temporal structure of the clock's behavior by indicating how the spatial distribution of its parts at one time is related to the distribution at a later time. However, this temporal structure cannot be deduced simply from the clock's spatial structure (or its "anatomy"), any more than its spatial structure at any given time can be derived from the general laws of mechanics. Accordingly, the question why a given anatomical structure is associated with specific functions may be irresolvable, not because it is beyond our capacities to answer it, but simply because the question in the sense in which it is intended asks for what is *logically* impossible. In short, anatomical structure does not *logically* determine function, though *as a matter of contingent fact* the specific anatomical structure possessed by an organism does set bounds to the kinds of activities in which the organism can engage. And conversely, the pattern of behavior exhibited by an organism does not *logically* imply a unique anatomical structure, though *in point of fact* an organism manifests specific modes of activity only when its parts possess a determinate anatomical structure of a definite kind.

It follows from these various considerations that the distinction between structure and function covers nothing that distinguishes biology from the physical sciences, or that necessitates the use of biology of a distinctive logic of explanation. It has not been the aim of the present discussion to deny the patent differences between biology and other natural sciences with respect to the role played by functional analyses. Nor has it been its aim to cast doubt on the legitimacy of such explanations in any domain in which they are appropriate because of the special character of the systems investigated. The objective of the discussion has been to show only that the prevalence of teleological explanations in biology does not constitute a pattern of explanation incomparably different from those current in the physical sciences, and that the use of such explanations in biology is not a sufficient reason for maintaining that this discipline requires a radically distinctive logic of inquiry.

# 19

# Functions

〜〜〜〜〜〜〜〜〜〜〜〜〜〜〜〜〜〜〜〜

## LARRY WRIGHT

THE NOTION OF FUNCTION IS not all there is to teleology, although it is sometimes treated as though it were. Function is not even the central, or paradigm, teleological concept. But it *is* interesting *and* important; and it is still not as well understood as it should be, considering the amount of serious scholarship devoted to it during the last decade or two. Let us hope this justifies my excursion into these murky waters.

Like nearly every word in English, "function" is multilaterally ambiguous. Consider:

1. $y = f(x)$ / The pressure of a gas is a function of its temperature.
2. The Apollonaut's banquet was a major state function.
3. I simply can't function when I've got a cold.
4. The heart functions in this way . . . (something about serial muscular contractions).
5. The function of the heart is pumping blood.
6. The function of the sweep-second hand on a watch is to make seconds eaiser to read.
7. Letting in light is one function of the windows of a house.
8. The wood box next to the fireplace currently functions as a dog's sleeping quarters.

It is interesting to notice that the word "function" has a spectrum of meanings even within the last six illustrations, which are the only ones at all relevant to a teleologically oriented study. Numbers 3, 4, and 8 are substantially different from one another, but they are each, from a teleological point of view, peripheral cases by comparison with 5, 6, and 7, which are the usual paradigms. And even these latter three are individually distinct in some respects, but much less profoundly than the others.

Quite obviously, making some systematic sense of the logical differentiation implicit in categorizing these cases as peripheral and paradigmatic is a major task of this paper. But a clue that we are on the right track here can be found in a symptomatic grammatical distinction present in the last six illustrations: in the peripheral cases the word "function" is itself the the verb, whereas in the more central cases "function" is a noun, used with the verb "to be." And since the controversy revolves around what *the function* of something *is,* the grammatical role of "function" in 5, 6, and 7 makes them heavy favorites for the logical place of honor in this discussion.

### SOME RUDIMENTARY DISTINCTIONS

1. *Functions v. goals.* There seems to be a strong temptation to treat functions as representative of the set of central teleological concepts which cluster around goal-directedness. However, even a cursory examination of the usual sorts of examples reveals a very important distinction. Goal-directedness is a behavioral predicate. The *direction* is the direction of behavior. When we do speak of objects (homing torpedoes) or individuals (General MacArthur) as being goal-directed, we are speaking indirectly of their behavior. We would argue against the claim that they are goal-directed by appeal to their behavior (for example, the torpedo, or the General, did not *change course* at the appropriate time, and so forth). On the other hand, many things have *functions* (for example, chairs and windpipes) which do not behave *at all,* much less goal-directedly. And behavior can have a function without being goal-directed—for example, pacing the floor or blinking your eye. But even when goal-directed behavior has a function, very often its function is quite different from the achievement of its *goal.* For ex-

ample, some fresh-water plankton diurnally vary their distance below the surface. The *goal* of this behavior is to keep light intensity in their environment relatively constant. This can be determined by experimenting with artificial light sources. The *function* of this behavior, on the other hand, is keeping constant the oxygen supply, which normally varies with sunlight intensity. There are many instances to be found in the study of organisms in which the function of a certain goal-directed activity is not some further goal of that activity, as it usually is in human behavior, but rather some natural concomitant or consequence of the immediate goal. Other examples are food-gathering, nest-making, and copulation. Clearly function and goal-directedness are not congruent concepts. There is an important sense in which they are wholly distinct. In any case, the relationship between functions and goals is a complicated and tenuous one; and becoming clearer about the nature of that relationship is one aim of this essay.

2. *A function v. the function.* Recent analyses of function, including all those treated here, have tended to focus on *a* function of something, by contrast with *the* function of something. This tendency is understandable; for any analysis of this sort aims at generality, and "a function" would seem intrinsically more general than "the function" because it avoids one obvious restriction. This generality, however, is superficial: the notion of *a* function is derivable from the notion of *the* function (more than one thing meets the criteria) just as easily as the reverse (only one thing meets the criteria). Furthermore, the notion of

*a* function is much more easily confused with certain peripheral, quasi-functional ascriptions which are examined below. In short, the discussion of this paper is concerned with *a* function of *X* only in so far as it is the sort of thing which would be *the* function of *X* if *X* had no others. Accordingly, I take the definite-article formulation as paradigmatic and will deal primarily with it, adding comments in terms of the indefinite-article formulation parenthetically, where appropriate.

3. *Function v. accident.* Very likely the central distinction of this analysis is that between the *function* of something and other things it does which are *not* its function (or one of its functions). The function of a telephone is effecting rapid, convenient communication. But there are many other things telephones do: take up space on my desk, disturb me at night, absorb and reflect light, and so forth. The function of the heart is pumping blood, not producing a thumping noise or making wiggly lines on electrocardiograms, which are also things it does. This is sometimes put as the distinction between a function, and something done merely "by accident." Explaining the propriety of this way of speaking—that is, making sense of the function/accident distinction—is another aim, perhaps the *primary* aim of the following analysis.

4. *Conscious v. natural functions.* The notion of accident will raise some interesting and important questions across another rudimentary distinction: the distinction between natural functions and consciously designed ones. Natural functions are the common organismic ones such as the function of the heart, mentioned above. Other examples are the function of the kidneys to remove metabolic wastes from the bloodstream, and the function of the lens of the human eye to focus an image on the retina. Consciously designed functions commonly (though not necessarily) involve artifacts, such as the telephone and the watch's sweep hand mentioned previously. Other examples of this type would be the function of a door knob, a headlight dimmer switch, the circumferential grooves in a pneumatic tire tread. or a police force. Richard Sorabji has argued[1] that "designed" is too strong as a description of this category, and that less elaborate conscious effort would be adequate to give something a function of this sort. I think he is right. I have used the stronger version only to overdraw the distinction hyperbolically. In deference to his point I will drop the term "designed" and talk of the distinction as between natural and conscious functions.

Of the two, natural functions are philosophically the more problematic. Several schools of thought, for different reasons, want to deny that there are natural functions, as opposed to conscious ones. Or, what comes to the same thing, they want to deny that natural functions are functions in anything like the same sense that conscious functions are. Some theologians want to say that the organs of organisms get their functions through God's conscious design, and hence these things *have* functions, but not natural functions *as opposed to* conscious ones. Some

1. Richard Sorabji, "Function," *Philosophical Quarterly,* 14 (1964), 290.

scientists, like B. F. Skinner, would *deny* that organs and organismic activity have functions *because* there is no conscious effort or design involved.

Now it seems to me that the notion of an organ having a function—both in everyday conversation and in biology—has no strong theological commitments. Specifically, it seems to me consistent, appropriate, and even common for an atheist to say that the function of the kidney is elimination of metabolic wastes. Furthermore, it seems clear that conscious and natural functions are functions in the same sense, despite their obvious differences. Functional ascriptions of either sort have a profoundly similar ring. Compare "the function of that cover is to keep the distributor dry" with "the function of the epiglottis is to keep food out of the windpipe." It is even more difficult to detect a difference in what is being requested: "What is the function of the human windpipe?" versus "What is the function of a car's exhaust pipe?" Certainly no analysis should begin by supposing that the two sorts are wildly different, or that only one is really legitimate. That is a possible *conclusion* of an analysis, not a reasonable presupposition. Accordingly, the final major aim of this analysis will be to make sense of natural functions, both as functions in the same sense as consciously contrived ones, and as functions independent of any theological presuppositions—that is, independent of conscious purpose. It follows that this

analysis is committed to finding a way of stating what it is to be a function—even in the conscious cases —that does not rely on an appeal to consciousness. If no formulation of this kind can be found despite an honest search, only then should we begin to take seriously the view that we actually mean something quite different by "function" in these two contexts.

## SOME ANALYSES OF FUNCTION

The analysis of function for which I wish to argue grew out of a detailed critical examination of several recent attempts in the literature to produce such an analysis, and it is best understood in that context. For this reason, and because it will help clarify the aims I have sketched above, I will begin by presenting the kernel of that critical examination.

The first analysis I want to consider is an early one by Morton Beckner.[2] Here Beckner contends that to say something $s$ has function $F'$ in system $s'$ is to say that "There is a set of circumstances in which: $F'$ occurs when $s'$ has $s$, AND $F'$ does not occur when $s'$ does not have s" (p. 113).[3] For example, "the human heart has the function of circulating blood" means that there is a set of circumstances in which circulation occurs in humans when they have a heart, and does not when they do not. Translated into the familiar jargon, $s$ has function $F'$ in $s'$ if and only if there is a set of circumstances containing $s$ which are sufficient for the occurrence of $F'$

---

2. Morton Beckner, *The Biological Way of Thought* (New York, 1959), chap. 6.

3. Beckner gives an alternative formulation in which we can speak of *activities* as having functions, instead of *things*. I have abbreviated it here for convenience and clarity. The logical points are the same.

and which also require $s$ in order to be sufficient for $F$. Now it is not clear whether the "requirement" here is necessity or merely nonredundancy. If it is necessity, then under the most natural interpretation of "circumstances" (environment), it is simply mistaken. There are *no* circumstances in which, for example, the heart is absolutely irreplaceable: we could always pump blood in some other way. On the other hand, if the requirement here is only nonredundancy, the mistake is more subtle.

In this case Beckner's formula would hold for cases in which $s$ merely *does* $F'$, but in which $F'$ is not the function of $s$. For example, the heart is a nonredundant member of a set of conditions or circumstances which are sufficient for a throbbing noise. But making a throbbing noise is not a function of the heart; it is just something it does—accidentally. In fact, there are even dysfunctional cases which fit the formula: in some circumstances, livers are nonredundant for cirrhosis, but cirrhotic debilitation could not conceivably be the (or a) function of the liver. So this analysis fails on the functional/accidental distinction: it includes too much.

After first considering a view essentially similar to this one, John Canfield has offered a more elaborate analysis.[4] According to Canfield:

A function of $I$ (in $S$) is to do $C$ *means* $I$ does $C$ and that $C$ is done is useful to $S$. For example, "(In vertebrates) a function of the liver is to secrete bile" means "the liver secretes bile, and that

bile is secreted in vertebrates is useful to them." (p. 290)

Canfield recognizes that natural functions are the problematic ones, but he devotes his attention solely to those cases. He treats only the organs and parts of organisms studied by biology, to the exclusion of the consciously designed functions of artifacts. As a result of this emphasis, his analysis is, without modification, almost impossible to apply to conscious functions. But even with appropriate modifications, it turns out to be inadequate to the characterization of either conscious or natural function.

In the conscious cases, there is an enormous problem in identifying the system $S$, *in* which $I$ is functioning, and *to* which it must be useful. The function of the sweep-second hand of a watch is to make seconds easier to read. It would be most natural to say that the system *in which* the sweep hand is functioning —by analogy with the organismic cases—is the watch itself; but it is hard to make sense of the easier reading being useful to the mechanism. On the other hand, the best candidate for the system *to which* the easier reading is useful is the person wearing the watch; but this does not seem to make sense as the system *in which* the sweep hand is functioning.

The crucial difficulty of Canfield's analysis begins to appear at this point: no matter what modifications we make in his formula to avoid the problem of identifying the system $S$, we must retain the requirement that

4. John Canfield, "Teleological Explanations in Biology," *The British Journal for the Philosophy of Science,* 14 (1964).

*C* be useful. This is really the major contribution of his analysis, and to abandon it is to abandon the analysis. The difficulty with this is that, for example in the watch case, it is clearly not necessary that easily read seconds be useful to the watch-wearer—or anyone else—in order that making seconds easier to read be the function of the sweep hand of that wearer's watch. My watch has a sweep-second hand, and I occasionally use it to time things to the degree of accuracy it allows: it is useful to me. Now suppose I were to lose interest in reading time to that degree of accuracy. Suppose my life changed radically so that nothing I ever did could require that sort of chronological precision. Would that mean the sweep hand on my particular watch no longer has the function of making seconds easier to read? Clearly not. If someone were to ask what the sweep hand's function was ("What's it do?," "What's it there for?") I would still have to say it made seconds easier to read, although I might yawningly append an autobiographical note about my utter lack of interest in that feature. Similarly, the function of that button on my dashboard is to activate the windshield washer, even if all it does is make the mess on the windshield worse, and hence is not useful at all. That would be its *function* even it I never took my car out of the garage —or if I broke the windshield.

It is natural at this point to attempt to patch up the analysis by reducing the requirement that *C* be useful to the requirement that *C usually* be useful. But this will not do either, because it is easy to think of cases in which we would talk of something's having a function even though doing

that thing was quite *generally* of no use to anybody. For example, a machine whose function was to count Pepsi Cola bottle caps at the city dump; or M.I.T.'s ultimate machine of a few years back, whose only function was to turn itself off. The source of the difficulty in all of these cases is that what the thing in question (watch, washer button, counting machine) was *designed* to do has been left out of the calculation. And, of course, in these cases, if something is designed to do *X,* then doing *X* is its function even if doing *X* is generally useless, silly, or even harmful. In fact, intention is so central here that it allows us to say the function of *I* is to do *C,* even when *I* cannot even *do C.* If the windshield washer switch comes from the factory defective, and is never repaired, we would still say that its *function* is to activate the washer system; which is to say: that is what it was *designed* to do.

It might appear that this commits us to the view that natural and consciously contrived functions cannot possibly be the same sort of function. If conscious intent is what *determines* the function an artifact has got, there is no parallel in natural functions. I take this to be mistaken, and will show why later. For now it is only important to show, from this unique vantage, the nature of the most formidable obstacle to be overcome in unifying natural and conscious functions.

The argument thus far has shown that meeting Canfield's criteria is not necessary for something to be a function. It can easily be shown that meeting them is also not sufficient. We are always hearing stories about the belt buckles of the Old West or on foreign battlefields which save their wearers' lives by deflecting bullets. From several points of view that is a very useful thing for them to do. But that does not make bullet deflection the function—or even *a* function—of belt buckles. The list of such cases is endless. Artifacts do all kinds of useful things which are not their functions. Blowouts cause you to miss flights that crash. Noisy wheel bearings cause you to have the front end checked over when you are normally too lazy. The sweep hand of a watch might brush the dust off the numbers, and so forth.

All this results from the inability of Canfield's analysis to handle what we took to be one of the fundamental distinctions of function talk: accidental versus nonaccidental. Something can do something useful purely by accident, but it cannot have, as its function, something it does only by accident. Something that *I* does by accident cannot be the function of *I*. The cases above allow us to begin to make some fairly clear sense of this notion of accident, at least for artifacts. Buckles stop bullets only by accident. Blowouts only accidentally keep us off doomed airplanes. Sweep hands only accidentally brush dust, if they do it at all. And this brings us back to the grammatical distinction I made at the outset when I divided the list of illustrations into "central" and "peripheral" ones. When something does something useful by accident rather than design, as in these examples, we signal the difference by a standard sort of "let's pretend" talk. Instead of using the verb "to be" or the verb "to have," and saying the thing in

question *has* such and such a function, or saying that *is* its function, we use the expression "functioning as." We might say the belt buckle *functioned as* a bullet shield, or the blowout *functioned as* divine intervention, or the sweep hand *functions as* a dust brush. Canfield's analysis does not embrace this distinction at all.

So far I have shown only that Canfield's formula fails to handle conscious functions. This means it is incapable of showing natural functions to be functions in the same full-blooded sense as conscious ones, which is indeed serious; but that, it might be argued, really misses the point of his analysis. Canfield is not interested in conscious functions. He would be happy just to handle natural functions. For the reasons set down above, however, I am looking for an analysis which will *unify* conscious and natural functions, and it is important to see why Canfield's analysis cannot produce that unification. Furthermore, Canfield's analysis has difficulties in handling natural functions that closely parallel the difficulties it has with conscious functions; which is just what we should expect if the two are functions in the same sense.

For example, it is absurd to say with Pangloss that the function of the human nose is to support eyeglasses. It is absurd to suggest that the support of eyeglasses is even one of its functions. The function of the nose has something to do with keeping the air we breathe (and smell) warm and dry. But supporting a pince-nez, just as displaying rings and warpaint, is something the human nose does, and is useful to the system having the nose: so it fits Canfield's formula. Even the heart throb, our paradigm of non-function, fits the formula: the sound made by the heart is an enormously useful diagnostic aid, not only as to the condition of the heart, but also for certain respiratory and neurological conditions. More bizarre instances are conceivable. If surgeons began attaching cardiac pacemakers to the sixth rib of heart patients, or implanting microphones in the wrists of C.I.A. agents, we could then say that these were useful things for the sixth rib and the wrist (respectively) to do. But that would not make pacemaker-hanging a function of the sixth rib, or microphone concealment a function of the human wrist.

There seems to be the same distinction here that we saw in conscious functions. It makes perfectly good sense to say the nose *functions as* an eyeglass support; the heart, through its thump, *functions as* a diagnostic aid; the sixth rib *functions as* a pacemaker hook in the circumstances described above. This, it seems to me, is precisely the distinction we make when we say, for example, that the sweep-second hand *functions as* a dust brush, while denying that brushing dust is one of the sweep hand's functions. And it is here that we can make sense of the notion of accident in the case of natural functions: it is merely fortuitous that the nose supports eyeglasses; it is happy chance that the heart throb is diagnostically significant; it would be the merest serendipity if the sixth rib were to be a particularly good pacemaker hook. It is (would be) only *accidental* that (if) these things turned out to be useful in these ways.

Accordingly, we have already drawn a much stronger parallel between natural functions and conscious functions than Canfield's analysis will allow.

Thus far I have ignored Canfield's analysis of usefulness:

[In plants and animals other than man, that *C* is done is useful to *S* means] if, *ceteris paribus, C* were not done in *S*, then the probability of that *S* surviving or having descendants would be smaller than the probability of an *S* in which *C* is done surviving or having descendants. [p. 292]

I have ignored it because its explicit and implicit restrictions make it even more difficult to work this analysis into the unifying one I am trying to produce. Even within its restrictions (natural functions in plants and animals other than man), however, the extended analysis fails for reasons very like the ones we have already examined. Hanging a pacemaker on the sixth rib of a cardiovascularly inept lynx would be useful to that cat in precisely Canfield's sense of "useful": it would make it more likely that the cat would survive and/or have descendants. Obviously the same can be said for the diagnostic value of an animal's heart sounds. So usefulness—even in this very restricted sense—does not make the right function/accident distinction: some things do useful things which are not their functions, or even one of their functions.

The third analysis I wish to examine is a more recent one by Morton Beckner.[5] This analysis is particularly

5. Morton Beckner, "Function and Teleology," *Journal of the History of Biology,* 2 (1969).

interesting for two reasons. First, Beckner is openly (p. 160) trying to accommodate both natural and conscious functions under one description. Second, he wants to avoid saying things like (to use his examples) "A function of the heart is to make heart sounds" and "A function of the Earth is to intercept passing meteorites." So his aims are very like the ones I have argued for: to produce a unifying analysis, and one which distinguishes between functions and things done by accident. And since the heart sound is useful, and intercepting metoeorites could be (perhaps already is), Beckner would probably agree in principle with the above criticism of Canfield.

Beckner's formulation is quite elaborate, so I will present it in eight distinct parts, clarify the individual parts, and then offer an illustration before going on to raise difficulties with them collectively as an analysis of the concept of function. That formulation is:

*P* has function *F* in *S* if and only if [6]

1. *P* is a part of *S* (in the normal sense of "part").
2. *P* contributes to *F*. (*P*'s being part of *S* makes the occurrence of *F* more likely.)
3. *F* is an activity in or of the system *S*.
4. *S* is structured in such a way that a significant number of its parts contribute to the activities of other parts, and of the system itself.
5. The parts of *S* and their mutual contributions are identified by the same conceptual scheme which is employed in the statement that *P* has function *F* in system *S*.
6. A significant number of critical parts (of *S*) and their activities definitionally contribute to one or more activities of the whole system *S*.
7. *F* is or contributes to an activity *A* of the whole system *S*.[7]
8. *A* is one of those activities of *S* to which a significant number of critical parts and their activities definitionally contribute.

Two points of clarification must be made at once. First, the notion of "the same conceptual scheme" in number 5 is obscure in some respects, and the considerable attention devoted to it by Beckner does not help very much. In general all one can say is that *P, F,* and the other parts and activities of *S* must be *systematically* related to one another. But in practice the point is easier to make. For example, if we wish to speak of removing metabolic wastes as the function of the human kidney, the relevant conceptual scheme contains other human organs, life, and perhaps ecology in general, but not atoms, molecular bonds, and force fields. The second point concerns the "definitional contribution" in number

6. As before, Beckner gives an alternative formulation so that we can speak either of a thing or of an activity having a function. My treatment will be limited to things, but again the logical points are the same.

7. Beckner seems to suggest (p. 160, top) that *F* must *be* an activity of the whole system *S*, which, of course, would conflict with part of 3. But his illustration, reproduced below, suggests the phrasing I have used here.

6. A part (or activity) makes a definitional contribution to an activity if that contribution is part of what we mean by the word which refers to that part (or activity). For example, part of what we mean by "heart" in a biological or medical context is "something which pumps blood": we would allow considerable variation in structure or appearance and still call something a heart if it served that function. Beckner illustrates how all these steps work together, once again using the heart.

> It is true that a function of the heart is to pump blood. The heart does pump blood; the body is a complex system of parts that by definition aid in certain activities of the whole body, such as locomotion, self-maintenance, copulation; the concepts "heart" and "blood" are recognizably components of the scheme we employ in describing this complex system; and blood-pumping does contribute to activities of the whole organism to which many of its organs, tissues and other parts definitionally contribute. (p. 160)

There are several difficulties with this analysis. They appear below, roughly in order of increasing severity.

First, Beckner's problems with the system $S$ are in some ways worse than Canfield's; for Beckner explicitly wants to include artifacts, and in addition he says much more definite things about the relationship among $P$, $F$, and $S$. So in this case, when we say the function of a watch's sweep hand is making seconds easier to read, we must not only find a system *of* which the sweep hand is a part, and *in* or *of* which "making seconds easier to read" is an activity, but this activity must be or contribute to one to which a number of the system's critical parts definitionally contribute. In the case of natural functions of the organs and other parts of organisms, the system $S$ is typically a natural unit, easy to subdivide from the environment: the organism itself. But for the conscious functions of artifacts, such systems, if they can be found at all, must be hacked out of the environment rather arbitrarily. With no more of a guide than Beckner has given us, there is nothing like a guarantee that we can always find such a system. Accordingly, when our minds boggle—as I take it they do in trying to conceive of "making seconds easier to read" being an activity at all, much less one meeting all of the other conditions of this analysis—we have to say that the analysis is at best too obscure to be applicable to such cases, and is perhaps just mistaken.

A second difficulty stems directly from the first. It is not at all clear that functions—even natural functions—have to be activities at all, let alone activities of the sort required by Beckner. Making seconds easier to read is an example, but there are many others: preventing skids in wet weather, keeping your pants up, or propping open my office door. All of these things are legitimate functions (of tire treads, belts, and doorstops, respectively); none are activities in any recognizable sense.

Thirdly, we noticed in our discussion of Canfield that something could do a useful thing by accident, in the appropriate sense of "accident." Similarly, a part of a system meeting all of Beckner's criteria might easily make a contribution to an activity of that system also quite by accident. For example, an internal-combustion engine is a system satisfying Beckner's criteria for *S*. If a small nut were to work itself loose and fall under the valve-adjustment screw in such a way as to adjust properly a poorly adjusted valve, it would make an accidental contribution to the smooth running of that engine. We would never call the maintenance of proper valve adjustment the *function* of the nut. If it got the adjustment right it was just an accident. But on Beckner's formulation, we would have to call that its function. The nut does keep the valve adjusted; the engine is a complex system of parts that by definition aid in certain activities of the whole body, such as generation of torque and self-maintenance (lubrication, heat dissipation); the concepts "nut," "valve," and "valve adjustment" are components of the scheme we employ in describing this complex system; and proper valve adjustment does contribute to the smooth running of the (whole) engine, which is an activity to which many of the other parts of the engine definitionally contribute (flywheel, connecting rod, exhaust ports).

The final difficulty is also related to one we raised for Canfield's analysis. There we noticed that if an artifact was explicitly designed to do something, *that* usually *determines* its function, irrespective of how well or badly it does the thing it was supposed to do. An analogous point can be made here. If *X* was designed to do *Y*, then *Y* is *X*'s function regardless of what contributions *X* does in fact make or fail to make. For example, the *function* of the federal automotive safety regulations is to make driving and riding in a car safer. And this is so even if they actually have just the opposite effect, through some psychodynamic or automotive quirk.

So in spite of their enormous differences, this analysis and Canfield's fail for very similar reasons: problems with the notion of system *S*, failure to rule out some accidental cases, and general inability to account for the obvious role of design.

There have been several other interesting attempts in the recent literature to provide an analysis of function. Most notable are those by Carl Hempel,[8] Hugh Lehman,[9] Richard Sorabji,[10] Francisco Ayala,[11] and Michael Ruse.[12] The last two of these do a somewhat better job on the function/accident distinction than the ones we have examined. But other than that, a discussion of these analyses would be largely redundant on the discussions of Beckner and Canfield. So I think we have gone

8. Carl Hempel, "The Logic of Functional Analyses," in L. Gross, ed., *Symposium on Sociological Theory* (New York, 1959).

9. Hugh Lehman, "Functional Explanations in Biology," *Philosophy of Science*, vol. 32 (1965).

10. Sorabji, *op. cit.*

11. Francisco J. Ayala, "Teleological Explanation in Evolutionary Biology," *Philosophy of Science*, vol. 37 (1970).

12. Michael E. Ruse, "Function Statements in Biology," *Philosophy of Science*, vol. 38 (1971).

far enough in clarifying the issues to begin constructing an alternative analysis.

## AN ALTERNATIVE VIEW

The treatments we have so far considered have overlooked, ignored, or at any rate failed to make, one important observation: that functional ascriptions are—intrinsically, if you will—explanatory. Merely saying of something, *X,* that it has a certain function, is to offer an important kind of explanation of *X.* The failure to consider this, or at least take it seriously, is, I think, responsible for the systematic failure of these analyses to provide an accurate account of functions.

There are two related considerations which urge this observation upon us. First, the "in order to" in functional ascriptions is a teleological "in order to." Its role in functional ascriptions (the heart beats in order to circulate blood) is quite parallel to the role of "in order to" in goal ascriptions (the rabbit is running in order to escape from the dog). Accordingly, we should expect functional ascriptions to be explanatory in something like the same way as goal ascriptions.[13] When we say that the rabbit is running in order to escape from the dog, we are explaining *why* the rabbit is running. If we say that John got up early in order to study, we are offering an explanation of his getting up early. Similarly in the functional cases. When we say that the distributor has that cover in order to keep the rain out, we are explaining *why* the distributor has that cover. And when we say the heart beats in order to pump blood, we are ordinarily taken to be offering an explanation of why the heart beats. This last sort of case represents the most troublesome problem in the logic of function, but it must be faced squarely, and, once faced, I think its solution is fairly straightforward.

The second consideration which recommends holding out for the explanatory status of functional ascriptions is the contextual equivalence of several sorts of requests. Consider:

1. What is the function of *X*?
2. Why do *C*'s have *X*'s?
3. Why do *X*'s do *Y*?

In the appropriate context, each of these is asking for the function of *X* "What is the function of the heart?," "Why do humans have a heart?," "Why does the heart beat?" All are answered by saying, "To pump blood," in the context we are considering. Questions of the second and third sort, being "Why?" questions, are undisguised requests for explanations. So in this context functional attributions are presumed to be explanatory. And why-form function requests are by no means bizarre or esoteric ways of asking for a function.

13. This is not to abandon, or even modify, the previous distinction between functions and goals: the point can be made in this form only *given* the distinction. Nevertheless, support is provided for the analysis I am presenting here by the fact that the "in order to" of goal-directedness can be afforded a parallel treatment. For that parallel treatment see my paper "Explanation and Teleology," in the June 1972 issue of *Philosophy of Science.*

Consider:

> Why do porcupines have sharp quills?
> Why do (some) watches have a sweep-second hand?
> Why do ducks have webbed feet?
> Why do headlight bulbs have two filaments?

These are rather ordinary ways of asking for a function. And if that is so, then it is ordinarily supposed that a function explains why each of these things is the case. The function of the quills is why porcupines *have* them, and so forth.

Moreover, the kind of explanatory role suggested by both of these considerations is not the anemic "What's it good for?" sort of thing often imputed to functional explanations. It is rather something more substantial than that. If to specify the function of quills is to explain why porcupines *have* them, then the function must be the reason they *have* them. That is, the ascription of a function must be explanatory in a rather strong sense. To choose the weaker interpretation, as Canfield does,[14] is once again to run afoul of the function-accident distinction. For, to use his example, if "Why do animals have livers?" is a request for a function, it cannot be rendered "What is the liver good for?" Livers are good for many things which are not their functions, just like anything else. Noses are good for supporting eyeglasses, fountain pens are good for cleaning your fingernails, and livers are good for dinner with onions.

No, the *function* of the liver is that *particular* thing it is good for which explains why animals have them.

Putting the matter in this way suggests that functional ascription-explanations are in some sense etiological, concern the causal background of the phenomenon under consideration. And this is indeed what I wish to argue: functional explanations, although plainly not causal in the usual, restricted sense, do concern how the thing with the function *got there.* Hence they *are* etiological, which is to say "causal" in an extended sense. But this is still a very contentious view. Functional and teleological explanations are usually *contrasted with* causal ones, and we should not abandon that contrast lightly: we should be driven to it.

What drives us to this position is the specific difficulty the best-looking alternative accounts have in making the function/accident distinction. We have seen that no matter how useful it is for $X$ to do $Z$, or what contribution $X$'s doing $Z$ makes within a complex system,[15] these sorts of consideration are never sufficient for saying that the function of $X$ is $Z$. It could still turn out that $X$ did $Z$ only by accident. But all of the accident counterexamples can be avoided if we include as part of the analysis something about how $X$ came to be there (wherever): namely, that it is there *because it does* $Z$—with an etiological "because." The buckle, the heart, the nose, the engine nut, and so forth were not there *because* they stop bullets, throb, support glasses, adjust the valve, and all the other things

14. Canfield, p. 295.

15. It is sometimes urged that this sort of thing is all a teleological explanation is asserting; this is all "why?" asks in these contexts.

which were falsely attributed as functions, respectively. Those pseudo functions could not be called upon to explain how those things *got* there. This seems to be what was missing in each of those cases.

In other words, saying that the function of $X$ is $Z$ is saying at least that

(1) $X$ is there *because* it does $Z$.
    or
    Doing $Z$ is the *reason* $X$ is there.
    or
    That $X$ does $Z$ is *why* $X$ is there.

where "because," "reason," and "why" have an etiological force. And it turns out that "$X$ is there because it does $Z$,"[16] with the proper understanding of "because," "does," and "is there" provides us with not only a necessary condition for the standard cases of functions, but also the kernel of an adequate analysis. Let us look briefly at those key terms.

"Because" is, of course, to be understood in its explanatory rather than evidential sense. It is not the "because" in "It is hot because it is red." More importantly, "because" is to be taken (as it ordinarily is anyway) to be indifferent to the philosophical reasons/causes distinction. The "because" in "He did not go to class because he wanted to study" and in "It exploded because it got too hot" are both etiological in the

appropriate way.[17] And finally, it is worth pointing out here that in this sense "$A$ because $B$" does not require that $B$ be either necessary or sufficient for $A$. Racing cars have airfoils because they generate a downforce (negative lift) which augments traction. But their generation of negative lift is neither necessary nor sufficient for racing cars to have wings: they could be there merely for aesthetic reasons, or they could be forbidden by the rules. Nevertheless, if you want to know why they are there, it is because they produce negative lift. All of this comes to saying "because" here is to be taken in its ordinary, conversational, causal-explanatory sense.

Complications arise with respect to "does" primarily because on the above condition "$Z$ is the function of $X$" is reasonably taken to entail "$X$ does $Z$." Although in most cases there is no question at all about what it is for $X$ to do $Z$, the matter is highly context-dependent and so perhaps I should mention an extreme case, if only as notice that we should include it. In some contexts we will allow that $X$ does $Z$ even though $Z$ never occurs. For example, the button on the dashboard activates the windshield washer system (that is what it does, I can tell by the circuit diagram) even though it never has and never will. An unused organic or organismic emergency reaction might have the same status. All that seems to be required is that $X$ be

---

16. I take the other forms to be essentially equivalent and subject, *mutatis mutandis,* to the same explication.

17. Of course, it follows that the notion of a *reason* offered in one of the alternative formulations is the standard conversational one as well: the reason it exploded was that it got too hot.

*able* to do Z under the appropriate conditions; for example, when the button is pushed or in the presence of a threat to safety.

The vagueness of "is there" is probably what Beckner and Canfield were trying to avoid by introducing the system S into their formulations. It is much more difficult, however, to avoid the difficulties with the system S than to clarify adequately this more general placemarker. "Is there" is straightforward and unproblematic in most contexts, but some illustrations of importantly different ways in which it can be rendered might be helpful. It can mean something like "is where it is," as in "keeping food out of the windpipe is the reason the epiglottis is where it is." It can mean "C's have them," as in "animals have hearts because they pump blood." Or it can mean merely "exists (at all)," as in "keeping snow from drifting across roads (and so forth) is why there are snow fences."

Now, saying that (I), understood in this way, should be construed as a necessary condition for taking Z to be the function of X, is merely to put in precise terms the moral of our examination of the function/accident distinction. We saw above that the accident counterexamples could not meet this requirement. On the other hand, this condition *is* met in all of the center-of-the-page cases. This is quite easy to show in the conscious cases. When we say the function of X is Z in these cases, we are saying that at least some effort was made to get X (sweep hand, button on dashboard) where it is precisely because it does Z (whatever). Doing Z is the reason X is there.

*That* is why the effort was made. The reason the sweep-second hand is there is that it makes seconds easier to read. It is there *because* it does that. Similarly, rifles have safeties because they prevent accidental discharge.

It is only slightly less obvious how natural functions can satisfy (I): We can say that the natural function of something—say, an organ in an organism—is the reason the organ is there by invoking natural selection. If an organ has been naturally differentially selected-for by virtue of something it does, we can say that the reason the organ is there is that it does that something. Hence we can say animals have kidneys *because* they eliminate metabolic wastes from the bloodstream; porcupines have quills *because* they protect them from predatory enemies; plants have chlorophyll *because* chlorophyll enables plants to accomplish photosynthesis; the heart beats *because* its beating pumps blood. And each of these can be rather mechanically put in the "reason that" form. The reason porcupines have quills is that they protect them from predatory enemies, and so forth.

It is easy to show that this formula does not represent a sufficient condition for being a function, which is to say there is something more to be said about precisely what it is to be a function. The most easily generable set of cases to be excluded is of this kind: oxygen combines readily with hemoglobin, and that is the (etiological) reason it is found in human bloodstreams. But there is something colossally fatuous in maintaining that the function of that oxygen is to combine with hemoglobin, even though it is there because it does that. The function of the oxygen in human bloodstreams is providing energy in oxidation reactions, not combining with hemoglobin. Combining with hemoglobin is only a means to that end. This is a useful example. It points to a contrast in the notion of "because" employed here which is easy to overlook and crucial to an elucidation of functions.

As I pointed out above, if producing energy is the function of the oxygen, then oxygen must be there (in the blood) because it produces energy. But the "because" in "It is there because it produces energy" is importantly different from the "because" in "It is there because it combines with hemoglobin." They suggest different *sorts* of etiologies. If carbon monoxide, which we know to combine readily with hemoglobin, were suddenly to become able to produce energy by appropriate (nonlethal) reactions in our cells and, further, the atmosphere were suddenly (!) to become filled with CO, we could properly say that the reason CO was in our bloodstreams was that it combines readily with hemoglobin. We could not properly say, however, that CO was there because it produces *energy*. And that is precisely what we could say about oxygen, on purely evolutionary-etiological grounds.

All of this indicates that it is the nature of the etiology itself which determines the propriety of a functional explanation; there must be specifically functional etiologies. When we say the function of X is Z (to do Z) we are saying that X is there because it does Z, but with a further qualification. We are explaining how X came to be there, but only certain kinds of explanations of how X came to be there will do. The causal/functional distinction is a distinction *among* etiologies; it is not a contrast between etiologies and something else.

This distinction can be displayed using the notion of a causal consequence.[18] When we give a functional explanation of X by appeal to Z ("X does Z"), Z is always a consequence or result of X's being there (in the sense of "is there" sketched above).[19] So when we say that Z is the function of X, we are not only saying that X is there because it does Z, we are also saying that Z is (or happens as) a result or consequence of X's being there. Not only is chlorophyll in plants *because* it allows them to perform photosynthesis, photosynthesis is a *consequence* of the chlorophyll's being there. Not only is the valve-adjusting screw there *because* it allows the clearance to be easily adjusted, the possibility of easy adjustment is a *consequence* of the screw's being there. Quite obviously, "consequence of" here does not mean "guaranteed by."

18. The qualification "causal" here serves merely to indicate that this is not the purely inferential sense of "consequence." I am not talking about the result or consequence of an argument—e.g., necessary conditions for the truth of a set of premises. The precise construction of "consequence" appropriate here will become clear below.

19. It is worth recalling here that "is there" can only sometimes, but not usually, be rendered "exists (at all)." So, contrary to many accounts, what is being explained, and what Z is the result of, can very often *not* be characterized as "that X exists" *simpliciter*.

"$Z$ is a consequence of $X$," very much like "$X$ does $Z$" earlier, must be consistent with $Z$'s not occurring. When we say that photosynthesis is a consequence of chlorophyll, we allow that some green plants may never be exposed to light, and that all green plants may at some time or other not be exposed to light. Furthermore, this consequence relationship does not mean that whenever $Z$ *does* occur, happen, obtain, exist, and so forth, it is as a consequence of $X$. There is room for a multiplicity of sufficient conditions, overdetermined or otherwise. Other things besides the adjusting screw may provide easy adjustment of the clearance. This (the inferential) aspect of consequence, as that notion is used here, can be roughly captured by saying that there are circumstances (of recognizable propriety) in which $X$ is non-redundant for $Z$. The aspect of "consequence" of central importance here, however, is its asymmetry. "$A$ is a consequence of $B$" is in virtually every context incompatible with "$B$ is a consequence of $A$." The source of this asymmetry is difficult to specify, and I shall not try.[20] It is enough that it be clearly present in the specific cases.

Accordingly, if we understand the key terms as they have been explicated here, we can conveniently summarize this analysis as follows:

The function of $X$ is $Z$ *means*

(2) (*a*) $X$ is there because it does $Z$,
(*b*) $Z$ is a consequence (or result) of $X$'s being there.

The first part, (*a*), displays the etiological form of functional ascription-explanations, and the second part, (*b*), describes the convolution which distinguishes functional etiologies from the rest. It is the second part of course which distinguishes the combining with hemoglobin from the producing of energy in the oxygen-respiration example. Its combining with hemoglobin is emphatically not a consequence of oxygen's being in our blood; just the reverse is true. On the other hand, its producing energy *is* a result of its being there.

The very best evidence that this analysis is on the right track is that it seems to include the entire array of standard cases we have been considering, while at the same time avoiding several very persistent classes of counter-examples. In addition to this, however, there are some more general considerations which urge this position upon us.[21] First, and perhaps most impressive, this analysis shows what it is about functions that is teleological. It provides an etiological rationale for the functional "in order to," just as recent discussions have for other teleological concepts. The role of the consequences of $X$ in its own etiology provide functional ascription-explanations with a convoluted forward orientation which precisely

---

20. It is often claimed that the asymmetry is temporal, but there are many difficulties with this view. Douglas Gasking, in "Causation and Recipes," *Mind* (October 1955), attempts to account for it in terms of manipulability with some success. But manipulability is even less generally applicable than time order, so, as far as I know, the problem remains.

21. The following considerations are intended primarily as support for the entire analysis considered as a whole. Since (*a*) has already been examined extensively, however, I have biased the argument slightly to emphasize (*b*).

parallels that found by recent analyses in ascription-explanations employing the concepts goal and intention.[22] In a functional explanation, the consequences of $X$'s being there (where it is, and so forth) must be invoked to explain why $X$ is there (exists, and so forth). Functional characterizations, by their very nature, license these explanatory appeals. Furthermore, as I hinted earlier, (b) is often simply implicit in the "because" of (a). When this is so, the "because" is the specifically teleological one sometimes identified as peculiarly appropriate in functional contexts. The peculiarly functional "because" is the normal etiological one, except that it is limited to consequences in this way. The request for an explanation as well will very often contain this implicit restriction, hence limiting the appropriate replies to something in terms of this "because"—that is, to functional explanations. "Why is it there?" in some contexts, and "What does it do?" in most, unpack into "What consequences does it have that account for its being there?"

The second general consideration which recommends this analysis is that it both accounts for the propriety of, and at the same time elucidates the notion of, natural selection. To make this clear, it is important first to say something about the unqualified notion of selection, from which natural selection is derived. According to the standard view, which I will accept for expository purposes, the paradigm cases of selection involve conscious

choice, perhaps even deliberation. We can then understand other uses of "select" and "selection" as extensions of this use: drawing attention to specific individual *features* of the paradigm which occur in subconscious or nonconscious cases. Of course, the range of extensions arrays itself into a spectrum from more or less literal to openly metaphorical. Now, there is an important distinction within the paradigmatic, conscious cases. I can say I selected something, $X$, even though I cannot give a reason for having chosen it: I am asked to select a ball from among those on the table in front of me. I choose the blue one and am asked why I did. I may say something like "I don't know; it just struck me, I guess." Alternately, I could without adding much give something which has the form of a reason: "Because it is blue. Yes, I'm sure it was the color." In both of these cases I want to refer to the selection as "mere discrimination," for reasons which will become apparent below. On the other hand, there are a number of contexts in which another, more elaborate reply is possible and natural. I could say something of the form "I selected $X$ because it does $Z$," where $Z$ would be some possibility opened by, some advantage that would accrue from, or some other result of having (using, and so forth) $X$. "I chose American Airlines because its five-across seating allows me to stretch out." Or "They selected DuPont Nomex because of the superior protection it affords in a fire."[23] Let me refer to

---

22. The primary discussions of this sort I have in mind are those in Charles Taylor's *Explanation of Behavior* and the literature to which it has given rise.

23. Of course the advantage is not always stated explicitly: "I chose American because of its five-across seating." But for it to be selection of the sort described here, as opposed to mere discrimination, something like an advantage must be at least implicit.

selection by virtue of resultant advantage of this sort as "consequence-selection." Plainly, it is this kind of selection, as opposed to mere discrimination, that lies behind conscious functions: the consequence *is* the function. Equally plainly, it is specifically this kind of selection of which *natural* selection represents an extension.

But the parallel between natural selection and conscious consequence-selection is much more striking than is sometimes thought. True, the presence or absence of volition is an important difference, at least in some contexts. We might want to say that *natural* selection is really *self*-selection, nothing is *doing* the selecting; given the nature of X, Z, and the environment, X will *automatically* be selected. Quite so. But here the above distinction between kinds of conscious selection becomes crucial. For consequence-selection, by contrast with mere discrimination, de-emphasizes volition in just such a way as to blur its distinction from natural selection on precisely this point. Given our criteria, we might well say that X *does* select itself in conscious consequence-selection. By the very nature of X, Z, and our criteria (the implementation of which may be considered the environment), X will automatically be selected.[24] The cases are very close indeed.

Let us now see how this analysis squares with the desiderata we have developed. First, it is quite clearly a unifying analysis: the formula applies to natural and conscious functions indifferently. Both natural and conscious functions are functions by

virtue of their being the reason the thing with the function "is there," subject to the above restrictions. The differentiating feature is merely the *sort* of reason appropriate in either case: specifically, whether a conscious agent was involved or no. But in the functional-explanatory context which we are examining, the difference is minimal. When we explain the presence or existence of X by appeal to a consequence Z, the overriding consideration is that Z must be or create conditions conducive to the survival or maintenance of X. The exact *nature* of the conditions is inessential to the possibility of this form of explanation: it can be looked upon as a matter of mere etiological detail, nothing in the essential form of the explanation. In any given case something could conceivably get a function through either sort of consideration. Accordingly, this analysis begs no theological questions. The organs of organisms could logically possibly get their functions through God's conscious design; but we can also make perfectly good sense of their functions in the absence of divine intervention. And in either case they would be functions in precisely the same sense. This of course was accomplished only by disallowing explicit mention of intent or purpose in accounting for conscious functions. Nevertheless, the above formula can account for the very close relationship between design and function which the previous analyses could not. For, excepting bizarre circumstances, in virtually all of the usual contexts, X was designed to do Z simply entails that X is there because it results in Z.

Second, this analysis makes a clear

24. This is a version of the old problem about the tension between rationality and freedom.

and cogent distinction between function and accident. The things $X$ can be said to do by accident are the things it results in which cannot explain how it came to be there. And we have seen that this circumvents the accident counterexamples brought to bear on the other analyses. It is merely accidental that the chlorophyll in plants freshens breath. But what it does for plants when the sun shines is no accident—that is why it is there. Furthermore, in this sense, "$X$ did $Z$ accidentally" is obviously consistent with $X$'s doing $Z$ having well-defined causal antecedents, just like the normal cases of other sorts of accident (automobile accidents, accidental meetings, and so forth). Given enough data it could even have been predictable that the belt buckle would deflect the bullet. But such deflection was still in the appropriate sense accidental: that is not why the buckle was there.

Furthermore, it is worth noting that something can get a function—either conscious or natural—*as the result of* an accident of this sort. Organismic mutations are paradigmatically accidental in this sense. But that only disqualifies an organ from functionhood for the first—or the first few—generations. If it survives by dint of its doing something, then that something becomes its function on this analysis. Similarly for artifacts.

For example, if an earthquake shifted the rollers of a transistor production-line conveyor belt, causing the belt to ripple in just such a way that defective transistors would not pass over the ripple, while good transistors would, we could say that the ripple was *functioning as* a quality control sorter. But it would be incorrect to say that the ripple *had* the function of quality control sorting. It does not *have* a function at all. It is there only by accident. Sorting can, however, *become* its function if its sorting ability ever becomes a reason for preserving the ripple: if, for example, the company decides against repairing the conveyor belt *for that reason.* This accords nicely with Richard Sorabji's comment that in conscious cases, saying the function of $X$ is $Z$ requires at least "that some efforts are or would if necessary be made" to obtain $Z$ from $X$.[25]

Third, the notion of something having more than one function is derivative. It is obtained by substituting something like "partly because"[26] for "because" in the formula. Brushing dust off the numbers is one of the functions of the watch's sweep-second hand if that feature is *one* of the (restricted, etiological) reasons the sweep hand is there. Similarly in the case of natural functions. If two or three things that livers do all contribute to the survival of organisms which have livers, we must appeal

25. Sorabji, *op.cit.*, p. 290.

26. Again, it is worth pointing out that "partly" here does not indicate that "because," when *not* so qualified, represents a sufficient condition relationship. It merely serves to indicate that more than one thing plays an explanatorily relevant role in this particular case. More than one thing must be mentioned to answer adequately the functional "why?" question in this context. But that answer, as usual, need not provide a sufficient condition for the occurrence of $X$.

to all three in an evolutionary account of why those organisms have livers. Hence the liver would have more than one function in such organisms: we would have to say that each one was *a* function of the liver.

Happily, the analysis I am here proposing also accounts for the undoubted attractiveness of the other analyses we have examined. Beckner's first analysis is virtually included in this one under the rubric "$X$ does $Z$." The rest of the formula can be thought of as a qualification to avoid some rather straightforward counterexamples which Beckner himself is concerned to circumvent in his more recent attempt. Canfield's "usefulness" is even easier to accomodate: the usefulness of something, $Z$, which $X$ does is *very usually* an informative way of characterizing why $X$ has survived in an evolutionary process, or the reason $X$ was consciously constructed. The important point to notice is that this is only *usually* the case, not necessarily: not all useful $Z$'s can explain survival and some things are constructed to do wholly useless things. As for Beckner's most recent analysis, the complex, mutually contributory relationship among parts central to it is precisely the sort of thing often responsible for the survival and reproduction of organisms on one hand, and for the construction of complex mechanisms on the other. Again the valuable features of that analysis are incorporated in this one.

There is still one sort of case in which we clearly want to be able to speak of a function, but which offends the letter of this analysis as it stands. In several contexts, some of which we have already examined, we want to be able to say that $X$ has the function $Z$, even though $X$ cannot be said to do $Z$. $X$ is not even *able* to do $Z$ under the requisite conditions. In the cases of this sort I have already mentioned (the defective washer switch and ineffective governmental safety regulations), it has seemed necessary to italicize (emphasize, underline) the word "function" in order to make its use plausible and appropriate. This is a logical flag: it signals that a special or peculiar contrast is being made, that the case departs from the paradigms in a systematic but intelligible way. Accordingly, an analysis has to make sense of such a case as a variant.

On the present analysis, the italic type signals the dropping of the (usually presumed) second condition. $X$ does *not* result in $Z$, although, paradoxically, doing $Z$ *is* the reason $X$ is there. Of course, in the abstract, this sounds fatuous. But we have already seen cases in which it is natural and appropriate. That *is* the reason $X$ (switch, safety regulations) is there. And a slightly more defensive formulation of (2) will include them directly: a functional ascription-explanation accounts for $X$'s being there by appeal to $X$'s resulting in $Z$. These cases *do* appeal to $X$'s resulting in $Z$ to explain the occurrence of $X$, even though $X$ does *not* result in $Z$. So the form of the explanation is functional even in these peculiar cases.

Interestingly, this account even handles the exotic fact that these italicized functions of $X$ can cease being even italicized functions without dispensing with or directly altering $X$. (Something that $X$ did not do can stop being its function!) For example, if the ineffective safety regulations were superseded by another set, and were merely left on the books through legislative sloth or expediency, we would no longer even say they had the (italicized) *function* of making driving less dangerous. But, of course, that would no longer be the reason they were there. The explanation would then have to appeal to legislative sloth or expediency. This is usually done with verb tenses: that *was* its function, but is not any longer; that was why it was there at one time, but is not why it is still there. A similar treatment can be given vestigial organs, such as the vermiform appendix in humans.

# 20

# Wright on Functions

## CHRISTOPHER BOORSE

ATTRIBUTING FUNCTIONS TO features of organisms is a favorite activity of biologists. The problem of analyzing these function statements has generated a lively controversy, to which the most carefully defended contribution to date is surely that of Larry Wright.[1] Wright also offers a comprehensive critique of rival views. Although this critique seems largely successful, there are reasons for thinking that his positive proposal remains unsatisfactory. Wright argues that a certain sort of explanatory force shown by function statements is the central element in their meaning. In particular, he holds that the function of a trait is that one among its effects by which its presence may be explained. I wish to argue that this etiological approach is inferior to a simple articulation of an older idea: that a function is a contribution to a goal.

In the first section I will discuss Wright's proposal; in the second I will defend the competing goal analysis.

I

Wright's main contention is that to attribute a function to a trait is to say something about its etiology, or causal history. Specifically, he suggests:

The function of $X$ is $Z$ *means*

(2) (*a*) $X$ is there because it does $Z$.

(*b*) $Z$ is a consequence (or result) of $X$'s being there [p. 161].

The function of gills in fish, for example, is respiration. According to Wright, that is because the historical development of gills took place by natural selection in favor of their respiratory effect. Respiration is the function of the gills because it is that

Support during the preparation of this paper from the Delaware Institute for Medical Education and Research and the National Institute of Mental Health (PHS grant 1 RO3 MH 24621) is gratefully acknowledged.

1. Larry Wright, "Functions," *Philosophical Review*, 82 (1973), 139–168. Parenthetical references in the text are to this article.

effect of their presence by which their presence may be causally explained.

Wright argues that this etiological analysis satisfies three crucial conditions of adequacy. First, it provides for the function-accident distinction (pp. 141, 165). Many of the effects of organs and processes, however beneficial, do not seem to be their functions. Although it is convenient that noses can support eyeglasses, that is not the function of noses (p. 148). According to Wright, this is because the supporting of eyeglasses played no role in the evolutionary etiology of the nose: the nose is not there because it supports eyeglasses. Second, the analysis applies univocally to organisms and artifacts (pp. 143, 164), as competing analyses frequently do not. We may say that the function of a sweep-second hand on a watch is to make seconds easier to read. This means, Wright thinks, that the sweep-second hand got into the design by virtue of having this effect (p. 158). With both organisms and artifacts, a functional part is there because of the function it performs. Finally, Wright's analysis shows why function statements have inherent explanatory force (pp. 154–156). Since they involve the claim that $X$ is there because it does $Z$, they constitute answers to the question "Why is $X$ there?"

These three adequacy conditions are extremely plausible ones, and Wright's explicit statement of them is a real advance. But the analysis he proposes to satisfy them raises serious difficulties.

We may begin by noting its trivializing effect upon a thesis of evolutionary biology. Consider the claim that in the evolution of organisms, functional traits tend to persist and nonfunctional ones tend to disappear. On Wright's analysis the statement:

$X$ is there because it has a function

receives the translation:

$X$ is there because it has a consequence because of which it is there.

This translation is not strictly a tautology; logically speaking, $X$ could get into an organism and remain without doing anything at all. But Wright's rendering is much closer to a tautology than one expects from an apparently substantive remark. From a different angle, the difficulty is that "$X$ is there because it has a certain function" has been made synonymous with "$X$ is there because it has a certain effect." This result is peculiar if, as Wright insists, "function" and "effect" are not synonymous. On his account there is clear content in a specific statement like "The heart is there because it pumps blood" or "The stomach is there because it digests food." But there is almost no content in the generalization that the presence of an organic character may be explained by its function. For this generalization reduces to the statement that organic characters may be explained by things they do by which they may be explained. One could wish that explanatory force might be accorded function statements at a lower price.

A second difficulty with Wright's analysis is that it is clearly incomplete as it stands. His formulation of the etiology clause—"$X$ is there because it does $Z$"—is quite general, as is dictated by the aim of univocality between organisms and artifacts. The only restriction on the clause is that "because" is to be taken "in its ordinary . . . causal-explanatory sense" (p. 157). Thus one would assume that regardless of whether organisms or artifacts are in question, any ordinary sort of etiological explanation of $X$ by $X$'s effects will support a function statement (cf. p. 164). But this is not the case. The fact fails to emerge in Wright's discussion largely because all his organic examples involve one pattern of etiological explanation and all his mechanical examples another. When organisms are in question, all cases are of an evolutionary sort: the trait $X$ arises in the first place by chance and then survives by virtue of doing $Z$. With artifacts, however, Wright considers only etiological explanations that appeal to the intentions of the designer. As soon as one examines cases where these pairings are reversed, it becomes clear that the restrictions to specific etiology patterns must be regarded as part of the analysis itself.

Consider first a counterpart to evolutionary etiology for an artifact. Suppose that a scientist builds a laser which is connected by a rubber hose to a source of gaseous chlorine. After turning on the machine, he notices a break in the hose, but before he can correct it he inhales the escaping gas and falls unconscious. According to Wright's explicit proposal, one must say that the function of the break in the hose is to release the gas. The release of the gas is a result of the break in the hose; and the break is there—that is, as in natural selection, it continues to be there—because it releases the gas. If it did not do so, the scientist would correct it. This and similar examples suggest that Wright will have to insist on the intention interpretation for artifacts. Conversely, consider an intentional

etiology for a trait of an organism. A man who is irritated with a barking dog kicks it, breaking one leg, with the intention of causing the animal pain. The dog's pain is a result of the fracture, and the fracture is there because its creator intends it to have that result. Yet I doubt whether Wright would wish to say that the function of the fracture is to cause the dog pain. So in parallel fashion one suspects that only the evolutionary interpretation of "$X$ is there because it does $Z$" is supposed to be relevant to organisms.

These examples suggest that Wright's analysis ought to be revised into a disjunction of specific etiology clauses for organisms and artifacts:

The function of $X$ is $Z$ *means*
  (*a*) (1) $X$ occurs in an organism and the presence of $X$ may be explained by its doing $Z$ according to pattern I, or
       (2) $X$ occurs in an artifact and the presence of $X$ may be explained by its doing $Z$ according to pattern II; and
  (*b*)  $Z$ is a consequence or result of $X$'s being there.

Exactly what patterns I and II should be is not clear from Wright's discussion and would have to be worked out in detail. Such a revision makes it hard to call the analysis univocal between organisms and artifacts. But a more important objection is that there are counterexamples within each domain to the view that any such etiology is what gives a trait its function.

Let us take the case of artifacts first. Wright makes it clear that he regards the intentions of the designer as central to determining the functions of an artifact's parts. In fact, these intentions are not even required to be successful, Wright says of a broken windshield-washer button that its function is to activate the washer even though the washer never gets activated (p. 158).[2] To see why this curious ruling is unsatisfactory, one need not dispute whether $X$ can "do $Z$" under these conditions in more than a Pickwickian sense. For if pure intention is sufficient, the analysis becomes flatly equivocal. Effects that are only intended cannot figure in any etiologies at all. But this defect could easily be remedied by adding the requirement that the intentions in question be successful and that their success induce the designer to continue the part in the machine. The revised thesis would be that the function of a part of an artifact is that actual result of its presence which the designer intended. In point of fact, however, such intentions are certainly not necessary and perhaps not even sufficient for $Z$ to be the function of $X$.

Such intentions are not necessary because parts of artifacts may have functions wholly unknown to their makers. Many ancient mechanisms achieved their desired goals without being understood by the people who built them. The following quotation from an article on brewing in no way strains the ordinary usage of "function":

In the succeeding twenty years [1838–58] the recognition of yeast as a living organism became more widespread, but its exact function in alcoholic fermentation remained a matter of controversy.[3]

As currently understood, the actual function of yeast in fermentation is to produce enzymes which catalyze the conversion of sugar to carbon dioxide and alcohol. Presumably, then, that has always been the function of yeast in brewing devices. It did not suddenly acquire this function with the advent of chemical theory. But brewers with no knowledge of enzymes cannot intend their yeast to produce them. Similarly, the function of wood ash was unknown to ancient soapmakers and the function of the glass in a Leyden jar to Franklin. Yet each of these components had a definite function. Wright's analysis can handle these cases only by saying that brewing, for instance, would not have been popular enough to continue in existence if yeast did not have catalytic effects, and that in this sense the yeast is there because of its catalysis. But such an appeal to the survival of a part via unintended effects is exactly what gave the wrong answer in the case of the chlorine laser. A different solution will be suggested in the next section: that secreting enzymes is the function of the yeast because that is their actual

---

2. Wright's comments on this example suggest that he sees the designer's original intentions alone as determining the function of a part, regardless of what later intentions people may have about it during the rest of its history. But his treatment of the safety-regulations case (pp. 167–168) makes later intentions relevant, too. A precise version of the clause for artifacts would have to state whose intentions count—e.g., whether the designer's, the user's, or both—and at what times.

3. W. E. Trevelyan, "Fermentation," *Encyclopedia Britannica,* 9 (1971), 187.

contribution, known or unknown, to the brewer's goals.

Conversely, is a designer's intention sufficient to give part of his artifact a function? Consider first the position adopted by Wright that pure intentions can make $Z$ the function of $X$ independently of whether $Z$ is actually produced. On this interpretation a part which once acquires a function can never lose it after the designer finishes his work. This ruling creates a peculiar divergence in the treatment of artifacts and organisms. Let us imagine that an air conditioner is designed with a special filter for removing a certain atmospheric pollutant $A$, and that eventually this pollutant vanishes from the air. Even many years after there is no $A$ in the air, Wright would apparently say that the function of the filter is to filter out the $A$. It seems more natural to hold that although the filter used to have a function, it is currently nonfunctional. That is exactly what we do say, and what Wright thinks we should say, about the appendix (p. 168). This divergence would be eliminated by our additional requirement that functional parts of artifacts must succeed in performing their intended functions. But it is unclear that even successful intentions confer functions on their objects. One often hears the hood ornaments on luxury cars described as nonfunctional. This statement is difficult to explain on Wright's analysis. A hood ornament does do something the designer intends it to do—namely, create an impression of opulence—and it is there because it does so in exactly the same sense as any other part of the car. Again one is inclined to adopt a goal analysis: hood ornaments are nonfunctional in that they make no

contribution to a contextually definite goal—that is, transportation—associated with the use of the car. It is difficult to see how else to distinguish intended effects which are functional from those which are nonfunctional except by appealing to an implicit goal. At any rate, we have seen that intentions about the parts of artifacts are not necessary for them to have functions, and so Wright's analysis is unsatisfactory even in the modified form.

Let us now turn from artifacts to organisms. My first criticism of the etiological interpretation of biological function statements is that it is historically implausible. The modern theory of evolution is of recent vintage; talk of functions had been going on for a long time before it appeared. When Harvey, say, claimed that the function of the heart is to circulate the blood, he did not have natural selection in mind. Nor does this mean that pre-evolutionary physiologists must therefore have believed in a divine designer. The fact is that in talking of physiological functions, they did not mean to be making historical claims at all. They were simply describing the organization of a species as they found it. This approach to physiology is still the standard one. Even today physiological function statements are not usually supported by, or regarded as refutable by, evolutionary evidence. Suppose we discovered, for example, that at some point the lion species simply sprang into existence by an unparalleled saltation. One would not regard this discovery as invalidating all functional claims about lions; it would show that in at least one case an intricate functional organization was created by chance. Given a little knowledge about what happens inside mammals, it is obvious that the function of the heart is to circulate the blood. That is what the heart contributes to the organism's overall goals, rather than its weight or its noise. But it cannot be obvious in any strict sense that the heart had an etiology in which this effect rather than the others played a role. Nothing about the etiology of the heart is obvious on inspection at all.

The reason Wright and others miss this point, I think. is that they assume that the point of a function statement is always to give a certain kind of explanation discussed by philosophers of science. Function statements do often provide an answer to the question "Why is $X$ there?," and it is this explanation pattern that Hempel called "functional explanation."[4] There is, however, another sort of explanation using function statements that has an equal claim to the name. This sort answers the question "How does $S$ work?," where $S$ is the goal-directed system in which $X$ appears. With some misgivings, I shall call such an explanation an "operational explanation." It is operational explanation, not the evolutionary sort, with which physiology has traditionally been concerned. Confronted with inordinately complex organisms that achieve certain goals in a remarkable range of conditions, physiologists set out to study them as one would study an alien artifact; that is, by determining what contribution, if any, each part makes to the operation of the whole. Any possibility of accounting for the origin of these parts is a very recent one. Wright is surely justified in insisting that an adequate analysis of function statements must exhibit their relation to the explanations they support. But it is also an adequacy condition that their support of both sorts of explanation be allowed for, and his analysis does little to show why function statements are of use in operational explanation.[5]

Actually Wright's account has the appearance of biological plausibility only in so far as some very specific references to evolutionary mechanisms are thought of as written into it. Consider some counterexamples to the formulation he explicitly presents. A hornet buzzing in a woodshed so frightens a farmer that he repeatedly shrinks from going in and killing it. Nothing in Wright's essay blocks the conclusion that the function of the buzzing, or even of the hornet, is to frighten the farmer. The farmer's fright is a result of the hornet's presence, and the hornet's presence continues because it has this result. By failing to make any requirement that functions benefit their bearers, the analysis also creates unwanted functions of the following sort. Obesity in a man of meager motivation can prevent him from exercising.

---

4. Carl G. Hempel, "The Logic of Functional Analysis," in *Aspects of Scientific Explanation* (New York, 1965), pp. 297-330.

5. Wright mentions and rejects the idea that function statements answer the question "What's it good for?" (p. 155). This question is not the one I have in mind. To ask what a thing is good for is different from asking how it contributes to the goals of the system in which it appears. In any case, Wright's apparent presumption that there is only one kind of functional explanation is unwarranted.

Although failure to exercise is a result of the obesity, and the obesity continues because of this result, it is unlikely that prevention of exercise is its function. These cases suggest that Wright must say explicitly that the function of $X$ is $Z$ only when the presence of $X$ in a species has resulted from selection pressure deriving from $Z$. But this alone is not enough. An organ $X$ once established by selection pressure deriving from its effect $Z$ may cease to be functional, as did the appendix, if a change in the rest of the organism or in the environment renders $Z$ useless. To cope with such cases, one must insist that functional traits create selection pressure in the present as well as the past.

Instead of attempting to find some precise etiology clause for organisms to cover all problematic cases, one would do better to realize that function statements in physiology do not carry evolutionary content at all. As something like a knockdown argument for this view I propose the following hypothetical case. Suppose that the penile urethra had evolved first as a sperm conductor, the evacuation of urine being accomplished by other means. In time, the rest of the organism then evolved in such a way that urine came regularly to be evacuated through the urethra as it is in our species. Here we have a case where a fully established organ acquires a new function without altering its features. And in such a case there is no sense at all in which the organ is there because it performs the new function. The urethra was established in its final form before the new function devolved upon it, and would have survived in this form had it never acquired the function of evacuating urine. Now Wright says that given his analysis of "The function of $X$ is $Z$," the proper analysis of "A function of $X$ is $Z$" would be that $X$ is there *partly* because it does $Z$ (p. 166). It is indeed hard to see what other suggestion would be plausible. But in our hypothetical case the urethra is not there even partly because it evacuates urine. Nevertheless there is no doubt that this remains a function of the urethra. Such multiple functions are no problem for a goal analysis: regardless of the evolutionary details, which physiologists do not usually know anyway, the urethra has both of these functions because both are standard causal contributions to the organism's over-all goals.

We may now conclude our critique of Wright by suggesting a further test for analyses of function statements which his account fails to meet. As he points out (p. 139), functional terminology occurs in various kinds of statements besides "The function of $X$ is $Z$" and "A function of $X$ is $Z$." One may say that the nose *functions as* an eyeglass support, or that it *performs* this function, while denying that supporting eyeglasses is *the* or *a* function of the nose. Wright calls these and other usages "peripheral" (p. 139) or "quasifunctional" (p. 141). It seems less tendentious to call call them simply "weak function statements." Other things being equal, it is surely a point in favor of any analysis of strong function statements if it lets us exhibit a connection between them and the weak varieties. Whatever the meaning of strong function statements is, one may expect it to support some explanation of why the term "function" seems appropriate in talk of "performing a function" as well.

Wright's comments on weak function statements do not show how his analysis can meet this test. He says:

> When something does something useful by accident rather than design, . . . we signal the difference by a standard sort of "let's pretend" talk. Instead of using the verb "to be" or the verb "to have," and saying the thing in question *has* such and such a function, or saying that *is* its function, we use the expresson "functioning as." We might say the belt buckle *functioned as* a bullet shield, or the blowout *functioned as* divine intervention, or the sweep hand *functions as* a dust brush. Canfield's analysis does not embrace this distinction at all [p. 147].

What is puzzling about this explanation is that the idea of usefulness nowhere appears in Wright's analysis, although it does in Canfield's. For Wright the essence of functionality is a certain sort of etiology. But in the "functioning as" examples he gives, nothing whatever about the etiology of the belt buckle, or the blowout, or the sweep hand is relevant to their useful effects. It remains obscure why in the world we might nevertheless wish to "pretend" that their etiology was of the right sort to justify functional language. If the pretense is made only where a trait has useful effects, that is some argument for a connection between functions and usefulness. Let us now see how a goal analysis can explain this feature of weak function statements without obliterating the distinction between the weak and the strong.

II

One of the unquestionable virtues of Wright's analysis is that his root concept, "causal history," offers some hope of covering the vast range of contexts in which function statements appear. Besides the organisms and mechanisms discussed by the literature, we also speak of the functions of words in sentences, signatures on wills, figures in paintings, and so on. But such breadth can also be achieved in another and less costly way by appealing to the notion of a goal. In every context where functional talk is appropriate, one has also to do with the goals of some goal-directed system. It would be a mistake to restrict the analysis to some particular goal like "reproductive fitness" (Ruse) or "usefulness" (Canfield) or "the good" (Sorabji).[6] Any goal pursued or intended by a goal-directed system may serve to generate a function statement. Functions are, purely and simply, contributions to goals.

Since the objects of functional ascription we have been discussing—that is, organisms and artifacts—are mechanisms, it is causal contributions to goals with which I will be concerned.

6. See Michael Ruse, "Functional Statements in Biology," *Philosophy of Science,* 38 (1971), 87–95; John Canfield, "Teleological Explanations in Biology," *British Journal for the Philosophy of Science,* 14 (1964), 285–295; and Richard Sorabji, "Function," *Philosophical Quarterly,* 14 (1964), 289–302. Ruse argues that function statements do not entail anything about goal-directedness. His argument, however, uses a function statement about a goal-directed system (a dog). A more convincing argument would be to present a function which does not contribute to some system's goals.

This notion of a causal contribution, or contributory cause, has come to seem unnecessarily obscure since Nagel's unhappy references to necessary and sufficient conditions.[7] It is true that contributory causes are not only insufficient but need not even be necessary for their effects. The pumping of the heart may be a contributory cause to the circulation of the blood without being essential to it, since circulation can occur by artificial means. But this does not mean the idea of a contributory cause should be abandoned, any more than any other ordinary causal concept for which one lacks a precise analysis. For purposes of discussing teleology, we are clear enough what it means to say that the heart is helping to cause the circulation of the blood, even if a heart-lung machine is ready to switch on at a moment's notice. We know well enough when a house is running off the power company rather than its emergency generator. With Ruse,[8] one may say that heart action is contributing to circulation when circulation is occurring by, or via, heart action, and let it go at that.

On the other hand, the notion of goal-directed behavior, when applied outside the realm of intentional action, looks more like a theoretical concept of biology to be explicated according to convenience. Different accounts of it will yield different analyses of functionality that agree extensionally only on clear cases of functions. At any rate, the notion of goal-directedness I shall employ is a slight revision of Sommerhoff's.[9] Sommerhoff's complicated account, like those of Braithwaite and Nagel,[10] is guided by the following idea. To say that an action or process *A* is directed to the goal *G* is to say not only that *A* is what is required for *G*, but also that within some range of environmental variation *A would have been modified* in whatever way was required for *G*. For example, capturing a bird may be the goal of a cat's behavior in so far as this behavior not only is appropriate for capturing a bird but would also have been appropriately

7. Ernest Nagel, *The Structure of Science* (New York, 1961), p. 403.

8. Ruse, *op. cit.*, p. 88.

9. The most extended discussion is G. Sommerhoff, *Analytical Biology* (London, 1950). A helpful supplement is "The Abstract Characteristics of Living Organisms," in F. E. Emery (ed.), *Systems Thinking* (Harmondsworth, 1959). Sommerhoff does not encounter the problem of the exploding fish which Lehman raises against Nagel (Hugh Lehman, "Functional Explanation in Biology," *Philosophy of Science*, 32, 1965, 5). The case involves a deep-sea fish that, once displaced upward, undergoes an involuntary expansion which causes it to continue rising until it bursts. At no stage of this process, according to Sommerhoff's definitions, is the fish goal-directed toward bursting (cf. pp. 93–94 and 54–55 of *Analytical Biology*). For it is never true that had the external pressure been different at some level, the "behavior" of the fish would have shown a modification required for the "goal" of bursting. Further defense against this example is provided by the revision of Sommerhoff I suggest in n. 15. My solution to the problem of heart sounds (Lehman, *op. cit.*, pp. 5–6) will become clear near the end of this paper; it is that the sounds make no regular contribution to the goals considered by physiology.

10. R. B. Braithwaite, *Scientific Explanation* (New York, 1960), pp. 328 ff.; Nagel, *op. cit.*, pp. 547–548.

modified if the bird had behaved differently.

Two difficulties mentioned by Scheffler[11] require changes in this account. Presumably a cat which waits by an empty mouse hole may have the goal of catching a mouse; but it is hard to see how any behavior can literally be required for catching a nonexistent mouse. The cat's behavior can, however, fairly be called appropriate to catching a mouse: it is, for instance, the kind of behavior that leads to catching mice when they are there. And this answer seems sufficient except for the second difficulty Scheffler mentions, the problem of multiple goals. In the presence or absence of a hidden mouse, the cat's behavior may be equally appropriate to catching a bowl of cream—grant this for the sake of argument. Furthermore, the cat may be so conditioned that in environments including a bowl of cream it would take appropriate steps to get it. Yet intuitively it seems to make sense to say that mice and not cream are the cat's goal.

I see no way to solve this second problem without abandoning Sommerhoff's ideal of a black-box analysis of goal-directedness. One might think that the cat's goal is mice rather than cream because one believed the cat has an idea of a mouse rather than an idea of cream. But not everyone attributes concepts to cats, and anyway it is not clear that having an idea of a mouse can be analyzed independently of tendencies to goal-directed activities involving mice. What seems to be required for the possibility that mice are the cat's real goal is something of which the idea of a mouse is only a special case—namely, an internal mechanism which standardly guides mouse-catching but not cream-catching. When a process appropriate to several ends at once has a true goal, I suggest it is because the process is produced by an internal mechanism which standardly guides pursuit of that goal but not the others. This principle, if correct, has an important application in biology. Since organisms contain no separate mechanisms that distinguish among the various goals that biological processes achieve, there is no way of finding a unique goal in relation to which traits of organisms have functions. I shall return to this point in due course.

With Sommerhoff as amended, then, one can say that organisms are centers of activity which is objectively directed at various goals—for example, survival and reproduction. Artifacts, by contrast, may or may not be goal-directed in and of themselves. Thermostats and guided missiles are; chairs and fountain pens are not. But we do attribute functions to chairs and fountain pens and their parts, and I think we do so by taking the artifact together with its purposive human user as a goal-directed system. Chairs have functions because they contribute to the goal-directed human activity of sitting, fountain pens because they contribute to the goal-directed human activity of writing. Such objects, which lack the appropriate organization to be independent centers of teleology, must inherit their functional features from people's use of them. In this way we may explain why, as Wright empha-

11. Israel Scheffler, "Thoughts on Teleology," *British Journal for the Philosophy of Science,* 9 (1959), 265–284.

sizes, human intentions are so central in fixing the functions of artifacts. But at the same time we avoid the false consequence that the function of a functional part must be known to the user. For the role of intention is merely to determine the over-all goal of the use of the artifact. Once that is established, the function of a part is its actual contribution to this goal regardless of whether its mode of contributing is understood or not.

With these preliminaries out of the way, we may now state some analyses and apply them to examples. I begin with what is perhaps the weakest of all functional attributions:

$X$ is performing the function $Z$ in the $G$-ing of $S$ at $t$, *means*

At $t$, $X$ is $Z$-ing and the $Z$-ing of $X$ is making a causal contribution to the goal $G$ of the goal-directed system $S$.

Two features of this proposal require comment. First, it contains five variables rather than two because all function statements, weak and strong, seem to me implicitly relative to system, goal, and time; of this more later. Second, I use the progressive tense rather than "performs" because the latter carries the unwanted implication that the function is performed repeatedly or over an appreciable time. Clearly functions may be performed only once and by accident. The Bible in the soldier's pocket, for example, may perform the function of stopping a bullet only once. Here the system $S$ is soldier plus Bible, and bullet-stopping contributes to its goals by saving his life. It sounds false to say: "This Bible performs the function of stopping bullets," but the reason is merely

that the simple present suggests that the function gets performed more often than it has been. This situation is worth noting. Although we give contrary verdicts on the two statements, the contrast depends on a feature of English verb tenses rather than on anything about the concept of a function. A good deal of the variation in our attitudes to function statements seems similarly irrelevant to the function concept. In particular, I shall argue that the contrast between weak and strong function statements is solely an effect of the English articles "a" and "the."

Most of Wright's examples of useful effects of things which are said not to be their functions are still cases of functions being performed at certain times. The stray nut which falls under the valve-adjustment screw and thereby adjusts a poorly adjusted valve (p. 152) does perform a useful function in so doing. It performs the function of adjusting the valve—that is, it contributes in that way to the goals of our activities with the car. This fact, of course, has nothing to do with the intentions of the owner or manufacturer concerning the nut. As in the brewery case, intentions about parts are unnecessary given intentions about the whole. Lehman's bullet in the brain[12] or the flat feet of a draftee are likewise cases of the performance of functions. It is easy to imagine a doctor saying: "This bullet has been performing a useful function for you all these years, Mr. Jones . . . ." On the other hand, not just any accidental effect is the performance of a function. The accidental ripple in Wright's conveyor belt (p. 165) performs no

12. Lehman, *op. cit.*, p. 14.

function if it passes only defective transistors. But it does perform a function if it passes only good ones, and that is because it promotes the goals for which we employ the machine.

Now it would be unreasonable to claim against Wright that the useful effects of all these items are in fact their functions—that is, "the functions" of them. Nevertheless I suggest that the distinction between weak and strong function statements is illusory to this extent: there is no important conceptual constituent of the idea of "the function" or "a function" which is missing in "performing the function." To accept our analysis of performing a function is to settle the question of what sort of thing a function is—namely, a contribution to a goal. Whatever "the" or "a" function of a thing is, then, it must be at least a contribution to a goal. And I shall now try to show that what more is required for a function performed by $X$ to be among "the functions" of $X$ is not any fixed property but instead varies from context to context.

Initial support for this view comes from the difficulty of dividing cases that justify weak statements from those that justify strong ones. Suppose I say that although I was never assigned any particular job on the tenure committee, my function in the group has been largely to mediate between opposing factions. Is this a function that I have merely been performing? Or is it actually *the* function of Boorse in the group, a function that I have? It certainly sounds right to say that *the* function of Boorse *has been* largely . . . and so forth. Similarly, if a battle-weary veteran looks back over a long career in which he was shot thirteen times in the pocket Bible, can he truly say that stopping bullets has been one of the functions of the book? Contrary to Wright's analysis, even if he carried the Bible for religious reasons alone, it is unclear that the answer is no. And is holding down papers one of the many functions of my telephone? Again the answer does not seem to depend on my intentions in acquiring the telephone, but in some vague way simply on how often I use it to do so.

To remove the uncertainty in these cases, what we need is not to discover that the items in question have some fixed general property which transforms functions performed into functions possessed. What converts a function $X$ performs into "the function of $X$" is our background interests in the context in which the function statement is made. Philosophers influenced by Russell tend to forget how often the articles "a" and "the" work in a context-sensitive way. Phrases like "the man" and "the book" do not, of course, assert or presuppose that there is exactly one man or book. Which of the many men and books is under reference is determined by the context of utterance. Similar comments apply to the indefinite article. Suppose I say: "When I walked into the department meeting, a man was quoting from Chairman Mao." This outrageous fabrication does not become true if it happens that at the moment I entered the room in Newark, a meeting of the central committee began in Peking. If we apply this point to function statements[13] the moral is obvious. "The function of $X$" will be simply that one among

13. Not all uses of the definite or indefinite article are sensitive to context. What I suggest

all the functions performed by $X$ which satisfies whatever relevance conditions are imposed by the context of utterance. In other words:

"The function of $X$ is $Z$" means that in some contextually definite system $S$ with contextually definite goal(s) $G$, during some contextually definite time interval $t$, the $Z$-ing of $X$ is the sole member of a contextually circumscribed class of functions being performed during $t$ by $X$ in the $G$-ing of $S$—that is, causal contributions to $G$.

Similarly, something like this will explain "A function of $X$ is $Z$":

"A function of $X$ is $Z$" means that in some contextually definite goal-directed system $S$, during some contextually definite time interval $t$, the $Z$-ing of $X$ falls within some contextually circumscribed class of functions being performed by $X$ during $t$—that is, causal contributions to a goal $G$ of $S$.

These statements are not analyses in the sense of two-place synonymy relations, but for reasons just given I doubt that such analyses are possible.

To reinforce the claim that these formulations are an adequate general account of strong function statements, we may mention some of the contextual limitations by which we determine in practice what "the function of $X$" or "a function of $X$" is. With artifacts, overwhelmingly the most common limitation is the one Wright

discusses: the intentions of the designer. If one of the functions $X$ does or could perform is intended by the designer, the question "What is the function of $X$?" is almost inevitably heard as a request for this function. But that is not because we are trying to trace the etiology of, say, our automobile. It is because we want to know what good $X$ is for making our car run, and the designer is usually the best authority on how the artifact works. Etiology seems relevant only in so far as it is expected to show how the part can advance the goals we pursue with the car. And as one would suppose from this explanation, when the presumption that the designer understands his mechanism fails, and none of the functions performed are intended by him, a quite different contextual limitation comes into play. In that case "the function" of a part will be simply that function, if any, which it both regularly and importantly performs. This situation was illustrated earlier by brewer's yeast. If ingenious mechanisms like cars usually resulted from lucky accidents on the part of ignorant manufacturers, we might disregard the intention of designers altogether. In asking our mechanic for the function of the carburetor, we would then be asking simply for whatever it standardly contributes to our goals of driving. So Wright's intentional convention is no part of the meaning of functional talk. It is a generally convenient contextual device reflecting part of the normal background of our discussions of artifacts—namely, the insight of their

is that "The function of $X$ is $Z$" is not like "The color of daffodils is yellow" or "The capital of Maine is Augusta," but instead like "The qualifications of the candidate are exceptional." Colors and capital of things at any one time are fixed properties of them; the qualifications of a candidate vary with the employment under discussion in the context.

designers—and would disappear if that background changed.

Although much more could be said about the functions of parts of artifacts,[14] I shall now turn to the real payoff of the contextual view—its implications for biology. For our analysis discloses an ambiguity in biological function statements which illuminates some recent controversy in the field. This ambiguity arises partly because biologists see teleology on so many different time scales and levels of organization. At one level, individual organisms are goal-directed systems whose behavior contributes to various goals. By eating insects, say, a bird contributes to its own survival and also (a fortiori) to the survival of the species, but further to the survival of its genes, the equilibrium of the insect population, and so on. More-

over, unlike the cat stalking the mouse, the bird contains no separate mechanisms reserved for each general goal. One can say only that some of its behavior serves one goal, some another, and most serves all at once. This means that there is no sense in asking which goal $G$ is *the* goal at which the bird's behavior aims, and in respect to which its parts have functions, except in so far as this goal is clear from the context of discourse. And the individual bird is only one system $S$. At the level of a population or species, new goal-directed processes seem to appear—for example, the maintenance of protective coloration in a changing environment[15]—and one can ascend still higher by viewing a whole ecosystem as a teleological unit.

Given this latitude for choice, what

14. The fact that a part may fail to perform its function deserves some explanation. In Wright's example, one may say that the function of a certain switch is to activate the windshield washer, even though the switch is defective and the washer does not get activated. This kind of remark can be understood in at least two ways. One might mean that activating the washer is the intended function of the switch—i.e., its intended contribution to our goals in using the car. Here the intentional convention selects from all possible functions performable by $X$ that one which the designer had in mind. Or one might mean that activating the washer is the function performed by this (kind of) switch in most cars of this model. In either of these senses a defective switch still has a function; but at the same time one could rightly deny that it has any *current* function. This explanation seems less mysterious than Wright's view that a purely intentional "doing" of $Z$ by $X$ can play a role in $X$'s creation.

15. Sommerhoff cites protective coloration as an example of directive correlation on the level of a species. It seems to me, however, that his definitions ought to have excluded it. Protective coloration is certainly adaptive—i.e., functional—for each individual organism that has it. To this extent biologists are right to view it teleologically. But a species is no more goal-directed toward maintaining this function than an ornamental hedge is goal-directed toward the shape in the gardener's mind. The hedge also fits Sommerhoff's definition perfectly: for a range of values of the gardener's ideal (the coenetic variable) at $t_0$, the hedge (by progressive trimming) would come to match this ideal at $t_1$. Intuitively, the hedge is not goal-directed because it in no way *modifies* its behavior between $t_0$ and $t_1$ in response to the coenetic variable. Each part of it simply keeps growing and is or is not trimmed off by the gardener. But the same is true of the species. Each member simply reproduces as usual while caterpillars of unsuitable color are trimmed by a hungry environment. Since space prevents any explication of what it is for a system to "modify its behavior," I will merely say this: if, as I believe, it is wrong to call cryptic coloration or the maintenance of an eco-

has happened is that various sub-fields of biology have carved out various slices of the teleological pie and interpreted their function statements accordingly. In physiology the goal-directed system $S$ is the individual organism and the relevant goals its own survival and reproduction. Whatever contributes to these goals reliably, throughout a species or other reference class, is assigned a physiological function.[16] Thus, although physiological statements are claims about a whole class, the goals in question are states of individual organisms. From this standpoint a trait which contributes only to other goals like the survival of the species—for example, the immolation response of some ants to fires—will be omitted from physiological theory. But such traits may very well be assigned functions by other biologists whose system $S$ is the species itself. Ecologists create yet another distinct context of functional talk by considering an ecosystem rather than a species as their unit. In discussing an analysis very similar to our own, Beckner says:

the ecosystem of a mountain lake . . . is directively organized with respect to the biomass ratio of predator and prey fishes. But we would not say that a function of the trout is to eat the bluegills, although this does play a role in the regulation of the ratio.[17]

That, however, is exactly the sort of thing ecologists do say.[18] Our analysis explains why it seems natural to them to say it. As long as a biologist's task is to provide, for some genuinely goal-directed natural system $S$, an *operational* explanation—that is, an explanation of how $S$ works—he cannot be faulted for using functional talk. One must simply be clear about the variables suppressed in his function statements: the system $S$, the goals $G$, and the time interval $t$.

Having thus placed the functional vocabulary in its broader setting, we can once again agree with Wright to a limited degree. When the biologist, like a good natural scientist, is interested not only in how a system works but also in how it got that way, there will almost certainly be one or

system goal-directed, then some of the contexts for function statements mentioned in the text may reflect confusion among biologists.

16. It is worth noting that since standing states may be causally relevant to the achievement of goals, relatively inert parts of organisms (e.g., hair or exoskeletons) may have functions. Thus, as Wright mentions (p. 152), functions need not be activities. It is also worth noting that a functional trait must make a net contribution to physiological goals. That is why breaks in blood vessels which lead to their own repair (Lehman, *op. cit.*, pp. 14–15), though they activate a function, are not themselves functional.

A brief discussion of the relativity of physiological function statements to a reference class is contained in my "Health as a Theoretical Concept," *Philosophy of Science*, 44 (1977): 542–573.

17. Morton Beckner, "Function and Teleology," *Journal of the History of Biology*, 2 (1969), 156.

18. E.g., George L. Clarke, *Elements of Ecology* (New York, 1954), defines an ecological niche as "the function of [an] organism in the community" (p. 468). Clifford B. Knight, *Basic Concepts of Ecology* (New York, 1965), calls it "the functional role an organism plays" (p. 171).

more functions of a trait with a unique claim to be called *the* functions of the trait. These functions will be the ones that explain, via evolutionary theory, the trait's development. Furthermore, if George Williams is right, such functions are invariably contributions to one particular goal—namely, the survival of the genes that govern them. Williams argues that the only effective evolutionary force is the natural selection of alternative alleles in a Mendelian population; group selection is impotent.[19] If this is true, then any group- or ecosystem-related functions of a trait may be completely worthless in explaining why it is there. For this reason among others Williams explicitly proposes a "terminological convention" about function statements: "The designation of something as the *means* or *mechanism* for a certain *goal* or *function* or *purpose* will imply that the machinery involved was fashioned by selection for the goal attributed to it."[20]

When the question at issue is evolutionary explanation, the appropriateness of this Wrightian convention is apparent. But it is equally apparent that such a convention is inappropriate to those other biological contexts where the aim is an explanation of how some system currently achieves the goals it does. Contrary to Wright and Williams, there is no reason to force the physiologist to use "function" in such a way that his statement that urine evacuation is a function of the male urethra could be overthrown by new evolutionary evidence. Especially in human physiology, the focus of inquiry is not evolution at all, but how the mechan-

ism currently operates and how to to keep it in shape. Similarly, the aim of an ecologist may be simply to understand how an equilibrium is in fact maintained so that we can avoid disrupting it. In these contexts talk of functions has a clear and legitimate use without any etiological implications.

One can see, then, why Wright's analysis so often seems ideal and yet sustains so many counterexamples. To ask for "the function" of a thing is, indeed, often to ask for that function which explains its presence. The utility of this convention is easy to understand. With either organisms or artifacts, mechanisms well adapted to definite goals do generally arise not by chance but by selection. Any important functions of a part are quite likely to have played some role in its origin. Furthermore, etiological explanation is the only thing that interests some biologists. In contexts where these considerations operate, Wright's analysis seems ideal. But this link between functions and etiology is a feature of "the" rather than of "function." By missing this point, Wright misses the possibility of alternative contextual conventions on "the" as well as the real content of the function concept. Consequently, his analysis suffers two sorts of counterexamples. It fails to account for those contexts—for example, ordinary physiology or a discussion of brewing—where the explanatory force of function statements is not etiological. And it falsely attributes functions —for example, to hornets, gas leaks, and obesity—where there is no contribution to a goal.

---

19. George C. Williams, *Adaptation and Natural Selection* (Princeton, 1966).
20. Williams, *op. cit.*, p. 9.

Besides preserving Wright's successes and avoiding his difficulties, our context-sensitive goal analysis meets his tests of adequacy. As regards the term "function" itself, it is completely univocal between organisms and artifacts, thus passing his first test. It also distinguishes between functions and accidents: all Wright's cases of accidental effects of things which are not their function either fail to contribute to a goal or violate relevance conditions imposed by the context of utterance. Finally, our analysis allows for the kind of functional explanation Wright has in mind wherever such explanations are possible. Unlike Wright's, however, it accounts for the operational variety of functional explanation as well and exhibits a clear connection between strong and weak function statements. Wright has made an essential contribution in stating explicit desiderata for an analysis of function statements and in showing that previous attempts fail to meet them. But there seems to be less promise in his own account than in our rehabilitation of the goal analysis.

# 21

# Functional Analysis

*∽∿∽∿∽∿∽∿∽∿∽∿∽∿∽∿∽∿∽∿∽∿∽∿∽*

## ROBERT CUMMINS

I

A survey of the recent philosophical literature on the nature of functional analysis and explanation, beginning with the classic essays of Hempel in 1959 and Nagel in 1961, reveals that philosophical research on this topic has almost without exception proceeded under the following assumptions.[1]

(A) The point of functional characterization in science is to explain the presence of the item (organ, mechanism, process, or whatever) that is functionally characterized.

(B) For something to perform its function is for it to have certain effects on a containing system, which effects contribute to the performance of some activity of, or the maintenance of some condition in, that containing system.

Putting these two assumptions together, we have: a function-ascribing statement explains the presence of the functionally characterized item $i$ in a system $s$ by pointing out that $i$ is present in $s$ because it has certain effects on $s$. Give or take a nicety, this fusion of (A) and (B) constitutes the core of almost every recent attempt

1. Cf. Carl Hempel, The logic of functional analysis, in *Aspects of Scientific Explanation,* New York, Free Press, 1965, reprinted from Llewellyn Gross, ed., *Symposium on Sociological Theory,* New York, Harper and Row, 1959; and Ernest Nagel, *The Structure of Science,* New York, Harcourt, Brace and World, 1961, chapter 12, section I. The assumptions, of course, predate Hempel's 1959 essay. See, for instance, Richard Braithwaite, *Scientific Explanation,* Cambridge, Cambridge University Press, 1955, chapter X; and Israel Scheffler, Thoughts on teleology, *British Journal for the Philosophy of Science,* 11 (1958). More recent examples include Francisco Ayala, Teleological explanations in evolutionary biology, *Philosophy of Science,* 37 (1970); Hugh Lehman, Functional explanations in biology, *Philosophy of Science,* 32 (1965); Richard Sorabji, Function, *Philosophical Quarterly,* 14 (1964); and Larry Wright, Functions, *Philosophical Review,* 82 (1973).

to give an account of functional analysis and explanation. Yet these assumptions are just that: assumptions. They have never been systematically defended; generally they are not defended at all. I think there are reasons to suspect that adherence to (A) and (B) has crippled the most serious attempts to analyze functional statements and explanation, as I will argue in sections II and III below. In section IV, I will briefly develop an alternative approach to the problem. This alternative is recommended largely by the fact that it emerges as the obvious approach once we take care to understand why accounts involving (A) and (B) go wrong.

II

I begin this section with a critique of Hempel and Nagel. The objections are familiar for the most part, but it will be well to have them fresh in our minds as they form the backdrop against which I stage my attack on (A) and (B).

Hempel's treatment of functional analysis and explanation is a classic example of the fusion of (A) and (B). He begins by considering the following singular function-ascribing statement.

(1) The heartbeat in vertebrates has the function of circulating the blood through the organism.

He rejects the suggestion that "function" can *simply* be replaced by "effect" on the grounds that, although the heartbeat has the effect of producing heartsounds, this is not its function. Presuming (B) from the start, Hempel takes the problem to be

how one effect—the having of which is the function of the heartbeat (circulation)—is to be distinguished from other effects of the heartbeat (e.g., heartsounds). His answer is that circulation, but not heartsounds, ensures a necessary condition for the "proper working of the organism." Thus Hempel proposes (2) as an analysis of (1).

(2) The heartbeat in vertebrates has the effect of circulating the blood, and this ensures the satisfaction of certain conditions (supply of nutriment and removal of waste) which are necessary for the proper working of the organism.

As Hempel sees the matter, the main problem with this analysis is that functional statements so construed appear to have no explanatory force. Since he assumes (A), the problem for Hempel is to see whether (2) can be construed as a deductive nomological explanans for the presence of the heartbeat in vertebrates and, in general, to see whether statements having the form of (2) can be construed as deductive nomological explananda for the presence in a system of some trait or item that is functionally characterized.

Suppose, then, that we are interested in explaining the occurrence of a trait $i$ in a system $s$ (at a certain time $t$), and that the following functional analysis is offered:

(a) At $t$, $s$ functions adequately in a setting of kind $c$ (characterized by specific internal and external conditions).

(b) $s$ functions adequately in a

setting of kind *c* only if a certain necessary condition, *n,* is satisfied.

(c) If trait *i* were present in *s* then, as an effect, condition *n* would be satisfied.

(d) Hence, at *t*, trait *i* is present in *s*. [2]

(d), of course, does not follow from (a) – (c), since some trait *i'* different from *i* might well suffice for the satisfaction of condition *n*. The argument can be patched up by changing (c) to (c'): "condition *n* would be satisfied in *s* only if trait *i* were present in *s*," but Hempel rightly rejects this avenue on the grounds that instances of the resulting schema would typically be false. It is false, for example, that the heart is a necessary condition for circulation in vertebrates, since artificial pumps can be, and are, used to maintain the flow of blood. We are thus left with a dilemma. If the original schema is correct, then functional explanation is invalid. If the schema is revised so as to ensure the validity of the explanation, the explanation will typically be unsound, having a false third premise.

Ernest Nagel offers a defense of what is substantially Hempel's schema with (c) replaced by (c').

. . . a teleological statement of the form, "The function of *A* in a system *S* with organization *C* is to enable *S* in the environment *E* to engage in process *P*," can be formulated more explicitly by: every system *S* with organization *C* and in environment *E* engages

in process P; if *S* with organization *C* and in environment *E* does not have *A,* then *S* does not engage in *P;* hence, *S* with organization *C* must have *A*. [3]

Thus he suggests that (3) is to be rendered as (4):

(3) The function of chlorophyll in plants is to enable them to perform photosynthesis.

(4) A necessary condition of the occurrence of photosynthesis in plants is the presence of chlorophyll.

So Nagel must face the second horn of Hempel's dilemma: (3) is presumably true, while (4) may well be false. Nagel is, of course, aware of this objection. His rather curious response is that, as far as we know, chlorophyll *is* necessary for photosynthesis in the green plants. [4] This may be so, but the response will not survive a change of example. Hearts are *not* necessary for circulation, artificial pumps having actually been incorporated into the circulatory systems of vertebrates in such a way as to preserve circulation and life.

A more promising defense of Nagel might run as follows. While it is true that the presence of a working heart is not a necessary condition of circulation in vertebrates under all circumstances, still, under *normal* circumstances—most circumstances, in fact—a working heart is necessary for circulation. Thus it is perhaps true that, at the present stage of evolution, a vertebrate that

---

2. Hempel, p. 310.          3. Nagel, p. 403.          4. Ibid., p. 404.

has not been tampered with surgically would exhibit circulation only if it were to contain a heart. If these circumstances are specifically included in the explanans, perhaps we can avoid Hempel's dilemma. Thus instead of (4) we should have:

(4) At the present stage of evolution, a necessary condition for circulation in vertebrates that have not been surgically tampered with is the operation of a heart (properly incorporated into the circulatory system).

(4'), in conjuction with statements asserting that a given vertebrate exhibits circulation and has not been surgically tampered with and is at the present stage of evolution, will logically imply that that vertebrate has a heart. It seems, then, that the Hempelian objection could be overcome if it were possible, given a true function-ascribing statement like (1) or (3), to specify "normal circumstances" in such a way as to make it true that, in those circumstances, the presence of the item in question is a necessary condition for the performance of the function ascribed to it.

This defense has some plausibility as long as we stick to the usual examples drawn from biology. But if we widen our view a bit, even within biology, I think it can be shown that this defense of Nagel's position will

not suffice. Consider the kidneys. The function of the kidneys is to eliminate wastes from the blood. In particular, the function of my left kidney is to eliminate waste from my blood. Yet the presence of my left kidney is not, in normal circumstances, a necessary condition for the removal of the relevant wastes. Only if something seriously abnormal should befall my right kidney would the operation of my left kidney become necessary, and this only on the assumption that I am not hooked up to a kidney machine.[5]

A less obvious counterexample derives from the well-attested fact of hemispherical redundancy in the brain. No doubt it is in principle possible to specify conditions under which a particular duplicated mechanism would be necessary for normal functioning of the organism, but (a) in most cases we are not in a position to actually do this, though we are in a position to make well-confirmed statements about the functions of some of these mechanisms, and (b) these circumstances are by no means the normal circumstances. Indeed, given the fact that each individual nervous system develops somewhat differently owing to differing environmental factors, the circumstances in question might well be different for each individual, or for the same individual at different times.

Apparently Nagel was pursuing the wrong strategy in attempting to analyze functional ascriptions in terms

5. It might be objected here that although it is the function of the kidneys to eliminate waste, that is not the function of a particular kidney unless operation of that kidney *is* necessary for removal of wastes. But suppose scientists had initially been aware of the existence of the left kidney only. Then, on the account being considered, anything they had said about the function of that organ would have been false, since, on that account, *it has no function in organisms having two kidneys!*

of necessary conditions. Indeed, we are still faced with the dilemma noticed by Hempel: an analysis in terms of necessary conditions yields a valid but unsound explanatory schema; analysis in terms of sufficient conditions along the lines proposed by Hempel yields a schema with true premises, but validity is sacrificed.

Something has gone wrong, and it is not too difficult to locate the problem. An attempt to explain the presence of something by appeal to what it does—its function—is bound to leave unexplained why something else that does the same thing—a functional equivalent—isn't there instead. In itself, this is not a serious matter. But the accounts we have been considering assume that explanation is a species of deductive inference, and one cannot deduce hearts from circulation. This is what underlies the dilemma we have been considering. At best, one can deduce circulators from circulation. If we make this amendment, however, we are left with a functionally tainted analysis; 'the function of the heart is to circulate the blood' is rendered 'a blood circulator is a (necessary/sufficient) condition of circulation, and *the heart is a blood circulator.*' The expression in italics is surely as much in need of analysis as the analyzed expression. The problem, however, runs much deeper than the fact that the performance of a certain function does not determine how that function is performed. The

problem is rather that to "explain" the presence of the heart in vertebrates by appeal to what the heart *does* is to "explain" its presence by appeal to factors which are causally irrelevant to its presence. Even if it were possible, as Nagel claimed, to *deduce* the presence of chlorophyll from the occurrence of photosynthesis, this would fail to explain the presence of chlorophyll in green plants in just the way deducing the presence and height of a building from the existence and length of its shadow would fail to explain why the building is there and has the height it does. This is not because all explanation is causal explanation: it is not. But to explain the presence of a naturally occurring structure or physical process—to explain why it is there, why such a thing exists in the place (system, context) it does—this does require specifying factors which causally determine the appearance of that structure or process.[6]

There is, of course, a sense in which the question "Why is $x$ there?" is answered by giving $x$'s function. Consider the following exchange. X asks Y, "Why is that thing there (pointing to the gnomon of a sundial)?" Y answers, "Because it casts a shadow on the dial beneath, thereby indicating the time of day." It is exchanges of this sort that most philosophers have had in mind when they speak of functional explanation. But it seems to me that, although such exchanges

---

6. Even in the case of a designed artifact, it is at most the designer's *belief* that $x$ will perform $f$ in $s$ which is causally relevant to $x$'s presence in $s$, not $x$'s actually performing $f$ in $s$. The nearest I can come to describing a situation in which $x$ performing $f$ in $s$ is causally relevant to $x$'s presence in $s$ is this: the designer of $s$ notices a thing like $x$ performing $f$ in a system like $s$, and this leads to belief that $x$ will perform $f$ in $s$, and this in turn leads the designer to put $x$ in $s$.

do represent genuine explanations, the use of functional language in this sort of explanation is quite distinct from its explanatory use in science. In section IV below I will sketch what I think *is* the central explanatory use of functional language in science. Meanwhile, if I am right, the evident propriety of exchanges like that imagined between X and Y has led to premature acceptance of (A), hence to concentration on what is, from the point of view of scientific explanation, an irrelevant use of functional language. For it seems to me that the question "Why is $x$ there?" can be answered by specifying $x$'s function only if $x$ is or is part of an artifact. Y's answer, I think, explains the presence of the gnomon because it rationalizes the action of the agent who put it there by supplying a *reason* for putting it there. In general, when we are dealing with the result of a deliberate action, we may explain the result by explaining the action, and we may explain a deliberate action by supplying the agent's reason for doing it. Thus when we look at a sundial, we assume we *know* in a general way how the gnomon came to be there: someone deliberately put it there. But we may wish to know *why* it was put there. Specifying the gnomon's function allows us to formulate what we suppose to be the unknown agent's reason for putting it there, viz., a belief that it would cast a shadow such that . . . , and so

on. When we do this, we are elaborating on what we assume is the crucial causal factor in determining the gnomon's presence, namely a certain deliberate action.

If this is on the right track, then the viability of the sort of explanation in question should depend on the assumption that the thing functionally characterized is there as the result of deliberate action. If that assumption is evidently false, specifying the thing's function will not answer the question. Suppose it emerges that the sundial is not, as such, an artifact. When the ancient building was ruined, a large stone fragment fell on a kind of zodiac mosaic and embedded itself there. Since no sign of the roof remains, Y has mistakenly supposed the thing was designed as a sundial. As it happens, the local people have been using the thing to tell time for centuries, so Y is right about the function of the thing X pointed to.[7] But it is simply false that the thing is there because it casts a shadow, for there is no agent who put it there "because it casts a shadow." Again, the function of a bowl-like depression in a huge stone may be to hold holy water, but we cannot explain why it is there by appeal to its function if we know it was left there by prehistoric glacial activity.

If this is right, then (A) will lead us to focus on a type of explanation which will not apply to natural systems: chlorophyll and hearts are not

---

7. *Is* casting a shadow the function of this fragment? Standard use may confer a function on something: if I standardly use a certain stone to sharpen knives, then that is its function, or if I standardly use a certain block of wood as a door stop, then the function of that block is to hold my door open. If non-artifacts *ever* have functions, appeals to those functions cannot explain their presence. The things functionally characterized in science are typically not artifacts.

"there" as the result of any deliberate action; hence the essential presupposition of the explanatory move in question is missing. Once this becomes clear, to continue to insist that there *must* be *some* sense in which specifying the function of chlorophyll explains its presence is an act of desperation born of thinking there is no other explanatory use of functional characterization in science.

Why have philosophers identified functional explanation exclusively with the appeal to something's function in explaining why it is there? One reason, I suspect, is a failure to distinguish teleological explanation from functional explanation, perhaps because functional concepts do loom large in "explanations" having a teleological form. Someone who fails to make this distinction, but who senses that there is an important and legitimate use of functional characterization in scientific explanation, will see the problem as one of finding a legitimate explanatory role for functional characterization within the teleological form. Once we leave artifacts and go to natural systems, however, this approach is doomed to failure, as critics of teleology have seen for some time.

This mistake probably would have sorted itself out in time were it not the case that we do reason from the performance of a function to the presence of certain specific processes and structures, e.g., from photosynthesis to chlorophyll, or from coordinated activity to nerve tissue. This is perfectly legitimate reasoning: it is a species of inference to the best explanation. Our best (only) explana-

tion of photosynthesis requires chlorophyll, and our best explanation of coordinated activity requires nerve tissue. But once we see what makes this reasoning legitimate, we see immediately that inference *to* an explanation has been mistaken for an explanation itself. Once this becomes clear, it becomes equally clear that (A) has matters reversed: given that photosynthesis is occurring in a particular plant, we may legitimately infer that chlorophyll is present in that plant precisely because chlorophyll enters into our best (only) explanation of photosynthesis, and given coordinated activity on the part of some animal, we may legitimately infer that nerve tissue is present precisely because nerve tissue enters into our best explanation of coordinated activity in animals.

To attempt to explain the heart's presence in vertebrates by appealing to its function in vertebrates is to attempt to explain the occurrence of hearts in vertebrates by appealing to factors which are causally irrelevant to its presence in vertebrates. This fact has given "functional explanation" a bad name. But it is (A) that deserves the blame. Once we see (A) as an undefended philosophical hypothesis about how to construe functional explanations rather than as a statement of the philosophical problem, the correct alternative is obvious: what we can and do explain by appeal to what something does is the behavior of a containing system.[8]

A much more promising suggestion in the light of these considerations is that (1) is appealed to in explaining *circulation.* If we reject (A) and adopt

---

8. A confused perception of this fact no doubt underlies (B), but the fact that (B) is nearly inseparable from (A) in the literature shows how confused this perception is.

this suggestion, a simple deductive-nomological explanation with circulation as the explicandum turns out to be a sound argument.

(5) a. Vertebrates incorporating a beating heart in the usual way (in the way $s$ does) exhibit circulation.
   b. Vertebrate $s$ incorporates a beating heart in the usual way.
   c. Hence, $s$, exhibits circulation.

Though by no means flawless, (5) has several virtues, not the least of which is that it does not have biologists passing by an obvious application of evolution or genetics in favor of an invalid or unsound "functional" explanation of the presence of hearts. Also, the redundancy examples are easily handled, e.g., the removal of wastes is deduced in the kidney case.

The implausibility of (A) is obscured in examples taken from biology by the fact that there are two distinct uses of function statements in biology. Consider the following statements.

(a) The function of the contractile vacuole in protozoans is elimination of excess water from the the organism.
(b) The function of the neurofibrils in the ciliates is coordination of the activity of the cilia.

These statements can be understood in either of two ways. (i) They are generally used in explaining how the organism in question comes to exhibit certain characteristics or behavior. Thus (a) explains how excess water, accumulated in the organism by osmosis, is eliminated from the organism; (b) explains how it happens that the activity of the cilia in paramecium, for instance, is coordinated. (ii) They may be used in explaining the continued survival of certain organisms incorporating structures of the sort in question by indicating the survival value which would accrue to such organisms in virtue of having structures of that sort. Thus (a) allows us to infer that incorporation of a contractile vacuole makes it possible for the organism to be surrounded by a semi-permeable membrane, allowing the passage of oxygen into, and the passage of wastes out of, the organism. Relatively free osmosis of this sort is obviously advantageous, and this is made possible by a structure which solves the excess water problem. Similarly, ciliates incorporating neurofibrils will be capable of fairly efficient locomotion, the survival value of which is obvious.[9]

The second sort of use occurs as part of an account which, if we are not careful, can easily be mistaken for an explanation of the presence of the sort of item functionally characterized, and this has perhaps encouraged philosophers to accept (A). For it might seem that natural selection provides the missing causal link between what something does in a certain type of organism and its presence in that type of organism. By performing their respective

9. Notice that the second use is parasitic on the first. It is only because the neurofibrils explain the coordinated activity of the cilia that we can assign a survival value to neurofibrils: the survival value of a structure $s$ hangs on what capacities of the organism, if any, are explicable by appeal to the functioning of $s$.

functions the contractile vacuole and the neurofibrils help species incorporating them to survive, and thereby contribute to their own continued presence in organisms of those species, and this might seem to explain the presence of those structures in the organisms incorporating them.

Plausible as this sounds, it involves a subtle yet fundamental misunderstanding of evolutionary theory. A clue to the mistake is found in the fact that the contractile vacuole occurs in marine protozoans which have no excess water problem but the reverse problem. Thus the function and effect on survival of this structure is not the same in all protozoans. Yet the explanation of its presence in marine and fresh-water species is almost certainly the same. This fact reminds us that the processes actually responsible for the occurrence of contractile vacuoles in protozoans are totally insensitive to what that structure does. Failure to appreciate this point not only lends spurious plausibility to (A) as applied to biological examples, but seriously distorts our understanding of evolutionary theory. Whether an organism $o$ incorporates $s$ depends on whether $s$ is "specified" by the genetic "plan" which $o$ inherits and which, at a certain level of abstraction, is characteristic of $o$'s species. Alterations in the plan are not the effects of the presence or exercise of the structures the plan specifies. This is most obvious when the genetic change is the result of random mutation. Though not all genetic change is due to random mutation, some certainly is, and that fact is enough to show that specifying the function of a biological structure cannot, in general, explain the

presence of that structure. If a plan is altered so that it specifies $s'$ rather than $s$, then the organisms inheriting this plan will incorporate $s'$ regardless of the function or survival value of $s'$ in those organisms. If the alteration is advantageous, the number of organisms inheriting that plan may increase, and, if it is disadvantageous, their number may decrease. But this typically has no effect on the plan, and therefore no effect on the occurrence of $s'$ in the organisms in question.

One sometimes hears it said that natural selection is an instance of negative feedback. If this is meant to imply that the relative success or failure of organisms of a certain type can affect their inherited characteristics, it is simply a mistake: the characteristics of organisms which determine their relative success are determined by their genetic plan, and the characteristics of these plans are typically independent of the relative success of organisms having them. Of course, if $s$ is very disadvantageous to organisms having a plan specifying $s$, then organisms having such plans may disappear altogether, and $s$ will no longer occur. We could, therefore, think of natural selection as reacting on the *set* of plans generated by weeding out the bad plans: natural selection cannot alter a plan, but it can trim the set. Thus we may be able to explain why a given plan is not a failure by appeal to the functions of the structures it specifies. Perhaps this is what some writers have had in mind. But this is not to explain why, e.g., contractile vacuoles occur in certain protozoans; it is to explain why the sort of protozoan incorporating contractile vacuoles occurs. Since we cannot appeal to the relative success or failure of these organisms to explain why their genetic plan specifies contractile vacuoles, we cannot appeal to the relative success or failure of these organisms to explain why they incorporate contractile vacuoles.

Once we are clear about the explanatory role of functions in evolutionary theory, it emerges that the function of an organ or process (or whatever) is appealed to in order to explain the biological capacities of the organism containing it, and from these capacities conclusions are drawn concerning the chances of survival for organisms of that type. For instance, appeal to the function of the contractile vacuole in certain protozoans explains how these organisms are able to keep from exploding in fresh-water. Thus evolutionary biology does not provide support for (A), but for the idea instanced in (5): identifying the function of something helps to explain the capacities of a containing system.[10]

(A) misconstrues functional explanation by misidentifying what is explained. Let us abandon (A), then, in favor of the view that functions are appealed to in explaining the capacities of containing systems, and turn our attention to (B).

Whereas (A) is a thesis about functional explanation, (B) is a thesis about the analysis of function-ascribing statements. Perhaps when divorced from (A), as it is in (5), it will fare better than it does in the accounts of Hempel and Nagel.

### III

In spite of the evident virtues of (5), (5a) has serious shortcomings as an analysis of (1). In fact it is subject to the same objection Hempel brings to the analysis which simply replaces 'function' by 'effect': vertebrates incorporating a working heart in the usual way exhibit the production of heartsounds, yet the production of heartsounds is not a function of hearts in vertebrates. The problem is that whereas the production of certain effects is essential to the heart's performing its function, there are some effects the production of which is irrelevant to the functioning of the heart. This problem is bound to infect any "selected effects" theory, i.e., any theory built on (B).

What is needed to establish a selected effects theory is a general formula which identifies the appropriate effects.[11] Both Hempel and Nagel attempt to solve this problem by identifying the function of

10. In addition to the misunderstanding about evolutionary theory just discussed, biological examples have probably suggested (A) because biology was the *locus classicus* of teleological explanation. This has perhaps encouraged a confusion between the teleological *form* of explanation, incorporated in (A), with the explanatory role of functional ascriptions. Function-ascribing statements do occur in explanations having a teleological form, and when they do, their interest is vitiated by the incoherence of that form of explanation. It is the legitimate use of function-ascribing statements that needs examination, i.e., their contribution to nonteleological theories such as the theory of evolution.

11. Larry Wright (op. cit.) is aware of this problem but does not, to my mind, make much progress with it. Wright's analysis rules out "The function of the heart is to produce heartsounds," on the ground that the heart is not there because it produces heartsounds. I

something with just those effects which contribute to the maintenance of some special condition of, or the performance of some special activity of, some containing system. If this sort of solution is to be viable, there must be some principled way of selecting the relevant activities or conditions of containing systems. For no matter which effects of something you happen to name, there will be some activity of the containing system to which just those effects contribute, or some condition of the containing system which is maintained with the help of just those effects. Heart activity, for example, keeps the circulatory system from being entirely quiet, and the appendix keeps people vulnerable to appendicitis.[12]

Hempel suggests that, in general, the crucial feature of a containing system, contribution to which is to count as the functioning of a contained part, is that the system be maintained in "adequate, or effective, or proper working order."[13] Hempel explicitly declines to discuss what constitutes proper working order, presumably because he rightly thinks that there are more serious problems with the analysis he is discussing than those introduced by this phrase. But it seems clear that for something to be in working order is just for it to be capable of performing its functions, and for it to be in adequate or effective or proper working order is just for it to be capable of performing its functions adequately or effectively or properly. Hempel seems to realize this himself, for in setting forth a deductive schema for functional explanation, he glosses the phrase in question as 'functions adequately.'[14] More generally, if we identify the function of something $x$ with those effects of $x$ which contribute to the performance of some activity $a$ or to the maintenance of some condition $c$ of a containing system $s$, then we must be prepared to say as well that a function of $s$ is to perform $a$ or to maintain $c$. This suggests the following formulation of "selected effects" theories.

(6) The function of an $F$ in a $G$ is $f$ just in case (the capacity for) $f$ is an effect of an $F$ incorporated in a $G$ in the usual way (or: in the way *this F* is

---

agree. But neither is it there because it pumps blood. Or if, as Wright maintains, there is a sense of "because" in which the heart *is* there because it pumps blood and not because it produces heartsounds, then this sense of "because" is as much in need of analysis as "function." Wright does not attempt to provide such an analysis, but depends on the fact that, in many cases, we are able to use the word in the required way. But we are also able to use "function" correctly in a variety of cases. Indeed, if Wright is right, the words are simply interchangeable with a little grammatical maneuvering. The problem is to make the conditions of correct use explicit. Failure to do this means that Wright's analysis provides no insight into the problem of how functional theories are confirmed, or whence they derive their explanatory force.

12. Surprisingly, when Nagel comes to formulate his general schema of functional attribution, he simply ignores this problem and thus leaves himself open to the trivialization just suggested. Cf. Nagel, p. 403.

13. Hempel, p. 306.

14. Ibid., p. 310.

incorporated in this $G$), and that effect contributes to the performance of a function of the containing $G$.

It seems that any theory based on (B) —what I have been calling "selected effects" theories—must ultimately amount to something like (6).[15] Yet (6) cannot be the whole story about functional ascriptions.

Suppose we follow (6) in rendering "The function of the contractile vacuole in protozoans is elimination of excess water from the organism." The result is (7).

(7) Elimination of excess water from the organism is an effect of a contractile vacuole incorporated in the usual way in a protozoan, and that effect contributes to the performance of a function of a protozoan.

In order to test (7) we should have to know a statement of the form "$f$ is a function of a protozoan." Perhaps protozoans have no functions. If not, (7) is just a mistake. If they do, then presumably we shall have to appeal to (6) for an analysis of the statement attributing such a function and this will leave us with another unanalyzed functional ascription. Either we are launched on a regress, or the analysis breaks down at some level for lack of functions, or perhaps for lack of a plausible candidate for containing system. If we do not wish to simply acquiesce in the autonomy of functional ascriptions, it must be possible to analyze at least some functional ascriptions without appeal-ing to functions of containing systems. If (6) can be shown to be the only plausible formulation of thories based on (B), then no such theory can be the whole story.

Our question, then, is whether a thing's function can plausibly be identified with those of its effects contributing to production of some activity of, or maintenance of some condition of, a containing system, where performance of the activity in question is not a function of the containing system. Let us begin by considering Hempel's suggestion that functions are to be identified with the production of effects contributing to the proper working order of a containing system. I claimed earlier that to say something is in proper working order is just to say that it properly performs its functions. This is fairly obvious in cases of artifacts or tools. To make a decision about which sort of behavior counts as working amounts to deciding about the thing's function. To say something is working, though not behaving or disposed to behave in a way having anything to do with its function, is to be open, at the very least, to the charge of arbitrariness.

When we are dealing with a living organism, or a society of living organisms, the situation is less clear. If we say, "The function of the contractile vacuole in protozoans is elimination of excess water from the organism," we do make reference to a containing organism, but not, apparently, to its function (if any). However, since contractile vacuoles do a number of things having nothing to do with their function, there must be some implicit

---

15. Hugh Lehman (op. cit.) gives an analysis that appears to be essentially like (6).

principle of selection at work. Hempel's suggestion is that, in this context, to be in "proper working order" is simply to be alive and healthy. This works reasonably well for certain standard examples, e.g. (1) and (3): circulation does contribute to health and survival in vertebrates, and photosynthesis does contribute to health and survival in green plants.[16] But once again, the principle will not stand a change of example, even within the life sciences. First, there are cases in which proper functioning is actually inimical to health and life: functioning of the sex organs results in the death of individuals of many species (e.g., certain salmon). Second, a certain process in an organism may have effects which contribute to health and survival but which are not to be confused with the function of that process: secretion of adrenalin speeds metabolism and thereby contributes to elimination of harmful fat deposits in overweight humans, but this is not a function of adrenalin secretion in overweight humans.

A more plausible suggestion along these lines in the special context of evolutionary biology is this:

> (8) The functions of a part or process in an organism are to be identified with those of its effects contributing to activities or conditions of the organism which sustain or increase the the organism's capacity to contribute to survival of the species.

Give or take a nicety, (8) doubtless does capture a great many uses of functional language in biology. For instance, it correctly picks out elimination of excess water as the function of the contractile vacuole in fresh water protozoans only, and correctly identifies the function of sexual organs in species in which the exercise of these organs results in the death of the individual.[17]

In spite of these virtues, however, (8) is seriously misleading and extremely limited in applicability even within biology. Evidently, what contributes to an organism's capacity to maintain its species in one sort of environment may undermine that capacity in another. When this happens, we might say that the organ (or whatever) has lost its function. This is probably what we would say about the contractile vacuole if fresh-water protozoans were successfully introduced into salt water, for in this case the capacity explained would no longer be exercised. But if the capacity explained by appeal to the function of a certain structure continued to be exercised in the new environment, though now to the individual's detriment, we would not say that that structure had lost its function. If, for some reason, flying ceased to contribute to the capacity of pigeons to maintain their species, or even under-

16. Even these applications have their problems. Frankfurt and Poole, Functional explanations in biology, *British Journal for the Philosophy of Science,* 17 (1966), point out that heartsounds contribute to health and survival via their usefulness in diagnosis.

17. Michael Ruse has argued for a formulation like (8). See his Function statements in biology, *Philosophy of Science,* 38 (1971), and *The Philosophy of Biology,* London, Hutchinson, 1973.

mined that capacity to some extent,[18] we would still say that a function of the wings in pigeons is to enable them to fly. Only if the wings ceased to function as wings, as in the penguins or ostriches, would we cease to analyze skeletal structure and the like functionally with an eye to explaining flight. Flight is a capacity which cries out for explanation in terms of anatomical functions regardless of its contribution to the capacity to maintain the species.

What this example shows is that functional analysis can properly be carried on in biology quite independently of evolutionary considerations: a complex capacity of an organism (or one of its parts or systems) may be explained by appeal to a functional analysis regardless of how it relates to the organism's capacity to maintain the species. At best, then, (8) picks out those effects which will be called functions when what is in the offing is an application of evolutionary theory. As we shall see in the next section, (8) is misleading as well in that it is not *which* effects are explained but the style of explanation that makes it appropriate to speak of functions. (8) simply identifies effects which, as it happens, are typically explained in that style.

We have not quite exhausted the lessons to be learned from (8). The plausibility of (8) rests on the plausibility of the claim that, for certain purposes, we may assume that a function of an organism is to contribute to the survival of its species. What (8) does, in effect, is identify a function of an important class of (uncontained) containing systems without providing an analysis of the claim that a function of an organism is to contribute to the survival of its species.

Of course, an advocate of (8) might insist that it is no part of his theory to claim that maintenance of the species is a function of an organism. But then the defense of (8) would have to be simply that it describes actual usage, i.e., that it is in fact effects contributing to an organism's capacity to maintain its species which evolutionary biologists single out as functions. Construed in this way, (8) would, at most, tell us *which* effects are picked out as functions; it would provide no hint as to *why* these effects are picked out *as functions*. We know why evolutionary biologists are interested in effects contributing to an organism's capacity to maintain its species, but why call them functions? This is precisely the sort of question a philosophical account of function-ascribing statements should answer. Either (8) is defended as an instance of (6)—maintenance of the species is declared a function of organisms—or it is defended as descriptive of usage. In neither case is any philosophical analysis provided. For in the first case (8) relies on an unanalyzed (and undefended) function-ascribing statement, and in the second it fails to give any hint as to the point of identifying certain effects as functions.

The failings of (8) are I think bound to cripple any theory which identifies a thing's functions with effects contributing to some antecedently speci-

18. Perhaps, in the absence of serious predators, with a readily available food supply, and with no need to migrate, flying simply wastes energy.

fied type of condition or behavior of a containing system. If the theory is an instance of (6), it launches a regress or terminates in an unanalyzed functional ascription; if it is not an instance of (6), then it is bound to leave open the very question at issue, viz., why are the selected effects seen as functions?

IV

In this section I will sketch briefly an account of functional explanation which takes seriously the intuition that it is a genuinely distinctive style of explanation. The assumptions (A) and (B) form the core of approaches which seek to minimize the differences between functional explanations and explanations not formulated in functional terms. Such approaches have not given much attention to the characterization of the special explanatory strategy science employs in using functional language, for the problem as it was conceived in such approaches was to show that functional explanation is not really different in essentials from other kinds of scientific explanation. Once the problem is conceived in this way,

one is almost certain to miss the distinctive features of functional explanation, and hence to miss the point of functional description. The account of this section reverses this tendency by placing primary emphasis on the kind of problem which is solved by appeal to functions.

### 1. Functions and Dispositions

Something may be capable of pumping even though it does not function as a pump (ever) and even though pumping is not its function. On the other hand, if something functions as a pump in a system $s$, or if the function of something in a system $s$ is to pump, then it must be capable of pumping in $s$.[19] Thus function-ascribing statements imply disposition statements; to attribute a function to something is, in part, to attribute a disposition to it. If the function of $x$ in $s$ is to $\phi$, then $x$ has a disposition to $\phi$ in $s$. For instance, if the function of the contractile vacuole in freshwater protozoans is to eliminate excess water from the organism, then there must be circumstances under which the contractile vacuole would actually manifest a disposition to eliminate excess water

19. Throughout this section I am discounting appeals to the intentions of designers or users. $x$ may be intended to prevent accidents without actually being capable of doing so. With reference to this intention, it *would* be proper in certain contexts to say, "$x$'s function is to prevent accidents, though it is not actually capable of doing so."

There can be no doubt that a thing's function is often identified with what it is typically or "standardly" used to do, or with what it was designed to do. But the sorts of things for which it is an important scientific problem to provide functional analyses—brains, organisms, societies, social institutions—either do not have designers or standard or regular uses at all, or it would be inappropriate to appeal to these in constructing and defending a scientific theory because the designer or use is not known—brains, devices dug up by archaeologists— or because there is some likelihood that real and intended functions diverge—social institutions, complex computers. Functional talk may have originated in contexts in which reference to intentions and purposes loomed large, but reference to intentions and purposes does not figure at all in the sort of functional analysis favored by contemporary natural scientists.

from the protozoan which incorporates it.

To attribute a disposition $d$ to an object $a$ is to assert that the behavior of $a$ is subject to (exhibits or would exhibit) a certain law-like regularity: to say $a$ has $d$ is to say that $a$ would manifest $d$ (shatter, dissolve) were any of a certain range of events to occur ($a$ is put in water, $a$ is struck sharply). The regularity associated with a disposition—call it the dispositional regularity—is a regularity which is special to the behavior of a certain kind of object and obtains in virtue of some special fact(s) about that kind of object. Not everything is water-soluble: such things behave in a special way in virtue of certain (structural) features special to water-soluble things. Thus it is that dispositions require explanation: if $x$ has $d$, then $x$ is subject to a regularity in behavior special to things having $d$, and such a fact needs to be explained.

To explain a dispositional regularity is to explain how manifestations of the disposition are brought about given the requisite precipitating conditions. In what follows I will describe two distinct strategies for accomplishing this. It is my contention that the appropriateness of function-ascribing statements corresponds to the appropriateness of the second of these two strategies. This, I think, explains the intuition that functional explanation is a special *kind* of explanation.

## 2. Two Explanatory Strategies[20]

(i) *The Instantiation strategy.* Since dispositions are properties, not events to explain a disposition requires explaining how it is instantiated. To explain an event, we cite its cause, and to explain an event type requires a recipe (law) for constructing causal explanations of its tokens. But dispositions, being properties, not events, are not explicable as effects. The *acquisition* of a property is an event, but explaining the acquisition of a property is quite distinct from explaining the property itself. One can explain why/ how a thing became fragile without thereby explaining fragility, and one can explain why/how something changed properties—e.g., why something changed temperature—without thereby explaining the property that changed. To explain a property one must show how that property is instantiated in the things that have it.

Simple dispositions are explained by exhibiting their instantiations: water solubility is instantiated as a certain kind of molecular structure, temperature as (average) kinetic energy of molecules, flammability as a kind of subatomic structure (allowing for bonding with oxygen at relatively low temperatures). When we understand how a disposition is instantiated, we are in a position to understand why the dispositional regularity holds of the disposed objects.

20. For a detailed discussion of the two explanatory strategies sketched here, see Cummins, *The Nature of Psychological Explanation*, Bradford Books/M.I.T. Press, Cambridge, 1983. In the original version of this paper, I called the two strategies the Subsumption Strategy and the Analytical Strategy. I have retained the latter term, but the former I have replaced. What I was calling the subsumption strategy in 1975 was simply a confusion; a conflation of causal subsumption of events, and the nomic derivation of a property via the facts of its instantiation. Since functions are dispositions and dispositions are properties, only the latter is relevant here.

Brian O'Shaughnessy has provided an example that allows a particularly simple illustration of this strategy.[21] Consider the disposition he calls elevancy: the tendency of an object to rise in water of its own accord. To explain elevancy, we must explain why freeing a submerged elevant object causes it to rise.[22] This we may do as follows. In every case, the ratio of an elevant object's mass to its nonpermeable volume is less than the density (mass per unit volume) of water: that is how elevancy is instantiated. Once we know this, we may apply Archimedes' Principle, which tells us that water exerts an upward force on a submerged object equal to the weight of the water displaced. In the case of an elevant object, this force evidently exceeds the weight of the object by some amount $f$. Freeing the object changes the net force on it from zero to a net force of magnitude $f$ in the direction of the surface, and the object rises accordingly. Here we subsume the connection between freeings and risings under a general law connecting changes in net force with changes in motion by citing a feature of elevant objects which allows us (via Archi-

medes' Principle) to represent freeing them under water as an instance of introducing a net force in the direction of the surface.

(ii) *The analytical strategy*. Rather than deriving the dispositional regularity that specifies $d$ (in $a$) from the facts of $d$'s instantiation (in $a$), the analytical strategy proceeds by analyzing a disposition $d$ of $a$ into a number of other dispositions $d_1 \ldots d_n$ had by $a$ or components of $a$ such that programmed manifestation of the $d_i$ results in or amounts to a manifestation of $d$.[23] The two strategies will fit together into a unified account if the analyzing dispositions (the $d_i$) can be made to yield to the instantiation strategy.

When the analytical strategy is in the offing one is apt to speak of capacities (or abilities) rather than of dispositions. This shift in terminology will put a more familiar face on the analytical strategy,[24] for we often explain capacities by analyzing them. Assembly-line production provides a transparent example of what I mean. Production is broken down into a number of distinct tasks. Each point on the line is responsible for a certain task, and it is the function of the components at that point to complete

21. Brian O'Shaughnessy, The powerlessness of dispositions, *Analysis*, October (1970). See also my discussion of this example in Dispositions, states and causes, *Analysis*, June (1974).

22. Also, we must explain why submerging a free elevant object causes it to rise, and why a free submerged object's becoming elevant causes it to rise. One of the convenient features of elevancy is that the same considerations dispose of all these cases. This does not hold generally: gentle rubbing, a sharp blow, or a sudden change in temperature may each cause a glass to manifest a disposition to shatter, but the explanations in each case are significantly different.

23. By "programmed" I simply mean organized in a way that could be specified in a program or flow chart: each instruction (box) specifies manifestation of one of the $d_i$ such that if the program is executed (the chart followed), $a$ manifests $d$.

24. Some might want to distinguish between dispositions and capacities, and argue that to ascribe a function to $x$ is in part to ascribe a *capacity* to $x$, not a disposition as I have claimed. Certainly (1) is strained in a way (2) is not.          (*continued on page 403*)

that task. If the line has the capacity to produce the product, it has it in virtue of the fact that the components have the capacities to perform their designated tasks, and in virtue of the fact that when these tasks are performed in a certain organized way—according to a certain program—the finished product results. Here we can explain the line's capacity to produce the product—i.e., explain how it is able to produce the product—by appeal to certain capacities of the components and their organization into an assembly line. Against this background we may pick out a certain capacity of an individual component the exercise of which is its function on the line. Of the many things it does and can do, its function on the line is doing whatever it is that we appeal to in explaining the capacity of the line as a whole. If the line produces several products—i.e., if it has several capacities—then, although a certain capacity $c$ of a component is irrelevant to one capacity of the line, exercise of $c$ by that component may be its function with respect to another capacity of the line as a whole.

Schematic diagrams in electronics provide another obvious illustration. Since each symbol represents any physical object whatever having a certain capacity, a schematic diagram of a complex device constitutes an analysis of the electronic capacities of the device as a whole into the capacities of its components. Such an analysis allows us to explain how the device as a whole exercises the analyzed capacity, for it allows us to see exercises of the analyzed capacity as programmed exercise of the analyzing capacities. In this case the "program" is given by the lines indicating how the components are hooked up. (Of course, the lines are themselves function-symbols.)

Functional analysis in biology is essentially similar. The biologically significant capacities of an entire organism are explained by analyzing the organism into a number of "systems" —the circulatory system, the digestive system, the nervous system, etc.— each of which has its characteristic capacities.[25] These capacities are in turn analyzed into capacities of component organs and structures. Ideally, this strategy is pressed until physiology takes over—i.e., until the analyzing capacities are amenable to the instantiation strategy. We can easily imagine biologists expressing their analyses in a form analogous to the schematic diagrams of electrical engineering, with special symbols for pumps, pipes, filters, and so on. Indeed, analyses of even simple cognitive capacities are typically expressed in flow-charts or programs, forms designed specifically to represent analyses of information-processing capabilities generally.

Perhaps the most extensive use of

---

(1) Hearts are disposed to pump.
    Hearts have a disposition to pump.
    Sugar is capable of dissolving.
    Sugar has a capacity to dissolve.
(2) Hearts are capable of pumping.
    Hearts have a capacity to pump.
    Sugar is disposed to dissolve.
    Sugar has a disposition to dissolve.

25. Indeed, what makes something part of, e.g., the nervous system is that its capacities

the analytical strategy in science occurs in psychology, for a large part of the psychologist's job is to explain how the complex behavioral capacities of organisms are acquired and how they are exercised. Both goals are greatly facilitated by analysis of the capacities in question, for then acquisition of the analyzed capacity resolves itself into acquisition of the analyzing capacities and the requisite organization, and the problem of performance resolves itself into the problem of how the analyzing capacities are exercised. This sort of strategy has dominated psychology ever since Watson attempted to explain such complex capacities as the ability to run a maze by analyzing the performance into a series of conditioned responses, the stimulus for each response being the previous response, or something encountered as the result of the previous response.[26] Acquisition of the complex capacity is resolved into a number of distinct cases of simple conditioning—i.e., the ability to learn the maze is resolved into the capacity for stimulus substitution, and the capacity to run the maze is resolved into abilities to respond in certain simple ways to simple stimuli. Watson's analysis proved to be of

limited value, but the analytic strategy remains the dominant mode of explanation in behavioral psychology.[27]

### 3. Functions and Functional Analysis

In the context of an application of the analytical strategy, exercise of an analyzing capacity emerges as a function: it will be appropriate to say that x functions as a $\phi$ in s, or that the function of x in s is $\phi$-ing, when we are speaking against the background of an analytical explanation of some capacity of s which appeals to the fact that x has a capacity to $\phi$ in s. It is appropriate to say that the heart functions as a pump against the background of an analysis of the circulatory system's capacity to transport food, oxygen, wastes, and so on, which appeals to the fact that the heart is capable of pumping. Since this is the usual background, it goes without saying, and this accounts for the fact that "The heart functions as a pump" sounds right, and "The heart functions as a noise-maker" sounds wrong, in some context-free sense. This effect is strengthened by the absence of any actual application of the analytical strategy which makes use of the fact that the heart makes noise.[28]

---

figure in an analysis of the capacity to respond to external stimuli, coordinate movement, etc. Thus there is no question that the glial cells are part of the brain, but there is some question whether they are part of the nervous system or merely auxiliary to it.

26. John B. Watson, *Behaviorism*, New York, W. W. Norton, 1930, chapters IX and XI.

27. Writers on the philosophy of psychology, especially Jerry Fodor, have grasped the connection between functional characterization and the analytical strategy in psychological theorizing but have not applied the lesson to the problem of functional explanation generally. The clearest statement occurs in J. A. Fodor, The appeal to tacit knowledge in psycological explanation, *Journal of Philosophy*, 65 (1968), 627–640.

28. It is sometimes suggested that heartsounds do have a psychological function. In the context of an analysis of a psychological disposition appealing to the heart's noise-making capacity, "The heart functions as a noise-maker" (e.g., as a producer of regular thumps) would not even *sound* odd.

We can capture this implicit dependence on an analytical context by entering an explicit relativization in our regimented reconstruction of function-ascribing statements.

(9) $x$ functions as a $\phi$ in $s$ (or: the function of $x$ in $s$ is to $\phi$) relative to an analytical account $A$ of $s$'s capacity to $\psi$ just in case $x$ is capable of $\phi$-ing in $s$ and $A$ appropriately and adequately accounts for $s$'s capacity to $\psi$ by, in part, appealing to the capacity of $x$ to $\phi$ in $s$.

Sometimes we explain a capacity of $s$ by analyzing it into other capacities of $s$, as when we explain how someone ignorant of cookery is able to bake cakes by pointing out that he/she followed a recipe each instruction of which requires no special capacities for its execution. Here we don't speak of, e.g., stirring as a function of the cook, but rather of the function of stirring. Since stirring has different functions in different recipes, and at different points in the same recipe, a statement like "The function of stirring the mixture is to keep it from burning to the bottom of the pan" is implicitly relativized to a certain (perhaps somewhat vague) recipe. To take account of this sort of case, we need a slightly different schema: where $e$ is an activity or behavior of a system $s$ (as a whole), the function of $e$ in $s$ is to $\phi$ relative to an analytical account $A$ of $s$'s capacity to $\psi$ just in case $A$ appropriately and adequately accounts for $s$'s capacity to $\psi$ by, in part,

appealing to $s$'s capacity to engage in $e$.

(9) explains the intuition behind the regress-ridden (6): functional ascriptions do require relativization to a "functional fact" about a containing system—i.e., to the fact that a certain capacity of a containing system is appropriately explained by appeal to a certain functional analysis. And, like (6), (9) makes no provision for speaking of the function of an organism except against a background analysis of a containing system (the hive, the corporation, the eco-system). Once we see that functions are appealed to in explaining the capacities of containing systems, and indeed that it is the applicability of a certain strategy for explaining these capacities that makes talk of functions appropriate, we see immediately why we do not speak of the functions of uncontained containers. What (6) fails to capture is the fact that uncontained containers can be functionally analyzed, and the way in which function-analytical explanation mediates the connection between functional ascriptions ($x$ functions as a $\phi$, the function of $x$ is to $\phi$) and the capacities of the containers.

### 4. Function-analytical explanation

If the account I have been sketching is to draw any distinctions, the availability and appropriateness of analytical explanations must be a nontrivial matter.[29] So let us examine an obviously trivial application of the analytical strategy with an eye to determining whether it can be dismissed on principled grounds.

29. Of course, it might be that only arbitrary distinctions are to be drawn. Perhaps (9) describes usage, and usage is arbitrary, but I am unable to take this possibility seriously.

(10) Each part of the mammalian circulatory system makes its own distinctive sound, and makes it continuously. These sounds combine to form the "circulatory noise" characteristic of all mammals. The mammalian circulatory system is capable of producing this sound at various volumes and various tempos. The heartbeat is responsible for the throbbing character of the sound, and it is the capacity of the heart to beat at various rates that explains the capacity of the circulatory system to produce a variously tempoed sound.

Everything in (10) is, presumably, true. The question is whether it allows us to say that the function of the heart is to produce a variously tempoed throbbing sound.[30] To answer this question we must, I think, get clear about the motivation for applying the analytical strategy. For my contention will be that the analytical strategy is most significantly applied in cases very unlike that envisaged in (10).

The explanatory interest of an analytical account is roughly proportional to (i) the extent to which the analyzing capacities are less sophisticated than the analyzed capacities, (ii) the extent to which the analyzing capacities are different in type from the analyzed capacities, and (iii) the relative sophistication of the program appealed to, i.e., the relative complexity of the organization of

component parts/processes which is attributed to the system. (iii) is correlative with (i) and (ii): the greater the gap in sophistication and type between analyzing capacities and analyzed capacities, the more sophisticated the program must be to close the gap.

It is precisely the width of these gaps which, for instance, makes automata theory so interesting in its application to psychology. Automata theory supplies us with extremely powerful techniques for constructing diverse analyses of very sophisticated tasks into very unsophisticated tasks. This allows us to see how, in principle, a mechanism such as the brain, consisting of physiologically unsophisticated components (relatively speaking), can acquire very sophisticated capacities. It is the prospect of promoting the capacity to store ones and zeros into the capacity to solve problems of logic and recognize patterns that makes the analytical strategy so appealing in cognitive psychology.

As the program absorbs more and more of the explanatory burden, the physical facts underlying the analyzing capacities become less and less special to the analyzed system. This is why it is plausible to suppose that the capacity of a person and a machine to solve a certain problem might have substantially the same explanation, while it is not plausible to suppose that the capacities of a synthesizer and a bell to make similar sounds have substantially similar explanations. There is no work for a sophisticated hypothesis about the organization of various capacities to

30. The issue is not whether (10) forces us, via (9), to say something false. Relative to *some* analytical explanation, it may be true that the function of the heart is to produce a variously tempoed throbbing. But the availability of (10) should not support such a claim.

do in the case of the bell. Conversely, the less weight borne by the program, the less point to analysis. At this end of the scale we have cases like (10) in which the analyzed and analyzing capacities differ little if at all in type and sophistication. Here we could apply the instantiation strategy without significant loss, and thus talk of functions is comparatively strained and pointless. It must be admitted, however, that there is no black-white distinction here, but a case of more-or-less. As the role of organization becomes less and less significant, the analytical strategy becomes less and less appropriate, and talk of functions makes less and less sense. This may be philosophically disappointing, but there is no help for it.

### CONCLUSION

Almost without exception, philosophical accounts of function-ascribing statements and of functional explanation have been crippled by adoption of assumptions (A) and (B). Though there has been widespread agreement that extant accounts are not satisfactory, (A) and (B) have escaped critical scrutiny, perhaps because they were thought of as somehow setting the problem rather than as part of proffered solutions. Once the problem is properly diagnosed, however, it becomes possible to give a more satisfactory and more illuminating account in terms of the explanatory strategy which provides the motivation and forms the context of function-ascribing statements. To ascribe a function to something is to ascribe a capacity to it which is singled out by its role in an analysis of some capacity of a containing system. When a capacity of a containing system is appropriately explained by analyzing it into a number of other capacities whose programmed exercise yields a manifestation of the analyzed capacity, the analyzing capacities emerge as functions. Since the appropriateness of this sort of explanatory strategy is a matter of degree, so is the appropriateness of function-ascribing statements.

# VI
# THE REDUCTION OF MENDELIAN GENETICS TO MOLECULAR BIOLOGY

# 22

# The Standpoint of Organismic Biology

## ERNEST NAGEL

VITALISM OF THE SUBSTANTIVE type advocated by Driesch and other biologists during the preceding century and the earlier decades of the present one is now almost entirely a dead issue in the philosophy of biology. The issue has ceased to be focal, perhaps less as a consequence of the methodological and philosophical criticisms to which vitalism has been subjected than because of the sterility of vitalism as a guide in biological research and the superior heuristic value of other approaches to the study of vital phenomena. Nevertheless, the historically influential Cartesian conception of biology as simply a chapter of physics continues to meet resistance. Many outstanding biologists who find no merit in vitalism are equally dubious about the validity of the Cartesian program; and they sometimes advance what they believe are conclusive reasons for affirming the irreducibility of biology to physics and the intrinsic autonomy of biological method. The standpoint from which this antivitalistic and yet antimechanistic thesis is currently advanced commonly carries the label of "organismic biology." The label covers a variety of special biological doctrines that are not always mutually compatible. Nonetheless, the doctrines falling under it generally share the common premise that explanations of the "mechanistic" type are not appropriate for vital phenomena. We shall now examine the main contentions of organismic biology.

1. Although organismic biologists deny the suitability if not always the possibility of "mechanistic theories" for vital processes, it is frequently not clear what it is they are protesting against. But such unclarity can undoubtedly be matched by the ambiguity that often marks the statements of aims and programs by professed "mechanists" in biology. As we had occasion to note in an earlier chapter,* the word "mechanism" has a variety of meanings, and "mechanists" in biology as well as their opponents take few pains to

*Of Nagel's *The Structure of Science,* New York, Harcourt Brace and World, 1961. All chapter references in this selection refer to chapters of that work [Editor].

make explicit the sense in which they employ it. There are biologists who profess themselves to be mechanists simply in the broad sense that they believe that vital phenomena occur in determinate orders and that the conditions for their occurrence are spatiotemporal structures of bodies. But such a view is compatible with the outlook of all schools in biology, with the exception of the vitalists and radical indeterminists; and in any case, when mechanism in biology is so understood, no issue divides those who profess it from most organismic biologists. There have also been biologists who proclaimed themselves to be mechanists in the sense that they maintained that all vital phenomena were explicable exclusively in terms of the science of mechanics (more specifically, in terms of either pure or unitary mechanical theories in the sense of Chapter 7, and who therefore believed living things to be "machines" in the original meaning of this word. It is doubtful, however, whether any biologists today are mechanists in this sense. Physicists themselves have long since abandoned the seventeenth-century hope that a universal science of nature could be developed within the framework of the fundamental ideas of classical mechanics. And it is safe to say that no contemporary biologist subscribes literally to the Cartesian program of reducing biology to the science of mechanics, and especially to the mechanics of contact action.

In any event, most biologists today who call themselves mechanists profess a view that is at once much more specific than the general thesis of causal determinism, and much less restrictive than the one which identifies a mechanistic explanation with an explanation in terms of the science of mechanics. A mechanist in biology, we shall assume, is one who believes, as did Jacques Loeb, that all living processes "can be unequivocally explained in physicochemical terms,"[1] that is, in terms of theories and laws which by common consent are classified as belonging to physics and chemistry. However, biological mechanism so understood must not be taken to deny that living bodies have highly complex organizations. On the contrary, most biologists who adopt such a standpoint usually note quite emphatically that the activities of living bodies are not explicable by analyzing "merely" their physical and chemical compositions without taking into account their "ordered structures or organization." Thus, Loeb's characterization of a living body as a "chemical machine" is an obvious recognition of such organization. It is recognized even more explicitly by E. B. Wilson, who declares, after defining the "development" of germ plasm as the totality of operations by which the germ gives rise to its typical product, that the particular course of this development

is determined (given the normal conditions) by the specific 'organization' of the germ-cells which form its starting-point. As yet we have no adequate conception of this organization, though we know

1. Jacques Loeb, *The Mechanistic Conception of Life,* Chicago, 1912.

that a very important part of it is represented by the nucleus. ... Its nature constitutes one of the major unsolved problems of nature. ... Nevertheless the only available path toward its exploration lies in the mechanistic conception that somehow the organization of the germ-cell must be traceable to the physico-chemical properties of its component substances and the specific configurations which they may assume.[2]

If such is the content of current biological mechanism, and if organismic biologists, like mechanists, reject the postulation of non-material "vitalistic" agents whose operations are to explain vital processes, in what way do the approach and content of organismic biology differ from those of mechanism? The main points of difference, as noted by organismic biologists themselves, appear to be the following:

a. It is a mistake to suppose that the sole alternative to vitalism is mechanism. There are sectors of biological inquiry in which physico-chemical explanations play little or no role at present, and a number of biological theories have been successfully exploited which are not physicochemical in character. For example, there is available an impressive body of experimental knowledge concerning embryological processes, though few of the regularities that have been discovered can be explained at present in exclusively physicochemical terms; and neither the theory of evolution even in its current forms, nor the gene theory of heredity, is based on any definite physicochemical assumptions concerning vital processes. It is certainly not inevitable that mechanistic explanations will eventually prevail in these domains; and, since in any event these domains are now being fruitfully explored without any necessary commitment to the mechanistic thesis, organismic biologists possess at least some ground for their doubts concerning the ultimate triumph of that thesis in all sectors of biology. For just as physicists may be warranted in holding that some branch of physics (e.g., electromagnetic theory) is not reducible to some other branch of science (e.g., to mechanics), so an organismic biologist may be warranted in espousing an analogous view with respect to the relation of biology to the physical sciences. Thus there is a genuine alternative in biology to both vitalism and mechanism—namely, the development of systems of explanation that employ concepts and assert relations neither defined in nor derived from the physical sciences.

b. However, organismic biologists generally claim far more than this. Many of them also maintain that the analytic methods of the physicochemical sciences are intrinsically unsuited to the study of living organisms; that the central problems connected with vital processes require a distinctive mode of approach; and that, since biology is inherently

2. Edmund B. Wilson, *The Cell* (New York, 1925), p. 1037, quoted by kind permission of the Macmillan Company, New York.

irreducible to the physical sciences, mechanistic explanations must be rejected as the ultimate goal of biological research. One reason commonly advanced for this more radical thesis is the "organic" nature of biological systems. Indeed, perhaps the dominant theme upon which the writings of organismic biologists play so many variations is the "integrated," "holistic," and "unified" character of a living thing and its activities. Living creatures, in contrast to nonliving systems, are not loosely jointed structures of independent and separable parts, are not assemblages of tissues and organs standing in merely external relations to one another. Living creatures are "wholes" and must be studied as "wholes"; they are not mere "sums" of isolable parts, and their activities cannot be understood or explained if they are assumed to be such "sums." But mechanistic explanations construe living organisms as "machines" possessing independent parts, and thereby adopt an "additive" point of view in analyzing vital phenomena. Accordingly, since the action of the whole organism "has a certain unifiedness and completeness" which is left out of account in the course of analyzing it into its elementary processes, E. S. Russell concludes that "the activities of the organism as a whole are to be regarded as of a different order from physico-chemical relations, both in themselves and for the purposes of our understanding"[3] Therefore biology must observe two "cardinal laws of method": "The activity of the whole cannot be fully explained in terms of the activities of the parts isolated by analysis"; and "No part of any living entity and no single process of any complex organic unity can be fully understood in isolation from the structure and activities of the organism as a whole."[4]

c. An additional though closely related point which organismic biologists stress is the "hierarchical organization" of living bodies and processes. Thus, a cell is known to be a structure of various constituents, such as the nucleus, the Golgi bodies, and the membranes, each of which may be analyzable into other parts and these in turn into still others, so that the analysis presumably terminates in molecules, atoms, and their "ultimate" parts. But in multicellular organisms the cell is also only an element in the organization of a tissue, the tissue is a part of some organ, the organ a member of an organ system, and the organ system a constituent in the integrated organism. It is patent that these various "parts" do not occur at the same "level" of organization. In consequence, organismic biologists place great stress on the fact that an animate body is not a system of parts

---

3. E. S. Russell, *The Interpretation of Development and Heredity,* Oxford, 1930, pp. 171–72.

4. *Ibid.,* pp. 146–47. Similar statements of the central tenet of organismic biology will be found in Russell's *Directiveness of Organic Activities,* Cambridge, England, 1945, esp. chaps. 1 and 7; Ludwig von Bertalanffy, *Theoretische Biologie,* Berlin, 1932, chap. 2; his *Modern Theories of Development,* Oxford, 1933, chap. 2; and his *Problems of Life,* New York and London, 1952, chaps. 1 and 2; and W. E. Agar, *The Theory of the Living Organism,* Melbourne and London, 1943.

homogeneous in complexity of organization, but that on the contrary the "parts" into which an organism is analyzed must be distinguished according to the different levels of some particular type of hierarchical structure (there may be several such types) to which the parts belong. Now organismic biologists do not deny that physicochemical explanations are possible for the activities of parts on the "lower" levels of a hierarchy. Nor do they deny that the physicochemical properties of the parts on lower levels "condition" or "limit" in various ways the occurrence and modes of action of higher levels of organization. They do deny, on the other hand, that the processes found at higher levels of a hierarchy are "caused" by, or are fully explicable in terms of, lower-level properties. Biochemistry is acknowledged to be the study of the "conditions" under which cells and organisms act the way they do. Organismic biology, on the other hand, investigates the activities of the whole organism "regarded as conditioned by, but irreducible to, the modes of action of lower unities."[5]

We must now examine these alleged differences between the organismic and the mechanistic approaches to biology, and attempt to assess the claim that the mechanistic approach is generally inadequate to biological subject matter.

2. At first blush, the sole issues raised by organismic biology are those we have already discussed in connection with the doctrine of emergence and the reduction of one science to another. In point of fact, other questions are also involved. But to the extent that the issues are those of reduction, we can dispose of them quite rapidly.

Let us remind ourselves of the two formal conditions, examined at some length in the preceding chapter, that are necessary and suffcient for the reduction of one science to another. When stated with special reference to biology and physico-chemistry, they are as follows:

a. *The condition of connectability*. All terms in a biological law that do not belong to the primary science (such as 'cell,' 'mitosis,' or 'heredity') must be "connected" with expressions constructed out of the theoretical vocabulary of physics and chemistry (out of terms such as 'length,' 'electric charge,' 'free energy,' and the like). These connections may be of several kinds. The meanings of the biological expressions may be analyzable, and perhaps even explicitly definable, in terms of physicochemical ones, so that in the limiting case the biological expressions are

---

5. Russell, *The Interpretation of Development and Heredity,* p. 187. For an analogous view, cf. Ludwig von Bertalanffy and Alex B. Novikoff, "The Conception of Integrative Levels and Biology," *Science,* 101 (1945), pp. 209–15, and the discussion of this article in the same volume, pp. 582–85 and in Vol. 102 (1945), pp. 405–06. A careful and sober analysis of the nature of hierarchical organization in biology and of its import for the possibility of mechanistic explanation is given in J. H. Woodger, *Biological Principles,* New York, 1929, chap. 6, and in Woodger's "The 'Concept of Organism' and the Relation between Embryology and Genetics," *Quarterly Review of Biology,* 5 (1930), and 6 (1931).

eliminable in favor of the physico-chemical terms. An alternative mode of connection is that biological expressions are associated with physicochemical ones by some type of coordinating definition, so that the connections have the logical status of conventions. Finally, and this is the more frequent case, the biological terms may be connected with physicochemical ones on the strength of empirical assumptions, so that the sufficient conditions (and possibly the necessary ones as well) for the occurrence of whatever is designated by the biological terms can be stated by means of the physicochemical expressions. Thus, if the term 'chromosome' can be associated in neither of the first two ways with some expression constructed out of the theoretical vocabulary of physics and chemistry, then it must be possible to state in the light of an assumed law the truth-conditions for a sentence of the form '$x$ is a chromosome' entirely by means of a sentence constructed out of that vocabulary.

b. *The condition of derivability.* Every biological law, whether theoretical or experimental, must be logically derivable from a class of statements belonging to physics and chemistry. The premises in these deductions will contain an appropriate selection from the theoretical assumptions of the primary discipline, as well as statements formulating the associations between biological and physicochemical terms required by the condition of connectability. In general, some of the premises will state in the vocabulary of the primary science the boundary conditions or specialized spatiotemporal configurations under which the theo-retical assumptions are being applied.

As was shown in the preceding chapter, the condition of derivability cannot be fulfilled unless the condition of connectability is satisfied. It is beyond dispute, however, that the task of satisfying the first of these conditions for biology is still far from completed. We do not know at present, for example, the detailed chemical composition of chromosomes in living cells. We are therefore unable to state in exclusively physicochemical terms the conditions for the occurrence of those organic parts, and hence to state in such terms the truth-conditions for the application of the word 'chromosome.' And a fortiori we are not able at present to formulate in physicochemical language the structure of any of the systems, such as cell nucleus, cell, or tissue, of which chromosomes are themselves parts. Accordingly, in the current state of biological knowledge it is logically impossible to deduce the totality of biological laws and theories from purely physicochemical assumptions. In short, biology is not at present simply a chapter of physics and chemistry.

Organismic biologists are therefore on firm ground in maintaining that mechanistic explanations of all biological phenomena are currently impossible, and will remain impossible until the descriptive and theoretical terms of biology can be shown to satisfy the first condition for the reduction of that science to physics and chemistry—that is, until the composition of every part or process of living things, and the distribution and arrangement of their parts at any time, can be exhaustively specified in physicochemical terms. Moreover, even if this condition were realized, the triumph of the mechanistic standpoint would not thereby be assured. For as we have already shown, the satisfaction of the condition of connectability is a necessary but in general not a sufficient requirement for the absorption of biology into physics and chemistry. Although the connectability condition might be fulfilled, there would still remain the question whether all biological laws are deducible from the current theoretical assumptions of these physical sciences. The answer to this question is conceivably in the negative, since physicochemical theory in its present form may be insufficiently powerful to permit the derivation of various biological laws, even if these laws were to contain only terms properly linked with expressions belonging to those primary disciplines. It should also be noted that, even if both formal conditions for the reducibility of biology were satisfied, the reduction might nevertheless have little if any scientific importance, for the reason that some of the conditions previously labeled "nonformal" might not be adequately realized.

On the other hand, the facts cited and the argument thus far examined do not warrant the conclusion that biology is *in principle* irreducible to the physical sciences. The task facing such a proposed reduction is admittedly a most difficult one; and it undoubtedly impresses many students as one which, if not utterly hopeless, is at present not worth pursuing. However, no *logical* contradiction has yet been exhibited to the supposition that both the formal and nonformal conditions for the

reduction of biology may some day be fulfilled. We can therefore terminate this part of the discussion with the conclusion that the question whether biology is reducible to physicochemistry is an open one, that it cannot be settled by a priori argument, and that an answer to it can be provided only by further experimental and logical inquiry.

3. Let us next turn to the argument for the inherent "autonomy" of biology based on the fact that living systems are hierarchically organized. The burden of the argument, as we have seen, is that properties and modes of behavior occurring on a higher level of such a hierarchy cannot in general be explained as the resultants of properties and behaviors exhibited by isolable parts belonging to lower levels of an organism's structure.

There is no serious dispute among biologists over the thesis that the parts and processes into which living organisms are analyzable can be classified in terms of their respective loci into hierarchies of various types, such as the essentially spatial hierarchy mentioned earlier. Nor is there disagreement over the contention that the parts of an organism belonging to one level of a hierarchy frequently exhibit forms of relatedness and of activity not manifested by organic parts belonging to another level. Thus, a cat can stalk and catch mice; but though the continued beating of its heart is a necessary condition for these activities, the cat's heart cannot perform these feats. Again, the heart can pump blood by contracting and expanding its muscular tissues, although no single tissue can keep the blood in circulation; and no tissue is able to divide by fission, even though its constituent cells may have this property. Such examples suffice to establish the claim that modes of behavior appearing at higher levels of a hierarchically organized system are not explained by merely listing each of the various lower-level parts and processes of the system as an aggregate of isolated and unrelated elements. Organismic biologists do not deny that the occurrence of higher-level traits in hierarchically structured living organisms is contingent upon the occurrence, at different levels of the hierarchy, of various component parts related in definite ways. But they do deny, and with apparent good reason, that statements formulating the traits exhibited by components of an organism, when the components are not parts of an actually living organism, can adequately explain the behavior of the living system that contains those components as parts related in complex ways to other elements in a hierarchically structured whole.

But do these admitted facts establish the contention that mechanistic explanations are either impossible or unsuitable for biological subject matter? It should be noted that various forms of hierarchical organization are exhibited by the materials of physics and chemistry, and not only by those of biology. Our current theories of matter assume atoms to be structures of electric charges, molecules to be organizations of atoms, solids and liquids to be complex systems of molecules. Moreover, the available

evidence indicates that elements at different levels of this hierarchy exhibit traits which their component parts do not invariably possess. However, these facts have not stood in the way of establishing comprehensive theories for the more elementary physical particles and processes, in terms of which it has been possible to account for some if not all of the physicochemical properties exhibited by objects having a more complex organization. To be sure, we do not possess at present a comprehensive and unified theory competent to explain the full range even of purely physicochemical phenomena occurring on various levels of organization. Whether such a theory will ever be achieved is certainly an open question. It is also relevant to note in this connection that biological organisms are "open systems," never in a state of "true equilibrium" but at best only in a steady state of "dynamic equilibrium" with their environment, because they continually exchange material components and not only energies with the latter.[6] In this respect, living organisms are unlike the "closed systems" usually studied in current physical science. Indeed, an adequate theory for physicochemcal processes in open systems—for example, a thermodynamics competent to deal with systems in nonequilibrium as well as equilibrium states—is at present only in an early stage of development. Nevertheless, the circumstance remains that we can now account for some characteristics of fairly complex systems with the help of theories formulated in terms of

6. L. von Bertalanffy, *Problems of Life,* chap. 4.

relations between relatively more simple ones—for example, the specific heats of solids in terms of quantum theory, or the changes in phase of compounds in terms of the thermodynamics of mixtures. This circumstance must make us pause in accepting the conclusion that the hierarchical organization of living systems by itself precludes a mechanistic explanation for their traits.

Let us, however, examine in greater detail some of the organismic arguments on this issue. One of them has been persuasively stated by J. H. Woodger, whose careful but sympathetic analyses of organismic notions are important contributions to the philosophy of biology. Woodger maintains that it is essential to distinguish between chemical *entities* and chemical *concepts;* he believes that if the distinction is kept in mind, it no longer appears plausible to assume that a thing can be satisfactorily described in terms of chemical concepts exclusively, merely because the thing is held to be composed of chemical entities. "A lump of iron," Woodger declares, "is a chemical entity, and the word 'iron' stands for a chemical concept. But suppose that the iron has the form of a poker or a padlock, then although the iron is still chemically analyzable in the same way as before it cannot still be fully described in terms of chemical concepts. It now has an organization above the chemical level."[7]

Now there is no doubt that many of the uses to which iron pokers or padlocks may be put are not, and may never be, described in purely physicochemical terms. But does the fact that a piece of iron has the form of a poker or of a padlock stand in the way of explaining an extensive class of its properties and modes of behavior in exclusively physicochemical terms? The rigidity, tensile strength, and thermal properties of the poker, or the mechanism and the qualities of endurance of the padlock, are certainly explicable in such terms, even if it may not be necessary or convenient to invoke a microscopic physical theory to account for all these traits. Accordingly, the mere fact that a piece of iron has a certain organization does not preclude the possibility of a physicochemical explanation for some of the characteristics it exhibits as an organized object.

Some organismic biologists maintain that, even if we were able to describe in minute detail the physicochemical composition of a fertilized egg, we would nevertheless still be unable to explain mechanistically the fact that such an egg normally segments. In the view of E. S. Russell, for example, we might be able on the stated supposition to formulate the physicochemical conditions for segmentation, but we would be unable to "explain the course which development takes."[8]

This claim raises some of the

---

7. J. H. Woodger, *Biological Principles,* p. 263. Woodger continues, "In the same way an organism is a physical entity in the sense that it is one of the things we become aware of by means of the senses, and is a chemical entity in the sense that it is capable of chemical analysis just as is the case with any other physical entity, but it does not follow from this that it can be fully and satisfactorily described in chemical terms."

8. E. S. Russell, *The Interpretation of Development and Heredity,* p. 186.

previously discussed issues associated with the distinction between structure and function. But quite apart from these issues, the claim appears to rest on a misunderstanding if not on a confusion. It is cogent to maintain that a knowledge of the physicochemical composition of a biological organism does not suffice to explain mechanistically its modes of action—any more than an enumeration of the parts of a clock together with a description of their spatial distribution and arrangement suffices to explain or to predict the behavior of the timepiece. To make such an explanation, we must also assume some theory or set of laws (in the case of the lock, the theory of mechanics) which formulates the way certain elements act when they occur in some initial distribution and arrangement, and which permits the calculation (and hence the prediction) of the subsequent development of that organized system of elements. Moreover, it is conceivable that, despite our assumed ability at some given stage of scientific knowledge to describe in full detail the physicochemical composition of a living thing, we might nevertheless be unable to deduce from the established physicochemical theories of the day the course of the organism's development. In short, it is conceivable that the first but not the second formal condition for the reducibility of one science to another is satisfied at a given time. It is a misunderstanding, however, to suppose that a fully codified explanation in the natural sciences can consist only of instantial premises formulating initial and boundary conditions but containing no statements of law

or theory. It is an elementary blunder to claim that, because some one physicochemical theory (or some class of such theories) is not competent to explain certain vital phenomena, it is *in principle* impossible to construct and establish a mechanistic theory that can do so.

On the other hand, it would be foolish to underestimate the enormity of the task facing the mechanistic program in biology because of the intricate hierarchical organization of living things. Nor should we dismiss as pointless the protests of organismic biologists against versions of the mechanistic thesis that appear to ignore the fact of such organization. As biologists of all schools have often observed, there is no such thing as a homogeneous and structurally undifferentiated "living substance," analogous to "copper substance." There have nevertheless been mechanists who in their statements on biological method, if not in their actual practice as biological investigators, have in effect asserted the contrary. It is therefore worth stressing that the subject matters of their inquiry have compelled biologists to recognize not just a single type of hierarchical organization in living things but several types, and that a central problem in the analysis of organic developmental processes is the discovery of the precise interrelations between such hierarchies.

The hierarchy most frequently cited is generated by the relation of spatial inclusion, as in the case of cell parts, cells, organs, and organisms. However, on any reasonable criterion for distinguishing between various "levels" of such a hierarchy, it turns out that there are bodily parts in most organisms (such as the blood plasma) which cannot be fitted into it. Furthermore, there are types of hierarchy that are not primarily spatial. Thus, there is a "division hierarchy," with cells as elements, which is generated by the division of a zygote and of its cell descendants. Biologists also recognize a "hierarchy of processes": the hierarchy of physicochemical processes in a muscle, the contraction of the muscle, the reaction of a system of muscles, the reaction of the animal organism as a whole; and other types which could be added to this brief list. In any event, it should be noted that in embryological development the spatial hierarchy changes, since in this process new spatial parts are elaborated. This fact can be expressed by saying that, when the division hierarchy of an embryo is compared at different times, its spatial hierarchy at a later time contains elements that did not exist at earlier times. Accordingly, organismic biologists are obviously correct in claiming that to a large extent biological research is concerned with establishing relations of interdependence between various hierarchical structures in living bodies.[9]

Let us now, however, state briefly the schematic form of a hierarchical organization (not necessarily a spatial hierarchy), with a view to assessing in general terms one component in the organismic critique of biological mechanism. Suppose $S$ is some biological system which is analyzable into three major constituents $A$, $B$, and $C$, so that $S$ can be conceived as the relational complex $R(A, B, C)$,

where $R$ is some relation. Assume further that each major constituent is in turn analyzable into subordinate constituents $(a_1, a_2, \ldots, a_i)$, $(b_1, b_2, \ldots, b_j)$, and $(c_1, c_2, \ldots, c_k)$, respectively, so that the major constituents of $S$ can be represented as the relational complexes $R_A(a_1, \ldots, a_i)$, $R_B(b_1, \ldots, b_j)$, and $R_C(c_1, \ldots, c_k)$. The $a$'s, $b$'s, and $c$'s may be analyzable still further, but for the sake of simplicity we shall assume only two levels for the hierarchical organization of $S$. We also stipulate that some of the $a$'s (and similarly some of the $b$'s and $c$'s) stand to each other in various special relations, subject to the condition that all of them are related by $R_A$ to constitute $A$ (with analogous conditions for the $b$'s and $c$'s). Moreover, we assume that some of the $a$'s may stand in certain other special relations to some of the $b$'s and $c$'s, subject to the condition that the complexes $A$, $B$, and $C$ are related by $R$ to constitute $S$. If $S$ is such a hierarchy, one aim of research on $S$ will be to discover its various constituents, and to ascertain the regularities in the relations connecting them with $S$ and with constituents on the same or on different levels.

The pursuit of this aim will in general require the resolution of many serious difficulties. To discover just what the presence of $A$, for example, contributes to the traits manifested by $S$ taken as a whole, it may be necessary to establish what $S$ would be like in the absence of $A$, as well as how $A$ behaves when it is not a constitutive part of $S$. There

---

9. Cf. the writings of Woodger cited above, as well as his *Axiomatic Method in Biology*, Cambridge, England, 1937; also L. von Bertalanffy, *Problems of Life*, chap. 2.

may be grave experimental problems in attempting to isolate and identify such causal influences. But quite apart from these, the fundamental question must at some point be faced whether the study of $A$, when it is placed in an environment differing in various ways from the environment provided by $S$ itself, can yield pertinent information about the behavior of $A$ when it occurs as an actual constituent of $S$. Suppose, however, that we possess a theory $T$ about the components $a$ of $A$, such that if the $a$'s are assumed to be in the relation $R_A$ when they occur in an environment $E$, it is possible to show with the help of $T$ just what traits characterize $A$ in that environment. On this supposition it may not be necessary to experiment upon $A$ in isolation from $S$. The above crucial question will nevertheless continue to be unresolved unless the theory $T$ permits conclusions to be drawn not only when the $a$'s are in the relation $R_A$ in some artificial environment $E$, but also when they are in that relation in the particular environment that contains the $b$'s and $c$'s all jointly organized by the relations $R_B$, $R_C$, and $R$. Without such a theory, it will generally be the case that the only way of ascertaining just what role $A$ plays in $S$ is to study $A$ as an actual component in the relational complex $R(A,B,C)$.

Accordingly, organismic biologists are right in insisting on the general principle that "an entity having the hierarchical type of organization such as we find in the organism requires investigation at all levels, and in-vestigation of one level cannot replace the necessity for investigation of levels higher up in the hierarchy."[10] On the other hand, this principle does not entail the impossibility of mechanistic explanations for vital phenomena, though organismic biologists sometimes appear to believe that it does. In particular, if the $a$'s, $b$'s, and $c$'s in the above schematism are the submicroscopic entities of physics and chemistry, $S$ is a biological organism, and $T$ is a physicochemical theory, it is not impossible that the conditions for the occurrence of the relational complexes $A$, $B$, $C$, and $S$ can be specified in terms of the fundamental concepts of $T$, and that furthermore the laws concerning the behaviors of $A$, $B$, $C$, and $S$ can be deduced from $T$. But, as has been argued in the preceding chapter, whether in point of fact one science (such as biology) is reducible to some primary science (such as physico-chemistry), is contingent on the character of the particular theory employed in the primary discipline at the time the question is put.

4. We must finally turn to what appears to be the main reason for the negative attitude of organismic biologists toward mechanistic explanations of vital phenomena, namely, the alleged "organic unity" of living things and the consequent impossibility of analyzing biological wholes as "sums" of independent parts. Whether there is merit in this reason obviously depends on what senses are attached to the crucial expressions 'organic unity' and 'sum.' Organismic biologists have done little to clarify the

10. J. H. Woodger, *Biological Principles*, p. 316.

meanings of these terms, but at least a partial clarification has been attempted in the preceding and present chapters of this book. In the light of these earlier discussions the issue now under examination can be disposed of with relative brevity.

Let us assume, as do organismic biologists, that a living thing possesses an "organic unity," in the sense that it is a teleological system exhibiting a hierarchical organization of parts and processes, so that the various parts stand to each other in complex relations of causal interdependence. Suppose also that the particles and processes of physics and chemistry constitute the elements at the lowest level of this hierarchical system, and that $T$ is the current body of physicochemical theory. Finally, let us associate with the word 'sum' in the statement 'A living organism is not the sum of its physicochemical parts,' the "reducibility" sense of the word distinguished in the preceding chapter. The statement will then be understood to assert that, even when suitable physicochemical initial and boundary conditions are supplied, it is not possible to deduce from $T$ the class of laws and other statements about living things commonly regarded as belonging to the province of biology.

Subject to an important reservation, the statement construed in this manner may very well be true, and probably represents the opinion of most students of vital phenomena, whether or not they are organismic biologists. The statement is widely held, despite the fact that in many cases physicochemical conditions for biological processes have been ascertained. Thus, an unfertilized egg of the sea urchin does not normally develop into an embryo. However, experiments have shown that, if such eggs are first placed for about two minutes in sea water to which a certain quantity of acetic acid has been added and are then transferred to normal sea water, the eggs presently begin to segment and to develop into larvae. But, although this fact certainly counts as impressive evidence for the physicochemical character of biological processes, the fact has thus far not been fully explained, in the strict sense of 'explain,' in physicochemical terms. For no one has yet shown that the statement that sea urchin eggs are capable of artificial parthenogenesis under the indicated conditions is *deducible* from the purely physicochemical assumptions $T$. Accordingly, if organismic biologists are making only the *de facto* claim that no systems possessing the organic unity of living things have thus far been proved to be sums (in the reducibility sense) of their physicochemical constituents, the claim is undoubtedly well founded.

On the other hand, in the prevailing circumstances of our knowledge there should be no cause for surprise that the fact about the artificial parthenogenesis of sea urchin eggs is not deducible from $T$. The deduction is not possible, if only because the elementary logical requirements for performing it are currently not satisfied. No theory can explain the operations of any concretely given system unless a complete set of initial and boundary conditions for the application of the theory is stated in a manner consonant with the specific notions employed in the theory. For example, it is not possible to deduce the distribution of electric charges on a given insulated conductor merely from the fundamental equations of electrostatic theory. Additional instantial information must be supplied in a form prescribed by the character of the theory—in this instance, information about the shape and size of the conductor, the magnitudes and distribution of electric charges in the neighborhood of the conductor, and the value of the dielectric constant of the medium in which the conductor is embedded. In the case of the sea urchin eggs, however, although the physicochemical composition of the environment in which the unfertilized eggs develop into embryos is presumably known, the physicochemical composition of the eggs themselves is still unknown, and cannot be formulated for inclusion in the indispensable instantial conditions for the application of $T$. More generally, we do not know at present the detailed physicochemical composition of any living organism, nor the forces that may be acting between the elements on the lowest level of its hierarchical organization. We are therefore currently unable to state in exclusively physicochemical terms the initial and boundary conditions requisite for the application of $T$ to vital systems. Until we can do this, we are in principle precluded from deducing biological laws from mechanistic theory. Accordingly, although it may indeed be true that a living organism is not the sum of its physicochemical parts, the available evidence does not warrant the assertion either of the truth or of the falsity of this dictum.

Although the point just stressed is elementary, organismic biologists often appear to neglect it. They sometimes argue that, while mechanistic explanations may be possible for traits of organic parts when these parts are studied in "abstraction" (or isolation) from the organism as a whole, such explanations are not possible when the parts function conjointly in mutual dependence as actual constituents of a living thing. But this claim ignores the crucial fact that the initial conditions required for a mechanistic explanation of the traits of organic parts manifested when the parts exist *in vitro* are generally insufficient for accounting mechanistically for the conjoint functioning of the parts in a biological organism. For it is evident that when a part is isolated from the rest of the organism it is placed in an environment which is usually different from its normal environment, where it stands in relations of mutual dependence with other parts of the organism. It therefore follows that the initial conditions for using a given theory to explain the behavior of a part in isolation will also be different from the initial conditions for using that theory to explain behavior in the normal environment. Accordingly, although it may indeed be beyond our actual competence at present or in the foreseeable future to specify the instantial conditions requisite for a mechanistic explanation of the functioning of organic parts *in situ,* there is nothing in the logic of the situation that limits such explanations in principle to the behavior of organic parts *in vitro.*

One final comment must be added. It is important to distinguish the question of whether mechanistic explanations of vital phenomena are possible from the quite different though related problem of whether living organisms can be effectively synthesized in a laboratory out of nonliving materials. Many biologists seem to deny the first possibility because of their skepticism concerning the second. In point of fact, however, the two issues are logically independent. In particular, although it may never become possible to manufacture living organisms artificially, it does not follow that vital phenomena are therefore incapable of being explained mechanistically. A glance at the achievements of the physical sciences will suffice to establish this claim. We do not possess the power to manufacture nebulae or solar systems, despite the fact that we do possess physicochemical theories in terms of which nebulae and planetary systems are tolerably well understood. Moreover, while modern physics and chemistry provide competent explanations for various properties of chemical elements in terms of the electronic structure of the atoms, there are no compelling reasons for believing that, for example, men will some day be capable of manufacturing hydrogen by putting together artificially the subatomic components of the substance. On the other hand, the human race possessed skills (e.g., in the construction of dwellings, in the manufacture of alloys, and in the preparation of foods) long before adequate explanations for the traits of the artifically constructed articles were available.

Nonetheless, organismic biologists often develop their critique of the mechanistic program in biology as if its realization were equivalent to the acquisition of techniques for literally taking apart living things and then overtly reconstituting the original organisms out of their dismembered and independent parts. However, the conditions for achieving mechanistic explanations for vital phenomena are quite different from the requirements for the artificial manufacture of living organisms. The former task is contingent on the construction of factually warranted theories of physicochemical materials, and on the invention of effective techniques for combining and controlling them. It is perhaps unlikely that living organisms will ever be synthesized in the laboratory except with the help of mechanistic theories of vital processes; in the absence of such theories, the artificial manufacture of living things, were this ever accomplished, would be the outcome of a fortunate but improbable accident. But in any event, the conditions for achieving these patently different tasks are not identical, and either may some day be realized without the other. Accordingly, a denial of the possibility of mechanistic explanations in biology on the tacit supposition that these conditions do coincide, is not a cogently reasoned thesis.

The main conclusion of this discussion is that organismic biologists have not established the absolute autonomy of biology or the inherent impossibility of physicochemical explanations of vital phenomena. Nevertheless, the stress they place on the hierarchical organization of living things and on the mutual dependence of organic parts is not a misplaced one. For, although organismic biology has not convincingly secured all its claims, it has demonstrated the important point that the pursuit of mechanistic explanations for vital processes is not a *sine qua non* for valuable and fruitful study of such processes. There is no more reason for rejecting a biological theory (e.g., the gene theory of heredity) because it is not a mechanistic one (in the sense of "mechanistic" we have been employing) than there is for discarding some physical theory (e.g., modern quantum theory) on the ground that it is not reducible to a theory in another branch of physical science (e.g., to classical mechanics). A wise strategy of research may indeed require that a given discipline be cultivated as a relatively independent branch of science, at least during a certain period of its development, rather than as an appendage to some other discipline, even if the theories of the latter are more inclusive and better established than are the explanatory principles of the former. The protest of organismic biology against the dogmatism often associated with the mechanistic standpoint in biology is salutary.

There is, however, an obverse side to the organismic critique of that dogmatism. Organismic biologists sometimes write as if any analysis of vital processes into the operation of distinguishable parts of living things entails a seriously distorted view of these processes. For example, E. S. Russell has maintained that in analyzing the activities of an organism into elementary processes "something is

lost, for the action of the whole has a certain unifiedness and completeness which is left out of account in the process of analysis."[11] Analogously, J. S. Haldane claimed that we cannot apply mathematical reasoning to vital processes, since a mathematical treatment assumes a separability of events in space "which does not exist for life as such. We are dealing with an indivisible whole when we are dealing with life."[12] And H. Wildon Carr, a professional philosopher who subscribed to the organismic standpoint and wrote as one of its exponents, declared that "Life is individual; it exists only in living beings, and each living being is indivisible, a whole not constituted of parts."[13]

Such pronouncements exhibit an intellectual temper that is as much an obstacle to the advancement of biological inquiry as is the dogmatism of intransigent mechanists. In biology as in other branches of science knowledge is acquired only by analysis or the use of the so-called "abstractive method"—by concentrating on a limited set of properties things possess and ignoring (at least for a time) others, and by investigating the traits selected for study under controlled conditions. Organismic biologists also proceed in this way, despite what they may say, for there is no effective alternative to it. For example, although J. S. Haldane formally proclaimed the "indivisible unity" of living things, his studies on respiration and the chemistry of the blood were not conducted by considering the body as an indivisible whole. His researches

11. E. S. Russell, *The Interpretation of Development and Heredity,* p. 171.
12. J. S. Haldane, *The Philosophical Basis of Biology,* London, 1931, p. 14.
13. Quoted in L. Hogben, *The Nature of Living Matter,* London, 1930, p. 226.

involved the examination of relations between the behavior of one part of the body (e.g., the quantity of carbon dioxide taken in by the lungs) and the behavior of another part (the chemical action of the red blood cells). Like everyone else who contributes to the advance of knowledge, organismic biologists must be abstractive and analytical in their research procedures. They must study the operations of various prescinded parts of living organisms under selected and often artificially instituted conditions—on pain of mistaking enlightening statements liberally studded with locutions like 'wholeness,' 'unifiedness,' and 'indivisible unity' for expressions of genuine knowledge.

# 23

# Reduction in Biology: Prospects and Problems*

## KENNETH SCHAFFNER

### I. INTRODUCTION

Explications of what it means for one science to be reduced to another have been the focus of considerable interest in recent years, and when the reducing science is physics (and chemistry) and the reduced science is biology, additional concerns seem to arise. In this paper I wish to represent a model for theory reduction which has occupied my attention for some years now, and consider its applicability in the area of the biological sciences.[1]

I should state in advance that the model to be discussed represents an ideal standard for accomplished reductions, and does not characterize the research programmes of molecular biologists. Recently I have argued, in point of fact, that a research programme which might be generated from the tenets of the reduction model to be outlined in this paper

would not represent some of the most significant advances in molecular biology, such as the development of the Watson-Crick model of DNA or the *genesis* of the Jacob-Monod operon theory.

However, even though I would subscribe to what I have termed a 'peripherality thesis' as regards reductionism as the primary *aim* of molecular biology, I believe that molecular biology is *resulting in* at least *partial* reductions of biology to physics and chemistry, and that even these partial instances are in accord with the ideals proposed as part of the general reduction model to be considered below. In any unfinished and rapidly advancing science which is resulting in reductions, however, it is possible to find aspects which can be interpreted as difficulties for such a model as I wish to present and defend. I shall

*Essays 23–26 were originally presented as a Philosophy of Science Association symposium (Cohen *et al.*, 1976). Authors referring to "other symposium contributions" are referring to these selections [Editor].

1. See Schaffner (1967a) and (1969a), and (1974b). The present paper relies extensively on some of the arguments developed in this latter paper.

mention some of these difficulties in this paper; the other symposiasts develop these and related problems in considerably more detail.

## II. A GENERAL MODEL OF
## THEORY REDUCTION

The model to be given in this paper is but a slight modification of a model which I sketched several years ago, and which is indebted to various earlier forms which have existed in the literature of the philosophy of science since the pioneering work of Ernest Nagel in 1949.[2] Recently, the model I shall outline has been criticized, I believe mistakenly, by several philosophers including the present symposiasts.[3] It will be appropriate to respond to some of those criticisms as I think that an answer to those attempted refutations both clarifies and deepens the original model.

Following standard practice, I shall term a theory to be reduced the reduced theory, and the theory which does the reducing the reducing theory. Theories can be constructed as attempts to capture the essentials of the subject areas, or domains to use

Dudley Shapere's term,[4] which they explain, and thus a reduction of a theory (assuming its adequacy) is a reduction of the area of application of a theory. For the sake of precision as well as to insure the adequacy of a reduction, I shall assume that the basic principles of the reducing and reduced theories have been codified and that the primitive terms of both theories have been determined.[5] We can thus assume that we know the fundamental entities, processes, and generalizations (or laws) that constitute the reducing and reduced theories and their subject areas.

The conditions for the adequate reduction of a theory have both formal and informal aspects.[6] On the formal side, terms, e.g., 'gene' and 'phenotype' which appear in the reduced theory but not in the reducing theory, must be somehow associated with terms in the reducing theory. The simplest and least question-begging way to do this seems to be to require (i) that the entity terms (e.g., 'gene' or 'DNA') to be associated, be construed as extensionally referring to the same entity, even though that

2. Nagel (1949); also see Nagel's more developed ideas in his (1961).

3. See M. Ruse (1971a) and (1971b) and D. Hull (1972), (1973) and (1974).

4. D. Shapere (1974).

5. A formalization, say in first order logic or in set theoretical language, is not necessary on this view, though an axiomatization (in English, for example) is very desirable to insure this 'codification' and the determination of the primitive terms and relations. The requirement is a logical one and requires fulfillment for a determination of the logical 'effectiveness' of a reduction. Such an axiomatization may not be a necessary condition of a de facto reduction since molecular biologists themselves probably can proceed informally, identifying, for example, a particular problem area which has been recalcitrant to physicochemical characterization and explanation. They thus demarcate areas which have been reduced from those which have not. In their research, their description of the area in terms of 'biological' terminology would amount to a provisional codification of the basic principles characterizing the subject area.

6. This distinction between formal and informal aspects of a reduction is not the same as Nagel's similar distinction between formal and nonformal conditions made in his (1961), pp. 358ff.

entity is described in different ways and (ii) that predicates and relations (e.g., 'dominant') in the two theories, be interpreted as referring to the same state of affairs, characterized with the help of the entity relations spoken above. (This point will be developed more extensively later.) Thus, for example, the term 'gene' can be understood to refer to the same entity which is named by a sequence of nucleotides of DNA (or RNA in some special cases involving viruses). Sentences which formulate such extensional references are best construed as *synthetic* identities, i.e., identities which at least initially require empirical support for their warrant.[7] (Genes were not discovered to be DNA via the analysis of *meaning;* important and difficult empirical research was required to make such an identification.) I am going to term such sentences 'reduction functions' since they have values which exhaust the universe of the reduced theory, e.g., gene, for arguments in the universe of the reducing theory, e.g., DNA.[8] Structural relations between chemical constituent parts, e.g., DNA *sequences,* appear in the chemical side of the reduction functions.

The reduction functions thus contain part of the 'organization' which characterizes biological systems. This last point is relevant to some of Michael Polanyi's claims concerning the impossibility of the reduction of biology.[9]

Again, continuing to comment on the formal aspects of theory reduction, in order to insure that a reduced theory is *in principle eliminable,* i.e., that the reducing theory can do all that the reduced theory accomplished in terms of explanation and systematization of data, we stipulate that the reduced theory must be *derivable* from the reducing theory when the reducing theory is supplemented with the reduction functions mentioned above. The reduced theory thus stands to the reducing theory somewhat as a set of *theorems* in geometry stands to the basic *axioms* of geometry. The reduced theory is derivative, and its ontology and empirical content are more limited aspects of the reducing theory's ontology and basic assertions about the world.

Formalizable relations between the reducing and reduced theory do not, however, exhaust the relationship

7. See H. Feigl (1958), p. 370, and L. Sklar, (1964) for a discussion of synthetic identities in reduction. For an argument similar in spirit to Feigl's point that only synthetic identities are satisfactory in that they are not 'nomological danglers' see R. Causey, (1972).

8. The locution 'reduction function' is patterned on Quine's (1964) 'proxy function' in his article on ontological reduction.

9. Other aspects of the organization of chemical systems will appear in the section explaining chemical mechanisms (see text below). M. Polanyi has offered an intriguing 'vitalistic' interpretation of the need to accept organization in chemical systems in his (1968). His arguments are not successful, however, inasmuch as (1) structural relations embodying chemical organization are statable in chemical terms and explicable in their workings by normal chemical theories, and (2) the existence of these structures or 'boundary conditions' as Polanyi terms them, is explicable in principle by a chemical evolutionary theory. See my (1976b), (1969b), for more specific details. For a recent lucid introduction to the issues of the chemical origin of life and chemical evolution, see L. E. Orgel, (1973). For a completely sequenced replicating molecule which tells us much about Darwinian evolution in chemical terms, see D. R. Mills, F. R. Kramer, and S. Spiegelman (1973).

between these theories. As has been stressed over the last decade by Paul Feyerabend,[10] reducing theories often contradict or are incommensurable with aspects of the purportedly reduced theory. The reduction functions of which I spoke above are thus often very difficult to formulate in such a way that they are not fundamentally misleading, confusing, and incoherent. Feyerabend has offered a number of examples from the physical sciences to support his claim. Consider the concept of temperature in classical phenomenological thermodynamics.[11] This notion of temperature, definable in terms of a reversible Carnot cycle and heat exchanges, can be shown to be inextricably associated with the strict, i.e., nonstatistical, form of the second law of thermodynamics. Kinetic theory, Feyerabend notes, cannot "give us such a concept". The terms in both theories cannot be associated by a reduction function without contradiction (if the function is conceived to be analytic) or without falsity (if the function is thought to be a physical hypothesis). The term 'temperature' is thus only a homonym for Feyerabend, and reduction of the macro-theory, thermodynamics, to the micro-theory, statistical

mechanics, is impossible. The theories are incommensurable. At best, statistical mechanics replaces phenomenological thermodynamics. (Feyerabend's views are quite similar to Thomas Kuhn's, which are probably more widely known.)[12]

Without attempting to analyze Feyerabend's most interesting arguments in this paper, what I wish to do is to consider the *possibility* that something similar is at work in the relationship of classical genetics to molecular genetics, and to consider what its import may be for the reduction of classical genetics by molecular biology.[13]

It is clear that some modification of the concepts of classical genetics has occurred since 1950 both as a consequence of more sophisticated genetic techniques, e.g., the use of bacteria and phage, and the development of chemical models of genetic processes. Biological dictionaries in the early 1950's used to characterize the gene as the unit of mutation, recombination, *and* function (in the sense of being a necessary condition for a phenotypic characteristic). With the advent of fine structure genetics, Benzer and other geneticists saw the necessity to redefine the notion of

10. Feyerabend (1962) and (1965).

11. The example is from Feyerabend's (1962). Also see Feyerabend's (1965), p. 223, for a later discussion of his views, and also p. 226 for a presentation of a modified thermodynamics with fluctuation (due to Leo Szilard) which is relevant for the argument given in the text below concerning a 'corrected' $T_2{}^*$.

12. T. S. Kuhn (1962).

13. The need for redefinitions of basic terms in the reduced science in order to insure that adequate reduction functions are formulatable was discussed in my (1967a). The prospect that the reducing theory might have to be altered to accomplish a reduction was considered in my (1969a) esp. pp. 331–332. Recently David Hull has discussed a number of problems which are associated with the issue of formulating specific reduction functions between classical genetics and molecular genetics, sharply raising the Feyerabend type of problem for this example of reduction. See text below and notes 19 and 20.

the classical gene and to distinguish between these heretofore extensionally equivalent concepts. The early 1960's saw the introduction of new classes of genes, e.g., regulator genes such as repressor-making genes and operator genes. A more careful analysis of phenotypic characteristics via biochemical analysis allowed for a much more precise characterization of genetic effects.[14]

Classical genetics of the type understood by Morgan and Muller in the 1920's, say, has evolved and become much richer even when presented without reference to underlying biochemical details. (Fine structure genetics and regulatory genetics can, though they usually are not, be presented in a roughly phenomenological manner without much stress on the biochemical foundations.)[15] Let us indicate that an evolution of classical genetics has taken place by saying that classical genetics has been modified or corrected to a more adequate form of classical genetics, say neo-classical genetics or classical genetics*. This correction has in part been caused by the successes of

research into the chemical nature of the gene and into chemical characterizations of phenotypes. One can, nonetheless, speak in *biological* terminology and use expressions which are characteristically *nonchemical* in presenting classical genetics*, e.g., we could talk about genetic interaction or the blocking effect of the inductibility gene ($i^+$) on the lactose metabolizing gene ($z^+$) in the *lac* region of *E. coli,* without knowing that the *i+ DNA region* makes a special *protein* which prohibits transcription of the *$z^+$ DNA region,* thus ultimately preventing the synthesis of the *enzyme* β-galactosidase.[16] (It should be noted here however that because the partial reductions of genetics thus far achieved have resulted in a kind of 'intertwining' of chemistry and genetics,[17] such a separation as is analytically desired for our purposes is not explicitly found in textbooks and research papers.)

The point of the foregoing discussion is as follows: I believe that we can take a modification of classical genetics and explain that *modification* by physics and chemistry, i.e., by

14. The changing definition of a gene is not simply a 'philosophical' problem; the need for a more precise conceptual analysis of the notion of a 'gene' occasioned by new discoveries in molecular genetics has been argued for in the editorial pages of the distinguished British scientific journal *Nature New Biology* 230 (1971), 194.

15. To obtain a very rough example of what genetics without the underlying biochemical foundations might look like it is instructive to compare the first edition of R. P. Wagner's and H. K. Mitchell's *Genetics and Metabolism* (New York: Wiley, 1955), with the 1964 second edition, which contains more molecular material, and to compare both of these with the manner of presenting genetics in J. D. Watson's *Molecular Biology of the Gene,* 2nd ed. (New York: Benjamin, 1970).

16. This case of genetic interaction effects in the lac operon is discussed extensively in my (1974a) and (1974b), section 2. In point of historical fact, the genetic interactions were discovered prior to a determination of much of the underlying biochemistry.

17. The felicitious term 'intertwining' as used to characterize what happens in actual reductions is, I believe, due to Sidney Morgenbesser.

General Reduction Model

$T_1$—the reducing theory.
$T_2$—the original reduced theory.
$T_2^*$—the 'corrected' reduced theory.

---

Reduction occurs if and only if:

(1) All primitive terms of $T_2^*$ are associated with one or more of the terms of $T_1$ such that:

   (a) $T_2^*$(entities)=function$[T_1$ (entities)$]$.
   (b) $T_2^*$ (predicates)=function $[T_1$ (predicates)$]$.

   (This is the condition of referential identity.)

(2) Given fulfillment of condition (1), that $T_2^*$ be derivable from $T_1$ supplemented with 1(a) and 1(b) functions. (Condition of derivability.)

(3) $T_2^*$ corrects $T_2$, i.e., $T_2^*$ makes more accurate predictions.

(4) $T_2$ is explained by $T_1$ in that $T_2$ and $T_2^*$ are strongly analogous, and $T_1$ indicated why $T_2$ worked as well as it did historically.

---

Figure 1. The general reduction model.

molecular genetics, even if reduction functions associating an *unmodified* classical genetics with physics and chemistry cannot be formulated. In this way I hope to be able to outflank the Feyerabend type of objection. In such a way we can, assuming the validity of the model sketched in this paper, obtain a formally precise characterization for the reduction of classical genetics* by physics and chemistry, but only an *informal* relationship between either (i) the physics and chemistry and the *uncorrected* form of classical genetics, or (ii) the corrected and the uncorrected forms of classical genetics. To say, however, that we have reduced classical genetics and not some *totally* different theory to physics and chemistry, we require that the corrected reduced theory bear a strong analogy with the *uncorrected* reduced theory, and, that the reducing theory explain why the uncorrected reduced theory worked as well as it did historically, e.g., by pointing out that the theories lead to approximately equal laws, and the experimental results will agree very closely except in special extreme circumstances. These relations of approximate equality, close agreement, and analogy have yet to find formally precise characterizations, and to date represent informal aspects of a reduction. These elements in the reduction should not, however, be taken as implying that the relation between the reducing theory and reduced theory, in its corrected form, is vague or imprecise.[18] The vagueness lies in the historical relation of strong similarity between the uncorrected and the corrected reduced theory. Such vagueness is in fact logically demanded since the corrected theory is thought to be more adequate, and perhaps even completely true, whereas the uncorrected theory is inadequate and in part false, and it would be most unusual if a formal relation that could be codified in logical terms existed between such earlier and later versions of the reduced theory.

A summary of the conditions for an adequate reduction of theory $T_2$* by the reducing theory $T_1$ is given in Figure 1.

### III. CRITICISMS OF THE MODEL

Criticisms of the model sketched above have tended to focus on three issues: first, the nature of the reduction functions and whether these can be specified. David Hull in particular has questioned the possibility of being able to state these functions and has also suggested the conditions that in point of fact exist between classical and molecular genetics are so complex that we would encounter a bizarre many-many set of relations if we attempted to articulate the reduction functions.[19] Secondly, both Hull and Michael Ruse have queried whether the distinction between reduction and *replacement* is not blurred to the point where what I term a reduction is not in fact a

replacement of classical genetics by molecular genetics.[20] Finally, Tom Nickles has recently suggested that the model presented here characterizes but one of several different types of reduction, and that emphasis on the derivational or deductive relation between $T_1$ and $T_2$* diverts our attention from construing intertheoretic relations under alternative useful non-deductive rubrics.[21] I believe that Bill Wimsatt is in agreement with Nickles' suggestions here.[22]

In the scope of this paper I cannot develop a full set of replies to each of these criticisms. It will, however, be useful to examine David Hull's criticisms in some detail and touch very briefly on the replacement and non-deductive relation issues.

Let us begin by looking at Hull's claims, reiterated in the present group of papers, that argues for nonconnectability of classical and molecular genetics. The essence of Hull's view is that connections between classical genetics and molecular genetics are many-many, i.e., that to one Mendelian term (say the predicate 'dominant') there will correspond many diverse molecular mechanisms which result in dominance. Further, Hull adds, any one molecular mechanism, say an enzyme-synthesizing system, will be responsible for many different types of Mendelian predicates: dominance in some cases, recessiveness in others. With the increase in precision which molecular methods allow, furthermore, major

18. See for example G. Massey's comment in his (1973) p. 208.
19. See Hull (1972), (1973), and (1974).
20. Hull (1974), (1975) and Ruse (1971a and b).
21. Nickles (1973).
22. Wimsatt (1975).

reclassifications of genetic traits are likely, and when all of these difficulties are sorted out, any corrected form of genetics that can be reduced is accordingly likely to be very different from the uncorrected form of genetics. To Hull, "given our pre-analytic intuitions about reduction [the transition from Mendelian to molecular genetics] is a case of reduction, a paradigm case. [However on] . . . the logical empiricist analysis of reduction [i.e., on the basis of a type of reduction model introduced earlier] . . . Mendelian genetics cannot be reduced to molecular genetics. The long awaited reduction of a biological theory to physics and chemistry turns out not to be a case of 'reduction' . . . , but an example of replacement".

It will be useful, in an attempt to present Hull's position as fairly as possible, to allow him to give his arguments in his own terms. Hull contends:

> Numerous phenomena are now explained in Mendelian genetics in terms of recessive epistasis [i.e., a certain form of gene interaction]; e.g., color coat in mice, feather color in Plymouth Rock and Leghorn chickens, a certain type of deaf-mutism in man, feathered shanks in chickens, and so on. There is little likelihood that all of these phenomena are produced by a single molecular mechanism. At best, several alternative mechanisms are involved. Conversely, there is little likelihood that these various molecular mechanisms always produce phenomena which are appropriately characterized by the Mendelian predicate term 'recessive epistasis'.

> . . . For instance, one might expect dominant and recessive epistasis to be produced by a combination of those molecular mechanisms that produce epistasis with those that produce dominance and recessiveness.

One does not have to look very deeply into the relation between Mendelian and molecular genetics to discover how naive the preceding expectations actually are. Even if all gross phenotypic traits are translated into molecularly characterized traits, the relation between Mendelian and molecular predicate terms express prohibitively complex, many-many relations. Phenomena characterized by a single Mendelian predicate term can be produced by several different types of molecular mechanisms. Hence, any possible reduction will be complex. Conversely, the same types of molecular mechanism can produce phenomena that must be characterized by different Mendelian predicate terms. Hence, reduction is impossible. Perhaps these latter ambiguities can be eliminated by further 'correcting' Mendelian genetics but then we are presented with the problem of justifying the claim that the end result of all this reformulation is reduction and not replacement. Perhaps something is being reduced to molecular genetics once all these changes have been made in Mendelian genetics, but it is not Mendelian genetics.

When one combines all of the complexities mentioned thus far with the requirement that the relations between the original Mendelian predicate terms be retained, one begins to appreciate the scope of the task confronting a serious proponent of reduction in genetics. Not only must he list all the mechanisms which eventuate in recessive epistasis as well as those that eventuate in dominant epistasis, but also these two lists of molecular mechanisms must be symmetrical in the same way in which traits which are simple dominant or recessive are symmetrical. According to the current analysis of reduction supplied by the logical empiricists and their contemporary descendents, a certain amount of reshuffling and reclassification of the phenomena of the reduced theory is to be expected, but in this instance it seems as if the vast scaffolding of Mendelian genetics must either be ignored or else dismantled and reassembled in a drastically different form.[23]

I believe that Hull has misconstrued the application of the general model of reduction which I outlined earlier in several ways, and that he thus sees logical problems where they do not exist. Empirical problems do exist, namely what mechanisms are involved in the cytoplasmic interactions that result in genetic epistasis of the dominant and recessive varieties, but these ought not to be confused with the logical problems of the type of connections which have to be established and with how a corrected form of classical genetics is to be derived from molecular genetics. Let me argue for this position in some detail.

Hull's primary concern is with the reduction functions for Mendelian predicates. Let us briefly recall what predicates are prior to sketching the manner in which they can be replaced in a reduction. In earlier characterizations of the general model of theory reduction as outlined above, I have followed modern logicians such as Quine in construing predicates in an extensional manner. From this perspective, a predicate is simply a class of those things or entities possessing the predicate. This nominalistic position can be put another way by using standard logical terminology and understanding predicates to be open sentences which become true sentences when the free variable(s) in the open sentence are replaced by entities which possess the predicate characterized by the open sentence. Thus the open sentence '$x$ is odd' where $x$ ranges over positive integers, say, introduces the predicate 'odd' (or 'is-odd'). When the free variable $x$ is replaced by a named object, then the sentence becomes closed and can be said to be true or false depending on whether the named object in fact possesses the predicate. Relations can be similarly introduced by open sentences containing two free variables, e.g., the relation of being a brother, by '$x$ is the brother of $y$'. Following Quine, we may then say that "the extension of a closed one place predicate is the class of all things of which the predicate is true; the extension of the closed two place predicate is the class of all the pairs of which the predicate is true; and so on."[24] The extensional

23. Hull (1974) pp. 40, 39, 41.
24. Quine (1959), p. 136.

characterization of predicates and relations is important because reduction functions relate entities and predicates of reduced and reducing theories extensionally, via an imputed relation of synthetic identity.

Let us examine the predicate 'dominant' or 'is-dominant' from this point of view.

First it will be essential to unpack the notion. Though the predicate is one which is ascribed to genes and might prima facie appear to be a simple one-place predicate,[25] in fact the manner in which it is used in both traditional classical genetics and molecular genetics indicates that the predicate is a rather complex relation. The predicate is defined in classical genetics as a relation between the phenotype produced by an allele of a gene in conjunction with another different allele of the same locus (a heterozygous condition), compared with the phenotype produced by each of these alleles in a double dose of the same allele (a homozygous condition). The predicate 'epistatic,' which Hull believes to be a particularly troublesome one for the general reduction model, is similarly unpackable as a rather complex relation between different genes and not simply between different alleles of the same gene.

More specifically, suppose that the phenotype of a gene $a$ when the gene is in a homozygous situation, i.e., $aa$ is $\mathbf{a}$.[26] Similarly suppose that genotype $AA$ biologically produces phenotype $A$. (The causal relation of biological production will be represented in the following pages by a single arrow: $\rightarrow$.) If $a$ and $A$ are alleles of the same genetic locus, and the phenotype produced by genotype $aA$ is $\mathbf{a}$, then we say that gene $a$ is dominant (with respect to gene $A$).[27] The predicate 'dominant' therefore is in reality a

25. David Hull has suggested, in his comments on a version of the present paper (delivered at a meeting of a Chicago seminar on reduction on 17 July 1973) that the cellular environment ought to be explicitly included in a characterization of dominance, since the environment can affect the dominance relation. In reply I noted that (i) in the case considered the environment is assumed, for reason of simplicity and logical clarity, to be constant and (ii) expanding the relation provided from a two- to a three-place predicate does not, in any event, interestingly alter the logic of the predicate reduction functions.

26. This notation is a slight change from that used in my (1974a) and is due to suggestions by David Hull and William Wimsatt.

27. In my (1967a), which outlined the tenets of the general reduction model (or 'paradigm'), I suggested that the model would be better comprehended if a simple example clarifying the dual (entity and predicate) nature of the requisite reduction functions was provided to illustrate the rather abstract form of the model. The example I chose was from genetics, with the entity in the reduced science being a 'gene' and the predicate being 'dominant'. I suggested that the predicate dominant could be interpreted molecularly as "the ability to make an active enzyme". This is a useful first approximation and, in point of historical fact, is not too different from the early and oversimplified 'presence and absence' theory of dominance proposed by Bateson in 1906. (See E. A. Carlson's (1966), Ch. 8, for a discussion of this theory and its weaknesses.) Hull has rightly criticized this interpretation as an inadequate explication at the molecular level of the property of dominance, though he has misinterpreted the function of the example in the 1967 article. In my (1969a), which was explicitly on reduction in biology, I indicated more precisely the complexity of predicate reduction functions. Essentially the same logic proposed there is offered in

relation and is extensionally characterized by those pairs of alleles of which the relationship $(aa \to \textbf{a})$ & $(AA \to \textbf{A})$ & $(aA \to \textbf{a}) =_{df} a$ is dominant (with respect to $A$), is true.

Epistasis can be similarly unpacked. If two non-allelic genes, $a$ and $c$ say, are such that $aa \to \textbf{a}$ and $cc \to \textbf{c}$, when each genotype is functioning in isolation from the other, but that the genotype $aacc$ biologically produces $\textbf{a}$, we say that $a$ 'hides' $c$ or 'is epistatic to' $c$. The gene $a$ may so relate to $c$ that $a$ in its dominant allelic form obscures $c$. (This is 'dominant epistasis'.) The gene $a$ may also relate to $c$ such that $a$'s absence (or recessive allele of $a$) prevents the appearance of $c$'s phenotype. (This is 'recessive epistasis'). $\textbf{a}$ and $\textbf{c}$ here are contrasting traits.

The various interactions between genes, of which the two forms of epistasis mentioned above are but a subset, can be operationally characterized since they result in a modification of the noninteractive independence assortment ratio of 9:3:3:1 for two genes affecting the phenotypes of the organisms. (For a survey of these interactions and the specific modifications of the ratios which they produce the reader must be referred to other sources.)[28]

I have now sufficiently analyzed some of the predicates which appear in classical genetics so as to return to the problem of specifying reduction functions for such predicates.

Specification of reduction functions, which are like dictionary entries that aid a translator, can be looked at in two ways analogous to the two directions of translation: (1) From the context of the theory to be reduced one must be able to take any of the terms (entity terms or predicate–including relations–terms) and univocally replace these by terms or combinations of terms drawn from the reducing theory. (2) From the context of the reducing theory one must be able to pick out combinations of terms which will unambiguously yield equivalents of those terms in the reduced theory which do not appear in the reducing theory. Such sentences characterizing replacements and combination selections must, together with the axioms and initial conditions of the reducing theory, entail the axioms of the reduced theory if the reduction is to be effected.

To obtain reduction functions for Mendelian predicates we proceed in the following manner. (1) the predicate is characterized extensionally by indicating what class of biological entities or pairs of entities, etc., represent its extension. This was done for dominance and epistasis above in terms of genotypes and phenotypes. (2) In place of each occurrence of a biological entity term, a chemical term or combination of terms is inserted in accord with the reduction functions for entity terms. (3) Finally, in place of the biological production arrow which represents the looser or

the present paper, along with a more detailed but still schematic explication of dominance and together with some suggestions for reducing still more complex predicates, such as 'epistatic', which Hull feels are particularly troublesome in reduction.

28. For survey of the different types of ratios produced by genetic interaction, see R. C. King (1965), pp. 89ff, and R. P. Wagner and H. K. Mitchell (1964), pp. 392–438.

less detailed generalizations of biological causation, we write a double-lined arrow ($\Rightarrow$), which represents a promissory note of a chemically causal account, i.e., a specification of various chemical mechanisms which would yield the chemically characterized consequent on the basis of the chemically characterized antecedent. Thus we could do the following: Let the reduction function for gene $a$, say, be 'gene $a$ = DNA sequence $\alpha$' and 'gene $A$ = DNA sequence $\beta$'. We also suppose phenotype **a** possesses a reduction function 'phenotype **a** = amino acid sequence א ', and phenotype A = amino sequence ב. Then if we represented the Mendelian predicate for dominance as $(aa \rightarrow a)$ & $(AA \rightarrow A)$ & $(aA \rightarrow a)$ we would obtain:

Allele $a$ is dominant (with respect to $A$) =
(DNA sequence $\alpha$, DNA sequence $\alpha \Rightarrow$ amino acid sequence א ) & (DNA sequence $\beta$, DNA sequence $\beta \Rightarrow$ amino acid sequence ב ) & (DNA sequence $\alpha$, DNA sequence $\beta \Rightarrow$ amino acid sequence א ).

This reduction function for dominance is, as was the characterization of dominance introduced earlier, somewhat oversimplified and for complete adequacy ought to be broadened to include Muller's more sensitive classification and quantitative account of dominance.[29] This need not be done, however, for the purposes of illustrating the logic of reduction

functions for predicates. I have deliberately left the interpretation of the double-lined arrow in the right-hand side of the above reduction function unspecified. As noted, it essentially represents a telescoped chemical mechanism of production of the amino acid sequence by the DNA. Such a mechanism must be specified if a chemically adequate account of biological production or causation is to be given. Such a complete account need not function explicitly in the reduction function per se, even though it is needed in any complete account of reduction. Such a reduction function, and similar though more complex ones, could be constructed in accordance with the above steps for epistasis in its various forms, and allows the translation from the biological language into the chemical and vice versa. Notice that the relations among biological entities which yield the various forms of dominance and epistasis, for example, can easily be preserved via such a construal of reduction functions. This is the case simply because the interesting predicates can be unpacked as relations among biological entities, and these selfsame entities identified with chemical entities. The extensional characterization of predicates and relations as classes and (ordered) pairs of entities accordingly allows an extensional interpretation of reduction functions for predicates.[30]

Assuming that we can identify any

---

29. Muller's original article suggesting his analysis (or, better, reanalysis) of dominant and recessive genes into amorph, hypomorph, and hypermorph classes appears in his (1932), p. 213. A more recent account of this quantitative theory of dominance appears in Wagner and Mitchell (1964).

30. The synthetic identity interpretation of entity reduction functions and the exten-

gene with a specific DNA sequence and that, as Francis Crick says, "the amino acid sequences of the proteins of an organism . . . are the most delicate expression possible of the phenotype of an organism,"[31] we have a program for constructing reduction functions for the primitive entities of a corrected classical genetics. (A detailed treatment would require that different types of genes in the corrected genetic theory, e.g., regulator genes, operator genes, structural genes, and the recon and muton subconstituents of all these, be identified with specific DNA sequences.)

Substitution of the identified entity terms in the relations which constitute the set of selected primitive predicates, which yields reduction functions for predicates, then, allows for the two-way translation, i.e., from classical genetics to molecular genetics and vice versa. Vis-à-vis David Hull, molecular mechanisms do not appear in connection with these reduction functions *per se,* rather they constitute part of the reducing theory. Chemically characterized relations of dominance can be and are explained by a variety of mechanisms. The only condition for adequacy which is relevant here is that chemical mechanisms yield predictions which are unequivocal when the results of the mechanism(s) working on the available DNA and yielding chemically characterized phenotypes are translated back into the terms of neoclassical genetics.

What Hull has done is to misconstrue the locus where molecular mechanisms function, logically, in a reduction, and he has thus unnecessarily complicated the logical aspects of reduction. Different molecular mechanisms can appropriately be appealed to account for the same genetically characterized relation, as the genetics is less sensitive. The same molecular mechanism can also be appealed to account for different genetic relations, but only if there are further differences at the molecular level. It would be appropriate to label these differences as being associated with different initial conditions in the reducing theory.[32]

---

sional construal of $n$-ary predicates as ordered $n$-tuples of entities suggests, although we provisionally distinguish between reduction functions for entities and predicates, that ultimately all reduction functions connect reduced entities and reducing entities. It may also be useful to point out here that we can, for similar heuristic reasons, relax the requirement that the reduction functions for entities only range over (ordered $n$-tuples of) the reducing entities and allow predicates of the reducing theory to appear on the right-hand side of the entity reduction functions. Such a relaxation would very probably allow for an initially easier statement of the reduction functions, e.g., one could not incorporate properties of the DNA in reduction functions for genes. Again, though, if we ultimately construe predicates as ordered $n$-tuples of entities, the distinction formally collapses.

31. See F. H. Crick (1958). In addition to primary structure, specification of secondary, tertiary and quaternary structures are probably also needed, given the current state of folding theories.

32. This point is important and is one of the main areas of continued disagreement between Hull and myself. Hull seems to believe that the list of 'initial conditions' is likely to be impossibly long. I, on the other hand, believe that molecular biologists are able to identify a small set of crucial parameters as initial conditions in their elucidation of fundamental

There are several appeals which I believe can be made to contemporary genetics which will support my position over Hull's. First, traditionally discovered gene interactions, not only epistasis but also suppression and modification, are in fact currently being treated from a molecular point of view. Several textbooks and research papers can be cited in support of this.[33] The logic of their treatment is not to drastically reclassify the traditionally discovered gene interactions, but rather to identify the genes with DNA, and the phenotype results of the genes with the chemically characterized biosynthetic pathways, and to regard gene interactions as due to modifications of the chemical environments within the cells. Traditional genetics is not drastically reanalyzed, it is corrected and enriched and then explained or reduced, at least in part.

Second, I would note that it is sometimes the practice of molecular geneticists to consider what types of genetic effects would be the result of different types of molecular mechanisms. A most interesting case in point is the predictions by Jacob and Monod of what types of genetic interaction effects in mutants would be observable on the basis of different

hypotheses concerning the operator region in the lac system of *E. coli.*[34] In brief, they were able to reason from the chemical interactions of the repressor and the different conjectures about the operator, and infer whether the operator gene would be dominant or recessive, and pleiotropic or not. The fact that in such a case molecular mechanisms yielded unequivocal translations into a corrected genetic vocabulary is a strong argument against Hull's concerns.

Finally, I should add that if the relation between modified classical genetics and molecular genetics were one of replacement, then it is very likely that considerably more controversy about the relations of classical to molecular genetics would have infused the literature. Further, it is also likely that those persons trained in classical genetics would not have made the transition as easily as they clearly have; nor would they comfortably be able to continue to utilize classical methods without developing, in a conscious manner, an elaborate philosophical rationale about the pragmatic character of classical genetics. Clearly such was the case with the type of historical transitions which are termed replacements and not reductions, e.g., the replacement of the

biochemical mechanisms. These initial conditions are usually discussed explicitly in the body of their research articles (sometimes in the 'materials and methods' sections). The specification of the initial conditions, e.g., the temperature at which a specific strain of organisms was incubated and the specific length of time, is absolutely necessary to allow repetition of the experiment in other laboratories whose scientists know the experiment only through the research article. The fact that mechanisms such as the Krebs cycle and the *lac* operon are formulatable in a relatively small set of sentences, with a finite list of relevant initial conditions specifiable, indicates to me that Hull is too pessimistic.

33. See Wagner and Mitchell (1964) pp. 394–401, for a discussion of a biochemical explanation of gene interaction.

34. See my (1974c) for a discussion of this strategy in connection with the development of the operon theory of Jacob and Monod.

theories of Aristotle by Newton's or of Ptolemy's by Copernicus' and Kepler's.

I believe that these three considerations count heavily against a replacement interpretation of the relationship between classical genetics and classical genetics* (or between classical genetics and molecular genetics). I am gratified to see that Ruse, who originally seemed disposed to accept a replacement interpretation, has recently come to accept more of a close-analogy or a reduction relationship.[35] He has also in his recent paper provided an additional example from a contemporary dispute in population genetics to support the reduction interpretation.

Finally I would briefly like to consider Nickles' views concerning nondeductive accounts of reduction. Nickles believes that a deductive relationship between reduced and reducing theories is too narrow and that the Nagel type of relation, which he terms reduction$_1$, has to be supplemented with an additional concept of reduction, a reduction$_2$. This latter concept arises out of the senses of reduction which refer to "being led back from one thing to another . . . and to the related notion of transforming something into a different form by performing an operation on it" . . . [such as] the "reduction of ores to their metals, the reduction of wood to pulp by pounding it . . . , and indeed, the reduction of $m_0 v / \sqrt{(1 - v^2/c^2)}$ to $m_0 v$ by taking the limit as $v$ goes to zero."[36]

Reduction$_1$ is 'essentially *derivational reduction*' or a deductive explanation, whereas reduction$_2$ is neither a theoretical explanation nor an ontological reduction.[37] Reduction$_2$ is rather 'justificatory and heuristic'. The general idea is that a later reducing$_2$ theory $T_L$ yields a limiting consequence C which is apparently prima facie equivalent to some important constituent of an earlier theory $T_E$ (if not prima facie equivalent to the entire $T_E$). Reduction$_2$, Nickles contends, characterizes the relation in which $T_L$ reduces in *different ways to* $T_E$ as different limits are taken, whereas in reduction$_1$, there is only one deductive relation betweeen $T_L$ and $T_E$. Finally, Nickles suggests that reduction$_2$ is not very sensitive to the problem of meaning change and logical compatability since its function is heuristic and justificatory.

With this distinction between two concepts of reduction in view, Nickles explicitly raises some problems with the general reduction model sketched above.

Nickles points out that the analogy condition (condition (4) in Figure 1 above) is problematical in that the notion of analogy has, as I had previously pointed out,[38] not yet received an adequate philosophical treatment. In place of this, Nickles suggests:

> . . . the concept of reduction$_2$ may be of some help. Presumably limit

35. See Ruse (1976), p. 633.
36. Nickles (1973), p. 184.
37. Nickles (1973), p. 185.
38. See my (1967a), p. 146.

relationships and the like will sometimes be useful in spelling out the analogy of $T_2$ and $T_2$* in particular cases. That is, in cases of approximative reduction, we can regard $T_2$* as a (fictitious) successor theory to the historical $T_2$, and in many of these cases reduction$_2$ techniques should be useful in spelling out the analogy. There is no reason to think that reductions$_2$ can in all cases take over the work of Schaffner's analogy relation, but when it can we shall have $T_1$ reducing $T_2$ approximately by reducing $T_2$*, which in turn reduces$_2$ to $T_2$.[39]

Reduction$_2$ can be stated in more general terms by employing the notion of a set of 'intertheoretic operations' performed on $T_1$ which yield $T_2$. Nickles has proposed the following schema:

Instead of a single reduction relation (the logical-consequence relation), we now recognize a set of intertheoretic operations $O_1$ (or corresponding relations) which we might write $O_2[O_1(T_1)] \rightarrow T_2$ or simply $O_2 O_1(T_1) \rightarrow T_2$, signifying that by performing operations $O_1$ and $O_2$ on theory $T_1$ we get $T_2$. No attempt will be made to build this notational scheme into a general account of reduction$_2$.[40]

Nickles goes on to argue that one of the problems is that the approximate nature of the relation between $T_2$ and $T_2$* characterized in the general reduction model is still basically *derivational,* and that instead of attempting to conceive of approximative reduction as a *failure* to achieve a cleaner non-approximative reduction, approximate reductions ought to be construed under an entirely different rubric, as instances of reduction$_2$.

Let us briefly consider some of the issues arising out of the reduction of classical genetics ($T_2$) and classical genetics* ($T_2$*) by molecular genetics ($T_1$).[41] I have argued that the relation between $T_1$ (with added reduction functions) and $T_2$* is deductive or 'derivational' in Nickles' terminology. Presumably the issue on which Nickles would favor his alternative construal is whether there is a relation of reduction$_2$ between $T_1$ (with reduction functions) and $T_2$.[42]

The problem that I see with Nickles' interesting suggestion as applied to the genetics case is that the intertheoretic operations which are not limit notions are difficult to specify and are deficient compared with the more intuitively appealing notion of a strong analogy. Indeed, as Nickles is willing to admit, it is not difficult to trivialize the notion of reduction$_2$ unless constraints on the types of operations allowed are articulated,

---

39. Nickles (1973), p. 195.

40. Nickles (1973), p. 197.

41. Nickles does not specifically address this case so I am perhaps unfairly characterizing his views without sufficient reason.

42. Nickles has also raised certain problems about identification, but I do not think these come up forcefully in the context of the present case. See his comments in Nickels (1973). p. 194.

and Nickles does not see any general way to do this. Though there may be disagreements about the details of the strength of analogies involved in the genetics case, I think that it can be said that there is general agreement among the participants in the present symposium (with the possible exception of Hull) that there is a very close analogy between $T_2$ and $T_2{}^*$. I would predict also that most members of the community of geneticists would also agree with this position. We are thus dealing with an unanalyzed or primitive relation, but with one which is testable.

Using a set of unspecified operations with no general restrictions would not, in my view, clarify the relations involved in the genetics case as much as the analogy condition does. As I perceive the situation, it would be unwise to dispense with the comparatively clearer analogy condition and with the paradigmatically clear relation of 'deductive consequence', to replace it with a set of unspecified intertheoretic relations characterizing (possibly vacuously) a concept of reduction.[2]

## REFERENCES

Carlson, E. A. 1966, *The Gene: A Critical History,* Philadelphia, Saunders.

Causey, R., 1972, *J. Phil.,* 69: 407.

Cohen, R. S., C. A. Hooker, A. C. Michalos, and J. van Evra (eds.), 1976, *PSA 1974,* Dordrecht, Reidel.

Crick, F. H. C., 1958, *Symp. Soc. Exp. Biol.,* 12: 138.

Feigl, H., 1958, in *Minnesota Studies in the Philosophy of Science, II,* (ed. by H. Feigl, M. Scriven, and G. Maxwell), Minneapolis, University of Minnesota Press.

Feyerabend, P. K., 1962, in *Minnesota Studies in the Philosophy of Science, III,* (ed. by H. Feigl and G. Maxwell), Minneapolis, University of Minnesota Press.

——, 1965, in *Boston Studies in the Philosophy of Science, II,* (ed. by R. S. Cohen and M. W. Wartofsky), Humanities Press, New York.

Hull, D., 1972, *Phil. Sci.* 39: 491.

——, 1973, in *Logic, Methodology, and Philosophy of Science, IV: Methodology and Philosophy of Biological Science* (ed. by P. Suppes *et al.*), Amsterdam, North Holland.

——, 1974, *Philosophy of Biological Science,* Englewood Cliffs, N. J., Prentice-Hall.

——, 1976, 'Informal Aspects of Theory Reduction', in Cohen et al. (1976), p. 653.

King, R. C., 1965, *Genetics,* Oxford University Press, New York.

Kuhn, T. S., 1962, *The Structure of Scientific Revolutions,* Chicago, University of Chicago Press.

Massey, G., 1973, *Ann. Japan Assn. Phil. Sci.,* 4: 203.

Mills, D. R., Kramer, F. R., and Spiegelman, S., 1973, *Science,* 180: 916.

Muller, H. J., 1932, *Proc. Intern. Congr. Genet.,* Ithaca, N. Y., I, 213.

Nagel, E., 1949, in *Science and Civilization* (ed. by R. C. Stauffer), Madison, University of Wisconsin Press.

——, 1961, *The Structure of Science,* New York, Harcourt, Brace, and World.

Nickles, T., 1973, *J. Phil.,* 70: 181.

Orgel, L. E., 1973, *The Origins of Life: Molecules and Natural Selection,* New York, Wiley.

Polanyi, M., 1968, *Science,* 160: 1308.

Quine, W. V. O., 1959, *Methods of Logic,* New York, Holt, Rinehart, and Winston.

——, 1964, *J. Phil.,* 51: 209.

Ruse, M., 1971a, *Dialectica,* 25: 17.

——, 1971b, *Dialectica,* 25: 39.

Ruse, M., 1976, 'Reduction in Genetics', in Cohen *et al.* (1976), p. 633.

Schaffner, K., 1967a, *Phil. Sci.,* 34: 137.

——, 1967b, *Science,* 157: 644.

——, 1969a, *Brit. J. Phil. Sci.,* 20: 325.

——, 1969b, *Amer. Sci.,* 57: 410.

——, 1974a, in R. Cohen and R. J. Seeger (ed.), *Boston Studies in the Philosophy of Science, II,* Reidel, Dordrecht, p. 207.

——, 1974b, *J. Hist. Biol.,* 7: 111.

——, 1974c, *Stud. Hist. Phil. Sci.,* 4: 349.

Shapere, D., 1974, in *The Structure of Scientific Theories.* (ed. by F. Suppe), Urbana, University of Illinois Press, p. 518.

Sklar, L., 1964, 'Intertheoretic Reduction in the Natural Sciences', unpubl. diss., Princeton University.

Wagner, R. P. and Mitchell, H. K., 1955, *Genetics and Metabolism,* New York Wiley.

——, 1964, *Genetics and Metabolism,* 2nd ed., New York, Wiley.

Watson, J. D., 1970, *Molecular Biology of the Gene,* 2nd ed., New York, Benjamin.

Wimsatt, W., 1976, 'Reductive Explanation: A Functional Account', in Cohen *et al.* (1976), p. 671.

# 24

# Reduction in Genetics

## MICHAEL RUSE

THERE IS A DISAGREEMENT BETWEEN Kenneth Schaffner and David Hull about the relationship between the biological theory of Mendelian genetics and the physico-chemical theory of molecular genetics.[1] Schaffner believes that the logical-empiricist thesis about theory-reduction is, in important respects, applicable to this relationship and illuminating when so applied. Hull denies its applicability and its illumination—indeed, he has gone as far as to say that "I find the logical empiricist analysis of reduction inadequate at best, wrong-headed at worst" (Hull, 1974a, p. 12). And he adds that "the conclusion seems inescapable that the logical empiricist analysis of reduction is not very instructive in the case of genetics. For my own part, I found that it hindered rather than facilitated understanding the relationship between Mendelian and molecular genetics" (1974a, p. 44).

More precisely, the disagreement between Schaffner and Hull seems to be this. Schaffner follows Ernest Nagel (1961) in arguing that a theory-reduction has occurred when all the claims[2] of one theory $T_1$ (the reduced theory) can be shown to be deductive consequences of the claims of another theory $T_2$ (the reducing theory), or, when the two theories talk in different languages, of $T_2$ together with the principles which allow one to translate from the language of one theory to the language of the other. However, Schaffner goes beyond Nagel in arguing that we can legitimately weaken the above criteria of theory-reduction and still meaningfully talk of reduction (of $T_1$ to $T_2$) if, although a theory $T_1$ cannot be deduced from $T_2$, a corrected version

---

1. Pertinent references are Schaffner 1967, 1969, 1974a, 1974b; Hull 1972, 1973, 1974a, 1974b.

2. Particularly the axioms.

of $T_1$, $T_1$*, can be deduced from $T_2$. How much correction is permissible Schaffner leaves open, but he does specify that there must be 'strong analogy' between $T_1$ and $T_1$*.

Applying this thesis about theory-reduction to biology, Schaffner agrees that traditional Mendelian genetics cannot in fact be deduced from molecular genetics (even when the latter is supplemented with appropriate translation principles), if for no other reason than that the two theories make conflicting claims. Mendelian genetics denies that the basic heritable unit of biological function (the Mendelian gene) is divisible by crossing-over—molecular genetics allows for such a division in its unit of function (the molecular gene, which is in fact a strip of DNA). However, argues Schaffner, Mendelian genetics can easily be corrected, and this corrected version can be deduced from molecular genetics. Indeed, this correction has taken place—we now have the biological theory of transmission genetics, where the unit of function, the cistron, is divisible by crossing-over. Moreover, argues Schaffner, there is strong analogy between traditional Mendelian genetics and transmission genetics. Hence, he concludes that we can meaningfully speak of theory-reduction in the case of genetics—"reduction" in the weaker, modified sense outlined above.

Hull's disagreements seem to be three. In the first place, he does not think that transmission genetics has, in fact, been deduced from molecular genetics and he is not convinced, at least not convinced *a priori,* that this can, in fact, be done. In the second place, even if such a deduction were done, he feels that there would be so great a difference between the deduced transmission genetics and traditional Mendelian genetics that one could not properly talk of reduction, even in Schaffner's weaker sense. In the third place, perhaps most importantly, Hull feels that the demand for a reduction is misleading—it does not tell us much about present real science, and it leads to little of new scientific or philosophical interest.

Since, after some initial dithering, I have committed myself fairly firmly to Schaffner's side of the argument,[3] it behooves me to try to answer Hull's objections. This I shall now do.

1. Hull points out rightly that as things stand at the moment, transmission genetics has not been deduced from molecular genetics—apart from anything else such deduction requires that the two theories be in hypothetico-deductive form and both are in a far looser state than that. Hull has also done great service, to me at least, in pointing out what an immense amount of work such a deduction would still require. However, Hull wants to go farther that this, and it is here that I start to feel uncomfortable. As Hull notes correctly, there are four possible relations between the concepts of transmission and molecular genetics—one to one (i.e. the entities[4]

---

3. See Ruse, 1973; but see also Ruse, 1971.

4. I mention here only entities—there is also the question of properties, but they introduce no new principles into the discussion.

referred to by a concept in one theory are the very same entities referred to by a concept in the other theory), one to many (a concept in one theory refers to entities referred to by two or more concepts in the other theory), many to one, and many to many. Now, if we have a one-to-one relationship, a deduction is at least possible. Letting small Greek letters stand for concepts in molecular genetics, and capital Arabic letters stand for concepts in transmission genetics, then if we have two laws $\alpha \to \beta$, and $A \to B$, and one to one correspondences $A = \alpha$ and $\beta = B$, then we can deduce $A \to B$ from $\alpha \to \beta$.[5] If we have one to many correspondences from transmissions genetics to molecular genetics, reduction is still possible. Suppose we have transmission genetical law $A \to B$, molecular genetical laws $\alpha_1 \to \beta_1$, $\alpha_2 \to \beta_2$, and correspondences $A = (\alpha_1 \lor \alpha_2)$, $B = (\beta_1 \lor \beta_2)$, then we can deduce $A \to B$ from the molecular laws. But if we have many-one or many-many correspondences, deduction is impossible. Thus, for instance, from $\alpha \to \beta$, $(A_1 \lor A_2) = \alpha$, and $(B_1 \lor B_2) = \beta$ we can get neither $A_1 \to B_1$, nor $A_2 \to B_2$. All we can get is the weaker $(A_1 \lor A_2) \to (B_1 \lor B_2)$.

Hull seems to feel that in an important sense we are stuck with many-one or many-many correspondences of this kind just described. Having pointed out (truly) that we often get one-many correspondences (from transmission to molecular genetics), he writes:

> Conversely, the same molecular mechanisms can produce different phenotypic effects. For example,

in a heterozygote, one allele may be completely operative, the other completely inoperative. Depending on the nature of the alleles and the biosynthetic pathways in which they are functioning the effect can vary. It may be the case that the single allele can produce all the product necessary to maintain the reaction at full capacity. If so, the phenomenon would be termed "dominant." Or it might be the case that the presence of the single operative allele decreases the rate of the reaction only slightly or perhaps cuts it in half. In such cases, the phenomenon would be termed "incomplete dominance." Hence, it would seem that even at the molecular level the relation between Mendelian predicate terms and molecular mechanisms is many-many. (Hull, 1974a, 40–41)

Clearly, if this is all that can be said on the matter a deduction (and hence a reduction) is impossible. But there is a fairly obvious reply to this objection. Water is $H_2O$ whether it be at 20°C or 80°C; but being of the same chemical constitution at the two temperatures hardly rules out a molecular explanation of the phenomenological differences of water at the two temperatures. More information must be fed in at the molecular level—for example, information about the differences in energy of the molecules at the two temperatures. Similarly, what seems to follow from the genetical case is not the conclusion that a deduction is in principle impossible, but that more information

---

5. '$\to$' stands for some kind of causal connexion; '$=$' stands for some kind of sameness–

is required at the molecular level showing why the various phenotypic effects occur. Then, we can get away from such correspondences as $\beta = (B_1 \vee B_2)$, replace them by such as $\beta_1 = B_1$, and $\beta_2 = B_2$, and hence deduction is once again possible.

Hull is not unaware of this counter-move.

At this point, one might object that perhaps the same molecular mechanism can result in different phenotypic effects, but that is because relevant factors are being left out—the temperature, hydrogen-ion concentration, and so on. Once all these relevant factors have been included, the one-many relation from the molecular to the phenotypic level will have been converted to a one-one relation. (Hull, 1974a, 42)

However, Hull seems unimpressed by this move. It is, we are told, "a covert restatement of the principle of deterministic causality. The same cause always produces the same effect. If the effects are different, then the causes must be different" (Hull, 1974a, p. 42). It is certainly not Hull's intention to deny the principle of causality—that would be a bit like attacking motherhood—but he feels that appeals to causality (perhaps like appeals to motherhood) are really so sweeping as to be practically uninformative. We have been offered some rather strong claims by the logical empiricist about the relationship between molecular and biological genetics, and suddenly these collapse into truisms about causality.

Although there is certainly truth in this position, I feel Hull's conclusion is nevertheless an overstatement. Had the logical empiricist nothing more to go on than a belief in the principle of causality, then it could hardly be claimed that there is a reduction—even of the weaker kind suggested by Schaffner—between molecular and Mendelian genetics. However, there is more than this—there is evidence which I think makes the logical empiricist's position far more plausible. In particular, there are cases where, given a general molecular mechanism which could lead to various Mendelian effects, by specifying the peculiarities of a particular case, biologists have acutally followed through to a uniquely determined effect. In other words, logical empiricists have more to rely on than general beliefs in causality—they can point to actual cases where the kinds of connections they believe ought in principle to hold have in practice been shown to hold.

Take, for example, a molecular mechanism which Hull cites as implying a many-many relationship between Mendelian and molecular genetics, namely molecular heterozygosity. This arises when one has different allelic molecular genes, and Hull's claim is that all kinds of biological phenotypes are possible—hence a reduction is impossible (or one has to resort to general claims about

---

Schaffner suggests synthetic identity. The concepts in these laws and identities might well refer to combinations of entities and properties.

6. A general discussion of this phenomenon can be found in Ruse, 1973. The specific details I am about to give now can be found in Winchester, 1972, Strickberger, 1968, Luzzatto, et al., 1970, and references.

causality). An instance of this heterozygosity occurs in the case of sickle-
cell anaemia in humans.[6] This disease,
endemic in certain African tribes
(and thus to be found in American
negroes), normally leads to death in
early childhood. The disease is known
to be caused by the mutant allele of
a human hemoglobin gene—homozygotes for the so-called 'sickle-cell'
gene suffer from the disease. Heterozygotes however do not suffer in any
appreciable way from anaemia, although the 'wild-type' allele is not
entirely dominant over the sickle-cell
allele. Where the heterozygote differs
from the normal homozygote is that,
apparently because of the heterozygosity, it has an increased resistance to malaria. Thus the sickle-
cell gene is maintained in malarially infected districts in a balanced situation.

Now, given the way Hull presents
the case, one would hardly expect
that the geneticist could go from the
molecular heterozygosity for the
sickle-cell allele to increased malarial
resistance in the heterozygote. We
have a one-many relationship, and if
we assume more then all we are
doing is committing ourselves to a
"covert restatement of the principle
of deterministic causality." However,
in fact, a great deal is known about
the molecular mechanics of the
sickle-cell case, so much so in fact
that geneticists can practically give
an entire molecular explanation of
the (biological) phenotypic effects,
without need for blind faith in principles of causality. In particular,
hemoglobin which consists essentially
of two pairs of polypeptide chains
($\alpha$ and $\beta$) is thought to be the product
of two (non-allelic) genes. The sickle-
cell mutant causes the substitution
of one amino acid in the $\beta$-chain

for another amino acid (valine for
glutamic acid). This substitution leads
to 'stacking' of the molecules in a
uniform alignment, which increases
the viscosity of the hemoglobin, which
leads in turn to the characteristic sickle-
shaped distortion of the cell. In the
heterozygote, because of the existence
of some normal hemoglobin one gets
less stacking in a cell, which can thus
function, but still causes increased viscosity. Parasitical infection by the
falciparum malarial organism of such a
heterozygote cell increases sickling because the parasite uses oxygen, thus
further increasing viscosity. It is then
believed that the infected, much sick-
led cell is removed by the body by pha-
gocytosis. Thus the malarial parasite
sows the seed of its own destruction.

It seems to me that the conclusion
that all one has in this case is a general
molecular mechanism and a belief
in the principle of causality is unwarranted. Knowledge of the general
molecular situation and of the specific
molecular peculiarities of the sickle-
cell case points in a fairly unambiguous way to the biological genetical
consequences—as the logical empiricist
suggests they should. In short, given
that this example is but one of many
that could be cited, whilst I agree
entirely with Hull that an actual reduction might be a lot more complex
than some of us have supposed—and
much credit is due to Hull for under-
lining this fact—I do not find troublesome his doubts about the possibility
of a reduction even in principle.

In his most recent writings on this
subject, Hull has a variant on the argument being considered. He suggests
that even if one's assumption of the
principle of causality bore fruit and
one were to link up molecular and
biological mechanisms entirely, one

might still have no reduction. So detailed an analysis may be required, that at best one could link individual instances with individual instances—something falling short of a reduction which is between theories, which (by definition) go beyond the individual to the general. Hull writes:

It may well be true that a particular molecular mechanism of some kind or other can be found for each instance of a particular pattern of Mendelian genetics on a case by case basis, but more than that is required in reduction. Reduction functions relate *kinds* of entities and predicates, not particulars. The problem is to discover natural kinds in the two theories which can be related systematically in reduction functions. (Hull, 1974b, p. 20)

Again I find Hull's objection not to be devastating in theory (although again I think he points to an important practical question). Clearly Hull is right in thinking that we are going to need a terrific number of connexions between molecular mechanisms and Mendelian (or transmission) genetical effects, so many so in fact that one might well question whether the whole enterprise of articulating a fully worked-out reduction is worthwhile. But is the result really going to be that *every* instance of a Mendelian effect will have to be treated separately? Surely, for example, one is going to have similar instances between members of the same species. Take for example the M and N blood groups in man, which in the Mendelian (or transmissions) world are believed a function of the segregation of two alleles. Is Hull really suggesting that every case of

a human homozygous for the M-gene (and thus with blood which works in a particular way in agglutination tests) is going to call for a different molecular mechanism? If he is, then more argument by him is needed. Or take for example the sickle-cell phenomenon just discussed. Is it Hull's claim that geneticists are wrong in dealing with the heterozygotes (and homozygotes) on a collective basis, rather than on an individual by individual basis? If so, then yet more argument is needed—argument which I think Hull might find rather embarrassing for he must show that scientists ought not be behaving as they do, whereas the thrust of much of his criticism of the logical empiricists is that logical empiricists are too ready to prescribe the ideal course of science, when they should be describing its actual course.

In short, it seems clear that when dealing with members of the same species (or sub-groups within a species) generalities can be made between molecular and biological levels. A priori it seems plausible also to suggest that in some cases such generalities might go beyond one species—particularly to members of closely related species. Thus, required generalities do seem available. Moreover, although one cannot deny that the available generalities might be a lot more restricted than most reductionists have supposed, Hull has really given no arguments ruling out general similarities between widely different organisms linking molecular mechanisms and biological effects—general similarities which can be applied to specific cases. Thus take for example the biological phenomenon of epistasis, where the phenotypic effects of alleles at one locus seem to be a function in part of alleles at another locus. In such a situation, although different combinations of alleles $a_1$ and $a_2$ at one locus might cause phenotypes $A_1$ and $A_2$ under normal circumstances, it may be that neither phenotype will occur unless one has at least one instance of allele $b_1$ at another locus. Hull rightly points out that epistasis may have several molecular mechanisms—nevertheless there are some general models which seem applicable to widely differing cases. Thus, for example, the most obvious model supposes that we have a cellular product which is being produced sequentially by enzymes caused by the two sets of alleles.

enzyme $\beta$                    enzyme $\alpha$

substance $p$ —→ substance $q$ —→ substance $r$

Using the above terminology, we might suppose that enzyme $\beta$ is produced when and only when allele $b_1$ is present, that enzyme $\alpha$ is produced when and only when allele $a_1$ is present, that phenotype $A_1$ occurs when and only when substance $r$ is produced, that phenotype $A_2$ occurs when and only when substance $r$ is absent but substance $q$ is present, and that when neither substances $q$ nor $r$ are present a different phenotype (say $A_3$) occurs. This model is known to fit many cases of epistasis. One instance is epistatic expression of hydrocyanic acid in white clover, something which leads to vigorous growth (Burns, 1972, pp. 81–82). Taken alone, one pair of alleles in clover segregate 3:1 for HCN: cyanogenic glucoside (*i.e.* the HCN-producing allele is dominant). However, the segregation of this pair of genes is in

turn controlled epistatically by another pair which segregate 3:1 for the production of cyanogenic glucoside. And it is known moreover that this state of epistasis fits the above model.

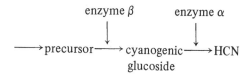

enzyme $\beta$                    enzyme $\alpha$

—→ precursor —→ cyanogenic —→ HCN
                        glucoside

There are many other cases of epistasis known to fit this model, particularly those affecting color. (Strickberger, 1968, gives examples of epistasis involving colour, and indeed, uses the above model to explain them.) Hence, although Hull is no doubt right in doubting that one molecular mechanism will fit all cases of epistasis, it seems wrong to conclude that something completely new must be found for every case. Some generalities do apply.

2. Hull's second criticism of Schaffner's position is that were one to deduce a biological genetics from molecular genetics, the deduced biological genetics would be so different from Mendelian genetics that talk of "reduction" would still be inappropriate. There would be no strong analogy between traditional Mendelian genetics and the corrected biological genetics ("transmission genetics"). "The amount of reconstruction necessary to permit the deduction of transmission genetics from molecular genetics would seem to preclude the existence of any strong analogy between classical Mendelian genetics and reconstructed transmission genetics" (Hull, 1974a, p. 43).

Of course, as Hull rightly points out, part of the problem here is that to date no one has been prepared to give a thorough explication of what might be meant by "strong analogy," and it is clear that if the debate about reduction in genetics is to continue, someone soon must turn to this task. At present one must work by example and rough intuition—Kepler's astronomy and that which can be deduced from Newtonian axioms do seem strongly analogous, Darwin's theory of evolution through natural selection and Richard Owen's Platonic theory of organic origins seem to be miles apart. But even though we have to work with such crude guidelines as these, I would suggest that Hull's conclusion is not well taken.

It is agreed by all that Mendelian genetics and transmission genetics are not identical—if they were, then there would be no argument. Mendelian genetics rolls together the units of function, mutation, and crossing-over, whereas transmission genetics separates these out, allowing mutation of just part of the unit of function and crossing-over within the unit of function. But where else are there essential differences? It seems to me that just about everything else is retained from the old biological genetics in the new biological genetics, and what differences there are do not make themselves much felt in a lot of cases. Take for example the work of someone like Th. Dobzhansky. The third and final edition of his classic *Genetics and the Origin of Species* appeared in 1951, before the work of Watson, Crick, or Benzer. In it he relied on the Mendelian law of segregation, its generalization to large groups (the Hardy-Weinberg law), and so on. The revised and retitled edition of

his work *Genetics of the Evolutionary Process* appeared in 1970. The early chapters discuss, as accepted scientific fact, the major findings of molecular biology, for instance the Watson-Crick model of DNA. But then what do we find? All the old favorites like the Hardy-Weinberg law make their reappearance and play just as great a role in Dobzhansky's theorizing as they ever did. Hence, not much change between the old and new seems to have occurred.

Of course, one might argue that Dobzhansky is holding to and working with two contradictory theories—traditional Mendelian genetics and modern molecular genetics. But I see no reason why one should assume this. He identifies explicitly his unit of function (which he normally calls the "gene") with the unit of function of transmission genetics, the cistron. But, having made this change, he can immediately make use of just about all his old theory, because for the kinds of problems he tackles the modern sophisticated analysis (replacing the old Mendelian gene) is not needed. Basically, Dobzhansky is untroubled by crossing-over within the unit of function, because he works at a rather cruder level analysis. Hence, it seems to me true to say that the modern *biological* geneticist incorporates into his new biological genetics much of the old Mendelian genetics.

To conclude this section let me look briefly at an argument Hull puts forward in support of his position that there can be no strong analogy between traditional Mendelian genetics and the corrected biological genetics derivable from molecular genetics (what I have been calling 'transmission genetics'). Hull's argument is that any biological genetics derived from molecular genetics must reflect differences at the molecular level, even though there may be no such differences at the biological level. But if there are no differences at the biological level, they will not presently be mentioned in biological genetics (either Mendelian or transmission). Hence, argues Hull, a great deal of correction in this respect will be required before Mendelian genetics can be converted to a derivable biological genetics—therefore, strong analogy seems out of the question. Hull writes:

> Transmission geneticists claim no interest in the variety of biochemical mechanisms that eventuate in the characters whose transmission they follow, unless these differences are reflected in different Mendelian ratios. If not, this additional information about biochemical pathways is of no use to them. However, if reduction functions are to be established that associate the terms of transmission and molecular genetics, the important biochemical differences must be read back into transmission genetics. If the same molecular structure is produced in two different ways, this difference must be reflected in the relevant reduction functions and Mendelian genetics changed accordingly. (Hull, 1974a, p. 43)

Put simply, this argument seems fallacious. As Hull himself points out, a many-one relationship from the molecular to the biological does not prevent a reduction. But this kind of many-one relationship is what is being supposed here. Suppose in a species a molecular gene product $\alpha$ which leads to phenotype $A$ can be produced in two ways, as follows. ($a$ is the biological gene, $G$ the molecular gene, $\alpha$, $\beta$, $\gamma$, $\delta$ are molecular cell products, the alleles $B_1$, $B_2$, $C_1$, $C_2$ are producing enzymes driving the changes, alleles at $B$ and $C$ are supposed always linked, perhaps through suppression of crossing-over). We have here the kind of situation envisioned by Hull—biological theory could be quite ignorant of the differ-

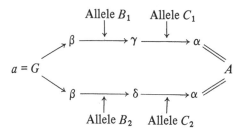

ent pathways—but a reduction is possible. From, $G \rightarrow \beta$, $\beta \rightarrow \gamma$, $\beta \rightarrow \delta$, $\gamma \rightarrow \alpha$, $\delta \rightarrow \alpha$, $a = G$, $\alpha = A$, we can get $a \rightarrow A$. We can also get $a \rightarrow A$ if the two pathways lead to slightly different molecular end products ($\alpha_1$ and $\alpha_2$), if, as supposed, the molecular difference makes no difference at the gross phenotypic level. There is certainly no need, as Hull suggests, to reconstruct our biological theory (say from $A$ to $A_1$ and $A_2$) to reflect the differences at the molecular level.

3. Hull's third criticism is probably his most important. This is that there is something really rather irrelevant about the logical empiricist thesis of reduction. Somehow it directs us away from an understanding of what is happening in genetics, rather than towards it. Hull writes:

To my knowledge, no biologist is currently engaged in the attempt to reconstruct Mendelian and molecular genetics so that Mendelian genetics can be derived from molecular genetics. What is more, I can think of no reason to encourage a biologist to do so. This state of affairs strikes me as strange. . . . Either this reduction is a peculiar case or else whatever it is that makes the pre-analytic notion of reduction seem important has dropped out of the logical empiricist analysis of reduction. (Hull, 1973, p. 622).

Somewhat paradoxically Schaffner (and Ruse) agree fully with Hull about the actual place reduction seems to play in geneticists' research programmes. Schaffner concedes that the really exciting advances in molecular genetics have not come through the attempt to spell out a reduction, and he writes:

I wish to propose that the aim of reductionism is *peripheral* to molecular biology, and that an attempt to construe the development of molecular biology as exemplifying a research program or set of research programs whose conscious and constant *intent* is reductionism is both inappropriate and historically misleading. (Schaffner, 1974a, p. 111)

In a similar vein, Ruse thought it prudent to add the following footnote to his most recent discussion of reduction: "It is perhaps important to emphasize that no actual deduction from a purely molecular theory to a purely biological theory seems yet in existence. Indeed, reading the works of biologists one gets the feeling that the whole question of reduction is more of a philosopher's problem than a scientist's" (Ruse, 1973, p. 207n). What more damning comment could one make than that! Perhaps Hull is right and we should find better things to do with our time.

In defence of the logical empiricist, I offer the following two arguments. First, although some of us may secretly have a hope that our most recent attack on the covering-law model will some day win us a Nobel prize in science, we are at present trying to do philosophy not science. Our job is not to make brilliant new scientific discoveries, but to analyse and understand the way science works. In John Locke's great words "it is ambition enough to be employed as an under-laborer in clearing the ground a little, and removing some of the rubbish that lies in the way to knowledge" (Locke, 1959, p. 14). Hence, that scientists are not pushing full-speed ahead on a reduction does not mean that philosophers ought not use the formal, idealized analyses of logical empiricism to see if they can throw light on the relationship between molecular and biological genetics. Scientists have their job to do, we have ours, and ours is certainly not to follow slavishly in the footsteps of science—nor is it necessarily a mark that we have done our job badly if scientists do not immediately rush to put flesh on our formal analyses.

Since this point I am trying to make is rather important, it may perhaps be worthwhile to consider for a moment a closely related example. No one could deny that in its present form evolutionary theory taken as a whole is not a rigorously formulated hypothetico-deductive system. But it does not necessarily follow that the hypothetico-deductive ideal is irrelevant to evolutionary studies— that the philosopher ought not use the hypothetico-deductive model in his attempt to understand what the evolutionist is doing. Certain parts of evolutionary science, particularly those concerned with the spread of genes in populations, do approximate to a hypothetico-deductive theory and the philosopher can therefore use his model directly to throw light on these parts and on how these parts relate to other parts of evolutionary studies (like organic geographical distribution). Moreover, those places where the theory fails to fit the model can guide the philosopher to an understanding of the problems facing the evolutionist—essential data irretrievably lost, vast timespans, and so on. And although it is certainly true that logical empiricists would (or at least should) feel most uncomfortable were there no parts of evolutionary theory in any sense hypothetico-deductive or were there no sense in which evolutionary studies had progressed towards a greater manifestation of the hypothetico-deductive ideal, there is nothing in their espousal of this ideal which insists that scientists must put this ideal before all else. It is fully recognized that things like missing data might make a significant complete hypothetico-deductive evolutionary theory a practical impossibility, and that scientists might feel the technical details of filling out such a theory with full rigour to be rather boring and not something

which would lead to dramatic new insights—as in mathematical proofs we often see how a problem can be solved but might not wish to fill in every last step.

An exactly analogous situation seems to me to prevail with respect to theory reduction and ·genetics. Because of their understanding of DNA, the genetic code, and so on, biologists can see in broad outline how transmission genetics follows from molecular genetics (in the manner suggested by logical empiricists), and as we have seen in the sickle-cell case, when it is in their interests they can often trace through the details of some particular case with precision. But for various reasons, particularly because it does not seem very exciting and would seem only to involve filling out details of a general picture already broadly understood, biologists are not trying to set up a fully articulated deduction of transmission genetics from molecular genetics. The logical empiricist, however, can and does recognize this fact. It is the broad outline which interests him—the general relationship between molecular and non-molecular genetics—and the way in which particular details are filled in when there is a specific need or interest. Hence, I would suggest that Hull is wrong when he writes that:

The crucial observation is that no geneticists to my knowledge are attempting to derive the principles of transmission genetics from those of molecular genetics. But according to the logical empiricist analysis of reduction, this is precisely what they should be doing. (Hull, 1974a, p. 44)

But, in a sense this is all rather tangential. Hull's main claim is not so much that the logical empiricist thesis on reduction is scientifically irrelevant (although he thinks it is), but that it is philosophically harmful. It "hindered" his understanding of the true situation in genetics, rather than facilitated it (Hull, 1974a, p. 44). It is here that I want to make my second argument, for I think Hull is wrong in claiming this. I think rather that the logical empiricist analysis does help us towards an understanding of what is really happening in genetics.

In order to defend the logical empiricist analysis, let me first ask of Hull the question—What, if the logical empiricist account is incorrect, is indeed the true situation? At times Hull seems to have no answer to this question, and seems almost in despair to give up hope of finding such an analysis. Rather, somewhat wearily, even he seems prepared to go along with the logical empiricist account.

If my estimation of the situation in genetics turns out to be accurate, then even the latest versions of the logical empiricist analysis of reduction are inadequate and must be improved. This conclusion, however, should not be taken as being too damning of the logical empiricist analysis of reduction. In spite of its shortcomings, it is currently the best analysis which we have of reduction. In fact, it is the only analysis which we have. (Hull, 1973, p. 634)

However, at other times he is more positive in his opposition. The logical empiricist sees the relationship between the two theories as one of logical, deductive connexion. The one theory, in some sense, 'contains' the other. Hull suggests however that the two theories belong in some important way to different worlds—at least, to incommensurable ways of viewing this world. He writes:

Most contemporary geneticists know both theories. They can operate successfully within the conceptual framework of each and even leap nimbly back and forth between the two disciplines, but they cannot specify how they accomplish this feat of conceptual gymnastics. Whatever connections there might be, they are subliminal. In a word, those geneticists who work both in Mendelian and molecular genetics are schizophrenic. The transitions which they make from one conceptual schema to the other are not so much inferences as *gestalt* shifts . . . (Hull, 1973, p. 626)

We have then two sharply different analyses of the situation in genetics. One sees the reduced theory $T_1$ in some sense contained in the reducing theory $T_2$, the other sees $T_1$ and $T_2$ as being logically quite distinct. To use N. R. Hanson's example of a drawing which looks, in one way, like a rabbit, and, in another way, like a duck, $T_1$ is a rabbit and $T_2$ is a duck. This is a somewhat extreme position I am ascribing to Hull, and it should in fairness be added that he does qualify the passage just quoted by saying "To be sure, these observations are psychological in nature, and reduction as set out by philosophers of science is a logical

|             | Two randomly chosen individuals |
|-------------|---------------------------------|

Classical    $\underline{+++m+\cdots++}$     $\underline{+++++\cdots m+}$

hypothesis   $+++++\cdots m+$     $+++++\cdots++$

(+ = wild-type allele; $m$ = deleterious mutant)

Balance      $\underline{A_3 B_2 C_2 D E_5 \cdots Z_2}$     $\underline{A_2 B_4 C_1 D E_2 \cdots Z_1}$

hypothesis   $A_1 B_7 C_2 D E_2 \cdots Z_3$     $A_3 B_5 C_2 D E_3 \cdots Z_1$

($A_1, A_2, \ldots$ = different alleles)

relation between rational reconstructions of scientific theories..." On the other hand he adds "but these psychological facts provide indirect evidence for the position I am about to urge concerning the logical relation." What is clear is that even if Hull would not accept all the implications that Hanson might draw from the situation, he believes that from a formal viewpoint molecular genetics is a new theory replacing Mendelian genetics, not one absorbing Mendelian genetics.

> If the logical empiricist analysis of reduction is correct, then Mendelian genetics cannot be reduced to molecular genetics. The long-awaited reduction of a biological theory to physics and chemistry turns out not to be a case of "reduction" after all, but an example of replacement. (Hull, 1974a, p. 44)

Hull is certainly right in pointing to the way in which geneticists switch blithely to and fro between the molecular and biological levels, without giving too much thought to the relationship between them. But the extreme position to which he pushes this observation clouds rather than clears the view to correct understanding. Consider for a moment a situation in genetics where geneticists do work at both molecular and biological levels, namely over the problem of the amount of genetically caused variation in natural populations.[7] There are, or rather were, two hypotheses about variation when the problem was considered entirely at the biological level. On the one hand, followers of H. J. Muller argued for the so-called "classical" hypothesis, namely that most organisms in a population are genetically similar, with just the occasional mutant allele, usually recessive, usually deleterious, deviating from the "wild-type." On the other hand, followers of Th. Dobzhansky supported the "balance" hypothesis—they saw selection primarily maintaining genetic variability in a population, for example through heterosis (this refers to a situation like the sickle-cell case, where the heterozygote for two alleles is fitter than either homozygote—at equilibrium both alleles are retained in a balance). Hence, for balance hypothesis supporters there is in an important sense no such thing as a wild-type—almost every genotype in a population is different. Diagramatically we can represent the differences between the two populations (see above).

7. Full details can be found in Lewontin 1974.

Now, at the biological level it is all but impossible to decide between these hypotheses. Balance hypothesis supporters argued that the phenotypic differences caused by allelic changes would normally be very slight (although such differences could have significant evolutionary implications). But the measurement of slight differences due to genetic change is incredibly difficult, mainly because environmental changes can also cause phenotypic differences and one can never be absolutely sure that one is keeping the environment constant (remember, the environment of an individual includes the other members of its population).

What geneticists have done in the past ten years, however, is switch the debate from the biological to the molecular level. Functional strips of DNA lead, via RNA, to polypeptide chains—strings of linked amino acids. Changes in the DNA lead (taking note of redundancies) to changes in the polypetide chains. Using a technique known as 'gel electrophoresis' (which is based on the fact that some amino acids have different electrostatic charges and that hence changes in polypeptide chains lead to changes in charge) geneticists can detect changes in polypeptide chains, which, in turn, they interpret as changes in the DNA structure. Both classical and balance hypothesis supporters think that the findings about polypeptide chains are relevant to their debate. In particular it is found that polypeptide chains (caused by DNA strips at the same locus) do vary greatly. Balance hypothesis supporters take this to be confirmation of their position, namely that selection maintains genetic variability in populations. Classical sup-

porters, whilst they cannot deny this variation, argue that the variation is non-adaptive, that it is due to drift not selection, and they can give both theoretical and empirical arguments to support their position.

Clearly the resort to the molecular level has not brought debate to an end—if anything it rages more fiercely than ever before. But what I want to ask is what is the most reasonable way to interpret geneticists' resort to the molecular level at all? Why did geneticists get so excited about gel electrophoresis and so on? On Hull's interpretation of the situation it is difficult to see why geneticists should have turned (in the instance we are considering) to the molecular level, thinking (as they obviously did) that the molecular level was going to break down some barriers. Given Hull's analysis, geneticists had a debate at the biological level. It was not getting untangled. So they gave up, made a gestalt switch, and turned to a different problem, a problem at the molecular level. Two levels, two problems, and never the twain shall meet. On the logical empiricist analysis however, all is readily explicable, including geneticists' enthusiasm for the molecular level. Mendelian genes (or cistrons) can be identified with strips of DNA, DNA causes polypeptide chains, and these either lead to or can be identified with gross phenotypic characteristics. Hence, inasmuch as one gathers information about genetically caused variation amongst polypeptide chains, one is throwing light on genetically caused variation at the gross phenotypic level—the matter at the heart of the debate in the first place.

Thus what we find in the case of

balance hypothesis supporters is the belief that much of the difference in the polypeptide chains reflects as difference at the gross phenotypic level—difference which they believe must be maintained by selection. And in support of their case they point to several facts which they believe explicable only on their hypothesis (at least, certainly not on the classical hypothesis). One thing in particular they believe proves their case, namely that polypeptide ratios between closely related groups are often very similar, pointing to a systematic cause, which they believe can only be selection and is certainly not drift.

But the position of classical hypothesis supporters is also readily explicable given the logical empiricist analysis. They cannot deny the differences in polypeptide chains; however, they argue that either different chains lead to the same gross phenotypic effects, or that different chains lead to different effects but that these effects are selectively neutral between themselves. In other words, they argue either for a many-one relationship (from molecular to biological) or for a one-one relationship, either of which, as we have seen, is permissible in a reduction. But either way, we see that they do believe that polypeptides have direct implications for gross phenotypic differences, which is what the logical empiricist account leads us to expect, and indeed demands.

In short, I suggest that the logical empiricist account throws much light on the current debate in population genetics, something which Hull's alternative does not. Thus I suggest that the logical empiricist account of reduction as applied to genetics is not as irrelevant or misleading as Hull claims.

## REFERENCES

Burns, G. W., 1972, *The Science of Genetics,* 2nd ed., New York, Macmillan.

Dobzhansky, Th., 1951, *Genetics and the Origin of Species,* 3rd ed., 1951, New York, Columbia University Press.

——, 1970, *Genetics of the Evolutionary Process,* New York, Columbia.

Hull, D. L., 1972, Reduction in genetics—Biology or Philosophy?, *Phil. Sci.,* 39: 491–99.

——, 1973, Reduction in genetics—doing the impossible. In P. Suppes *et al.* (eds.), *Logic, Methodology and Philosophy of Science,* IV: 619–35.

——, 1974a, *Philosophy of Biological Science,* Englewood Cliffs, Prentice-Hall.

——, 1974b, 'Informal Aspects of Theory Reduction.' Read at P.S.A. conference 1974.

Lewontin, R. C., 1974, *The Genetic Basis of Evolutionary Change,* New York, Columbia University Press.

Locke, J., 1959, A. C. Fraser (ed.), *An Essay Concerning Human Understanding,* New York, Dover.

Luzzatto, L., Nwachuku-Jarrett, E.S., and Reddy, S., 1970, Increased sickling of parasitised erythrocytes as mechanism of resistance against malaria in the sickle-cell trait. *Lancet,* I: 319–22.

Nagel, E., 1961, *The Structure of Science,* London, Routledge and Kegan Paul.

Ruse, M., 1971, Reduction, replacement, and molecular biology. *Dialectica,* 25: 39–72.

——, 1973, *The Philosophy of Biology,* London, Hutchinson University Library.

Schaffner, K. F., 1967, Approaches to reduction. *Phil. Sci.,* 34: 137–47.

——, 1969, The Watson-Crick Model and Reductionism, *Brit. J. Phil. Sci.,* 20: 325–48.

——, 1974a, The peripherality of reductionism in the development of molecular biology. *J. Hist. Biol.*, 7: 111–39.

——, 1974b, 'Reductionism in Biology: Prospects and Problems.' Read at P.S.A. conference 1974.

Strickberger, M. W., 1968, *Genetics,* New York, Macmillan.

# 25

# Informal Aspects of Theory Reduction

*ᵕᶜᵕᶜᵕᶜᵕᶜᵕᶜᵕᶜᵕᶜᵕᶜᵕᶜᵕᶜᵕᶜᵕᶜᵕᶜᵕᶜᵕᶜᵕᶜᵕᶜᵕᶜᵕ*

## DAVID L. HULL

THE ISSUES WHICH SEPARATE the members of this symposium* concern the nature of reduction and scientific theories. However, these issues involve primarily informal aspects of theory reduction. But more than this, they concern the nature of philosophy of science itself.

Kenneth Schaffner (1967) has set out what he terms a general reduction paradigm, a development of earlier efforts by such philosophers as Ernest Nagel, J. H. Woodger, and Carl Hempel. This model, according to Schaffner, "represents an ideal standard for *accomplished* reductions, and does not characterize the research programmes of molecular biologists." Following the lead of his predecessors, Schaffner does not intend for his model of theory reduction to characterize the ongoing process of science (Wimsatt's rational₁) but an abstract, formal relation between atemporal rational reconstructions of scientific theories (Wimsatt's rational₂).

Numerous objections have been raised to this way of doing philosophy of science. In this paper I will deal with three. First, very little has been said about the process of rational reconstruction. How does one go from scientific theories as scientists set them out (what might be termed "raw science") to the rational reconstructions of philosophers? By what criteria are alternative analyses to be judged? Second, there is little agreement about how explicit and complete these analyses have to be. Must philosophers actually axiomatize the relevant theories, set out the appropriate reduction functions, and then carry out the derivation, or are vague gestures about in principle possibilities good enough? Finally, even if a Nagel-type analysis of theory reduction adequately captures one aspect of theory reduction, mightn't there be more to it than that? For example, one might wish to provide an analysis of reduction as a temporal process. Why must philosophers studiously exclude all temporal considerations from their analyses?

*That is, the authors of selections 23–26 [Editor].

For all the importance which philosophers of science have placed on the process of extracting a scientific theory from the scientific literature and reconstructing it rationally, they have said precious little about how to do it. At times they seem to act as if scientific theories exist right out there in nature, as if there were some one thing called a "Mendelian theory of inheritance" and some one thing called a "molecular theory of inheritance." But such is clearly not the case. At any one time in the development of a particular theory, numerous partially incompatible versions of this theory can be found in the primary literature of science. When these clusters of theories are traced through time, the multiplicity only increases. As David M. Knight (1967, p. 2) has observed with respect to atomic theory, "There was no one classical, received atomic theory but rather a number of theories overlapping in their explanatory ranges." At the turn of the century, when Mendelian genetics was rediscovered, a half dozen geneticists set out differing versions of it. During the first ten years of the development of Mendelian genetics, these versions changed radically. About the only important substantive claim that remained unchanged was the "law of segregation" or, as it was also called, the principle of the purity of the gametes. According to this principle, genes do not contaminate each other in the heterozygote. During the next half century, additional changes were made in Mendelian genetics, including the discovery that genes as physical entities *do* contaminate each other. As both Ruse and Schaffner have pointed out, crossover occurs within genes as well

as between them, resulting in the physical mixing of alleles.

Traditional logical empiricist philosophers of science would surely reply, and several have, that all the preceding is irrelevant. An historian of science might be interested in this diversity but not a philosopher. The subject matter of philosophy of science is not raw science but the philosopher's own rational reconstructions of science. There are two problems with the preceding response. First, philosophers of science must get their rational reconstructions of science from somewhere. If it is not from the multifarious primary literature of science, then where? I suspect the answer to this question is, from college textbooks. Textbooks present just the sort of ahistorical, after the fact simplifications required by philosophers of science. Second, if the subject matter of philosophy of science is the rational reconstructions of philosophers of science and these reconstructions are produced in accordance with their own analyses of science, how then are these analyses to be evaluated? For example, the analyses of theory and theory reduction produced by the logical empiricists are interdependent. Theories are exactly the sorts of things that can be reduced, and reduction is exactly the sort of relation which can exist between theories. Obviously, reconstructing Mendelian and molecular genetics as theories on the logical empiricist analysis of "theory" stacks the deck in favor of reduction. The issue is whether or not the bias introduced at this step is sufficient to preclude independent evaluation of the resulting claims of a successful reduction.

Philosophers have exerted considerable effort on investigating the relationship of scientific theories, laws, and observation statements to the empirical world. "Snow is white" is true, if and only if snow is white. But very little time has been spent discussing the relation of philosophical analyses of science to science. Karl Popper's doctrine of falsificationism is true, if and only if, . . . ? Because so little has been said about this general issue, the participants in this symposium have been placed in the uncomfortable position of not having any generally acknowledged criteria to use in evaluating the particular example of reduction under investigation. The subject matter of philosophy of science may well be the philosopher's own rational reconstructions, but some external constraints must be placed on his freedom to reconstruct science, or else philosophy of science runs the risk of being as much a self-justifying activity as theology. The most likely candidate for these external constraints is the actual practice and productions of scientists. For this purpose, textbook expositions will not do.

Scientists do not produce theories in precisely the form advocated by logical empiricist philosophers. These philosophers reply that they should. However, if too many of the things which scientists produce as scientific theories depart too radically from the philosopher's notion of a scientific theory, and if scientists insist on producing their deviant theories even after they are made aware of philosophers' views on the subject, then philosophers just might be led to reconsider their original analyses. But how true to raw science must these rational reconstructions be? Only Imre Lakatos (1971, p. 107) has had the courage (or foolhardiness) to claim that the connection is as free as one might wish to make it. He authorizes philosophers to reconstruct history of science the way it should have happened and to "indicate *in the footnotes* how actual history 'misbehaved' in the light of its rational reconstruction."[1] Philosophy of science may be to some extent normative, but until the source of these normative powers are more fully understood, it would be wise for philosophers of science to exercise some restraint lest we find ourselves once again arguing that space is necessarily Euclidean and species necessarily immutable.

Ruse objects to my claim that the logical empiricist analysis of theory and theory reduction have been philosophically harmful, that they have hindered rather than facilitated understanding of the true situation in genetics (Hull, 1972, 1973, 1974). He himself cannot see what harm it has done. On the logical empiricist analysis, scientific theories are in-

1. Following Lakatos's suggestion, I have stated in the text what Lakatos should have said, and now, in a note, I indicate how he misbehaved. Lakatos distinguishes between internal history (history as it should have happened given a particular set of philosophical views about science) and external history (history as it actually happened regardless of any such philosophical views). The more internal history coincides with external history, Lakatos argues, the better the philosophy of science which gave rise to the internal history. However, Lakatos never tells us how one can write a factually accurate external history independent of any views about the nature of science.

dividuated on the basis of substantive content. Two theories are two different theories because they differ with respect to one or more substantive claims. From this perspective, the molecular theory of Watson and Crick is different from the biological theory of Mendel, but Watson and Crick's theory is also different from the molecular theories of Jacob and Monod, Lederberg, and Kornberg as is Mendel's theory from those of de Vries, Correns, Morgan, Dobzhansky and Muller. But all these theories are not equally different. One would expect them to form two clusters, one roughly molecular, the other Mendelian. But they do not if these clusters are to be formed on the basis of the substantive claims made in these theories. As Wimsatt has pointed out, theories evolve in a process which he terms "successional reduction." According to Wimsatt, successional reduction, like Schaffner's criterion of strong analogy, is an intransitive similarity relation. An early version of a theory can be similar to a later version of that same theory (that is why they are versions of the *same* theory) and this version similar to an even later version of that theory, without the earliest and latest versions being similar at all. They are all versions of the same theory because of continuous and unitary development.

I tend to agree with Wimsatt's analysis of successional reduction as far as it goes, but similarity in substantive content, even interpreted as a serial relation, is not sufficient for individuating scientific theories. The substantive content of science changes too rapidly and sporadically for that. Instead the continuing commitment on the part of scientists to certain procedures, goals, problems, and metaphysical presuppositions supply most of the continuity to be found in science. Early Mendelian geneticists abandoned or modified nearly all of the basic tenets of Mendelian genetics, but they were still Mendelian geneticists improving Mendelain theory, not opponents producing competing theories. Similar stories can be told for other episodes in the history of science. Jacques Loeb's disciples rejected all of Loeb's substantive claims about animal tropisms, yet were still disciples, not rebels. They were merely "developing" his mechanistic conception of life (Fleming, 1964). T. H. Huxley was a Darwinian, though he thought that evolution was usually saltative and that natural selection might well play a subsidiary role in evolution.

Previously I have characterized Mendelian genetics as a theory of hereditary transmission in which the inheritance of various phenotypic traits is followed from generation to generation, ratios discerned in the distribution of these traits, and the appropriate number and kind of genes postulated to account for the distributions. Although Mendelian geneticists assumed that some sort of causal chains connected genes with the resultant traits, Mendelian genetics was not a developmental theory. So far no one has objected to this characterization, yet the substantive content of Mendelian genetics plays almost no role in it. Ruse cannot see how the logical empiricist analysis of science could impede an adequate understanding of science. Here are but two ways. It ignores the temporal dimension to science and directs attention away from the chief means by which various stages of a scientific theory can be integrated to form a single theory.[2]

The final criticism I wish to make of the philosophical literature on reduction in genetics is that it is too sketchy and programmatic. None of the philosophers who have discussed the problem of reduction in genetics have actually axiomatized the two theories, set out the necessary reduction functions, and performed the derivation. Everyone comments that Mendelian genes are to be identified with some segment of DNA, possibly those segments delineated by the *cis-trans* test, and predicate terms of Mendelian genetics, like dominant, with various kinds of molecular mechanisms such as the production of an active enzyme. To date, this has been the sum total of all the efforts of philosophers to show how Nagel's formal conditions of theory reduction can be fulfilled in the case of genetics. Perhaps I expect too much of the philosophical notion of reduction. Perhaps my standards are unrealistically high. But on any standards, these efforts do not seem impressive. The problem is whether or not there is any point to fulfilling the requirements of the logical empiricists analysis of theory reduction. If not, why not? If so, who should be doing it?

Ruse says that scientists have their jobs to do, philosophers theirs, and among neither is the task of setting out reductions in any sort of detail. Schaffner subscribes to what he has termed a peripherality thesis to the effect that reductionism is not a primary aim of molecular biology. Even so, it "is *resulting in* at least *partial* reduction of biology to physics and chemistry." I find these views puzzling. If crucial experiments are as important in science as philosophers once claimed they were, then one has every right to expect scientists to be performing them. If not, then a philosopher is within his rights to urge them to do so. Similar observations seem to be warranted for other philosophical theses about science. To claim simultaneously that falsification is the chief distinguishing feature of science and that no scientist need ever attempt to falsify a scientific law or theory seems to me to be more than a little paradoxical.

In the middle of the nineteenth century, physicists such as Bernoulli, Joule, and Kronig suggested that the behavior of gases could be accounted for by considering the motion of their constituent molecules. Was thermodynamics thereby reduced to statistical mechanics? Later Clausius, Maxwell, Boltzman, and others developed a detailed theory of gases. With a few

---

2. Numerous philosophers have argued for an evolutionary interpretation of science, most recently Stephen Toulmin, *Human Understanding* (Princeton University Press, 1972). On this analysis, scientific theories are historical entities developing continuously in time; see my "Central Subjects and Historical Narratives," *History and Theory*, 1975, 14: 253–274. With respect to the particular example of reduction at issue, Michael Simon, *The Matter of Life* (Yale University Press, 1971) assumes an evolutionary viewpoint. Even though radical changes took place in the transition from Mendelian to molecular genetics, Simon argues that the process is one of successive correction and refinement, not replacement, because none of the changes took place abruptly.

modifications, several thermodynamic laws were actually derived from these statistical theories.[3] Boyle's law was explained in terms of the gas molecules striking the walls of the container, balancing the external pressure applied to the gas. If the volume of the gas is cut in half, each of the molecules strikes the walls twice as often, thereby doubling the pressure. In a similar manner the law of Charles and Gay-Lussac can also be explained. If the absolute temperature of a gas is doubled, the speed of the molecules is increased by a factor $\sqrt{2}$. This causes the molecules to make $\sqrt{2}$ times as many collisions as before, and each collision is increased in force by $\sqrt{2}$, so that the pressure itself is doubled by doubling the absolute temperature. Avogadro's law is also explained by the fact that the average kinetic energy is the same at a given temperature for all gases.

Where are comparable derivations for genetics? At the turn of the century, Garrod suggested that either genes were enzymes or else controlled enzymatic action. Was Mendelian genetics thereby reduced to molecular genetics? By the middle of the twentieth century, we knew that genes were made primarily out of DNA and proteins out of twenty or so amino acids. Was it then that Mendelian genetics was reduced to molecular genetics? I suggest not. What is needed is the derivation of the basic principles of Mendelian genetics from molecular biology. Ruse (1973, p. 203) concludes that the derivation of Mendel's

first law from molecular premises "looks very promising." According to this law, alleles do not contaminate each other while residing on homologous chromosomes in the heterozygote. If a Mendelian gene is defined as a "piece of DNA which is just enough to serve as the cause of a cellular product (i.e. a polypeptide chain)," as Ruse suggests, then Mendel's first law is false, because crossover takes place as commonly within such functional units as between them. Only single nucleotides remain pure in a physical sense. One must correct Mendelian genetics to accommodate these findings. As Ruse (1973, p. 206) says:

> Now, if we consider not the older classical Mendelian genetics but this new yet still entirely biological, fine-structure genetics, then the task of showing that one can have a Nagelian-type reductive relationship between molecular and nonmolecular genetics seems once again to be within the realm of possiblity. One links the (biological) cistron with the (nonbiological) molecular gene and the (biological) muton and the (biological) recon with a very few non-biological nucleotide pairs of the DNA molecule.

Ruse claims that the Mendelian gene is the basic unit of biological function and that these units of function at the level of DNA are distinguished by the *cis-trans* test. Yet Mendel's first law must be "corrected" to apply to the

---

3. At least, so the story goes. L. Sklar (1976) argues for a position with respect to the reduction of classical thermodynamics to statistical mechanics similar to the one that I am urging with respect to genetics.

recon, the smallest unit of recombination. I have several objections to this suggestion. First, like any operationally defined unit, the cistron cannot fulfill the functions of a theoretical entity. There *is,* for example, intracistronic complementation because two or more cistrons quite commonly cooperate to produce single polypeptide chains. The cistron is hardly *the* unit of biological function. Second, on Ruse's interpretation, Mendel's first law becomes a tautology. If a recon is defined as the smallest unit of recombination and Mendel's first law is reformulated to state that recons do not contaminate each other at meiosis by recombination, then Mendel's first law is converted from a false synthetic claim to one that is true by definition. Of course, given Mendel's original intention, his first law is not all that false. Mendel was concerned with explaining how recessive traits could reemerge from the heterozygote in as pure a form as when they went in. Finally, we still are left with no justification for calling this new, corrected biological theory "Mendelian genetics." *And* I'd like to see a few sample derivations.

Who is going to do all this? Schaffner seems to think that molecular biologists are laying the groundwork for some such derivation, but only as a peripheral part of their endeavors. Ruse admits that scientists are not rushing to flesh out the formal analyses of theory reduction presented by philosophers. I do not see them even strolling in that general direction. Perhaps, then, philosophers might attempt to flesh out their own analyses. By this, I do not mean that they should proceed to carry out empirical investigations but that on the basis of the primary scientific literature, they should attempt to reconstruct the two theories, set out representative reduction functions, and perform representative derivations—just to show that it can be done. In the past, some philosophers of science *have* occupied themselves with tasks such as these and without any intent or hope of winning a Nobel prize. Woodger (1937), for example, actually tried to axiomatize part of Mendelian genetics. (Unfortunately, his axiomatization is extremely deficient; see Kyburg, 1968.) In doing so, such formalistically-inclined philosophers are performing one of the traditional tasks of philosophy of science.

Given the turn which the controversy over reduction in genetics has taken recently, a formal treatment would seem to be more important than ever. Unless one goes off in a completely new direction, as Wimsatt suggests, the current dispute can be resolved only if the reductionist thesis is presented in a more coherent and unified manner than it has been in the past. Piecemeal observations are no longer good enough. Everyone now seems to agree that Mendelian genetics is not just a special case of molecular genetics. Both theories must be modified if reduction in any formal sense is to take place. How similar must the modified and unmodified versions of these two theories be in order to term the relation beween the modified versions "reduction"?

In his early writings, Ruse (1971) found the differences between Mendelian and molecular genetics too extensive to permit reduction in any strict, formal sense. His current intuitions are that the differences are not all that great. From the first, Schaffner has felt that a strong analogy existed between corrected and uncorrected Mendelian genetics, even though no one had presented these theories explicitly or had explicated the notion of strong analogy. My intuitive impression continues to be that the differences between the corrected and uncorrected versions of these theories are too numerous and too fundamental to consider the relationship between the two corrected theories reduction in the formal sense of the term. Pre-analytically, the relation between Mendelian and molecular genetics is a paradigm case of theory reduction, but from the point of view of the logical empiricist analysis of theory reduction, it looks more like replacement. However, I do not see how our continuing to trade intuitions is going to get us anywhere. If this dispute is to be resolved, the two theories will have to be set out in greater detail and the necessary changes specified with greater precision than they have been in the past.

The absence of any explicit formulation of the two theories also serves to blunt a second objection which has been raised to the straightforward application of the logical empiricist analysis of theory reduction to the situation in genetics. Schaffner and I do not disagree about the discrepancies which exist between historically accurate rational reconstructions of Mendelian and molecular genetics. Our major point of contention is how much slippage exists between the two and where to put it. Should it be reflected in the reduction functions, incorporated into the theories, or buried in the boundary conditions, auxiliary hypotheses, background knowledge, and the like? Until a reasonably complete and explicit analysis of Mendelian and molecular genetics has been presented, including a specification of which elements are to be put into each of these categories, any objection can be evaded by shunting it somewhere else. If a reduction function becomes too complicated, then get rid of the complexity by some vague reference to differences in the environment. I am not claiming that such evasive maneuvers are inherent in the position currently espoused by Schaffner and Ruse, only that in the absence of a more formal analysis, they are all but impossible to avoid.[4]

Before turning to issues more

---

4. Schaffner at least should not complain of my request for a more explicit and complete explication of the two theories and the necessary reduction functions, because he raised a similar objection to an argument set out by Bentley Glass (The relation of the physical sciences to biology—Indeterminacy and causality, in B. Baumrin, ed., *Philosophy of Science: The Delaware Seminar,* Interscience, New York, 1963). In this paper, Glass argued that the statistical laws found for phenomena at one level of organization are not derivable from the statistical laws at a lower level of organization. Schaffner responded that in order to eliminate the possibility of interlevel reduction between statistical laws, "Glass would have to present an appropriate axiomatization of a true probabilistic theory in biology and demonstrate that the identification of biological entities with physiochemical entities and explanation of the biological entities' behavior on the basis of either causal or statistical laws involving physiochemical terms would entail a contradiction" (Antireductionism and molecular biology, *Science,* 157, 1967: 645). If Glass must do all Schaffner claims to substantiate his antireductionist thesis, there is no good reason to expect Schaffner to do less if he is to substantiate his reductionist thesis.

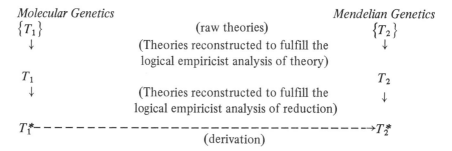

Fig. 1. Diagram of the logical empiricist analysis of theory reduction.

intimately connected to the specific example of reduction under investigation, let me summarize in diagrammatic form the message of the preceding pages. Let $\{T_1\}$ stand for all the various versions of molecular genetics to be found in the primary literature of that discipline and $\{T_2\}$ stand for all the various versions of Mendelian genetics to be found in the literature of that field. Let $T_1$ and $T_2$ stand for historically accurate rational reconstructions of $\{T_1\}$ and $\{T_2\}$ respectively, corrected only to the extent necessary to fulfill the logical empiricist analysis of a scientific theory. I doubt that anyone has ever supposed that $\{T_2\}$ could be derived from $\{T_1\}$, but someone might be tempted to think that $T_2$ is derivable from $T_1$. But as Schaffner (1967, 1969) admits, both theories need further modification, some of it for the sole purpose of fulfilling the logical empiricist analysis of theory reduction. Following the symbols introduced by Schaffner, let $T_1^*$ and $T_2^*$ stand for rational reconstructions of molecular and Mendelian genetics respectively, further modified to take such corrections into account. Thus, when a logical empiricist says that Mendelian genetics is reducible to molecular genetics, all he is claiming

is that a modified, corrected rational reconstruction of Mendelian genetics, $T_2^*$, is derivable from a modified, corrected rational reconstruction of molecular genetics, $T_1^*$.

In Figure 1, the broken arrow represents derivation. If derivation is taken to be deduction, then a highly formal analysis of it is available. The solid arrows represent rational reconstruction, modification, and correction. So far very little has been said about these processes. In the absence of any explicit analysis of these currently informal aspects of theory reduction, the claim that Mendelian genetics can or cannot be reduced to molecular genetics seems to be premature. With enough ingenuity (and lack of intellectual integrity), the phlogiston theory could probably be reduced to statistical mechanics, once the two theories had been sufficiently "modified" and "corrected."

Previously, I (Hull, 1972, 1973, 1974) have argued that the reduction functions necessary for the reduction of anything which might legitimately be called Mendelian genetics to molecular genetics are prohibitively complex many-many relations. Many-one relations from molecular to Mendelian genetics make the reduction complicated, so complicated that specific derivations would rarely be carried out in practise. Like three-body problems in physics, it is nice to know that they can be solved, but no one is tempted to do so very often. One-many relations from molecular to Mendelian genetics make reduction impossible. Assume for the moment that the goal of reduction functions is to identify the thing terms of Mendelian genetics (e.g., genes and traits) with some sort of entity terms at the molecular level (e.g., DNA and proteins) and the predicate terms of Mendelian genetics (e.g. dominant and epistatic) with various molecular mechanisms (e.g., production of an active enzyme or a necessary substrate). This is what all of the philosophers involved in the dispute seem to have had in mind in their early writings (Schaffner, 1967, 1969; Hull, 1972, 1973, 1974; Ruse, 1971, 1973; Simon, 1971). Thus, the one situation which is precluded, if reduction is to be possible, is a single molecular entity or mechanism being identified with two or more Mendelian entities, properties, or relations.

In order to decide whether or not Mendelian genetics is reducible to molecular genetics, it is not enough to correlate textbook versions of these two theories. The actual complexities of both types of phenomena must be recognized and taken into account. For example, in an early publication, Schaffner (1967, p. 144) set out a sketch of what a reduction function for "dominant" should look like. According to Schaffner, $gene_1$ is dominant $\equiv$ DNA $segment_1$ is capable of directing the synthesis of an active enzyme. But like the definition of "gene" in terms of the production

of polypeptide chains, this reduction function for "dominant" required considerable refinement. Not all dominant genes function in the production of enzymes, active or otherwise. But of greater significance, "dominant" is a relational term. One allele can be dominant to another of its alleles but recessive to a third. In addition, one allele can be dominant to another with respect to one trait which it controls but recessive with respect to a second. The age of the organism must also be taken into account. One allele can be dominant to another allele with respect to a particular trait early on in the ontogenetic development of an organism; recessive later. For example, red shell color in a particular species of snail is dominant to yellow early on but then gradually changes to recessive.

Both the genetic and external environments also matter. A change in a gene in one place in the genome can affect the functioning of genes elsewhere. Thresholds also exist such that below a particular level one trait develops; above that level another trait. Sometimes such switches are only stochastic, determining only the probability of certain phenotypes. For example, the castes in some social insects are genetically similar and are determined by the quality and quantity of food fed to the larvae. Flowering in many plants and diapause in many insects are induced by the length of daylight and darkness. In the marine worm *Bonellia,* sex is determined by the environment: larvae growing on a female's proboscis become males; anywhere else females. And I'm sure no one will object to my adding at this point, "And so on."

A comparable story can be told for

molecular genetics. There seems to be a reasonably close correlation between molecular and Mendelian genes, especially if the genome is divided into genes on the basis of those segments of DNA which produce single molecules of RNA. The correlation between specific DNA segments and specific molecules of RNA is pretty much one-one, though some exceptions do exist. But once these molecules of RNA begin to function in the production of proteins, the situation becomes a great deal more complicated. As Schaffner (1969, p. 411) has observed:

> The requirements for protein synthesis include about 60 transfer RNA molecules, at least 20 activating enzymes which attach the amino acids to the t-RNAs, special enzymes for initiation, propagation, and termination of peptide synthesis, ribosomes containing three different structural RNAs, and as many as 50 different structural proteins, messenger RNA, ATP, GTP, and $Mg^{+2}$.

And other factors matter as well, including temperature, pH, ionic strength, and so on.

I have not listed all the preceding complexities just for the sake of obscurantism. Perhaps a few, many, or all of them can be taken into account. The point is that some hint must be given as to which of these factors are to be included in the relevant theory, which in the environment and in what sense "environment," and how these theories and environments are to be correlated. As Robert Causey (1972, p. 194) observes in his discussion of uniform microreductions, the reducing theory and its primitive nonlogical predicates will have to make reference to environmental conditions and to state some relations between the elements of the domain of the reducing theory and the relevant aspects of the environment. That these factors be stated *explicitly* and correlated *systematically* does not seem to be asking too much. Neither vague promises nor a case by case treatment is good enough.

Ruse has responded to my claim that the relation between particular *kinds* of molecular mechanisms and *patterns* of Mendelian inheritance is one-many by arguing that "given a general mechanism which could lead to various Mendelian effects, by specifying the peculiarities of a particular case, one can follow through to a uniquely determined effect." He even toys with the idea that I might be claiming that "*every* instance of a Mendelian effect will have to be treated separately." Instead, Ruse counters, "there are some general models which seem applicable to widely differing cases." As I have noted previously (Hull, 1974, p. 41), one likely mechanism for epistasis is the production by the epistatic gene of a necessary precursor for the reaction controlled by the hypostatic gene. As Ruse observes, this "model is known to fit many cases." True, but which ones? After all, reduction functions are supposed to by synthetic *identity* statements, not statements of loose correlations. But Ruse replies, if one supplements the description of the general model with a specification of the *particular circumstances,* then one can infer unerringly which pattern of Mendelian inheritance should result. I take this statement to be fairly safe, because the only way that it could be false is by miraculous intervention of a supernatural power. The issue is the relative importance of the general model and the particular circumstances in such inferences. As it turns out in the case of genetics, knowledge of the general model does not contribute much to the inference. Such models are too ubiquitous for that. Instead it is the peculiarities of the particular case which are decisive. For example, sometimes the total inactivation of a particular allele which codes for an enzyme will result in no enzyme being produced, sometimes in the continued production of the enzyme at full capacity, and sometimes any of the intermediary states. Given knowledge of the general mechanism, no inference can be made to any sort of general Mendelian pattern of inheritance.

I realize that from a general law one can infer to the particular case only by introducing statements of particular circumstances. In order to infer the path of the earth around the sun, one must know the masses, positions, and velocities of these two bodies. Newton's laws alone will not suffice. What is needed in the case of the reduction functions in genetics is a specification *in advance* of the molecular mechanisms (what factors are to be included, which excluded, and how these factors are related) and what *sorts* of things are part of the particular circumstances. To be told that in each case some molecular mechanism or other combined with whatever particular circumstances turn out to be relevant are responsible for the observed Mendelian pattern of

inheritance is just a covert statement of determinism. I have no quarrel with macroscopic determinism, but the issue is reduction, not determinism. The logical empiricist analysis of theory reduction concerns the inferential relationship between *theories*. In order for the thesis to be significant, a *systematic* relationship must exist between the key elements in the two theories, in this case between various *kinds* of molecular mechanisms and various *patterns* of Mendelian inheritance.

Although Schaffner's original suggestion for a reduction function for the Mendelian notion of dominance initiated the recent dispute about the actual relationship between molecular mechanisms and Mendelian patterns of inheritance, Schaffner in the interim has taken a completely different tack. On Schaffner's current analysis, all one needs to know in order to reduce Mendelian genetics to molecular genetics is that genes are made out of DNA and proteins out of amino acids. This is the total empirical content of the necessary reduction functions. If we let $A$ and $a$ stand for two alleles, ④ and ⑳ stand for the resulting phenotypic traits, and a single arrow stand for "Mendelianly produce," then the Mendelian claim that allele $A$ is dominant to allele $a$ with respect to character states ④ and ⑳ can be diagrammed as follows:

$$AA \rightarrow ④, \text{ and}$$
$$Aa \rightarrow ④, \text{ and}$$
$$aa \rightarrow ⑳.$$

Now if we identify Mendelian alleles $A$ and $a$ with DNA sequences $\alpha$ and $\beta$ respectively, and the phenotypic traits ④ and ⑳ with amino acid sequences א and ב respectively, and let a double arrow stand for "molecularly produce," then the preceding Mendelian claim can be re-written as follows:

$$\alpha\alpha \Rightarrow \aleph, \text{ and}$$
$$\alpha\beta \Rightarrow \aleph, \text{ and}$$
$$\beta\beta \Rightarrow \beth.$$

As the reader can see simply by comparing these two formulations, all Schaffner has done is substitute particular DNA sequences for particular Mendelian genes, an arrow standing for "molecularly produce" for an arrow standing for "Mendelianly produce," and amino acid sequences for phenotypic traits. At first one might have expected the formulation of reduction functions for Mendelian predicate terms to be quite difficult and require considerable knowledge of molecular genetics. On Schaffner's current analysis, the sum total of the empirical content of his *synthetic* identity statements is the identification of Mendelian genes with segments of DNA and phenotypic traits with amino acids. All the rest of the empirical content of molecular genetics is buried in the double arrow. Schaffner justifies this move by noting that the double arrow "essentially represents a telescoped chemical mechanism of production of the amino acid sequence by the DNA. Such a mechanism must be specified if a *chemically adequate* account of biological production or causation is to be given. Such a complete account need not function explicitly in the *reduction function per se,* even though it is needed in any account of reduction."

Schaffner has chosen one possible way to represent the reduction of Mendelian to molecular genetics. In his characterization, the reduction functions for "gene" and "phenotypic trait" are empirically significant. Those for the large array of Mendelian predicate terms follow automatically from these initial two reduction functions and contribute nothing empirically significant to the reduction. I must admit that I find Schaffner's move disheartening. I had thought that Nagel, Woodger, Hempel *et al.,* were attempting to analyze an important notion of reduction. Whatever it was that made it important either philosophically or scientifically seems to have dropped out along the way. Defenders of the logical empiricist analysis of theory reduction seem to be willing to trivialize their analysis in order to salvage it. As with similar disputes over the existence of God, I personally much prefer to retain a significant notion of the concept at issue and admit that it does not apply rather than trivialize it to the extent necessary to insure its application. "Does God exist?" Certainly! God is the Universe, and only lunatics doubt the existence of the universe. "Ah yes, then physicists are theologians."

In summary, both Ruse and Schaffner have defended the possibility of providing the reduction functions necessary to derive Mendelian from molecular genetics. They both agree that these reduction functions are *synthetic identity* statements. To put my objections to this thesis succinctly, I have argued that Ruse's reduction functions are not *identities* and Schaffner's reduction functions for the predicate terms of Mendelian genetics are not *synthetic.* In both of their accounts, all empirical difficulties are shunted into some area whose elements happily never need to be stated explicitly. Schaffner's leaving the causal relation between Mendelian genes and the phenotypic traits which they control unspecified is prefectly legitimate. Although most Mendelian geneticists assumed that some sort of causal chains connected genes with phenotypic traits, the specification of these causal chains played no role in transmission genetics. The same cannot be said for molecular genetics. Molecular genetics sets out in detail the molecular structure of genes and proteins and the mechanisms by which DNA replicates itself and produces molecules of RNA which cooperate in the synthesis of proteins. Leaving the double arrow unexplicated as a "promissory note for a *chemically* causal account" to be specified later is to leave the reduction itself largely promissory, to be supplied later. Nor is the burying of all empirical difficulties in the theory, particular circumstances, auxiliary hypotheses, and what have you just to keep the reduction functions simple a very propitious maneuver. The complexity of the relationship between molecular and Mendelian genetics must be measured in terms of *all* the operative elements, not just in terms of the simplicity of the reduction functions. The decision to keep the reduction functions simple merely requires that the complexities which I have mentioned be accommodated elsewhere. They lie in wait for anyone attempting to set out a complete analysis of theory reduction in genetics.

One of the major impediments to the straightforward application of the logical empiricist analysis of theory reduction to the case in genetics is that Mendelian genetics is almost exclusively a transmission theory. Mechanisms are irrelevant. Whereas molecular genetics is concerned first and foremost with mechanisms. In fact, it is difficult to set out molecular genetics in the form of a "theory," in the usual logical empiricist sense of the term. Wimsatt has questioned whether reduction is best construed as a relation between the theories at all:

At least in biology, most scientists see their work as explaining types of phenomena by discovering mechanisms, rather than explaining theories by deriving them from or reducing them to other theories, and *this* is seen by them as reduction, or as integrally tied to it.

I suggest that the explanation of phenomena at one level of analysis by setting out the mechanisms which produce it at a lower level is the important sense of reduction which has gradually been eliminated as the logical empiricist analysis of theory reduction has been reduced to its current sad state. Little by little the notion of reduction has been so trivialized that it is difficult to imagine a situation which could not be made to fit it.

## REFERENCES

Causey, R. L., 1972, Uniform microreductions. *Synthese*, 25: 176–218.

Fleming, Donald (ed.), 1964, *The Mechanistic Conception of Life,* Cambridge, Mass., The Belknap Press.

Hull, D. L., 1972, Reduction in genetics—biology or philosophy? *Philosophy of Science,* 39: 491–499.

——, 1973, Reduction in genetics—doing the impossible. In P. Suppes (ed.), *Logic, Methodology and Philosophy of Science,* vol. IV: 619–635, North-Holland Publishing Co.

——, 1974, *Philosophy of the Biological Sciences,* Englewood Cliffs, Prentice-Hall.

Knight, David, M., 1967, *Atoms and Elements,* London, Hutchinson University Library.

Kyburg, H., 1968, *Philosophy of Science,* New York, Macmillan.

Lakatos, I., 1971, History of science and its rational reconstruction. *Boston Studies in the Philosophy of Science,* vol. 8: 91–136.

Ruse, Michael, 1971, Reduction, replacement and molecular biology. *Dialectica,* 25: 39–72.

——, 1973, *The Philosophy of Biology,* London, Hutchinson University Library.

Schaffner, Kenneth, 1967, Approaches to reduction. *Philosophy of Science,* 34: 137–147.

——, 1969, The Watson-Crick model and reductionism, *The British Journal for the Philosophy of Science,* 20: 325–348.

Simon, Michael, 1971, *The Matter of Life,* New Haven, Yale University Press.

Sklar, L., 1976, Thermodynamics, statistical mechanics, and the complexity of reductions. In Cohen *et al.* (1976), *PSA 1974,* Reidel, Dordrecht, pp. 15–32.

Woodger, J., 1937, *The Axiomatic Method in Biology,* Cambridge, Cambridge University Press.

# 26

# Reductive Explanation: A Functional Account

*~~~~~~~~~~~~~~~~~~~~~~~~~~~~~~~~~~*

## WILLIAM WIMSATT

I

Philosophical discussions of reduction seem at odds or unsettled on a number of questions:

(i) Is it a relation between real or between reconstructed theories, and if the latter, how much reconstruction is appropriate?[1] Or is reduction best construed as a relation between theories at all?[2]

(ii) Is it primarily connected with theory succession, with theoretical explanation, or with both?[3]

(iii) Is translatability *in principle* sufficient, or must we have the translations in hand, and if the former, how do we judge the possibility of translation when we don't have one?[4]

(iv) What is the point of defending the formal model of reduction if it doesn't actually happen (Hull[5], Ruse[6]), or if the defense has the consequence that if reductions occur, they are trivial and uninformative (Hull[7]), or merely incidental consequences of the purposeful activity of the scientist *qua* scientist in devising explanations (Schaffner[8])?

Furthermore:

(v) At least in biology, most scientists see their work as explaining types of phenomena by discovering mechanisms, rather than explaining theories by deriving them from or reducing them to other theories, and *this* is seen by them as reduction, or as integrally tied to it.[9]

(vi) None of the symposiasts* present are suggesting inadequacies in the kinds[10] of mechanisms postulated by molecular geneticists for the explanation of more macroscopic genetic phenomena.

(vii) Nonetheless, two of them (Ruse in his earlier work, though no longer, and Hull) seem to suggest that there is no reduction (only a replacement) and the third (Schaffner) suggests that a reduction is occurring, but is a merely incidental consequence of the activity of these scientists.

What possibly can explain this wide disagreement between scientists who appear to take reductive explanation seriously and to regard it as an—indeed, as perhaps *the*—important consequence of their work, and philosophers

*That is, the authors of selections 23–26 [Editor].

who are attempting to faithfully characterize their activity and its rationale? Can reduction be as unimportant (or nonexistent) in science as these philosophers seem to suggest? I think the answer must be 'no', and that there are four main factors which are responsible for the present philosophical confusion on this point:

(1) Philosophers have taken the 'linguistic turn' and talk about relations between linguistic entities whereas biologists are more frequently unabashed (or sometimes abashed) realists, and talk about mechanisms, causal relations and phenomena. Though not necessarily vicious, I think that the linguistic move has led philosophers astray. I will here defend a realistic account of reduction.[11]

(2) While virtually everyone agrees that a philosopher by the nature of his task must be interested in doing some rational reconstruction, doing so serves different ends in different contexts. A failure to distinguish these ends and how they may be served contributes to the apparent defensibility of the formal model of reduction.

(3) No real competitor to the formalistic (or more generally structuralist) account of reduction has been forthcoming. Therefore there has been a tendency to regard 'informal' reductions (Ruse)[12] as either nonreductions or as *deficient* reductions, which can be remedied by becoming formalized. I will outline some aspects of a functional account of reduction which suggest that *'informal reductions'* are the proper end of scientific analyses aiming at reductive explanations.

(4) An emphasis on structural (deductive, formal, logical) similarities has led to a lumping of cases of theory succession with cases of theoretical explanation, with the result that discussions of reduction, replacement, identification and explanation (which have radically different significances in the two contexts) have become thoroughly muddled.[13] A functional account of these activities yields important clarifications of their nature.

I wish to say something about (2), before turning to my analysis of reduction, which concerns primarily (3) and (4). The first point enters mainly by implication.

## II. TWO KINDS OF RATIONAL RECONSTRUCTION

There are at least two (and probably more) contexts where talk of rational reconstruction seems appropriate in connection with plausible and useful activities of philosophers of science:

### Rational₁ —An Optimal Strategy

One might want to abstract from the often irrelevant details and sometimes mistaken moves of the actual practice of science to reconstruct the significant patterns of scientific activity.[14] Insofar as these patterns can be claimed to be a relatively efficient, or even an *optimal* way of achieving or trying to acheive the ends of such activity, the reconstruction could claim to be a rational reconstruction in the sense of rational decision theory—that it represented the way one ought to do that activity. As such the philosopher of science is a *therapist with respect to scientific strategy.*

*Rational₂ —A Canon of Logical Rigor*

A physicist (and nowadays with increasing frequency, a biologist) might ask a mathematician for "formal" help. He might wish to prove a mathematical conjecture whose truth or falsity he is uncertain of and which has important implications for his work. Or he may have an argument which he can formulate more informally, but desires more rigor either to buttress the argument or to determine more precisely the conditions under which it holds. As such, a mathematician is a *therapist with respect to formal argument,* logic, and "critical thinking", and these are also roles which could legitimately and usefully be played by a philosopher of science.

In either case the philosopher of science would be analyzing or criticizing an acitvity in terms of how well it served the ends of the scientist, and in each case, the activity itself and the analysis of it further these ends.

Note that the functions of the philosopher of science in these two cases are, at least *prima facie* not equivalent. It is not at all clear that improvements in rigor, *per se,* are a rational *qua* efficient way to do science—say, for finding explanations —nor even that the ultimate end state of science will be to improve the rigor of theories *which are otherwise adequate*—i.e., after their other problems have been solved. Improvements in rigor are sometimes useful, but not always. Philosophers of science have sometimes talked as if improvement in rigor is a scientific-end-in-itself, but no one here is doing so. I believe that the sort of confirma-

tion and troubleshooting suggested above is the main function of rigorous argument in science, and that rigor is not a scientific-end-in-itself.

One effect of logical empiricism (with its emphasis on the 'logical structure' of laws, theories, explanations, predictions, and experiments) has been to blur—even to obliterate —the distinction between these two senses of 'rational reconstruction'. This conflation has had a disastrous effect upon the analysis of reduction— proceeding as it has in terms of the formal model. Schaffner's thesis of the 'peripherality' of reduction suggests that any successful defense of the formal model would win a pyrrhic victory. In terms of the above distinctions, I would describe this 'peripherality' of the formal model as follows: It is not rational₁ to view formal (i.e., rational₂) reduction as a scientific-end-in-itself because science then becomes an inefficient and ineffective way of pursuing known scientific ends (such as explanation). And although the formal model of reduction is by definition a rational₂ model, it is not even an effective *means* to some end because it is not the answer to a request for formal (i.e. rational₂) assistance which anyone has made or would be likely to make! Thus, although early discussions of formal reduction seemed to hold out the hope that it would perform the functions of both kinds of rational criticism, it is my impression that more recent sophisticated discussions (such as Schaffner's) have given up on both claims. But these claims are not peripheral and readily dispensable. They represent one of the major motivations for pursuing either a formalistic or a reductionistic strategy in science. If they must be given up, one's claim to be analyzing reduction as that concept is *used* in science must be suspect.

Paradoxically, if a non-formal (or perhaps "partially formal") account of reduction is allowed, it can be seen to be a rational activity in both senses: It is an efficient (rational$_1$) way in which to proceed, and it proceeds by using logical instruments for the critical (rational$_2$) evaluation of theoretical and observational claims. Because it is a partially formal model, the use of formal methods (as discussed by Schaffner and Ruse) is to be expected on this model also, and it derives confirmation from the cases they adduce to support the formal model. It does not require total systematization, however, which has *not* been exemplified in any of the cases they discuss and which formal reduction requires (See, e.g., Schaffner, 1976, p. 614).

How do we get such an alternative to the formal model of reduction? Just as a characterization of logical *structure* (a rational$_2$ reconstruction) suggests and is suggested by a formal model of reduction, the view of scientific activity as purposive suggests a *functional* analysis and characterization—a rational$_1$ reconstruction—of reduction. Such an analysis distinguishes activities which may, in some respects have similar structure,[15] and may point to and explain further structural differences which have been ignored on the formal approach. Most importantly, I believe that a functionalist approach shows why the research aims of the scientist *contribute to* (in the sense of moving in the direction of) fulfilling the aims of the formal model, but are in fact *different from* and even, *inconsistent with,* actually getting there. Then a stronger version of Schaffner's (1974b) "peripherality" thesis is justified:

(P1) Not only is progress toward formal reduction incidental, but

(P2) It also seems to be epiphenomenal, since this progress towards formal reduction appears to have no *further* consequences.

(P3) Finally, if (as I believe) getting there is inconsistent with the real aims of science, this 'progress' is bound to remain incomplete.

## III. SUCCESSIONAL VS. INTER-LEVEL REDUCTIONS

The functional viewpoint is perhaps best explicated by expanding upon and modifying Schaffner's model, which has many useful features, although the end result will be quite different. (See Figures 1 and 3.) Most importantly, Schaffner distinguishes between and includes both a derivability condition between the reducing theory ($T_1$), and a corrected version of the reduced theory ($T_2^*$), and a condition of strong analogy between $T_2^*$ and its uncorrected predecessor, $T_2$. These two relations are prototypic of two distinct relationships, each of which has been called 'reduction'.

Schaffner's condition of strong analogy is closely related to Nickles' "reduction$_2$" (Nickles, 1973, p. 194ff.) and to what I elsewhere (Wimsatt, 1975) and below call "successional" or "intra-level" reduction. Nickles' account, emphasizing transformational and possibly non-deductive relations between successive competing theories affords an important partial

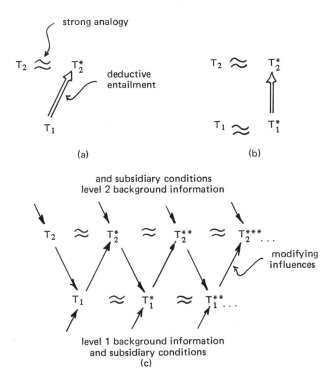

Figure 1. (*a*) Theory Reduction: Schaffner (1967). $T_2$: reduced theory; $T_2^*$: corrected reduced theory; $T_1$: reducing theory. (*b*) Theory Reduction: Schaffner (1969). $T_1^*$: modified (corrected?) reducing theory. (*c*) 'Coevolution of theories at Different Levles': Wimsatt (1973) [an early draft of (1975)].

explication of 'strong analogy'. A functional account of this activity explains many of the structural features Nickles proposes, and others which he does not mention.

What is not clear on Schaffner's model, but implicit in Nickles', is that 'reduction₂' (which is a kind of 'pattern matching' problem and could also be regarded as *demonstrating and analyzing* the 'strong analogy' between $T_2$ and $T_2^*$)[16] is neither automatic nor self evident. It has a point, involves work, and is performed for reasons separate from the functions of the 'other' reductive relation. Nickles suggests that reduction₂ performs heuristic and justificatory

functions *vis-a-vis* the uncorrected older $T_2$.[17]

I believe that reduction₂ is fundamentally connected with theory succession (of $T_2$ by $T_2^*$) and performs rather more functions than Nickles makes out. *It is most immediately a transformational operation whose function is to localize and analyze the similarities and differences between $T_2$ and $T_2^*$* which in turn serve a variety of further functions. Most interestingly, because none of these functions are served by making comparisons other than between $T_2^*$ and its immediate predecessor, $T_2$, and in any case, similarities and differences become

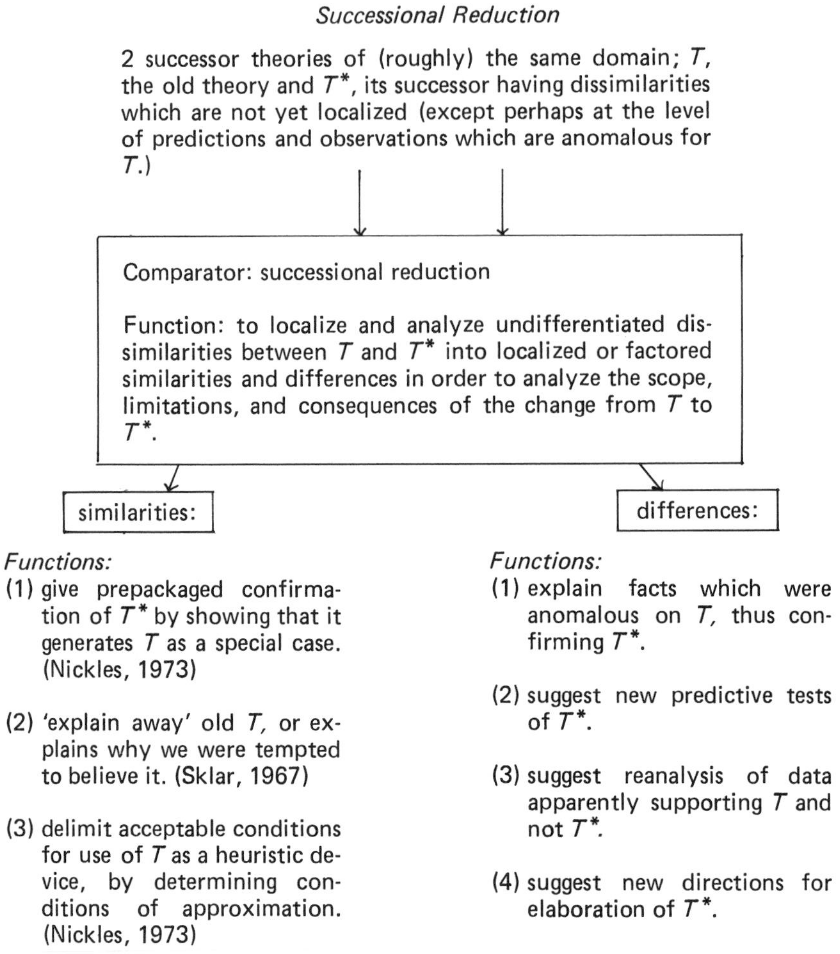

## Successional Reduction

2 successor theories of (roughly) the same domain; $T$, the old theory and $T^*$, its successor having dissimilarities which are not yet localized (except perhaps at the level of predictions and observations which are anomalous for $T$.)

Comparator: successional reduction

Function: to localize and analyze undifferentiated dissimilarities between $T$ and $T^*$ into localized or factored similarities and differences in order to analyze the scope, limitations, and consequences of the change from $T$ to $T^*$.

similarities:

Functions:
(1) give prepackaged confirmation of $T^*$ by showing that it generates $T$ as a special case. (Nickles, 1973)

(2) 'explain away' old $T$, or explains why we were tempted to believe it. (Sklar, 1967)

(3) delimit acceptable conditions for use of $T$ as a heuristic device, by determining conditions of approximation. (Nickles, 1973)

differences:

Functions:
(1) explain facts which were anomalous on $T$, thus confirming $T^*$.

(2) suggest new predictive tests of $T^*$.

(3) suggest reanalysis of data apparently supporting $T$ and not $T^*$.

(4) suggest new directions for elaboration of $T^*$.

Figure 2.

*less* localizeable as changes accumulate, successional reduction would be expected to be *intransitive,* and to behave as a similarity relation.[18] *Thus the intransitivity of successional reduction is an explicable feature, not a given, on the functional account of this activity.*

For further analysis of the specific uses made of these localized similarities and differences between $T_2$ and $T_2^*$ and diagrammed in Figure 2, I refer you to part II of Wimsatt (1975). The following contrasts between "suc-cessional" and "explanatory" reductions should be noted here, however.

(1) *Successional reduction is* and must be *a relation between theories* (since it is these which exhibit the similarities and differences), unlike *explanatory reduction* which *is not,* in any but degenerately simple cases.

(2) *Replacement* occurs only with the *failure* of successional reduction —failure to localize similarities and differences among successive competing theories. Replacement and suc-cessional reduction are opposites.

But for explanatory reductions, replace*ability* is closer to and is by many treated as a *synonym* for reduce*ability*. A failure of $T_1$ to reduce $T_2$ (perhaps derivatively, by reducing $T_2^*$) would make $T_2$ and its successors *emergent* and *irreplaceable* relative to $T_1$. *Replacement obviously has two different meanings here.*

(3) *Successional reductions are intransitive.* A number of them "add up" to a replacement. *Explanatory reductions are transitive.* [It is this last fact which raised the hopes among advocates of "unity of science" for great ontological economies through reduction, about which I have more to say (1975 and below).]

(4) Talk about elimination might be appropriate for the posited entities of corrected and replaced theories if the new theory is sufficiently different that there is no significant continuity between old and new entities. But such talk is frequently illegitimately extended to contexts of *explanatory reduction.* This is often motivated by talk of ontological or postulational simplicity in the light of supposed translateability and deduceability, (discussed further below), but in at least some cases looks suspiciously like treating reduction and replacement as opposites. Thus, in arguing that the formal model of

reduction doesn't fit the relation of Mendelian to molecular genetics, Hull and Ruse[19] each suggest that it looks more like a case of replacement. As I suggested in (2) above, the opposition between reduction and replacement is appropriate for successional reduction, but *not* for inter-level or explanatory reduction. Their claim is thus misplaced if it concerns the relation between $T_1$ and $T_2$. Though intelligible if construed as concerning the relation between $T_2$ and $T_2^*$, I would disagree on the facts of the case, and agree with Schaffner (1976) and Ruse's (1976) most recent view that there is no replacement, but a reduction. To explain why, I must say a great deal more about explanatory reductions. In what follows, I will be talking about them unless otherwise indicated.

## IV. LEVELS OF ORGANIZATION AND THE CO-EVOLUTION AND DEVELOPMENT OF INTER-LEVEL THEORIES

Rather than talking directly about reductive relations between theories, the approach I have taken (Wimsatt, 1975) is the realistic one of regarding levels of organization—features of the world—as primary, and defined in such a way that it is natural that

## NOTES

(1) Nickles (1973) also suggests that 'reductions$_2$' may be done in a variety of ways. This is understandable if the point of the transformation is how best to factor out similarities and differences.

(2) Successional reductions may be possible 'locally' (for parts of theories) even when not possible globally (for the whole theory.).

(3) Differences in meanings of the key terms may be regarded as irrelevant as long as they are localizable in a way that allows fixing praise and blame on specific components of $T$ and $T^*$ in comparatively evaluating them. (see also Glymour, 1975). Thus the 'meaning change' objection is avoidable.

theories should be about entities at these levels of organization. The notion of a level implies a partial ordering, such that higher level entities are composed of lower level entities, and, in a universe where reductionism is a good research strategy, the properties of higher level entities are predominanatly best explained in terms of the properties and interrelations of lower level entities.

But I argue further that levels of organization are primarily characterized as local maxima of regularity and predictability in the phase space of different modes of organization of matter. Given this, selection forces (and at lower levels, the stability considerations into which these shade) suggest that the majority of readily defineable entities will be found in the (phase space) neighborhood of levels of organization, and that the simplest and most powerful theories will be about entities at these levels.[20]

Nothing in this approach entails that levels defined as local maxima of regularity and predictability must always be well-defined and delineated, or strictly linearly orderable (although they usually are for simpler systems), and in fact certain conditions can be suggested (in *this* world) where these assumptions are false (see Wimsatt, 1974 and 1975, part III). These are conditions where neat composition relations cannot be specified for all (or perhaps even for any) of the entities in these different "perspectives." (Level talk *requires* the possibility of specifying composition relations, so I talk about "perspectives" when this condition is not met.) This failure of orderability leads to the "intertwining" of theories mentioned by Schaffner (1974b) in

discussing the operon model (see also his 1974a), in support of his thesis of the "peripherality of reduction," and to the much more extreme situation suggested by Roth (1974) in her penetrating analysis of the same case—which she sees as the development of an inter-(multiple) level theory rather than as the tying or merging together of preexisting theories.

These sorts of complexities have been ignored in discussions of the standard model of reduction, and Hull's discussions of the difficulties of translation just begin to characterize one of their major effects. Nor is this problem limited to genetics. Fodor's recent (1974) discussion supports the view that it is of substantially greater scope and provides a careful analysis of problems that arise for the standard ("type reduction") account of reduction in these areas. But the standard model just looks so right that it is hard to see how it *could* be wrong. In this light, claims like those of Hull and Fodor look almost counterintuitive, and it becomes easy to give them short shrift. There are several sources of bias in favor of the "standard model" which contribute to this appearance:

(1) There is a general tendency to characterize the lower-level theory $(T_1)$ as "more general" and "more explanatory" than the upper-level theories $(T_2$ and $T_2^*)$, trading on our general reductionistic prejudices in favor of using compositional information (rather than, e.g., contextual information) in an explanation. This has complex sources which I have discussed elsewhere (Wimsatt, 1975), and has as one of its effects the tendency to assume that lower-level

theories correct upper-level theories, but not conversely.[21]

(2) Another important source of bias leading to this error is the distinction between contexts of justification and contexts of discovery, and the attention paid to the former at the expense of the latter. We primarily worry about justifying edifices—theoretical structures that have already undergone substantial revision and selection, and that we have begun to presuppose in a variety of other areas and are thus loath to revise in any substantial way. We discover and propose models—tentatively and usually without much commitment. We give them up or modify them easily because little else depends upon it. For reductions (or at least for those which look much as if they will come close to satisfying the formal model) the lower-level theory is already well into the edifice stage, and it is thus not surprising that lower-level corrections are less visible, having for the most part already occurred.

(3) A bias towards the 'standard model' is introduced *via* the view that explanations involve giving laws rather than citing causal factors or giving causal mechanisms. How this is introduced (laws suggest greater systematization than do causal factors) and avoided (by accepting Salmon's account, 1971, of statistical explanation) is discussed below in part V.

(4) Discussions of translatability tend to revolve around those cases where it looks easiest to give a translation. It is often easier for properties than for objects (which are characterized by a variety of theoretically relevant properties if they are important objects). It is easier for objects if they are not functionally defined (or are fallaciously *treated* as if they were not), since function makes features of the *context* highly relevant. (As linguists know, a context-dependent translation is an incomplete translation.) Functionally defined processes can be the most difficult, since they will often be associated with a number of objects which will also be involved in *other* functional processes (see Wimsatt, 1974), and can be realized in a variety of different ways.

Discussions of reduction in genetics have not even approached the translation of some of these terms. Terms from population genetics like "heterosis", "additive (multiplicative, nonadditive, non-multiplicative) interactions in fitness" (see Lewontin, 1974), and Lewontin's "coupling coefficient" (ibid., p. 294) represent things we look for and find mechanisms for, but general or context-independent translations at a molecular level seem absurd—both impossible and pointless. "Context-dependent translations" are easy to come by, of course. Discovering the mechanisms in specific cases *gives* us that. But that won't do for the formal model: for those purposes a *"context-dependent translation" is not a translation.*

What would a new view of interlevel reduction look like? Schaffner's later move (1969) in allowing modifications to $T_1$ in order to affect the reduction (Figure 1b) is a step towards the picture I would draw: *Theoretical conceptions of entities at different levels coevolve and are mutually elaborated* (particularly at places where they "touch"—where we come closest to having inter-level translations)[22] *under the pressure of one another and "outside" influences* (see Figure 1c).

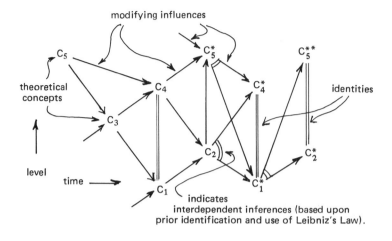

(a)

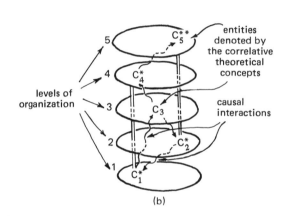

(b)

Figure 3. (*a*) "Inference Structure of the Development of the Theory." (*b*) "Resultant Causal Structure of the Mechanism According to the Theory." "Development of an Inter-level Theory." An extension of the model of Roth (1974) involving the use of identities as proposed in Wimsatt (1975) in the coevolution of concepts in the development of an inter-level theory of the operation of a causal mechanism. Strong analogy between concepts and their descendants ( $C_m^{n*}$, $C_m^{n+*}$ ) is assumed generally (but not necessarily universally) to hold, but to simplify the diagram is not represented here.

In this picture, both successional reductions (or replacements) and explanatory reductions are occurring in an intricately interwoven fashion. Very roughly, all corrections in theory get packed into a "successional" component, (because Leibniz's Law applied to inter-level identities ferrets them out of the other component) and all unfalsified explanatory and compositional statements get packed into the "explanatory reduction" component. Theory at different levels progresses by piecemeal modification, in a manner paradigmatically exemplified by Roth's discussion of the operon theory (1974, ch. II); see Figure 3.

Three things should be noticed about these modifications:

(1) Their form may well be deductive or quasi-deductive, but if so, the arguments are usually both enthymematic and riddled with *Ceteris paribus* assumptions. Typically, it is decided that a $T_1$-level mechanism cannot accommodate a $T_2$-level phenomenon without modification to $T_1^*$, in which case inferential failure of $T_1$ is the source of the change; or from $T_1$ and appropriate boundary conditions, we infer, predict, or deduce that a phenomenon which is incompatible with $T_2$, but not with a $T_2^*$ and observed results, should occur, in which case an inferential success of $T_1$ and its associated mechanisms is the source of the change.

(2) The modification occurs without a total deductive systematization, or often even an informal recodification of the theories. The new theories are characterized in terms of the changes from the preceding theories, but since they were similarly characterized, there is hardly ever a thorough systematization.

(3) The important difference of this picture from Schaffner's is that it is primarily the *changes* in theories which result from deductive arguments. Seldom if ever is any even sizable fragment of a theory deduced wholesale from another, and seldom if ever is even a single theory sufficiently systematized to meet the conditions for applying the formal model. Furthermore, it is so clearly unnecessary and irrelevant to the search for explanations to do so.

Schaffner's own accounts (1974a, 1974b) and that of Roth (1974) are beautiful confirmation of this highly efficient, but formally, highly confusing strategy of theory evolution. These suggest that the vertical arrows *not* be interpreted as total entailments between theories (or reductions, where upwards arrows are concerned), but as single rough deductions or inferences from attempts to match the structure of causal mechanisms as described at different levels resulting in changes in *parts* of theories. There is, to be sure, use of deductive argument, and lower-level explanation of upper level phenomena. The examples of Ruse (1976) (hemoglobin and sickle cell anemia), Roth (1974), and Schaffner (1974a, 1974b) are marvelous. But as Hull points out, they do *not* touch the issue of whether a total deductive systematization is occurring, since such cases would also be expected on the view of reduction advanced here. But if this is all that happens, why should one bother to attempt to characterize reduction along the lines of the formal model? There just seems to be too big a gap between principle and practice for the principle to be very interesting.

Aside from philosophical predilections of an "eliminative" sort, there seem to be two reasons for holding onto the formal model of reduction:

(a) The belief that as the "fit" gets better between upper- and lower-level theories, their relationship asymptotically approaches the conditions of the formal model of reduction.

(b) The belief that even if the "fit" never asymptotes, or if it does, doesn't converge on the formal model, the latter represents an aim of scientists.

While Schaffner (1974b) has questioned whether trying to accomplish the reductionistic program per se is a good scientific strategy, I suspect that he (and perhaps many scientists) believe that it is at least a secret hope or end. I want to examine the grounds for this latter belief and suggest an alternative interpretation that is more consistent with scientists' actual behavior. This interpretation also raises serious questions about the first assumption.

Finally, the formal model would not be nearly as tempting if there were not, for each philosopher talking about "translating away" upper-level vocabulary, a scientist talking about "analyzing away" upper-level entities. It thus looks as if a claim about words can be "cashed in" for a claim about entities, and a claim about entities which many scientists appear to accept. The formal model thus appears to have direct support in the talk of many scientists of the "nothing more than" persuasion. But of what are they persuaded? Are the translations or analyses like those promised by Schaffner *immediately* forthcoming? Usually not. No one actually ground them all out, but that's said to be just a *practical* difficulty. It is *in principle* possible. But "in principle" claims have been failing, only to be replaced by new ones, since the time of Democritus. Given their history, such *in principle* claims could not plausibly be treated as self-warranting. But then what else warrants them? How can we evaluate these *in principle* claims, to distinguish good ones from bad ones? Or perhaps these *in principle* claims are not the claims they seem to be to knowledge the claimant cannot have: I suggest rather that they are important tools in the task of looking for explanations. Before discussing this (in Section VIII) I must talk about explanation.

## V: TWO VIEWS OF EXPLANATION: MAJOR FACTORS AND MECHANISMS VS. LAWS AND DEDUCTIVE COMPLETENESS

I accept Salmon's (1971) account of explanation as a successful search for "statistically relevant" partitions of the reference class of the event being explained, with two provisos: First, I will make some modifications (to be explained below and in the appendix) to bring it into line with a view of science as an acitivity conducted according to cost-benefit considerations. Secondly, I assume that in finding "statistically relevant" partitions, we are doing so with the aim of partitioning the reference class into *kinds of mechanisms,* or kinds of cases involving a given mechanism. (I am thus giving a realist interpretation to his model.) In a *reductive* explanation, these mechanisms or factors are at a lower level of organization than that of the phenomenon being explained.

One of the intriguing features of Salmon's account is his move from constructing (statistical) *laws* to a search for statistically relevant *factors.* Laws suggest the need for a complete account of the conditions under which they apply and are correct, and the connection of explanation with laws thus naturally suggests the sort of exhaustive search for factors and conditions that would go along with a complete translation of terms or a complete deductive reduction. By contrast, a search for factors (especially a search for the *major* factors—enter cost-benefit considerations!) ties in naturally with a view of explanation as a search for the mechanisms which produce a given phenomenon, and as an account of how they do it. This search stops short of an exhaustive deductive account by sticking much of the initial and boundary conditions and many background assumptions into a *ceteris paribus* qualifier on the explanation *because they are too unimportant or insufficiently general to be accounted part of the "mechanism".*

The deductivist or formal account *can* give superficial recognition to such differences of importance by different labeling (laws, boundary conditions, initial conditions, etc.) of different parts of the deductive basis. However, in looking first for a valid deduction, the formal account treats all such information as if it were fundamentally alike because it is all equally necessary for the deduction to go through. It thus rides roughshod over realistic intuitions as to differences in the roles and importance of these different kinds of information. Hull is sensitive to this in arguing that a single molecular mechanism can lead to different Mendelian traits, for which he has been criticized by Ruse (1976) and Schaffner (1976). Neither Hull nor I nor the scientists who would agree with us are anti-reductionists or anti-determinists. We are simply responding to widespread and reproducible intuitions as to when a change in the total state-description is counted as a change in the "mechanism", and when it is not.

This judgment and its reproducibility are explicable on a combination of realistic, evolutionary, and cost-benefit considerations about the nature of scientific theorizing: A mechanism is a "kind," and cost-benefit considerations on the complexity of the theory introduce a "crossover point" beyond which a phenomenon or state is too infrequent or unimportant in a theory to be reified as a kind. There will thus be cases involving the same 'mechanism' with different outcomes which will be attributed to differences in the (more viable and less central) initial or boundary conditions, or to violation of the nebulous *ceteris paribus* clause.

The deductivist also makes and must make such judgments of relative importance, but the baggage of having to construct a valid deduction and of having to treat the correspondences between lower and upper levels as "translations" leads to dangerous misdescriptions of what is going on in several respects:

(1) It is only too easy to assume that variations in the boundary conditions are predictively negligible because they are treated as of negligible or lesser general explanatory importance. A failure to include them as part of the "mechanism" as Hull has done indicates the latter, but in no way implies either that the same mechanism always produces the same output, or that this failure indicates that the same total state of the system is on different occasions yielding different outcomes. These are mistaken interpretations which become tempting when Hull's discussion of mechanisms is read as if it were about state-descriptions, and when the only differences of importance are assumed to be differences of deducibility or predictability.

(2) Schaffner's claim (1976, pp.

624–625) that Hull's discussion of mechanisms misconstrues the logic of the formal model is double-edged. He would in effect substitute talk about state-descriptions. But if the scientists are interested in *mechanisms* and Hull's point is defensible in terms of the way we investigate and reason about mechanisms, (as I think are so), of what relevance is Schaffner's probably correct claim that the formal model is defensible if we translate from talk about mechanisms to talk about state-descriptions? If scientists aren't interested in state-descriptions, Schaffner has apparently defended the formal correctness of his model at the cost of showing its irrelevance to how scientists talk and reason about reduction. Schaffner's claim about the peripherality of reduction begins to look more and more as if it applies more modestly and correctly *to the formal model of reduction.*

(3) An equally dangerous move accompanies Schaffner's account of the relation between micro- and macro-descriptions as "translation." Schaffner (1975, n. 25) *assumes* the constancy of the environment and unstated initial and boundary conditions over a range of different cases in constructing his "translation" for the dominance relation. This is done "for reasons of simplicity and logical clarity" (ibid.). But while this is an appropriate defense of simplifying assumptions in a model or idealization, it is not an appropriate move in defense of a "translation" which is to be used in the way that his are. Thus *one thing his assumption does is to mask the real context-dependence of his 'translation' by artificially assuming that the context in constant!*

But if one is trying to establish that context-independent translations can be given (a necessary move if one is to use these translations as general premises in a deduction over range of cases in which the context changes), this move is to beg the question. It is to hide deductive incompleteness by trading it for translational incorrectness or equivocation. Schaffner *cannot* do so. (See his chapter, below, pp. 428–445.)

Schaffner would not assume this constancy if it were admitted or discovered that there were an important variable (or "part of the mechanism") contained in that set of things assumed constant. He would then attempt to delineate that variable, and include it in the "translation." Thus the boundary between what is in the "translation" and what is "assumed constant" is fixed by the same judgments of importance used in delineating "mechanism" from "background" on the model which I (and I believe Hull) would defend. But what is not in the translation (or mechanism) is not thereby "constant." It is quite variable in fact, and *its very variability is one of the reasons for not including a detailed specification for it in the general theoretical account.* Its variability makes it unimportant for theory construction, and often for selection as well,[23] though it can often produce divergent predictive results, and frustrate attempts at "translation."

Although Salmon is probably not to be considered a scientific realist, his account of scientific explanation is thus a natural ally of realistic accounts of science because of its natural structural affinities for such explanations in terms of major factors and mechanisms, in general, and lower-level mechanisms in the case of reductive explanations. (See Shimony, 1971; Boyd, 1973, 1974; Campbell, 1974a, 1974b; and Wimsatt, 1975.)[24]

## VI. LEVELS OF ORGANIZATION AND EXPLANATORY COSTS AND BENEFITS

Suppose that the primary aim of science and inter-level reduction is explanation. We wish to be able to explain every phenomenon under every informative description by showing, first if possible, how it is a product of causal interactions at its own level, but barring that, how it is a product of causal interactions at lower levels (a micro-level or reductive explanation), or least probably and desirably in our reductionistic conceptual scheme, (but absolutely unavoidably in a world of evolution driven by selection processes), how it is a product of causal interactions at higher levels (most commonly, a functional explanation).

This order of priorities in the search for an explanation follows naturally from the account of levels as local maxima of regularity and predictability, together with acceptance of a weakly but generically reductionistic world view, and the assumption that the *search* for explanatory factors is also conducted according to some sort of efficiency optimizing or cost-benefit considerations. The rationale for this is discussed more fully in Wimsatt (1975) and is roughly as follows:

(1) The characterization of levels of organization as local maxima of regularity and predictability implies that most entities will most probably interact most strongly with (and most phenomena will be most probably explained in terms of) other entities and phenomena at the same level.

(2) A reductionist's conceptual scheme (or world) is at least one in which, when explanations are not forthcoming in terms of other same-level entities and phenomena, one is more likely to look for (or find) an explanation in terms of lower-level phenomena and entities than in terms of higher-level phenomena and entities.

(3) If a search for explanatory factors is conducted along some such principle as "Look in the most likely place first, and then in other places in the order of their likelihoods of yielding an explanation," then the above order of priorities is established.[25]

Salmon's account of explanation will be generally presupposed here, but with a "cost-benefit" clause added to it: not only are "statistically irrelevant" partitions products of a choice of explanatorily irrelevant variables, (as he points out), but "statistically negligible" partitions are similarly products of explanatory negligible variables. This change is consonant with the remarks of the preceding section on recognizing the different roles and importance of mechanisms, boundary conditions, and the like in an explanation, but also has some extremely important further ramifications. The most important of these is that the intuitive sense of what it is for one variable to "screen off" another changes (in a manner described in the appendix), from the account of Salmon, with consequences to be explored below.

The idea that there can be explanatorily negligible partitions of the reference class of the event or phenomenon being explained suggests an asymmetry of explanatory strategy for cases which do and cases which do not meet macroscopic regularities or laws. When a macro-regularity has relatively few exceptions, redescribing a phenomenon that *meets* the macro-regularity in terms of an *exact* micro-regularity provides no (or negligibly) further explanation. All (or most) of the explanatory power of the lower level description is 'screened off' (Salmon, 1971, p. 55, but see also appendix, below) by the success of the macro-regularity. The situation is different, however, for cases that are anomalies for or exceptions to the upper-level regularities. Since an anomaly does not meet the macro-regularity, the macro-regularity *cannot* "screen off" the micro-level variables. If the class of macro-level cases within which exceptions occur is significantly non-homogenous when described in micro-level terms, *then* going to a lower-level description can be significantly explantory, in that it may be possible to find a micro-level description partitioning the cases into exceptional and nonexceptional ones at the macro-level. We would then have a micro-explanation for the deviant phenomenon.

Thus, for example, the ideal gas law (or its corrected phenomenological successor), as a relationship between macroscopic causal factors, is explanation enough for occasions when gases obey it. Going to the micro-level in such a case is not (or negligibly) more explanatory. Of course, if all of the molecules go to one corner of the container, the

micro-level must be invoked, since the macro-level law does *not* apply, and in *that* case partitions in terms of micro-variables will be statistically relevant.

I have discussed one main reason to look for information at lower levels: to explain exceptional cases at the upper level. The other main reason is to explain upper-level regularities. But part of explaining exceptional cases involves explaining why they are exceptional in a way that is consistent with the patterns found in the motley of cases explained by the upper level law (*qua* set of interrelated causal factors.) This usually involves explaining exceptional and motley cases in terms of a single class of mechanisms or micro-variables. This requires that the relevant kinds of micro-descriptions necessary to explain the exceptional cases *also* be usable in generating the upper law as a "special case" or "limiting" or "approximate" description. It thus leads to an explanation of a revised version of the upper level law.[26]

But what is a law, and why bother to explain one if, as I have argued, mechanisms and major factors bear the primary role in explanations of events that laws have been thought to do? The answer that suggests itself in the cases I have looked at where laws are being explained in terms of lower-level factors and mechanisms is that *laws are regularities involving distributions of cases characterized at the macro-level.* They are explained as the product of the interaction of the mechanisms and major factors invoked at the micro-level with the micro-level distributions of initial and boundary conditions. They are not *mere* regularities (or "accidental generalizations", as Nagel, 1961, characterizes the infirm statement of law-like form), because they are exhibited as the product of *causal* interactions of micro-level mechanisms, factors, and initial and boundary conditions. Such law-statements thus support the appropriate counterfactual and sub-junctive conditionals. Indeed, when a macro-regularity is explained in this manner, an understanding of the micro-level mechanisms and conditions which generate the macro-level distri-bution and how they do so give a much richer structure of counter-factuals expressable in terms of micro-descriptions than before.

I am not sure whether this characterization of a law is generalizable. It might seem limited to cases where the phenomena of a law admit of meaningful redescription at a lower level. But, at least in those cases where this characterization applies, (and this would appear to cover all cases of (inter-level) reductive ex-planation) a law should be explicable in the same general way as an event. The only difference would be that instead of talking about individual constellations of mechanisms, factors and conditions, we are talking about assumed *distributions* of the above.

The reduction of thermodynamics to statistical mechanics would provide useful examples of explanations of this sort. (See, e.g., the much dis-cussed explanations of the second law of thermodynamics.) But so also would the history of the assumption of the "purity of the gametes in the heterozygote" which Hull (1974, 1976) makes much of in arguing that molecular genetics replaces, rather than reduces Mendelian genetics. I believe that Hull is incorrect in his

conclusion, and that an illustration of how this "law" is explained reduc-tively helps us to see how much real continuity there is between Men-delian and molecular genetics.

## VII. AN EXAMPLE: THE ASSUMPTION OF "THE PURITY OF THE GAMETES" IN THE HETEROZYGOTE

This assumption began life as Mendel's "law of segregation"—to explain the fact that some apparently lost charac-ters ("recessives") reappeared ap-parently in successive generations. Mendel's explanation was that in the company of certain alleles ("domi-nants"), the factors did not express themselves as characters, *but were transmitted to offspring unchanged (by their allelic factors or anything else)* to express themselves in future genotypes in which they were homo-zygous or dominant.[27]

In the Mendelism that Castle at-tacked, with his belief that the allelic genes "contaminated" one another in the heterozygous state, it was accepted that genes affecting a given character came in pairs (were alleles), but Mendel's other law—of "indepen-dent assortment" (that non-allelic genes assorted independently of one another in the offspring)—was being challenged, both experimentally and theoretically, by Bateson and others, including Morgan and his students.

The "linear linkage" model of the Morgan school explained some of Castle's results (gradual changes in coat color conformation in rats) by the gradual accumulation through selection of so-called "modifier" genes at *other* loci (presumably linked on the same chromosome) which modified the *effect* of the genes identified as producing coat color, *without modifying the allelic genes themselves*. There was thus no need to suppose (in this case) that allelic genes "contaminated" one another in the heterozygous state. Castle's supporting claim that these modifications were irreversible were successfully contested experimentally.

The Morgan model supposed that the genes were linearly arranged on chromosomes, with allelic genes on corresponding places on the homologous paired chromosomes. According to this model, homologous chromosomes would, at a certain part of the cell cycle, wind around one another, forming "chiasmata," break, and exchange segments. This was called crossing-over and recombination. A central feature of the model was that genes on the same chromosome would tend to assort together, constituting linkage groups. This was in contradiction to Mendel's law of independent assortment. A prediction of the linear model and the mechanisms of recombination was that the probability of recombination between two points along the chromosome was a monotonic increasing function of the distance between points, (being approximately linear for small distances and approaching 50 percent (or random assortment) for large distances.)[28] These also were experimentally confirmed. Furthermore, *in the absence of any "atomistic" assumptions* (placing a lower bound on minimum distance between recombinations), *this model would predict a finite frequency for crossing-over within genes of any finite size.*

A gene has a size, and this was recognized by members of the Morgan school, though different ways of estimating it.[29] produced different results. Although it was usually assumed that the genes behaved like "beads-on-a-string" (or independent atoms) as far as recombination was concerned, Muller, a Morgan student, questioned whether these "atoms" were the same for both recombinational and mutational events. Other observed phenomena (like "position effect") also raised questions about the "beads-on-a-string" model. It also was generally supposed that genes had an underlying molecular nature, though it was unknown what this was, and how it produced the properties manifested by genes, so the idea that genes had a molecular infrastructure was not new. Indeed, the "atomicity" of the genes was clearly believed, to the extent that it was only with respect to the genetic or biological properties of the genes.

The details of how the molecular account of the gene explains "position effect" and the possibility of differences between recombinational, functional and mutational criteria for individuating genes are well known (see, e.g., Hull, 1974, or any modern genetics text) and uncontroversial here. All of these have the effect of compromising the view of genes as monolithic, monadic "atoms" with respect to some of their biological properties. If there are any "atomic" units of DNA, it is the individual base pair—again not because smaller changes are impossible, but because if they occur, they are not counted as *genetic* changes. But while this would show that there were no "atomic" genes of the size Morgan and his school had assumed, and that their different criteria of individuation picked out *different* larger compound assemblages of bases as genes, it is not necessarily a disproof of their genetic "atomism." It could just as well be taken as a demonstration that their "atoms" were smaller than they had thought, (see below, note 32) and (being of at that time unknown constitution) had some unexpected properties which explained others that the genes had been thought to have.

How does the assumption of the "purity of the genes in the heterozygote" fare? This becomes a question of the possibility of intra-genic recombination—but not a simple question: we must ask not only what happens, but also what an experiment detects.

We can now explain in terms of the design (I nearly said "logic"!) of the recombination experiment why, even if they should occur readily, it was very difficult to find intragenic cross-overs and recombinations. We can do this in terms of the molecularly characterized gene, *but there is no need to do so*. Morgan could have done so himself, for it is an obvious consequence of the "classical model" of the genome.

(1) On this model, there were a large number of genes on each chromosome. Muller estimated in 1919 that there were at least 500 genes on the *X* chromosome in *Drosophila,* and we now know that to have been at least a fourfold underestimate.[30]

(2) It was taken as a given then, as now, that any individual gene has a very high stability, which would have applied either to intra-genic recombination or to any other mutational event.

(3) The design of a recombination experiment involved looking at a small number of "marker" genes spaced along the chromosome in order to see how frequently they (or, more accurately, the traits which signal their presence) stay together in offspring. The usual number of marker genes was 2, though 3 and 4 were occasionally used to detect multiple crossing-over by Sturtevant.[31] Supposing even that one could detect any intra-genic recombination occurring in any of the marker genes (see 4), the very small fraction of the genome being used as marker genes renders it very probable that recombinational events will not occur in any of the markers, but will occur elsewhere along the chromosome, separating whole the marker genes on either side of the break.

(4) We now know that intra-genic recombination would produce a nonfunctioning gene. This would have been scored by the Morgan school as a "loss" or "mutation" of a gene, rather than as an intra-genic recombination, so they probably did *not* detect any such events that did occur. (Only with the later work on intra-cistronic complementation were the classical techniques sufficiently refined to detect such intra-genic events. But it is worth emphasizing that the problem was a technical one, and not a conceptual one for the classical approach.)

The net effect of this is twofold:

(a) The classical model itself predicts that if genes are as small and as numerous as they had to be (and they were smaller and more numerous), intra-genic recombination would be hard or impossible to detect, even if virtually all recombinational events were intra-genic.

(b) What was *seen* in recombination

experiments was whole (marker) genes separating from one another untouched.

The first fact might have produced caution. It did not. The second observation led to an extrapolated assumption that recombination occurred *between genes, generally,* rather than just *between the observed genes.* But the first fact means that the new molecular picture is *not* that different from the old model. By analogy with the old model:

(19) Crossing-over should be a monotonic increasing function of the length of the DNA involved.

(2) The probability of crossing-over should be very near 0 for lengths of DNA of the order of functional genes—e.g., cistrons.

(3) Individual base pairs, *at least,* still have the "atomistic" status of the beadlike genes of the old model, since crossing-over cannot meaningfully be said to occur within a base.

(4) The linear arrangement of the genes on chromosomes (preserved in the linearity of the primary structure of the DNA molecule) is unchanged in the modern account and plays a central role in accounting for the high stability of the genes, the high reliability of the segregation mechanisms (without which genetics would be impossible), and the low frequency of "contamination" in the heterozygote.

But intra-genic recombination *is* assumed to be possible on the molecular account[32] and not on the beads-on-a-string model. Does this make the molecular theory a "neocontaminationist" theory rather than a neoclassical one?

Castle had no well worked out mechanism, but only a set of experiments which purported to show that classical (pre-Morganian) Mendelism did not work. There was little in Castle's work from which "neo-contaminationists" could claim descent. The purported phenomena of Castle's experiments for "contamination" turned out to be nonexistent or to admit of Morganian explanations. His explanations had no important connections with the explanations a molecular "neocontaminationist" would give for his "neocontamination" phenomena, but Morgan's did. Thus without a theory, a mechanism, or a set of phenomena persisting through time to call their own, there is no "Castilian genetics," and there are no molecular "neocontaminationists."

The kinds of connections between the two accounts clearly support the claim that the mechanism of the Morganian and molecular theories (especially when looked at with the time and size scale appropriate to the Morganian account—a move appropriate to showing that $T_2$ and $T_2^*$ are strongly analogous) are indeed strongly analogous. I thus agree with Schaffner and Ruse on this issue.

Indeed, there has been so little change, and what has changed has done so with such continuity that it is tempting to describe this as not a case of successional reduction at all. It is very tempting to say that Morgan's gene *is* the molecular gene, at a different level of description, and conversely. But to make this identification in the same breath with a claim of strong analogy is to invite confusion of identity by descent of concepts in successive theories (which is a similarity relation) with referential identity of different level descriptions of the same object (which is an identity relation.) The former notion requires no further attention now, but the latter concept and its role in reductive explanations and analyses is radically different on this account from that suggested by the formal model. Furthermore, the much better fit of this account of the role and uses of identity hypotheses with actual scientific practice is one of the strongest arguments for this account and against that of the formal model.

## VIII. IDENTIFICATORY HYPOTHESES AS TOOLS IN THE SEARCH FOR EXPLANATIONS

In its earlier formulations, the classical model of reduction had nothing to say about the role of identifications in reduction. Thus Nagel (1961) suggests that bridge laws or correspondence rules might be grounded in definitions, conventions, or empirically discovered correlations or hypothesized identifications, as if one were as good as another. The widespread instrumentalism and mistrust of identifications as metaphysical and as going beyond the evidence has perhaps led many writers away from asking why scientists might prefer to make one claim rather than another. In the one area where this has been hotly debated (and where postulating identities or postulating correspondences is seen as making a metaphysical difference which bears immediately on matters of importance), philosophers of mind appear to believe almost universally that identity claims are a solely metaphysical and evidentially unsupportable extension beyond the evidence of observable correspondences. (See Kim, 1966, for a representative and influential view.) Only recently (see, e.g., Causey, 1972) have philosophers of science found a necessary role for identities in reduction. I wish to suggest a heretofore unexplored and absolutely central role for hypothesized identifications as tools in the search for explanations which, among other things, explains a number of features concerning their use which have been considered to be unjustified, unjustifiable, or otherwise anomalous. (I have discussed some aspects of this analysis more fully in Wimsatt, 1975, 1976).

I will assume that we are faced with some upper-level explanatory problem: some phenomenon for which we have no micro-level explanation, or perhaps something which lower-level accounts would lead us to expect at the upper level, but which has not been observed. Such an explanatory failure suggests inaccurate compositional information, or none. How do we discover the source of these inaccuracies, of the locus of our incomplete information? An identity claim with its subsequent application of Leibniz's Law provides the most rigorous detector of possible error or of a failure of fit of applicable descriptions at different levels: *Two things are identical if and only if any property of either is a property of the other.* If there are properties apparently had by one but not by the other, then either the identity claim is false (as many are) or else *there are as yet undiscovered translations between descriptions at the different levels* which show that the relevant properties are indeed shared.

Thus, *in principle,* translatability

(or analyzability) is a corollary to, and the cutting edge of, an identity claim. The identity claim is in turn a tool to ferret out the source of explanatory failures which, by its transitivity, allows one to delve an arbitrary number of levels lower if need be to pinpoint the mismatch, or by its scope, to any properties—however diffuse and relational—to detect a relevant but ignored interaction. (For this reason, I do not share the view of some writers that Leibniz's law should be weakened in all sorts of ways for intensional contexts, and the like.)

Several interesting features follow from this account:

(1) It would be expected that identity claims and claims of translatability should be honored more in the breach than in the observance. They function primarily as templates which help us to locate and to focus upon *relevant* differences—differences which can help us to solve explanatory problems—in order to remove these differences and thereby to make more accurate identity claims. Thus the warrant for claims of *in principle* translatability, which was questioned in section IV, is the same as that for making the identity claim from which it flows.

(2) The warrant for this claim is in part the warrant for using a good tool appropriately: that its employment at this time and in this place may help us to discover a description or suggest a redescription which will allow us to explain some heretofore unexplained phenomenon. There is *no* warrant for using the claim if it is *known* to be false. The strength of the claim, which makes it such a sensitive template, renders it easily

falsified, and like any strong claim, its negation carries no or little significant information. Thus, if one of the standard defeating conditions for identification, such as causal relation or failure of spatiotemporal coincidence is known to obtain, the claim is dropped—though perhaps in favor of a correspondence claim (see Wimsatt, 1975, part II).

(3) This kind of warrant can, however, apply early in the stages of an investigation, and explains behavior which seems irrational and unjustifiable on a more inductivist account of the making of identity claims. Identity claims are often made on the basis of correspondences between or explanations of only two or three properties, often together with some subsidiary background information of a noncorrelational nature. I have argued (in Wimsatt, 1976) that this was, in fact, true for the early identifications, by Boveri and by Sutton, of Mendel's "factors" with the chromosomes. To the inductivist this would look like a wildly irresponsible claim: a projection from two or three properties of a pair of entities to *all* properties of those entities. Moreover, to add insult to injury, the burden of proof after the making of such a claim is not upon its maker (as one would expect on an inductivist account), but upon those who *doubt* the claim, who must come up with a counterinstance. Only then is the maker obligated to respond to the putative counterinstance, either by elaborating and defending his claim or by giving it up, as the case seems to demand. Sutton and Boveri proposed a number of new correspondences on the basis of their identifications, and these were later observed, though

subsequent conceptual modifications and clarifications led to an elaboration of the identification claims by Morgan and his students, and the generation of many new predicted correspondences (Wimsatt, 1976). The early stages at which identities are proposed; the fact that they seem to provide the basis for, rather than be made on the basis of, claims of correspondence; and the location of the burden of proof after the making of an identity claim—all support this account of the role of identity claims against the inductivist, who should expect the opposite in each case.

(4) The fragility and falsifiability of identity claims are hidden by the "open texture" of our concepts (Waismann, 1951), and in more severe cases by the same tendency to claim identity by descent of our concepts that makes successional reduction possible. With successional reduction, the similarities *and* differences in the successive theories are analyzed critically and used. Only afterwards is the similarity implied by the possibility of performing a successional reduction invoked to maximize the apparent continuity in this identity-by-descent of theoretical concepts. Similarly, with interlevel identifications, the similarities are used critically to ferret out the differences, and only afterwards are the newly assimilated differences reified after the fact into the original identification. The fact that it has become more specific, more detailed, and sometimes has undergone outright changes is hidden from us, so that we see only the continuity of "identity by descent" in our concept of the specific identifications we have made.

(5) This analysis suggests that scientists should prefer identity claims to claims of correspondence when there is no specific reason (such as the violation of one of the identity conditions mentioned in section 2, above) to prefer correspondence. They should do so because they prefer the stronger tool, and not for reasons of "ontological simplicity" (or whatever), as suggested by Kim (1966). From a specific identification, after all, one can generate all necessary correspondences, including new ones which might arise as new properties and relationships are discovered at one level or another. But from the set of correspondences one might derive from an identification, given what is known at a given time, one could *not* (without covert reintroduction of the identification) know how to generate new correspondences to fit the new information as it comes in. Identifications are an effective guide to theory elaboration. Correspondences are not. Thus one can understand not only why identity claims might be made early in the course of an investigation, but also why the metaphysically more conservative strategy of making correspondence claims instead will not work. In a static view of science, identity claims and corresponding claims of correspondence only may be empirically indistinguishable. But in a dynamic view of science, only identity claims can effectively move science forward.

The analysis of reduction and of correlative activities proposed here has differed from most extant analyses in two important respects: it has been primarily functional, with the aim of deriving and explaining salient structural features (including some not explained by the standard model) in terms of their functioning in efficiently promoting the aims of science; most notably, explanation. Secondly, it has aimed at a dynamical account of science, in which optimally efficient change and elaboration are the primary process, and in which stasis is an artificial construct, or a temporary blockage which must be explained, or an end state which we are not likely to reach in the foreseeable future. I believe further that it supports realistic conceptions of the nature of theoretical entities, and of the functions and roles of scientific theory, and does so while being truer to the ways in which scientists *actually*

behave than the extant analyses of these activities deriving from the structuralist, static, and often instrumentalist logical empiricist tradition. Finally, it fits into a broader generically evolutionary account of man and his activities, and encourages me to believe that biology may soon be a source for paradigms and analyses which will inform philosophy and philosophy of science generally, rather than being little more than the backwards field for the brushfire skirmish in which philosophical imperialists moving out from the "hard" sciences stop to try their weapons. The latter time is now fast receding into the past, but it is not yet so far that almost all of us cannot remember it.

## APPENDIX I. MODIFICATIONS APPROPRIATE TO A COST–BENEFIT VERSION OF SALMON'S ACCOUNT OF EXPLANATION

Salmon (1971, p. 55) defines what it is for one variable to "screen off" another as follows:

> ...$D$ screens off $C$ from $B$ in reference class $A$ if and only if:

(i)      $P(B/A.C.D) = P(B/A.D)$                    [$C$ adds nothing to $D$.]

(ii)     $P(B/A.C.D) \neq P(B/A.C)$                   [$D$ adds something to $C$.]

Thus, on this interpretation, microstate description $D$ in statistical thermodynamics *screens off* the macrostate description $C$ from $B$ (a macrostate in accordance with a phenomenological macro-law) in $A$ (a macroscopically characterized assumed-ideal gas). This is so because of those fluctuations from the equilibrium state predictable from $D$, but not predictable from $C$, which generates the inequality in (ii).

Note how this definition handles an upper-level anomaly (say, a macroscopically unpredictable fluctuation). Since it would be true that:

(1)      $P(B*/A.C.D) = P(B*/A.D)$

(2)      $P(B*/A.C.D) \neq P(B*/A.C)$

where all is as before except that $B*$ is a macrostate violating phenomenological macro-laws, it is clear that according to the above definition, $D$ screens off $B*$ from $C$ in $A$.

It is the consequence and intent of Salmon's definition that any strict improvement in information requires saying that the variables generating the improvement screen off any other set of variables which they represent this sort of improvement upon. *This so no matter how small the improvement and how great the cost resulting from adopting the new set of variables.* It is another consequence of accepting a view of scientific method appropriate to Laplacean demons.

I think that scientific practice and good sense suggest the value of a different notion of "screening off," which, because of its obvious connections with cost-benefit analysis might be called the "effectively screens off" relation:

*C effectively screens off D* from $\dot{B}$ in reference class *A* if (and perhaps not only if):

(a)   $P(B/A.C.D) = P(B/A.D)$

(b)   $P(B/A.C.D) \simeq P(B/A.C)$  [*D* improves the characterization only a little.]

(c)   $C(D) \gg C(C)$  [*D* is enormously more expensive information to get than *C*.]

(c')   *D* is a *compositional redescription* of *C*. .

Some comments are in order about conditions (c) and (c'), which are probably alternatives, or nearly so. The second condition comes closer to capturing the intended application of the effective screening-off relationship in the present context, since I am here considering inter-level explanatory reductions, where the lower level is a compositional redescription of the upper level. Furthermore, at least empirically, the truth of (c') appears to guarantee the truth of (c), at least for those kinds of cases we are likely to regard as interesting compositional redescriptions, and thus for all of those cases where we are likely to find any room for debate in the matter of inter-level reduction. Indeed, I am inclined to feel that the proposed "upper level" is not a distinct level unless at least most of the compositional redescriptions of upper-level phenomena in terms of lower-level entities meet condition (c), which would, in turn, guarantee that any inter-level reduction would be nontrivial.

Condition (c) gives explicitly the cost part of the cost-benefit condition, whereas the approximate equality in (b) guarantees that the benefits, if any, of using redescription *D* are small. Obviously, the deviation from strict equality in (b) and the cost-ratio in (c) required for the effective screening-off relation to hold are interdependent, and are in turn both dependent upon outside factors which determine the importance of additional information and level of acceptable costs. These may vary with the purposes for which the theory is being used, and with any other factors (such as the current explosion in the development of computers and computational facilities) that may radically affect these costs or importances.

The situation where the approximate equality in (b) is in fact an inequality is by far the most interesting one, for *under these circumstances, D screens off C* (according to Salmon's definition) *but C effectively screens off D* (according to my characterization.) Thus, in this case, the two criteria would pick out different factors to include in an explanation of phenomenon *B*.

Condition (a) was also included for the same reason: it is the same as condition (i) in Salmon's definition of the screening off relation, and thus points directly to a class of cases in which $X$ screens off $Y$ but $Y$ effectively screens off $X$. Condition (a) would presumably be met in any case in which a successful and total theory reduction (along deductivist lines outlined by Nagel and Schaffner) holds between two theories, such that $D$ is a description imbedded in the reducing theory and $C$ is a description imbedded in the reduced theory. (I would guess that this should be provable as a theorem in the probability calculus from the characteristics of their model of reduction.)

I am not sure, however, how or even whether this result would be provable for reduction as I have characterized that relation. I rather suspect that it is not. Furthermore, in cases where no reduction or only a partial reduction has been accomplished, it would at least be true that condition (a) would not be known to be met for at least some descriptions $C$ in the upper-level theory (and further, that on a subjectivist notion of probability, condition (a) would almost certainly *not* be met for these cases).

In fact, I see no reason why condition (a) should not be dropped for the effective screening off relation, since conditions (b) and (c) (or (c')) seem to include all that is necessary—namely, the cost-benefit conditions. I have included it for the time being because it heightens the contrast between the screening off and effective screening off relations, and because I think that substantial further work is necessary to see what if any other modifications and applications seem desirable in developing a cost-benefit model of explanation. The need for at least one further clarification should be immediately obvious: since Salmon (1971, p. 105) points out that his screening-off rule follows from his characterization of explantion, if I believe that the effective screening-off relation says something fundamental about the notion of explanation (as I do), it is necessary for me to produce an appropriately modified concept of explanation. This is better left to some future date.

An important consequence of adopting the effective screening-off relation rather than the screening-off relation was assumed in the text. This was that although upper-level descriptions meeting upper-level laws would effectively screen off lower level redescriptions, upper-level anomalies—upper-level descriptions which failed to meet upper-level laws—would *fail* to effectively screen off lower-level redescriptions. This introduced an important asymmetry between cases which met upper-level laws (and thus which were acceptably explained at the upper level) and cases which were upper-level anomalies (and which thus had to be explained at the lower level). On Salmon's screening-off relation, there is no asymmetry of course, since both cases which meet and cases which fail to meet upper-level laws are explained at the lower level, because lower level variables screen off upper level variables in either case.

This asymmetry arises in the following way for the effective screening-off relation. Suppose as before that $B^*$ represents an upper-level description which is anomalous for upper-level theory. Presumably, then:

(a)     $P(B^*/A.C.D) = P(B^*/A.D)$
(b)     $P(B^*/A.C.D) \neq P(B^*/A.C)$

The failure of condition (b) occurs because if $B^*$ is an anomaly, then $P(B^*/A.C)$ must either equal zero or be very low, and much lower, for example, than the probability of states that are held to be explained by the upper-level theory under similar circumstances. On the other hand, if $B^*$ is to be explicable by an account in terms of lower-level variables, it must be that there exists an appropriate description of $B^*$ such that $P(B^*/A.D)$ is appreciably greater than zero—and in general of the order that similar phenomena held to be explicable on the lower-level theory would exhibit. Thus the failure of condition (b) means that the benefits of redescribing $B^*$ at a lower level are not negligible, and in general justify the greater costs implied by conditions (c) or (c').

## NOTES

The major portion of this paper was written while I was a visiting research fellow in Humanities, Science and Technology at Cornell University. I wish to thank the program and especially Max Black and Stuart Brown for their support.

1. See Ruse (1971), and Hull (1974).

2. See Roth (1974), and Wimsatt (1975).

3. See Nickles (1973), and Wimsatt (1975).

4. On the point of *in principle* translatability, see Boyd (1972) for a masterful discussion and doubts of a more general and pervasive nature.

5. Hull (1974).

6. Ruse (1971).

7. Hull (1974).

8. Schaffner (1974b).

9. Boyd (1973, 1974, unpublished manuscript), and especially Kauffman (1970).

10. It is naturally important to distinguish between disputes over details of particular mechanisms from objections—e.g., like those of Haldane (1914) or Elsasser (1965)—which challenge the adequacy of an entire approach.

11. See Boyd (1974), and Wimsatt (1975), parts II and III. Boyd tends to locate the primary difficutly in verificationism and in acceptance of the Humean account of causation, but as a realist would also agree with the views advanced here and in my (1975).

12. Ruse has (in this symposium) retreated from his earlier attack on the formal model and attempt to characterize "informal reduction." I am more in sympathy with his earlier views.

13. This line of criticism was initiated by Nickles (1973). See also Wimsatt (1975), part II and below.

14. All of us believe that some reconstruction is necessary. Hull and I appear to believe that less is necessary (me) or appropriate (both of us) than Schaffner or Ruse. See Hull's discussion in this symposium (1976), which, however, does *not* mention the specific alternative discussed here: reconstruction of reduction as an efficient end-directed activity.

15. This approach is explicit in Kim's (1964) analysis of the deductive-nomological (or D–N) model for explanation and prediction, though Kim advances this as a defense of the D–N model, (by suggesting that the differences are pragmatic and epistemological rather than structural), and I am using it as an attack on the formal model (by suggesting that the structural similarities are more superficial than the functional differences).

Nickles (1973) individuates two types of reduction on both functional and structural grounds, but concentrates on what I call "successional" or "intra-level" reductions, largely accepts the formal model for the other kind (which is most relevant here) and does not draw the close links between functional and structural characterizations that can be made for each of the two types. Schaffner's (1974b) argument for the peripherality of (formal) reduction in the development of molecular biology invokes Bayesian arguments for choosing scientific research strategies, which presupposes a purposive account of scientific activity, but he has not attempted a functional analysis of reduction or of other related activities.

16. Schaffner (1976, pp. 626–628) appears to regard Nickles' "reduction$_2$" and the correlative notion of a transformation as a competitor to his condition of strong analogy, and criticizes it—uncharitably, I think—for being too open ended, in that there seems (says Nickles) to be no general way to characterize what kinds of transformations should be allowable and what should not. He claims, by contrast, that the notion of "strong analogy" can be applied with general agreement (3 out of 4, at least—*pace* Hull!) in the case of genetics and thus, though it is unanalyzed and primitive, it is at least testable. But surely Nickles could claim as much for the notion of an allowable transformation. I suspect that there would be general agreement in any given case on what transformations would be allowable in constructing a "reduction$_2$." I believe that Nickles despaired of finding something which I don't think exists: a general theory-independent criterion which would determine the allowable transformations. This is impossible for the same reason that a theory independent notion of "strong analogy" would be impossible: what transformations are allowable (or even interesting) and what features of an analogy are salient depend upon usually quite general and important features of theory in that area. And on these, there would usually be general agreement. Further, the notion of a transformation is mathematically an extremely powerful and suggestive one, and is less tied down to intuitive notions of similarity than analogy. For three relevant examples which are very different in terms of allowable transformations, but for each of which there would be agreement on what transformations would be allowable, see Minsky and Papert's applications of linear transformations to the analysis of the data-manipulating capabilitites of certain classes of neural networks (1969); the "law of similitude" and its use in building scale models of ships and aircraft for testing in wind tunnels and towing tanks; and the continuous deformations allowable in the applications of conformal mapping to 2–dimensional airfoil theory (see Prandtl and Tietjens, 1957) and in D'Arcy Thompson's application of his (1961) theory of transformations to problems of development and allometric growth. Indeed, none of these has been seen as involving anything like reduction, and it is one of the more provocative aspects of Nickles' analysis that it suggests the possibility of seeing them in a new light.

17. Nickles gives a more complete account of theory succession and elaboration in his paper in this volume (Nickles, 1976). His paper suggests and may require modifications to the account of successional reduction adumbrated here, but seems to lend further support to the general functionalist approach. His more recent account seems to show some of the features of both intra- and inter-level reduction, but this is to be expected in the analysis of any multi-level historical case, which should involve both components of change. Further, those ways in which his new account differs from his earlier one, or from the view advanced here, should be of no comfort to advocates of the "standard model."

18. Ruse (1971) suggests that reducibility is a similarity relation, but gives different reasons (which I do not accept) for saying so.

19. In his symposium paper (Ruse, 1976), Ruse gives up this view and attacks Hull for holding it. In this matter, I agree with Ruse (and Schaffner), though in virtually all other respects I agree with Hull.

20. For the earliest statement of a closely related view, see Simon (1969, chap. 4). See also Bronowski (1970). For my general approval of and some dissatisfactions with Simon's view see my (1974), and for a thorough discussion of levels see my (1975), part III.

21. Thus, e.g., in his first (1967) presentation of his general reduction paradigm, Schaff-

ner made provision for upper-level modifications or corrections, but not for lower-level ones, a matter which he corrected later (in his (1969)).

22. This picture is in this respect very close to that drawn by Friedrich Waismann in his penetrating essays "Verifiability" and "Language strata" (Waismann, 1951, 1953), though he put more weight on the language and less on the underlying structure of the world than I would. In particular, Waismann suggests that different language strata might not fit exactly, but would permit nearly exact translations at some points and none, or only very rough and partial ones, at others. This is roughly what I believe to be true for the languages which best describe phenomena and entities at different levels of organization.

23. Traits which are highly variable in irregular ways are unusually difficult to select for in most cases, so one might argue that it would be highly unlikely that they would be included as part of a *functional* mechanism. But all or virtually all mechanisms which are of interest in biological organisms are functional. Thus highly variable things are not likely to be included as parts of biological mechanisms. No less an ecologist than G. E. Hutchinson used this elegantly to argue (Hutchinson, 1964) that certain trace materials probably could not be utilized by organisms to perform any characteristic functions because they were present in amounts of less than about $10^4$ atoms per cell, which Hutchinson suggests as a rough stochastic threshold below which fluctuation phenomena rendered their presence too unreliable to be used by selection in any biological processes. Unfortunately, this reasoning does not apply symmetrically to allow one to assume (as Dinman, 1972, does) that lower concentrations of trace elements could not *disrupt* functional processes.

24. This realism may look superficially very much like a kind of instrumentalism, because our perceptual apparatus, senses, cognitive apparatus, and theories are all treated as instruments designed by biological psychological and social selection processes according to cost-benefit constraints which naturally introduce biases. But the biases are taken seriously as deviations from a correct portrayal of the real world. We regard the biases of the senses, theories, etc. as leading to *false* judgements which we try to correct when appropriate. That a good theory is a useful *instrument* for getting around in the world is a product of the fact that it contains a good deal of *truth*. This is no form of instrumentalism.

25. I do not think that this is the *best* way to argue for this conclusion, primarily because I believe that judgments as to where one should look for an explanation of a phenomenon are made on other grounds, which determine whether a standard causal, micro-level, or functional explanation is appropriate, and that the judgments of relative likelihood follow from these in any given case. Nonetheless, at least globally (not in specific cases), I think that the likelihoods are assumed to be as they are in the argument in the text, and the matter is clearly worth further study.

26. If it were to turn out that there were a single micro-variable which partitioned the macroscopic reference class into exceptions and non-exceptions to the macro-law, this micro-variable would give the relevant lower-level type-descriptions for a reduction. The force of Hull's complaint concerning the complexity of reduction functions is that there isn't even a small number of such variables. The force of ergodic theory is to suggest that the same problem affects statistical mechanics, but that the number of "pathological" states involved is so small (of measure 0) that we nonetheless treat it as a reduction. (See Sklar, 1974.) The number of "pathological" states in the case of genetics is *not* likely to be of measure 0 however.

27. I am not in all respects using Mendel's terminology (or even his assumptions) in this description, but the respects in which it is thereby distorted do not affect the present argument.

28. I am here talking about the possibility of a single break, so the complications of "interference" and multiple crossing over do not arise. But even this ignores the complica-

cation that breakage strength may vary along the chromosome. All of these factors were recognized and discussed by the Morgan school.

29. See Carlson (1966), pp. 83, 85, 158ff.

30. The underestimates in the number of genes was a crucial factor in overestimating their size. This was one area in which further progress raised questions about the classical model, such as Muller's doubts that the unit of mutation was the same as the unit of recombination.

31. There are a variety of reasons why it becomes experimentally more difficult to handle a large number of markers in a given experiment, and the largest number ever followed at once to my knowledge was 6, by Muller, and that for a very special kind of test of the linearity hypothesis. (See Muller, 1920, especially table II, for discussion of why smaller numbers of marker genes were usually followed.)

32. Indeed, this may exaggerate the difference. Evidence is accumulating in *Neurospora* (a bread mold widely used in genetic experiments) that there is a strong or even an absolute bias against intra-genic recombination at a molecular level. This is a product of site specificities where the "nickases" (enzymes which nick open the DNA to allow recombination) will act. If this phenomenon is veridical and generalizable, then the "beads on a string" view of the genome is inappropriate only for suggesting a macro-mechanical metaphor rather than a chemical or a micro-mechanical one. (See Whitehouse, 1973, pp. 367–369, for relevant discussion.) I thank Thomas Kass for helpful discussion of this and other related points.

## REFERENCES

Boyd, Richard, 1972, Determinism, laws, and predictability in principle. *Philosophy of Science,* 39:431–450, No. 4 (December).

——, 1973, Realism, underdetermination, and a causal theory of evidence. *Nous,* 7: 1–12, No. 1, (March).

——, 1974, Materialism without reductionism: Non-humean causation and the evidence for physicalism. Mimeographed draft, 140 pp.

Bronowski, Jakob, 1970, New concepts in the evolution of complexity: Stratified stability and unbounded plans. *Synthese,* 21: 228–246.

Campbell, Donald, T., 1974a, Evolutionary epistemology. In P. A. Schilpp, ed., *The Philosophy of Karl Popper,* v. 1: 413–463 (LaSalle Illinois: Open Court).

——, 1974b, 'Downwards causation' in hierarchically organized biological systems. In F. J. Ayala and T. Dobzhansky, eds., *Studies in the Philosophy of Biology,* pp. 179–186 (Berkeley, University of California Press.)

Carlson, Elof, A., 1966, *The Gene: A Critical History,* Philadelphia, Saunders.

Causey, R. W., 1972, Attribute-identities in micro-reductions. *Journal of Philosophy,* 69: 407–422, No. 14 (August 3).

Dinman, Bertram, D., 1972, 'Non-concept' of 'no-threshold' chemicals in the environment. *Science,* 175: 495–497 (February 4).

Elsasser, Walter, M., 1965, *Atom and Organism,* Princeton, Princeton University Press.

Fodor, Jerry, A., 1974, Special sciences (Or: the disunity of science as a working hypothesis). *Synthese,* 28: 97–115.

Glymour, Clark, 1975, Relevant evidence. *Journal of Philosophy,* 72: 403–425, (August 14).

Haldane, J., S., 1914, *Mechanism, Life and Personality,* New York, Dutton.

Hull, David, L., 1972, Reduction in genetics biology or philosophy?, *Philosophy of Science,* 39: 491–499 (December).

——, 1974, *Philosophy of Biological Science,* Englewood Cliffs, Prentice-Hall.

——, 1976, Informal aspects of theory reduction. This volume, pp. 462–476.

Hutchinson, G. E., 1964, The influence of

the environment. *Proceedings of the National Academy of Sciences,* 51: 930–934.

Kauffman, Stuart, A., 1972, Articulation of parts explanation in biology and the rational search for them. In *PSA–1970,* R. C. Buck and R. S. Cohen, eds., *Boston Studies in the Philosophy of Science,* 8: 257–272.

Kim, Jaegwon, 1964, Inference, explanation and prediction. *Journal of Philosophy,* 61: 360–368, No. 12 (July 11).

——, 1966, On the psycho-physical identity thesis. *American Philosophical Quarterly,* 3: 227–235.

Lewontin, Richard, C., 1974, *The Genetic Basis of Evolutionary Change,* New York, Columbia University Press.

Maull, 1974, see Roth, 1974.

Minsky, Marvin, and Papert, Seymour, 1969, *Perceptrons: A Study in Computational Geometry,* Cambridge, M.I.T. University Press.

Muller, Hermann, J., 1920, Are the factors of heredity arranged in a line? *American Naturalist,* 54: 97–121 (March–April).

Nagel, Ernest, 1961, *The Structure of Science,* New York, Harcourt.

Nickles, Thomas, 1973, Two concepts of inter-theoretic reduction. *Journal of Philosophy,* 70: 181–201 (April 12).

——, 1976, Theory generalization, problem reduction, and the unity of science. In R. S. Cohen *et al.,* eds., *PSA 1974,* Dordrecht, Reidel.

Prandtl, Ludwig, and Tietjens, O. G., 1957, *Fundamentals of Aero- and Hydromechanics* (New York, Dover) (reprint of original volume published in 1934 by McGraw-Hill).

Roth, Nancy Maull, 1974, Progress in Modern Biology: An Alternative to Reduction, Ph.D. dissertation, Committee on Conceptual Foundations of Science, University of Chicago.

Ruse, Michael, 1971, Reduction, replacement, and molecular biology. *Dialectica,* 25: 39–72.

——, 1973, *The Philosophy of Biology,* London, Hutchinson University Library.

——, 1976, Reduction in genetics. This volume, pp. 446–461.

Salmon, Wesley, C., 1971, *Statistical Explanation and Statistical Relevance,* Pittsburgh, University of Pittsburgh Press.

Schaffner, Kenneth, F., 1967, Approaches to reduction. *Philosophy of Science,* 34: 137–147, (June).

——, 1969, The Watson-Crick model and reductionism. *British Journal for the Philosophy of Science,* 20: 325–348.

——, 1974a, Logic of discovery and justification in regulatory genetics. *Studies in History and Philosophy of Science,* 4: 349–385, No. 4.

——, 1974b, The peripherality of reductionism in the development of molecular biology. *Journal of the History of Biology,* 7: 111–139, No. 1 (Spring).

——, 1976, Reductionism in biology: Prospects and problems. This volume, pp. 428–445.

Shimony, Abner, 1971, Perception from an evolutionary point of view. *Journal of Philosophy,* 68: 571–583, No. 19 (October 7).

Simon, Herbert, A., 1969, *The Sciences of the Artificial,* Cambridge, M.I.T. University Press.

Sklar, Lawrence, 1967, Types of inter-theoretic reduction. *British Journal for the Philosophy of Science,* 18: 106–124, No. 2, (August).

——, 1973, Statistical explanation and ergodic theory. *Philosophy of Science,* 40: 194–212, (June).

Thompson, D'Arcy, W., 1961, *On Growth and Form,* abridged edition, edited with commentary by J. T. Bonner, London, Cambridge University Press.

Waismann, Friedrich, 1951, Verifiability. In A. G. N. Flew, ed., *Logic and Language* (first series), London, Blackwell, pp. 117–144.

——, 1953, Language strata. In A. G. N. Flew, ed., *Logic and Language* (second series), London, Blackwell, pp. 11–31.

Whitehouse, H. L. K., 1973, *Towards an Understanding of the Mechanisms of Heredity,* third revised edition, New York, St. Martin's Press.

Wimsatt, William, C., 1974, Complexity and organization. In K. F. Schaffner and R. S. Cohen, eds., *PSA–1972, Boston Studies in the Philosophy of Science,* 20: 67–86, Dordrecht, Reidel.

——, 1975, Reductionism, levels of organization, and the mind-body problem. In *Consciousness and the Brain,* edited by G. G. Globus, G. Maxwell, and I. Savodnik, New York, Plenum, pp. 205–267.

——, 1976, Correspondence versus identity and the problem of spatiality in the localization of the genome and determining the configuration of the mental realm. Invited address, Section VIII (Foundations of Biology), *5th International Congress on Logic, Methodology and Philosophy of Science,* London, Ontario, August 31, 1975. To be published in the proceedings, edited by Jaako Hintikka, by D. Reidel, Dordrecht.

# 27

# Unifying Science without Reduction

*ᗡᖾᗡᖾᗡᖾᗡᖾᗡᖾᗡᖾᗡᖾᗡᖾᗡᖾᗡᖾᗡᖾᗡᖾᗡᖾᗡ*

## NANCY MAULL

### INTRODUCTION

Surely, doubts that scientists once entertained about the "ontological status" of classical genes have now been laid to rest: classical genes really are parts of intracellular molecules. Likewise, it is no longer profitable to doubt that the molecules of living things function in accord with the same physico-chemical laws that hold for all molecules.[1] The decisive discoveries about the relationship between the "biological" and the "physico-chemical" came about as the result of developments in different branches of science—for example, in genetics, biochemistry, and physical chemistry. Yet none of these areas of investigation developed "autonomously"—that is to say, unaffected by neighboring fields of research. Rather, each of these branches of science has, as part of its own evolution, generated important conceptual, theoretical, and problem-solving links with other branches. Indeed, and this

is my main point, the very genesis of links between different areas of investigation is an important though neglected type of scientific change. Such processes of linkage, in fact, contribute to a gradual "unification" between biological and physical sciences. Until quite recently, of course, a widespread methodological commitment to the logical "unity of science" has led philosophers to ignore the historical processes through which such "unification" takes place. This paper, in contrast, focuses on precisely these processes. It is thereby meant as a contribution to the genetic and developmental reinterpretation of the "unity" of science.

Every attempt to understand the interaction between different branches of science soon confronts a cluster of problems called "reduction." Philosophers anticipating a logical unity among sciences commonly interpreted reduction as the deductive-nomological explanation of one theory by another theory. Yet the approach

called *derivational reduction* leaves
unanswered—not to say unasked—
many important questions about
(1) interactions between branches
of science, (2) intermediate or linking
theories, and (3) the way the "unity"
of science comes about.[2]

1. What differences, if any, are to
be found between theories and
branches of research? Recall that
derivational reduction is explicitly
conceived as an intertheoretic relation. But can interactions between
branches of science be adequately
reconstructed as intertheoretic relations?

2. How can we understand the
generation and function of hypotheses
that join theories or branches of
research? Almost all derivational reductions, as we shall see, require
"connections" between the terms of
the reduced and the reducing theories.
Connections between terms, however,
are not established by virtue of the
meanings of those terms. Rather,
connections are discovered in processes
of research; they are hypotheses.
But under what circumstances are
such linking hypotheses generated?
And what relation do these hypotheses
bear to the parts of science that they
connect?

3. Finally, what do the fairly recent changes in the relationship between the biological and the physical
sciences tell us about their unification?
The derivational reduction model mis-
leadingly identifies the unity of science with the cumulative reduction
of theories, established "in principle"
if not in fact. But what relation does
such an account bear to the historical
development and goals of science?

I am concerned with precisely these
questions, questions that are seldom

raised because of the widespread influence of the derivational reduction
model. I shall not, of course be able
to give an exhaustive answer to all
of them, though that is the direction
in which I want to move. In "Derivational Reduction" I explain why
reductionists have never been able
to deal intelligently with such questions. And in the subsequent two
sections on descriptive levels and
problem shifts, I propose a general
answer based on my own historical
investigations of the relationship between the biological and physical
sciences. This alternative to derivational reduction begins by drawing
attention to the way a vocabulary
can be "shared" by different areas
of research. Such a "shared" vocabulary, it turns out, can be used to
identify a very special sort of problem, one that, although it arises within
one branch of inquiry, can be solved
only with the aid of another science.
As I shall argue, this type of "shared
problem" is solved by a theory but
not by a reduction of theories. Thus
in the final two sections of the paper
(on interlevel theories and on the unification of science) I return to the
failure of reduction, using that failure
to explain the relevance of my own
approach for answering the pivotal
question: What kind of unity can
reasonably be attributed to the
development of science itself?

## DERIVATIONAL REDUCTION

Models of derivational reduction require that two formal conditions be
met, conditions which Ernest Nagel
calls *connectability and derivability*.[3]
The first of these, the connectability
condition, stipulates that all primitive

terms of the secondary theory ($T_2$, the theory to be reduced) be associated with one or more terms of the primary theory ($T_1$, the reducing theory). These connections are necessary because in some cases, perhaps the most interesting ones, the secondary theory employs a number of terms not included among those of the primary theory. The second and stronger requirement for derivational reduction is that the secondary theory be *derivable* from the primary theory plus the requisite connections between terms. Although the construction of connections between the terms of what Nagel calls "heterogeneous" theories is a necessary condition, it is not a sufficient condition for derivational reduction. A deductive relationship between the propositional components of theories, on the other hand, *is* sufficient for reduction. Deductive relationships, of course, have the notable advantage of being truth-preserving. In order to secure this advantage, philosophers like Nagel do not hesitate to conceive derivational reduction as a relation which holds between fully articulated and well-supported theories. Indeed, one reason why the deductive model is so attractive is that it provides what initially appears to be a transparent and simple way of talking about the unity of science in terms of a deductive hierarchy of secure theories.

Following Nagel, philosophers interested in the relations between different areas or branches of science hold the derivational reduction of theories to be a model or idealization of the most interesting of these relations. Not surprisingly, this model is limited in at least two crucial ways.

First it confines our attention to relations between theories and, as a consequence, reinforces the tendency to blur theories and branches of science. (Nagel talks about the "reduction" of a branch of science as if it were simply a matter of reducing the comprehensive theory of that branch![4]) This conflation seems arbitrary, however, for a distinction between theories on the one hand and broader aspects of science on the other is a conspicuous element in our ordinary thinking about the history of science. For example, although we do not usually think of theories as including techniques and problems, we *do* associate these elements with areas of science—that is to say, with disciplines or branches of inquiry. Of course, the case for such a distinction, or rather for its fruitfulness, does not rest with ordinary or historiographical usage. Still, philosophy cannot simply ignore conceptual distinctions which are well entrenched in the historical discussion of research. We must remain open to the likelihood that relations between branches of science can be distinguished philosophically from relations between theories.

Second, in Nagel's account, derivation is taken to be the essential feature of relations between theories. The elaboration of connections between terms is understood as a mere means to derivations. As we shall see, this is unfortunate, since connections between terms (*either* terms of theories *or* terms of different branches of science) are far more important than derivational accounts indicate. Indeed, as I said, such connections are actually hypotheses about the relationships between different areas of investigation.

In short, then, derivational reduction is a model for the unity of science that uncritically confines our attention to deductive relations between theories. Moreover, derivational reduction is not only supposed to give an exhaustive account of the interesting and important relations between theories and (by implication) between branches of science. It is sometimes presented as a description or, perhaps, a rational reconstruction of the actual course of scientific developments. To be sure, the relationship between a reduction model and particular developments in research is not always obvious. When such reduction models fail to be descriptive, however—and this is what provokes suspicion—they are rebaptized "idealizations" of science. Thus Kenneth Schaffner recently described his own reduction model as "peripheral" to the research strategies and goals of some of the important developments at the interface of genetics and biochemistry.[5] As a consequence of such evasive maneuvers, however, no particular development in science can stand as evidence for or against a model of reduction.

Not only does the model fail as an adequate description of particular research developments, but it also fails as a description of long-term tendencies in science. Notice that successful derivational reduction is always said to depend upon the *greater generality* of the reducing theory (an assumption that drew Nagel's attention to the analysis of explanatory generality).[6] Otherwise the reducing theory would be nothing more than a "translation" of the reduced theory, and the asymmetry of reduction would be lost. Thus if

derivational reduction has *any* significance for a discussion of scientific change and progress, and especially of the "unity of science," it must stand or fall with this claim: science exhibits (or *should* exhibit) a trend toward the greater generality of theories.

Unfortunately for Nagel and other proponents of derivational reduction, the biology of the last century has *not* been marked by a trend toward greater generality of theories. To this extent it cannot be compared to Newtonian mechanics, which indeed was characterized by an advance in explanatory generality over Kepler's laws. What has happened in biology, in contrast, can more readily be compared with the development—from Galileo to Newton—of a unified treatment for terrestrial and celestial phenomena. Thus the idea of "special biological laws" (like that of "special supra-lunar laws") was gradually abandoned. This shift, however, has nothing to do with *reducing* biological theories to "more general" theories in chemistry and physics. To the extent that biochemistry shares a comprehensive theory with chemistry proper and with physics, there is now a *unified treatment* of the organic and the inorganic. Nonetheless, the collaborative relationship between these neighboring areas of research cannot be understood in terms of a deductive hierarchy of generality.

Interestingly enough, as Schaffner argues, a research strategy along the lines of a derivational model is not even to be recommended to science.[7] We cannot even suggest that scientists *should* develop increasingly general theories as a long-term strategy. In sum, although we might expect a derivational reduction analysis to be helpful for understanding the relationship between biological and physical sciences, and although these sciences have recently exhibited a fruitful and promising unification of theory, the type of greater generality which deductive-nomological explanation attributes to one theory over another is patently missing. As a result, we need to find a new way to understand the interaction of the biological and the physical sciences.

### DESCRIPTIVE LEVELS

The development of an alternative to derivational reduction can begin, I think, with the observation that many important developments, especially in modern biology, have involved interactions not between two *theories,* but between two or more branches of science, or, as I shall now call them (using the concept developed by Lindley Darden), *fields.*[8] Examples of fields between which important interactions occur are ecology and population genetics, cytology and transmission genetics, transmission genetics and biochemistry, and biochemistry and physical chemistry. It is the last two of these pairs that I shall discuss. Of course, it is difficult to characterize either a theory or a field, and the difficulties are easy enough to anticipate. But we do speak, rather naturally, of genetics, biochemistry, and physical chemistry as fields (or as sciences, disciplines, areas of research, and so forth), and not as theories. Genetics, biochemistry, and physical chemistry are not called "theories," although we do speak of "theories *of* heredity," "theories *of* macromolecules," and "theories *of* chemical binding." Moreover it seems natural to view these as theories within genetics, within biochemistry, or within physical chemistry. To speak of *a* theory of genetics, for example, suggests the possibility of competitor theories. But we do not think of fields as competing, at least not in the sense that theories compete. In other words, we already understand the elementary difference between a theory and a field and by simply refining this preliminary understanding we can begin to develop an alternative to the reductionist account of the relation between biological and physical sciences.

A field can be specified by reference to a focal problem, a domain consisting of "facts" related to that problem, explanatory goals providing expectations as to how the problem is to be solved, special methods and techniques, and sometimes, but not always, laws and theories.[9] Theodosius Dobzhansky characterizes Mendelian genetics in just this way: "Mendelian genetics is concerned with gene differences; the operation employed to discover a gene is hybridization: parents differing in some trait are crossed and the distribution of the trait in hybrid progeny is observed."[10] While classical or Mendelian genetics used gross character differences as traits (*e.g.,* eye color), modern transmission genetics focuses on molecular differences (for example, differences in proteins). Modern transmission genetics (like its classical Mendelian predecessor) can be distinguished as a field which investigates the problem of gene differences (or differences in the hereditary determinants) using the technique of hybridization and the method of inferring the gene difference from observations of the distribution of a trait in hybrid progeny. Biochemistry, another field with which I shall be concerned, can also be characterized in relation to a domain, a set of problems, explanatory goals, techniques, and methods; for example, a central problem for biochemistry lies in determining the network of interactions between the molecules of cellular systems and their molecular environments. Physical chemistry, in contrast, is now concerned with the determination of interactions between the parts of molecules relative to one another, under varying conditions, and thus with the structure and three-dimensional configuration of molecules.[11]

In order to develop an apparatus for handling the interactions between fields, I shall focus attention on the *special vocabulary* associated with each field and then on the relationships between special vocabularies of different fields.

The special vocabulary of a field, let us say, consists of the *proper terms* of that field. A term can become proper to a field in either of two ways. First, terms can originate within the historical development of the field (e.g. 'epistasis' and 'mutation' with genetics, and 'ligase' and 'transcription' with biochemistry). Some of these are "new," "coined" terms, and others are terms already used "outside" the field that are subsequently appropriated for very new uses within the field. The appropriation of terms can be illustrated by taking, as an example, the term 'mutation' and its introduction into the field of genetics. 'Mutation' was

used in evolutionary contexts in the late nineteenth century to refer to successional subspecies, that is, to morphologically different groups that occur within a phylogenetic lineage without the branching of that lineage.[12] De Vries, in his *Mutation Theory* of 1903, used the term differently, associating it with changes in the state of a pangen (a hereditary determinant). Some of de Vries' mutations could *result* in successional subspecies; but with de Vries, mutation is no longer a change in species, but a change in pangens which could result in a change in species.[13] The appropriation of 'mutation' for genetics is thus characterized by an implicit rejection of the knowledge claims previously associated with its use ("Mutations are successional subspecies"), and the replacement of such claims with others ("Mutations are changes in the state of a pangen").

There is, however, a second way in which terms can become proper to a field. In fact, if we examine the terms used in more than one scientific field, a "sharing of terms" becomes quite apparent. A term may be taken from another scientific field and, while the (unrefuted) knowledge claims previously associated with its use are retained, new claims are generated and justified within the new field. When this happens, the term is, I shall say, *transformed* and becomes a proper term of both fields. Indeed, this transformation can be illustrated by completing the history of the term 'mutation.' First a proper term of genetics, it eventually became a proper term of biochemistry as well.

De Vries' mutation theory, in fact, was rejected; he thought that some mutations could result in the formation of new species in one generation. But the cases in which de Vries thought that speciation occurred were subsequently shown to be changes due to recombination (or rearrangement among the hereditary determinants) and not due to alteration in a single determinant.[14] Thus, de Vries made some claims about alteration in the hereditary determinants; some claims were later justified, some not, and others were reinterpreted in the light of further developments in genetics.

By 1913 Morgan and his colleagues were using 'mutation' in a different way: to indicate alterations which produce new, usually small differences in traits, the inheritance of which is always Mendelian. The changes are said to be in "factors" (or, later, "genes"). With Morgan and throughout the subsequent development of genetics, mutations are understood as heritable alterations in the genotype of an organism.[15]

However, 'mutation' was to become a proper term of biochemistry by the process which I am calling transformation. Mutation was initially heritable alteration in the genotype of an organism, but we now know it also to be heritable alteration in the base sequence of the DNA of an organism. Major developments, both in genetics and in biochemistry, led to the biochemical understanding of mutation as heritable alteration in base sequence. There was early and continuing speculation as to the material nature of the hereditary determinants as to the nature of change in these determinants. The chromosome theory of heredity allowed identification of a linear order of genes with the chromosomes. And as early as 1916, Morgan was thinking of the chromosome as composed of a chain of chemical particles; any number of rearrangements would be possible within each particle and any *could,* he thought, give rise to a mutant character.[16]

By 1920, the view that genes were proteins and, more specifically enzymes, had gained popularity. Even in the famous 1941 paper of Beadle and Tatum, in which gene action was associated with enzyme activity in the "one gene–one enzyme" formula, the question whether genes *are* enzymes or *make* enzymes remained open.[17] Of course, in a series of developments beginning in 1928 with Griffith (in the transformation of pneumococcus) and culminating in the elaboration of the structure of DNA in 1953 by Watson and Crick, the hereditary material was found to be nucleic acid. On the basis of these discoveries it was then possible to investigate and understand the hereditary code, the biochemical pathway for protein synthesis, and to develop a theory of mutagenesis. The use of chemical mutagens (analogues of the bases of DNA) permitted the development of a theory of mutagenesis: individual mutant sites are changes in purines and pyrimidines, the DNA bases.[18]

I have summarized this development in order to point out how specific research findings (both new observations and new hypotheses) led to the realization that heritable alteration in the genotype was heritable alteration in the base sequence of DNA—that is to say, how 'mutation$_g$', a proper term of genetics, was transformed into 'mutation$_b$' a proper term of biochemistry.

Notice that the transition from 'mutation$_g$' to 'mutation$_b$' might have been described (far less effectively, I think) as an example of "meaning change." However, it is striking that questions of meaning change seem largely irrelevant to an examination of the development of fields and of interactions between fields. Philosophers, of course, have often overlooked this point. Paul Feyerabend's critique of derivational reduction, for example, depends on a continuing commitment to the importance of the concept "meaning" for our understanding of scientific change.[19] A study of the relations between genetics, biochemistry, and physical chemistry, however, need not await an adequate "theory of meaning," but is, rather, a matter of assessing the *problems* seen to require solutions, the methodological and technical access to those problems, and the acceptability of the claims made (or the claims that should have been made) on the basis of the "good reasons" available.

Of course, in setting out part of the history of the term 'mutation' as an example of transformation, I have advanced a hypothesis about the generation and justification of knowledge claims within genetics and within biochemistry: claims about mutation from genetics were retained and biochemical claims added. The hypothesis rests on substantial historical evidence for the following:

(i) The claims made about mutation within genetics (the claims that arose as a result of solutions to its characteristic problems and within the limits of its techniques and methods) posed problems that could not be solved within genetics. For example:

What is the physical nature of these alterations in the hereditary determinants? Then, what structural changes in the genes result in the functional differences seen in mutants? Or, what are the biochemical causes of the inheritance of an altered phenotype? Such a problem shift is characteristic of transformation. An examination of other transformations, I am convinced, would reveal that 'mutation' is typical in this regard.

(ii) Biochemistry, by contributing to the solution of such problems, generated and justified additional claims about mutation. (Such solutions would, in turn, provide new problems for biochemistry, but also for physical chemistry.)

(iii) The previous claims about mutation from genetics were, with very few exceptions (and these amounted to "corrections") not rejected as a result of the new biochemical information.

Of course, a proper term from one scientific field is often used in another field *without* the characteristic problem-shift just described in the transformation of proper terms. In this case, terms are *imported,* that is, used in a new field context, but not in such a way as to be associated with problems originating in the field from which they are imported. For example, in modern genetic contexts, proper terms of biochemistry are often used. Here is a sample of the importation of biochemical terms into genetics. In summarizing his work, Joshua Lederberg writes:

Data are given in detail of the segregation of factors involved in the biosynthesis of biotin, methionine, threonine, leucine or thiamin, in the fermentation of lactose, and in resistance to bacterial viruses T1 and T6. On the basis of these data a tentative 8 point genetic map of the chromosome of *E. coli* is presented.[20]

Notice that the proper terms of biochemistry like 'biosynthesis,' 'leucine,' and 'fermentation' are used in association with a problem and with a methodology that are characteristic of genetics, namely, the mapping of genes by hybridization studies, and *not* with a problem that arose within biochemistry, only to be solved within genetics. For this reason, the biochemical terms remain proper biochemical terms.

Remember that one of the flaws I attributed to derivational reduction accounts was a tendency to minimize the importance of connections between terms. Thus the development of analytic tools for talking about connections, such as the notion of the proper terms of a field and (especially those terms proper to more than one field—the transformed proper terms), is certainly desirable. It could allow us to discuss connections without falling back into talk about the logical relations between theories or between theories and what they explain (e.g., "correspondence rules," "bridge laws").

Indeed, the transformation and importation of terms must be essential features in any alternative program to that of derivational reduction. For the reduction program, connection (and derivation) between *theories* is required: for an alternative analysis, connections between the special vocabularies of fields are required for the ordering of special vocabularies as *descriptive levels*. The possibility depends on an asymmetry or "directionality" in importation when, for example, terms are more often imported from biochemistry to genetics than *vice versa*. Directionality is also characteristic of transformation, although the "direction" is precisely the contrary to that found in importation; genetic terms are more commonly transformed into biochemical ones than *vice versa*. (An instance of such directionality is the transformation of 'mutation$_g$' to 'mutation$_b$'.) Naturally, this observation of directionality in importation and transformation is based on the study of a number of terms in genetics and biochemistry. (The same kind of directionality can be observed between the special vocabularies of biochemistry and physical chemistry.) I also suspect that it holds as a generalization for other fields having relations similar to those exhibited by genetics and biochemistry. The directionality of importation and transformation, moreover, can be used to order the special vocabularies as descriptive levels:

I. If a proper term is imported from special vocabulary *B* to special vocabulary *A*, then a condition for *B* being a deeper descriptive level than *A* is fulfilled.

II. If a proper term is transformed from *A* to *B*, then a condition for *B* being a deeper descriptive level than *A* is fulfilled.

III. Conditions 1 and 2 will not order special vocabularies as descriptive levels if there are more than a few terms in *B* which are transformed in *A*. Jointly, these conditions define the binary relation ' . . . is a deeper descriptive level than . . .' This relation is identified with a set of ordered pairs, an example of which is <biochemistry, genetics>, with biochemistry being the deeper level. (Another pair is <physical chemistry, biochemistry> with physical chemistry the deeper level.)

Notice that descriptive levels are the ordered special vocabularies of fields. Notice, too, that the problem of different levels of scientific inquiry is usually approached as the problem of ordering *phenomena*. The ordering of phenomena might be undertaken by utilizing any number of criteria; for example, entities might be ordered according to their spatial dimensions, once a decision has been made as to how entities are to be individuated. The application of criteria for ordering often results in *structural levels,* that is, in levels exhibiting part/ whole relationships. Lower level entities are then parts of higher level wholes. For example, Paul Oppenheim and Hilary Putnam have discussed six "reductive levels" in elaborating a notion of the unity of science: social groups; (multicellular) living things; cells; molecules; atoms; and elementary particles.[21]

The point to be made here is that the descriptive levels which I have been discussing are *not* structural levels in this sense. Whatever fundamental relations are characteristic of descriptive levels, they are *not* part/ whole orderings (just as they are *not* deductive hierarchies of generality). In fact, my way of characterizing descriptive levels of science is desirable precisely because the different domains of scientific fields cannot always be ordered according to part/whole relationships between their

items. Oppenheim and Putnam say that we can assign a branch of science (which is to say, all its "accepted theories") to one and only one reductive level, on the assumption that the universe of discourse of the branch is uniquely assignable. However, the actual assignment of branches to levels is not always possible, and, in the end, Oppenheim's and Putnam's initial assumption is at fault. For example, where does genetics (or its theories) fall within the postulated hierarchy of reductive levels? Hereditary determinants, of course, constitute the domain of genetics. Genes are parts of macromolecules which in turn "contain" atoms and elementary particles (although this is not usually of primary interest to geneticists); genes are located within cells and organisms and are distributed within social groups. If we were forced to assign genetics to a single reductive level, we would certainly select Putnam's and Oppenheim's molecular level, since, after all, genes are molecules. Indeed, the fact that genetics is emphatically *not* a molecular science should convince us of the final irrelevance of the "structural level" model for understanding the relations between genetics and its neighboring fields.

Some of the difficulty encountered in assigning theories or fields to structural levels can be explained by appealing to the phenomena that underlie the transformation of proper terms; for the *same* parts and wholes are often investigated at *different* levels and designated *differently* at each level. The genes (cistrons) of genetics, for instance, are also the DNA sequences of biochemistry. Likewise, many of the entities and processes of interest to biochemistry and physical chemistry are the "same", even though they may be designated in different ways. In physical chemistry, the conformation of molecules is the focal problem for investigation while in biochemistry, function is the pivotal problem. The "same" parts of molecules are important for both areas of investigation. In evolutionary or developmental fields, the problems of assigning a single structural level are perhaps even greater. The selection of molecules, genes, cells, species and social groups is investigated within the field of evolutionary biology (including population genetics). In developmental biology, on the other hand, the ontogenesis of organisms and parts of organisms is studied. Thus, the "same" wholes and parts are investigated in different fields according to distinct (though related) *problems*.

In a given domain, characteristic problems cannot be solved solely by the determination of part/whole relationships. The determination of causal and functional relationships is also required. The discovery and elaboration of part/whole relations, in other words, does not exhaust scientific investigation. This helps explain why the branches of science— or rather their universes of discourse —cannot be satisfactorily characterized in terms of different structural levels. The special vocabulary associated with a given branch of science is not a set of predicates uniquely assigned to a single structural level. Of course, any discussion of the reduction of branches or fields of science which (like Oppenheim's and Putnam's) bears little resemblance to the historical processes of unification among these branches, is doomed at

the outset. A more fruitful approach to the unification of science is promised by a study of problems and of the interaction between neighboring problem constellations.

## PROBLEM SHIFTS

Between different scientific fields, then, a shared vocabulary often emerges. Although this is an undeniable fact about different areas of investigation, it is not a finding that has engaged the curiosity of philosophers of science. The sharing of vocabularies *should* interest philosophers, however, for it is symptomatic of other fundamental relationships between fields of research, relationships that will help us explain how a problem can shift from one area of investigation to another.

The transformation of a term (its inclusion in a deeper descriptive level) comes about because the field set at this deeper level can provide a new kind of information. New information can be provided in one or more of the following three ways:

(a) In the transformation of a term, the "deeper" field may provide *specification* of the physical nature (for example, DNA) of an entity (for example, gene) or process (for example, the alteration of base sequence). Such specification, in turn, can provide details about physical components, that is, information about the relation of whole to part.

(b) In the transformation of a term, the "deeper" field may provide information about the *structure* of entities or processes, the *function* of which is investigated within the field from which the transformed term originates. Thus, physical chemistry elucidates the structure of molecules, while the function of these molecules can be characterized biochemically.

(c) With transformation, finally, the "deeper" field may provide causal information. That is, effects detectable within one field may be causally explained by new information provided within a "deeper" field. For example, in genetics, mutation is known by its effects, namely, the predictable inheritance of an altered phenotype. New information from biochemistry (that mutation is alteration of the base sequence of DNA) provides the cause of inherited changes in the phenotype.

In short, transformations of terms between two descriptive levels can be analysed as providing new information about (a) the physical nature of an entity or process, (b) the structures associated with a particular function, and (c) the cause of particular effects (Notice that the part/whole relation, sometimes involved in specifying the physical nature of an entity or process, is one way—though not the only way—in which fields may be interrelated by the transformation of terms.)

Even more helpful for dispelling reductionist errors is the idea that a problem shift accompanies all transformations. It is possible for problems to *arise* within a field even though they cannot be *solved* within that field. Their solutions may well require the concepts and techniques of another field. In this case, we say that the problem "shifts". Let us return to our example. The discovery that mutation (mutation$_g$) is heritable alteration in the genotype posed the following problems for biochemistry

as soon as genes were localized on chromsomes: What is the physical nature of the alterations in hereditary determinants? What structural changes in the genes result in the functional differences characteristic of some mutants? What are the biochemical causes of the inheritance of an altered phenotype?

By contrast, the discovery that mutation is also alteration in base sequence, although significant, did not pose new or major problems for genetics. (To be sure, this discovery did clear up a few difficulties within genetics and was useful as background information in generating new lines of research.) However, the same biochemical finding did pose a problem for physical chemistry: What is the causal sequence (or sequences) that results in the alteration of bases in DNA? I take this example as evidence for the directionality or asymmetry of those problem shifts which accompany the transformation of terms from one special vocabulary to another.

## INTERLEVEL THEORIES

To talk about a problem shift from one field to another is to describe an especially interesting way in which fields may share a problem. Thus, we can describe the solution to a problem that shifts from one field to another as the solution to a shared problem. What, then, are the chief characteristics of solutions to a problem that is shared by two fields—supposing one field to have generated a problem that it, alone, cannot solve? One important fact about such shared problems, and a starting point for an examination of their eventual solutions, is that the concepts and techniques of both fields are available for the discovery of solutions. Of course, as I said, the concepts and techniques of the field that generates the problem will be seen to be insufficient for its solution. And moreover, the field that subsequently adopts such a problem is very unlikely to have generated it independently, just because the concepts and techniques of this field are also limited. For the same reason, this new field is even less liable to solve the problem without the aid of information from the field of origin.

Solutions to the type of problem I have described often take the form of a causal sketch. These causal sketches, I want to claim, are really interlevel theories, For example, a causal sketch (or interlevel theory) tells us how the heritable alteration in base sequence causes the predictable inheritance of an altered phenotype. The following major discoveries and developments in both genetics and biochemistry have provided the basis of our understanding of this causal relationship: the development of the concept of hereditary determinants as discrete units, the chromosome theory of heredity, the physical, chemical, and biochemical characterization of the gene, the genetic code and the biochemical pathway for protein synthesis, and a biochemical account of mutagenesis. The cause of the predictable inheritance of an altered phenotype can thus be identified with the biochemical cause (the heritable alteration in the base sequence of DNA); in other words, the causal sketch warrants the identification of 'mutation$_g$' and 'mutation$_b$'. The causal sketch is not a detailed biochemical account giving

*sufficient* conditions for the inheritance of an altered phenotype. The sketch has gaps, any number of important "initial conditions" are ignored, and it is formulated in both genetic and biochemical terms.

Other causal sketches (solutions to shared problems) are more narrowly circumscribed than the one just mentioned. The operon theory of Jacob and Monod and the Monod-Wyman-Changeux theory of allosteric regulation are both interlevel theories that postulate causal links between descriptive levels. Both are theories of metabolic control and so have a common relation to biochemistry. The operon theory bridges genetics and biochemistry; it explains how protein levels are controlled by genes. The theory of allosteric regulation, by contrast, bridges biochemistry and physical chemistry; it is a theory of the control of protein activity by conformational change. I shall briefly describe both theories.

The problem for which the operon theory provides a solution is that of how gene expression is regulated. (You can see this as a problem immediately; after all, every cell of an organism has all the organism's genes, but only some genes are expressed in each cell—in short, differentiation occurs.) In genetics, the problem of regulated gene expression originates in this form: How can the genetic apparatus cause the regular alteration of the phenotype without mutation? (The problem of gene expression was encountered very early in transmission studies; the terms 'dominant,' 'recessive,' and 'position effect' were introduced to describe states of gene expression.) For biochemistry, the problem of regulated gene expression is reinterpreted as that of understanding the biochemical interactions that affect regulated protein synthesis. In the operon theory both genetic and biochemical aspects of the problem are solved, for the interaction between structural and regulatory genes and gene products accounts for phenotype alteration without mutation, and at the same time for the regulation of protein synthesis, a biochemical pathway. The operon theory warrants connections between the genetic and biochemical levels in much the same way that the causal sketch (linking the genetic effects of mutation with a biochemical cause) warrants the identity between the genetic and biochemical designations. The theory postulates certain causal links between the levels; regulatory genes and regulatory gene products control the expression of certain structural genes (that is, genes coding for enzymes or structural proteins) in response to the availability of certain small molecules.[22]

Similarly, the theory of allosteric regulation bridges biochemistry and physical chemistry. The theory postulates that certain conformational features are common to proteins having a certain pattern of activity (sigmoid activity curves and inhibition or activation by specific small molecules). The theory is also a causal sketch of the reversible sequence of conformational change resulting in changes in protein activity. The pattern of biochemical activity of the protein is the effect of conformational changes caused by the interaction of small molecules and their stereospecific sites on the molecule.[23]

## THE UNIFICATION OF SCIENCE

I take the relations between the field of genetics and the field of biochemistry as exemplary for the kind of "unity" which can reasonably be attributed to science. Like biochemistry and physical chemistry, these fields are ordered descriptive levels. This ordering relation contrasts sharply with derivational reduction. First of all, the *relata* are not theories but fields (which, of course, may have associated theories but are certainly not resolvable into them). Derivational reduction, as I have said, places a disproportionate emphasis on the relations between theories. Thus reductionists often make the mistake of assuming that a postulated "reduction" of theories implies a corollary "reduction" of domains, areas of inquiry, or fields. This is a rather crude error, though an explicable one. Often enough, such mistakes rest on the idea that every field has a comprehensive theory or even that a field *is* its comprehensive theory. Indeed, one of the most persuasive counterarguments to such brisk simplifications is provided by an inspection of the field of biochemistry —a field which conspicuously lacks a distinctive comprehensive theory. In fact, the comprehensive theory of biochemistry is surely quantum mechanics, a theory obviously shared with physics and chemistry proper. This sharing of a comprehensive theory, however, does nothing to jeopardize the distinctiveness of biochemistry as a field of research. Thus, it is simply a confusion to conflate a theory (even a "comprehensive theory") with a field. As a matter of fact, by keeping these concepts distinct we can understand correctly the usual claim that biochemistry is "nothing more" than physics and chemistry (for biochemistry shares, but cannot be reduced to, their comprehensive theory) and still hold that biochemistry is a field (with a distinguishable domain, special problems, methodologies and techniques) distinct from chemistry and physics. Furthermore, by successfully disentangling (and thus relating) the concepts of theory and field, we can also begin to explore the role of interlevel theories in the gradual "unification" of science.

Recall that, until recently, the formal concept of reduction (derivational reduction) was used to characterize quite dissimilar kinds of scientific change. The work of Nickles, Sklar, and Wimsatt has been very important in this regard, for it has shown that reduction may be of different types and fulfill different functions.[24] It is not very surprising, moreover, that the fruits of such detailed analyses of scientific change, analyses responsive to developmental and functional differences among theories, should yield results very far removed from the original derivational concept of reduction. Indeed, the results of these recent analyses also suggest that talk of "reduction," because of derivational connotations, is no longer appropriate for the study of scientific change. "Reduction talk" seems irrelevant for analyses which, like the one I have tried to present, no longer rely on the "structural" features of sentences or systems of sentences, but rather on developmental and functional features to pick out the significant units of scientific change.

As I have emphasized throughout this paper, these two alternative ways of approaching the unity of science lead to quite incongruous results. According to derivational reduction, as I have said, the reducing theory explains the reduced theory. According to the present analysis, in contrast, an interlevel theory bridges two fields by establishing, explaining, and warranting the connections between descriptive levels. Historically, special vocabularies simply are not just theoretical vocabularies and thus the connections between the vocabularies of neighboring fields are far more intricate and extensive than those postulated by derivational reductionists.[25] According to these old accounts, in fact, a distinction between theoretical and observational terms is accompanied by a requirement that all the theoretical terms of the theory to be reduced (and only these terms) be connected with those of the reducing theory. By contrast, our examination of the subtler working relationship between terms of neighboring fields shows that some terms are transformed in a new field. We have seen that transformed terms are embedded in interlevel theories. Further, we have seen that terms can be imported from one special vocabulary to another and that, in a similar fashion, theories and laws may be shared by fields.

In fact, the present analysis provides a new set of concepts for understanding the unity of science and its relation to scientific change. For example, we have seen that causal sketches, like the operon theory and the theory of allosteric regulation, are one kind of interlevel theory. But interlevel theories only constitute a subset of the *interfield* theories previously investigated by Darden and me.[26] While an interfield theory, the more general type, can be said to explain connections between fields, an interlevel theory explains the connections between fields ordered as descriptive levels. While the generation of an interfield theory is associated with a shared problem, the genesis of an interlevel theory is associated with a problem-shift characteristic of fields ordered as descriptive levels.

This latter feature of interlevel theories (the directionality of problem shifts) means that interlevel theories can be readily misinterpreted in terms of derivational reductions. Derivational reduction, too, has a directional component, though it is usually characterized as movement from lesser to greater generality. Interestingly enough, derivational reduction not only fails in its analysis of scientific change for the reasons already given. It also fails because it requires an extreme distortion of actual science in order to pose as an explanation of the scientific change which I have shown to be brought about by means of interlevel theories. Two examples suggesting why interlevel theories can never be adequately explained in reductionist terms are the following: (1) The operon theory cannot be viewed as a partial reduction of genetics to biochemistry (nor, for that matter, can the theory of allosteric regulation be a partial reduction of biochemistry to physical chemistry). Viewed as an instance of derivational reduction, the operon theory would have to be *two* theories, one genetic and the other biochemical. The genetic operon theory would have to be derivable from the biochemical

operon theory, given the biochemical-genetic connections. In short, reduction requires two theories while we have only one. (2) Alternatively, theories like the operon theory and the theory of allosteric regulation might be thought to supply the connections between terms required for a derivational reduction. In a reduction account, the connections and the reducing theory are premises while the reduced theory is the consequence. According to the view which I have been developing, however, connections are in turn explained by interlevel theories. But this points out another crippling inadequacy in the standard reduction accounts. It is always assumed in these accounts that connections are sufficiently warranted by the evidence for the reduced and the reducing theories, and not by some additional theories. If the connections are embedded in an interlevel theory distinct from the reducing and reduced theories, however, the reduction is bogus. Consider, for example, the deductive-nomological explanation of a genetic theory by a biochemical theory *with the help of a third theory, both genetic and biochemical.* I suspect that something similar is always the case with pretended derivational reductions—that additional interlevel theories must be invoked to justify the connections between terms.

Before we can meaningfully ask whether a theory of one field is derivable from a theory of another, that is to say, before the question of reduction can even arise, extensive unification between fields must already have taken place. Connections between terms of the fields must already have been established, explained, and warranted by an inter-level theory. In other words, before philosophers could begin to axiomatize and deductively order theories, a theoretical bridge between fields had to be constructed within the problem-solving strategies of research itself. I am convinced that this bridging process underlies whatever sort of unification we can reasonably attribute to science. With an interlevel theory at hand, moreover, what can be gained by the *reduction* of one theory to another? In contrast to the reductionists, in sum, I would suggest the following as a working hypothesis: the unity which science is capable of exhibiting at any stage in its development is always the result of specific and identifiable inter-level or (more broadly speaking) interfield theories.

## NOTES

1. Of course, positivistic doubts about the status of classical genes were overcome before DNA was identified as the genetic material and long before the structure of the DNA molecule had been elaborated. (How scientists developed an understanding of genes as discrete, material hereditary units has not yet, I think, received full and adequate historical treatment.) By contrast, as late as 1944, Erwin Schrödinger was still suggesting, now on the basis of Max Delbrück's speculations, that the study of living things would reveal "other laws of physics, hitherto unknown." See Erwin Schrödinger, *What is Life? The Physical Aspect of the Living Cell,* Cambridge, Cambridge University Press, 1944, p. 73.

2. Thomas Nickles says that derivational reduction, the type described by Nagel, is helpful for understanding what Nickles calls "domain-combining reductions," although he admits that derivational reduction does *not* help us understand the reduction of predecessor theories by successor theories ("domain-preserving reductions"). My claim, in contrast, is that derivational reduction fails even as an account of the relations between domains. See Nickles, Two concepts of intertheoretic reduction, *Journal of Philosophy,* 70, no. 7 (1973), 181–201.

3. Ernest Nagel, *The Structure of Science,* New York, Harcourt, Brace, and World, 1961, p. 354.

4. According to Nagel, the reduction of one *science* to another "is effected when the experimental laws of the secondary science (and if it has an adequate theory, its theory as well) are shown to be the logical consequences of the theoretical assumptions (inclusive of the coordinating definitions) of the primary science" (ibid., p. 352).

5. Kenneth Schaffner, The peripherality of reduction in the development of molecular biology, *Journal of the History of Biology,* 7 (1964), 111–139.

6. Nagel, *The Structure of Science,* pp. 37–42. According to Nagel, any adequate explanation of a law (and presumably a theory) must contain at least one explanatory premise that is "more general" than the law (or theory) that it explains. However, Nagel does not explicitly require that reductions satisfy the conditions for adequate explanation even though he says, "Reduction . . . is the explanation of a theory or a set of experimental laws established in one area of inquiry, by a theory usually though not invariably formulated for some other domain" (*The Structure of Science,* p. 338).

7. Schaffner, The peripherality of reduction.

8. Darden first saw that the concept of a field might be helpful in my analysis. She has examined the emergence of the field of genetics in "Reasoning in Scientific Change: The Field of Genetics at Its Beginnings," Ph.D. dissertation, University of Chicago, 1974. Darden and I discuss fields and their relationships in "Interfield Theories," forthcoming in *Philosophy of Science,* 1 (1977), 43–64.

9. "Domain" is used here in the sense analyzed by Dudley Shapere in Scientific theories and their domains, in F. Suppe, ed., *The Structure of Scientific Theories,* Urbana, University of Illinois Press, 1974, pp. 518-565.

10. Theodosius Dobzhansky, *Genetics of the Evolutionary Process,* New York, Columbia University Press, 1970, pp. 221-222.

11. At the turn of the century, however, physical chemistry was under the influence of Wilhelm Ostwald, who was interested only in the energy relations in biological systems. By contrast, organic chemists like Emil Fisher were interested in the structural analysis of molecules. By 1910 even Ostwald, who had been impressed by Ernst Mach's positivism, admitted that molecules exist, thus opening the way to the use of kinetic techniques in the discovery of molecular structure. For an excellent account of the interaction between physical chemistry and organic chemistry, see Joseph S. Fruton, *Molecules and Life: Historical Essays on the Interplay of Chemistry and Biology,* New York, Wiley Interscience, 1972.

12. To the best of my knowledge, the term "mutation" first appeared in Wilhelm Heinrich Waagen's Die Formenreihe des Ammonites subradiatus, *Geognostische-paläontologische Beiträge,* E. W. Beneke, 1868, pp. 179-257.

13. Most important for de Vries' theory was a hypothesis concerning the formation of new pangens that remained latent for a period of time and then became active—that is, were finally expressed as a new visible character. This form of mutation (progressive mutation) was thought by De Vries to be important in the formation of a new species. While he claimed that other altered pangens assorted according to the law of segregation, he did not believe that this law was applicable to the pangens that had undergone progressive mutation. For commentary on Hugo de Vries' *Die Mutationstheorie* (Leipzig, Viet, 1901-1903), see Garland Allen, Hugo de Vries and the reception of the "mutation theory," *Journal of the History of Biology,* 2 (1969), 55-87.

14. See Hermann J. Muller, Genetic variability, twin hybrids and constant hybrids, in a case of balanced lethal factors, *Genetics,* 2 (1913), 422-499.

15. The striking stability in the use of the term "mutation" in genetics can be illustrated by comparing the definitions given by Muller in 1923, Sturtevant and Beadle in 1939, and Hayes in 1968. See Hermann J. Muller, Mutation, *Eugenics, Genetics and the Family,* 1 (1923), quoted in Elof Axel Carlson, *The Gene: A Critical History,* Philadelphia, W. B. Saunders, 1966, p. 87; A. H. Sturtevant and G. W. Beadle. *An Introduction to Genetics,* Philadelphia, W. B. Saunders, 1939, and New York, Dover Publications, 1962, p. 209; and William Hayes, *The Genetics of Bacteria and Their Viruses: Studies in Basic Genetics and Molecular Biology,* New York, John Wiley, 1968, p. 40.

16. T. H. Morgan and C. B. Bridges, Sex-Linked inheritance in *Drosophila, Carnegie Institute of Washington Publications,* n. 237 (1916), quoted in Carlson, *The Gene,* p. 75.

17. G. W. Beadle and E. L. Tatum, Genetic control of biochemical reactions in *Neurospora, Proceedings of the National Academy of Science, U. S. A.,* 27 (1941), 499-506.

18. These developments are reviewed by H. L. K. Whitehouse in *Towards an Understanding of the Mechanism of Heredity,* London, Edward Arnold, 1969.

19. See Paul Feyerabend, Explanation, reduction, and empiricism, in H. Feigl and G. Maxwell, eds., *Scientific Explanation, Space, and Time,* Minnesota Studies in the Philosophy of Science, Vol. 3, Minneapolis: University of Minnesota Press, 1962, pp. 28-97. The debate over meaning change initiated by Feyerabend provides a wealth of examples, all difficult to judge. I am often at a loss, when confronted by a particular case, to say whether meaning has changed or not. And I assume that answers to questions of meaning in particular cases depend on an "adequate" theory of meaning—the theory we all await. However, with the necessary historical footwork, it is possible to determine whether, in a particular case of the transference of terms, knowledge claims are accepted, rejected, or modified.

20. Joshua Lederberg, Gene recombinations and linked segregations in *Escherichia coli*, *Genetics*, 32 (1947), 524. Lederberg did recombination mapping of some genes, of *E. coli*, a bacterium. That is, by hybridization studies, he attempted to order the genes. In this study he found that certain genetic markers (the biosynthesis of biotin, methionine, and so forth) behaved as if they were the results of a system of linked genes. Furthermore, he obtained some evidence for a linear order of the genes.

21. Paul Oppenheim and Hilary Putnam, Unity of science as a working hypothesis, in H. Feigl, M. Scriven, and G. Maxwell, eds., *Concepts, Theories, and the Mind-Body Problem*, Minnesota Studies in the Philosophy of Science, Vol. 2, Minneapolis, University of Minnesota Press, 1968, pp. 3–36. For Oppenheim and Putnam the unity of science is the cumulative reduction of branches of science. The term "unity of science" is used in their paper "in two senses, to refer, first to an ideal *state* of science, and, second, to a pervasive *trend* within science, seeking the attainment of that ideal" (Unity of science, p. 4).

22. François Jacob and Jacques Monod, Genetic regulatory mechanisms in the synthesis of proteins, *Journal of Molecular Biology*, 3 (1961), 318–356.

23; Jacques Monod, Jeffries Wyman, and Jean-Pierre Changeux, On the nature of allosteric transitions: A plausible model, *Journal of Molecular Biology*, 12 (1965), 88–118.

24. Nickles, Two concepts of intertheoretic reduction; Lawrence Sklar, Types of intertheoretic reduction, *British Journal for the Philosophy of Science*, 18, no. 2 (1967), 109–124; and William Wimsatt, "Reductionism, levels of organization, and the mind-body problem," in G. Globus, G. Maxwell, and I. Savodnick, eds., *Consciousness and the Brain: A Scientific and Philosophical Inquiry*, New York, Plenum, 1976, pp. 205–267.

25. To employ a rough distinction, some proper terms of a field are introduced in order to characterize parts of the domain (for example, "bar eye," in genetics), others in order to fomulate problems (for instance, "position effect"), and still others to solve problems ("mutation" is a good example). Only some of the proper terms employed in a problem solution are obviously "theoretical" (in *contrast* to "observational"), and even those can cease to function as "theoretical" when they characterize an item in the domain at some later stage in the development of the field.

26. Lindley Darden and Nancy Maull, "Interfield Theories," *Philosophy of Science*, 1 (1977), 43–64.

# VII
# THE NATURE OF SPECIES

# 28

# Species Concepts
# and Their Application

ERNST MAYR

DARWIN'S CHOICE OF TITLE FOR his great evolutionary classic, *On the Origin of Species,* was no accident. The origin of new "varieties" within species had been taken for granted since the time of the Greeks. Likewise the occurrence of gradations, of "scales of perfection" among "higher" and "lower" organisms, was a familiar concept, though usually interpreted in a strictly static manner. The species remained the great fortress of stability, and this stability was the crux of the anti-evolutionist argument. "Descent with modification," true biological evolution, could be proved only by demonstrating that one species could originate from another. It is a familiar and often-told story how Darwin succeeded in convincing the world of the occurrence of evolution and how—in natural selection—he found the mechanism that is responsible for evolutionary change and adaptation. It is not nearly so widely recognized that Darwin failed to solve the problem indicated by the title of his work. Although he demon-strated the modification of species in the time dimension, he never seriously attempted a rigorous analysis of the problem of the multiplication of species, of the splitting of one species into two. I have examined the reasons for this failure (Mayr, 1959a) and found that foremost among them was Darwin's uncertainty about the nature of species. The same can be said of those authors who attempted to solve the problem of speciation by saltation or other heterodox hypotheses. They all failed to find solutions that are workable in the light of the modern appreciation of the population structure of species. An understanding of the nature of species, then, is an indispensable prerequisite for the understanding of the evolutionary process.

## SPECIES CONCEPTS

The term *species* is frequently used to designate a class of similar things to which a name has been attached. Most often this term is applied to living organisms, such as birds, fishes, flowers, or trees, but it has also been used for inanimate objects and even for human artifacts. Mineralogists speak of species of minerals, physicists of nuclear species; interior decorators consider tables and chairs species of furniture. The application of the same term both to organisms and to inanimate objects has led to much confusion and an almost endless number of species definitions (Mayr, 1963, 1969); these, however, can be reduced to three basic species concepts. The first two, mainly applicable to inanimate objects, have considerable historical significance, because their advocacy was the cause of much past confusion. The third is the species concept now prevailing in biology.

### 1. The Typological Species Concept

The typological species concept, going back to the philosophies of Plato and Aristotle (and thus sometimes called the essentialist concept), was the species concept of Linnaeus and his followers (Cain, 1958). According to this concept, the observed diversity of the universe reflects the existence of a limited number of underlying "universals" or types (*eidos* of Plato). Individuals do not stand in any special relation to one another, being merely expressions of the same type. Variation is the result of imperfect manifestations of the idea implicit in each species. The presence of the same underlying essence is inferred from similarity, and morphological similarity is, therefore, the species criterion for the essentialist. This is the so-called morphological species concept. Morphological characteristics do pro-

vide valuable clues for the determination of species status. However, using degree of morphological difference as the primary criterion for species status is completely different from utilizing morphological evidence together with various other kinds of evidence in order to determine whether or not a population deserves species rank under the biological species concept. Degree of morphological difference is not the decisive criterion in the ranking of taxa as species. This is quite apparent from the difficulties into which a morphological-typological species concept leads in taxonomic practice. Indeed, its own adherents abandon the typological species concept whenever they discover that they have named as a separate species something that is merely an individual variant.

### 2. The Nominalistic Species Concept

The nominalists (Occam and his followers) deny the existence of "real" universals. For them only individuals exist; species are man-made abstractions. (When they have to deal with a species, they treat it as an individual on a higher plane.) The nominalistic species concept was popular in France in the eighteenth century and still has adherents today. Bessey (1908) expressed this viewpoint particularly well: "Nature produces individuals and nothing more . . . species have no actual existence in nature. They are mental concepts and nothing more . . . species have been invented in order that we may refer to great numbers of individuals collectively."

Any naturalist, whether a primitive native or a trained population geneticist, knows that this is simply not true. Species of animals are not human constructs, nor are they types in the sense of Plato and Aristotle; but they are something for which there is no equivalent in the realm of inanimate objects.

From the middle of the eighteenth century on, the inapplicability of these two medieval species concepts (1 and 2 above) to biological species became increasingly apparent. An entirely new concept, applicable only to species of organisms, began to emerge in the later writings of Buffon and of many other naturalists and taxonomists of the nineteenth century (Mayr, 1968).

### 3. The Biological Species Concept

This concept stresses the fact that species consist of populations and that species have reality and an internal genetic cohesion owing to the historically evolved genetic program that is shared by all members of the species. According to this concept, then, the members of a species constitute (1) *a reproductive community*. The individuals of a species of animals respond to one another as potential mates and seek one another for the purpose of reproduction. A multitude of devices ensures intraspecific reproduction in all organisms. The species is also (2) *an ecological unit* that, regardless of the individuals composing it, interacts as a unit with other species with which it shares the environment. The species, finally, is (3) *a genetic unit* consisting of a large intercommunicating gene pool, whereas an individual is merely a temporary vessel holding a small portion of the contents of the gene pool for a short period of time. These three properties

raise the species above the typological interpretation of a "class of objects" (Mayr, 1963, p. 21). The species definition that results from this theoretical species concept is: *Species are groups of interbreeding natural populations that are reproductively isolated from other such groups.*

The development of the biological concept of the species is one of the earliest manifestations of the emancipation of biology from an inappropriate philosophy based on the phenomena of inanimate nature. The species concept is called biological not because it deals with biological taxa, but because the definition is biological. It utilizes criteria that are meaningless as far as the inanimate world is concerned.

When difficulties are encountered, it is important to focus on the basic biological meaning of the species: A species is a protected gene pool. It is a Mendelian population that has its own devices (called isolating mechanisms) to protect it from harmful gene flow from other gene pools. Genes of the same gene pool form harmonious combinations because they have become coadapted by natural selection. Mixing the genes of two different species leads to a high frequency of disharmonious gene combinations; mechanisms that prevent this are therefore favored by selection. Thus it is quite clear that the word "species" in biology is a relational term. $A$ is a species in relation to $B$ or $C$ because it is reproductively isolated from them. The biological species concept has its primary significance with respect to sympatric and synchronic populations (existing at a single locality and at the same time), and these—the "nondimensional species"—are precisely the ones where the application of the concept faces the fewest difficulties. The more distant two populations are in space and time, the more difficult it becomes to test their species status in relation to each other, but also the more irrelevant biologically this becomes.

The biological species concept also solves the paradox caused by the conflict between the fixity of the species of the naturalist and the fluidity of the species of the evolutionist. It was this conflict that made Linnaeus deny evolution and Darwin the reality of species (Mayr, 1957). The biological species combines the discreteness of the local species at a given time with an evolutionary potential for continuing change.

## THE SPECIES CATEGORY AND SPECIES TAXA

The advocacy of three different species concepts has been one of the two major reasons for the "species problem." The second is that many authors have failed to make a distinction between the definition of the species category and the delimitation of species taxa (for fuller discussion see Mayr, 1969).

A *category* designates a given rank or level in a hierarchic classification. Such terms as "species," "genus," "family," and "order" designate categories. A category, thus, is an abstract term, a class name, while the organisms placed in these categories are concrete zoological objects.

Organisms, in turn, are classified not as individuals, but as groups of organisms. Words like "bluebirds," "thrushes," "songbirds," or "vertebrates" refer to such groups. These are the concrete objects of classification. Any such group of populations is called a *taxon* if it is considered sufficiently distinct to be worthy of being formally assigned to a definite category in the hierarchic classification. *A taxon is a taxonomic group of any rank that is sufficiently distinct to be worthy of being assigned to a definite category.*

Two aspects of the taxon must be stressed. A taxon always refers to specified organisms. Thus *the* species is not a taxon, but any given species, such as the Robin (*Turdus migratorius*) is. Second, the taxon must be formally recognized as such, by being described under a designated name.

Categories, which designate a rank in a hierarchy, and taxa, which designate named groupings of organisms, are thus two very different kinds of phenomena. A somewhat analogous situation exists in our human affairs. Fred Smith is a concrete person, but "captain" or "professor" is his rank in a hierarchy of levels.

## THE ASSIGNMENT OF TAXA TO THE SPECIES CATEGORY

Much of the task of the taxonomist consists of assigning taxa to the appropriate categorical rank. In this procedure there is a drastic difference between the species taxon and the higher taxa. Higher taxa are defined by intrinsic characteristics. Birds is the class of feathered vertebrates. Any and all species that satisfy the definition of "feathered vertebrates" belong to the class of birds. An essentialist (typological) definition is satisfactory and sufficient at the level of the higher taxa. It is, however, irrelevant and misleading to define species in an essentialistic way because the species is not defined by intrinsic, but by *relational* properties.

Let me explain this. There are certain words that indicate a relational property, like the word "brother." Being a brother is not an inherent property of an individual, as hardness is a property of a stone. An individual is a brother only with respect to someone else. The word "species" likewise designates such a relational property. A population is a species with respect to all other populations with which it exhibits the relationship of reproductive isolation—noninterbreeding. If only a single population existed in the entire world, it would be meaningless to call it a species.

Noninterbreeding between populations is manifested by a gap. It is this gap between populations that coexist (are sympatric) at a single locality at a given time which delimits the species recognized by the local naturalist. Whether one studies birds, mammals, butterflies, or snails near one's home town, one finds each species clearly delimited and sharply separated from all other species. This demarcation is sometimes referred to as the species delimitation *in a nondimensional system* (a system without the dimensions of space and time).

Anyone can test the reality of these discontinuities for himself, even where the morphological differences are slight. In eastern North America, for instance, there are four similar species of the thrush genus *Catharus* (Table 1), the Veery (*C. fuscescens*), the Hermit Thrush (*C. guttatus*), the Olive-Backed or Swainson's Thrush

(*C. ustulatus*), and the Gray-Cheeked Thrush (*C. minimus*). These four species are sufficiently similar visually to confuse not only the human observer, but also silent males of the other species. The species-specific songs and call notes, however, permit easy species discrimination, as observationally substantiated by Dilger (1956). Rarely do more than two species breed in the same area, and the overlapping species, $f + g$, $g + u$, and $u + m$, usually differ considerably in their foraging habits and niche preference, so that competition is minimized with each other and with two other thrushes, the Robin (*Turdus migratorius*) and the Wood Thrush (*Hylocichla mustelina*), with which they share their geographic range and many ecological requirements. In connection with their different foraging and migratory habits the four species differ from one another (and from other thrushes) in the relative length of wing and leg elements and in the shape of the bill. There are thus many small differences between these at first sight very similar species. Most important, no hybrids or intermediates among these four species have ever been found. Each is a separate genetic, behavioral, and ecological system, separated from the others by a complete biological discontinuity, a gap.

## DIFFICULTIES IN THE APPLICATION OF THE BIOLOGICAL SPECIES CONCEPT

The practicing taxonomist often has difficulties when he endeavors to assign populations to the correct rank. Sometimes the difficulty is caused by a lack of information concerning the degree of variability of the species

TABLE 1. CHARACTERISTICS OF FOUR EASTERN
NORTH AMERICAN SPECIES OF *CATHARUS* (from DILGER 1956)

| Characteristic compared | C. fuscescens | C. guttatus | C. ustulatus | C. minimus |
|---|---|---|---|---|
| Breeding range | Southernmost | More northerly | Boreal | Arctic |
| Wintering area | No. South America | So. United States | C. America to Argentina | No. South America |
| Breeding habitat | Bottomland woods with lush undergrowth | Coniferous woods mixed with deciduous | Mixed or pure tall coniferous forests | Stunted northern fir and spruce forests |
| Foraging | Ground and arboreal (forest interior) | Ground (inner forest edges) | Largely arboreal (forest interior) | Ground (forest interior) |
| Nest | Ground | Ground | Trees | Trees |
| Spotting on eggs | Rare | Rare | Always | Always |
| Relative wing length | Medium | Short | Very long | Medium |
| Hostile call | *veer* *pheu* | *chuck* *seeeep* | *peep* *chuck-burr* | *beer* |
| Song | Very distinct | Very distinct | Very distinct | Very distinct |
| Flight song | Absent | Absent | Absent | Present |

with which he is dealing. Helpful hints on the solution of such practical difficulties are given in the technical taxonomic literature (Mayr, 1969).

More interesting to the evolutionist are the difficulties that are introduced when the dimensions of time and space are added. Most species taxa do not consist merely of a single local population but are an aggregate of numerous local populations that exchange genes with each other to a greater or lesser degree. The more distant that two populations are from each other, the more likely they are to differ in a number of characteristics. I show elsewhere (Mayr, 1963, ch. 10 and 11) that some of these populations are incipient species, having acquired some but not all characteristics of species. One or another of the three most characteristic properties of species taxa—reproductive isolation, ecological difference, and morphological distinguishability—is in such cases only incompletely developed. The application of the species concept to such incompletely speciated populations raises considerable difficulties. There are six wholly different situations that may cause difficulties.

(1) *Evolutionary continuity in space and time.* Widespread species may have terminal populations that behave toward each other as distinct species even though they are connected by a chain of interbreeding populations. Cases of reproductive isolation among

geographically distant populations of a single species are discussed in Mayr, 1963, ch. 16.

(2) *Acquisition of reproductive isolation without corresponding morphological change.* When the reconstruction of the genotype in an isolated population has resulted in the acquisition of reproductive isolation, such a population must be considered a biological species. If the correlated morphological change is very slight or unnoticeable, such a species is called a sibling species (Mayr, 1963, ch. 3).

(3) *Morphological differentiation without acquisition of reproductive isolation.* Isolated populations sometimes acquire a degree of morphological divergence one would ordinarily expect only in a different species. Yet some such populations, although as different morphologically as good species, interbreed indiscriminately where they come in contact. The West Indian snail genus *Cerion* illustrates this situation particularly well (Fig. 1).

(4) *Reproductive isolation dependent on habitat isolation.* Numerous cases have been described in the literature in which natural populations acted toward each other like good species (in areas of contact) as long as their habitats were undisturbed. Yet the reproductive isolation broke down as soon as the characteristics of these habitats were changed, usually by the interference of man. Such cases of secondary breakdown of isolation are discussed in Mayr, 1963, ch. 6.

(5) *Incompleteness of isolating mechanisms.* Very few isolating mechanisms are all-or-none devices (see Mayr, 1963, ch. 5). They are built up step by step, and most isolating mechanisms of an incipient species are imperfect and incomplete. Species level is reached when the process of speciation has become irreversible, even if some of the secondary isolating mechanisms have not yet reached perfection (see Mayr, 1963, ch. 17).

(6) *Attainment of different levels of speciation in different local populations.* The perfecting of isolating mechanisms may proceed in different populations of a polytypic species (one having several subspecies) at different rates. Two widely overlapping species may, as a consequence, be completely distinct at certain localities but may freely hybridize at others. Many cases of sympatric hybridization discussed in Mayr, 1963, ch. 6, fit this characterization (see Mayr, 1969, for advice on handling such situations).

These six types of phenomena are consequences of the gradual nature of the ordinary process of speciation (excluding polyploidy; see Mayr, 1963, p. 254). Determination of species status of a given population is difficult or arbitrary in many of these cases.

## DIFFICULTIES POSED BY UNIPARENTAL REPRODUCTION

The task of assembling individuals into populations and species taxa is very difficult in most cases involving uniparental (asexual) reproduction. Self-fertilization, parthenogenesis, pseudogamy, and vegetative reproduction are forms of uniparental reproduction. The biological species concept, which is based on the presence or absence of interbreeding between natural populations, cannot be applied to groups with obligatory asexual reproduction because interbreeding of

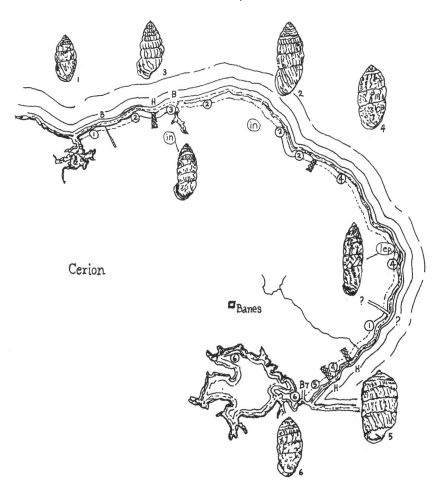

Fig. 1. The distribution pattern of populations of the halophilous land snail *Cerion* on the Banes Peninsula in eastern Cuba. Numbers refer to distinctive races or "species." Where two populations come in contact (with one exception) they hybridize (*H*), regardless of degree of difference. In other cases contact is prevented by a barrier (*B*). *In* = isolated inland population.

populations is nonexistent in these groups. The nature of this dilemma is discussed in more detail elsewhere (Mayr, 1963, 1969). Fortunately, there seem to be rather well-defined discontinuities among most kinds of uniparentally reproducing organisms. These discontinuities are apparently produced by natural selection from the various mutations that occur in the asexual lines (clones). It is customary to utilize the existence of such discontinuities and the amount of morphological difference between them to delimit species among uniparentally reproducing types.

## THE IMPORTANCE OF A NONARBITRARY DEFINITION OF SPECIES

The clarification of the species concept has led to a clarification of many evolutionary problems as well as, often, to a simplification of practical problems in taxonomy. The correct classification of the many different kinds of varieties (phena), of polymorphism (Mayr, 1963, ch. 7), of polytypic species (ibid. ch. 12), and of biological races (ibid. ch. 15) would be impossible without the arranging of natural populations and phenotypes into biological species. It was impossible to solve, indeed even to state precisely, the problem of the multiplication of species until the biological species concept had been developed. The genetics of speciation, the role of species in large-scale evolutionary trends, and other major evolutionary problems could not be discussed profitably until the species problem was settled. It is evident then that the species problem is of great importance in evolutionary biology and that the growing agreement on the concept of the biological species has resulted in a uniformity of standards and a precision that is beneficial for practical as well as theoretical reasons.

## THE BIOLOGICAL MEANING OF SPECIES

The fact that the organic world is organized into species seems so fundamental that one usually forgets to ask why there are species, what their meaning is in the scheme of things.

There is no better way of answering these questions than to try to conceive of a world without species. Let us think, for instance, of a world in which there are only individuals, all belonging to a single interbreeding community. Each individual is in varying degrees different from every other one, and each individual is capable of mating with those others that are most similar to it. In such a world, each individual would be, so to speak, the center of a series of concentric rings of increasingly more different individuals. Any two mates would be on the average rather different from each other and would produce a vast array of genetically different types among their offspring. Now let us assume that one of these recombinations is particularly well adapted for one of the available niches. It is prosperous in this niche, but when the time for mating comes, this superior genotype will inevitably be broken up. There is no mechanism that would prevent such a destruction of superior gene combinations, and there is, therefore, no possibility of the gradual improvement of gene combinations. The significance of the species now becomes evident. The reproductive isolation of a species is a protective device that guards against the breaking up of its well-integrated, coadapted gene system. Organizing organic diversity into species creates a system that permits genetic diversification and the accumulation of favorable genes and gene combinations without the danger of destruction of the basic gene complex. There are definite limits to the amount of genetic variability that can be accommodated in a single gene pool without producing too high a proportion of inviable recombinants. Organizing genetic diversity into protected gene pools—that is, species—guarantees that these limits are not overstepped. This is the biological meaning of species.

## REFERENCES

Bessey, C. E., 1908, The taxonomic aspect of the species. *American Naturalist,* 42: 218–224.

Cain, A. J., 1958, Logic and memory in Linnaeus's system of taxonomy. *Proc. Linn. Soc.,* London, 169: 144–163.

Mayr, E., 1957, Species concepts and definitions. *Amer. Assoc. Adv. Sci.,* Publ. No. 50: 1–22, Washington, D.C.

——, 1959a, Darwin and the evolutionary theory in biology. In *Evolution and Anthropology: A Centennial Approach,* Anthropological Society of America, Washington, D. C.

——, 1963, *Animal Species and Evolution,* Cambridge, Harvard University Press.

——, 1968, Illiger and the biological species concept. *J. Hist. Biol.,* 1: 163–178.

——, 1969, *Principles of Systematic Zoology,* New York, McGraw-Hill.

# 29

# The Biological Species Concept:
# A Critical Evaluation

## ROBERT R. SOKAL AND THEODORE J. CROVELLO

### I. INTRODUCTION

A species concept has been a central tenet of biological belief since the early origins of biology as a science. The implications of this term have changed over the years: the fixed, immutable, and sharply distinct entities of the Linnaean period gave way to the more variable and intergrading units of the post-Darwinian era. For many taxonomists before and after Darwin, the species has simply implied the recognition of groups of morphologically similar individuals that differ from other such groups.

Through much of biological history there has been controversy regarding the existence of species in nature. Are species real units in nature? Can the species category be defined objectively? Given an affirmative answer to the above two questions, can real organisms be assigned to one of the nonoverlapping species so delimited? Darwin's work contributed to the recognition of species as real entities.

The very title of his book, *On the Origin of Species,* stressed this category. But as Mayr (1959) has pointed out, Darwin himself was so impressed by the variability and intergradation in the material he studied that he considered the term "species" to be arbitrary, not differing in essential features from "variety." Argument regarding these questions has persisted through changing concepts of the biological universe and the increasing insights into the genetic and ecological mechanisms governing the behavior of individuals and populations. The history of these ideas and controversies is reviewed by Mayr (1957), and we shall not enlarge upon it here. Some have considered species as man-made, arbitrary units either because of their philosophical orientation or because of the difficulty of interpreting variable material from widely ranging organisms as consisting of one or more species. These arguments have been countered by evidence of the common-sense recognition of discontinuities in nature even by lay observers (see Mayr, 1963, p. 17, for an account of species recognition by New Guinea natives, but see Berlin, Breedlove, and Raven, 1966, for a contrary view) and also of species recognition, presumably instinctive, by other organisms. Such discontinuities are most easily noted by naturalists who study local faunas and floras, and the species concept derived from such situations has been called the "nondimensional species concept" by Mayr (1963). But in some taxa, such as in willows, groups generally assigned generic or sectional rank are more easily recognized by local naturalists than are the species.

The apparent necessity to accommodate within one species concept several aspects of organisms led to the development of the so-called *biological species concept* (hereafter abbreviated BSC). These aspects include the variation of characteristics over large geographic areas, changes in these characteristics as populations adapt to environmental challenges or interact with other populations, and the integration of individuals into populations to form gene pools through direct processes, as well as indirectly through their ecological interactions. We shall not trace the development of the concept during the 1930's. Ernst Mayr, recognized as its foremost advocate, has called the BSC a "multidimensional concept" (Mayr, 1963), because it deals with populations that are distributed through time and space, interrelated through mutual interbreeding, and distinguished from others by reproductive barriers.

Since its formulation, there have been objections to the BSC from a variety of sources and motives. Many taxonomists have ignored it for practical reasons. Some workers (e.g., Blackwelder, 1962; Sokal, 1962) have charged that the employment of the BSC is misleading in that it imbues species described by conventional morphological criteria with a false aura of evolutionary distinctness and with unwarranted biosystematic implications. In fairness we point out that some supporters of the BSC (e.g., Simpson, 1961, p. 149) state clearly the difficulties of correlating phenetic and genetic species criteria even in the same taxonomic group but especially across diverse taxa. Nevertheless, such caveats do not generally affect either taxonomic practice or teaching as it filters down to the level of the introductory courses. These critics also point out that the actual procedures employed even by systematists with a modern outlook are quite different from those implied or required by the BSC. Recent trends toward quantification in the biological sciences and especially emphasis on operationalism in systematic and taxonomic procedures (Ehrlich and Holm, 1962; Ehrlich and Raven, 1960; Sokal, 1964; Sokal and Camin, 1965; Sokal and Sneath, 1963) have raised fundamental questions about the BSC to discover whether it is operational, useful, and/or heuristic with relation to an understanding of organic evolution.

The general purposes of this paper are: (1) to show, by means of a detailed flow chart, that the BSC is largely a phenetic concept; (2) given the above, to show that the BSC should be at least as arbitrary as phenetic taxonomic procedure; and (3) to explore the value of the BSC to evolution by posing a set of specific questions. Specifically, we shall first review the definition of the BSC and enumerate those of its attributes that require extended discussion and analysis. Next, we shall discuss three operations required for making decisions about actual populations with respect to these attributes of the BSC. Armed with an understanding of these operations, we shall then consider a flow chart of the detailed steps necessary to determine which of a set of organisms under study can be considered to form a biological species.

As a next step we shall note the difficulties of applying the BSC even in the optimal case of complete knowledge regarding the material under study, and examine how problems multiply as knowledge of the organisms diminishes.

Finally, given the difficulties of the BSC as a workable concept for the practicing taxonomist, we shall briefly examine the necessity for such a concept in evolutionary theory, its heuristic value, and the evidence for the existence of biological species in spite of the difficulty of their recognition and definition.

Although our philosophical attitude in systematics is that of empiricism—and, consequently, we are not committed to the existence of biological species—we have approached our task with minds as open as possible. We recognize, as must any observer of nature, that there are discontinuities in the spectrum of phenetic variation. The question we have asked ourselves, one which we believe must be asked by every biologist concerned with problems of systematics and of evolution, is whether there is a special class of these discontinuities that

delimits units (the biological species) whose definition and description should be attempted because they play an especially significant role in the process of evolution or help in understanding it.

## II. THE BIOLOGICAL SPECIES CONCEPT

The number of species definitions that have been proposed since the advent of the New Systematics, and that fall within the general purlieus of the BSC, is very large, but an extended review and discussion of these definitions would serve little useful purpose here. Many are but minor variants of the one to be discussed below, and they share in most ways the problems that we shall encounter with it. We shall employ the classical definition of biological species as restated by Mayr (1963, p. 19) in his definitive treatise. The definition is:

| | |
|---|---|
| Groups of | (1) |
| actually | (2) |
| or potentially | (3) |
| interbreeding | (4) |
| populations, | (5) |
| which are repro- | |
| ductively isolated | (6) |
| from other groups | (7) |

We have deliberately arranged the definition in the above manner to emphasize those terms or phrases which make separate and important contributions to the overall definition. Let us briefly go through these. We are dealing with *populations* (line 5) whose members *interbreed* (line 4) *actually* (line 2) *or potentially* (line 3). The difficulties of the latter term will be taken up in the next section.

There usually is more than one such population (line 1). This group of populations will not exchange genes (line 6) with other interbreeding groups (line 7). This phenomenon is referred to as *reproductive isolation.*

According to Mayr (1963, p. 20) there are three aspects of the BSC: "(1) Species are defined by distinctness rather than by difference." By this he means reproductive gaps rather than phenetic differences (Mayr, personal communication). "(2) Species consist of populations rather than of unconnected individuals; and (3) species are more unequivocally defined by their relation to nonconspecific populations ('isolation') than by the relation of conspecific individuals to each other. The decisive criterion is not the fertility of individuals but the reproductive isolation of populations."

Thus to discover whether a given set of individuals is a biological species in the sense of the above definition we must have information about three essential components of the BSC: (1) that some individuals lack distinctness (*sensu* Mayr) from other individuals and join these in comprising biological populations of interbreeding individuals (this is the meaning by implication of the term "population" in the definition of the BSC); (2) that there is a group of such populations among which interbreeding does, or could, take place (this follows from the "actually or potentially interbreeding" clause of this definition); (3) that this group lacks gene flow with other groups of populations (this covers the "reproductively isolated" portion of the definition). These three aspects of the biological species are worked into the flow chart (see Fig. 1, page 550).

## III. FUNDAMENTAL OPERATIONS

To ascertain whether a given assemblage of organisms belongs to one or more biological species, three types of operations for grouping organisms and population samples will be found necessary (although only the third is directly implied by the definition given above). The first operation groups organisms by geographic contiguity; the second, by phenetic relationships; and the third, by reproductive relationships. In all these cases there will be some difference in the procedure when the initial grouping is of individuals into subsets (populations), and when these subsets are the basic units being grouped into more inclusive sets (species).

All grouping procedures will, of necessity, be based on samples of organisms and populations. Only in a minuscule number of instances will we have knowledge of all the individuals about which inferences are being made. This is not necessarily an unsatisfactory state of affairs, but it is important to specify the size of the samples required to estimate parameters of the populations with a desired level of confidence. Also, the use of samples necessitates that some assumptions be made about the spatiotemporal distribution of individuals and populations.

The grouping operations will frequently refer to the idea of *connectedness.* We shall consider two operational taxonomic units (OTU's; see Sokal and Sneath, 1963, p. 121—individuals or population samples in this context) to be connected if there exists some definable relation between them (for example, geographic contiguity, phenetic similarity, or interfertility). *Minimally* connected sets of such

OTU's have at least as many such relations as permit any two OTU's to be connected via any other members of the set. *Fully* connected sets have relations between every pair of members of the set. We use these terms by analogy with their employment in graph theory (Busacker and Saaty, 1965).

We shall take up the three types of operations below in the order in which they were introduced.

The first operation groups by *geographic contiguity.* In order to belong to one population, organisms must be within reach of some others, that is, have the possibility of encountering for reproductive purposes other organisms within the same spatiotemporal framework. A first prerequisite for individuals to belong to the same population is that they come from sites which would enable them to be within reach of each other, considering the normal vagility of these organisms or of their propagules. In many cases we can simply assume this when we have samples from one site containing numerous individuals such as are obtained by seining, light traps, or botanical mass collecting. In other cases (especially with large organisms) where single individuals are found at specific sites, we have to be reasonably certain that individuals from separate sites presumed to be within the same local population have intersecting home ranges. In developing a criterion of geographic connectedness among local populations we need to be concerned with the probability of members of one locality visiting members of another one to permit the necessary gene flow required by the model. Again, this will be a function of the distance between localities, the vagility of organisms, and the ecological conditions that obtain between points. Various techniques of locational analysis (see Haggett, 1966) can be used for establishing these linkages. We note in passing that the essential information required for this operation is lacking for most taxa. For example, the pollen and seed ranges for most flowering plant taxa are unknown (Harper, 1966).

A second operation is the establishment of *phenetic similarity* between individuals within population samples and between such samples from various areas. While the definition of the BSC does not invoke phenetic considerations, it will be shown in the next section that any attempt to apply the definition to an actual sample of organisms will need to resort to phenetics in practice. In the initial stages of a study it may be that sufficient estimates of phenetic similarity can be determined by visual inspection of the specimens. Clearly, when the material is very heterogeneous such an initial sorting of the material into putatively conspecific assemblages can be profitable. When more refined analysis is indicated, a quantitative phenetic approach is necessary. Here again we need not concern ourselves with the technical details, which are by now well established through the techniques of numerical taxonomy (Sokal and Sneath, 1963).

The third operation involves grouping *interbreeding individuals* into population samples and grouping *interbreeding population samples* into larger assemblages. Before discussing this in detail, a semantic digression is necessary. In most relevant texts the term "interbreeding" is not defined precisely or distinguished clearly from intercrossing, interfertility, mating, and similar terms. Recourse to a dictionary is not enlightening. The reader is aware that the very act of mating (i.e., copulation in animals with or without insemination, or pollination in plants, to name only two of the more common mechanisms of sexual reproduction) does not of itself insure the production of viable offspring and especially of fertile offspring. Clearly, the act of mating or the transfer of male gametes toward a female gamete is the single necessary precondition for successful interbreeding, but it does not in itself insure fertile offspring. We shall use the term "interbreeding" to mean crossing between individuals resulting in the production of fertile offspring, but we shall occasionally use the terms "interfertility" or simply "mating" in a similar context.

The only unequivocal, direct basis for forming interbreeding groups is to observe organisms interbreeding in nature. If we wanted to make the definition absurdly rigorous, we would wish to insist that an interbreeding population sample be one where a sufficient number of females from the local population sample is mated with a sufficient number of males in the same sample to insure reproductive connectedness to the required degree. Fertile offspring would have to result from all of these unions. Obviously such observations are unlikely. Even if we were to turn to experiments to answer the question, we could not insist on so complete a test of interfertility, both because the number of experiments would be far too great and because, in most cases such crosses would be impossible,

since the biological nature of the organisms precludes more than a single mating (e.g., longevity of mating individuals, incompatibility toward further mates by an already mated female, developmental period of the young, etc.).

Thus, as noted earlier, we shall have to resort to samples of field observations or of crossing experiments. The latter raise the often discussed issue of whether laboratory tests of interbreeding should be considered as evidence when contrasted with field observations. Clearly, first consideration must be given to observations of nature as it is. Success in crossing experiments might indicate "potential" interbreeding. In designing crossing experiments as criteria of infertility, clear instructions must be given on what role these experiments will play and whether the definition to be tested will be satisfied by laboratory crossing experiments or whether field observations are required.

Added to these difficulties is the fact that most of the material systematists deal with is already dead at the time of study and cannot be brought into the laboratory or experimental garden for crossing purposes. Thus, extensive interbreeding tests are impractical, and one needs to resort to partial or circumstantial evidence on crossing for inference on interfertility. As direct evidence on interbreeding diminishes, the methods become increasingly phenetic. Phenetic information is of value in ascertaining interbreeding relationships only insofar as one may assume that phenetic similarity is directly related to ease of interbreeding. Yet we know that phenetics is an imperfect reflection of interfertility between organisms. In fact, this has been one of the main criticisms of numerical taxonomy by evolutionists.

The above arguments should not be interpreted as insistence on our part for "complete" knowledge of reproductive relationships. Just as one samples in phenetic studies to obtain estimates of phenetic structure of a larger population, so it is entirely justified to test reproductive relationships among only a sample of individuals and make inferences about a larger population. However, both sampling procedures are based on prior phenetic sorting out of specimens and populations. Thus we test reproductive relationships only among organisms likely to be interfertile, and the only way we can recognize these is on a phenetic basis. Therefore, except for the absurdly extreme reproductive test of each organism against every other one—biologically and experimentally infeasible, as well as destructive of the original taxa if it were possible to carry out such a test—reproductive tests based on samples reflect phenetic considerations in choosing the individuals to be tested. Furthermore, we must stress that even if we carried out some crossing experiments we would still need to employ phenetic inference to reason from the results of our limited number of crosses to the larger population sample, to the entire local population living today, and to the entire local population both living and dead.

Depending on the set of reproductive properties chosen by a given scientist, interbreeding will range continuously from complete interbreeding through intermediate stages to total lack of interbreeding. The two properties most often considered are connectedness and success of reproduction. If every individual in a group could interbreed with every other one of the opposite sex, *connectedness* would be complete. But the total number of possible combinations will likely be reduced; that is, some pairs may not be able to interbreed. This could be so for a variety of reasons, directly and indirectly genetic, such as sterility genes, reproductive incompatibilities, behavioral differences, seasonal isolation, etc. We are prepared to accept a sample as connected within itself if each individual is capable of interbreeding with one or more of the opposite sex in such a way that the reproductive relationships would yield a minimally connected graph (Busacker and Saaty, 1965) (with $n + m - 1$ edges, where $n$ is the number of one sex and $m$ that of the other), with terminal members being connected to one mate only. Such a minimal interbreeding relationship is unlikely in a large biological sample, because it would imply a very complex system of mating types and intersterilities; yet even such a system practiced over many generations would ensure genetic connectedness among its members. A sample whose reproductive relations are less than a minimal connected set should be separated into those subsets which are connected.

But the ability to mate is clearly not enough. Fertile offspring, which have a nonzero probability of survival and of leaving new offspring, must result from such a union. This consideration leads us directly to the second property characterizing interbreeding.

*Success of reproduction* can be expressed as the percentage of fertile offspring resulting from a given mating measured in terms of percentage of eggs hatched, percentage of seed set, litter size, and similar criteria in the $F_1$ or later generations. The standards set for such criteria and acceptable levels of success will vary with the investigator.

Therefore, members of a local population sample may be considered to interbreed either if they are completely interfertile as defined above or if they are partially interfertile. In the latter case, only samples whose members show at least minimal connectedness and whose average success of reproduction is greater than an arbitrarily established value would qualify.

If organisms are apomicts or obligate selfers, then by their very nature they cannot form biological species (as has indeed been pointed out by proponents of the BSC, e.g., Simpson, 1961, p. 161, or Mayr, 1963, p. 27). If these biological facts are not known to us, they might be suggested by all individuals forming a disjoint set in this step (i.e., no individuals will reproduce with any other individual in the sample). Technically, we should no longer process such samples through the flow chart. However, a useful classification could be arrived at if we ran the individuals of each local sample through the phenetic pathways of the flow chart. We infer this because taxonomists have had no apparent difficulty in describing species by conventional methods in these forms.

Once it has been demonstrated that the individuals *within* each local population sample interbreed, we need only show that there is some gene flow among the samples studied in order to establish interbreeding among them. Once genes from population $A$ enter population $B$, (and those from $B$ enter $A$), interbreeding among the members of $A$ and $B$ provides an opportunity for the establishment of the new genes in both populations.

We can conceive of several partially interfertile population samples as a connected set. It would follow that in order to be considered actually interbreeding the several population samples would have to represent at least a minimally connected set of reproductive relationships. Therefore, not every population sample needs to be directly reproductively connected to every other population sample in the study. A *Rassenkreis* is an example of such a situation. These relationships may be somewhat difficult to represent because the paths of connection will have to pass through either the offspring or parents of mates in a zigzag fashion. However, in populations among which there is substantial gene flow, it should be possible to make a chain of connection between any two organisms by going through relatively few ancestral and descendant generations.

The term "potentially interbreeding," which is included in some definitions of the biological species, has never really been defined, let alone defined operationally. It appears to us that the only possible answer one could get to the question of whether two samples are potentially interbreeding is "don't know." At best, one would be reduced to inferences about potential interfertility from phenetic evidence (and we have already seen that this is not too reliable). It is interesting to note that in his latest work, Mayr (1969) has dropped "potentially interbreeding" from his biological species definition.

## IV. FLOW CHART FOR RECOGNIZING BIOLOGICAL SPECIES

The actual flow chart is shown in Figure 1. The various steps in this figure are listed in this section, each followed by an explanatory account of the reasons for the step, the manner in which it could be carried out, and inherent difficulties.

1. *Assemble phenetically similar individuals.* This preliminary step is important because unless the individuals used for the study are "relatively" similar, it is not reasonable to suppose that they interbreed. Lacking such a procedure, one would be forced to carry out a vast amount of fruitless testing for interfertility. Cottonwoods, aphids, and field mice could all be obtained in samples from the same locality, and while the subsequent logic of the flow chart should ensure their separation into independent biological species (if we can determine that they are not actually or potentially interbreeding populations), a large amount of unnecessary and most likely impractical work would have to be done to test for fertility between cottonwoods and field mice, for example.

Systematists have appropriately decided not to trouble about this point but to use the relatively great phenetic dissimilarity of such groups of organisms to infer that they would be intersterile if an attempt at artificial crossing were made. Substantial evidence is available, especially from plants, that individuals allocated to

different orders, families, or genera are usually intersterile. However, in the vast majority of organisms we may state with certainty that decisions about the presumptive intersterility of two dissimilar individuals or populations are based on phenetics alone. But since phenetic similarity is a continuous variable (as is reproductive interrelationship), it is difficult to designate anything but arbitrary similarity levels above which individuals and populations are potentially interbreeding and hence potentially conspecific and therefore need further testing and below which they are phenetically so different that the likelihood of interbreeding (hence of conspecificity) is small enough to be neglected. In the absurdly extreme instance of cottonwoods and field mice, this phenetic comparison is made instantaneously by the taxonomist without the need for more precise and sophisticated phenetic methods. This step is stressed here mainly to make the logic of the flow chart complete. When assembling similar individuals, dimorphisms and polymorphisms may give rise to practical difficulties, and relational criteria based on knowledge of the biology of the organisms involved may be invoked. Thus, knowing that given caterpillars give rise to given butterflies, we shall associate them, and in cases of marked sexual dimorphism we would wish to associate males and females that appear to form sexual pairs. This can sometimes be done by refined biometric techniques, but where previous knowledge or simple observations suffice these should surely be preferred.

A second point is that step 1 should not be carried out so finely that

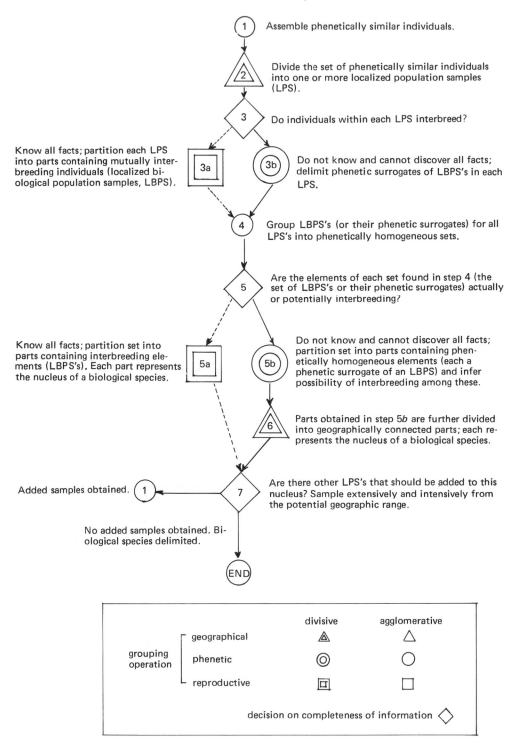

Figure 1. Flow chart for determining biological species. For explanation, see text.

potential candidates for conspecific status are excluded. Thus the grouping should err by inclusion rather than by exclusion. Otherwise, since the flow chart will not cycle through the original sample again, some of the initial sample of organisms that also belong to the same biological species would be excluded.

2. *Divide the set of phenetically similar individuals into one or more localized population samples.* The procedure leading to localized population samples is that of grouping by geographical contiguity as defined in section III. Since we are at the moment concerned with the grouping of individuals to form population samples, we would be unlikely to encounter fixed geographic points from which we can create an interconnected network. Rather, we are likely to obtain a scattering of incidental collection sites from which we must draw inferences about the potential for geographic overlap of the lifetime movement ranges of the individuals concerned.

We shall define a *localized population sample* (LPS) in terms of the natural vagility of the organism. The need for such a definition stems from the biological attributes of populations as integrated gene pools which require that the members of a population be within the geographic range making such integration possible. We use the term "localized," following the conventions of the statistical geographers (Haggett, 1966), rather than the more common "local" population, which has certain biological, genetic, and ecological connotations that, although hard to define, nevertheless are generally invoked in the minds of systematists. By localized population sample we mean to

imply only connection by an external relationship, largely spatial but also temporal and ecological. Unless otherwise qualified, this does not necessarily imply genetic or phenetic similarity among its members.

Gametes or propagules will differ in the distances they travel. A distribution of such distances, if known, could serve as a measure of vagility. The ninety-ninth percentile, $V$, gives a near upper limit to the distances travelled. If the largest observed distance between any two members of a cluster being formed is less that $kV$, where $k$ is an arbitrary constant, we may define this cluster as a localized population sample. Problems might arise with uniformly spaced individuals, but such instances invite arbitrary decisions by any procedure. Also, while the samples are likely to be phenetically similar following step 1, we have no assurance that each sample represents one and the same species. Hence the vagilities of the individuals within each sample are potentially heterogeneous as well. Percentiles other than the ninety-ninth might be employed.

Many times one will not have a distribution of exact locations at which individual specimens have been obtained because the sample will have been collected at one spot or because the collection records for the entire sample refer to one spot or to a broad area. In the former case we are clearly dealing with a sample from a localized population; in the latter we have to make a judicious definition of the area sampled. For instance, if a botanist furnishes only county records and the sample may be from anywhere within a county, the maximum straight-line distance within the county will have to stand for the greatest distance between any two members of the sample.

Of course, in most instances we will not know enough about the biology of the organisms studied to make a useful estimate of $V$. We therefore may have to guess at this value by analogy with known similar organisms.

The definition for localized samples given above should perhaps also include other criteria, such as time and ecological factors. The biological species definition as generally stated does not specifically refer to synchronous populations; yet, as has been repeatedly pointed out, the delimitation of species becomes much more complicated if chronistic aspects are also considered. One might very well impose an analogous criterion of chronistic connectedness on the definition and obtain samples localized in both space and time. Restriction to a given general habitat such as crowns of trees or leaf litter could also be imposed to restrict the possibility further, but we do not pursue this subject here.

Each resulting LPS is not necessarily a local population in the conventional biological sense. To be that it would have to be connected not only in the geographical sense but also by interbreeding relationships. The next step in the flow chart will impose this added constraint.

3. *Is each localized population sample, defined by geographic contiguity in step 2 above, also interbreeding within itself; that is, do its individual members interbreed among themselves?* Localized populations that are not interfertile within themselves cannot make up the elements of a biological species population. In its

rigid interpretation, we would have to ascertain whether there is either actual or potential interbreeding within each localized population sample. We have two choices in answering this question: we can either claim to know or hope to find out what the actual interbreeding relations among the organisms are—this leads us to step 3a—or we may decide that the question cannot be answered fully or at all in terms of interbreeding relationships and proceed to make inferences about these from other evidence, usually phenetics (step 3b).

3a) Knowing all the facts about interbreeding interrelationships within each LPS, we may partition it into parts containing mutually interbreeding elements. Each such part represents a *localized biological population sample* (LBPS).

The general criteria for recognizing interbreeding have been given in the previous section and will not be repeated here. The difficulties of testing even within a limited sample the interbreeding of a sufficient number of members are considerable, and in fact step 3a is, for all intents and purposes, impracticable. Even if all necessary crosses were feasible in theory, we have seen that sampling based on phenetics will be required sooner or later for inferences about potential interfertility of some untested members of each LPS and of the larger local population. For this reason a broken arrow leads to and from the grouping operation based strictly on reproductive criteria, 3a, to indicate that this is *not* the usual path.

3b) When we do not know and cannot discover all the facts regarding interbreeding, we have to delimit at least some phenetically homogeneous subsets in each LPS and infer interbreeding of members of each subset (we may call these subsets phenetic surrogates of an LBPS).

We may assume that markedly dissimilar organisms have already been eliminated in step 1. When eliminating grossly different organisms, one should also take care that polymorphic forms representing sexes, genetic polymorphs, or different ontogenetic or cyclomorphic stages are not excluded. If no obvious differences are present, the establishment of the homogeneity of the individuals within the sample may require sophisticated biometrical analysis. Even then, homogeneity cannot be proved. It can only be established that for the set of characters which has been measured the individuals appear to be homogeneous. If a heterogeneity is discovered, as, for example, in the form of a bimodality of a given character or a constriction or discontinuity in character hyperspace, we need to allocate the sampled individuals to the two or more subpopulations thus defined. Intergrades will be troublesome in this context, and final decisions on boundaries of the phenetic groups are bound to be arbitrary.

Another method for grouping the subsamples within the original LPS would be to cluster the organisms by one of the methods of numerical taxonomy. As before, such a procedure is quite arbitrary in terms of the choice of a criterion of homogeneity.

4. *Group the LBPS's (or their phenetic surrogates) for all LPS's into phenetically homogeneous sets.* This is an agglomerative phenetic grouping procedure and is necessary as a preliminary to the rigorous test by the

defined criteria for the BSC. This is so because, following the strict guidelines of the biological species definition of "actually or potentially interbreeding populations," one would have to test all samples obtained at the various localities for mutual interbreeding. As will be seen later, this is a formidable, if not impossible, task even when the samples are homogeneous within and among LBPS's, so that one may presume that they all belong to the same species. However, at this point in our procedure for determining whether a group of populations constitutes a biological species, we do not as yet know that the separate subsets in the various LBPS's defined in step 3 are similar to such a degree. All we know is that they are homogeneous *within* LBPS's. This does not necessarily mean that they are homogeneous *among* LBPS's.

Markedly different populations will already have been eliminated in step 1 of the flow chart. Thus we would no longer find one LPBS of drosophila and another LBPS of field mice. However, there might well be several species of drosophila, from the same locality, each in a single LBPS formed by a partition of one original LPS during step 3. We now must take all LBPS's (subsets from different LPS's) and combine them to form one or more sets whose elements are phenetically closely related LBPS's, regardless of the LBP from which they originated. It should be understood here that LBPS's in this step include not only those samples defined by step 2a, when this is possible, but must often include their phenetic surrogates established in step 3b.

When the LBPS's comprise two or

more phenetically closely related but reproductively isolated groups of samples, this admixture becomes a problem. In most cases, techniques like numerical taxonomy should be able to cluster the populations correctly into those that are phenetically alike and therefore candidates for becoming a biological species, subject to further tests in this flow chart. The criterion of phenetic similarity to be employed is necessarily arbitrary, and for this step to be operational we have to establish phenetic limits. One situation where such an analysis might result in clusters undesirable for the present purpose is with marked geographic variation, possibly related to adaptation to ecological differences. Suppose there were two sibling species distributed over the area. It may well be that samples from reproductively isolated populations showing parallel ecological adaptations may cluster before joining with freely interbreeding samples from ecologically different areas. In such cases, some other form of multivariate analysis that removed the effect of ecological differences from a series of morphological variables would reveal the correct situation.

In summary, in most instances of testing for biological species the preliminary test (step 1) is carried out automatically, often already by the collector who does not bother to pick up animals other than those of the species group he is interested in. Nevertheless, it must be clearly recognized that unless the *phenetic* decisions of steps 1 and 4 are taken, one cannot in practice proceed with the determination of the specific status of these populations.

5. *Are the elements of each set found in step 4, the set of LBPS's or their phenetic surrogates, actually or potentially interbreeding among themselves?* This question refers to the most important criterion of the BSC. In its rigid interpretation in terms of the definition, we would have to ascertain whether there is either actual or potential interbreeding among individuals of all the population samples obtained for our study. We have two choices in answering this question: we can either claim to know or hope to find out what the actual interbreeding relations among the organisms are—this leads us to step 5a— or we may decide that the question cannot be answered fully or at all in terms of fertility relationships and proceed to make inferences about these from other evidence, usually phenetics (step 5b).

5a) Knowing all the facts about interbreeding interrelationships among elements of this set, we may partition it into parts containing mutually interbreeding elements (LBPS's). Each such part represents the nucleus of a biological species.

As has been pointed out repeatedly by proponents as well as opponents of the biological species definition, it is impracticable to ascertain these facts in most real situations. The difficulties encountered are of many kinds. The only kind of evidence that would unequivocally answer the question posed is direct observations of marked individuals and of their dispersal (or that of their gametes or offspring), plus observations on mating and success of the progeny in the field. Laboratory experiments on interfertility could be carried out but would indicate neither whether such interbreeding would take place in the field nor whether the offspring of such unions would be viable and reproduce under field conditions.

Even if we were to admit the evidence of laboratory tests, or of crossing experiments by botanists in experimental gardens, the number of crosses required would be formidable. With only two reciprocal crosses for any pair of population samples, we would need $a^2$ tests for $a$ samples (including controls within samples). Thus, for 10 local populations (a far from adequate number in most modern studies of speciation), 100 crosses would have to be made. Yet, we have no assurance that a single representative of each local sample would suffice to establish the necessary facts. After all, if an incomplete sterility barrier exists between these populations, then certain genotypes representing the population might not be able to cross while others would do so successfully. Doubtless, a more representative subsample of each population sample is needed to arrive at a decision on this matter.

On the other hand, since it was demonstrated—or inferred—in step 3 that the individuals within each sample interbreed, we have already stressed in section III that we need only show that there is *some* gene flow among the samples being compared in order to establish interbreeding. Again, we need to distinguish between complete interbreeding, which would mean total panmixia or swamping among all population samples (an unlikely occurrence if the samples are reasonably far apart), and partial interbreeding. The latter, again, could depend on *connectedness* between some individuals in different

LBPS's, which will govern the amount of gene flow, and *success of reproduction,* which refers to the percentage of fertile offspring from such crosses and the success of these offspring, evaluated by some standard. Arbitrary levels for these parameters must be designated to make the definition operational. We shall not suggest such levels here. In any event, the amount of experimental work and of field observations necessary to obtain answers for step 5a would become staggering and is clearly not practical. Sampling and inferences for the larger population are again phenetically based. For this reason there is once more (as in step 3a) a broken arrow leading to and from this operation.

5b) We do not know and cannot discover reproductive relationships among all of the elements (LBPS's) of this set. We therefore partition it into parts containing phenetically homogeneous elements (phenetic surrogates of LBPS's) and infer the possibility of interfertility among these.

This is a phenetic grouping procedure. The type of phenetic connectedness that should reflect whether samples (LBPS's) are actually interbreeding includes a high degree of overall phenetic similarity or the presence of intermediaries (introgression). Both kinds of phenetic evidence are subject to the same arbitrariness associated with the degree to which isolating mechanisms must be present before one can call two samples the same biological species. Here we have to decide what degree of phenetic similarity must be present before considering two samples members of the same biological species. This will vary, of course, with the particular group under study and most of all with the characters chosen for analysis. As Davis and Heywood (1963) as well as critics of numerical taxonomy (e.g., Stebbins, 1963), point out, morphological similarity is not a very accurate reflection of the evolutionary status of biological species. Also, overall similarity may not be the most critical phenetic relationship to be established. Phenetic evidence of introgression may be considered a more important criterion. We shall not discuss the possible procedures in detail here, since our main point is to point out the necessity of inference from phenetic evidence.

Had we been able to follow through on step 5a and define parts containing mutually interfertile elements, we could have bypassed step 6 below because we would have met the requirements of the biological species definition. Since we could not rigidly proceed by step 5a and had to resort to phenetic evidence in step 5b, we should strengthen our inferences by determining the geographical connectedness of these elements as shown in step 6.

6. *The parts of homogeneous sets of phenetic surrogates of LBPS's obtained in step 5b are further divided into one or more parts by geographic contiguity.* This is done to increase our accuracy in the delimitation of biological species. Criteria of geographic proximity should reflect the likelihood of gene flow occurring between any two populations. Thus localities will be considered connected if some members of one LBPS at one locality have an opportunity to join a similar LBPS at the other locality. Geographical distances in such a model would be modified into ecological distances expressing the probability of propagules from one population entering the other population. We are now in a position to make joint judgments about the biological status of the resulting parts, which are phenetically homogeneous and geographically connected sets, constructed by a technique analogous to that of Gabriel and Sokal (1969) for geographic variation analysis.

It will be obvious that, since the level of phenetic homogeneity designated for assigning LBPS's to the same biological species is arbitrary, as is the accepted degree of geographic connectedness, decisions on membership in a biological species are arbitrary as well. That is, we may occasionally decide to include within the same biological species phenetically homogeneous populations that are not fully geographically connected; and, conversely, we may include populations that are phenetically distinct but seem to be fully geographically connected. Since these criteria do not, in any case, meet the formal definition of the BSC, their exact interpretation is not at issue here, unless we wish to infer "potential interbreeding" from them.

Following steps 5b and 6 we obtain the intersection of the parts resulting from these procedures. We infer that the elements in such an intersection (LBPS's or their phenetic surrogates) can represent the nucleus of a biological species. If we are prepared to accept the concept of potentially interbreeding populations, then we may simply use phenetic similarity as a criterion and bypass step 6, which implies actual gene flow in the geographic connections defined by its operations. To avoid confusion, this alternative is not shown in Figure 1.

We now must ask ourselves whether the delimitation of this particular biological species can be extended to include other local populations. This is done by the final step, which follows.

7. *Are there other LPS's that should be added to the above nucleus?* This step tests the adequacy of sampling. The question can be answered by further sampling of organisms from newly studied LPS's, starting with step 1 and repeating the entire procedure.

We define two kinds of additional sampling. *Extensive sampling* gathers further samples beyond the spatial limits of previous samples. *Intensive sampling* seeks to sample areas within the spatial limits of previous samples that have not been sampled before. This step will involve phenetic and geographic criteria, since it would be even more impractical to employ fertility criteria here as well. There is little point in going back through the flow chart, since the same information (phenetics and contiguity in distribution) will be used. In this step, as in steps 1, 3b, 4, and 5b, phenetic considerations will in the end largely delimit the biological species.

## V. PHENETIC BOTTLENECKS

We now can examine the flow chart as a whole and imagine ourselves running some organisms through it to determine into how many biological species they should be divided. Let us design the optimal case for the systematic study of these organisms by the BSC criteria. Therefore we assume un-

limited quantities of live material available from suitably positioned locations throughout the range of the organisms. Since tests of fertility would still require an enormous amount of experimentation, we shall imagine ourselves equipped with an all-knowing computer of unlimited capacity which will provide correct answers for meaningful questions asked of it, obviating experimental tests for interbreeding between pairs of individuals within and between locality samples. To make the situation correspond more closely to the real world about which we wish to make inferences, we shall restrict the computer's performance as follows. It cannot be queried simultaneously about the interbreeding of all individuals of interest, but it will provide correct replies to sequential questions about relationships between each and every pair of individuals.

Given the above (and assuming that we have agreed on a criterion of interbreeding as discussed in section III), we should be able to eliminate all steps in the flow chart except those that make critical tests of interbreeding, namely, 3a and 5a. But we would find that even our phenomenal computer would soon be running overtime providing answers to the millions of questions about interbreeding results of the possible combinations of individuals which we would have to ask. Hence, even in this utopian case we would wish to avail ourselves of steps 1 and 4 for purposes of grouping individuals and populations initially by phenetic likeness so as to cut down on the number of questions about interbreeding that need to be asked. (Thus we shall avoid asking whether an individual cottonwood would cross with an individual aphid.) However, even this timesaving device would not be sufficient. We would still have so many questions to ask about interbreeding, that our patience, if not that of a computer, would soon be exhausted, and we would take certain shortcuts, that is, ask questions about interbreeding of some of the individuals while resorting to phenetic similarities of these with other untested individuals for conclusions about the entire sample. But, having made this concession (i.e., having taken the path of the solid arrows in Figure 1), we are back at steps 3*b* and 5*b*, which we call the *phenetic bottlenecks* because limitations of time will force all studies, even the imaginary optimal study just discussed, into these operations.

Hence, while the definition of the BSC does not involve phenetics, the actual determination of a biological species always will do so, even in the optimal case. As soon as we permit less favorable (and more realistic) conditions to obtain, such as more limited material and no omniscient computer but a hard-working scientist with limited resources and facilities, establishment of biological species from fertility characteristics is entirely quixotic. We are left with what is essentially a phenetic criterion of homogeneous groups that show definite aspects of geographic connectedness and in which we have any evidence at all on interbreeding in only a minuscule proportion of cases.

The above is true for all animal organisms and for most plant organisms, as well. But even in those plant groups where crossing tests (the so-called experimental taxonomy) have been applied, the basic definition of the species is of necessity phenetic because the statements that are made rest on phenetic inferences from the relatively few crosses that have actually been carried out in these groups.

Phylogenetically oriented systematists have pointed out in the past that there are practical difficulties in determining the potentiality of interbreeding in given cases. But, as we have shown here, the concept cannot be used even under optimal circumstances. Simpson (1961, p. 150) has called this a pseudo-problem. He feels that the difficulty of ascertaining whether the definition is met in a given case with a sufficient degree of probability is different from the validity of the concept as such. Yet, as will be discussed below, there is serious question that the concept is evolutionarily meaningful.

## VI. DISCUSSION

*The BSC is imprecise in its formulation and inapplicable in practice.* An obvious conclusion from the flow chart and analysis is that in practice phenetics plays an essential role at several crucial points in the delimitation of a biological species. This leads to the critical question of the degree to which phenetics reflects interbreeding among individuals and populations. But many examples are known (see Davis and Heywood, 1963) where phenetics can only mislead the biosystematist who is seeking the biological species. This ranges from simple polyploidy without phenotypic change and cryptic species, on the one hand, to problems of reactions to the environment, on the other. For example, small flowers are a result of dryness but can also be produced by mutation (Grant, 1954). Without subjecting his material to experimental analysis the practicing systematist could not distinguish between these two causes. In other words, the inductive inference that is necessary here is often unwarranted.

Our study of the operations necessary to delimit a biological species revealed considerable arbitrariness in the application of the concept. This is in direct conflict with the claims of nonarbitrariness by proponents of the BSC. We use the terms "arbitrary" and "nonarbitrary" here in the sense of Simpson (1961, p. 115), where "a group is nonarbitrary as to inclusion if all its members are continuous by an appropriate criterion, and nonarbitrary as to exclusion if it is discontinuous from any other group by the same criterion. It is arbitrary as to inclusion if it has internal discontinuities and as to exclusion if it has an external continuity." The degree of sterility required in any given cross and the number of fertile crosses between members of populations, not to mention the necessarily arbitrary decisions proper to the hidden phenetic components of the BSC, make this concept no less arbitrary than a purely phenetic species concept, and perhaps even more so, since phenetics is but one of its components.

Relevant at this point is a contradiction in the use of the BSC regarding hybridization. This is a confusing term because at one extreme some authors call successful crosses between members of two strains a hybrid, while at the other extreme only crosses between members of two species, or between two genera, are hybrids. If a hybrid is produced in nature from two species and there is *any* backcrossing at all, then by a strict application of the BSC the two parents should belong to the same species, even if such hybrids appear in only a small part of the range of the species. But such an application is not usually made, since the investigator has some arbitrary level of frequency of crossing that he will tolerate before assigning the parents to the same species.

One of the prime complaints of the opponents of a phenetic taxonomy has been that it is typological (Inger, 1958; Mayr, 1965; Simpson, 1961). Whether empirical or statistical typology is an undesirable approach for a classificatory procedure is not at issue here. This question is discussed in some detail by Sokal (1962). In his most recent work on systematics, Mayr (1969, p. 67) describes essentialist ideology as synonymous with typology in the following terms: "This philosophy, when applied to the classification of organic diversity, attempts to assign the variability of nature to a fixed number of basic types at various levels. It postulates that all members of a taxon reflect the same essential nature, or in other words that they conform to the same type. . . . The constancy of taxa and the sharpness of the gaps separating them tend to be exaggerated by [the typologist]. The fatal flaw of essentialism is that there is no way of determining what the essential properties of an organism are." However, it should be pointed out that whether this is desirable or not, the BSC as advanced by its proponents is in itself a typological concept in the above sense. It is typological because it is defined by strict genetic criteria which are rarely tested, and which may not be met by its members (individuals or local populations). We shall examine below the question of whether populations in nature correspond to the biological species type erected by the new systematists. It may well be that the BSC does not reflect a widespread phenomenon in nature but rather represents a theoretical ideal to which existing situations are forced to fit as closely as possible.

It might be claimed that other variants of the biological species definition than the one employed by us could have been shown not to involve unwarranted inferences. However, a careful study of a great variety of such definitions shows this not to be the case. The definition by Emerson (1945)—"evolved (and probably evolving), genetically distinctive, reproductively isolated, natural population" —and that by Grant (1957)—"a community of cross-fertilizing individuals linked together by bonds of mating and isolated reproductively from other species by barriers to mating" —are both prone to the same difficulties. Simpson (1961, p. 153) defined "evolutionary species" as "a linkage (an ancestral-descendant sequence of populations) evolving separately from others and with its own unitary evolutionary role and tendencies." This is so vague as to make any attempt at operational definition foredoomed to failure.

Some plant biosystematists consider the BSC definition we have chosen to be genetic, and not necessarily evolutionary. Some, for example, would maintain that two populations belong to two biological species if they differ in at least one

qualitative character and if there exists a certain amount of sterility between them. But this and similar definitions contain the same drawbacks of necessary phenetic inferences and arbitrariness as the concept we have discussed. It still is based in large part on phenetic inferences that may be unwarranted, and it still distorts relationships among populations by lumping them into a smaller number of biological species. The same comments apply to the definition of a biological species as a set of individuals sharing a common gene pool. This last definition may appear to have one advantage over previous ones. It does not demand that local populations be erected during the process of species delimitation. In terms of our flow chart, steps 2 and 3 would be deleted and subsequent steps reworded. Although this has the "advantage" of reducing the number of necessary steps in the process, this is more than outweighed by the increased amount of inference about gene-pool membership that now must be made from only phenetic evidence, as opposed to inferences made previously from both phenetic and geographic information.

*Some essential questions about the BSC.* From the above conclusions drawn about the BSC, we see that only in rare instances, such as a species consisting entirely of one small endemic population, is the concept even partly operational in practice. But a nonoperational concept may still be of value. For example, it may be used to generate hypotheses of evolutionary importance. We shall examine several relevant questions for systematists and evolutionists concerned with the BSC. At this time we can do little more than to ask the questions and to suggest possible answers.

1. Is the BSC necessary for practical taxonomy? By practical taxonomy we mean the straightforward description of the patterns of variation in nature for the purpose of ordering knowledge. This is phenetic taxonomy, or perhaps simply taxonomy as Blackwelder (1967) sees it. The BSC is not a necessary part of the theory of practical taxonomy, although the category "species" is. The answer to question one is no.

2. Is the BSC necessary (or useful) for evolutionary taxonomy? This is a more difficult question to answer, since different workers attach different meanings to the term "evolutionary taxonomy." It may mean the relatively less complex task of putting all members believed to be derived from the same ancestral stock into the same taxon, say at the genus, or family, level. Or it may involve detailed (usually phenetically inferred) description of cladistic relationships among taxa at some categorical level. The property of interbreeding may or may not be possessed by all members of the group currently under study. Most evidence for decisions in evolutionary taxonomy (and all evidence above the level of classification where crossing is not possible, e.g., between members of two families) is based not on interbreeding but on phenetics and homologies, whether they are morphological, behavioral, physiological, serological, or DNA homologies. Most work to date, especially on DNA homologies, has involved very dissimilar taxa, such as wheat, corn, pigs, monkeys, and man. Since the biological species does not play an essential role in any of the above work, the answer to question 2 would appear to be no.

3. Is the BSC valuable as a unique, heuristic concept from which hypotheses valuable for evolutionary theory can be generated at a high rate? It would appear that any evolutionary hypothesis generated in terms of the BSC can also be generated in terms of the less abstract localized population and perhaps generated more easily. Significantly, population genetics, both theoretical and practical, in nature and in the laboratory, concerns itself with the localized population, or a small number of adjacent localized biological populations. There are few if any insights supposedly obtained from species that cannot be better interpreted at the population level. In fact, some would say that they can be interpreted only at the population level. Nothing is gained by additional abstraction to the species level (except perhaps in efficiency of names), but much is lost, namely, accuracy, for no two localized biological populations are alike. By forcing a large series of them into one biological species we lose the resolution of their differences. The answer to question 3 appears to be no.

4. Is the BSC necessary (or useful) for evolutionary theory? That is, does the general theory of evolution, or any particular evolutionary process, require, or use, the BSC? With respect to the general theory, the answer appears to be no. If we examine the evolutionary situation within some ecosystem, we can generate the same theory based on localized biological populations without grouping sets of interbreeding populations into more

abstract biological species. Parenthetically, we may point out that what are probably the most important and progressive books on evolutionary theory that have been published within the last year or so essentially do not refer to the biological species at all. MacArthur and Wilson (1967) in their study of island biogeography, Wallace (1968) in his analysis of evolutionary mechanisms, and Levins (1968) in his theory of evolution in changing environments base their entire discussions on Mendelian populations and hardly mention the BSC. Williams (1966, p. 252) believes that the species is "a key taxonomic and evolutionary concept but [it] has no special significance for the study of adaptation. It is not an adapted unit and there are no mechanisms that function for the survival of the species."

Let us turn to evolution over geological time and consider the birth and death of a presumed biological species. Assume that a certain phenetic form appeared at time $i$ in the fossil record, subsequently became abundant, and then became extinct at time $j$. What does this mean? It means only that certain populations that possessed the given phenotype were able to survive from time $i$ to time $j$. Ignoring polytopic origins, this means that this favorable character combination was transmitted among several localized biological populations. Nowhere does such a process demand that this set of populations be put into one group and that it be called a biological species. This can be done, but it is not essential to evolutionary theory. Of course, it is done for convenience of reference. It orders our knowledge

in a certain way, as does grouping organisms into taxonomic species, then into genera, then families, etc. Thus it would seem to us that the biological species is an arbitrary category, which may be useful in given situations but is not a fundamental unit of evolution, except possibly in a case in which there is only one local biological population, and therefore the biological species as a class has only one member.

Furthermore, if we assume a priori that all organisms can be put into some biological species, then we of necessity concentrate on finding such classes. Could it be that the occurrence of well-circumscribed biological species is *not* the rule but the exception, in biology? Although Stebbins (1963) says that 70 to 80 percent of higher plant species conform well to the BSC, other evolutionists, upon the accumulation of more and more evidence (e.g., Grant, 1963, pp. 343 ff.) recognize the frequent occurrence of borderline situations.

We do not in any of the above statements imply that reproductive barriers are either nonexistent or unimportant in evolution. Quite clearly they are of fundamental significance. But we do question whether they can be employed to define species and whether emphasis in evolutionary theory should be based on phenomena (including reproductive barriers) pertaining to the species category or to a lower category, the local population.

The answer to question 4 appears to be unclear at best.

*Conclusions.* If our contention that the BSC is neither operational nor necessary for evolutionary theory is granted, what consequences result for general evolutionary theory? There would be few changes if any in terms of our understanding of speciational mechanisms. For example, the numerous important principles outlined by Mayr (1963) in his treatise on the species would still be relevant even if the term "species" as such were removed and replaced by others referring to phenetically different populations, or reproductively isolated populations, or populations with both properties. The positive aspect of such a procedure would be that evolutionary theory and research would concern themselves more with discovering and describing mechanisms bringing about population changes than with trying to bring organic diversity into an order conforming to an abstract ideal. The emphasis would be on unbiased description of the variety of evolutionary patterns that actually exist among organisms in nature, and of the types of processes bringing about the different varieties of population structure. We believe that in the long run this approach would lead to greater and newer insights into the mechanisms of evolution. Fundamentally this would be so because such an approach would free hypothesis construction in evolution from the language-bound constraint imposed by the species concept. (See Kraus, 1968, for a lucid exposition of some of these issues and especially the role of the Whorfian hypothesis.) Even if the Whorfian hypothesis is only partially correct, the very fact that we need no longer put our major emphasis on species definition and description would have a liberating effect on evolutionary thinking. By not tying the variation of individuals and populations to abstract ideals or relating it to a

one-dimensional nomenclatural system incapable of handling the higher dimensionality of the variation pattern, we would be led to new ways of looking at nature and evolution.

Having decided that the BSC is neither operational nor heuristic nor of practical value, we conclude that the phenetic species as normally described and whose definition may be improved by numerical taxonomy is the appropriate concept to be associated with the taxonomic category "species," while the local population may be the most useful unit for evolutionary study.

In advocating a phenetic species concept we should stress that, in concert with most numerical taxonomists, we conceive of phenetics in a very wide sense. All observable properties of organisms and populations are considered in estimating phenetic similarities between pairs of OTU's. These would include not only traditional morphological similarity but also physiological, biochemical, behavioral similarity, DNA homologies (Reich, et al., 1966), similarities in amino acid sequences in proteins (Eck and Dayhoff, 1966; Fitch and Margoliash, 1967), ecological properties (Fujii, 1969), and even intercrossability (Morishima, 1969). Critics of a phenetic taxonomy have claimed that such a wide definition of phenetics makes the term meaningless, since all possible relationships among organisms are then by definition phenetic. But this is not necessarily so. Similarities over the set of all known properties are surely different from similarities based solely on the ability to produce fertile offspring.

Insistence on a phenetic species

concept leads inevitably to a conceptualization of species as dense regions within a hyperdimensional environmental space in the sense of Hutchinson (1957, 1969). Current trends in evolutionary thinking do, in fact, consider this approach to species definition as a more useful and heuristic concept, and, as already mentioned, the existence of apparently "good" asexual species supports this view. However, the establishment of such an environmentally bounded species concept, an idea whose germs can be found in numerous recent papers, is beyond the scope of the present article, which limits itself to pointing out the weaknesses of the generally promulgated BSC.

### SUMMARY

The term "species" has been a central tenet of biological belief since the early days of biology. But the concepts attached to the term have varied and often were not defined rigorously. The purpose of this paper is to investigate the biological species concept (BSC): to consider its theoretical aspects, how one would actually delimit a biological species in nature, whether such species exist in nature, and whether the concept is of any unique value to the study of evolution.

The classical definition of the BSC is partitioned into its essential components, and some of their aspects and problems are discussed. Three fundamental operations necessary for the delimitation of biological species in nature are described in detail. These are operations based on criteria of: (1) geographic contiguity, (2) phenetic similarity, and (3) interbreeding.

Two properties of interbreeding, connectedness and success of reproduction, are defined and discussed.

A flow chart for recognizing biological species is constructed from the definition as given by Mayr. Each step involves one of the three operations mentioned above. Reasons are given for including each step, as well as the inherent difficulties of each. It can be seen that most steps are either largely or entirely phenetic, even in theory. The necessary phenetic steps are termed "phenetic bottlenecks." To test the flow chart, we assume the unrealistic but optimal situation of total knowledge about the interbreeding relations among sampled organisms. The phenetic bottlenecks remain in this optimal case, and the degree of reliance on phenetic information for the delimitation of biological species increases as we depart from the optimal situation and make it more realistic.

The BSC is found to be arbitrary (*sensu* Simpson) when attempts are made to apply it to actual data in nature, and not only because arbitrary phenetic decisions are a necessary part of the delimitation of biological species in nature.

On asking some essential questions about the value of the BSC to taxonomy and evolution, we find that the BSC is not necessary for practical taxonomy, is neither necessary nor especially useful for evolutionary taxonomy, nor is it a unique or heuristic concept necessary for generating hypotheses in evolutionary theory. Most of the important evolutionary principles commonly associated with the BSC could just as easily be applied to localized biological populations, often resulting in deeper insight into evolution.

Having decided that the BSC is neither operational nor heuristic nor of any practical value, we conclude that the phenetic species as normally described is the desirable species concept to be associated with the taxonomic category "species," and that the localized biological population may be the most useful unit for evolutionary study.

### ACKNOWLEDGMENTS

We are fortunate to have benefited from a critical reading of an earlier draft of this paper by several esteemed colleagues, representing considerable diversity in their attitudes to the "species problem." Paul R. Ehrlich, Richard W. Holm, and John A. Hendrickson, Jr., of Stanford University, James S. Farris of the State University of New York at Stony Brook, David L. Hull of the University of Wisconsin at Milwaukee, and Arnold G. Kluge of the University of Michigan contributed much constructive criticism and helped us remove numerous ambiguities and obscurities. A similar function was performed by many members of the Biosystematics Luncheon Group at the University of Kansas. We are much in the debt of all of these individuals, even in those rare instances where we have chosen not to follow their advice.

Collaboration leading to this paper was made possible by grant no. GB-4927 from the National Science Foundation and by a Research Career Award (no. 5-KO3-GM22021) from the National Institute of General Medical Sciences, both to Robert R. Sokal.

### REFERENCES

Berlin, B., D. E. Breedlove, and P. H. Raven, 1966, Folk taxonomies and biological classification. *Science,* 154: 273–275.

Blackwelder, R. E., 1962, Animal taxonomy and the new systematics. *Survey Biol. Progress,* 4: 1–57.

Blackwelder, R. E., 1967, *Taxonomy,* New York, Wiley, 698 p.

Busacker, R. G., and T. L. Saaty, 1965, *Finite Graphs and Networks: An Introduction with Applications,* New York, McGraw-Hill, 294 p.

Davis, P. H., and V. H. Heywood, 1963, *Principles of Angiosperm Taxonomy,* London, Oliver & Boyd, 558 p.

Eck, R. B., and M. O. Dayhoff, 1966, *Atlas of Protein Sequence and Structure,* Silver Spring, Md., Nat. Biomed. Res. Found., 215 p.

Ehrlich, P. R., and R. W. Holm, 1962, Patterns and populations. *Science,* 137: 652–657.

——, and P. H. Raven, 1969, Differentiation of populations. *Science,* 165: 1228–1232.

Emerson, A. E., 1945, Taxonomic categories and population genetics. *Entomol. News,* 56: 14–19.

Fitch, W. M., and E. Margoliash, 1967, Construction of phylogenetic trees. *Science,* 155: 279–284.

Fujii, K., 1969, Numerical taxonomy of ecological characteristics and the niche concept. *Syst. Zool.* 18: 151–153.

Gabriel, K. R., and R. R. Sokal, 1969, A new statistical approach to geographic variation analysis. *Syst. Zool.,* 18: 259–278.

Grant, V. E., 1954, Genetic and taxonomic studies in *Gilia.* IV. *Gilia achilleaefolia. Aliso,* 3: 1–18.

——, 1957, The plant species in theory and practice, p. 39–80. In E. Mayr, ed., *The*

*Species Problem,* Amer. Assoc. Advance. Sci. Publ., 50.

———, 1963, *The Origin of Adaptions,* New York, Columbia Univ. Press, 606 p.

Haggett, P., 1966, *Locational Analysis in Human Geography,* New York, St. Martin's, 310 p.

Harper, J. L., 1966, The reproductive biology of the British poppies, p. 26–39. In J. G. Hawkes, ed., *Reproductive Biology and Taxonomy of Vascular Plants,* New York, Pergamon.

Hutchinson, G. E., 1957, Concluding remarks. *Cold Spring Harbor Symp. Quant. Biol.,* 22: 415–427.

———, 1969, When are species necessary? p. 177–186. In R. C. Lewontin, ed., *Population Biology and Evolution,* Syracuse, N.Y., Syracuse Univ. Press.

Inger, R. F., 1958, Comments on the definition of genera. *Evolution,* 12: 370–384.

Kraus, R. M., 1968, Language as a symbolic process in communication. *Amer. Sci.,* 56: 265–278.

Levins, R., 1968, *Evolution in Changing Environments,* Princeton, N.J., Princeton Univ. Press, 120 p.

MacArthur, R. H., and E. O. Wilson, 1967, *The Theory of Island Biogeography,* Princeton, N.J., Princeton Univ. Press, 203 p.

Mayr, E., 1957, Species concepts and definitions, p. 1-22. In E. Mayr, ed., *The Species Problem,* Amer. Assoc. Advance. Sci. Publ., 50.

———, 1959, Isolation as an evolutionary factor. *Amer. Phil. Soc., Proc.* 103: 221–230.

———, 1963, *Animal Species and Evolution,* Cambridge, Mass., Harvard Univ. Press 797 p.

———, 1965, Numerical phenetics and taxonomic theory. *Syst. Zool.,* 14: 73–97.

———, 1969, *Principles of Systematic Zoology,* New York, McGraw-Hill, 428 p.

Morishima, H., 1969, Phenetic similarity and phylogenetic relationships among strains of *Oryza perennis,* estimated by methods of numerical taxonomy. *Evolution,* 23: 429–443.

Reich, P. R., N. L. Somerson, C. J. Hybner, R. M. Chanock, and S. M. Weissman, 1966, Genetic differentiation by nucleic acid homology. I. Relationships among *Mycoplasma* species of man. *J. Bacteriol.,* 92: 302–310.

Simpson, G. G., 1961, *Principles of Animal Taxonomy,* New York, Columbia Univ. Press, 237 p.

Sokal, R. R., 1962, Typology and empiricism in taxonomy. *J. Theoretical Biol.,* 3: 230–267.

———, 1964, The future systematics, p. 33–48. In C. A. Leone, ed., *Taxonomic Biochemistry and Serology,* New York, Ronald.

———, and J. H. Camin, 1965, The two taxonomies: Areas of agreement and conflict. *Syst. Zool.,* 14: 176–195.

———, and P. H. A. Sneath, 1963, *Principles of Numerical Taxonomy,* San Francisco, Freeman, 359 p.

Stebbins, G. L., 1963, Perspectives. I. *Amer. Sci.,* 51: 362–370.

Wallace, B., 1968, *Topics in Population Genetics,* New York, Norton, 481 p.

Williams, G. C., 1966, *Adaptation and Natural Selection,* Princeton, N.J., Princeton Univ. Press, 307 p.

# 30

# Contemporary Systematic Philosophies

## DAVID L. HULL

DURING THE PAST DECADE, taxonomists have been engaged in a controversy over the proper methods and foundations of biological classification. Although methodologically inclined taxonomists had been discussing these issues for years, the emergence of an energetic and vocal school of taxonomists, headed by Sokal and Sneath, increased the urgency of the dispute. This phenetic school of taxonomy had its origins in a series of papers in which several workers attempted to quantify the processes and procedures used by taxonomists to classify organisms. Of special interest was the process of weighting. These early papers give the impression that the primary motivation for the movement was the desire to make taxonomy sufficiently explicit and precise to permit quantification and, hence, the utilization of computers as aids in classification (22, 23, 41, 91, 106, 107, 111, 112). The initial conclusion that these authors seemed to come to was that taxonomy, as it was then being practiced, was too vague, intuitive, and diffuse to permit quantification. Hence, the procedures and foundations of biological classification had to be changed.

The central issue in this dispute, however, has not been quantification but the extremely empirical philosophy of taxonomy which the founders of phenetic taxonomy seemed to be propounding (54, 79). The pheneticists' position on these issues is not easy to characterize because it has undergone extensive development in the last few years. The words have remained the same. Pheneticists still maintain that organisms should be classified according to overall similarity without any a priori weighting. But the intent of these words has changed. However, one thing seems fairly certain. Pheneticists believed that there was something fundamentally wrong with taxonomy as it was being practiced, especially as set out by such evolutionists as Dobzhansky, Mayr, and Simpson. Later, a third group of taxonomists, led by Hennig, Brundin, and Kiriakoff, entered the dispute, appropriating the name phylogenetic school for themselves. The evolutionists and the phylogeneticists agree that evolutionary theory must play a central role in taxonomy and that biological classification must have a systematic relation to phylogeny. They disagree only over the precise nature of this relation. For the purpose of this paper *evolutionary taxonomy* will refer to the views of the Dobzhansky-Mayr-Simpson school and *phylogenetic taxonomy* will refer to the views of Hennig, Brundin, and Kiriakoff. Together, these two schools will be referred to as *phyleticists* in contrast to the pheneticists.

Although the emphasis of this paper will be on contemporary systematic philosophies and not on the role of quantification in taxonomy, some of the resistance which phenetic taxonomy met was due to a blanket distaste on the part of some taxonomists for mathematical techniques as such and, in particular, for the pheneticists' attempt to quantify taxonomic judgment (104, 105). When Huxley called for "more measurement" in the *New Systematics* (68), he did not have in mind the processes by which taxonomists judge affinity. It is easy to sympathize with both sides, with the biologists who were less than elated over the prospect of learning all the new, high-powered notations and techniques that were beginning to flood the literature and with the pheneticists whose work was rejected on occasion, not because the particular mathematical techniques suggested were inadequate, but because they were mathematical. Hap-

pily, this aspect of the conflict has largely abated, although pockets of resistance still remain. The question is no longer whether or not to quantify but which are the best methods for quantifying.[1]

Recognition should also be made of the majority of taxonomists who, though they consider themselves mildly evolutionary in outlook, feel that all such disputes over foundations and methodology are idle chatter. Taxonomy is not the kind of thing one has to talk about. One just does it. The closest approximation to a spokesman for this group is R. E. Blackwelder, but he is atypical of the majority for which he speaks since he still advocates essentialism in almost its pristine, Aristotelian form (7-13), 15, 121-123). The inadequacy of essentialism as a philosophical foundation for biological classification has been discussed so extensively that nothing more needs to be said here (63, 65, 83, 86).

Not only will this paper be limited to the philosophical aspects of the phenetic-phyletic controversy, but also, of the various issues which have been raised, it will deal with only two—the relation of phylogeny to classification and the species problem. Many of the objections raised against evolutionary taxonomy are actually criticisms of the synthetic theory of evolution, rather than of the classifications built upon it. Nor are these criticisms of recent origin. Every objection raised by the pheneticists to evolutionary theory and evolutionary taxonomy can be found in the work of earlier biologists,

1. For those interested in a review of the numerical aspects of the phenetic-phyletic controversy, I recommend Johnson (73).

usually in the writings of the evolutionists themselves. The difference is that the evolutionists are optimistic about the eventual resolution of these difficulties, whereas the pheneticists, in the early years of the school, believed that they were insoluble. When viewed in the context of the development of biology during the past thirty years, phenetic taxonomy does not appear so much a recent insurrection as the culmination of long-standing grievances.

Soon after the turn of the century, both taxonomy and evolutionary theory had reached a low ebb in the esteem of the rest of the scientific community. Taxonomists seemed to be engaged in a frenzy of splitting and were viewed as nit-picking, skin-sorters, more as quarrelsome old librarians than scientists. Evolutionists had indulged themselves in reconstructing phylogenies in far greater detail and scope than the data and theory warranted and were looked upon as uncritical speculators, more authors of science fiction than science. Among evolutionists themselves, there were controversies. Were the laws of macroevolution different from those of microevolution? Was there such a thing as orthogenesis and, if so, what were the mechanisms for it? At this critical period, Mendel's laws were rediscovered, but instead of clarifying the situation, the birth of modern genetics confused it even further. A whole series of prejudices, conceptual confusions, and peculiarly pernicious terminologies made it seem as if the new genetics conflicted with evolutionary theory. Adding to the

intensity of the controversy was the fact that evolutionists tended to be museum and field workers, whereas geneticists were, by and large, experimentalists at home in the laboratory. It was in this setting that the synthetic theory of evolution and the New Systematics had their inception.

The initial impetus for the rebirth of evolutionary theory was Fisher's *The Genetical Theory of Natural Selection* (49), followed by similar works by Haldane (58) and Wright (130, 131). In these works it was shown that a mathematical model of evolutionary theory could be constructed in which the genetic mechanisms of Mendelian genetics meshed perfectly with the selective mechanisms of evolutionary theory. Evolutionary theory and, hence, evolutionary taxonomy had become respectable again. However, the models supplied by Fisher, Haldane, and Wright were highly restrictive and very far removed from any situation a naturalist was likely to encounter in nature. Using the techniques of idealization which had proved so successful in physics, they showed that in certain overly simple, ideal cases, natural selection working on mutations which obeyed the laws of Mendelian genetics could result in the gradual evolution and splitting of species. The task still remained of showing how the insights gained in these idealizations could be applied to real situations in nature.[2] The classic works on this are those by Dobzhansky (37), Mayr (80, 85, 86), Huxley (68, 69), Simpson (99, 100, 102, 103), Rensch (94, 95),

---

2. For a more realistic, formal axiomatization of evolutionary theory, see Williams (129).

Stebbins (120), Hennig (60, 61), and Remane (93). In the following discussion, the earliest works of these authorities will be cited as freely as their later works because the basic features of the synthetic theory of evolution and evolutionary taxonomy have changed very little during this period.

## PHYLOGENY AND CLASSIFICATION

One of the most persistent problems in biology has been the quest for a natural classification. Prior to Darwin a natural classification was one based on the essential natures of the organisms under study. Of the possible patterns that could be recognized in nature, a taxonomist would settle on one, partly because of his own peculiar psychological make-up and partly because of the scientific theories he held. Of course, another taxonomist with a different psychological make-up, perhaps holding different theoretical views, frequently recognized a different pattern. The controversies that ensued were usually settled by force of authority. The case of Cuvier and his disciples Owen and Agassiz is typical in this respect (83). There are four basic plans in the animal kingdom, no more, no less!

Evolutionary theory promised to put an end to all this dogmatic haggling. After Darwin a natural classification would be one that was genealogical. No longer would biologists have to search fruitlessly for some ideal plan but would need only to discover the genealogical relationships among the organisms being studied and record this information

in their classifications. The alacrity with which many biologists adopted Darwin's suggestion stemmed in part from two illegitimate sources—an inherent vagueness in the proposal and a misconception of the relation which any system of indented, discontinuous words can have to something as continuous and complex as phylogeny. As Darwin (34) observed of Naudin's simile of a tree and classification, "He cannot, I think, have reflected much on the subject, otherwise he would see that genealogy by itself does not give classification." Nearly a century later Gilmour (56) was still forced to remark that he doubted "whether the real significance of the term 'phylogenetic relationship' is yet fully understood."

The purpose of this section will be to investigate the relationships which phylogeny can have to biological classification—assuming that phylogeny can be known with sufficient certainty. The major criticism of evolutionary taxonomy by pheneticists has been that such reconstructions are too often impossible to make. Discussion of this criticism will be postponed until the next section.

No term in taxonomy seems immune to ambiguity and misunderstanding; this includes the term *classification*. Mayr (86) has already pointed out the process-product distinction between the process of classifying and the end product of this enterprise—a classification. But even the words *a classification* are open to misunderstanding. At one extreme, a classification is nothing but a list of taxa names indented to indicate category levels. Others would also include all the characters and the

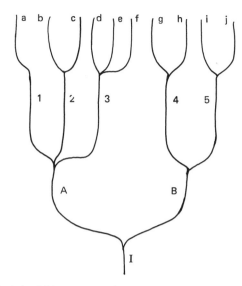

Figure 1. A dendritic representation of a hierarchical classification.

taxonomic principles used to construct a classification as part of the classification. At the other extreme, some authors use the words *a classification* to refer to the entire taxonomic monograph. Unless otherwise stipulated, *a classification* in the following pages will be used in the first, restricted sense.

The simplest view of the relation of a biological classification to phylogeny is that, given a classification, one can infer the phylogeny from which it was derived. One source of this misconception is a naive yet pervasive misconstrual of the relation between a hierarchical classification and a dendritic representation of phylogeny. According to this mistaken view, the classification of Order I sketched below

    Order I
       Family A
          Genus 1
             species a

Genus 2
   species b
   species c
Genus 3
   species d
   species e
   species f
Family B
   Genus 4
      species g
      species h
   Genus 5
      species i
      species j

corresponds to the phylogenetic tree in Figure 1. However, Figure 1 is not a dendritic representation of a possible phylogeny. Rather it is merely a representation of the hierarchic indentations of the classification in a dendritic form. A true dendrogram of the possible phylogenetic development of the organisms involved would consist only of the species listed in the classification. One possible phy-

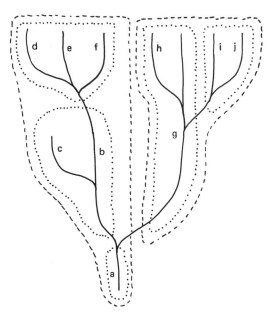

Figure 2. A phylogenetic tree subdivided into taxa.

logeny from which the classification of Order I could have been derived is shown in Figure 2.

In this section we assume the phylogenetic development of the groups under discussion to be completely known. Hence, all the ancestral species are included in the classification along with extant species. In actual classifications, of course, not all ancestral species are known, but at least some are. At least sometimes, biological classifications contain reference to extinct forms. Hence, the interpretation of I in Figure 1 as the unknown stem species which gave rise to Order I, of *A* and *B* as the unknown stem species which gave rise to Families *A* and *B* respectively, and so on, cannot be carried through consistently. On occasion, at least, an-

cestral species will be known and will be included in the classification. The mistake is to confuse the inclusion relations in the taxonomic hierarchy with species splitting (103). An order does not split into genera nor genera into species.

A second impediment to seeing clearly the relation between phylogeny and classification has been a failure to distinguish cladistic from patristic relations (2, 5, 23, 75, 76, 84, 89, 115, 116). The primary difference between the phylogenetic school of Hennig, Kiriakoff, and Brundin and the evolutionary school of Dobzhansky, Mayr, and Simpson is that the former want classification to reflect only cladistic affinity, whereas the latter feel that the classification should also reflect such factors as degree of divergence, amount of diversification, or in general, patristic affinity.

Hennig's principles of classification are extremely straightforward (60, 61). The stem species of every single higher taxon must be included in that taxon and must be indicated as the stem species by not being included in any of the other subgroups of that taxon. Splitting is the only mechanism of species formation that is recognized. Even though a group may evolve progressively until later members are extremely divergent from their ancestors, if no splitting has taken place, all the individuals are considered members of the same species. Upon splitting, the parental species is always considered to be extinct, even though individuals may persist which are morphologically identical to members of the parent species. As far as ranking is concerned, sister groups must always be given coordinate ranks. In addition, Hennig is predisposed to Bigelow's (2–5) observation that in a truly phylogenetic classification recency of common ancestry must be considered a criterion for ranking. Taxa that evolved earlier should be given a higher taxonomic rank than those that evolved later.

The major consequences of the adoption of these principles of classification is precisely the one intended by Hennig. Given a strictly phylogenetic classification, cladistic development can be read off directly. That is, given a classification and Hennig's principles, a dendrogram could be constructed which would accurately represent the cladistic relations of the groups classified. Hennig's principles of classification have something esthetically satisfying

about them. They are straightforward and exceptionless. But this satisfaction is purchased at a price higher than many biologists are willing to pay. Early groups, even if they immediately became extinct without leaving descendants, would have to be recognized as separate phyla, equivalent to highly diversified, persistent groups. Hence, if it could be shown that a species split off in the Precambrian but gave rise to no other species, it nevertheless would have to be classed as a phylum. The resulting classification would be exceedingly monotypic. Our increasing ignorance as phylogeny is traced further back through the geological strata saves phylogeneticists from actually having to introduce such extreme asymmetries into their classification, but even so, enough is known so that classifications erected on the purely cladistic principles of the Hennig school would be much more asymmetrical than those now commonly accepted. Evolutionists also complain that the Hennig school is too narrow since it limits itself just to cladistic affinity. Patristic affinity is also important. Thus, for both practical and theoretical reasons, the evolutionists feel that Hennig's solution to the problem of the relation between phylogeny and classification is unacceptable.

When we turn our attention to evolutionary taxonomy, the situation is not so straightforward. The principles of evolutionary taxonomy are extremely fluid and intricate. As Simpson (103) has said, the practice of evolutionary taxonomy requires a certain flair. There is an art to taxonomy. Vagueness as to the actual relation which evolutionary classification is

to have to phylogeny can be discerned in the earliest statements on
the subject. The main purpose of
Dobzhansky's *Genetics and the Origin
of Species* (37) was to reconcile the
differences between naturalists and
geneticists: to convince the naturalists that the geneticists' experimental
findings in the laboratory were relevant to their work in the field and
museum, and to convince the geneticists that their understanding of
evolutionary theory was grossly inadequate. Dobzhansky is not a systematist
and is not especially interested in
the problems of systematics. The little
that Dobzhansky (37) had to say
about classification can be quoted in
its entirety.

*A knowledge of the position of
an organism in an ideal natural system
would permit the formation of a
sufficient number of deductive propositions for its complete description.
Hence, a system based on the empirically existing discontinuities in the
materials to be classified, and following the hierarchical order of the discontinuous arrays, approaches most
closely to the ideal natural one. Every
subdivision made in such a system
conveys to the student the greatest
possible amount of information pertaining to the objects before him. The
modern classification of organisms
uses the principles on which an ideal
system could be built, although it
would be an exaggeration to think
that the two are consubstantial.*

*On the other hand, since the time
of Darwin and his immediate followers the term "natural classification" has meant in biology one based
on the hypothetical common descent
of organisms. The forms united*
*together in a species, genus, or phy-
lum were supposed to have descended
from a single common ancestor, or
from a group of very similar ancestors.
The lines of separation between the
systematic categories were, hence, adjusted, at least in theory, not so much
to the discontinuities in the observed
variations as to the branching of real
or assumed phylogenetic trees. And
yet the classification has continued
to be based chiefly on morphological
studies of the existing organisms rather than of the phylogenetic series of
fossils. The logical difficulty thus in-
curred is circumvented with the aid
of a hypothesis according to which
the similarity between the organisms
is a function of their descent. In
other words, it is believed that one
may safely base the classification on
studies on the structures and functions of the organisms existing at
one time level, in the assurance that
if such studies are made complete
enough, a picture of the phylogeny
will emerge automatically. This comfortably complacent theory has received some rude shocks from certain
palaeontological data that cast a grave
doubt on the proposition that similarity is always a function of descent.
Now, if similar organisms may, however rarely, develop from dissimilar
ancestors, a phylogenetic classification
must sometimes unite dissimilar, and
separate similar, forms. The resulting
system will be, at least in some of
its parts, neither natural in the sense
defined above nor convenient for
practical purposes.*

*Fortunately, the difficulty just
stated is more abstract than real. The
fact is that the classification of organisms that existed before the advent
of evolutionary theory has undergone*

*surprisingly little change in the times following it, and whatever changes have been made depended only to a trifling extent on the elucidation of the actual phylogenetic relationships through palaeontological evidence. The phylogenetic interpretation has been simply superimposed on the existing classifications; a rejection of the former fails to do any violence to the latter. The subdivisions of the animal and plant kingdoms established by Linnaeus are, with few exceptions, retained in the modern classification, and this despite the enormous number of new forms discovered since then. These new forms were either included in the Linnaean groups, or else new groups were created to accommodate them. There has been no necessity for a basic change in the classification. This fact is taken for granted by most systematists, and all too frequently overlooked by the representatives of other biological disciplines. Its connotations are worth considering. For the only inference that can be drawn from it is that the classification now adopted is not an arbitrary but a natural one, reflecting the objective state of things.* [3]

To begin with, the position of an organism in a hierarchical classification permits the inference of numerous propositions about it only if the characters used to classify the organisms are also listed. For example, knowledge that an organism is a chordate in conjunction with the defining characters of Chordata permits the inference that at some time in its ontogenetic development it has gill slits, a dorsal, hollow nerve cord, and probably a notochord. Knowledge that it is a vertebrate in conjunction with the defining characters of Vertebrata and the fact that Vertebrata is included in Chordata permits additional inferences and so on. A claim frequently made in the recent literature is that the best classification is the one with the highest information content; that is, the one which permits the greatest number of inferences. Colless (31), for example, says:

> The current conflict between the "phenetic" and "phylogenetic" approaches to taxonomy thus boils down to whether a classification should in some fashion act as a storage-and-retrieval system for information about the distribution of attributes over organisms, and thus as a theory that predicts unexamined parts of that distribution: or whether it should reflect, as closely as possible, the historical course of evolution of the organisms concerned.

What is being blurred in the preceding quotation is that a biological classification as such (whether phenetic or phyletic) permits little in the way of inferences. Only a classification in conjunction with the principles and characters used to construct it is sufficient to permit any extensive inferences about the organisms being classified. For example, given the phenon levels of a phenetic classification, it is possible to infer that

---

3. Dobzhansky condenses the preceding discussion to about half its length in the 3rd edition of his work, and Mayr quotes the final paragraph in his *Systematics and the Origin of Species* (80).

members of two taxa at a particular phenon level share a certain percentage of their characteristics, but it is not possible to infer which these may be. Similarly, from an evolutionary point of view, it would be reasonable to infer that two organisms classed together at the 40 phenon level are likely to have a more recent common ancestor than two organisms which are not classed together until the 10 phenon level—if one were given this information.[4] Only when the characteristics used to partition the organisms into taxa are included can specific predictions be made about which organisms are likely to have which characters. But with the addition of such information, we are rapidly approaching the point at which the word *classification* has become expanded to include the entire monograph. Classifications in the narrow sense are incapable of storing much in the way of specific information. Rather than being storage-and-retrieval systems themselves, they serve as indexes to such storage-and-retrieval systems. The information resides in the monograph, not in the classification (128). The classification merely provides a nested set of names which can be used to refer to the relevant taxa in as felicitous a manner as possible.

A second basic misunderstanding concerning the relation between a classification and a phylogeny has contributed to the belief that phylogeny can be inferred from an evolutionary classification. One commonly meets the assertion that proximity of names in a classification implies propinquity of descent. A glance back at Figure 2 shows that this belief is mistaken. If our knowledge of phylogeny were reasonably complete, every single higher taxon would contain at least one species which would be as closely related to a species in another taxon of the same rank as it is to its closest relative in its own taxon. For example, in the classification sketched on p. 571, species *a* is twice removed from species *b* (species *b* appears two lines below species *a*) and nine places removed from species *g* (species *g* appears nine lines below species *a*)— and yet both of these species are directly descended from *a* (see Figure 2).

When stated so baldly, the claim that inferences concerning propinquity of descent can actually be made from an evolutionary classification seems incredible; yet such a view is implicit in the writings of many phyleticists. From an evolutionary classification, even in conjunction with the stated criteria of classification, implications of cladistic relations are not possible. With a reconstructed phylogeny, indefinitely many classifications are possible. With any one of these classifications, an indefinite number of phylogenies are compatible. As Mayr has observed, "Even if we had a perfect understanding of phylogeny, it would be possible to convert it into many different classifications" (86; see also 14, 62, 103). Of course, one way to falsify these claims is to expand the meaning of *a classification* to include phylogenetic dendrograms.

---

4. One of the most persistent problems in taxonomy has been the explication of the notion of "similarity," which is to be some function of descent. An analysis of this concept must be postponed until the next section.

Then, in a trivial sense, phylogeny can be inferred from an evolutionary classification.

Note that not all classifications are acceptable to an evolutionist. For example, all taxa must be "monophyletic." Each taxon can "contain only the descendants of a common ancestor." Early in the history of evolutionary theory, this meant that all the members of a taxon had to be descended from at most a single individual or pair of individuals in an immediately ancestral taxon. As the emphasis in evolutionary theory shifted from individuals to populations and species, the principle of monophyly was expanded so that descent from a single immediately ancestral species was all that was necessary for a taxon to be monophyletic. Hennig and the phylogenetic school still retain this rather stringent notion of monophyly. Unfortunately, if this principle is adhered to, many well-known and easily recognizable taxa such as the mammals (à la Simpson) become polyphyletic. The compromise suggested by Simpson (101, 103), and Gilmour (56) before him, is that all the members of a taxon may be descended, not from a single immediately ancestral species, but from a single, immediately ancestral taxon of the same or lower rank. (For opposing views, see 32.) Thus, since all the species which contributed to the class Mammalia were in all likelihood therapsid reptiles, Mammalia is minimally monophyletic. As reasonable as this decision seems from the point of view of retaining well-marked groups and reflecting degree of divergence, its adoption further weakens the relation between classification and phylogeny. Not all classifications are compatible with a given phylogeny, but too many to permit any precise inferences.

Numerous authors before and after Dobzhansky (37) have observed that, from their classifications alone, "it is practically impossible to tell whether zoologists of the middle decades of the nineteenth century were evolutionists or not." Evolutionists have taken this fact to imply that preevolutionary taxonomists had been reflecting evolution in their classifications all along, though unwittingly. Pheneticists have argued for an additional factor. Classifications, before and after the introduction of evolutionary theory, are basically phenetic. Evolutionary theory, for all intents and purposes, is irrelevant to biological classification. It has been argued in this section that a third factor is actually responsible for the similarity between pre- and post- evolutionary classifications. Hierarchical classifications, in the absence of a rigid adherence to principles of classifications like those of Hennig, do not permit any extensive inferences—whether phyletic or phenetic. Hennig says that hierarchic classifications are completely adequate to indicate phylogeny because he has incorporated the requirements of hierarchic classification into his principles of classification. From a strictly phylogenetic classification (just a list of indented names of taxa) and Hennig's principles, cladistic relations can be deduced. To the extent that this is not done, to that extent the number and variety of phylogenetic inferences which can be drawn from a classification will be diminished (28, 62, 125, 126).

Thus, biologists who maintain that biological classifications should be genealogical are presented with a dilemma. If they adopted a system like Hennig's, in which cladistic development is inferable from a classification, they would have to put up with the loss of information about patristic affinities and the cumbersome classifications that would result. If they retained the more tractable classifications that result from the more pliant principles of evolutionary taxonomy, they would have to abandon the ideal that classifications imply anything very precise about phylogenetic development. Evolution and evolutionary theory would still influence evolutionary classifications, but mainly in decisions as to homologies and the basic units of classification. The way in which evolutionary theory influences estimations of homologies will be discussed in the next section. The relation between evolutionary theory and the basic units of classification will be treated in the last section.

## PHYLETIC INFERENCES AND PHENETIC TAXONOMY

In the preceding section, certain formal difficulties inherent in any attempt to establish a systematic relationship between classification and phylogeny were pointed out. The main thrust of the pheneticists' objections to evolutionary taxonomy, however, has been against permitting phylogeny to influence biological classification in the first place. The chief reasons that the pheneticists have given for excluding evolutionary considerations from biological classifications are as follows: 1. We cannot make use of phylogeny in classification since, in the vast majority of cases, phylogenies are unknown (3, 4, 6, 43, 119). 2. The methods which evolutionists use to reconstruct phylogeny, when not blatantly fallacious, are not sufficiently explicit and quantitative (6, 45, 115, 119). 3. With the help of techniques being developed by the pheneticists, it eventually may be possible to reconstruct reasonably accurate phylogenies for certain groups of organisms, but since phylogeny cannot be known with sufficient certainty for all groups, it should not be used in those few cases in which we do have good reconstructions (6, 24, 115, 119). 4. Even if the necessary evidence were available for all groups and the methods of reconstructing phylogeny were reformulated to make them completely acceptable, the resulting evolutionary classification would still be a special purpose classification and inadequate for biology as a whole; a general purpose classification would still be needed (55–57, 108–110, 119).

Like other criticisms of evolutionary taxonomy, these are not new. As early as 1874, Huxley (70), hardly an enemy of evolution, can be found saying, "Valuable and important as phylogenetic speculations are, as guides to, and suggestions of, investigation, they are pure hypotheses incapable of any objective test; and there is no little danger of introducing confusion into science by mixing up such hypotheses with Taxonomy, which should be a precise and logical arrangement of verifiable facts."

There is little that a philosopher can say about the first two objections to evolutionary taxonomy. After obvious inconsistencies have been removed and warnings about the type of certainty possible in empirical science duly entoned, the controversy becomes largely an empirical matter to be decided by scientists, not philosophers (64, 66). If extensive fossil evidence for a group is necessary for reconstructing the phylogeny of that group, then the phylogenetic development of a majority of plants and animals will never be known, but many biologists think that various laws (or rules of thumb, if you prefer) can be used to reconstruct tentative phylogenies even in the absence of more direct evidence (85, 103, 124).

An interesting development in the phyletic-phenetic controversy is that some numerically minded biologists are beginning to set out formalisms for inferring phylogeny which they feel fulfill the various criteria of objectivity, etc., which more traditional methods are reputed to lack (22–24, 27, 40, 47, 48, 77, 91, 112, 118, 126). Implicit in this endeavor is the conviction that attempts to reconstruct phylogeny even in the absence of fossil evidence are not inherently fallacious. Perhaps the practice of some evolutionists has been slipshod and certain reconstructions of the methods by which phylogenies are inferred have been mistaken, but the phyletic enterprise as such is not hopeless. For example, Colless (30) says, "I must stress at the outset that I am *not* denying that we can, and do, have available a body of reasonably credible phylogenies, which are probably fair reproductions of historical fact. I do, however, assert that some influential taxonomists have an erroneous view of the process by which such phylogenies are inferred; and, if my view is correct, such a situation clearly invites faulty inferences and sterile controversy."

Initially, phenetic and evolutionary taxonomy were treated by all those concerned as if they were in opposition to each other (84, 90, 108, 115, 116). Pheneticists argued that evolutionary classifications, based on a priori weighting, were limited in their uses because they were biased toward a single scientific theory. Phenetic classifications, on the other hand, were general purpose classifications, based on the total number of unweighted or equally weighted characters, and were equally useful to all scientists because they were biased toward no scientific theory whatsoever. Pheneticists like Cain (18–21) attributed the mistakes which early taxonomists like Aristotle, Linnaeus, and Cuvier made to their letting theoretical and philosophical beliefs affect their classifications. Evolutionists had carried on in this misbegotten tradition. To eliminate such errors, pheneticists argued that no theoretical considerations should enter into the initial stages of a purely phenetic classification. A pheneticist must classify as if he were completely ignorant of all the scientific achievements (and failures) which preceded him. Characters must be delineated, homologies established, and clusters derived without recourse to any preconceived ideas whatsoever. No character could be weighted more heavily than another because it proved to be a "good" character in previous studies (unless those studies themselves were phenetic) or because the studies were theoretically important according to current scientific theories. There must be no a priori weighting! Later, after several such purely phenetic studies had been run, certain characters would be found that tended to covary. They then could be weighted a posteriori. This a posteriori weighting would be, however, purely a function of the observed covariations of the characters being studied, not of any theoretical considerations. Finally, evolutionary interpretations could be placed on these purely phenetic classifications which would transform them into special purpose evolutionary classifications. In short, phenetic taxonomy was just look, see, code, cluster.

This initial sharp contrast between evolutionary and phenetic classification has been modified considerably in recent years. In their latest utterances, pheneticists tend to view phenetic taxonomy somewhat differently. Doubts are raised as to whether any pheneticist ever held the views described above. Purely phenetic studies are still considered necessary preliminaries to scientific endeavors of any kind, including the construction of evolutionary classifications, but these phenetic studies are no longer thought of as being performed in isolation from all scientific theories—just from evolutionary theory. Homologies are not established just by observation, but are inferred via relevant genetic, embryological, physiological, and other scientific theories. Prior to any phenetic study, decisions are made as to which characters are to be considered the same, and in what sense they are to be so considered. For example, two organs which are structurally very similar in adult forms might be considered different organs because they have decidedly different embryological developments. Phenetic taxonomy is a matter of look, see, infer, code, and cluster. The resulting phenetic classifications are general

purpose classifications because they have been constructed using all available knowledge, including all well-established scientific theories—except evolutionary theory. Finally, evolutionary interpretations can be placed on these phenetic classifications, but if the phenetic classification is properly constructed to begin with, it will actually be an evolutionary classification. Hence, phenetic and evolutionary classifications, when properly constructed, are equivalent to each other and are equally general purpose classifications.

It will be the purpose of this section to trace the change in phenetic taxonomy from its early, antitheory stage to its current state and to point out the fallacies in the early phenetic position which made it seem attractive and the reasons for changing it. It will be argued that purely phenetic classifications, as they were originally explicated, are impossible and that even if they were possible, they would be undesirable. To the question "Theory now or theory later?" only one answer is possible. The two processes of constructing classifications and of discovering scientific laws and formulating scientific theories must be carried on together. Neither can outstrip the other very far without engendering mutually injurious effects. The idea that an extensive and elaborate classification can be constructed in isolation from all scientific theories and then transformed only later into a theoretically significant classification is purely illusory. A priori weighting of the theoretical kind is not only desirable in taxonomy, it is necessary. The price one pays for theoretical significance is, obviously, that any change or abandonment of the theories which give rise to the classification will necessitate corresponding changes in the classification (52, 53, 67, 68, 69).

There is less to criticize in the latest versions of the phenetic position. One still must question why, of all scientific theories, evolutionary theory must be scrupulously excluded from the process of biological classification. There may be reasons for such a rejection, but the pheneticists have not been very articulate in stating them. Most of the objections which they have raised against evolutionary theory would count equally against any scientific theory and must be interpreted as utterances stemming from their early, antitheory stage of development. Now that pheneticists are willing to accept the role of theory in science, it would be helpful if they were to spell out exactly what faults they still find with evolutionary theory. A final question must also be asked before we turn to a detailed analysis of the evolution of phenetic taxonomy. What was all the controversy about? Except for a greater emphasis on making taxonomic practice explicit and perhaps even quantitative, how does phenetic taxonomy differ from evolutionary taxonomy? If patristic affinity is equivalent to some function of phenetics, chronistics, and cladistics, why all the acrimony? Has the phenetic-phyletic controversy been just one extended terminological confusion?

That the pheneticists actually held the early views attributed to them can easily be demonstrated. For example, as late as 1965, Sokal, Camin, Rohlf & Sneath (116) can be found saying, "Numerical taxonomists *do not disparage* interpretation or speculation or the inductive-deductive method in science. They simply feel that the process of constructing classification should be as free from such inferences as possible...," (See also 18–23, 29, 30, 39, 115, 116.) According to Colless (30), phenetic taxonomy makes reference "only to the observed properties of such entities, without any reference to inferences that may be drawn *a posteriori* from the patterns displayed. Such a classification can, and, to be strictly phenetic *must,* provide nothing more than a summary of observed facts." Even in their most recent publications, pheneticists can still be found making such extremely empirical claims; for example, Sokal (114) says that taxonomy is "the grouping of like organisms based on direct observation."

The key notion in the empiricist philosophy is the claim that, ideally, a priori weighting is to be completely expunged from taxonomic practice. What pheneticists have intended by such interdictions has been extremely equivocal. At one extreme they claim that homologies must be established on the basis of pure observation (as if there were such a thing). Two instances of a character are instances of the same character if they look, smell, taste, sound, and feel the same; otherwise not. Systematics "is a pure science of relation, unconcerned with time, space, or cause" (15). All operational homologies are observational homologies.

So far no pheneticist has produced anything like a strict phenetic classification as described above. Pheneticists make reference to things like wings, antennae, anal gills, dorsal nerve cords, enzymes, and nucleotides. These are hardly pure observation terms. They presuppose all sorts of previous knowledge of a highly theoretical kind. For example, a taxonomist working on brachiopods today describes his specimens and forgets that at one time considerable effort was expended to decide whether brachiopod valves were front and back, dorsal and ventral, or right and left and that the eventual decision reached was based on various theoretical beliefs concerning their ontogenetic and phylogenetic development (35). As Sneath (108) has observed, "Many taxonomic problems start part of the way along the classificatory process, and one is apt to forget what previous knowledge is assumed."

Pheneticists take this to be a fault with traditional taxonomy rather than a characteristic of all scientific undertakings, including their own. They think that, ideally, a purely descriptive, nontheoretical classification must be possible. The source of the persuasiveness of this view can be found in empiricist epistemology, according to which all empirical knowledge stems from sense impressions. Hence, all knowledge must be reducible to pure observation statements. Empiricists themselves have shown that such a reduction is impossible and, specifically, that scientific theories are not replaceable by sets of observation statements (59). There remains the metaphysical compulsion to believe that such a reduction must be possible, and with it, the notion of a purely phenetic classification.

At times pheneticists are a little more liberal in their interpretation of what is to count as a priori weighting. For example, Colless (29) says, "Of course, the simple act of observation of 'existing' entities involves inferences, but they are of a primitive nature and, I believe, can be clearly distinguished from those which I

am concerned to exclude." But how primitive is primitive enough? What criteria does Colless have for making this distinction? And why are primitive a priori weightings acceptable but sophisticated ones illegitimate?

There is a continuum between terms that are largely observational, like white precipitate, flammable fluids, and red appendage, and those that are more theoretical, like inertia, unit charge, and selection pressure. The reason why pheneticists want classifications to be constructed using those terms nearer the observational end of the scale is too apparent. Time and again Cain (18–21) has argued that the greatest source of error in early classifications is their reliance on scientific theories which we now know to be erroneous. Wouldn't the safest procedure be to classify neutrally? That way theories would come and go and the classification, nevertheless, remain unchanged. Such a procedure would assuredly be safe, but in the extreme it is impossible to accomplish and in moderation undesirable.

The basic fallacy underlying the phenetic position on a priori weighting is the confusion of the logical order of epistemological reconstructions with the temporal order in actual scientific investigations (50). Perhaps an analogous example from a different discipline will help to bring this fallacy into sharper focus. In the epistemological approach advocated by Sneath (108), a classification of inorganic substances must begin with purely phenetic studies in which samples are collected of a wide variety of inorganic substances, purely observational homologies established, and various clustering techniques used to group these substances into OTUs. Certain characters might then turn out to be good indicators of certain clusters and weighted more heavily for future runs. Eventually, a classification would emerge which would be equally useful for all purposes. Later, if one wished, this general purpose classification could be transformed into a special purpose classification by introducing atomic theory and weighting atomic number more heavily than all other characters put together.

The actual history of the construction of the periodic table does not, of course, read anything like this epistemological reconstruction. For example, gold was orginally recognized and defined in terms of color, malleability, weight, and so on—a characterization inadequate to distinguish gold from various alloys. Thus, Archimedes was presented with the problem of discovering a more important characteristic of gold. He hit upon specific density. What we tend to forget is that his selection of specific density rather than a host of other characters was his acceptance of the physics of his day in which the four elements were fire, air, earth, and water! Later, as physical theory developed, atomic weight replaced specific density as the key character in distinguishing inorganic substances. In the interim a new concept of element had evolved in the context of atomic theory. Not until atomic number replaced atomic weight could elements, in this new sense, be distinguished from each other and from compounds.

The analogy to the development of evolutionary theory and the species concept is obvious. The point is that a priori considerations were, not after-the-fact interpretations, but necessary factors in every step of the formation of the periodic table. Inorganic elements are distinguished from compounds and from each other on largely theoretical grounds. Incidentally, some very rough clusters of observable characters also accompany this theoretically significant classification. Atomic number, even if considered a phenetic character, was not treated as of equal weight to all other characters. Nor was its weight established a posteriori by discovering that numerous other characters tended to covary with it. The correlation between atomic number and the overall similarity of physical elements is about on the same order of magnitude as that observed by Dobzhansky between breeding habits and the overall similarity of living organisms.

Pheneticists might reply that perhaps this is how the periodic table was constructed, but it shouldn't have been. It should have been constructed by purely phenetic means, and to be justified it must be. This contention has yet to be proven. To do so, pheneticists would have to sample all inorganic substances. They could not limit themselves to just the elements, because that would presuppose that they knew which inorganic substances were elements, a blatant instance of a priori weighting. After establishing homologies purely on the basis of observation, pheneticists would then have to erect various alternative phenetic classifications. Atomic number could hardly appear as one of these phenetic characters, since electrons are observable in only the widest sense of the word. If electrons are observable, so is evolutionary development! If one of these phenetic classifications can distinguish between elements and compounds and can order the elements as they are ordered on the periodic table, then the pheneticists will have proved their case. If recourse to atomic theory is permitted in the early stages of the investigation and atomic number weighted more heavily than all other phenetic characters put together, then phenetic taxonomy, as it was originally explicated and as it is still propounded by many, has been abandoned. If it is to be abandoned, then the original criticisms of evolutionary classifications need to be re-evaluated.

Pheneticists seem to have gradually come to realize that the notion of a theoretically neutral phenetic classification is an illusion and have modified their position accordingly. Operational homologies are established utilizing any respectable scientific theory except evolutionary theory. The reasons given for permitting morphological, behavioral, physiological, serological, and DNA homologies, but forbidding evolutionary homologies, have all depended on repeated equivocations on the terms *phenetic character* and *operational homology*. Pheneticists claim that operational homologies are observed, whereas evolutionary homologies must be inferred. In the first place, only characters are observed. That two instances of a character are instances of the same character (i.e., that they are operationally homologous) must be inferred. Only if operational homologies are limited to observational homologies (i.e., if they both look blue then they are blue) will these inferences be made solely on the basis of observation. All other types of inferences to operational

homologies will make essential reference to a particular scientific theory, and with the introduction of theory the overly simplistic notion of observational homology must be abandoned. One cannot observe that two nucleotides are operationally homologous. Both the existence of the nucleotides and which of the nucleotides are homologous must be inferred from extremely indirect evidence in the context of current biochemical theories.

Colless (30) claims that there is a "phylogenetic fallacy"—the view that "in reconstructing phylogenies, we can employ something more than the observed attributes of individual specimens, plus some concept of 'overall resemblance' and some concept of 'attribute' of a set or class of such specimens." Scientists in general, not just evolutionists, do employ something more than observed attributes and some concept of overall resemblance. This something more is scientific theory. As Colless (30) himself says, "The codon elements thus employed as attributes must, surely, be the ultimate approximation to our notion of 'unit attributes' . . ." What is or is not a codon is determined in large measure by biochemical theory. Codons are certainly not observable. In this instance, pheneticists and not phylogeneticists are guilty of reasoning fallaciously. The phenetic fallacy is the belief that in reconstructing phylogenies, we employ anything less than all the data and all the scientific theories at our disposal. For example, even a theory as far removed from biology as quantum theory is used in the process of carbon dating.

Each of the various kinds of homology has its own special problems. For example, behavioral homologies cannot be obtained very readily for extinct species, nor are the results obtained for extant species by controlled experiments in the laboratory very reliable. Thus, the argument that evolutionary homologies should not be used for any group because we cannot obtain them for all groups cannot be cogent, since, if it were, it would count against all types of homologies. Even morphological homology, the most pervasive type of homology used in classification, has limited applicability. For example, individual viruses and bacteria have few morphological characters which can be used in classifying them. The likelihood of obtaining information about DNA homologies for more than an infinitesimally small percentage of species (and these all extant) is very slim, and yet no one would want to argue that this information should not be used when we do have it. Sokal and Camin (115) say, "Because phenetic classifications require only description, they are possible for all groups and are more likely to be obtained as a first stage in the taxonomic process." The preceding claim is true only if operational homologies are limited to observational homologies. If not, then phenetic classifications require more than description. They require the establishment of theoretically significant operational homologies.

However, the abandonment of the distinction between a priori and a posteriori weighting has certain ramifications for the notions of overall similarity and a general purpose classification. If it is admitted that the establishment of homologies presupposes various scientific theories, then the idea of a single parameter which might be termed *overall similarity* loses much of its plausibility and all classifications become special purpose classifications. As Edwards and Cavalli-Sforza (40) observed, "To say that the purpose of a classification is 'general' is, in our view, too vague to be of use in its construction." The idea of a general purpose classification is still another phenetic illusion. Pheneticists themselves have come to realize that too many parameters exist which have equal right to be termed measures of overall similarity and, hence, that there is no such thing as a general purpose classification. As the Ehrlichs (44) have said recently, "Theoretical considerations make it seem unlikely that the idea of 'overall similarity' has any validity. . . . *All* classifications are inherently special." They quickly add, however, that "no special classification is any more or less 'correct' than any other." (See also 45, 50–53, 73, 110, 115, 116.)

All actual biological classifications are mixed classifications; that is to say, they are affected to a greater or lesser degree by all current biological theories. No classification is purely evolutionary, purely embryological, and certainly none is purely phenetic. The justifications for this irregular mixing of these various considerations in a single classification are both practical and theoretical. In the current state of these theories, evolutionary considerations could no more be untwined from all other considerations and excluded from classification than could embryological or physiological considerations. They are too interconnected. They are interconnected because the theories from which they are partially derived are themselves partially interdependent. Of course, this situation need not be permanent. These various theories may gradually become more carefully and completely formulated, and the relevant derivations more distinct. When this happens, the ideal of providing a straightforward reconstruction of the inferences involved in biological classification can be more closely approximated. We must resist at all costs the tendency to superimpose a false simplicity on the exterior of science to hide incompletely formulated theoretical foundations.

### THE BIOLOGICAL SPECIES CONCEPT

Although Dobzhansky (36–38) first emphasized the biological species concept, it has received its most extensive development at the hands of Ernst Mayr. From his earliest to his most recent writings, Mayr (80–83, 85–87) has set himself the task of demolishing the typological species concept and replacing it with a species concept adequate for its role in evolutionary theory. According to the typological species concept, each species is distinguished by one set of essential characteristics. The possession of each essential character is necessary for membership in the species, and the possession of all the essential characters sufficient. On this view, either a character is essential or it is not. There is nothing intermediate. If a character is essential, it is all-important. If it is accidental, then it is of no importance (63, 65).

In taxonomy, the essentialist position is known as *typology,* a word with decidedly bad connotations. In the recent literature, every school of taxonomy has been called typological at one time or another. The phylogeneticists term the evolutionists typologists because they let degree of divergence take precedence over recency of common ancestry in their classifications (75, 76). The pheneticists call both evolutionists and phylogeneticists typologists because they claim to use criteria which are rarely tested and may not actually obtain (113, 116, 117). The pheneticists in turn are called typologists because their classifications are intended to reflect overall similarity (71, 98, 103). The pheneticists reply that they are typologists but without types and of a statistical variety (113). Their opponents reply that this is not typology but nominalism (84–86)! To put a nice edge on the dispute, some taxonomists openly claim the honor of being called typologists. "Now the great object of classification everywhere is the same. It is to group the objects of study in accordance with their essential natures." (See also 13, 15, 97, 121–123, 127.)

The key feature of essentialism is the claim that natural kinds have real essences which can be defined by a set of properties which are severally necessary and jointly sufficient for membership. Hence, strictly speaking, there can be no such things as statistical typology. Biologists were always aware that the characters which they used to distinguish species did not always universally covary, as the essentialist metaphysics which they tacitly assumed entailed, but not until evolutionary theory were they forced to admit that such variation was not an accidental feature of the organic world, but intrinsic to it. After evolutionary theory was accepted, variation was acknowledged as the rule, not the exception (63, 65). Instead of ignoring it, taxonomists had to take variation into account by describing it statistically. No one specimen could possibly be typical in any but a statistical sense. Species could no longer be viewed as homogeneous groups of individuals, but as polytypic groups, often with significant subdivisions. Polythetic definitions, in terms of statistically covarying properties, replaced essentialist definitions in terms of a single character or several universally covarying characters (1, 26, 33, 63, 92).

One of the accompanying characteristics of essentialism was the gradual insinuation of metaphysical properties and entities into taxonomy. Whenever naturalists attempted to define natural kinds in terms of observable attributes of the organisms being studied, exceptions always turned up. One way to reconcile this apparent contradiction was to dismiss all exceptions as monsters. Another way was to define the names of natural kinds in terms of unobservable attributes. However, two kinds of unobservables must be distinguished at this juncture—metaphysical entities and theoretical entities which, in the context of a particular scientific theory, are indirectly observable. The entities and attributes postulated by classical essentialists tended to be of the former type. The genetic criteria of the biological definition of species may be tested very rarely, but they

are testable and, hence, are not metaphysical. What the pheneticists have in common with typologists is a belief in the existence of natural units of overall similarity. They differ in that these units can be defined only polythetically.

Recognizing the existence of variation among contemporary forms as a necessary consequence of the synthetic theory of evolution is one thing; formulating a methodology in taxonomy sufficient to handle such variation is another. The history of the biological species concept is a story of successive attempts to define species so that the resulting groups are significant units in evolution, or in Simpson's (101, 103) words, an evolutionary species is an "ancestral-descendant sequence of populations ... evolving separately from others and with its own unitary evolutionary role and tendencies." Dobzhansky (36, 37) began by defining a species as that stage of the evolutionary process "at which the once actually or potentially interbreeding array of forms becomes segregated in two or more separate arrays which are physiologically incapable of interbreeding," and he emphasized the necessity of geographic isolation in species formation. "Species formation without isolation is impossible." Mayr concurred with Dobzhansky and distinguished with him between various isolating mechanisms, as such, and geographic and ecological isolation, since these latter are temporary and are readily removed. The species level is reached "when the process of speciation has become irreversible, even if some of the (component) isolating mechanisms have not yet reached perfection" (85). The classic formulation of the biological species definition is as follows:

> A species consists of a group of populations which replace each other geographically or ecologically and of which the neighboring ones intergrade or interbreed wherever they are in contact or which are potentially capable of doing so (with one or more of the populations) in those cases where contact is prevented by geographical or ecological barriers.

Or it may be defined more briefly:

> Species are groups of actually or potentially interbreeding natural populations, which are reproductively isolated from other such groups (80).

Special attention in the preceding definition must be paid to the fact that it is populations which are said to be actually or potentially interbreeding, reproductively isolated, and so on; not individuals. In ordinary discourse, the same terms are applied both to individuals and to groups of individuals—like populations. For example, both individuals and populations are frequently said to interbreed. In most cases, the use of two distinct senses of interbreed causes no confusion, especially since the notion of populations interbreeding is defined in terms of individuals interbreeding. Similarly, Mayr (85) says of isolating mechanisms that they are "biological properties of individuals that prevent the interbreeding of populations that are actually or potentially sympatric." By their very nature, claims about populations interbreeding, etc, are statistical notions derived from the corresponding actions and properties of individuals. Thus, complaints that evolutionists continue to consider two groups as separate species even though members of these groups occasionally cross and produce fertile offspring are misplaced. It is the amount of crossing and the degree of viability and fertility of the offspring that matter. Complaints that values for these variables are too often difficult or impossible to specify are obviously relevant.

Since there is a definitional interdependence between species and population, charges of circularity must be allayed before we proceed further. *Species* is defined in terms of interbreeding, potential interbreeding, and reproductive isolation. Populations are included in species. Hence, populations must at least fulfill all the requirements for species. Additional requirements are added for populations. *Populations* are defined in terms of geographic distribution, ecological continuity, and genetic exchange. A population is "the total sum of conspecific individuals of a particular locality comprising a single potential interbreeding unit" (85). The members of a population must not be separated from each other by ecological or geographic barriers. They must be actually interbreeding among themselves. As a unit, they are potentially interbreeding with other such units.

Throughout his long career, Mayr has continually opposed the typological species concept and essentialism, and yet on some interpretations, the biological species concept has itself been treated typologically, as if it provided both necessary and sufficient conditions for species status. Dobzhansky (37, 82), for example, has argued that individuals which never reproduce by interbreeding can form neither populations nor species because potential interbreeding is a necessary condition for the correct application of these terms. He even goes so far as to say that the terminal populations of a *Rasenkreis,* if intersterile, are to be included in separate species, even though these populations are exchanging genes through intermediary populations! Dobzhansky seems to be confusing the importance of a particular species criterion with the importance of the species concept. The crucial issue is not whether some one character is possessed, but whether the units function in evolution as species. As Mayr (87) has said, *"Species are the real units of evolution,* they are the entities which specialize, which become adapted, or which shift their adaption." Do asexual "species" specialize, become adapted, split, diverge, become extinct, invade new ecological niches, compete, etc.? If so, then from the point of view of evolutionary theory, they form species and criteria must be found to delimit them.

The three elements in the biological species definition are actual interbreeding, potential interbreeding, and reproductive isolation. As succinct as Mayr's shorter version of the biological species definition is, it nevertheless contains redundancies. Two or more populations are reproductively isolated from each other if, and only if, they are neither actually nor potentially interbreeding with each other. Thus, one or the other side of the equivalence could be omitted with no

loss of assertive content. Species are groups of natural populations which are not reproductively isolated from each other but which are reproductively isolated from other such groups. In his most recent publication, Mayr himself omits reference to potential interbreeding in his revised version of the biological species definition: "Species are groups of interbreeding natural populations that are reproductively isolated from other such groups" (86).

In his new biological species definition Mayr still retains reference to interbreeding to indicate that the definition is applicable only to populations whose members reproduce by interbreeding and because successful interbreeding is the most directly observable criterion for species status. Reference to potential interbreeding is omitted because anything that can be said in terms of potential interbreeding can be said in terms of reproductive isolation. Neither morphological similarity nor time is mentioned in any of the formulations of the biological species definition. Among synchronous populations, morphological similarity and difference are of no significance, as far as species status is concerned. Questions of inferring species status aside, they function only in distinguishing phena of the same population, subspecies, sibling species, etc. [See Mayr's (86) discrimination grid.]

The omission of any temporal dimension from the biological species definition is of greater significance. The application of the biological species definition successively in time would lead to the recognition of a series of biological species with minimal temporal dimensions. What is to integrate these successive time species? The answer, as Simpson (103) pointed out earlier, is descent. If species are to be significant evolutionary units, some reference to descent eventually must be made. It is also implicit in any definition of population, since males, females, young and adults, workers and asexual castes are all to be included in the same populations. Morphological similarity won't do, because the types of individuals listed are often morphologically quite dissimilar. However, once a temporal dimension is introduced into the species concept and speciation without splitting is permitted (contra Hennig), an additional criterion must be introduced to divide gradually evolving phyletic lineages into species. The only candidate for such a criterion is degree of divergence, as indicated by morphological and physiological similarity and difference. Thus in the discernment of biological species, morphological similarity and difference play a dual role, in most cases as the evidence by which the fulfillment of the other criteria is inferred and in some instances as criteria themselves. By now it should be readily apparent that any adequate definition of species as evolutionary units can no more be typological in form than can any definition of any theoretically significant term in science. As Julian Huxley (68) observed quite early in the development of the synthetic theory of evolution, "Species and other taxonomic categories may be of very different types and significance in different groups; and also . . . there is no single criterion of species."

The objections, however, which have been made most frequently by the pheneticists against the biological species concept are not those just enumerated, but the following: 1. As important as biological species may be in evolutionary theory, such theoretical considerations should not be allowed to intrude into biological classification, both because they are theoretical and because the presence or absence of reproductive isolation can seldom be inferred with sufficient certainty. 2. There may be fairly pervasive evolutionary units in nature, but reproductive isolation does not mark them. 3. There are no pervasive evolutionary units in nature, regardless of the criteria used to discern them.

As in the case of inferring phylogeny, the commonest complaint raised by extreme empiricists in general, and the pheneticists in particular, against the biological species concept is that too often reproductive isolation cannot be inferred with sufficient certainty to warrant its intrusion into classification. As early as the *New Systematics* (68) Hogben objected that biological species could not be determined often enough, and recently Mayr (85) has said that to "determine whether or not an incipient species has reached the point of irreversibility is often impossible." The problem is not distinguishing one taxon from another but deciding when one or more taxa have reached the level of evolutionary unity and distinctness required of species. If two groups are reproductively isolated from each other, then they are included in separate species; but how often and with what degree of certainty can the presence or absence of reproductive isolation be determined?

If just the two factors space and time are taken into account, four possible situations confront the taxonomist: In the ideal case, two populations, for a while at least, are synchronous and partially overlap. Here, in principle, it is possible to confirm species status by observation. In practice, the situation is not so ideal because the making of such observations is expensive, time consuming, and difficult—not to mention that decisions have to be made regarding the frequency of crossing, the degree of viability and fertility of the offspring, etc. In most cases, even under such optimal conditions, taxonomists depend heavily on inferences from morphological similarity to aid them in their decisions. In cases of synchronic but allopatric populations, the presence or absence of reproductive isolation must be inferred. The advantage here is that on occasion such inferences can be checked, both indirectly by fertility tests in the laboratory and directly, if the populations happen to meet in nature. Usually of course, species status is inferred via morphological similarity and difference. When two populations are separated by appreciable durations of time, inferences of species status are even more circumstantial and can never be checked by any of the more direct means. "Hence, while the definition of the BSC [biological species concept] does not involve phenetics, the actual determination of a biological species always will do so, even in the optimal case" (117).

Pheneticists have objected both to the failure of evolutionists to give phenetics its just due in the application of the biological species concept and to the deficiencies of phenetic similarity as an indicator of reproductive isolation. Since phenetics plays such a predominant role in species determination anyway and since inferences from phenetic similarity to interbreeding status are very shaky at best, they ask why one should not abandon oneself to phenetic taxonomy right from the start. The problem in replying to this question is in deciding precisely what phenetic taxonomy is. By a rigid interpretation, phenetic taxonomy, as it was originally set out, is something radically new, but by this interpretation it can be shown that there can be no such thing as phenetic taxonomy. By a more reasonable interpretation, phenetic taxonomy loses its originality, since it becomes by and large what traditional taxonomists have been doing all along. The jargon of phenetic taxonomy is different, and greater emphasis is placed on mathematical techniques of evaluation, but with such an interpretation phenetic taxonomy is not very revolutionary.

Sokal and Crovello (117) complain that since the words *potential interbreeding* have "never really been defined, let alone defined operationally, . . . it appears to us that the only possible answer one could get from the question whether or not two samples are potentially interbreeding is 'don't know.'" In the first place, potential interbreeding has been defined. If two populations are kept from interbreeding only by geographical or ecological barriers, then they are potentially interbreeding; otherwise not. It is another story, of course, whether or not ecologists and population biologists are in a position to make reasonable inferences on these matters. Sometimes, however, detailed analyses of particular situations have been provided and biologists are in a position to say more than "don't know." With equal justification, an evolutionist could say that since the words *phenetic similarity* have never really been defined, let alone defined operationally, the only possible answer one could get from the question whether or not two samples are phenetically similar is "don't know." Of course, for specific studies, when the OTU's, characters, and clustering method are specified, more specific decisions can be made, but the same is true for potential interbreeding claims. In both disciplines loose and specific questions can be asked.

As unflattering as the appellation may sound, *phenetic* has been a weasel word in phenetic taxonomy. Its meaning changes as the occasion demands. When the principles of other schools of taxonomy are being criticized, it is given a strict interpretation. Phenetic taxonomy is look, see, code, cluster. A methodologically sophisticated ignoramus could do it. But when the pheneticists turn to the elaboration of the methods and procedures of phenetic taxonomy, it takes on a whole spectrum of more significant meanings, heedless of the fact that under these various interpretations the original criticisms of other taxonomic schools lose much of their decisiveness.

For example, in the flow chart designed by Sokal and Crovello (117) for the recognition of biological species, they begin by grouping individuals into rough-and-ready samples. "In the initial stages of the study it may be that sufficient estimations of phenetic similarity can be determined by visual inspection of the specimens." But they go on to admit that such groupings are not the result of mindless look-see. "Knowledge of the biology of the organisms involved may be

invoked." Throughout this flow chart, *phenetically homogenous sets* must include all stages in the life cycle of the organism, various castes in social insects, males and females, etc., regardless of the polymorphisms involved (11, 16, 85, 88, 95, 96). They see this as a practical difficulty, when it is plainly a theoretical difficulty. The admission of such theoretical considerations in the initial stages of a phenetic study means that the pheneticists themselves are practicing a priori weighting, a practice which they have roundly condemned in others. Decisions to include males and females in the same taxon do not stem from earlier phenetic clustering but from previously accepted biological theories. Evolutionists emphasize reproductive isolation because they feel that it is of extreme importance in the phylogenetic development of species. They don't want to see evolutionary units broken up and scattered throughout the nomenclatural system. Similarly, biologists emphasize cellular continuity as a criterion for individuality because they feel that it is of extreme importance in the embryological development of the individual. They don't want to see embryological units broken up and scattered throughout the nomenclatural system. The theory of the individual, as Hennig calls it, may be so fundamental that it has become commonplace, but a biological theory does not cease to be a theory just because it has been around for a long time. As was argued earlier in the section on inferring phylogeny, pheneticists themselves admit theoretical (i.e., a priori) considerations in the initial stages of their studies—as well they should. The point in making this observation is not that pheneticists should be more rigorous in purging their procedures of such theoretical considerations—which are absolutely necessary—but that pheneticists should recognize them for what they are and modify their criticisms of evolutionary taxonomy accordingly.

What is a phenetic property, a phenetic classification, phenetic similarity? If a phenetic property is to be some minimal attribute analyzed in the absence of all scientific theories, regardless of how rudimentary, such characters will certainly be useless in any attempt to construct a scientifically meaningful classification. Arguments have even been set out that, in principle, such an analysis is impossible. If a phenetic property is to be some minimal unit analyzed in the context of some but not all scientific theories (and, in particular, not of evolutionary theory), then the criteria for deciding which scientific theories are legitimate and which illegitimate must be stated explicitly and defended. If some scientific theories are to be admitted even at the initial stages of a phenetic study, then the criticisms of comparable admissions of evolutionary theory must be re-evaluated. The establishment of evolutionary homologies on the basis of evolutionary theory may still be illegitimate, but not just because it is a scientific theory entering into the initial stages of a taxonomic study.

Sokal and Crovello (117) say that phenetic taxonomy is closely related to what Blackwelder calls practical taxonomy—"the straight-forward description of the patterns of variation in nature for the purpose of ordering knowledge." As efforts of the pheneticists have ably proved, there are indefinitely many ways of describing the patterns of variation in nature, and in each way there are indefinitely many patterns to be recognized. The problem is not so much that there is nothing which might be called overall phenetic similarity, but that there are too many things which might answer to this title. The question is whether or not some of these possible ways of ordering knowledge are perhaps more significant than others. The whole course of science attests to the reply that there are some preferable orderings—those which are most compatible with current scientific theories.

Evolutionists claim that their classifications, though they may be constructed in part by intuitive means, are objective, real, nonarbitrary, and so on, because they reflect something which really exists in nature. Pheneticists reply that character covariation also really exists in nature. As might be expected, this sort of exchange has done little to clarify the issues. The difference between evolutionary and phenetic taxonomy in this respect is that evolutionists have biologically significant reasons for making one decision rather than another while, by a strict interpretation, pheneticists do not. On purely phenetic criteria, any group of organisms can be arranged in indefinitely many OTU's with coefficients of similarity ranging from zero to unity. In contrast, evolutionists contend that biological species are important units in nature, more important than numerous other units which might be discernible. They are functioning as evolutionary units in evolution. Hence, from the point of view of evolutionary theory, there is good reason to pay special attention to these units and not to others.

If science were a theoretically neutral exercise, all decisions would be on a par. There would be no difference between the claims that it rains a lot in San Francisco and that all bodies attract each other with a force equal to the product of their masses divided by the square of their distances. As soon as scientific theory is allowed to intrude, certain alternatives are closed, certain decisions are preferable. This is the important sense of natural which has lurked behind the distinction between natural and artificial classifications from the beginning.

In the absence of any scientific theory, the only difference between a natural and an artificial classification is the number of characters used. A natural classification is constructed using a large number of characters, while an artificial classification is constructed using only a few (22, 55–57, 78, 115, 128). Biologists have tended to object to this characterization because it seemed to leave something out, but they have not been too articulate in describing this something. They have argued that a natural biological classification is one based on biologically relevant attributes—as many as possible. An artificial classification is one based on biologically irrelevant attributes—regardless of how many. The controversy has surrounded the sense in which attributes can be biologically relevant or irrelevant.

Taxonomists have tended to term an attribute relevant or taxonomically useful if it has served to cluster organisms into reasonably discrete groups. Thus, for future runs on a group, it would be given greater weight a posteriori. Pheneticists are in full agreement with this usage. But taxonomists also wish to extend their taxonomically useful attributes to cover additional, unstudied groups. This is the a priori weighting to which the pheneticists raised such vocal objections. The justification for such an extension, when it is justified, rests on the second and more important sense of biologically relevant. Certain concepts are central to biological theories; others are not. For example, canalization, geographic isolation, crossing over, epistatic interaction, and gene flow are important concepts in contemporary biological theory. Hence, a classification in which they were central would be natural in the above sense. Of course, gene flow is not used to define the name of a particular taxon, but it does serve two other functions. It plays an important part in the definition of species, and this definition, in turn, determines which taxa are classed at the species level and which are not. In addition, it might play a part in justifying the claim that an attribute which was taxonomically useful in group $A$ should also prove to be taxonomically useful in group $B$. To the extent that such claims are justified, they must be backed up by appropriate scientific laws.

An empiricist might object that all attributes of organisms are equally real. This is certainly true. The broken setae of an insect are as real as a mutation which permits it to produce double the number of offspring, but they hardly are equally important. Just as physical elements are classified on the basis of their atomic number—an attribute selected because of its theoretical significance—evolutionary elements are classified on the basis of their reproductive habits and for the same reasons. Evolutionists contend that if all the data were available, a high percentage of organisms which reproduce by interbreeding could be grouped for long periods of their duration into phylogenetically significant units by the biological species definition.

The pheneticists have attacked this contention on two fronts. First, they have argued that biological species, like phenetic species, are arbitrary units and, second, that biological species, even if they could be determined, would not form pervasive, significant units in evolution. At the heart of the first criticism is the evaluative term *arbitrary*. Claiming to use *arbitrary* in Simpson's sense, Sokal and Crovello (117) say, "Our study of the operations necessary to delimit a biological species revealed considerable arbitrariness in the application of the concept. This is in direct conflict with the claim of nonarbitrariness by proponents of the BSC. . . . The degree of sterility required in any given cross, the number of fertile crosses between members of populations, not to mention the necessarily arbitrary decisions proper to the hidden phenetic components of the BSC, make this concept no less arbitrary than a purely phenetic species concept, and perhaps even more so, since phenetics is one of its components."

Simpson's definition of *arbitrary* is hardly relevant to the issues at hand. According to Simpson (101, 103), when there is a criterion of classification and a classification, groups in this classification are nonarbitrary to the extent that they have actually been classified according to the criterion. For example, if species *A* is de-

fined in terms of property *f*, then the species is nonarbitrary if all of its members have *f;* otherwise, it is not. Simpson's definition is extraneous to this discussion since it assumes precisely what is at issue.

What then do Sokal and Crovello mean by *arbitrary?* Since they repeatedly designate decisions in phenetic taxonomy as arbitrary and since they are advocates of phenetic taxonomy, one might reasonably infer that they do not take it to be a term of condemnation. Yet in one place they talk of arbitrariness as being a drawback to various species definitions. *Arbitrary* is used in ordinary discourse in a host of different senses, and the pheneticists, in a manner not confined to themselves, seem to switch casually from one to another in their criticisms of evolutionary taxonomy. At one extreme, a decision is arbitrary if more than one choice is possible. This is unfortunate because in science more than one reasonable decision is always possible. Hence, all scientific decisions become arbitrary, and the term ceases to make a distinction. For example, should physicists retain Euclidean geometry and complicate their physical laws, or should they retain the simplicity of their laws and treat space as non-Euclidean? Either choice is possible, but physicists' decision for the latter is hardly arbitrary.

A more reasonable use of *arbitrary* is in the division of continua into segments. Biologists of all persuasions commonly admit that whenever an even gradation exists, any classificatory decision automatically becomes arbitrary (17, 103, 119). Here there are not just two or a few possible choices, but many, perhaps infinitely many. Hidden in this line of reasoning is the essentialist prejudice that the only distinctions that exist are sharp distinctions. Unless there is a complete, abrupt break in the distribution of the characters being used for classification, no meaningful decisions can be made. This prejudice was one of the primary motives for philosophers' refusing to countenance even the possibility of evolution by gradual variation and for many philosophers' and biologists' opting for evolution by saltation (65). But this prejudice runs counter to both the very nature of modern science and the methods being introduced by the pheneticists. Various statistical means exist for clustering elements, even when at least one element exists at every point in the distributional space. For example, there are reasons for dividing a bimodal curve at some points rather than at others. Darwin argued that species as well as varieties intergraded insensibly. He concluded, therefore, that they were equally arbitrary. Owing to the mathematical and philosophical prejudices of his day, Darwin's conclusion is understandable. There is no excuse for similar prejudices still persisting (25, 72).

All decisions in phenetic taxonomy are hardly arbitrary in any meaningful sense. If they were, then all the techniques of phenetic taxonomy could be replaced by the single expedient of flipping a coin. Similarly, all decisions as to the degree of crossing, the number of fertile offspring and their viability, etc., sufficient to assure the presence or absence of reproductive isolation are hardly arbitrary in any meaningful sense of the term. From all indications, various thresholds exist in the empirical world. The temperature of water can be varied continu-

ously, but it does not follow thereby that the attendant physical phenomena also vary continuously. At the boiling point, at the freezing point, and near absolute zero, a change of a single degree is accompanied by extremely discontinuous changes in the attendant physical phenomena. Similarly, for example, Simpson refers to quantum evolution, the burst of proliferation that follows a population managing to make its way through an adaptive valley to invade a new ecological niche (88, 101, 103; see also Lewontin's "The Units of Selection," *Annual Rev. Ecol. Syst.*, 1970, 1: 1–14.

There seems to be no question that such significant thresholds exist in evolution. Recently, however, pheneticists have contended that the biological species concept does not mark such a threshold (42, 45, 115). Of all the criticisms leveled at evolutionary taxonomy in the last ten years, this is the most serious. Most of the other criticisms have been largely methodological, resting uneasily on certain dubious philosophical positions, but this criticism is empirical. In a recent study by Ehrlich and Raven (46), evidence was adduced to show that selection is so overwhelmingly important in speciation that the occasional effects of gene flow can safely be ignored in the general evolutionary picture. If this contention is borne out by additional investigation, then the role of the biological definition of species will have been fatally undermined and the synthetic theory of evolution will have to be modified accordingly.

Sokal and Crovello (117), concurring with the position of Ehrlich and Raven, observe that "possibly concepts such as the BSC are more of a burden than a help in understanding evolution." They go on to conclude, however, that "the phenetic species as normally described and whose definition may be improved by numerical taxonomy is the desirable appropriate concept to be associated with the category, species, while the local population may be the most useful unit for evolutionary study." If it can be shown that biological species are not significant units in evolution, then from the point of view of evolutionary taxonomy, the role of the biological species has been fatally undermined. It does not follow, therefore, that the phenetic species, as normally described, should automatically replace them in biological classification, if for no other reason than that no description has been provided yet for the phenetic species. Instead, there are literally an infinite number of phenetic units, all of which have an equal right, on the principles of numerical taxonomy, to be called species.

## CONCLUSION

Numerous distinctions have been drawn in the preceding pages, but little notice has been taken of the most important distinction underlying the phenetic-phyletic controversy—the difference between explicit and implicit or intuitive taxonomy. Simpson (103) has argued that taxonomy, like many other sciences, is a combination of science and art. For example, tempering vertical with horizontal classification, dividing a gradually evolving lineage into species, deciding how much interbreeding is permissible before two populations are included in the same species, the assignment of category rank above

the species level, choices between alternative ways of classifying the same phylogeny, balancing splitting and lumping tendencies, and the inductive inferences by which phylogenies are inferred are all to some extent part of the art of taxonomy. The question is whether the intuitive element in taxonomy should be decreased and, if so, at what cost.

It has been assumed in this paper that decreasing the amount of art in taxonomy is desirable. Taxonomists can be trained to produce quite excellent classifications without being able to enunciate the principles by which they are classifying, just as pigeons can be trained to use the first-order functional calculus in logic. Human beings can be trained to be quite efficient classifying machines. They can scan complex and subtle data and produce estimates of similarity with an accuracy which far exceeds the capacity of current techniques of multivariate analysis. Taxonomists as classifying machines, however, have several undesirable qualities. Although taxonomists, once trained, tend to produce consistent, accurate classifications, the programs by which they are producing these classifications are unknown to other taxonomists and vary from worker to worker. In addition, just when a taxonomist is reaching the peak of his abilities, he tends to die. Only recently one of the most accomplished taxonomists passed away and with her, all the experience which she had accumulated during decades of doing taxonomy.

The resistance to making taxonomic practice and procedures explicit seems to have stemmed from two sources: one, an obscurantist obsession with the ultimate mystery of the human intellect; the other, a concern over how much theoretical significance one must sacrifice in order to make biological classification explicit. With respect to the first reservation, Kaplan (74) has distinguished between reconstructed logic and logic-in-use. Frequently, during the course of development of formal and empirical science, empirical scientists use certain modes of inference which are beyond the current formal reconstructions. There is the tendency to dismiss these modes of inference by attributing them to genius, imagination, and unanalyzable, fortuitous guesswork. Kaplan (74) views the intuition of great scientists, not as lucky guesswork, but as currently unreconstructed logic-in-use. Intuition is any logic-in-use which is preconscious and outside the inference schemata for which we have readily available reconstructions. "We speak of intuition, in short, when neither we nor the discoverer himself knows quite how he arrived at his discoveries, while the frequency or pattern of their occurrence makes us reluctant to ascribe them merely to chance."

The second reservation which tax-onomists have had about making taxonomy less intuitive and more explicit is less subtle, but equally important. In the early days of phenetic taxonomy, pheneticists seemed willing to dismiss the theoretical side of biological classification, since it seemed to make straightforward reconstructions extremely difficult, if not impossible. They tended to conflate the complexity of taxonomic inferences with taxonomists being muddle-headed. Certainly some of the complexity of traditional taxonomy may well have been due just to sloppy thinking, but instead of this evaluation being the immediate, initial response, it should have been the last resort. Traditional taxonomists and computer taxonomists are going to have to adapt to each other, but this adaptation cannot be purchased at the expense of the purposes of scientific investigation. These ends are better characterized by the words *theoretical significance* than by *usefulness.* An extremely accurate scientific theory of great scope will certainly be useful, but there are many things which are useful, though of little theoretical significance.

## ACKNOWLEDGMENTS

I wish to thank Donald H. Colless, Theodore J. Crovello, Michael T. Ghiselin, Ernst Mayr, and Robert R. Sokal for reading and criticizing this paper. The preparation of this paper was supported in part by NSF grant GS–1971.

# REFERENCES

1. Beckner, M., 1959, *The Biological Way of Thought*. New York, Columbia Univ. Press, 200 pp.

2. Bigelow, R. S., 1956, Monophyletic classification and evolution. *Syst. Zool.*, 5: 145–46.

3. ——, 1958, Classification and phylogeny. *Syst. Zool.*, 7: 49–59.

4. ——, 1959, Similarity, ancestry, and scientific principles. *Syst. Zool.*, 8: 165–68.

5. ——, 1961, Higher categories and phylogeny. *Syst. Zool.*, 10: 86–91.

6. Birch, L. C., Ehrlich, P. R., 1967, Evolutionary history and population biology. *Nature*, 214: 349–52.

7. Blackwelder, R. E., 1959, The present status of systematic zoology. *Syst. Zool.*, 8: 69–75.

8. ——, 1959, The functions and limitations of classification. *Syst. Zool.*, 8: 202–11.

9. ——, 1962, Animal taxonomy and the new systematics. *Surv. Biol. Progr.*, 4: 1–57.

10. ——, 1964, Phyletic and phenetic *versus* omnispective classification. In *Phenetic and Phylogenetic Classification*, ed. V. H. Heywood, J. McNeill, 17–28, London, Systematics Assoc., 164 pp.

11. ——, 1967, A critique of numerical taxonomy. *Syst. Zool.*, 16: 64–72.

12. ——, 1967, *Taxonomy*, New York, Wiley, 698 pp.

13. Blackwelder, R. E., Boyden, A., 1952, The nature of systematics. *Syst. Zool.*, 1: 26–33.

14. Bock, W. J., 1963, Evolution and phylogeny in morphologically uniform groups. *Am. Natur.*, 97: 265–85.

15. Borgmeier, T., 1957, Basic questions of systematics. *Syst. Zool.*, 6: 53–69.

16. Boyce, A. J., The value of some methods of numerical taxonomy with reference to hominoid classification. See Ref. 10: 47–65.

17. Burma, B. H., 1949, The species concept: a semantic review. *Evolution,* 3: 369–70.

18. Cain, A. J., 1958, Logic and memory in Linnaeus's system of taxonomy. *Proc. Linn. Soc. London,* 169: 144–63.

19. ——, 1959, Deductive and inductive methods in Post-Linnaen taxonomy. *Proc. Linn. Soc. London,* 170: 185–217.

20. ——, 1959, Taxonomic concepts. *Ibis*, 101: 302–18.

21. ——, 1962, Zoological classification. *Aslib Proc.*, 14: 226–30.

22. Cain, A. J., Harrison, G. A., 1958, An analysis of the taxonomist's judgment of affinity. *Proc. Zool. Soc., London,* 131: 85–98.

23. ——, 1960, Phyletic weighting. *Proc. Zool. Soc. London,* 135: 1–31.

24. Camin, J. H., Sokal, R. R., 1965. A method for deducing branching sequences in phylogeny. *Evolution,* 19: 311–26.

25. Cargile, J., 1969, The sorites paradox. *Brit. J. Phil. Sci.*, 20: 193–202.

26. Carmichael, J. W., George, J. A., Julius, R. S., 1968, Finding natural clusters. *Syst. Zool.*, 17: 144–50.

27. Cavalli-Sforza, L. L., Edwards, A. W. F., 1967, Phylogenetic analysis: models and estimation procedures. *Evolution,* 21: 550–70.

28. Clark, R. B., 1956, Species and systematics. *Syst. Zool.*, 5: 1–10.

29. Colless, D. H., 1967, An examination of certain concepts in phenetic taxonomy. *Syst. Zool.*, 16: 6–27.

30. ——, 1967, The phylogenetic fallacy. *Syst. Zool.*, 16: 289–95.

31. ——, 1970, The relationship of evolutionary theory to phenetic taxonomy. *Evolution.*

32. Crowson, R. A., 1965, Classification, statistics and phylogeny. *Syst. Zool.*, 14: 144–48.

33. Daly, H. V., 1961, Phenetic classification and typology. *Syst. Zool.*, 10: 176–79.

34. Darwin, F., 1959, *The Life and Letters of Charles Darwin,* New York, Basic Books, 2 vols., 558 pp. and 562 pp.

35. Dexter, R. W., 1966, Historical aspects of studies on the Brachipoda by E. E. Morse, *Syst. Zool.,* 15: 241–43.

36. Dobzhansky, T., 1935. A critique of the species concept in biology. *Phil. Sci.,* 2: 344–55.

37. ——, 1937, *Genetics and the Origin of Species,* New York, Columbia Univ. Press, 364 pp.

38. ——, 1940, Speciation as a stage in evolutionary divergence. *Am. Natur.,* 74: 312–21.

39. DuPraw, E. J., 1964, Non-Linnaean taxonomy. *Nature,* 202: 849–52.

40. Edwards, A. W. F., Cavalli-Sforza, L. L., Reconstruction of evolutionary trees. See Ref. 10, 67–76.

41. Ehrlich, P. R., 1958, Problems of higher classification. *Syst. Zool.,* 7: 180–84.

42. ——, 1961, Has the biological species concept outlived its usefulness? *Syst. Zool.,* 10: 167–76.

43. ——, 1964, Some axioms of taxonomy. *Syst. Zool.,* 13: 109–23.

44. Ehrlich, P. R., Ehrlich, A. H., 1967, The phenetic relationships of the butterflies. *Syst. Zool.,* 16: 301–27.

45. Ehrlich, P. R., Holm, R. W. 1962, Patterns and populations. *Science,* 137: 652–57.

46. Ehrlich, P. R., Raven, P. H., 1969, Differentation of populations. *Science,* 165: 1228–31.

47. Farris, J. S., 1967, The meaning of relationship and taxonomic procedure. *Syst. Zool.* 16: 44–51.

48. ——, 1968, Categorical rank and evolutionary taxa in numerical taxonomy. *Syst. Zool.,* 17: 151–59.

49. Fisher, R. A., 1930, *The Genetical Theory of Natural Selection,* Oxford: Clarendon, 272 pp.

50. Ghiselin, M. T., 1966, On psychologism in the logic of taxonomic principles. *Syst. Zool.,* 15: 207–15.

51. ——, 1967, Further remarks on logical errors in systematic theory. *Syst. Zool.,* 16: 347–48.

52. ——, 1969, *The Triumph of the Darwinian Method.,* Berkeley, Univ. California Press, 287 pp.

53. ——, 1969, The principles and concepts of systematic biology. In *Systematic Biology,* Publ. 1962, Nat. Acad. Sci., ed. C. G. Sibley, 45–55, 632 pp.

54. Gilmartein, A. J., 1967, Numerical taxonomy—an eclectic viewpoint. *Taxon,* 16: 8–12.

55. Gilmour, J. S. L., 1937, A taxonomic problem. *Nature,* 139: 1040–47.

56. ——, 1940, Taxonomy and philosophy. In *The New Systematics,* ed. J. Huxley, 461–74. London, Oxford Univ. Press, 583 pp.

57. Gilmour, J. S. L., Walters, S. M., 1964, Philosophy and classification. *Vistas Bot.,* 4: 1–22.

58. Haldane, J. B. S., 1932, *The Causes of Evolution,* London, Harpers, 234 pp.

59. Hempel, C. G., 1965, *Aspects of Scientific Explanation,* New York, Free Press, 505 pp.

60. Hennig, W., 1950, *Grundzüge einer Theorie der phylogenetischen Systematik.* Berlin, Deut. Zentralverlag, 370 pp.

61. ——, 1966, *Phylogenetic Systematics,* Chicago, Univ. Illinois Press, 263 pp.

62. Hull, D. L., 1964, Consistency and monophyly. *Syst. Zool.,* 13: 1–11.

63. ——, 1965, The effect of essentialism on taxonomy. *Brit. J. Phil. Sci.,* 15: 314–26, 16: 1–18.

64. ——, 1967, Certainty and circularity in evolutionary taxonomy. *Evolution,* 2: 174–89.

65. ——, 1967, The metaphysics of evolution. *Brit. J. Hist. Sci.,* 3: 309–37.

66. ——, 1968, The operational imperative—sense and nonsense in operationism. *Syst. Zool.,* 16: 438–57.

67. ——, 1969, The natural system and the species problem. In *Systematic Biology,* Publ. 1962, Nat. Acad. Sci., ed. C. G. Sibley, 56–61, 632 pp.

68. Huxley, J., Ed., 1940, *The New Sys-*

*tematics*, London, Oxford University Press, 583 pp.

69. ——, 1942, *Evolution: the Modern Synthesis*, London, Allen & Unwin, 645 pp.

70. Huxley, T. H., 1874, On the classification of the animal kingdom. *Nature*, 11: 101–2.

71. Inger, R. R., 1958, Comments on the definition of genera. *Evolution*, 12: 370–84.

72. James, M. T., 1963, Numerical vs. phylogenetic taxonomy. *Syst. Zool.*, 12: 91–93.

73. Johnson, L. A. S., 1968, Rainbow's end: the quest for an optimal taxonomy. *Proc. Linn. Soc. N.S.W.*, 93: 8–45.

74. Kaplan, A., 1964, *The Conduct of Inquiry*, San Francisco, Chandler, 428 pp.

75. Kiriakoff, S. G., 1959, Phylogenetic systematics versus typology. *Syst. Zool.*, 8: 117–18.

76. ——, 1965, Cladism and phylogeny, *Syst. Zool.*, 15: 91–93.

77. Kluge, A. G., Farris, J. S., 1969, Quantitative phyletics and the evolution of Anurans. *Syst. Zool.*, 18: 1–32.

78. Lorch, J., 1961, The natural system in biology. *Phil. Sci.*, 28: 282–95.

79. Mackin, J. H., 1963, Rational and empirical methods of investigation in geology. In *The Fabric of Geology*, ed. C. C. Albritton, 135–63, New York: Addison-Wesley, 372 pp.

80. Mayr, E., 1942, *Systematics and the Origin of Species*, New York, Columbia Univ. Press., 334 pp.

81. ——, 1949, The species concept. *Evolution*, 3: 371–72.

82. ——, 1957, *The Species Problem*, AAAS Publ. N. 50, Washington, 338 pp.

83. ——, 1959, Agassiz, Darwin, and evolution. *Harvard Libr. Bull.*, 13: 165–94.

84. ——, 1965, Numerical phenetics and taxonomic theory. *Syst. Zool.*, 14: 73–97.

85. ——, 1965, *Animal Species and Evolution*, Cambridge, Harvard Univ. Press, 797 pp.

86. ——, 1969, *Principles of Systematic Zoology*, New York, McGraw-Hill, 428 pp.

87. ——, 1969, The biological meaning of species. *Biol. J. Linn. Soc.*, 1: 311–20.

88. Megletsch, P. A., 1954, On the nature of the species. *Syst. Zool.*, 3: 49–65.

89. Michener, C. D., 1957, Some bases for higher categories in classification. *Syst. Zool.*, 6: 160–73.

90. ——, 1963, Some future developments in taxonomy. *Syst. Zool.*, 12: 151–72.

91. Michener, C. D., Sokal, R. R., 1957, A quantitative approach to a problem in classification. *Evolution*, 11: 130–62.

92. Minkoff, E. C., 1964, The present state of numerical taxonomy. *Syst. Zool.*, 13: 98–100.

93. Remane, A., 1952, *Die Grundlagen des natürlichen Systems, der vergleichenden Anatomie und der Phylogenetik*, Leipzig, Geest & Portig, 364 pp.

94. Rensch, B., 1929, *Das Prinzip geographischer Rassenkreise und das Problem der Artbildung*, Berlin, Borntraeger.

95. ——, 1947, *Neure Probleme der Abstammungslehre*, Stuttgart, Enke, 407 pp.

96. Rohlf, F. J., 1963, The consequence of larval and adult classification in Aedes. *Syst. Zool.*, 12: 97–117.

97. Sattler, R., 1963, Methodological problems in taxonomy. *Syst. Zool.*, 13: 19–27.

98. ——, 1963, Phenetic contra phyletic systems. *Syst. Zool.*, 12: 94–95.

99. Simpson, G. G., 1944, *Tempo and Mode in Evolution.*, New York, Columbia Univ. Press, 434 pp.

100. ——, 1945, The principles of classification and a classification of mammals. *Bull. Am. Mus. Natur. Hist.*, 85: 1–350.

101. ——, 1951, The species concept. *Evolution*, 5: 285–98.

102. ——, 1953, *The Major Features of Evolution*, New York, Columbia Univ. Press, 434 pp.

103. ——, 1961, *Principles of Animal Taxonomy,* New York, Columbia Univ. Press, 247 pp.

104. ——, 1964, Numerical taxonomy and biological classification, *Science,* 144: 712–13.

105. Simpson, G. G., Roe, A., Lewontin, R. C., 1960, *Quantitative Zoology,* New York, Harcourt, Brace & World, 440 pp.

106. Sneath, P. H. A., 1957, The application of computers to taxonomy. *J. Gen. Microbiol.,* 17: 201–26.

107. ——, 1958, Some aspects of Adansonian classification and of the taxonomic theory of correlated features. *Ann. Microbiol. Enzimol.,* 8: 261–68.

108. ——, 1961, Recent developments in theoretical and quantitative taxonomy. *Syst. Zool.,* 10: 118–39.

109. ——, Introduction. See Ref. 10, 43–45.

110. ——, 1968, International conference on numerical taxonomy. *Syst. Zool.,* 17: 88–92.

111. Sokal, R. R., 1959, Comments on quantitative systematics. *Evolution,* 13: 420–23.

112. ——, 1961, Distance as a measure of taxonomic similarity. *Syst. Zool.,* 10: 70–79.

113. ——, 1962, Typology and empiricism in taxonomy. *J. Theor. Biol.,* 3: 230–67.

114. ——, 1969, Review of Mayr's *Principles of Systematic Zoology. Quart. Rev. Biol.,* 44: 209–11.

115. Sokal, R. R., Camin, J. H., 1965, The two taxonomies: areas of agreement and conflict. *Syst. Zool.,* 14: 176–95.

116. Sokal, R. R., Camin, J. H., Rohlf, F. J., Sneath, P. H. A., 1965, Numerical taxonomy: some points of view. *Syst. Zool.,* 14: 237–43.

117. Sokal, R. R., Crovello, T. J., 1970, The biological species concept: a critical evaluation. *Am. Natur.,* 104: 127–53.

118. Sokal, R. R., Michener, C. D., 1958, A statistical method for evaluating systematic relationships. *Univ. Kansas Sci. Bull.,* 38: 1409–38.

119. Sokal, R. R., Sneath, P. H. A., 1963, *The Principles of Numerical Taxonomy,* San Francisco, Freeman, 359 pp.

120. Stebbins, G. L., 1950, *Variation and Evolution in Plants,* New York, Columbia Univ. Press, 643 pp.

121. Thompson, W. R., 1952, The philosophical foundations of systematics. *Can. Entomol.,* 84: 1–16.

122. ——, 1960, Systematics: the ideal and the reality. *Studio Entomol.,* 3: 493–99.

123. ——, 1962, Evolution and Taxonomy. *Studio Entomol.,* 5: 549–70.

124. Thorne, R. F., 1963, Some problems and guiding principles of Angiosperm phylogeny. *Am. Natur.,* 97: 287–305.

125. Throckmorton, L. H., 1965, Similarity *versus* relationship in *Drosophila. Syst. Zool.,* 14: 221–36.

126. ——, 1968, Concordance and discordance of taxonomic characters in *Drosophila* classification. *Syst. Zool.,* 17: 355–87.

127. Troll, W., 1944, Urbild und Ursache in der Biologie. *Bot. Arch.,* 45: 396–416.

128. Warburton, F. E., 1967, The purposes of classification. *Syst. Zool.,* 16: 241–45.

129. Williams, M. B., 1970, Deducing the consequences of evolution: a mathematical model. *J. Theor. Biol.,* 29: 343–385.

130. Wright, S., 1931, Evolution in Mendelian populations. *Genetics,* 16: 97–159.

131. ——, 1931, Statistical theory of evolution. *Am. Statist. J.,* March suppl., 201–8.

# 31

# Phylogenetic Systematics

## WILLI HENNIG

SINCE THE ADVENT OF THE THEORY of evolution, one of the tasks of biology has been to investigate the phylogenetic relationship between species. This task is especially important because all of the differences which exist between species, whether in morphology, physiology, or ecology, in ways of behavior or even in geographical distribution, have evolved, like the species themselves, in the course of phylogenesis. The present-day multiplicity of species and the structure of the differences between them first becomes intelligible when it is recognized that the differences have evolved in the course of phylogenesis; in other words, when the phylogenetic relationship of the species is understood.

Investigation of the phylogenetic relationship between all existing species and the expression of the results of this research, in a form which cannot be misunderstood, is the task of the phylogenetic systematics.

The problems and methods of this important province of biology can be understood only if three fundamental questions are posed and answered: what is phylogenetic relationship, how is it established, and how is knowledge of it expressed so that misunderstandings are excluded?

The definition of the concept "phylogenetic relationship" is based on the fact that reproduction is bisexual in the majority of organisms, and that it usually takes place only within the framework of confined reproductive communities which are genetically isolated from each other. This is especially true for the insects, with which this paper is mainly concerned. The reproductive communities which occur in nature we call species. New species originate exclusively because parts of existing reproductive communities have first become externally isolated from one another for such extended periods that genetic isolation mechanisms have developed which make reproductive relationships between these parts impossible when

The survey of the literature pertaining to this review was concluded in 1963.

the external barriers which have led to their isolation are removed. Thus, all species ( = reproductive communities) which exist together at a given time—e.g., the present—have originated by the splitting of older homogeneous reproductive communities. On this fact is based the definition of the concept "phylogenetic relationship": under this concept, species B is more nearly related to species C than to another species, A, when B has at least one ancestral species source in common with species C which is not the ancestral source of species A (Hennig, 8).

"Phylogenetic relationship" is thus a relative concept. It is pointless (since it is self-evident) to say, as is often said, that a species or species-group is "phylogenetically related" to another. The question is rather one of knowing whether a species or species-group is more or less closely related to another than to a third. The measurement of the degree of phylogenetic relationship is, as the definition of the concept shows, "recency of common ancestry" (Bigelow, 1). A phylogenetic relationship of varying degree exists between all living species, whether we know of it or not. The aim of research on phylogenetic systematics is to discover the appropriate degrees of phylogenetic relationship within a given group of organisms.

The degree of phylogenetic relationship which exists between different species, and thus also the results of research on phylogenetic systematics, can be represented in a visual form which is not open to misinterpretation, as is a so-called phylogeny tree (dendrogram). To be able to discuss this, not only the species

but also the monophyletic groups included in the diagram must be given names. "Monophyletic groups" are small or large species-groups whose member species can be considered to be more closely related to one another than to species which stand outside these groups (Hennig, 8). When a phylogeny diagram, conforming to this postulate, has been rendered suitable for discussion by the naming of all of the monophyletic groups, then the diagram can be discarded and its information may be expressed solely by ranking the names of the groups:

A. Myriopoda
B. Insecta
  B.1 Entognatha
    B.1a Diplura
    B.1b Ellipura
      B.1ba Protura
      B.1bb Collembola
  B.2 Ectognatha

Such arrangement of monophyletic groups of animals according to their degree of phylogenetic relationship is called, in the narrower sense, a phylogenetic system of the group in question. Such a system belongs to the type called a "hierarchical" system. Since "system" in the wider sense means every arrangement of elements according to a given principle, the phylogeny tree, too, can be termed a phylogenetic system. Phylogeny diagrams and arrangement of the names of monophyletic groups in a hierarchical sequence are merely different but closely comparable forms of presentation whose content is the same. Therefore, everything which can be said about the methods of phylogenetic systematics (see below) applies irrespective of whether the results sought by the use of these methods are expressed only as a phylogeny tree or as a phylogenetic system in the narrower sense, in a hierarchically arranged list of the names of monophyletic groups.

In some cases, a hierarchical arrangement of group names, that is, a phylogenetic system in the narrower sense, is to be preferred to a phylogeny tree. One can, for instance, in a catalogue or check-list of Nearctic Diptera, give expression to all that one thinks is known about the phylogenetic relationship of all Nearctic species of Diptera in a form which can in no way be misinterpreted, without using a single phylogeny tree.

However, considerable difficulties arise because systems of the hierarchical type have also been used in biology with intentions other than of expressing the phylogenetic relationship of species. Long before the advent of the theory of evolution, "systematics" existed as the branch of biological science which had adopted as its aim an orderly survey of the plurality of organisms. Naturally, the principle of classification in systematics could not then be the phylogenetic relationship of species, which was still unrecognized, but only a morphological resemblance between organisms. This morphological systematics also used the hierarchical type of system to express its results although Linnaeus already held the view that morphological resemblances between organisms corresponded to a multidimensional net. Numerous attempts have also been made to introduce other types of system, which differ from the hierarchical, into biological systematics (see Wilson and Doner, 21). But they have not been successful.

Today, there are still many authors who consider that the purpose of biological "systematics" is to classify organisms according to their morphological resemblance, and who use a system of the hierarchical type to this end. It is hardly surprising that misunderstandings and serious errors can be produced by this formal identity between morphological and phylogenetic systems.

The source of danger in the formal identity between systems based on such different principles of classification is that, in a hierarchical system, each group formation relates to a "beginner," which is linked in "one-many relations" with all of the members of that group and only those (Gregg, 3). In morphological systems the "beginner" that belongs to each group is a formal idealistic standard ("Archetype") whose connections with the other members of the group are likewise purely formal and idealistic. But in a phylogenetic system the "beginner" to which each group formation relates is a real reproductive community which has at some time in the past really existed as the ancestral species of the group in question, independently of the mind which conceives it, and which is linked by genealogical connections with the other members of the group and only with these. One could, without difficulty, adduce many examples from the literature in which the formal beginner ("Archetype") of a group, conceived according to the principles of morphological systematics, has been erroneously taken, with all of the consequences of such an error of logic, as the real beginner (ancestral species) of a monophyletic group.

This dangerous difference between

a formal morphological (typological) hierarchical system and the equally hierarchical system of phylogenetic systematics, would not arise if the degree of morphological resemblance were an exact measurement of the degree of phylogenetic relationship. But this is not the case. Furthermore, there is yet no definition of the concept of morphological resemblance which is not open to theoretical objection, nor any method which can be accepted as the one and only method which achieves a satisfactory determination of more than the threshold of morphological resemblance, that is, the degree of resemblance between relatively similar species which agree in very many characters.

In these circumstances, the dangers which arise from the formal identity of phylogenetic and morphological systems will be avoided if agreement can be reached on whether or not the branch of biological science known simply as systematics will, in the future, always try to express the morphological resemblance of organisms or their phylogenetic relationship in the system in which it works.

It has often been stated, in defense of a system of morphological resemblance, that this has historical primacy over endeavors to express phylogenetic relationship in a system, because the morphological system had already existed as the aim of "systematics" before the advent of the theory of evolution. Even today, this reasoning is often augmented with the argument that the theory of evolution was established with the help, among other things, of the system of graduated morphological resemblances between organisms, and that therefore one is prescribing a circle if, in reverse, one

wishes to take the theory of evolution and the notion of the phylogenetic relationship of organisms which follows from it as the theoretical starting-point of their classification in a system (Sokal, 17; Blackwelder, Alexander, and Blair, 2). This "ebenso halt-wie heillose Einwand" (Günther, in discussing the work of Sokal, 12) has already been so often refuted that one can only attribute, to authors who persist in asserting it today, a lack of information.

It is certainly correct that the classification of organisms according to their morphological resemblance has led to the theory of evolution. This was possible only because the morphological differences between organisms are the result of a historical (phylogenetic) development and because, at least in rough terms, very similar organisms, are, in fact, generally more closely related than are very different ones. It was therefore inevitable that the classification of organisms according to their morphological resemblance, in association with certain features of their ontogenetic development and their geographical distribution, would sooner or later lead to the discovery of their successive degrees of phylogenetic relationship and thus to the theory of evolution.

However, there are historical origins not only of the morphological differences between organisms in the narrower sense, but also differences in their physiological functions, their ways of behavior and, in addition to these physical ("holomorphological") attributes, differences in their distribution in geographical and ecological space. Since it has been recognized and, moreover, become widely known that there are not the same degrees of agreement and difference in the various holomorphological and chorological resemblances which connect organisms, the way is open for establishing the phylogenetic relationship itself of organisms as the principle of classification, instead of successive degrees of resemblance in a single category of characters; for only from the phylogenetic relationship is it possible to establish direct connections with all other thinkable kinds of agreement and difference between organisms. The demand for a phylogenetic system is thus not so much a renunciation of pre-phylogenetic resemblance, systematics, as its consequential further development.

The claim of the phylogenetic system to elevation into the universal reference system of biology has a logical, even if not historical, foundation, and arises because few areas of research can be conceived which do not bear fruit and lead to more profound conclusions through a knowledge of the phylogenetic relationship of its objects, and which cannot, in turn, lead to the discovery of hitherto unknown relationships in the course of mutual exchange of information. This is not true to the same extent for any other system built on any other principle of classification. Other systems may also have their value as knowledge; but this value is, in each case, restricted to answering particular questions.

The logical primacy of the phylogenetic system also arises because it alone provides all parts of the field studied by biological systematics with a common theoretical foundation (Kiriakoff, 14). It is true that phylogenetic relationship exists only between different species, and species are not the simplest elements of biological systematics. These are not even the "individuals," but the individuals in given short periods of their lifetime ("semaphoronts"). The first and basic task of systematics is to establish that different individuals, or rather "semaphoronts," belong to particular species. The difficulty within this task rests in the fact that the species, which exist in nature as real phenomena independent of the men who perceive them, are units which are not morphologically but genetically defined. They are communities of reproduction, not resemblance. Of course, the morphological resemblance between members of a species is not unimportant for the practical establishment of specific limits. But it has only the significance of an auxiliary criterion whose capabilities of use are limited. This is because the definition of the phylogenetic relationship between species, as well as the definition of the species-concept, is deduced from the fact that the reproduction of species generally takes place within the framework of defined communities which cannot be unqualified communities of resemblance if, in the demand for a phylogenetic system, biological systematics has acquired for all its spheres of activity a common aim, that is, the discovery and recording of the "hologenetical" connections which exist between all organisms. In contrast with this, morphological resemblance-systematics, though not denying the modern genetic species-concept, employs different principles of classification above and below the specific level.

It would, of course, be meaningless

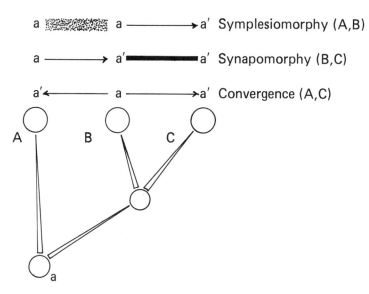

Figure 1. The three different categories of morphological resemblance $a$: plesiomorph. $a'$: apomorph expression of the morphological character $a$. Agreement may rest on symplesiomorphy ($a$–$a$), synapomorphy ($a'$–$a'$) or convergence ($a'$–$a'$).

to extol the need for a phylogenetic system, however well founded it might be theoretically, if this demand could not be put into practice. There is, in fact, a widespread notion that phylogenetic systematics, at least in those groups of animals for which no fossil finds are available, possesses no method of its own, but can only interpret the results of morphological systematics according to the principle that the degree of morphological resemblance equals the degree of phylogenetic relationship. This notion is false. The fundamental difference between the method of morphological and phylogenetic systematics is that the latter breaks up the simple concept of "resemblance" (Fig. 1).

It is a consequence of the theory of evolution that the differences between various organisms must have arisen through changes of characters in the course of a historical process. Therefore it is not the extent of re-

semblance or difference between various organisms that is of significance for research into phylogenetic relationship, but the connection of the agreeing or divergent characters with earlier conditions. It is valid to distinguish different categories of resemblance according to the nature of these connections.

The division of the concept of resemblance into various categories of resemblances probably began, in the history of systematics, with the introduction of the concept of convergence. Often this concept was linked with the distinction between analogous and homologous organs. Convergence is, in fact, commonly manifested by similar organs having arisen in adaptation to the same functions from different morphological foundations in different organisms. But there are also cases where virtually complete agreement in the form of homologous organs rests on convergence. "Convergence" means resemblance between the characters of different species which has evolved through the independent change of divergent earlier conditions of these characters. It shows how species which differed from one another are ancestors of species which have become similar to one another. If one associates in a group the species whose resemblance rests on convergence, then this is not a monophyletic but a polyphyletic group. There are few authors today who would specifically support the inclusion of demonstrably polyphyletic groups in a system. "Convergence" and "polyphyletic groups" are concepts which presuppose acceptance of the theory of evolution. Therefore, some systematists think they are already working with a "phylogenetic system" when, in their evaluation of morphological resemblance, they exclude convergence and thus polyphyletic groups from their system.

But even when purged of convergence, morphological resemblance is still not a satisfactory criterion for the degree of phylogenetic relationship between species. It still does not provide one with exclusively monophyletic groups, such as a phylogenetic system demands. This arises from the fact that characters can remain unchanged during a number of speciation processes. Therefore, it follows that the common possession of primitive ("plesiomorph") characters which remained unchanged cannot be evidence of the close relationship of their possessors.

Often, a given species can be phylogenetically more closely related to a species which possesses a particular character in a derivative ("apomorph") stage of expression than to species with which it agrees in the possession of the primitive ("plesiomorph") stage in the expression of this character. Therefore, a resemblance which rests on symplesiomorphy is of no more value in justifying a supposition of closer phylogenetic relationship than is a resemblance which has occurred through convergence. If, in a system, one associates in a group species whose agreement rests on convergence, a polyphyletic group is thereby formed, as has been established above and is generally recognized. If one associates species whose agreement rests on symplesiomorphy, then a paraphyletic group is formed (Fig. 2). Paraphyletic groups among insects are the "Apterygota" and Palaeoptilota ( = Palaeoptera), if one considers the closer relationship of the Odonata with the Neoptera as established. Paraphyletic vertebrate groups are the "Pisces" and the "Reptilia."

The supposition that two or more species are more closely related to one another than to any other species, and that together they form a monophyletic group, can be confirmed only by demonstrating their common possession of derivative characters ("synapomorphy"). When such characters have been demonstrated, then the supposition has been confirmed that they have been inherited from an ancestral species common only to the species showing these characters.

It must be recognized as a principle of inquiry for the practice of systematics that agreement in characters must be interpreted as synapomorphy as long as there are no grounds for suspecting its origin to be symplesiomorphy or convergence.

The method of phylogenetic systematics as that part of biological science

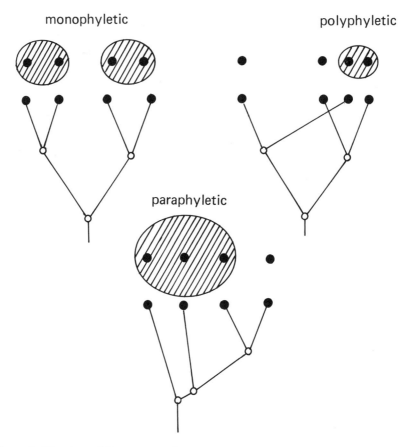

Figure 2. The three different categories of systematic group formations corresponding to the resemblance of their constituents resting on synapomorphy (monophyletic groups), convergence (polyphyletic groups), or symplesiomorphy (paraphyletic groups). For comparison with Figure 31-1.

whose aim is to investigate the degree of phylogenetic relationship between species and to express this in the system which it has designed thus has the following basis: that morphological resemblance between species cannot be considered simply as a criterion of phylogenetic relationship, but that this concept should be divided into the concepts of symplesiomorphy, convergence, and synapomorphy, and that only the last-named category of resemblance can be used to establish states of relationship.

The differences between the phylogenetic system and all other systems which likewise classify species on the basis of their morphological resemblance are as follows: (*A*) Systems which employ the simple criterion of morphological resemblance. Such systems include polyphyletic, paraphyletic, and monophyletic groups. (*B*) Systems which employ the criterion of morphological resemblance but fail to consider characters whose agreement rests on convergence. In such systems, polyphyletic groups are excluded but paraphyletic as well as monophyletic groups are admitted. (C) Phylogenetic system. Characters whose agreement rests on convergence or symplesiomorphy are not considered. Therefore, polyphyletic and paraphyletic groups are excluded and only monophyletic groups admitted.

The systems named under (B) have also often been termed phylogenetic systems in the literature (e.g., Stammer, 18; Verheyen, 20). But it is thereby overlooked that the paraphyletic groups admitted in these "pseudophylogenetic" or "cryptotypological" systems (Kiriakoff, 14) are similar in many respects to polyphyletic groups. No one would think of considering polyphyletic groups in studies concerned with the course and eventual rules of phylogenesis (zoogeographical studies, for instance, belong here), since they have no ancestors solely of their own and therefore no individual history. Exactly the same holds true, however, for paraphyletic groups. The sole common ancestors of all the so-called "Apterygota," for instance, were also the ancestors of the Pterygota, and the beginning of the history of the Apterygota was not the beginning of an individual history of this group, but the beginning of the individual history of the Insecta, which were at first Apterygota in the morphological-typological sense. Also, the concept of "extinction" is different in paraphyletic and monophyletic groups. Only monophyletic groups can become "extinct" in the sense that from a particular point in time no physical progeny of any member of the group have existed. But if, however, one says that a paraphyletic group has become "extinct," this can only mean that after a particular point of time no bearers of the morphological characters of this group have existed. But physical progeny of many of its members may, with changed characters, continue to live. Monophyletic and paraphyletic groups thus cannot be compared with each other in any question concerning their history. Failure to take account of this fact and invalid uncritical comparison of paraphyletic and monophyletic groups has led to some false conclusions in studies about the "Grossablauf der phylogenetischen Entwicklung" (Müller, 15) and the history of the distribution of animals.

From the premise that morphological agreement only confirms a supposition that the species concerned belong to a monophyletic group when it can be interpreted as synapomorphy, is derived for the practical work of the systematist, the "Argumentation plan of phylogenetic systematics" (Fig. 3). This plan shows that in a phylogenetic system which must contain only monophyletic groups, every group formation, irrespective of the rank to which it belongs, must be established by demonstration of derivative ("apomorph") characters in its ground plan. But it also shows clearly that in two monophyletic groups which together form a monophyletic group of higher rank and are therefore to be termed "sister-groups," one particular character must always occur in a more primitive (relatively plesiomorph) condition in one group than in its sister-group. For the latter, the same is true in respect to other characters. This mosaic-like distribution of relatively primitive and relatively derivative characters in related species and species-groups (Spezialisationskreuzungen, Heterobathmie der Merkmale:

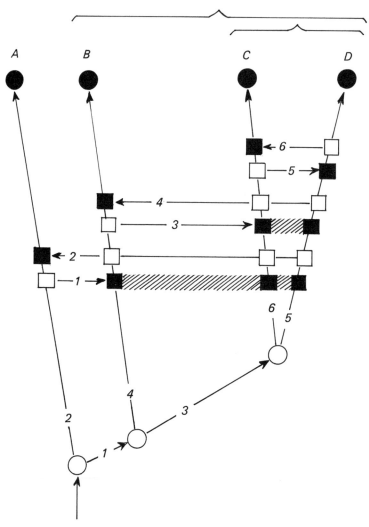

Figure 3. Argumentation plan of phylogenetic systematics. ☐plesiomorph, ■apomorph expression of characters. Equal numbers indicate how sister-group relations are established by the distribution of relatively plesiomorph (white) and relatively apomorph (black) characters ("heterobathmy of characters"). Adapted from Hennig (11).

Takhtajan, 19) has long been known. But one still finds it occasionally mentioned in the literature as a special peculiarity of some groups of animals that the classification of their constituent groups cannot be achieved in a definite sequence, because there are no solely primitive and no solely derivative species or species-groups. In a phylogenetic system there can indeed be no solely primitive and no solely derivative groups. The possession of at least one derivative (relatively apomorph) ground-plan character is a precondition for a group to be recognized at all as a monophyletic group. But it also follows from this that this same character in the nearest related group must be present in a more primitive (relatively plesiomorph) stage of expression. The exclusive presence of relatively plesiomorph characters is indicative of paraphyletic groupings: these are to be found only in pseudophyletic (see above under *B*) and purely morphological systems (see above under *A*), but not in phylogenetic systems. Heterobathmy of characters is therefore a precondition for the establishment of the phylogenetic relationship of species and hence a phylogenetic system.

It is sometimes said that the aims of phylogenetic systematics are not only practically but also theoretically unattainable, because the comparison of species living in a given time-horizon, such as the present, cannot in any way reveal their phylogenetic relationship, which refers to a completely different dimension. This view is false. Just as two stereoscopic views of a landscape, which themselves assume only a two-dimensional form, together contain exact information about the third spatial dimension, so the mosaic of heterobathmic characters in its distribution over a number of simultaneously living species contains reliable information about the sequence in which the species have evolved from common ancestors at different times. The study and use of the methods which serve to reveal this information needs, it is true, a far greater amount of knowledge and experience than some systematists are willing to employ. The theoretical foundation and refinement of these methods forms a special chapter in the theory of phylogenetic systematics which can only be touched upon in the present brief paper.

It is sometimes alleged that consideration of as many characters as possible which have so far not been studied is a prerequisite for the progress of phylogenetic systematics. In particular, the restriction of entomological systematics to comparatively easily recognizable characters of the external skeleton which lie open to view is often not highly regarded. This has some justification. The phenomena of convergence (particularly in its variant known under the name "parallel development"), reversed development of characters and paedomorphosis, which leads to pseudoplesiomorph conditions, make the establishment of true synapomorphy difficult. The more complex is the mosaic of heterobathmic characters which we have at our disposal in a chosen group of species, the more surely can their phylogenetic relationship be deduced from it.

Consideration of new and hitherto unobserved characters can, however, represent progress only if these are analyzed with the special methods of phylogenetic systematics. Thus, it is also necessary to distinguish between plesiomorph and apomorph expressions of characters in the internal anatomy and chemical structure, physiology, and serology and when considering different ways of behavior. Symplesiomorphy must be excluded just as much as convergence. If this is not observed, then consideration of however many characters leads, at best, only to a more precise determination of the overall similarity of the bearers of all these characters, but not to a more precise establishment of their degree of phylogenetic relationship.

This becomes particularly obvious in animal groups such as the insects, in which the life of the individual is subject to the phenomenon of metamorphosis. This is the cause of the incongruences which are so often discussed between larval, pupal, and imaginal classification in morphological and pseudo-phylogenetic systematics. A theoretically acceptable solution of such "incongruences" is possible only in phylogenetic systematics. It can indeed be the case that particular instances of synapomorphy, and therefore of monophyletic groups, can be recognized only in the larval or pupal stages and others only in the imaginal stage. But this is not a true incongruence, for the phylogenetic system does not try to classify organisms according to their degree of resemblances, but species according to their degree of phylogenetic relationship. It does not matter therefore which stage of development is used to establish relationship on the ground of synapomorphy. A monophyletic group remains such even if it can be established only with the characters of a single stage of development (for more detailed exposition see Hennig, 11).

The fact that not resemblance as such, but only agreement in a particular category of characters is significant for the study of phylogenetic relationship, also makes it possible for phylogenetic systematics to adduce for its purposes features other than physical (holomorphological) characters. Such nonholomorphological characters are the life history and geographical distribution of species. Phylogenetic systematics can, for instance, proceed from the plausible hypothesis that species which show a clearly derivative ("apooec") life history, and for which a certain

relationship is probable on other grounds, form a monophyletic group. This is, for instance, often true with parasites. However, hypotheses of this kind must always be verified by close morphological studies, for it is particularly with similar life histories that adaptive convergence is common.

A particularly great importance for phylogenetic systematics is presently often ascribed to parasites and to monophagous and oligophagous plant-feeders which are to be equated with them from the standpoint of phylogenetic method. The theoretical justification for this is supplied by the so-called parasitophyletic rules. Particularly important among these is the so-called "Fahrenholz rule," which supposes a marked parallelism between the phylogenetic development of parasitic groups of animals and their hosts in the majority of cases. If this is correct, then it might be concluded from the restriction of a monophyletic group of parasites to a particular group of host species that the latter, too, form a monophyletic group. But it can easily be shown that this conclusion would be correct only if one could assume that the ancestral species of the host group was attacked by one parasite species and that thereafter each process of speciation in the host group has been accompanied by one speciation process in the parasites. Clearly, this precondition is only rarely fulfilled, since the evolution of parasites often seems to be retarded in comparison with that of their hosts, in respect to both character changes and speciation. The result of this is that paraphyletic host groups can also be attacked by monophyletic groups of parasites. Moreover, it happens that parasites can transfer secondarily (without being passed from ancestors to progeny in the course of speciation) to host species which offer them similar conditions of life. This, too, is often seen as an indication of close phylogenetic relationship between host species which are exclusively attacked by particular parasite species or a monophyletic group of parasites. But this assumption would be valid only if one could assume that the "degree of resemblance" of different species and the "degree of their phylogenetic relationship" corresponded closely with each other. As has been shown, this is not the case. Resemblance can also be based, for instance, on symplesiomorphy, and this cannot be assumed to establish phylogenetic relationship. Since one cannot assume that parasites distinguish, in their choice of host range, the categories of resemblance connections (symplesiomorphy, synapomorphy and convergence) whose differences are important for phylogenetic systematics, the greatest care is necessary in attempting to draw conclusions about the phylogenetic relationship of their hosts from the occurrence of monophyletic groups of parasites. The importance of parasitology for phylogenetic systematics is considerable. But on the grounds given it is not so great as is sometimes supposed. In particular there is still no really satisfactory clarification of this whole complex of questions.

The geographical distribution of organisms is also of restricted though not to be underestimated importance for phylogenetic systematics. This can often proceed from the hypothesis that parts of a group which are restricted to a defined, more or less separated, part of the total range, whose ancestors may be assumed to have arrived from other regions, form a monophyletic group. This is particularly valid for the fauna of the marginal continents (Australia and South America), whose ease of accessibility has been different at different periods of the earth's history, and for some islands (e.g., Madagascar, New Zealand). One can, for instance, proceed on the working hypothesis that the Marsupialia of Australia form a monophyletic group, and then seek either to sustain or refute this hypothesis with the morphological methods of phylogenetic systematics. With groups of animals with disjunctive distribution, one may proceed on the hypothesis that both parts of the range (Australia and South America in the case of pouched mammals) have been settled by monophyletic subgroups and that between these a sister-group relationship exists. Extensive investigation of the phylogenetic development of animal groups (e.g., Hofer on the Marsupialia) often in themselves remain fruitless, since they do not proceed from a working hypothesis of this kind and as a result contain no statements which serve to answer the questions which first come clearly to light in such an hypothesis. This is often of even greater importance in studies of the history of the settlement of geographical space. Discussions about the earlier existence of direct land connections between now separate regions (Madagascar and the Oriental Region: Günther, 5; New Zealand and South America: Hennig, 12) have somewhat the same significance as have attempts to sustain or refute hypotheses about the monophyletic, paraphyletic, or poly-phyletic character of particular groups of animals. The inadequacy of mor-phological or pseudo-phyletic systems is shown here with particular clarity.

A special chapter in the theory of phylogenetic systematics which can only be touched upon here is the position of fossils in the system (Hen-nig, 9). Despite a widely held opinion, establishing the phylogenetic relation-ships of fossil animal forms is usually more difficult than that of recent spe-cies. The cause of this is that in fossil finds usually only a small, often ex-tremely small, section is available from the character structure of the whole organism. But since the meth-ods of phylogenetic systematics have a numerical character insofar as the cer-tainty of their conclusions grows as the number of characters at their dis-posal increases (see above), it follows necessarily that the reliability with which relationships can be established cannot usually be as great with fossils as with recent species. In the sphere of the lower categories of the system, the species and their subunits, palaeo-systematics is, in addition, at a de-cisive disadvantage because it can never observe its objects alive, and can therefore only solve its problems with the help of relatively unreliable morphological criteria. It is true that the systematics of recent organisms also satisfies itself mainly with mor-phological criteria to help it establish the limits of species. However, there is always the possibility, in principle, of testing in important cases that in-dividuals of similar or different ap-pearance actually belong to one or to different reproductive communi-ties by observation of their life in nature or by breeding and crossing

experiments. In species with seasonal and sexual dimorphism and those in which the life of the individual con-tains a metamorphosis, systematics de-pends upon such methods. But, in palaeontology, they cannot be em-ployed. Here systematics can establish the specific limits only with a much lower degree of accuracy than with recently known organisms. It would, however, be completely false to de-duce from this, as is sometimes done, that palaeontological systematics op-erates with other concepts (e.g., a different species-concept) and other methods. It differs from the syste-matics applicable to recent animal forms only in the lesser degree of cer-tainty and accuracy with which it is able to apply itself.

This applies to inquiry into specific limits just as it does to establishing the degree of the phylogenetic rela-tionship between species. If the pur-pose of systematics does not consist exclusively of conducting a survey of the animal forms which have existed on the earth at any time, then palae-ontology must also try to relate its objects to the phylogenetic system of recent organisms—that is, to in-clude them in this system. But this can be meaningful and fruitful only if the limits of the knowledge it can supply are known very precisely and are clearly expressed in each particu-lar case.

Subject to these conditions, the value of fossil finds lies in enabling one to interpret character agreements in recent species when this cannot be done solely from a knowledge of these recent forms. There are, in the recent fauna, monophyletic groups which agree in certainly derivative (apomorph) characters with other diverse groups which are just as surely monophyletic. Some of these agreements must therefore rest on convergence. But it is often impossible to decide with certainty which of these agreements are based on convergence and which are to be considered as true synapomorphy. The possibility of decision in such cases depends on a knowledge of the sequence in which the characters in question evolved. This is sometimes clarified by fossils. An example of this kind is supplied by the sea urchins (Echinoidea).

The Cidaroidea, which are shown to be a monophyletic group by their peculiar spine formation, agree completely with most other recent sea urchins in their possession of a rigid corona. The more primitive expression of this character, a flexible corona, is present only in the Echinothuridae. On the other hand, the Echinothuridae agree completely, in their possession of external gills, with the sea urchins which do not belong to the Cidaroidea. This is likewise a derivative character. This character distribution allows no decision on the question of whether the Cidaroidea or the Echinothuridae are more closely related to the bulk of recent sea urchins. One of the two derivative characters, the external gills or the rigid corona, must thus have evolved through convergence at least twice independently. The oldest fossil Cidaroidea, which are shown to belong to this group by their spine formation, possess a flexible corona. This is decisive evidence that the ridigity of the corona in recent Cidaroidean and in the remaining recent sea urchins (except the Echinothuridae) has evolved through convergence. Concerning the external gills, there are no reasons to suggest convergent evolution. Their presence in recent sea urchins which do not belong to the Cidaroidea may therefore be regarded as synapomorphy. However, it must also be said that they have often been lost secondarily. In other cases, only fossil finds make it possible to establish which expression of a character should be regarded as plesiomorph in a group and which as apomorph.

The importance of fossils thus lies, not so much in the fact that they reduce the morphological gap between different monophyletic groups of the recent fauna, but in that they help to make it possible to decide the categories of resemblance (symplesiomorphy, synapomorphy, or convergence) to which particular agreements of character belong.

Still greater is the value of fossils for determining the age of animal groups. But in this context it should be realized that age determinations have a meaning only in monophyletic groups, since only they have a history of their own (see above). It can be difficult, however, to demonstrate the relationship of a fossil to a given monophyletic group of animals. As has been shown above, heterobathmy of characters is characteristic for nearly related monophyletic groups. Therefore, it often happens that one of two sister-groups can be established as a monophyletic group only by a few apomorph characters which are difficult to verify or only present at a particular stage of metamorphosis. For the distinction of the two groups and the identification of the species belonging to them, this has no significance, because plesiomorph characters can also be employed for diagnosis, though they must be left out of consideration in establishing the monophyly of a group. One can, for instance, recognize at once that a recent arthropod species belongs to the Myriopoda from its possession of homonomous body segmentation with jointed appendages on more than three of its trunk segments, although both are plesiomorph characters and cannot be used to justify the supposition that the Myriopoda are monophyletic. But this is not the case with fossils. One cannot assume without qualification that fossils, especially from the early Palaeozoic, belong to the Myriopoda if they possess a homonomous segmentation and jointed appendages on more than three trunk segments. Both are plesiomorph characters which must also have been present in the common ancestors of the Insecta and Myriopoda. To demonstrate that fossils in fact belong to the Myriopoda, one must demonstrate in them those apomorphy characters in the ground plan of the group which suggest its monophyly—that is, the absence of ocelli and compound eyes. Such demonstration is often very difficult, since these characters are not preserved for us in the fossils. If, in this case, one proceeds uncritically, and classifies fossils on the basis of plesiomorph characters which suffice as diagnostic characters for the certain recognition of all recent species of a monophyletic group, then it can happen that the group will become a paraphyletic group solely through its acquisition of fossils. This can then become the source of all the errors which necessarily arise if one compares monophyletic and paraphyletic groups with one another in phylogenetic studies (see above).

When, however, it has been firmly established that a fossil belongs to a given monophyletic group, that fossil can then be of importance not only for determining the minimum age of the group to which it belongs, but also for determining the minimum age of related groups, of which no fossil finds are available. The existence of *Rhyniella praecursor* in the Devonian not only proves that Collembola, the group to which *Rhyniella* belongs, already occurred then, but from our relatively certain knowledge of the phylogenetic relationships of the principal monophyletic groups of insects it follows that at the same period the Protura, Diplura, and Ectognatha must also have existed, although, of course, not in the form of their present-day progeny.

In determining the age of animal groups, another factor should be considered as well. In the history of a monophyletic group of animals, there are two points of time which are especially important (see Fig. 4): one is at the time at which the group in question was separated from its sister-group by the splitting of their common ancestor (age of origin), and the other the time at which the last common ancestral species of all recent species of the group ceased to exist as a homogeneous reproductive community (age of division). The distinction between these two points of time is especially important in those groups whose recent species are distinguished from species of other groups by their agreement in a large number of derivative characters. One must assume that these characters were already present in the last common ancestral species to whose progeny they have been transmitted unchanged or, in part, further developed.

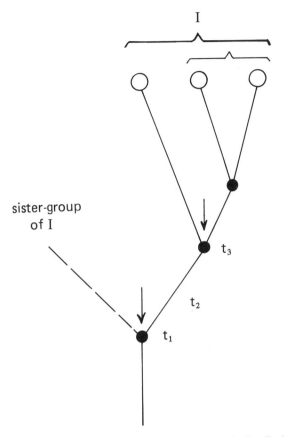

Figure 4. The three different meanings of questions about the "age" of an animal group.
$t_1$ age of origin (separation of group I from its sister-group),
$t_2$ first appearance of the "typical" characters of group I,
$t_3$ age of division (last common ancestor of all recent species of group I)

These characters must have evolved in the period between the two named points of time.

Speculation upon the age of a particular group of animals can have three appropriate but different meanings. The following may be intended: (*a*) When did the last common ancestral species of all the recent species of this group which have inherited their derivative characters from it, live? (question about the group's age of division). (*b*) When was the group separated from its sister group? (question about the group's age of origin).

(*c*) When, in the period between these two points of time, did species for the first time occur with the characters which justify their ascription to the "type" represented by the recent species?

It is seldom clear which of these three essentially different questions is intended when questions are asked about the age of fleas, lice, or other animal groups. This fact, in conjunction with the custom of seeing in phylogenesis mainly the emergence of particular "types" or "Baupläne" whose delimitation is dependent on subjective criteria, is the cause of endless and fruitless debate on the question of whether or not certain fossils should be considered "reptiles," "birds," "mammals," or "men," and when these groups evolved.

It might seem that questions about the age of animal groups lie outside the field of systematics. But this is not the case. The examples quoted should have shown that answering these questions has the same significance as systematically classifying fossils in particular groups, and that the meaning of an answer depends on the classificatory principle used in forming them.

The age of animal groups also has yet another significance for phylogenetic systematics, under some circumstances. It has been said above that the phylogeny diagram and the hierarchical system are closely corresponding kinds of presentation whose content is one and the same. The phylogeny tree presents, as the most important factor, the time dimension in which the degree of phylogenetic relationship between species or monophyletic groups of species is expressed by the sequence in which they evolved from each common ancestral species (i.e., recency of common ancestry); in a hierarchical system this is shown by the sequence of subordination in the group categories. It is a justifiable aim to perfect the phylogeny diagram by giving, not only the relative sequence of origin of the monophyletic groups, but also the actual time of their origin. This detail of a perfected phylogeny diagram can also be reproduced in a hierarchical system by means of the absolute rank of its group categories. In a hierarchical system, not only are the names of the monophyletic groups quoted but they are also given a specific absolute rank (class, order, family, etc.). Some clear-sighted authors (e.g., Simpson, 16) have quite correctly realized that the absolute rank which is attributed to a given group (e.g., family) does not generally mean that this group can be compared with any other of the same rank in any particular respect. Only within one and the same sequence of subordination is it true that the lower ranks show a higher degree of phylogenetic relationship than the higher. This situation can be accepted without injury to the basic principles of phylogenetic systematics. It could be changed, without injury to these principles, only if the absolute rank of categories was linked to their time of origin, just as in geology the sequence of strata in different continents is made comparable by its correlation with specific periods of the earth's history (e.g., Triassic, Jurassic, Cretaceous). Some authors (e.g., Stammer, 18) think that one must take into account, when according absolute rank to systematic groups, their different rates of evolution which have led to greater or lesser morphological "differentiation." But it needs little reflection to see that this is incompatible with the theoretical foundations of phylogenetic systematics and necessarily leads to pseudophylogenetic systems. This should already have been shown by the fact that sister-groups must have the same rank in a phylogenetic system, entirely without regard for the way in which this rank is established; for sister-groups can, of course, have morphologically unfolded (i.e., diverged from the form of their common ancestors) with completely different rates of evolution.

Biological systematics can no more do without a theoretical foundation for its work than can any other science. The theory of phylogenetic systematics is a comprehensive and complex edifice of thought, which here can only be touched upon lightly, even in its most important aspects. In this edifice there is, as always, a logical arrangement of individual problems. In critical expositions, this logical order must be observed. It is not permissible, as sometimes happens, to confuse the critique for answering logically subordinate questions with the critique concerning the principles of the phylogenetic system. From a thoroughgoing theory of phylogenetic systematics, there arise necessarily some unexpected demands on the practical work of the systematist. If the theory as such is accepted in principle it is not permissible to refuse these demands or leave them unconsidered merely because they conflict with certain customary methods obtaining at the time when systematics had no theory. There are many problems in biology whose solution presupposes knowledge of the phylogenetic relationship of one or many species—that is, a phylogenetic system of one or more groups of animals. To avoid false conclusions it is therefore especially important that every author of a system should make it easy to recognize whether, or rather to what extent, his system ought to meet the demands imposed by the theory of phylogenetic systematics. But even when these demands should be met in a system, according to the expressed wish of its author, there will always be differences of opinion over the actual relationships of some species or species groups. The person who requires a phylogenetic system as a premise for his own work will then have to decide on which side lie the better arguments; the criteria for this must again emerge from the theory of phylogenetic systematics. Differences of opinion on matters of fact are not, however, a special defect of phylogenetic systematics but the universal mark of every science.

It is impossible in a short paper to treat even sketchily the extensive field of phylogenetic systematics with all of the questions of detail which are important for the practical work of the systematist. A more comprehensive account in Spanish and another in English, with detailed bibliography, are in the course of preparation. Excellent introductions on its theoretical and methodological foundations with many critical comments on recent systematic works are given in the writings of Günther (4, 6). A valuable study on the philosophical foundations of biological systematics has very recently been published by Kiriakoff (14).

REFERENCES

1. Bigelow, R. S., 1956, Monophyletic classification and evolution. *System. Zool.*, 5: 145–46.

2. Blackwelder, R. A., Alexander, R. D., and Blair, W. F., 1962, The data of classification, a symposium. *System. Zool.*, 11: 49–84 [Critical review by Günther, K., in *Ber. Wiss. Biol.*, 1963].

3. Gregg, J. R., 1954, *The Language of Taxonomy. An Application of Symbolic Logics to the Study of Classificatory Systems*, New York, Columbia Univ. Press, 70 pp.

4. Günther, K., 1956, Systematik und Stammesgeschichte der Tiere 1939–1953. *Fortschr. Zool.* (N.F.), 10: 33–278.

5. ——, 1959, Die Tetrigidae von Madagaskar, mit einer Erörterung ihrer zoogeographischen Beziehungen und ihrer phylogenetischen Verwandtschaften. *Abhandl. Ber. Staatl. Mus. Tierkde Dresden*, 24: 3-56.

6. ——, 1962, Systematik und Stammesgeschichte der Tiere 1954–1959. *Fortschr. Zool.* (N.F.), 14: 268–547.

7. Hennig, W., 1950, *Grundzüge einer Theorie der Phylogenetischen Systematik*, Berlin, Deutscher Zentralverlag, 370 pp.

8. ——, 1953, Kritische Bemerkungen zum phylogenetischen System der Insekten. *Beitr. Entomol.*, 3: 1–85.

9. ——, 1954, Flügelgeäder und System der Dipteren, unter Berückschtigung der aus dem Mesozoikum beschriebenen Fossilien. *Beitr. Entomol.*, 4: 245–388.

10. ——, 1955, Meinungsverschiedenheiten über das System der niederen Insketen. *Zool. Ans.*, 155: 21–30.

11. ——, 1957, Systematik und Phylogenese. *Ber. Hundertjahrs. Deut. Entomol. Ges.* (Berlin, 1956), 50–71.

12. ——, 1960, Dipterenfauna von Neuseeland als systematisches und tiergeographisches Problem. *Beitr. Entomol.*, 10: 221–329.

13. ——, 1962, Veränderungen am phylogenetischen System der Insekten seit 1953. *Ber. Wandervers. Deut. Entomol.* (Berlin, 1951), 9: 29–42.

14. Kiriakoff, S. G., 1960, Les fondaments philosophiques de la systématique biologique. *Natuurw. Tijdschr.* (Ghent), 42: 35–57.

15. Müller, A. H., 1955, *Der Grossablaus der stammesgeschichtlichen Entwicklung* (Gustav Fisher, Jena), 50 pp.

16. Simpson, G. G., 1937, Supra-specific variation in nature and in classification from the view-point of paleontology. *Am. Naturalist*, 71: 236–67.

17. Sokal, R. R., 1963, Typology and empiricism in taxonomy *J. Theoret. Biol.*, 3: 230–67; critical review by Günther, K., in *Ber. Wiss. Biol., A*, 191, 70.

18. Stammer, H. J., 1961, Neue Wege der Insektensystematik. *Verhandl. Intern. Kongr. Entomol.*, Wien, 1950, 1: 1–7.

19. Takhtajan, A., 1959, *Die Evolution der Angiospermen* (Gustav Fischer, Jena).

20. Verheyen, R. A., 1961, A new classification for the non-passerine birds of the world. *Bull. Inst. Roy. Sci. Nat. Belg.*, 37 (27), 36 pp.

21. Wilson, H. F., and Doner, M. H., 1937, The historical development of insect classification. (Planographed by John S. Swift Co., Inc., St. Louis, Chicago, New York, Indianapolis.)

# 32

# A Matter of Individuality

DAVID L. HULL

*Biological species have been treated traditionally as spatiotemporally unrestricted classes. If they are to perform the function which they do in the evolutionary process, they must be spatiotemporally localized individuals, historical entities. Reinterpreting biological species as historical entities solves several important anomalies in biology, in philosophy of biology, and within philosophy itself. It also has important implications for any attempt to present an "evolutionary" analysis of science and for sciences such as anthropology which are devoted to the study of single species.*

## 1. INTRODUCTION

The terms "gene," "organism," and "species" have been used in a wide variety of ways in a wide variety of contexts. Anyone who attempts merely to map this diversity is presented with a massive and probably pointless task. In this paper I consciously ignore "the ordinary uses" of these terms, whatever they might be, and concentrate on their biological uses. Even within biology the variation and conflicts in meaning are suf-

ficiently extensive to immobilize all but the most ambitious ordinary language philosopher. Thus I have narrowed my focus even further to concentrate on the role which these terms play in evolutionary biology. In doing so, I do not mean to imply that this usage is primary or that all other biological uses which conflict with it are mistaken. Possibly evolutionary theory is *the* fundamental theory in biology, and all other biological theories must be brought into accord with it. Possibly all

The research for this paper was supported by NSF grant Soc 75 03535. I am indebted to the following people for reading and criticizing early versions of it; Michael Ghiselin, Stephen Gould, G. C. D. Griffiths, John Koethe, Ernst Mayr, Bella Selan, W. J. van der Steen, Gareth Nelson, Michael Perloff, Mark Ridley, Michael Ruse, Thomas Schopf, Paul Teller, Leigh Van Valen, Linda Wessels, Mary Williams, and William Wimsatt. Their advice and criticisms are much appreciated.

biological theories, including evolutionary theory, eventually will be reduced to physics and chemistry. But regardless of the answers to these global questions, at the very least various versions of evolutionary theory are sufficiently important in biology to warrant an investigation of the implications which they have for the biological entities which they concern.

Genes are the entities which are passed on in reproduction and which control the ontogenetic development of the organism. Organisms are the complex systems which anatomists, physiologists, embryologists, histologists, etc. analyze into their component parts. Species have been treated traditionally as the basic units of classification, the natural kinds of the living world, comparable to the physcal elements. But these entities also function in the evolutionary process. Evolution consists in two processes (mutation and selection) which eventuate in a third (evolution). Genes provide the heritable variation required by the evolutionary process. Traditionally organisms have been viewed as the primary focus of selection, although considerable disagreement currently exists over the levels at which selection takes place. Some biologists maintain that selection occurs exclusively at the level of genes; others that supragenic, even supraorganismic units can also be selected. As one might gather from the title of Darwin's book, species are the things which are supposed to evolve. Whether the relatively large units recognized by taxonomists as species evolve or whether much less extensive units such as populations are the effective units of evolution is an open

question. In this paper when I use the term "species," I intend to refer to those supraorganismic entities which evolve regardless of how extensive they might turn out to be.

The purpose of this paper is to explore the implications which evolutionary theory has for the ontological status of genes, organisms, and species. The only category distinction I discuss is between individuals and classes. By "individuals" I mean spatiotemporally localized cohesive and continuous entities (historical entities). By "classes" I intend spatiotemporal unrestricted classes, the sorts of things which can function in traditionally defined laws of nature. The contrast is between Mars and planets, the Weald and geological strata, Gargantua and organisms. The terms used to mark this distinction are not important; the distinction is. For example, one might distinguish two sorts of sets: those that are defined in terms of a spatiotemporal relation to a spatiotemporally localized focus, and those that are not. On this view, historical entities such as Gargantua become sets. But they are sets of a very special kind—sets defined in terms of a spatiotemporal relation to a spatiotemporally localized focus. Gargantua, for instance, would be the set of all cells descended from the zygote which gave rise to Gargantua.

The reason for distinguishing between historical entities and genuine classes is the differing roles which each plays in science according to traditional analyses of scientific laws. Scientific laws are supposed to be spatiotemporally unrestricted generalizations. No uneliminable reference can be made in a genuine law of nature to a spatiotemporally individuated entity. To be sure, the distinction between accidentally true generalizations (such as all terrestrial organisms using the same genetic code) and genuine laws of nature (such as those enshrined in contemporary versions of celestial mechanics) is not easy to make. Nor are matters helped much by the tremendous emphasis placed on laws in traditional philosophies of science, as if they were the be-all and end-all of science. Nevertheless, I find the distinction between those generalizations that are spatiotemporally unrestricted and those that are not fundamental to our current understanding of science. Whether one calls the former "laws" and the latter something else, or whether one terms both sorts of statements "laws" is of little consequence. The point I wish to argue is that genes, organisms *and* species, as they function in the evolutionary process, are necessarily spatiotemporally localized individuals. They could not perform the functions which they perform if they were not.

The argument presented in this paper is metaphysical, not epistemological. Epistemologically red light may be fundamentally different from infrared light and mammals from amoebae. Most human beings can see with red light and not infrared light. Most people can see mammals; few if any can see amoebae with the naked eye. Metaphysically they are no different. Scientists know as much about one as the other. Given our relative size, period of duration, and perceptual acuity, organisms appear to be historical entities, species appear to be classes of some sort, and genes cannot be seen at all. However, after acquainting oneself with the various entities which biologists count as organisms and the roles which organisms and species play in the evolutionary process, one realizes exactly how problematic our common-sense notions actually are. The distinction between an organism and a colony is not sharp. If an organism is the "total product of the development of the impregnated embryo," then as far back as 1899 T. H. Huxley was forced to conclude that the medusae set free from a hydrozoan "are as much organs of the latter as the multitudinous pinnules of a *Comatula,* with their genital glands, are organs of the Echinoderm. Morphologically, therefore, the equivalent of the individual *Comatula* is the Hydrozoic stock and all the Medusae which proceed from it" (24). More recently, Daniel Janzen (25) has remarked that the "study of dandelion ecology and evolution suffers from confusion of the layman's 'individual' with the 'individual' of evolutionary biology. The latter individual has 'reproductive fitness' and is the unit of selection in most evolutionary conceptualizations" (see also 2). According to evolutionists, units of selection, whether they be single genes, chromosomes, organisms, colonies, or kinship groups are individuals. In this paper I intend to extend this analysis to units of evolution.

If the ontological status of space-time in relativity theory is philosophically interesting in and of itself (and God knows enough philosophers have written on that topic), then the ontological status of species in evolutionary theory should also be sufficiently interesting philosophically to discuss without any additional justification. However, additional justification does exist. From Socrates and Plato to Kripke and Putnam, organisms have been paradigm examples of primary substances, particulars, and/or individuals, while species have served as paradigm examples of secondary substances, universals, and/or classes. I do not think that this paper has any necessary implications for various solutions to the problem of universals, identity, and the like. However, if my main contention is correct, if species are as much spatiotemporally localized individuals as organisms, then some of the confusion among philosophers over these issues is understandable. One of the commonest examples used in the philosophical literature is inappropriate. Regardless of whether one thinks that "Moses" is a proper name, a cluster concept, or a rigid designator, *"Homo sapiens"* must be treated in the same way.

## 2. THE EVOLUTIONARY JUSTIFICATION

Beginning with the highly original work of Michael Ghiselin (12, 13, 14), biologists in increasing numbers are beginning to argue that species as units of evolution are historical entities (15, 20, 21, 22, 23, 34, 38). The justification for such claims would be easier if there were one set of propositions (presented preferably in axiomatic form) which could be termed *the* theory of evolution. Unfortunately, there is not. Instead there are several, incomplete, partially incompatible versions of evolutionary theory currently extant. I do not take this state of affairs to be unusual, especially in periods of rapid theoretical change. In general the myth that some one set of propositions exists which can be designated unequivocally as Newtonian theory, relativity theory,

etc. is an artifact introduced by lack of attention to historical development and unconcern with the primary literature of science. The only place one can find *the* version of a theory is in a textbook written long after the theory has ceased being of any theoretical interest to scientists.

In this section I set out what it is about the evolutionary process which results in species being historical entities, not spatiotemporally unrestricted classes. In doing so I have not attempted to paper over the disagreements which currently divide biologists working on evolutionary theory. For example, some disagreement exists over how abruptly evolution can occur. Some biologists have argued that evolution takes place saltatively, in relatively large steps. Extreme saltationists once claimed that in the space of a single generation new species can arise which are so different from all other species that they have to be placed in new genera, families, classes, etc. No contemporary biologist to my knowledge currently holds this view. Extreme gradualists, on the other side, argue that speciation *always* occurs very slowly, over periods of hundreds of generations, either by means of a single species changing into a new species (phyletic evolution) or else by splitting into two large subgroups which gradually diverge (speciation). No contemporary biologist holds this view either. Even the most enthusiastic gradualists admit that new species can arise in a single generation, e.g., by means of polyploidy. In addition, Eldredge and Gould (11), building on Mayr's founder principle (36, 37), have recently argued that speciation typically involves small, peripheral isolates which develop quite rapidly into new species. Speciation is a process of "punctuated equilibria."

However, the major dispute among contemporary evolutionary theorists is the level (or levels) at which selection operates. Does selection occur *only* and *literally* at the level of genes? Does selection take place *exclusively* at the level of organisms, the selection of genes being only a consequence of the selection of organisms? Can selection also take place at levels of organization more inclusive than the individual organism, e.g., at the level of kinship groups, populations and possibly even entire species? Biologists can be found opting for every single permutation of the answers to the preceding questions. I do not propose to go through all the arguments which are presented to support these various conclusions. For my purposes it is sufficient to show that the points of dispute are precisely those which one might expect if species are being interpreted as historical entities, rather than as spatiotemporally unrestricted classes. Richard Dawkins puts the crucial issue as follows:

> Natural selection in its most general form means the differential survival of entities. Some entities live and others die but, in order for this selective death to have any impact on the world, an additional condition must

be met. Each entity must exist in the form of lots of copies, and at least some of the entities must be *potentially* capable of surviving—in the form of copies—for a significant period of evolutionary time.

The results of evolution by natural selection are *copies* of the entities being selected, not *sets*. Elements in a set must be characterized by one or more common characteristics. Even fuzzy sets must be characterized by at least a "cluster" of traits. Copies need not be.[1] A particular gene is a spatiotemporally localized individual which either may or may not replicate itself. In replication the DNA molecule splits down the middle producing two new molecules composed *physically* of half of the parent molecule while *largely* retaining its structure. In this way genes form lineages, ancestor-descendant copies of some original molecule. The relevant genetic units in evolution are not *sets* of genes defined in terms of structural similarity but lineages formed by the imperfect copying process of replication.[2] Genes can belong to the same lineage even though they are structurally different from other genes in that lineage. What is more, continued changes in structure can take place indefinitely. If evolution is to occur, not only *can* such indefinite structural variation take place within gene lineages, but it *must*. Single genes are historical

---

1. Once again I am excluding from the notion of class those "classes" defined by means of a spatiotemporal relation to a spatiotemporally localized individual. Needless to say, I am also excluding such constructions as "similar in origin" from the classes of similarities. I wish the need to state the obvious did not exist, but from past experience it does.

2. In population genetics the distinction between structurally similar genes forming a single lineage and those which do not is marked by the terms "identical" and "independent"; see (41), pp. 56–57.

entities, existing for short periods of time. The more important notion is that of a *gene lineage*. Gene lineages are also historical entities persisting while changing indefinitely through time. As Dawkins puts this point:

> Genes, like diamonds, are forever, but not quite in the same way as diamonds. It is an individual diamond crystal which lasts, as an unaltered pattern of atoms. DNA molecules don't have that kind of permanence. The life of any one physical DNA molecule is quite short—perhaps a matter of months, certainly not more than one lifetime. But a DNA molecule could theoretically live on in the form of *copies* of itself for a hundred million years. (8, p. 36)

Exactly the same observations can be made with respect to organisms. A particular organism is a spatiotemporally localized individual which either may or may not reproduce itself. In asexual reproduction, part of the parent organism buds off to produce new individuals. The division can be reasonably equitable, as in binary fission, or extremely inequitable, as in various forms of parthenogenesis. In sexual reproduction gametes are produced which unite to form new individuals. Like genes, organisms form lineages. The relevant organismal units in evolution are not sets of organisms defined in terms of structural similarity but lineages formed by the imperfect copying processes of reproduction. Organisms can belong to the same lineage even though they are structurally different from other organisms in that lineage. What is more, continued changes in structure can

take place indefinitely. If evolution is to occur, not only *can* such indefinite structural variation take place within organism lineages, but it *must*. Single organisms are historical entities, existing for short periods of time. Organism lineages are also historical entities persisting while changing indefinitely through time.

Both replication and reproduction are spatiotemporally localized processes. There is no replication or reproduction at a distance. Spatiotemporal continuity through time is required. Which entities at which levels of organization are sufficiently cohesive to function as units of selection is more problematic. Dawkins presents one view:

> In sexually reproducing species, the individual [the organism] is too large and too temporary a genetic unit to qualify as a significant unit of natural selection. The group of individuals is an even larger unit. Genetically speaking, individuals and groups are like clouds in the sky or dust-storms in the desert. They are temporary aggregates of federations. They are not stable through evolutionary time. Populations may last a long while, but they are constantly blending with other populations and so losing their identity. They are subject to evolutionary change from within. A population is not a discrete enough entity to be a unit of natural selection, not stable and unitary enough to be 'selected' in preference to another population. (8, p. 37)

From a common sense perspective, organisms are paradigms of tightly organized, hierarchically stratified systems. Kinship groups such as hives also seem to be internally cohesive entities. Populations and species are not. Dawkins argues that neither organisms (in sexually reproducing species) nor populations in any species are sufficiently permanent and cohesive to function as units in selection. In asexual species, organisms do not differ all that much from genes. They subdivide in much the same way that genes do, resulting in progeny which are identical (or nearly identical) with them. In sexual species, however, organisms must pool their genes to reproduce. The resulting progeny contain a combined sample of parental genes. Populations lack even this much cohesion.

Other biologists are willing to countenance selection at levels more inclusive than the individual gene, possibly parts of chromosomes, whole chromosomes, entire organisms or even kinship groups (32). The issues, both empirical and conceptual, are not simple. For example, G. C. Williams in his classic work (61) argues that selection occurs only at the level of individuals. By "individual," biologists usually mean "organism." However, when Williams is forced to admit that kinship groups can also function as units of selection, he promptly dubs them "individuals." One of the commonest objections to E. O. Wilson's (62) equally classic discussion of evolution is that he treats kin selection

as a special case of group selection. According to the group selectionists, entities more inclusive than kinship groups can also function as units of selection (63).[3] Matters are not improved much by vagueness over what is meant by "units of selection." Gene frequencies are certainly altered from generation to generation, but so are genotype frequencies. Genes cannot be selected in isolation. They depend on the success of the organism which contains them for survival. Most biologists admit that similar observations hold for certain kinship groups. Few are willing to extend this line of reasoning to include populations and entire species.

Although the dispute over the level(s) at which selection takes place is inconclusive, the points at issue are instructive. In arguing that neither organisms nor populations function as units of selection in the same sense that genes do, Dawkins does not complain that the cells in an organism or the organisms in a population are phenotypically quite diverse, though they frequently are. Rather he denigrates their cohesiveness and continuity through time, criteria which are relevant to individuating historical entities, not spatiotemporally unrestricted classes. Difficulties about the level(s) at which selection can operate to one side, the issue with which we are concerned is the ontological status of species. Even if entire species are not sufficiently well integrated to

3. Until recently even the most ardent group selectionists admitted that the circumstances under which selection can occur at the level of populations and/or entire species are so rare that group selection is unlikely to be a major force in the evolutionary process (30, 32, 33). Michael Wade (59), however, has presented a convincing argument to the effect that the apparent rarity of group selection may be the result of the assumptions commonly made in constructing mathematical models for group selection and not an accurate reflection of the actual state of nature. In his own research, the differential survival of entire populations has produced significant divergence.

function as units of selection, they are the entities which evolve as a result of selection at lower levels. The requirements of selection at these lower levels place constraints on the manner in which species can be conceptualized. Species as the results of selection are necessarily lineages, not sets of similar organisms. In order for differences in gene frequencies to build up in populations, continuity through time must be maintained. To some extent genes in sexual species are reassorted each generation, but the organisms which make up populations cannot be. To put the point in the opposite way, if such shuffling of organisms were to take place, selection would be impossible.

The preceding characteristic of species as evolutionary lineages by itself is sufficient to preclude species being conceptualized as spatiotemporally unrestricted sets or classes. However, if Eldredge and Gould are right, the case for interpreting species as historical entities is even stronger. They ask why species are so coherent, why groups of relatively independent local populations continue to display fairly consistent, recognizable phenotypes, and why reproductive isolation does not arise in every local population if gene flow is the only means of preventing differentiation:

> The answer probably lies in a view of species and individuals [organisms] as homeostatic systems—as amazingly well-buffered to resist change and maintain stability in the face of disturbing influences . . . In this view, the importance of peripheral isolates lies in their small size and the alien environment beyond the species border that they inhabit

—for only here are selective pressures strong enough and the inertia of large numbers sufficiently reduced to produce the "genetic revolution" (Mayr, 1963, p. 533) that overcomes homeostasis. The coherence of a species, therefore, is not maintained by interaction among its members (gene flow). It emerges, rather, as an historical consequence of the species' origin as a peripherally isolated population that acquired its own powerful homeostatic system. (11, p. 114)

Eldredge and Gould argue that, from a theoretical point of view, species appear so amorphous because of a combination of the gradualistic interpretation of speciation and the belief that gene exchange is the chief (or only) mechanism by which cohesion is maintained in natural populations. However, in the field, species of both sexual and asexual organisms seem amazingly coherent and unitary. If gene flow were the only mechanism for the maintenance of evolutionary unity, asexual species should be as diffuse as dust-storms in the desert. According to Eldredge and Gould, new species arise through the budding off of peripheral isolates which succeed in establishing new equilibria in novel environments. Thereafter they remain largely unchanged during the course of their existence and survive only as long as they maintain this equilibrium.

Another possibility is that evolutionary unity is maintained by both internal and external means. Gene flow and homeostasis within a species are internal mechanisms of evolutionary unity. Perhaps the external environment in the form of unitary selection pressures also contributes to the integrity of the entities which are evolving (10). For example, Jews have remained relatively distinct from the rest of humankind for centuries, in part by internal means (selective mating, social customs, etc.) but also in part by external means (discrimination, prejudice, laws, etc.). An ecological niche is a relation between a particular species and key environmental variables. A different species in conjunction with the same environmental variables could define quite a different niche. In the past biologists have tended to play down the integrating effect of the environment, attributing whatever unity and coherence which exists in nature to the integrating effect of gene complexes. At the very least, if the coherence of asexual species is not illusory, mechanisms other than gene flow must be capable of bringing about evolutionary unity.

### 3. INDIVIDUATING ORGANISMS AND SPECIES

By and large, the criteria which biologists use to individuate organisms are the same as those suggested by philosophers—spatiotemporal continuity, unity, and location. Differences between these two analyses have three sources: first, philosophers have been most interested in individuating persons, the hardest case of all, while biologists have been content to individuate organisms; second, when philosophers have discussed the individuation of organisms, they have usually limited themselves to adult mammals, while biologists have attempted to develop a notion of organism adequate to handle the wide variety of organisms which exist in nature; and finally, philosophers have felt free to

resort to hypothetical, science fiction examples to test their conceptions, while biologists rely on actual cases. In each instance, I prefer the biologists' strategy. A clear notion of an individual organism seems an absolute prerequisite for any adequate notion of a person, and this notion should be applicable to all organisms, not just a minuscule fraction. But most importantly, real examples tend to be much more detailed and bizarre than those made up by philosophers. Too often the example is constructed for the sole purpose of supporting the preconceived intuitions of the philosophers and has no life of its own. It cannot force the philosopher to improve his analysis the way that real examples can. Biologists are in the fortunate position of being able to test their analyses against a large stock of extremely difficult, extensively documented actual cases.

Phenotypic similarity is irrelevant in the individuation of organisms. Identical twins do not become one organism simply because they are phenotypically indistinguishable. Conversely, an organism can undergo massive phenotypic change while remaining the same organism. The stages in the life cycles of various species of organisms frequently are so different that biologists have placed them in different species, genera, families, and even classes—until the continuity of the organism was discovered. If a caterpillar develops into a butterfly, these apparently different organisms are stages in the life cycle of a single organism regardless of how dissimilar they might happen to be (see *Figure 1a*). In ontogenetic development, a single lineage is never divided successively in time into separate organisms;

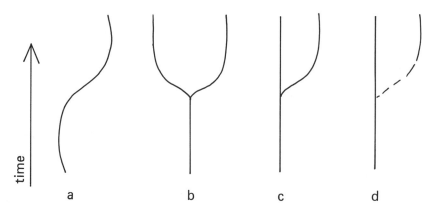

Figure 1. Diagrams which can be interpreted alternately as organisms undergoing onto-genetic change and the production of new organisms and as species undergoing phylogenetic change and speciation.

some sort of splitting is required. In certain cases, such as transverse fission in paramecia, a single organism splits equally into two new organisms (see *Figure 1b*). In such cases, the parent organism no longer exists, and the daughter organisms are two new individuals. Sometimes a single individual will bud off other individuals which are roughly its own size but somewhat different in appearance, e.g., strobilization in certain forms of Scyphozoa (see *Figure 1c*). At the other extreme, sometimes a small portion of the parent organism buds off to form a new individual, e.g., budding in Hydrozoa (see *Figure 1d*). In the latter two cases, the parent organism continues to exist while budding off new individuals. The relevant consideration is how much of the parent organism is lost and its internal organization disrupted.

Fusion also takes place at the level of individual organisms. For example, when presented with a prey too large for a single individual to digest, two amoebae will fuse cytoplasmically in order to engulf and digest it. However,

the nuclei remain distinct and the two organisms later separate, genetically unchanged. The commonest example of true fusion occurs when germ cells unite to form a zygote. In such cases, the germ cells as individuals cease to exist and are replaced by a new individual (see *Figure 2a*). Sometimes one organism will invade another and become part of it. Initially, these organisms, even when they become obligate parasites, are conceived of as separate organisms, but sometimes they can become genuine parts of the host organism. For example, one theory of the origin of certain cell organelles is that they began as parasites. Blood transfusions are an unproblematic case of part of one organism becoming part of another; conjugation is another (see *Figure 2b*). Sometimes parts of two different organisms can merge to form a third. Again, sexual reproduction is the commonest example of such an occurrence (see *Figure 2c*). In each of these cases, organisms are individuated on the basis of the amount of material involved and the effect of the change on the

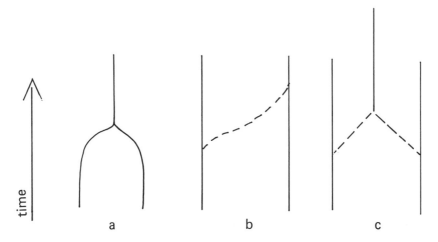

time

a          b          c

Figure 2. Diagrams which can be interpreted alternately as organisms merging totally or partially to give rise to new organisms and as species merging totally or partially to give rise to new species.

internal organization of the organisms. For example, after conjugation two paramecia are still two organisms and the same two organisms even though they have exchanged some of their genetic material.

If species are historical entities, then the same sorts of considerations which apply in the individuation of organisms should also apply to them, and they do (35). The only apparent discrepancy results from the fact that not all biologists have been totally successful in throwing off the old pre-evolutionary view of species as classes of similar organisms and replacing it with a truly evolutionary view. However, even these discrepancies are extremely instructive. For example, G. G. Simpson (50) maintains that a single lineage which changes extensively through time without speciating (splitting) should be divided into separate species (see *Figure 1a*). Willi Hennig (17) disagrees: new species should be recognized only upon splitting. This particular debate has been

involved, touching upon both conceptual and empirical issues. For example, how can a gradually evolving lineage be divided into discrete species in an objective, nonarbitrary way? Are later organisms considered to belong to different species from their ancestors because they are sufficiently dissimilar or because they can no longer interbreed with them even if they coexisted? Can such extensive change take place in the absence of speciation?

I cannot attempt to answer fully all of these questions here. Instead, I must limit myself to the remark that, on Simpson's view, species and organisms are quite different sorts of things. An organism undergoes limited change, constrained by its largely unchanging genotype. A single species is capable of indefinite, open-ended development. Although the course of a species' development is constrained from generation to generation by its gene pool, this gene pool is indefinitely modifiable. However, if Eldredge and Gould are right, species are more like organisms than anyone has previously supposed. Both are finite and can undergo only limited change before ceasing to exist. Significant evolutionary change can take place only through a series of successive species, not within the confines of a single species. Species lineages, not species, are the things which evolve. On this view, Hennig's refusal to divide a single lineage into two or more species is preferable to Simpson's alternative.

No disagreement exists between Simpson and Hennig over the situation depicted in *Figure 1b,* a single species splitting equally into two. Both agree that the ancestor species is extinct, having given rise to two new daughter species. However, this figure is drawn as if divergence always takes place upon speciation. When this diagram was interpreted as depicting the splitting of one organism into two, divergence was not presupposed. Two euglenae resulting from binary fission are two organisms and not one even though they may be phenotypically and genotypically identical. The same is true of species. Sometimes speciation takes place with no (or at least extremely minimal) divergence; e.g., sibling species are no less two species simply because they look alike. The assumption is, however, that in reproductively isolated species some divergence, at least in the mechanisms of reproduction, must have taken place, even if we cannot detect it. The role of similarity becomes controversial once again when speciation takes place and one species remains unchanged, while the other diverges from the parental type (see *Figure 1c*). According to Hennig (17), when

speciation occurs, the ancestor species must be considered extinct regardless of how similar it might be to one of its daughter species. Simpson (50) disagrees.

The factor which is causing the confusion in the preceding discussion is the role of similarity in the individuation of species. If species are classes defined by sets (or clusters) of traits, then similarity should be relevant. At one extreme, the pheneticists (54) argue that all that matters is phenetic similarity and dissimilarity, regardless of descent, reproduction, evolutionary cohesiveness, etc. Highly polytypic species such as dogs must be considered numerous different "species" because of the existence of so many reasonably discrete clusters. Sibling species must be considered a single "species" because they form a single cluster. At the other extreme, the Hennigians (commonly termed "cladists") concentrate solely on the splitting of phylogenetic lineages regardless of phenetic similarity. Polytypic species are single species because they form a single clade; sibling species are separate species because they form more than one clade. The evolutionists, represented by Simpson and Mayr, argue that somehow the two considerations must be balanced against each other.

However, on the historical entity interpretation, similarity is a red herring: it is not the issue at all. What really matters is how many organisms are involved and how much the internal organization of the species involved is disrupted. If speciation takes place when a small, peripheral isolate succeeds in bringing about a genetic revolution (see *Figure 1d*), then the parent species can still be said to persist unchanged. It has not lost significant numbers of organisms, nor has its internal organization been affected much. One Hennigian, at least, has come to this conclusion for precisely these reasons (60). If, however, the species is split into two or more relatively large subgroups, then it is difficult to see how the ancestral species can still be said to exist, unless one of these subgroups succeeds in retaining the same organization and internal cohesion of the ancestral species. Incidentally, it would also be phenetically similar to the ancestral species, but that would be irrelevant.

Fusion can also take place at the level of species. The breaking down of reproductive isolation sufficient to permit two entire species to merge into one is extremely unlikely (see *Figure 2a*). If it did occur, the consideration would be the same as those raised in connection with *Figure 1b*. However, introgression and speciation by polyploidy are common (see *Figures 2b* and *2c*). In such cases, a few organisms belonging to separate species mate and produce fertile offspring. Contrary to popular opinion, the production of an occasional fertile hybrid is not enough for biologists to consider two species one. What matters is how extensive the introgression becomes—exactly the right consideration if species are historical entities. As Dobzhansky remarks, "What matters is not whether hybrids can be obtained but whether the Mendelian populations do or do not exchange genes, and if they do whether at a rate which destroys the adaptive equilibrium of the populations concerned" (9, p. 586).

One final parallel between organisms and species warrants mentioning.

Organisms are unique. When an organism ceases to exist, numerically that same organism cannot come into existence again. For example, if a baby were born today who was identical in every respect to Adolf Hitler, including genetic makeup, he still would not be Adolf Hitler. He would be as distinct and separate a human being as ever existed because of his unique "insertion into history," to use Vendler's propitious phrase (58; see also 57). But the same observation can be made with respect to species. If a species evolved which was identical to a species of extinct pterodactyl save origin, it would still be a new, distinct species. Darwin himself notes, "When a species has once disappeared from the face of the earth, we have reason to believe that the same identical form never reappears" (7, p. 313). Darwin presents this point as if it were a contingent state of affairs, when actually it is conceptual. Species are segments of the phylogenetic tree. Once a segment is terminated, it cannot reappear somewhere else in the phylogenetic tree. As Griffiths observes, the "reference of an individual to a species is determined by its parentage, not by any morphological attribute" (15, p. 102).

If species were actually spatiotemporally unrestricted classes, this state of affairs would be strange. If all atoms with atomic number 79 ceased to exist, gold would cease to exist, although a slot would remain open in the periodic table. Later when atoms with the appropriate atomic number were generated, they would be atoms of gold regardless of their origins. But in the typical case, to *be* a horse one must be *born* of horse. Obviously, whether one is a gradualist or saltationist, there must have been instances in which non-horses (or borderline horses) gave rise to horses. The operative term is still "gave rise to." But what of the science fiction examples so beloved to philosophers? What if a scientist made a creature from scratch identical in every respect to a human being including consciousness, emtionality, a feeling of personhood, etc.? Wouldn't it be included in *Homo sapiens?* It all depends. If all the scientist did was to make such a creature and then destroy it, it was never part of our species. However, if it proceeded to mate with human beings born in the usual way and to produce offspring, introducing its genes into the human gene pool, then it would become part of our species. The criterion is precisely the same one used in cases of introgression. In the evolutionary world view, unlike the Aristotelian world view, an organism can change its species while remaining numerically the same individual (see 19).

One might complain that being born of human beings and/or mating with human beings are biological criteria, possibly good enough for individuating *Homo sapiens,* but inadequate for the humanistic notion of a human being. We are a social species. An entity which played the role of a human being in a society would *be* a "human being," even if it was not born of human beings or failed to mate with human beings. I'm not sure how one makes such decisions, but the conclusion is not totally incompatible with the position being presented in this paper. Species as they are commonly thought of are not the only things which evolve. Higher levels of organization also exist. Entities can belong to the same cultural system or ecosystem without belonging to the same biological species. As Eugene Odum has put it, "A human being, for example, is not only a hierarchical system composed of organs, cells, enzyme systems, and genes as subsystems, but is also a component of supraindividual hierarchical systems such as populations, cultural systems, and ecosystems" (44, p. 1289). If pets or computers function as human beings, then from certain perspectives they might well count as human beings even though they are not included in the biological species *Homo sapiens.*

## 4. BIOLOGICAL AND PHILOSOPHICAL CONSEQUENCES

Empirical evidence is usually too malleable to be very decisive in conceptual revolutions. The observation of stellar parallax, the evolution of new species right before our eyes, the red shift, etc. are the sorts of things which are pointed to as empirical reasons for accepting new scientific theories. However, all reasonable people had accepted the relevant theories in the absence of such observations. Initial acceptance of fundamentally new ideas leans more heavily on the increased coherence which the view brings to our general world picture. If the conceptual shift from species being classes to species being historical entities is to be successful, it must eliminate longstanding anomalies both within and about biology. In this section, I set out some of the implications of viewing species as historical entities, beginning with those that are most strictly biological, and gradually working my way toward those that are more philosophical.

The role of type specimens in biological systematics puzzles philosophers and biologists alike. As R. A. Crowson remarks, "The current convention that a single specimen, the Holotype, is the only satisfactory basic criterion for a species would be difficult to justify logically on any theory but Special Creation" (5, p. 29). According to all three codes of biological nomenclature, a particular organism, part of an organism, or trace of an organism is selected as the type specimen for each species. In addition, each genus must have its type species, and so on. Whatever else one does with this type and for whatever resson, the name goes with the type.[4] The puzzling aspect of the type method on the class interpretation is that the type need not be typical. In fact, it can be a monster. The following discussion by J. M. Schopf is representative:

> It has been emphasized repeatedly, for the benefit of plant taxonomists, at least, that the nomenclatural type (holotype) of a species is not to be confused or implicated in anyone's concept of what is "typical" for a taxon. A nomenclatural type is simply *the specimen,* or other element, with which a name is permanently associated. This element need not be "typical" in any sense; for organisms with a complicated life cycle, it is obvious that no single specimen could physically represent all the important characteristics, much less could it be taken to show many features near the mean of their range of variation. (See also 6, 39, 50, 51; 49, p. 1043.)

Species are polymorphic. Should the type specimen for *Homo sapiens,* for instance, be male or female? Species are also polytypic. What skin color, blood type, etc. should the type specimen for *Homo sapiens* have? Given the sort of variability characteristic of biological species, no one specimen could possibly be "typical" in even a statistical sense (37, p. 369). On the class interpretation, one would expect at the very least for a type specimen to have many or most of the more important traits characteristic of its species (16, p. 465–466), but on the historical entity interpretation, no such similarity is required. Just as a heart, kidneys and lungs are included in the same organism because they are part of the same ontogenetic whole, parents and their progeny are included in the same species because they are part of the same genealogical nexus, no matter how much they might differ phenotypically. The part/whole relation does not require similarity.

A taxonomist in the field sees a specimen of what he takes to be a new species. It may be the only specimen available or else perhaps one of a small

---

4. The three major codes of biological nomenclature are (1) the International Code of Botanical Nomenclature, 1966, International Bureau for Plant Taxonomy and Nomenclature, Utrecht; (2) the International Code of Nomenclature of Bacteria, 1966, *International Journal of Systematic Bacteriology,* 16: 459–490; and (3) the *International Code of Zoological Nomenclature,* 1964, International Trust for Zoological Nomenclature, London. In special circumstances the priority rule is waived, usually because the earlier name is discovered only long after a later name has become firmly and widely established.

sample which he gathers. The taxonomist could not possibly select a typical specimen, even if the notion made sense, because he has not begun to study the full range of the species' variation. He selects a specimen, any specimen, and names it. Thereafter, if he turns out to have been the first to name the species of which this specimen is part, that name will remain firmly attached to that species. A taxon has the name it has *in virtue of* the naming ceremony, not *in virtue of* any trait or traits it might have. If the way in which taxa are named sounds familiar, it should. It is the same way in which people are baptized.[5] They are named in the same way because they are the same sort of thing—historical entities (see Ghiselin, 13, 14).

But what, then, is the role of all those traits which taxonomists include in their monographs? For example, Article 13 of the Zoological Code of Nomenclature states that any name introduced after 1930 must be accompanied by a statement that "purports to give characteristics differentiating the taxon." Taxonomists distinguish between descriptions and diagnoses. A description is a lengthy characterization of the taxon, including reference to characteristics which are easily recognizable and comparable, to known variability within a population and from population to population, to various morphs, and to traits which can help in distinguishing sibling species. A diagnosis is a much shorter and selective list of traits chosen primarily to help differentiate a taxon from its nearest neighbors of the same rank. As important as the traits listed in diagnoses and descriptions may be for a variety of purposes, they are not definitions. Organisms could possess these traits and not be included in the taxon; conversely, organisms could lack one or more of these traits and be clear-cut instances of the taxon. They are, as the name implies, *descriptions.* As descriptions, they change through time as the entities which they describe change. Right now all specimens of *Cygnus olor* are white. No doubt the type specimen of this species of swan is also white. However, if a black variety were to arise, *Cygnus olor* would not on that account become a new species. Even if this variety were to become predominant, this species would remain the same species and the white type specimen would remain the type specimen. The species description would change but that is all. Organisms are not included in the same species *because* they are similar to the type specimen or to each other but *because* they are part

5. Although the position on the names of taxa argued for in this paper might sound as if it supported S. Kripke's (26) analysis of general terms, it does not. Taxa names are very much like "rigid designators," as they should be if taxa are historical entities. However, Kripke's analysis is controversial because it applies to *general* terms. It is instructive to note that during the extensive discussion of the applicability of Kripke's notion of a rigid designator to such terms as "tiger," no one saw fit to see how those scientists most intimately concerned actually designated tigers. According to Putnam's principle of the linguistic division of labor (47), they should have. If they had, they would have found rules explicitly formulated in the various codes of nomenclature which were in perfect accord with Kripke's analysis—but for the wrong reason. That no one bothered tells us something about the foundations of conceptual analysis.

of the same chunk of the genealogical nexus (Ghiselin, 13, 14).

On the class interpretation, the role of particular organisms as type specimens is anomalous. The role of lower taxa as types for higher taxa is even more anomalous. On the class interpretation, organisms are members of their taxa, while lower taxa are included in higher taxa (3). How could entities of two such decidedly different logical types play the same role? But on the historical entity interpretation, both organisms and taxa are of the same logical type. Just as organisms are part of their species, lower taxa are part of higher taxa. Once again, parts do not have to be similar, let alone typical, to be part of the same whole.

A second consequence of treating species as historical entities concerns the nature of biological laws. If species are actually spatiotemporally unrestricted classes, then they are the sorts of things which can function in laws. "All swans are white," if true, might be a law of nature, and generations of philosophers have treated it as such. If statements of the form "species $X$ has the property $Y$" were actually laws of nature, one might rightly expect biologists to be disturbed when they are proven false. To the contrary, biologists expect exceptions to exist. At any one time, a particular percentage of a species of crows will be non-black. No one expects this percentage to be universal or to remain fixed. Species may be classes, but they are not very important classes because their names function in no scientific laws. Given the traditional analyses of scientific laws, statements which refer to particular species do not count as scientific laws, as they

should not if species are spatiotemporally localized individuals (20, 21).

Hence, if biologists expect to find any evolutionary laws, they must look at levels of organization higher than particular taxa. Formulations of evolutionary theory will no more make explicit reference to *Bos bos* than celestial mechanics will refer to Mars. Predictions about these entities should be derivable from the appropriate theories but no uneliminable reference can be made to them. In point of fact, no purported evolutionary laws refer to particular species. One example of such a law is the claim that in diploid sexually reproducing organisms, homozygotes are more specialized in their adaptive properties than heterozygotes (31, p. 397). Evolutionary theory deals with the rise of individual homeostasis as an evolutionary mode, the waxings and wanings of sexuality, the constancy or variability of extinction rates, and so on. People are dismayed to discover that evolutionists can make no specific predictions about the future of humankind *qua* humankind. Since that's all they are interested in, they conclude that evolutionary theory is not good for much. But dismissing evolutionary theory because it cannot be used to predict the percent-*age of people who will have blue eyes* in the year 2000 is as misbegotten as dismissing celestial mechanics because it cannot be used to predict the physical make-up of Mars. Neither theory is designed to make such predictions.

The commonest objection raised by philosophers against evolutionary theory is that its subject matter—living creatures—are spatiotemporally localized (52, 53; see also 42). They exist here on earth and nowhere else. Even if the earth were the only place where life had arisen (and that is unlikely), this fact would not count in the least against the spatiotemporally unrestricted character of evolutionary theory. "Hitler" refers to a particular organism, a spatiotemporally localized individual. As such, Hitler is unique. But organisms are not. Things which biologists would recognize as organisms could develop (and probably have developed) elsewhere in the universe. "*Homo sapiens*" refers to a particular species, a spatiotemporally localized individual. As such it is unique. But species are not. Things which biologists would recognize as species could develop (and probably have developed) elsewhere in the universe. Evolutionary theory refers explicitly to organisms and species, not to Hitler and *Homo sapiens* (see 43, 48).

One advantage to biologists of the historical entity interpretation of species is that it frees them of any necessity of looking for any lawlike regularities at the level of particular species. Both "Richard Nixon has hair" and "most swans are white" may be true, but they are hardly laws of nature. It forces them to look for evolutionary laws at higher levels of analysis, at the level of *kinds* of species. It also can explain certain prevalent anomalies in philosophy. From the beginning, a completely satisfactory explication of the notion of a natural kind has eluded philosophers. One explanation for this failure is that the traditional examples of natural kinds were a mixed lot. The three commonest examples of natural kinds in the philosophical literature have been geometric figures, biological species, and the physical elements. By now it should be clear that all

three are very different sorts of things. No wonder a general anaylsis, applicable equally to all of them, has eluded us.

Some of the implications of treating species as historical entities are more philosophical in nature. For example, one of Ludwig Wittgenstein's most famous (or infamous) contributions to philosophy is that of family resemblances, a notion which itself has a family resemblance to cluster concepts and multivariate analysis (64). Such notions have found their most fertile ground in ethics, aesthetics, and the social sciences. Hence, critics have been able to claim that defining a word in terms of statistical covariation of traits merely results from ignorance and informality of context. If and when these areas become more rigorous, cluster concepts will give way to concepts defined in the traditional way. The names of biological species have been the chief counter-example to these objections. Not only are the methods of contemporary taxonomists rigorous, explicit, objective, etc., but also good reasons can be given for the claim that the names of species can never be defined in classical terms. They are inherently cluster concepts (18). On the analysis presented in this paper, advocates of cluster analysis lose their best example of a class term which is, nevertheless, a cluster concept. If "*Homo sapiens*" is or is not a cluster concept, it will be for the same reason that "Moses" is or (more likely) is not.

A second philosophical consequence of treating species as historical entities concerns the nature of scientific theories. Most contemporary philosophers view scientific theories as atemporal conceptual objects. A theory is a timeless set of axioms and that is that. Anyone who formulates a theory consisting of a particular set of axioms has formulated that theory period. Theories in this sense cannot change through time. Any change results in a new theory. Even if one decides to get reasonable and allow for some variation in axioms, one still must judge two versions of a theory to be versions of the "same" theory because of similarity of axioms. Actual causal connections are irrelevant. However, several philosophers have suggested that science might profitably be studied as an "evolutionary" phenomenon (4, 21, 27, 28, 29, 45, 46, 56). If one takes these claims seriously and accepts the analysis of biological species presented in this paper, then it follows that whatever conceptual entities are supposed to be analogous to species must also be historical entities. Theories seem to be the most likely analog to species. Because biological species cannot be characterized intelligibly in terms of timeless essences, it follows that theories can have no essences either. Like species, theories must be individuated in terms of some sort of descent and cohesiveness, not similarity.

The relative roles of similarity and descent in individuating scientific theories goes a long way in explaining the continuing battle between historians and philosophers of science. Philosophers individuate theories in terms of a set (or at least a cluster) of axioms. Historians tend to pay more attention to actual influence. For example, we all talk about contemporary Mendelian genetics. If theories are to be individuated in terms of a single set (or even cluster) of axioms, it is difficult

to see the justification of such an appellation. Mendel's paper contained three statements which he took to be basic. Two of these statements were rapidly abandoned at the turn of the century when Mendel's so-called "laws" were rediscovered. The third has been modified since. If overlap in substantive claims is what makes two formulations versions of the "same" theory, then it is difficult to see the justification for interpreting all the various things which have gone under the title of "Mendelian genetics" versions of the same theory. Similar observations are appropriate for other theories as well, including Darwin's theory of evolution. The theory that was widely accepted in Darwin's day differed markedly from the one he originally set out. Modern theories of evolution differ from his just as markedly. Yet some are "Darwinian" and others not.

When presented with comparable problems, biologists resort to the type specimen. One organism is selected as the type. Any organism related to it in the appropriate ways belongs to its species, regardless of how aberrant the type specimen might turn out to be or how dissimilar other organisms may be. Males and females belong to the same species even though they might not look anything like each other. A soldier termite belongs in the same species with its fertile congeners even though it cannot mate with them. One possible interpretation of Kuhn's notion of an exemplar (27) is that it is designed to function as a type specimen. Even though scientific change is extremely complicated and at times diffuse, one still might be able to designate particular theories by reference to "concrete problem-solutions," as long as one realizes that these exemplars have a temporal index and need not be in any sense typical.[6] Viewing theories as sets (or clusters) of axioms does considerable damage to our intuitions about scientific theories. On this interpretation, most examples of scientific theories degenerate into unrelated formulations. Viewing scientific theories as historical entities also results in significant departures from our usual modes of conception. Perhaps scientific theories really cannot be interpreted as historical entities. If so, then this is just one more way in which conceptual evolution differs from biological evolution. The more these disanalogies accumulate, the more doubtful the entire analogy becomes.

Finally, and most controversially, treating species as historical entities has certain implications for those sciences which are limited to the study of single species. For instance, if enough scientists were interested, one might devote an entire science to the study of *Orycteropus afer,* the African aardvark. Students of aardvarkology might discover all sorts of truths about aardvarks: that it is nocturnal, eats ants and termites, gives birth to its young alive, etc. Because aardvarks are highly monotypic, aardvarkologists might be able to discover sets of traits possessed by all and only extant aardvarks. But could they discover the essence of aardvarks, the traits which aardvarks must have necessarily to be aardvarks? Could there be scientific laws which govern aardvarks necessarily and exclusively? When these

questions are asked of aardvarks or any other nonhuman species, they sound frivolous, but they are exactly the questions that students of human nature treat with utmost seriousness. What is human nature and its laws?

Early in the history of learning theory, Edward L. Thorndike (55) claimed that learning performance in fishes, chickens, cats, dogs, and monkeys differed only quantitatively, not qualitatively. Recent work tends to contradict his claim (1). Regardless of who is right, why does it make a difference? Learning, like any other trait, has evolved. It may be universally distributed among all species of animals or limited to a few. It may be present in all organisms included in the same species or distributed less than universally. In either case, it may have evolved once or several times. If "learning" is defined in terms of its unique origin, if all instances of learning must be evolutionarily homologous, then "learning" is limited by definition to one segment of the phylogenetic tree. Any regularities which one discovers are necessarily descriptive. If, on the other hand, "learning" is defined so that it can apply to any organism (or machine) which behaves in appropriate ways, then it *may* be limited to one segment of the phylogenetic tree. It *need* not be. Any regularities which one discovers are at least candidates for laws of learning. What matters is whether the principles are generalizable. Learning may be species specific, but if learning theory is to be a genuine scientific theory, it cannot be limited *necessarily* to a

6. Kuhn himself (28) discusses taxa names such as *"Cygnus olor"* and the biological type specimen. Unfortunately, he thinks swans are swans because of the distribution of such traits as the color of feathers.

single species the way that Freud's and Piaget's theories seem to be. As important as descriptions are in science, they are not theories.

If species are interpreted as historical entities, then particular organisms belong in a particular species because they are part of that genealogical nexus, not because they possess any essential traits. No species has an essence in this sense. Hence there is no such thing as human nature. There may be characteristics which all and only extant human beings possess, but this state of affairs is contingent, depending on the current evolutionary state of *Homo sapiens*. Just as not all crows are black (even potentially), it may well be the case that not all people are rational (even potentially). On the historical entity interpretation, retarded people are just as much instances of *Homo sapiens* as are their brighter congeners. The same can be said for women, blacks, homosexuals, and human fetuses. Some people may be incapable of speaking or understanding a genuine language; perhaps bees can. It makes no difference. Bees and people remain biologically distinct species. On other, nonbiological interpretations of the human species, problems arise (and have arisen) with all of the groups mentioned. Possibly women and blacks are human beings but do not "participate fully" in human nature. Homosexuals, retardates, and fetuses are somehow less than human. And if bees use language, then it seems we run the danger of considering them human. The biological interpretation has much to say in its favor, even from the humanistic point of view.

1. Bitterman, M. E., 1975, The comparative analysis of learning. *Science,* 188: 699–709.
2. Boyden, A., 1954, The significance of asexual reproduction. *Systematic Zoology,* 3: 26–37.
3. Buck, R. C., and D. L. Hull, 1966, The logical structure of the Linnaean hierarchy. *Systematic Zoology,* 15: 97–111.
4. Burian, R. M., 1977, More than a marriage of convenience: On the inextricability of history and philosophy of science. *Philosophy of Science,* 44: 1–42.
5. Crowson, R. A., 1970, *Classification and Biology,* New York, Atherton Press.
6. Davis, P. H., and V. H. Heywood, 1963, *Principles of Angiosperm Taxonomy,* Princeton, Van Nostrand.
7. Darwin, C., 1966, *On the Origin of Species,* Cambridge, Mass., Harvard University Press.
8. Dawkins, R., 1976, *The Selfish Gene,* New York and Oxford, Oxford University Press.
9. Dobzhansky, T., 1951, Mendelian populations and their evolution. In L. C. Dunn (ed.) *Genetics in the 20th Century,* New York, Macmillan, pp. 573–589.
10. Ehrlich, P. R., and P. H. Raven, 1969, Differentiation of populations. *Science,* 165: 1228–1231.
11. Eldredge, N., and S. J. Gould, Punctuated equilibria: An alternative to phyletic Gradualism. In T. J. M. Schopf (ed.), *Models in Paleobiology,* San Francisco, Freeman, Cooper and Company, pp. 82–115.
12. Ghiselin, M. T., 1966, On psychologism in the logic of taxonomic controversies. *Systematic Zoology,* 15: 207–215.
13. ——, 1969, *The Triumph of the Darwinian Method,* Berkeley and London, University of California Press.
14. ——, 1974, A radical solution to the species problem. *Systematic Zoology,* 23: 536–544.
15. Griffiths, G. C. D., 1974, On the foun-

dations of Biological Systematics. *Acta Biotheoretica*, 23: 85–131.

16. Heise, H., and M. P. Starr, 1968, Nomenifers: Are they christened or classified? *Systematic Zoology*, 17: 458–467.

17. Hennig, W., 1966, *Phylogenetic Systematics*, Urbana, Illinois, University of Illinois Press.

18. Hull, D. L., 1965, 1966, The effect of essentialism on taxonomy. *The British Journal for the Philosophy of Science*, 15: 314–326; 16: 1–18.

19. ——, 1968, The conflict between spontaneous generation and Aristotle's metaphysics. *Proceedings of the Seventh Inter-American Congress of Philosophy*, Québec City, Les Presses de l'Université Laval, 2: 245–250.

20. ——, 1974, *Philosophy of Biological Science*, Englewood Cliffs, Prentice-Hall.

21. ——, 1975, Central subjects and historical narratives. *History and Theory*, 14: 253–274.

22. ——, 1976, Are species really individuals? *Systematic Zoology*, 25: 174–191.

23. ——, 1976, The ontological status of biological species. In R. Butts and J. Hintikka (eds.) *Boston Studies in the Philosophy of Science*, vol. 32, Dordrecht, D. Reidel, pp. 347–358.

24. Huxley, T. H., 1889, Biology. *Encyclopedia Britannica*.

25. Janzen, Daniel, 1977, What are dandelions and aphids? *American Naturalist*, 111: 586–589.

26. Kripke, S. S., 1972, Naming and necessity. In D. Davidson and H. Harman (eds.), *Semantics and Natural Language*, Dordrecht, Holland, D. Reidel, pp. 253–355.

27. Kuhn, T. S., 1969, *The Structure of Scientific Revolutions*, Chicago, The University of Chicago Press, 2nd ed.

28. ——, 1974, Second Thoughts on Paradigms. In F. Suppe (ed.), *The Structure of Scientific Theory*, Urbana, Illinois, University of Illinois Press.

29. Laudan, L., 1977, *Progress and Its Problems*, Berkeley and London, University of California Press, 1977.

30. Levins, R., 1968, *Evolution in Changing Environments*, Princeton, Princeton University Press.

31. Lewontin, R. C., 1961, Evolution and the theory of games. *Journal of Theoretical Biology*, 1: 382–403.

32. ——, 1970, The units of selection. *The Annual Review of Ecology and Systematics*, 1: 1–18.

33. ——, 1974, *The Genetic Basis of Evolutionary Change*, New York, Columbia University Press.

34. Löther, R., 1972, *Die Beherrschung der Mannigfaltigkeit*, Jena, Gustav Fisher.

35. Mayr, E., 1957 (ed.), *The Species Problem*, Washington, D.C., American Association for the Advancement of Science, Publication Number 50.

36. ——, 1959, Isolation as an evolutionary factor. *Proceedings of the American Philosophical Society*, 103: 221–230.

37. ——, 1963, *Animal Species and Evolution*, Cambridge, Mass., The Belknap Press of Harvard University Press.

38. ——, 1976, Is the species a class or an individual? *Systematic Zoology*, 25: 192.

39. Mayr, E., E. G. Linsley, and R. L. Usinger, 1953, *Methods and Principles of Systematic Zoology*, New York, McGraw-Hill Book Company.

40. Meglitsch, P. A., 1954, On the nature of species, *Systematic Zoology*, 3: 49–65.

41. Mettler, L. E., and T. G. Gregg, 1969, *Population Genetics and Evolution*, Englewood Cliffs, Prentice-Hall.

42. Monod, J. L., 1975, On the molecular theory of evolution. In R. Harré (ed.), *Problems of Scientific Revolution*, Oxford, Clarendon Press, pp. 11–24.

43. Munson, R., 1975, Is Biology a provincial science? *Philosophy of Science*, 42: 428–447.

44. Odum, E. P., 1977, The emergence of ecology as a new integrative discipline. *Science*, 195: 1289–1293.

45. Popper, K. R., 1972, *Objective Knowledge*, Oxford, Clarendon Press.

46. ——, 1975, The rationality of scientific revolutions. In R. Harré (ed.), *Problems*

of Scientific Revolution, Oxford, Clarendon Press, pp. 72-101.

47. Putnam, H., 1974, The meaning of meaning. In K. Gunderson (ed.), Minnesota Studies in the Philosophy of Science, vii, Minneapolis, University of Minnesota Press, pp. 131-193.

48. Ruse, M. J., 1973, The Philosophy of Biology, London, Hutchinson University Library.

49. Schopf, J. M., 1960, Emphasis on holotype. Science, 131: 1043.

50. Simspon, G. G., 1945, The principles of classification and a classification of mammals. Bulletin of the American Museum of Natural History, Vol. 85: 1-350.

51. ——, 1961, Principles of Animal Taxonomy, New York, Columbia University Press.

52. Smart, J. J. C., 1963, Philosophy and Scientific Realism, London, Routledge and Kegan Paul.

53. ——, 1968, Between Science and Philosophy, New York, Random House.

54. Sneath, P. H. A., and R. R. Sokal, 1973, Numerical Taxonomy, San Francisco, W. H. Freeman and Company.

55. Thorndike, E. L., 1911, Animal Intelligence, New York, Macmillan.

56. Toulmin, S., 1972, Human Understanding, Princeton, Princeton University Press.

57. Van Fraassen, Bas., 1972, Probabilities and the problem of individuation. In S. A. Luckenbach (ed.), Probabilities, Problems and Paradoxes, Encino, Calif., Dickinson Publishing Co., pp. 121-138.

58. Vendler, Z., 1976, On the possibility of possible worlds. Canadian Journal of Philosophy, 5: 57-72.

59. Wade, M. J., 1978, A critical review of the models of group selection. Quarterly Review of Biology, 1978, 53: 101-114.

60. Wiley, E. O., 1978, The evolutionary species concept reconsidered. Systematic Zoology, 27: 17-26.

61. Williams, G. C., 1966, Adaption and Natural Selection, Princeton, Princeton University Press.

62. Wilson, E. O., 1975, Sociobiology: The New Synthesis, Cambridge, Mass., The Belknap Press of Harvard University Press.

63. Wynne-Edwards, V. C., 1962, Animal Dispersion in Relation to Social Behaviour, Edinburgh and London, Oliver and Boyd.

64. Wittgenstein, L., 1953, Philosophical Investigations, New York, Macmillan.

# 33

# Biological Classification: Toward a Synthesis of Opposing Methodologies

ᶇᶇᶇᶇᶇᶇᶇᶇᶇᶇᶇᶇᶇᶇᶇᶇᶇᶇᶇᶇᶇᶇᶇᶇᶇᶇ

## ERNST MAYR

*Currently a controversy is raging as to which of three competing methodologies of biological classification is the best: phenetics, cladistics, or evolutionary classification. The merits and seeming deficiencies of the three approaches are analyzed. Since classifying is a multiple-step procedure, it is suggested that the best components of the three methods be used at each step. By such a synthetic approach, classifications can be constructed that are equally suited as the basis of generalizations and as an index to information storage and retrieval systems.*

For nearly a century after the publication of Darwin's *Origin* (1) no well-defined schools of classifiers were recognizable. There were no competing methodologies. Taxonomists were unanimous in their endeavor to establish classifications that would reflect "degree of relationship." What differences there were among competing classifications concerned the number and kinds of characters that were used, whether or not an author accepted the principle of recapitulation, whether he attempted to "base his classification on phylogeny," and to what extent he used the fossil record (2). As a result of a lack of methodology, radically different classifications were sometimes proposed for the same group of organisms; also new classifications were introduced without any adequate justification except for the claim that they were "better." Dissatisfaction with such arbitrariness and seeming absence of any carefully thought out methodology, led in the 1950's and 1960's to the establishment of two new schools of taxonomy, numerical phenetics and cladistics, and to a more explicit articulation of Darwin's methodology, now referred to as evolutionary classification.

## THE MAJOR SCHOOLS OF TAXONOMY

*Numerical phenetics.* From the earliest preliterary days, organisms were grouped into classes by their outward appearance, into grasses, birds, butterflies, snails, and others. Such grouping "by inspection" is the expressly stated or unspoken starting point of virtually all systems of classification. Any classification incorporating the method of grouping taxa by similarity is, to that extent, phenetic.

In the 1950's to 1960's several investigators went one step further and suggested that classifications be based exclusively on "overall similiarity." They also proposed, in order to make the method more objective, that every character be given equal weight, even though this would require the use of large numbers of characters (preferably well over a hundred). In order to reduce the values of so many characters to a single measure of "overall similarity," each character is to be recorded in numerical form. Finally, the clustering of species and their taxonomic distance from each other is to be calculated by the use of algorithms that operationally manipulate characters in certain ways, usually with the help of computers. The resulting diagram of relationship is called a phenogram. The calculated phenetic distances can be converted directly into a classification.

The fullest statement of this methodology and its underlying conceptualization was provided by Sokal and Sneath (3). They called their approach "numerical taxonomy," a somewhat misleading designation, since numerical methods, including numerical weighting, can be and have been applied to entirely different approaches to classification. The term numerical phenetics is now usually applied to this school. This has introduced some ambiguity since some authors have used the term phenetic broadly, applying it to any approach making use of the "similarity" of species and other taxa, while to the strict numerical pheneticists the term phenetic means the "theory-free" use of unweighted characters.

*Cladistics (or cladism).* This method of classification (4), the first comprehensive statement of which was published in 1950 by Hennig (5), bases classifications exclusively on genealogy, that is, on the branching pattern of phylogeny. For the cladist phylogeny consists of a sequence of dichotomies (6), each representing the splitting of a parental species into two daughter species; the ancestral species ceases to exist at the time of the dichotomy; sister groups must be given the same categorical rank; and the ancestral species together with all of its descendants must be included in a single "holophyletic" taxon.

*Evolutionary classification.* Phenetics and cladistics were proposed in the endeavor to replace the methodology of classification that had prevailed ever since Darwin and that was variously designated as the "traditional" or the "evolutionary" school, which bases its classifications on observed similarities and differences among groups of organisms, evaluated in the light of their inferred evolutionary history (7). The evolutionary school includes in the analysis all available attributes of these organisms, their correlations, ecological stations, and patterns of distributions and attempts to reflect both of the major evolutionary processes, branching and the subsequent diverging of the branches (clades). This school follows Darwin (and agrees in this point with the cladists) that classification must be based on genealogy and also agrees with Darwin (in contrast to the cladists) "that genealogy by itself does not give classification" (8).

The results of the evolutionary analysis are incorporated in a diagram, called a phylogram, which records both the branching points and the degrees of subsequent divergence. The method of inferring genealogical relationship with the help of taxonomic characters, as it was first carried out by Darwin, is an application of the hypothetico-deductive approach. Presumed relationships have to be tested again and again with the help of new characters, and the new evidence frequently leads to a revision of the influences on relationship. This method is not circular (9) as has sometimes been suggested.

### IS THERE A BEST WAY TO CLASSIFY?

Each of the three approaches to classification—phenetics, cladistics, and evolutionary classification—has virtues and weaknesses. The ideal classification would be one that would meet best as many as possible of the generally acknowledged objectives of a classification.

A biological classification, like any other, must serve as the basis of a convenient information storage and retrieval system. Since all three theories produce hierarchical systems, containing nested sets of subordinated taxa, they permit the following of information up and down the phyletic tree. But this is where the agreement among the three methods ends. Purely phenetic systems, derived from a single set of arbitrarily chosen characters, sometimes provide only low retrieval capacity as soon as other sets of characters are used. The effectiveness

of the phenetic method could be improved by careful choice of selected characters. However, the method would then no longer be "automatic," because any selection of characters amounts to weighting.

Cladists use only as much information for the construction of the classification as is contained in the cladogram. They convert cladograms, quite unaltered, into classifications, only when the cladograms are strictly dichotomous. Even though cladists lose much information by this simplistic approach, the information on lines of descent can be read off their classifications directly. However, a neglect of all ancestral-descendant information reduces the heuristic value of their classifications. By contrast, since evolutionary taxonomists incorporate a great deal more information in their classifications than do the cladists, they cannot express all of it directly in the names and ranking of the taxa in their classifications. Therefore, they consider a classification simply to be an ordered index that refers them to the information that is stored elsewhere (in the detailed taxonomic treatments).

A far more important function of a classification, even though largely compatible with the informational one, is that it establishes groupings about which generalizations can be made. To the extent that classifications are explicitly based on the theory of common descent with modification, they postulate that members of a taxon share a common heritage and thus will have many characteristics in common. Such classifications, therefore, have great heuristic value in all comparative studies. The validity of specific observations can be generalized by testing them against other taxa in the system or against other kinds of characters (10–12).

Pheneticists, as well as cladists, have claimed that their methods of constructing classifications are nonarbitrary, automatic, and repeatable. The criticisms of these methods over the last 15 years (13) have shown, however, that these claims cannot be substantiated. It is becoming increasingly evident that a one-sided methodology cannot achieve all the above-listed objectives of a good classification.

The silent assumption in the methodologies of phenetics and cladistics is that classification is essentially a single-step procedure: clustering by similarity in phenetics and establishment of branching patterns in cladistics. Actually a classification follows a sequence of steps, and different methods and concepts are pertinent at each of the consecutive steps. It seems to me that we might arrive at a less vulnerable methodology by developing the best method for each step consecutively. Perhaps the steps could eventually be combined in a single algorithm. In the meantime, their separate discussion contributes to the clarification of the various aspects of the classifying process.

## ESTABLISHMENT OF SIMILARITY CLASSES

The first step is the grouping of species and genera by "inspection." that is, by a phenetic procedure. (I use phenetic in the broadest sense, not in the narrow one of numerical phenetics.) All of classifying consists of, or at least begins with, the establishment of similarity classes, such as a preliminary grouping of plants into trees, shrubs, herbs, and grasses. The reason why the method is so often successful is simply that—other things being equal—descendants of a common ancestor tend to be more similar to each other than they are to species that do not share immediate common descent. The method is thus excellent in principle. Numerical phenetics has nevertheless proved to be largely unsuccessful because (i) claims, such as "results are objective and strictly repeatable," were not always justifiable since in practice different results are obtained when different characters are chosen or different programs of computation are used; (ii) the method was inconsistent in its claim of objectivity since subjective biological criteria were used in the assigning of variants (for example, sexes, age classes, and morphs) to "operational taxonomic units" (OTU's); and, most importantly, the method insisted on the equal weighting of all characters.

It is now evident that no computing method exists that can determine "true similarity" from a set of arbitrarily chosen characters. So-called similarity is a complex phenomenon that is not necessarily closely correlated with common descent, since similarity is often due to convergence. Most major improvements in plant and animal classifications have been due to the discovery of such convergence (14).

Different types of characters—morphological characters, chromosomal differences, enzyme genes, regulatory genes, and DNA matching—may lead to rather different grouping. Different stages in the life cycle may result in different groupings.

The ideal of phenetics has always been to discover a measure of total (overall) similarity. Since it is now evident that this cannot be achieved on the basis of a set of arbitrarily chosen characters, the question has been asked whether there is not a method to measure degrees of difference of the genotype as a whole. Improvements in the method of DNA hybridization offer hope that this method might give realistic classifications on a phenetic basis, at least up to the level of orders (15). The larger the fraction of the nonhybridizing DNA, the less reliable this method is, because it cannot be determined whether the nonmatching DNA is only slightly or drastically different.

### TESTING THE NATURALNESS OF TAXA

In the first step of the classifying procedure clusters of species were assembled that seemed to be more similar to each other than to species in other clusters. These clusters are the taxa we recognize tentatively (16). In order to make these clusters conform to evolutionary theory, two, operationally more or less inseparable, tests must be made: (i) determine for all species of a cluster (taxon) whether they are descendants of the nearest common ancestor and (ii) connect the taxa by a branching tree of common descent, that is, construct a cladogram. An indispensable preliminary of this testing is an analysis of the characters used to establish the similarity clusters.

*Character analysis.* A careful analysis shows almost invariably that some characters are better clues to relationship (have greater weight) than others. The fewer the number of available characters, the more carefully the weighting must be done. This weight-ing is one of the most controversial aspects of the classifying procedure. Investigators who come to systematics from the outside, say from mathematics, or who are beginners tend to demand objective or quantitative methods of weighting. There are such methods, principally ones based on the covariation of characters, but they are not nearly as informative as methods based on the biological evaluation of characters (17). But such an evaluation requires an understanding of many aspects of the to-be-classified group (that is, its life history, the inferred selection pressures to which it is exposed, and its evolutionary history) that may not be available to an outsider. This creates a genuine dilemma. If strictly taxometric methods were available that would produce satisfactory weighting, everyone would surely prefer them to weighting based on experience and biological knowledge. But so far such methods are still in their infancy.

The greatest difficulty for a purely phenetic method, indeed for any method of classification, is the discordance (noncongruence) of different sets of characters. Entirely different classifications may result from the use of characters of different stages of the life cycle as, for instance, larval versus adult characters. In a study of species of bees, Michener (18) obtained four different classifications when he sorted them into similarity classes on the basis of the characters of (i) larvae, (ii) pupae, (iii) the external morphology of the adults, and (iv) male genitalic structures. Phenetic delimitation of taxa unavoidably necessitates a great deal of decision-making on the use and weighting of characters. Often, when new sets of characters become available, their use may lead to a new delimitation of taxa or to a change in ranking.

*Determination of the genealogy.* Each group (taxon) tentatively established by the phenetic method is, so to speak, a hypothesis as to common descent, the validity of which must be tested. Is the delimited taxon truly monophyletic (19)? Are the species included in this taxon nearest relatives (descendants of the nearest common ancestor)? Have all species been excluded that are only superficially or convergently similar?

Methods to answer these questions have been in use since the days of Darwin, particularly the testing of the homology of critical characteristics of the included species. However, Hennig (5) was the first to articulate such methods explicitly, and these have been modified by some of his followers. These methods can be designated as the cladistic analysis.

Such an analysis involves first the partitioning of the joint characters of a group into ancestral ("plesiomorph" in Hennig's terminology) characters and derived ("apomorph") characters, that is, characters restricted to the descendants of the putative nearest common ancestor (20). The joint possession of homologous derived characters proves the common ancestry of a given set of species. A character is derived in relation to the ancestral condition of the character. The end product of such a cladistic character analysis is a cladogram, that is, a diagram (dendrogram) of the branching points of the phylogeny.

Although this procedure sounds simple, numerous practical difficulties have been pointed out (21, 22). Very often the branching points are inferred by way of single or very few characters and are affected by all the weaknesses of single character classifications. More serious are two other difficulties.

(1) *Polarity.* A derived character is often simpler or less specialized than the ancestral condition. For this reason it can be difficult to determine polarity in a transformation series of characters, that is, to determine which end of the series is ancestral. Tattersall and Eldredge (23) stressed that "in practice it is hard, even impossible, to marshall a strong, logical argument for a given polarity for many characters in a given group." Are they primitive (ancestral) or derived? Much of the controversy concerning the phylogeny of the invertebrates, for instance, is due to differences of opinion concerning polarity. Hennig tried to elaborate methods for determining polarity but, as others (24, 25) have shown, with rather indifferent success. Since characters come and go in phyletic lines and since there is much convergence, the problem of polarity can rarely be solved unequivocally. There are three best types of evidence for polarity reconstruction. First is the fossil record. Although primitiveness and apparent ancientness are not correlated in every case, nevertheless as Simpson (26) stressed, "for any group with even a fair fossil record there is seldom any doubt that characters usual or shared by older members are almost always more primitive than those of later members." Second is sequential constraints. Consecutive chromosomal inversions (as in *Drosophila*) or sets of amino acid replacements (and presumably certain other molecular events) form definite sequences. Which

end of the sequence is the beginning can usually not be read off from the sequence itself, but additional information (polarity of other character chains, geographical distribution, and the like) often permits an unequivocal determination of the polarity. Third is the reconstruction of the presumed evolutionary pathway. This can sometimes be done by studying evidence for adaptive shifts, the invasion of new competitors or the extinction of old ones, the behavior of correlated characters, and other biological evidence (see ref. 11, pp. 886–887; also ref. 24). Particular difficulties are posed when the polarity is reversed in the course of evolution, as documented in the fossil record.

(2) *Kinds of derived characters.* Two taxa may resemble each other in a given character for one of three reasons: because the character existed already in the ancestry of the two groups before the evolution of the nearest common ancestor (symplesiomorphy in Hennig's terminology), because it originated in the common ancestor and is shared by all of his descendants (homologous apomorphy or synapomorphy), or because it originated independently by convergence in several descendant groups (nonhomologous or convergent apomorphy) (27). Since, according to the cladistic method, sister groups are recognized by the possession of synapomorphies, convergence poses a major problem. How are we to distinguish between homologous and convergent apomorphies? Hennig was fully aware of the critical importance of this problem, but it has been quietly ignored by many of his followers. Both grebes and loons, two orders of diving birds, have a prominent spur on the knee

and were therefore called sister groups by one cladist. However, other anatomical and biochemical differences between the two taxa indicate that the shared derived feature was acquired by convergence. The reliability of the determination of monophyly of a group depends to a large extent on the care that is taken in discriminating between these two classes of shared apomorphy (11, pp. 880–890).

There is a third class of derived characters, so-called autapomorphies, which are characters that were acquired by and are restricted to a phyletic line after it branched off from its sister group.

The pheneticists do not undertake a character analysis. Cladists and evolutionary taxonomists agree with each other in principle on the importance of a careful character analysis. They disagree, however, fundamentally in how to use the findings of the character analysis in the construction of classifications, particularly the ranking procedure.

### THE CONSTRUCTION OF A CLASSIFICATION

*Cladistic classification.* Cladists convert the cladogram directly into a cladistic classification. In such a classification taxa are delimited exclusively by holophyly, that is, by the possession of a common ancestor, rather than by a combination of genealogy and degree of divergence (19). This results in such incongruous combinations as a taxon containing only crocodiles and birds, or one containing only lice and one family of Mallophaga.

Taxa based exclusively on genealogy are of limited use in most biological comparisons. Since, as Hull (28) pointed out, cladists really classify characters rather than organisms, they have to make the arbitrary assumption that new apomorph characters originate whenever a line branches from its sister line. This is unlikely in most cases. Surely the reptilian species that originated the avian lineage lacked any of the flight specializations characteristic of modern birds, except perhaps the feathers (29).

Two principles govern the conversion of a cladogram into a cladistic classification: (i) all branchings are bifurcations that give rise to two sister groups, and (ii) branchings are usually connected with a change in categorical rank. Cladistic classifications are only representations of branching patterns, with complete disregard of evolutionary divergence, ancestor-descendant relationships, and the information content of autapomorph characters. Because these aspects of evolutionary change are neglected, the cladistic method of classification "either results in lumping very similar forms (parasites and their relatives) or in recognizing a multitude of taxa (perhaps also of other categories) regardless of the extreme similarity of some of them. Such simplistic procedures do violence to most biological attributes other than the pattern of the cladistic branching system, as well as to the function of a classification for convenient information transmittal and storage." as Michener remarked (18).

These objections show that the methodology of cladistic classification is not satisfactory. Anyone familiar with the history of taxonomy is strangely reminded of the principles of Aristotelian logical division when encountering cladistic classifications with their rigid dichotomies, the mandate that every taxon must have a sister group, and the principle of a straight-line hierarchy.

There has been much argument over the relationship between classification and phylogeny (30). Both cladists and evolutionary taxonomists agree that all members of a taxon must have a common ancestor. A phylogenetic analysis, and in particular a clear separation of homologous apomorphies from convergences, is a necessary component of the classifying procedure. Classificatory analysis often leads to new inferences on phylogeny, and new insights on phylogeny may necessitate changes in classification. These interactions are not in the least circular (9).

It is quite unnecessary in most cases to know the exact species that was the common ancestor of two diverging phyletic lines. An inability to specify such an ancestral species has rarely impeded paleontological research (31, 32). For instance, it is of little importance whether *Archaeopteryx* was the first real ancestor of modern birds or some other similar species or genus. What is important to know is whether birds evolved from lizard-like, crocodile-like, or dinosaur-like ancestors. If a reasonably good fossil record is available, it is usually possible, by the backward tracing of evolutionary trends and by the backward projection of divergent phyletic lines, to reconstruct a reasonably convincing facsimile of the representative of a phyletic line at an earlier time.

Simpson (32) has provided us with cogent arguments about why it is not permissible to reject information from the fossil record under the pretext that it fails to give the phylogenetic connections between fossil and recent taxa with absolute certainty. Hence, there is no merit in the suggestion to construct separate classifications for recent and for fossil organisms. After all, fossil species belong to the same tree of descent as living species. Indeed, enough evidence usually becomes available through a careful character analysis to permit relatively robust inferences on the most probable phylogeny. A number of recent endeavors have been made to develop a cladistic methodology that is quantitative and automatic. New methods in this area are published in rapid succession and it would seem too early to determine which is most successful and freest of possible flaws (33).

*Evolutionary classification.* The taxonomic task of the cladist is completed with the cladistic character analysis. The genealogy gives him the classification directly, since for him classification is nothing but genealogy. The evolutionary taxonomist carries the analysis one step further. He is interested not only in branching, but, like Darwin, also in the subsequent fate of each branch. In particular, he undertakes a comparative study of the phyletic divergence of all evolutionary lineages, since the evolutionary history of sister groups is often strikingly different. Among two related groups, derived from the same nearest common ancestor, one may hardly differ from the ancestral group, while the other may have entered a new adaptive zone and evolved into a novel type. Even though they are sister groups in the terminology of cladistics, they may deserve different categorical rank, because their biological characteristics differ to such an extent as to affect any comparative study.

The importance of this consideration was stated by Darwin (1, p. 420): "I believe that the *arrangement* of the groups within each class, in due subordination and relation to the other groups, must be strictly genealogical in order to be natural, but that the *amount* of difference in the several branches or groups, though allied in the same degree in blood to their common progenitor, may differ greatly, being due to the different degrees of modification which they have undergone, and this is expressed by the forms being ranked under different genera, families, sections or orders." Darwin refers then to a diagram of three Silurian genera that have modern descendants; one has not even changed generically, but the other two have become distinct orders, one with three and the other with two families.

The question as to what extent an analysis of degrees of divergence is possible, is still debated. The cladist makes only "horizontal" comparisons, cataloging the synapomorphies of sister groups. The evolutionary taxonomist, however, also makes use of derived characters that are restricted to a single line of descent, so-called autapomorph characters (Fig. 1), which are apomorph characters restricted to a single sister group. The importance of autapomorphy is well illustrated by a comparison of birds with their sister group (34). Birds originated from that branch of the reptiles, the Archosauria, which also gave rise to the pterodactyls, dinosaurs, and crocodilians. The crocodilians are the sister group of the birds among living organisms: a stem group of archosaurians represents the common ancestry of birds and crocodilians. Although birds and croc-

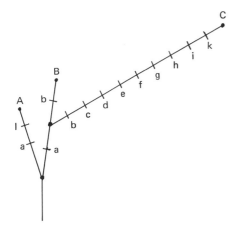

Figure 1. Cladogram of taxa A, B, and C. Cladists combine B and C into a single taxon because B and C share the synapomorph character b. Evolutionary taxonomists separate C from A and B, which they combine, because C differs by many (c through k) autapomorph characters from A and B and shares only one (b) synapomorph character with B.

odilians share a number of synapomorphies that originated after the archosaurian line had branched off from the other reptilian lines, nevertheless crocodilians are on the whole very similar to other reptiles, that is, they have developed relatively few autapomorph characters. They represent the reptilian "grade," as many morphologists call it. Birds, by contrast, have acquired a vast array of new autapomorph characters in connection with their shift to aerial living. Whenever a clade (phyletic lineage) enters a new adaptive zone that leads to a drastic reorganization of the clade, greater taxonomic weight may have to be assigned to the resulting transformation than to the proximity of joint ancestry. The cladist virtually ignores this ecological component of evolution.

The main difference between cladists and evolutionary taxonomists, thus, is in the treatment of autapomorph characters. Instead of automatically giving sister groups the same rank, the evolutionary taxonomist ranks them by considering the relative weight of their autapomorphies (Fig. 1). For instance, one of the striking autapomorphies of man (in comparison to his sister group, the chimpanzee) is the possession of Broca's center in the brain, a character that is closely correlated with man's speaking ability. This single character is for most taxonomists of greater weight than various synapomorphous similarities or even identities in man and the apes in certain macromolecules such as hemoglobins and cytochrome $c$. The particular importance of autapomorphies is that they reflect the occupation of new niches and new adaptive zones that may have greater biological significance than synapomorphies in some of the standard macromolecules.

I agree with Szalay (35) when he says: "The loss of biological knowledge when not using a scheme of ancestor-descendant relationship, I believe, is great. In fact, whereas a sister group relationship may . . . tell us a little, a postulated and investigated ancestor-descendant relationship may help explain a previously inexplicable character in terms of its origin and transformation, and subsequently its functional (mechanical) significance." In other words, the analysis of the ancestor-descendant relationships adds a great deal of information that cannot be supplied by the analysis of sister group relationships.

It is sometimes claimed that the analysis of ancestor-descendant relationships lacks the precision of cladistic sister group comparisons. However, as was shown above and as is also emphasized by Hull (36), the cladistic analysis is actually full of uncertainties. The slight possible loss of precision, caused by the use of autapomorphies, is a minor disadvantage in comparison with the advantage of the large amount of additional information thus made available.

The information on autapomorphies permits the conversion of the cladogram into a phylogram. The phylogram differs from the cladogram by the placement of sister groups at different distances from the joint common ancestry (branching point) and by the expression of degree of divergence by different angles. Both of these topological devices can be translated into the respective categorical ranking of sister groups. These methods (37) generally attempt to discover the shortest possible "tree" that is compatible with the data. Yet, anyone familiar with the frequency of evolutionary reversals and of evolutionary opportunism, realizes the improbability of the assumption that the tree constructed by this so-called "parsimony method" corresponds to the actual phylogenetic tree. "To regard [the shortest tree method] as parsimonious completely misconceives the intent and use of parsimony in science" (38).

It is not always immediately evident whether a tree construction algorithm is based on cladists' principles or on the methods of evolutionary classification. If the "special similarity" on which the trees are based are strictly synapomorphies, then the method is cladistic. If autapomorphies are also given strong weight, then the method falls under evolutionary classification.

The particular aspect of the method of evolutionary taxonomy found most unacceptable to cladists is the recognition of "paraphyletic" taxa. A paraphyletic taxon is a holophyletic group from which certain strikingly divergent members have been removed. For instance, the class Reptilia of the standard zoological literature is paraphyletic, because birds and mammals, two strikingly divergent descendants of the same common ancestor of all the Reptilia, are not included. Nevertheless, the traditional class Reptilia is monophyletic, because it consists exclusively of descendants from the common ancestor, even though it excludes birds and mammals owing to the high number of autapomorphies of these classes. The recognition of paraphyletic taxa is particularly useful whenever the recognition of definite grades of evolutionary change is important.

## THE RANKING OF TAXA

Once species have been grouped into taxa the next step in the process of biological classification is the construction of a hierarchy of these taxa, the so-called Linnaean hierarchy. The hierarchy is constructed by assigning a definite rank such as family or order to each taxon, subordinating the lower categories to the higher ones. It is a basic weakness of cladistics that it lacks a sensitive method of ranking and simply gives a new rank after each branching point. The evolutionary taxonomist, following Darwin, ranks taxa by the degree of divergence from the common ancestor, often assigning a different rank to sister groups. Rank determination is one of the most difficult and subjective decision processes in classification. One aspect of evolution that causes difficulties is mosaic evolution (39). Rates of divergence of different characters are often drastically different. Conventionally taxa, such as those of vertebrates, are described and delimited on the basis of external morphology and of the skeleton, particularly the locomotory system. When other sets of morphological characters are used (for example, sense organs, reproductive system, central nervous system, or chromosomes), the evidence they provide is sometimes conflicting. The situation can become worse, if molecular characters are also used. The anthropoid genus *Pan* (chimpanzee), for instance, is very similar to *Homo* in molecular characters, but man differs so much from the anthropoid apes in traditional characters (central nervous system and its capacities) and occupation of a highly distinct adaptive zone that Julian Huxley even proposed to raise him to the rank of a separate kingdom —Psychozoa.

It has been suggested that different classifications should be constructed for each kind of character, or at least for morphological and molecular characters. Yet there is already much evidence that the acceptance of several classifications based on different characters would lead to insurmountable complications. But taking all available data into consideration simultaneously, a classification can usually be constructed that can serve conveniently as an all-purpose classification or, as Hennig (5) called it, "a general reference system."

It is usually possible to derive more than one classification from a phylogram, because higher taxa are usually composed of several end points of the phylogram, and different investigators differ by the degree to which they lump such terminal branches into a single higher taxon (40). An example is the phylogram of the higher ferns on which, as Wagner (41) has shown, six different classifications have been founded (Fig. 2) and many more are possible. The extent to which investigators "split" or "lump" higher taxa, thus, is of considerable influence on the classifications they produce.

### COMPARISON OF THE THREE MAJOR SCHOOLS

Each school believes that its classification is the "best." Pheneticists as well as cladists claim that their respective methods have also the great merit of giving automatically nonarbitrary results. These claims cannot be substantiated. To be sure, grouping by phenetic characters and determination of holophyly by cladistic analysis are valuable components of the procedure of biological classification. The great deficiency of both phenetics and cladistics is the failure to reflect adequately the past evolutionary history of taxa.

What needs to be emphasized once more is the fact that groups of organisms are the product of evolution and that no classification can hope to be satisfactory that does not take this fact fully into consideration. Both pheneticists and cladists are ambiguous in their attitude toward the evolutionary theory. The pheneticists claim that their approach is completely theory-free, but they nevertheless assume that their method will produce a hierarchy of taxa that corresponds to descent with modification. On the basis of this assumption, they also claim to be "evolutionary taxonomists" (42),

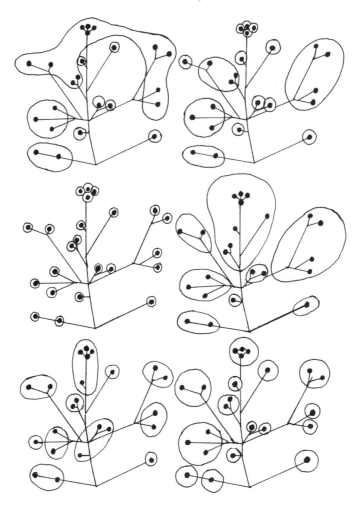

Figure 2. Six different possible classifications of ferns, based on the same dendrogram. Each filled circle is a genus, and each open circle is a family. The differences are due to which and how many genera are combined to make up the families. [From W. H. Wagner (41, figure 7)].

but the fact that different phenetic procedures may produce very different classifications and that their procedure is not influenced by evolutionary considerations refutes this assertion. The cladists exclude most of evolutionary theory (for example, inferences on selection pressures, shifts of adaptive zones, evolutionary rates, and rates of evolutionary divergence) from their consideration (43) and tend increasingly not to classify species and taxa, but only taxonomic characters (28) and their origin. The connection with evolutionary principles is exceedingly tenuous in many recent cladistic writings.

By contrast, the evolutionary taxonomists, as indicated by the name of their school and by well-articulated statements of some of its major representatives (7), expressly base their classifications on evolutionary theory. They aim to construct classifications that reflect both of the two major evolutionary processes, branching and divergence (cladogenesis and anagenesis). They make full use of information on shifts into new adaptive zones and rates of evolutionary change and believe that the resulting classifications are a key to a far richer information content.

Although the three schools still seem rather fundamentally in disagreement, as far as the basic principles of classification are concerned, the more moderate representatives have quietly incorporated some of the criteria of the opposing schools, so that the differences among them have been partially obliterated. For instance, Farris' (44) clustering of special similarities is a phenetic method based on the weighting of characters. The evolutionary school uses phenetic criteria to establish similarity classes and to construct a classification, and a cladistic criteria to test the naturalness of taxa. Comparing what McNeill (45) says in favor of phenetics (appropriately modified) and Farris (44) against it, we find that the gap has narrowed. I have no doubt that moderates will be able to develop an eclectic methodology, one that contains a proper balance of phenetics and cladistics that will produce far more "natural classifications" (16) than any one-sided approach that relies exclusively on a single criterion, whether it be overall similarity, parsimony of branching pattern, or what not. Evolutionary taxonomy, from Darwin on, has been characterized by the adoption of an eclectic approach that makes use of similarity, branching pattern, and degree of evolutionary divergence.

## CLASSIFICATION AND INFORMATION RETRIEVAL

Biological classifications have two major objectives: to serve as the basis of biological generalizations in all sorts of comparative studies and to serve as the key to an information storage system. Up to this point, I have concentrated on those aspects of classifying that help to secure a sound basis for generalizations. This leaves unanswered the question of whether achievement of this first objective is, or is not, reconcilable with achievement of the second objective. Is the classification that is soundest as a basis of generalizations also most convenient for information retrieval? This, indeed, seems to have been true in most cases I have encountered. However, we can also look at this problem from another side.

It is possible at nearly each of the three major steps in the making of a classification to make a choice between several alternatives. These choices may be scientifically equivalent, but some may be more convenient in aiding information retrieval than others. If we choose one of them, it is not necessarily because the alternatives were "falsified," but rather because the chosen method is "more practical." In this respect, biological classifications are not unique. Scientific theories are nearly always judged by criteria additional to truth or falsity, for instance, by their simplicity or, in mathematics, by their "elegance." Therefore, it can be asserted that convenience in the use of a classification, including its function as key to information retrieval, is not necessarily in conflict with its more purely scientific objectives (46–48).

## REFERENCES AND NOTES

1. C. Darwin, 1859, *On the Origin of Species,* London, Murray.
2. For an illuminating survey of the thinking of that period see F. A. Bather, *Proc. Geol. Soc. London,* 83, LXII (1927).
3. R. R. Sokal and P. H. A. Sneath, 1963, *Principles of Numerical Taxonomy,* San Francisco, Freeman. A drastically revised edition was published in 1973.
4. The method was first published under the misleading name phylogenetic systematics, but since it is based on only a single one (branching) of the various processes of phylogeny, the terms cladism or cladistics have been substituted and are now widely accepted.
5. W. Hennig's original statement is *Grundzüge einer Theorie der Phylogenetischen Systematik,* Deutscher Zentralverlag, Berlin (1950). A greatly revised second edition (reprinted in 1979) is *Phylogenetic Systematics,* D. D. Davis and R. Zangerl, eds., Univ. of Illinois Press, Urbana, 1966; see also W. Hennig (47). An independent phylogenetic analysis of characters was made by T. P. Maslin, *Syst. Zool.,* I (1952) 49. For an overview of the more significant recent literature see D. Hull (36) and J. S. Farris, *Syst. Zool.,* 28 (1979) 483.
6. Some cladists in recent years have relaxed the requirements of strict dichotomy and have permitted tri- and polyfurcations or have quietly abandoned dichotomy by admitting empty internodes in the cladogram. Polyfurcations can be translated into several alternate bifurcations [see J. Felsenstein, *Syst. Zool.,* 27 (1978) 27] and this makes the automatic conversion of the cladogram into a classification of sister groups impossible.
7. The classical statement of this theory is to be found in C. Darwin (1, pp. 411-

434). G. G. Simpson, *Principles of Animal Taxonomy,* New York, Columbia Univ. Press, 1961, and E. Mayr (48) provide comprehensive modern presentations of this theory. Several critical recent analyses are: W. Bock (11); C. D. Michener (18); *Syst. Zool.,* 27 (1978) 112; P. D. Ashlock (12).
8. F. Darwin, 1887, *Life and Letters of Charles Darwin,* London, Murray, vol. 2, p. 247.
9. D. Hull, *Evolution,* 21 (1967) 174; see also W. Bock (11).
10. F. E. Warburton, *Syst. Zool.,* 16 (1967) 241; W. Bock, *ibid.,* 22 (1973) 375.
11. W. Bock, 1977, in *Major Patterns in Vertebrate Evolution,* M. K. Hecht, P. C. Goody, B. M. Hecht, eds., NATO Advanced Study Institute Series, New York, Plenum, vol. 14: 851–895.
12. P. D. Ashlock, *Syst. Zool.,* 28 (1979) 441.
13. I shall not, at this time, recount the almost interminable controversies among the three schools. For critiques of phenetics see E. Mayr (48, pp. 203–211), L. A. S. Johnson, *Syst. Zool.,* 19 (1970) 203, and D. Hull, *Annu. Rev. Ecol. Syst.,* 1 (1970) 19. Some of the weaknesses pointed out by these early critics have been corrected in the 1973 edition of Sokal and Sneath (3) and by J. S. Farris (44). For critiques of cladistics see E. Mayr (21), R. R. Sokal (22), G. G. Simpson (32), D. Hull (36), P. D. Ashlock, *Annu. Rev. Ecol. Syst.,* 5 (1974) 81; and L. van Valen (49).
14. A particularly illuminating example is the breaking up of the plant group *Amentiferae,* which has been shown to consist of taxa secondarily adapted for wind pollination (R. F. Thorne, *Brittonia,* 25 (1973) 395). Examples among animals of radical reclassifications are the Rodentia, parasitic bees,

certain beetle families, and the turbellarians.

15. C. G. Sibley, in preparation.

16. There have been arguments since before the days of Linnaeus about how to determine whether or not a system, a classification, is "natural." William Whewell, at a time before Darwin had proclaimed his theory of common descent, expressed the then prevailing pragmatic consensus, "The maxim by which all systems professing to be natural must be tested is this: that the arrangement obtained from one set of characters coincides with the arrangement obtained from another set" (W. Whewell, *Philos. Inductive Sci.*, 1, 1840, 521). Interestingly, the covariance of characters is still perhaps the best practical test of the goodness of a classification. Since Darwin, of course, that classification is considered most natural that best reflects the inferred evolutionary history of the organisms involved.

17. For a tabulation and analysis of such qualitative methods of weighting, see E. Mayr (48, pp. 220–228).

18. C. D. Michener, *Syst. Zool.*, 26 (1977) 32.

19. I use the word monophyletic in its traditional sense, as a qualifying adjective of a taxon. Various definitions of monophyletic have been proposed but all of them for the same concept, a qualifying statement concerning a taxon. A taxon is monophyletic if all of its members are derived from the nearest common ancestor (E. Haeckel, 1868, *Natürliche Schöpsungsgeschichte*, Berlin, Reimer). Cladists have attempted to turn the situation upside down by placing all descendants of an ancestor into a taxon. Monophyletic thus becomes a qualifying adjective for descent, and a taxon is not recognized by its characteristics but only by its descent. The transfer of such a well-established term as monophyletic to an entirely different concept is as unscientific and unacceptable as if someone were to "redefine" mass, energy, or gravity by attaching these terms to entirely new concepts. P. D. Ashlock, *Syst. Zool.*, 20 (1971) 63, has proposed the term holophyletic for the assemblage of descendants of a common ancestor. See also P. D. Ashlock (12, p. 443).

20. Terms like apomorph, synapomorph, derived, ancestral, and so forth always refer to characters of taxa at all levels. A genus may have synapomorphies with another genus, and so may an order with another order. It is this applicability of the same criteria for taxa of all ranks that permits the construction of the Linnaean hierarchy.

21. E. Mayr, *Z. Zool. Syst. Evolutionsforsch.*, 12 (1974) 94; reprinted in E. Mayr, *Evolution and the Diversity of Life*, Cambridge, Mass., Harvard Univ. Press, 1976, pp. 433–478.

22. R. R. Sokal, *Syst. Zool.*, 24 (1975) 257.

23. J. Tattersall and N. Eldredge, *Am. Sci.*, 65 (1977) 204.

24. D. S. Peters and W. Gutmann, *Z. Zool. Syst. Evolutionsforsch.*, 9 (1971) 237.

25. O. Schindewolf, *Acta Biotheor.*, 18 (1968) 273; H. K. Erben, *Verh. Dtsch. Zool. Ges.*, 79 (1979) 116.

26. G. G. Simpson (32); see also L. van Valen (49).

27. For a diagram of these three categories of morphological resemblance see Figure 1 in W. Hennig (47).

28. "Cladistic classifications do not represent the order of branching of sister groups, but the order of emergence of unique derived characters." See D. Hull (36).

29. G. G. Simpson, 1953, *The Major Features of Evolution*, New York, Columbia Univ. Press, p. 348, discusses the fallacy of the cladist assumption.

30. Phylogeny is equated by cladists with cladogenesis (branching), while the evolutionary taxonomist subsumes both branching and evolutionary divergence (anagenesis) under phylogeny.

31. C. W. Harper, *J. Paleontol.*, 50 (1976) 180.

32. G. G. Simpson, 1975, in *Phylogeny of the Primates*, W. Pluckett and F. S.

Szalay, eds., New York, Plenum, pp. 3–19.

33. J. H. Camin and R. R. Sokal, *Evolution,* 19 (1965) 311; W. M. Fitch and E. Margoliash, *Science,* 155 (1967) 279; W. M. Fitch, in *Major Patterns of Vertebrate Evolution,* M. K. Hecht, P. C. Goody, B. M. Hecht, eds., NATO Advanced Study Institute Series, New York, Plenum, 1977, vol. 14: 169–204.

34. There are literally hundreds of cases to illustrate this situation. I use again the classical case of birds and crocodilians because even a nonbiologist will understand the situation if such familiar animals are used. The holophyletic classification of the lice (Anoplura) derived from one of the suborders of the Mallophaga is another particularly instructive example. K. C. Kim and H. W. Ludwig, *Ann. Entomol. Soc. Am.,* 71 (1978) 910.

35. F. S. Szalay, *Syst. Zool.,* 26 (1977) 12.

36. D. Hull, *ibid,* 28 (1979) 416.

37. J. W. Hardin, *Brittonia,* 9 (1957) 145; W. H. Wagner, in *Plant Taxonomy: Methods and Principles,* L. Benson, ed., Ronald, New York (1962), pp. 415–417; A. G. Kluge and J. S. Farris, *Syst. Zool.,* 18 (1969) 1; J. S. Farris, *Am. Nat.,* 106 (1972) 645.

38. L. H. Throckmorton, in *Biosystematics in Agriculture,* J. A. Romberger, ed., New York, Wiley, 1978, p. 237. Others who have questioned the validity of the so-called parsimony principle are M. Ghiselin, *Syst. Zool.,* 15 (1966) 214 and W. Bock (11).

39. Unequal rates of evolution for different structures or for any other components of phenotypes or genotypes are designated mosaic evolution.

40. See E. Mayr (48, pp, 238–241) on the differences between splitters and lumpers.

41. W. H. Wagner, 1969, The construction of a classification. In *Systematic Biology,* Publication 1692, National Academy of Sciences, Washington, D.C., pp. 67–90.

42. R. R. Sokal in (22). "I have yet to meet a nonevolutionary taxonomist."

43. Several leading cladists have recently published antiselectionist statements.

44. J. S. Farris, 1977, in *Major Patterns in Vertebrate Evolution,* M. K. Hecht, P. C. Goody, B. M. Hecht, eds., NATO Advanced Study Institute Series, New York, Plenum, vol. 14: 823–850.

45. J. McNeill, *Syst. Zool.,* 28 (1979) 468.

46. For criteria by which to judge the practical usefulness of biological classifications, see E. Mayr (48, pp. 229–242).

47. W. Hennig, *Annu. Rev. Entomol.,* 10 (1965) 97.

48. E. Mayr, 1969, *Principles of Systematic Zoology,* New York, McGraw-Hill.

49. L. van Valen, *Evol. Theory,* 3 (1978) 285.

50. Drafts were read by P. Ashlock, J. Beatty, W. Bock, W. Fink, C. G. Hempel, and D. Hull, to all of whom I am indebted for valuable suggestions and critical comments, not all of which was I able to accept.

# 34

# Cases in which Parsimony or Compatibility Methods Will Be Positively Misleading

## JOE FELSENSTEIN

*For some simple three- and four-species cases involving a character with two states, it is determined under what conditions several methods of phylogenetic inference will fail to converge to the true phylogeny as more and more data are accumulated. The methods are the Camin-Sokal parsimony method, the compatibility method, and Farris's unrooted Wagner tree parsimony method. In all cases the conditions for this failure (which is the failure to be statistically consistent) are essentially that parallel changes exceed informative, nonparallel changes. It is possible for these methods to be inconsistent even when change is improbable a priori, provided that evolutionary rates in different lineages are sufficiently unequal. It is by extension of this approach that we may provide a sound methodology for evaluating methods of phylogenetic inference.*

PARSIMONY OR MINIMUM EVOLUTION methods were first introduced into phylogenetic inference by Camin and Sokal (1965). This class of methods for inferring an evolutionary tree from discrete-character data involves making a reconstruction of the changes in a given set of characters on a given tree, counting the smallest number of times that a given kind of event need have happened, and using this as the measure of the adequacy of the evolutionary tree. (Alternatively, one can compute the weighted sum of the numbers of times several different kinds of events have occurred.) One attempts to find that evolutionary tree which requires the fewest of these evolutionary events to explain the observed data. Camin and Sokal treated

This report was prepared as an account of work sponsored by the United States Government. Neither the United States nor the United States Department of Energy, nor any of their employees, nor any of their contractors, subcontractors, or their employees, makes any warranty, express or implied, or assumes any liability or responsibility for the accuracy, completeness or usefulness of any information, apparatus, product or process disclosed, or represents that its use would not infringe privately owned rights.

the case of irreversible changes along a character state tree, minimizing the number of changes of character states required. A number of other parsimony methods have since appeared in the systematic literature (Kluge and Farris, 1969; Farris, 1969, 1970, 1972, 1977; Farris, Kluge, and Ekchardt, 1970) and parsimony methods have also found widespread use in studies of molecular evolution (Fitch and Margoliash, 1967, 1970; Dayhoff and Eck, 1968; see also Fitch, 1973). Cavalli-Sforza and Edwards (1967; Edwards and Cavalli-Sforza, 1964) earlier formulated a minimum evolution method for continuous character data.

An alternative methodology for phylogenetic inference is the compatability method, introduced by Le Quesne (1969, 1972). He suggested that phylogenetic inference be based on finding the largest possible set of characters which could simultaneously have all states be uniquely derived on the same tree. The estimate of the phylogeny is then taken to be that tree. While Le Quesne's specific suggestions as to how this might be done have been criticized by Farris (1969), his general approach, which is based on Camin and Sokal's (1965) concept of the compatibility of two characters, has been made rigorous and extended in a series of papers by G. F. Estabrook, C. S. Johnson, Jr., and F. R. McMorris (Estabrook, 1972; Estabrook, Johnson, and McMorris 1975, 1976a, 1976b; Estabrook and Landrum, 1975).

There has been relatively little examination of the properties of parsimony or compatibility methods as methods of statistical inference. Farris (1973, 1977) has shown that a number of different parsimony methods produce maximum likelihood estimates of an "evolutionary hypothesis" consisting of a phylogeny along with the reconstructed states of the characters in a large number of ancestral populations. However, when the object is to estimate only the phylogeny, the Camin-Sokal method has not been proven to give a maximum likelihood estimate except in the case when the probabilities of change in the character states are known to be small (Felsenstein, 1973).

For a given probabilistic model of evolution, one can construct a maximum likelihood estimate of the phylogeny, given the observed data on a set of discrete characters. Phylogenies constructed by the proper maximum likelihood method typically have the property of *consistency*. A statistical estimation method has the property of consistency when the estimate of a quantity is certain to converge to its true value as more and more data are accumulated. The purpose of this paper is to show that parsimony methods (as exemplified by the criterion of Camin and Sokal and by Farris's unrooted tree method) as well as compatibility methods do not possess the property of consistency in all cases. This is done by constructing a particular three-species case in which lack of consistency can be proven, a case in which parallel evolution is relatively probable. In finding such a case, we have thereby also shown that Farris's (1973) maximum likelihood estimate of the "evolutionary hypothesis" can give an inconsistent estimate of the phylogeny, since it always gives the same estimate as a parsimony method. Although it had been suspected that Farris's estimate of the phylogeny might be inconsistent, it was previously known only that it was not the same as direct maximum likelihood estimation of the phylogeny (Felsenstein, 1973), and no actual proof of its inconsistency had been made.

The result may be regarded as warning us of the weakness of parsimony and compatibility methods. Alternatively, the conditions which must hold in order to have lack of consistency may be regarded as so extreme that the result may be taken to be a validation of parsimony or compatibility approaches. Readers must decide for themselves. In either case the conclusion reached will have the merit of being based on an examination of the properties of phylogenetic methods when considered as methods of statistical inference. Systematists may be tempted to reject this sort of attempt to evaluate phylogenetic methods by the criteria of statistical inference, particularly in view of the oversimplified models of evolution used here. It would seem difficult to take such a reaction seriously if unaccompanied by an attempt to erect a more adequate set of criteria, or to use the present criteria to examine more realistic models of evolution.

To show that a parsimony or compatibility method does not yield a consistent estimate of the phylogeny, it is not sufficient simply to show that it does not yield a maximum likelihood estimate. There are many examples known in statistics of consistent estimation methods which are not maximum likelihood estimates. For example, in samples drawn independently from a normal distribution, the maximum likelihood estimate of the mean of the underlying normal distribution is the sample mean. But the sample median is also a consistent estimator of the true mean. As more and more points are collected, it too will approach the true mean. By analogy to this case it might be argued that, although parsimony and compatibility estimates of the phylogeny are not maximum likelihood estimates, they do provide consistent estimates of the phylogeny. While this will often be the case, we shall see that this conjecture is not always true.

## THE EXAMPLE

The example involves characters each of which has two states, 0 and 1. The ancestral state in each character is 0 and the derived state is 1. It is possible for the state of a population to change from 0 to 1, but not to revert from state 1 to state 0. Suppose that we have observed three species A, B, and C and that the (unknown) true phylogeny is as given in Fig. 1. Once a character is in state 1 at the beginning of a segment of the tree, it will not change thereafter, so that all we need to know for each segment is the probability that a character which is in state 0 at the beginning of the segment will have changed to state 1 by the end of the segment. These probabilities are assumed to be the same for all characters in this particular case; they are the quantities P, Q, and R shown in Fig. 1 next to the segments. In this particular case, the probabilities of change are assumed to be the same in segments II and IV of the tree, and the same in segments III and V. This is done purely to make the algebra easier: this assumption could be relaxed somewhat without

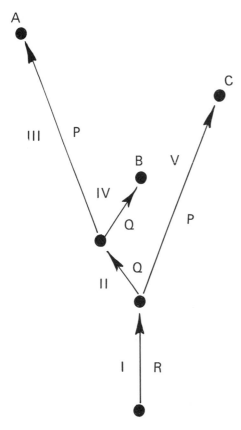

Figure 34-1. An evolutionary tree with three tip species. The segments of the tree are numbered I through V, and next to each is shown the probability of change from state 0 to state 1 in the segment.

ences from segment to segment of a sort which affects all characters. This amounts to the assumption that there are true differences in the overall rates of evolution of different lineages.

In Fig. 1 the segments of the tree are also numbered with Roman numerals. Knowing the probability of $0 \to 1$ change in each segment, we can easily obtain the probabilities of each of the possible combinations of states in the tips. For example, for the three tip species to be in states 1, 1, and 0 respectively, there must have been no change from state 0 in segments I and V. There may have been a $0 \to 1$ change in segment II, or else no change in that segment but $0 \to 1$ changes in both segments III and IV. The probability of observing states 1, 1, and 0 is thus

$$P_{110} = (1 - R)[Q + (1 - Q)PQ](1 - P) \tag{1}$$

Similarly, we can compute the probabilities of all eight possible configurations of character states:

$$P_{000} = (1 - P)^2 (1 - Q)^2 (1 - R) \tag{2a}$$
$$P_{001} = P(1 - P)(1 - Q)^2 (1 - R) \tag{2b}$$
$$P_{010} = (1 - P)^2 Q(1 - Q)(1 - R) \tag{2c}$$
$$P_{011} = P(1 - P)Q(1 - Q)(1 - R) \tag{2d}$$
$$P_{100} = P(1 - P)(1 - Q)^2(1 - R) \tag{2e}$$
$$P_{101} = P^2 (1 - Q)^2 (1 - R) \tag{2f}$$
$$P_{110} = (1 - P)[Q + (1 - Q)PQ](1 - R) \tag{2g}$$
$$P_{111} = PQ[P(1 - Q) + 1](1 - R) + R \tag{2h}$$

### RESULTS OF THE CAMIN-SOKAL PARSIMONY METHOD

If we examine N characters in these three species, we can count how many of the characters are in each of the eight possible combinations: 000, 001, 010, $\cdots$, 111. Let us call the resulting numbers of characters $n_{000}$,

altering the qualitative conclusions. It is important to realize that the constancy of P, Q, and R from character to character, and the differences between them from segment to segment, amount to strong assumptions about the biological situation. The differences in the probability of change may be due to the segments' being of different length in time (so that the tip species are not contemporaneous). Alternatively, they may be due to differences in the rate of evolution per unit time, differ-

$n_{001}, \cdots, n_{111}$. We can use these numbers to discover what will be the result of applying the Camin-Sokal parsimony method of these data. When a character has the configuration 000, then no matter which phylogeny we propose, no changes of character state will be required to explain the evolution of this character along that phylogeny. There are four other configurations of the data which will require only one character state change to be assumed, no matter what phylogeny is postulated. These are 001, 010, and 100, which require one character state change on the segment of the evolutionary tree leading to a single species, as well as 111, which requires a single change at the root of the tree.

The remaining three configurations, 110, 101, and 011, will require different numbers of changes of state on different phylogenies. Let us represent the three possible bifurcating phylogenies as (AB)C, A(BC), and (AC)B, placing parentheses around monophyletic groups. On the phylogeny (AB)C, the configuration 110 requires only one change while the others require two changes. If we let

$$S = n_{001} + n_{010} + n_{100} + n_{111}$$
$$+ 2(n_{110} + n_{101} + n_{011}), \qquad (3)$$

then (AB)C requires $S - n_{110}$ changes of state to be assumed. By similar logic, A(BC) requires $S - n_{011}$ changes, and (AC)B requires $S - n_{101}$ changes. Which tree we estimate depends on which requires us to assume the fewest changes of character state. We can immediately see that the Camin-Sokal parsimony method will estimate the correct phylogeny as (AB)C if and only if $n_{110} \geq n_{101}, n_{011}$. When $n_{011}$ is the greatest of these three numbers A(BC) will be the estimate, and when

$n_{101}$ is the greatest the estimate will be (AC)B. When there are ties for the greatest of $n_{110}$, $n_{101}$, and $n_{011}$, there will be two or more possible estimates.

## INCONSISTENCY OF THE RESULT

We assume that the N characters have evolved independently of one another, and have been chosen for study without regard to the configuration of their character states in these three species. Each character may be regarded as falling independently into one of the eight configurations 000, $\cdots$, 111 with probabilities $P_{000}, \cdots, P_{111}$. So the $n_{ijk}$ are drawn from a multinomial distribution with these probabilities.

In such a case, an elementary application of the Strong Law of Large Numbers (e.g., Feller, 1957:243–244) tells us that as we let N→∞, $n_{ijk}/N \to P_{ijk}$ for all configurations ijk. In particular, this implies that as we score more and more characters, $n_{110}$ will ultimately become larger and remain larger than either $n_{101}$ or $n_{011}$ if and only if $P_{110} > P_{101}, P_{011}$. Whichever of these three probabilities is largest determines which of the three bifurcating phylogenies is certain to be the ultimate estimate as we accumulate more and more characters. Thus the condition for the Camin-Sokal estimate to have the property of consistency is simple: that $P_{110}$ be greater than or equal to both $P_{101}$ and $P_{011}$. Note in particular that if this condition does not hold, the consequences are striking: if, say $P_{101} > P_{110}, P_{011}$, then as we accumulate more and more information the Camin-Sokal parsimony method is *increasingly certain to give the wrong answer*, in this case (AC)B.

We now examine the conditions on

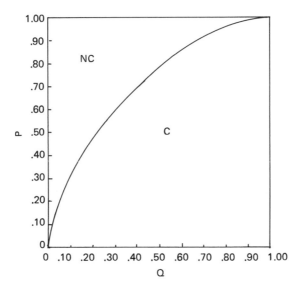

Figure 34-2. Values of P and Q for which the Camin-Sokal method fails to be consistent in the present case. C denotes the region of consistency, NC the region of inconsistency. Their boundary is the curve relating $P_1$ to Q.

P and Q which are required in order to have inconsistency of the Camin-Sokal parsimony methods. The three quantities $P_{110}$, $P_{101}$, and $P_{011}$ are given by the expressions (2g), (2f), and (2d). Note that all of these quantities contain a common factor of $(1 - R)$. Provided that $R < 1$ (which we assume), this factor can be dropped. The condition $P_{110} \geqslant P_{011}$ then becomes

$$(1 - P)[Q + (1 - Q)PQ] \geqslant P(1 - P)Q(1 - Q) \qquad (4)$$

which simplifies to
$$Q(1 - P) \geqslant 0. \qquad (5)$$

This will always hold, so in the present case it will always be true that $P_{110} \geqslant P_{011}$. Now we need only inquire whether $P_{110} \geqslant P_{101}$. This is the same as asking whether

$$(1 - P)[Q + (1 - Q)PQ] \geqslant P^2(1 - Q)^2, \qquad (6)$$

which is equivalent to requiring that
$$0 \geqslant P^2(1 - Q) + PQ^2 - Q. \qquad (7)$$

Let us view this as a quadratic equation in P whose coefficients depend on Q. Since $1 - Q > 0$ (which we assume), the quadratic in (7) has a minimum at $P = -Q/(1 - Q)$. Since this is never positive, the positive values of P for which (7) is satisfied are those values of P below the point where the quadratic function is zero:

$$P \leqslant P_1 = (-Q^2 + [Q^4 + 4Q(1 - Q)]^{1/2} /2(1 - Q) \qquad (8)$$

$P_1$ is always a real number, so no complications arise. Figure 2 shows $P_1$ plotted for values of Q between 0 and 1. $P_1$ rises from 0 to 1 as Q goes from 0 to 1. Above the $P_1$ curve is the region of values of P for which $P_{110} < P_{101}$.

This is the region in which the Camin-Sokal parsimony method is guaranteed to converge to the wrong estimate of the tree as we accumulate more and more data. Note that for every possible value of Q there is a range of values of P in which we will encounter this unpleasant behavior. A similar statement holds if we rearrange (7) to obtain limits on the values of Q as a function of P, so that for every value of P there is a range of Q values in which this unpleasant behavior occurs. Note that for small Q, the condition (8) is closely approximated by

$$P \leqslant Q^{1/2} \qquad (9)$$

and for Q near 1 it is closely approximated by

$$P \geqslant 1 - (1 - Q)^2 \qquad (10)$$

The effect of (8) is that the Camin-Sokal method will tend to fail when there is a sufficient disproportion between P and Q, which is the same as requiring that there be a sufficiently great disproportion between the lengths of the long and the short segments of the tree in Fig. 1. In a previous paper (Felsenstein, 1973), I showed that for sufficiently small probabilities of evolutionary change, the Camin-Sokal method yields a correct maximum likelihood estimate of the phylogeny, and hence would be consistent. This might appear to be contradicted by (8) and (9), since these show that the Camin-Sokal method can be inconsistent even when P and Q are small. But my earlier proof involved holding the lengths (in time) of the segments of the tree constant while letting the rate of change in the characters become small. This is equivalent to holding the ratio of P to Q constant while letting both approach zero. As will be apparent from dividing

both sides of (9) by Q, when this is done the values of P and Q enter the region of consistency for sufficiently small values of P and Q, no matter what the (constant) ratio P/Q. So in this sense, the Camin-Sokal method works for sufficiently small rates of character state change.

### COMPATIBILITY METHODS

It is a convenient fact that precisely the same three-species example also allows us to find conditions in which the compatibility methods yield an inconsistent estimate of the phylogeny. While the original approach of Le Quesne assumed that the direction of character state change was unknown, and could not be applied to a three-species case, the extensions of the compatiblility approach by Estabrook, Johnson, and McMorris do allow us to make inferences in a three-species case when the direction of change on the character-state trees is known. For example, if two binary characters have the states (1,1), (1,0), and (0,1) respectively in the three species, then it is impossible for the transition $0 \rightarrow 1$ to have taken place only once in each character on a branching phylogeny.

A pairwise consideration of all of the eight possible outcomes of the data will show that the outcomes 110, 101, and 011 are mutually incompatible, but that all other combinations are compatible. If we are trying to find the phylogenies suggested by the largest possible set of mutually compatible characters, these will include (AB)C if and only if $n_{110} \geqslant n_{101}$, $n_{011}$. Thus, the compatibility method for rooted binary character-state trees will give the same estimate as the Camin-Sokal method in the three-species case. We thus can apply all

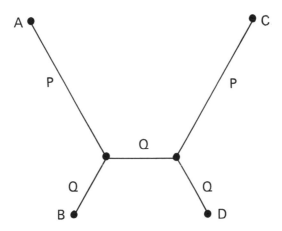

Figure 34-3. True unknown phylogeny (with root omitted) used to find cases in which unrooted Wagner tree parsimony methods will be inconsistent.

of the above conditions for inconsistency of the Camin-Sokal method to the compatibility approach. This allows the conclusion that consistency is not a general property of the compatibility methods, but must be proven for specific probability models of evolution if it is desired.

### UNROOTED WAGNER TREES

One of the most widely used parsimony methods has been Farris's (1970) method of inferring unrooted evolutionary trees under the assumption that character-state changes are reversible. The consistency of this method can be investigated by an extension of the present approach to a four-species case. This is necessary because there is only one possible unrooted tree in the three-species case, rendering it trivial. Figure 3 shows an unrooted tree with four species, A, B, C, and D. In order to more closely approximate the evolutionary model which underlies Farris's method, we assume that although the characters were originally in state 0, they have the same probability of reversion $1 \to 0$ once they are in state 1, as they have of origination $0 \to 1$ of state 1 when they are in state 0. Thus each segment of the evolutionary tree is characterized by a probability of character state change which applies equally to both forward change $0 \to 1$ and reversion $1 \to 0$. Once again, we assume for simplicity that characters are independently sampled and all have the same probabilities. There are 16 possible data outcomes, 0000 through 1111. Once again, the outcome of applying Farris's parsimony method will depend only on the numbers of characters $n_{0000}, \cdots, n_{1111}$ having each outcome.

It is easy to show, along the same lines as before, that whether the unrooted tree obtained is of form (AB)(CD), (AC)(BD), or (AD)(BC) is determined by which of the three numbers $n_{1100} + n_{0011}$, $n_{1010} + n_{0101}$, and $n_{1001} + n_{0110}$ is largest. It is not difficult to demonstrate that the exact placement of the root of the true tree will affect only the relative probabilities of obtaining 1100 and 0011, but will leave the total probability $P_{1100} + P_{0011}$ unchanged, and similarly for $P_{1010} + P_{0101}$ and $P_{1001} + P_{0110}$. Therefore, we need not specify the placement of the root on the (unknown) true tree to compute the probabilities which determine the outcome of this parsimony method. Suppose that the true phylogeny is one whose unrooted form is given in Fig. 3. We may as well assume that the root is at the left-hand end of the central segment, and that all characters start there in state 0, as these assumptions do not affect $P_{1100} + P_{0011}$ and the other relevant probabilities.

Considering the two possible character states at the right-hand end of the central segment, we find that

$$P_{1100} + P_{0011} = PQ[1 - Q]^2(1 - P) + Q^2 P]$$
$$+ (1 - P)(1 - Q)$$
$$\cdot [Q(1 - Q)(1 - P)$$
$$+ Q(1 - Q)P] \qquad (11)$$

with analogous expressions for the other two relevant probabilities:

$$P_{1010} + P_{0101} = P(1 - Q)$$
$$\cdot [Q^2(1 - P) + (1 - Q)^2 P]$$
$$+ (1 - P)$$
$$\cdot Q[Q(1 - Q)P$$
$$+ Q(1 - Q)(1 - P)]$$
$$\qquad (12)$$

and

$$P_{1001} + P_{0110} = P(1 - Q)$$
$$\cdot [Q(1 - Q)P$$
$$+ Q(1 - Q)(1 - P)]$$
$$+ Q(1 - P)$$
$$\cdot [(1 - Q)^2 P + Q^2(1 - P)].$$
$$\qquad (13)$$

After some elementary but tedious algebra it can be shown from (12) and (13) that provided that $Q \leqslant \frac{1}{2}$, which we assume,

$$P_{1010} + P_{0101} \geqslant P_{1001} + P_{0110}. \qquad (14)$$

This establishes that when the true tree is as shown in Fig. 3, our estimate of the unrooted tree topology may converge to either (AB)(CD) or to (AC)(BD), but never to (AD)(BC) as we collect more and more characters. So to establish the consistency of the estimation of unrooted tree topology, we need only enquire whether

$$P_{1100} + P_{0011} \geqslant P_{1010} + P_{0101}, \qquad (15)$$

which will be the condition for consistency. Using (11) and (12) we find after further tedious algebra that (15) is simply,

$$2P^2 Q - P^2 + 2Q^3 - 3Q^2 + Q \geqslant 0 \quad (16)$$

which is

$$(2Q - 1)(P^2 + Q(Q - 1)) \geqslant 0. \qquad (17)$$

Since $Q \leqslant \frac{1}{2}$, (17) is simply

$$P^2 \leqslant Q(1 - Q), \qquad (18)$$

a considerably simpler condition than (8). Note that when $Q$ is small, (18) reduces to (9). Thus, all the statements about consistency in the Camin-Sokal case when $P$ and $Q$ are small are also correct in the case of unrooted Wagner trees.

## DISCUSSION

We have seen that there are circumstances under which three different estimation methods are not statistically consistent, these being the Camin-Sokal parsimony method, the Estabrook-Johnson-McMorris compatibility method, and Farris's parsimony method for estimating unrooted Wagner trees. For small values of $P$ and $Q$, the condition for inconsistency amounts to requiring that simultaneous changes on two long segments of the tree be more probable a priori than one change on a short segment. This amounts to requiring that parallelism of changes be more probable than unique and unreversed change in an informative part of the tree (e.g., that simultaneous changes in segments III and V of the tree in Fig. 1 be more probable than a single change in segment II). This certainly seems like a reasonably intuitive condition for inconsistency. The advantage of the argument presented here lies not in leading to a particularly surprising conclusion, one that will cause abandonment of these parsimony and compatibility methods, but as a formal investigation of one of the statistical properties of phylogenetic inference methods.

The models employed here certainly have severe limitations: it will hardly ever be the case that we sample characters independently, with all of the characters following the same probability model of evolutionary change. Extending this analysis to more realistic evolutionary models will be difficult. Yet the task must be undertaken: if inconsistency of a parsimony or compatibility technique is suspected, it does little good simply to point out that the evolutionary models employed here do not apply to the type of data being encountered in practice. That amounts to a confession of ignorance rather than validation of the inference method in question.

### LIKELIHOOD METHODS

Methods of phylogenetic inference which entirely avoid the problem of statistical inconsistency are already known. Maximum likelihood estimation of the phylogeny is one of them.

I have outlined elsewhere (Felsenstein, 1973) how this may be done. In the three species cases maximum likelihood estimation methods can easily be developed. The likelihood of a tree will simply be

$$L = \prod_{ijk} P_{ijk}{}^{n_{ijk}}, \qquad (19)$$

where $P_{ijk}$ is the probability of data configuration ijk and $n_{ijk}$ is the number of characters having that configuration. Estimation is carried out by maximizing (19) over the unknown parameters of the evolutionary model (such as P and Q in equations [2]). This is done for each tree topology, and the final estimate consists of the topology and the evolutionary parameters which yield the highest likelihood. (Note that despite the connotations of the term, the likelihood of a tree is *not* the probability that it is the correct tree.) When there are larger numbers of species, the number of possible data configurations (the numbers of terms $n_{ijk}$) in each character becomes so large that it is impractical to use equation (19). I have presented eleswhere (Felsenstein, 1973) an algorithm for evaluating the likelihood of a tree which avoids this difficulty.

Maximum likelihood estimates are not desirable in themselves, but because they have desirable statistical properties such as consistency and asymptotic efficiency. In the case of discrete multistate characters under the sorts of evolutionary model considered here, it can be shown quite generally that the maximum likelihood estimation procedure has the property of consistency. In particular, in the case of the tree shown in Fig. 1, it will be a consistent method whatever the values of P, Q, and R.

The reader familiar with the paper of Farris (1973), which establishes a general correspondence between parsimony methods and maximum likelihood methods may be puzzled at this stage: if parsimony methods are maximum likelihood methods, why have the two been described here as separate methods? Why is one sometimes not consistent while the other is always consistent? This paradox is resolved once one recalls that the maximum likelihood methods used by Farris are different from those described in Felsenstein (1973) and here. Farris used the maximum likelihood method to estimate not only the parameters of the evolutionary tree, but also the states of the characters in a large number of ancestral populations. When this latter kind of maximum likelihood estimate is made, the number of parameters being estimated rises without limit as more characters are examined.

From the point of view of estimating the phylogeny, these extra parameters are "nuisance" parameters. As a result of their presence, the ratio between the number of data items and the number of parameters does not increase indefinitely as more characters are added. It is in situations such as this that maximum likelihood methods are particularly prone to lack of consistency, as I have previously pointed out (Felsenstein, 1973). Indeed, the present results establish that there are conditions under which Farris's likelihood method (giving the same results as a parsimony method) fails to be consistent.

## PERSPECTIVE

The weakness of the maximum likelihood approach is that it requires us to have a probabilistic model of character evolution which we can believe. The uncertainties of interpretation of characters in systematics are so great that this will hardly ever be the case. We might prefer to have methods which, while not statistically optimal for any one evolutionary model, were robust in that they had reasonable statistical properties such as consistency for a wide variety of evolutionary models. The present results establish that parsimony and compatibility methods can fail to be consistent if parallelism is expected to occur frequently. This helps establish that they do not yield maximum likelihood estimates. However, they pass the test of consistency when parallelism is rare. This leaves them as viable candidates for robust methods. Establishing that robustness (or disproving it) by examining a wider range of models is a daunting task, but it must be undertaken. If phylogenetic inference is to be a science, we must consider its methods guilty until proven innocent.

## ACKNOWLEDGMENTS

I wish to thank a reviewer for penetrating comments which were of great assistance. This work was supported by Department of Energy Contract No. EY–76–S–06–2225 5 with the University of Washington.

## REFERENCES

Camin, J. H., and R. R. Sokal, 1965, A method for deducing branching sequences in phylogeny. *Evolution,* 19: 311–326.

Cavalli-Sforza, L. L., and A. W. F., Edwards, 1967, Phylogenetic analysis: Models and estimation procedures. *Evolution,* 32: 550–570 (also in *Amer. J. Human Genetics,* 19: 233–257).

Dayhoff, M. O., and R. V. Eck, 1968, Atlas of protein sequence and structure 1967–1968. National Biomedical Research Foundation, Silver Spring, Md.

Edwards, A. W. F., and L. L. Cavalli-Sforza, 1964, Reconstruction of evolutionary trees. In Heywood, V. H., and J. McNeill (eds.), *Phenetic and Phylogenetic Classification,* Systematics Association, Publication No. 6, London, pp. 67–76.

Estabrook, G. F., 1972, Cladistic methodology: a discussion of the theoretical basis for the induction of evolutionary history. *Ann. Rev. Ecol. Syst.,* 3: 427–456.

Estabrook, G. F., C. S. Johnson, Jr., and F. R. McMorris, 1975, An idealized concept of the true cladistic character. *Math. Biosci.,* 23: 263–272.

Estabrook, G. F., and L. Landrum, 1975, A simple test for the possible simultaneous evolutionary divergence of two amino acid positions. *Taxon,* 24: 609–613.

Estabrook, G. F., C. S. Johnson, Jr., and F. R. McMorris, 1976a, A mathematical foundation for the analysis of character compatibility. *Math. Biosci.,* 29: 181–187.

——, 1976b, An algebraic analysis of cladistic characters. *Discrete Math.,* 16: 141–147.

Farris, J. S., 1969, A successive approximations approach to character weighting. *Syst. Zool.,* 18: 374–385.

——, 1970, Methods for computing Wagner trees. *Syst. Zool.,* 19: 83–92.

Farris, J. S., A. G. Kluge, and M. J. Eckhardt, 1970, A numerical approach to phylogenetic systematics. *Syst. Zool.,* 19: 172–189.

Farris, J. S., 1972, Estimating phylogenetic trees from distance matrices. *Amer. Nat.* 106: 645–668.

——, 1973, On the use of the parsimony criterion for inferring evolutionary trees. *Syst. Zool.,* 22: 250–256.

——, 1977a, Phylogenetic analysis under Dollo's Law. *Syst. Zool.,* 26: 77–88.

——, 1977b, Some further comments on Le Quesne's methods. *Syst. Zool.,* 26: 220–223.

Feller, W., 1957, An introduction to probability theory and its applications. Volume I. Second edition. Wiley, New York.

Felsenstein, J., 1973, Maximum likelihood and minimum-steps methods for estimating evolutionary trees from data on discrete characters. *Syst. Zool.,* 22: 240–249.

Fitch, W. M., and E. Margoliash, 1967, Construction of phylogenetic trees. *Science,* 155: 279–284.

Fitch, W. M., 1973, Aspects of molecular evolution. *Ann. Rev. Gen.,* 7: 343–380.

Kluge, A. G., and J. S. Farris, 1969, Quantitative phyletics and evolution of anurans. *Syst. Zool.,* 18: 1–32.

Le Quesne, W. J., 1969, A method of selection of characters in numerical taxonomy. *Syst. Zool.,* 18: 201–205.

——, 1972, Further studies based on the uniquely derived character concept. *Syst. Zool.,* 21: 281–288.

Morris, F. R., 1977, On the compatibility of binary qualitative taxonomic characters. *Bull. Math. Biol.,* 39: 133–138.

# 35

# The Logical Basis
# of Phylogenetic Analysis

## JAMES FARRIS

### INTRODUCTION

Phylogeneticists hold that the study of phylogeny ought to be an empirical science, that putative synapomorphies provide evidence on genealogical relationship, and that (aside possibly from direct observation of descent) those synapomorphies constitute the only available evidence on genealogy. Opponents of phylogenetic systematics maintain variously that genealogies cannot (aside from direct observation) be studied empirically, that synapomorphies are not evidence of kinship because of the possibility of homoplasy, or that raw similarities also provide evidence on genealogy. Most phylogeneticists recognize that inferring genealogy rests on the principle of parsimony—that is, choosing genealogical hypotheses so as to minimize requirements for ad hoc hypotheses of homoplasy. But other criteria as well have been proposed for phylogenetic analysis, and some workers believe that parsimony is unnecessary for that purpose. Others contend that

that principle is not truly "parsimonious," or that its application depends crucially on the false supposition that homoplasy is rare in evolution. Authors of all these criticisms have in common the view that phylogenetic systematics as it is now practiced may be dismissed as futile or at best defective. Phylogeneticists must refute that view, but accomplishing that goal seems complicated both by the apparent multiplicity of phylogenetic methods and by the diversity of the objections. I shall show here that the complexity of this problem is superficial. An analysis of parsimony will not only provide a resolution of the objections to that criterion, but will supply as well an understanding of the relationship of genealogical hypotheses to evidence, and with it a means of deciding among methods of phylogenetic inference.

### AD HOC HYPOTHESES

I share Popper's disdain for arguing definitions as such, but it is important to make intended meanings clear, and so I shall first dismiss terminological objections to the parsimony criterion. These all come to the idea that parsimonious phylogenetic reconstructions are so primarily by misnomer: the word might equally well refer to any of several other qualities. The meanings of "parsimony" would surely take volumes to discuss, but doing so would be quite pointless. Whether the word is used in the same way by all has no bearing on whether the phylogenetic usage names a desirable quality. I shall use the term in the sense I have already mentioned: most parsimonious genealogical hypotheses are those that minimize requirements for ad hoc hypotheses of homoplasy. If minimizing ad hoc hypotheses is not the only connotation of "parsimony" in general usage, at least it is scarcely novel. Both Hennig (1966) and Wiley (1975) have advanced ideas closely related to my usage. Hennig defends phylogenetic analysis on grounds of his auxiliary principle, which states that homology should be presumed in the absence of evidence to the contrary. This amounts to the precept that homoplasy should not be postulated beyond necessity, that is to say, parsimony. Wiley discusses parsimony in a Popperian context, characterizing most parsimonious genealogies as those that are least falsified on available evidence. In his treatment, contradictory character distributions provide putative falsifiers of genealogies. As I shall discuss below, any such falsifier engenders a requirement for an ad hoc hypothesis of homoplasy to defend the genealogy. Wiley's concept is then equivalent to mine.

Cartmill (1981) has effectively objected to that last equivalence, claiming

that neither phylogenetic analysis nor parsimony can be scientific in Popper's sense. His argument is superficially technical, but his principal conclusion is in fact based on a terminological confusion, and so I shall discuss his ideas here.

Cartmill cites Gaffney (1979) to the effect that character distributions are falsifiers of genealogical hypotheses, and that it is possible that every conceivable genealogy will be falsified at least once. From the first of these admissions he "deduces" that Gaffney must have relied on the "theorem" that any genealogy contradicted by a character distribution is false. Cartmill then reasons: Some genealogy must be true. "Gaffney's" theorem, together with a falsifier for every genealogy, implies that every genealogy is false. Therefore, Gaffney's claim that character distributions are falsifiers is false.

Cartmill's argument rests directly and entirely on a misrepresentation of the Popperian meaning of "falsifier:" a test statement that, if true, allows a hypothesis to be rejected. There is a great difference between "falsify" in Popper's sense and "prove false." The relationship between a theory and its falsifiers is purely logical; Popper never claimed that proof of falsity could literally be achieved empirically. "Observing" a falsifier of a theory does not prove that the theory is false; it simply implies that either the theory or the observation is erroneous. It is then seen that the only implication that can be derived from falsification of every genealogy is that some of the falsifiers are errors—homoplasies. It is thus seen as well that Cartmill's "syllogism" is nothing other than an equivocation.

So much for the claim that characters cannot be Popperian falsifiers, but is phylogenetic parsimony Popperian? Cartmill admits that phylogeneticists hold that the least falsified genealogy is to be preferred. The reason for this preference is that each falsifier of any accepted genealogy imposes a requirement for an ad hoc hypothesis to dispose of the falsifier. According to Popper—as Cartmill also cites—ad hoc hypotheses must be minimized in scientific investigation. Cartmill never attempts to argue that conflicts between characters and genealogies do not require hypotheses of homoplasy, and so none of his claims can serve to question the connection between parsimony and Popper's ideas.

## PARSIMONY AND SYNAPOMORPHY

The objection that parsimony requires rarity of homoplasy in evolution is usually taken to be just that: a criticism of parsimony. It might seem that the problem posed by that objection could be avoided simply by using some other criterion for phylogenetic analysis. Some quite non-phylogenetic proposals, such as grouping according to raw similarity, have been made along those lines, and I shall discuss those eventually. Of more immediate interest is the question whether grouping by putative synapomorphy can do without the parsimony criterion.

Watrous and Wheeler (1981) suggest that parsimony is needed only when characters conflict, with the implication that a set of congruent characters can be analyzed without avoiding ad hoc hypotheses of homoplasy. A similar idea would appear to underlie advocacy by Estabrook and others (reviewed by Farris and Kluge,

1979) of techniques ("clique" methods) that "resolve" character conflicts by discarding as many characters as necessary so that those surviving (the clique) are mutually congruent. The surviving characters are then used to construct a tree. Proponents of such methods maintain that the tree so arrived at rests on a basis different from parsimony.

The character selection process itself may well have a distinctive premise, a possibility that I shall discuss below. To claim that the interpretation of the characters selected rests on a basis other than parsimony, however, seems not to be defensible. The tree constructed from a suite of congruent characters by a clique method is chosen to avoid homoplasy in any of those characters, the possibility of doing so being assured by the selection. (Selection aside, Watrous and Wheeler proceed likewise.) It seems accurate, then, to describe that construction as minimizing requirements for ad hoc hypotheses of homoplasy for the characters within the congruent suite, but, more particularly, there seems to be no other sensible rationale for the construction. No one seems to have suggested any such principle, aside from the obvious: that if the characters were free of homoplasy (were "true" as it is often put), then the tree would follow. But the characters comprising a congruent suite are hardly observed to be free of homoplasy. At the most it might be said that the selected characters seem to suggest no genealogy other than the obvious one.

Of course it is what data suggest, or how they do it, that is at issue. If a suite of congruent characters is interpreted by avoiding unnecessary postulates of homoplasy, then the interpretation embodies parsimony. But the only apparent motivation for concentrating just on congruent characters is to avoid reliance on parsimony. That avoidance would seem sensible only on the supposition that parsimony is ill founded, and the only apparent reason for that supposition is the charge that parsimony depends crucially on unrealistic assumptions about nature. If that charge means anything at all, it must mean that taking conditions of nature realistically into account would lead to preference for a less parsimonious arrangement over a more parsimonious one. But if that charge were correct, then it would be—to say the least—less than obvious why the implications of those natural conditions would be expected to change simply because any characters incongruent with those chosen had been ignored.

If avoiding ad hoc hypotheses of homoplasy is unjustified, then neither Watrous and Wheeler nor clique advocates are entitled to the inferences on phylogeny that they draw, but the significance of parsimony for Hennigian methods is much more general than that. Watrous and Wheeler probably thought that they had no need of anything so questionable as parsimony, because they were simply applying Hennig's well established principle of grouping according to synapomorphy. Just how did that principle come to be well established? It is usually explained by taking note of the logical relationship between monophyletic groups and true synapomorphies, but that leaves open the question of how genealogies are related to observed features. It might well be questioned whether the logical con-

struct can legitimately be extended into a principle to guide interpretation of available characters. That question has in fact often been raised, and almost always in the form of the suggestion that putative synapomorphies are not evidence of kinship because they might well be homoplasies. Hennig's (1966) own reply to that objection was his auxiliary principle, which, as I have already observed, is a formulation of the parsimony criterion.

Hennig's defense of the synapomorphy principle by recourse to parsimony is not accidental, but necessary. The analytic relationship of correct synapomorphies to phylogeny is just that a property that evolved once and is never lost must characterize a monophyletic group. Synapomorphies are converted into a genealogy, that is, by identifying the tree that allows a unique origin for each derived condition. A phylogeny based on observed features is parsimonious to the degree that it avoids requirements for homoplasies—multiple origins of like features. Secondary plesiomorphies aside, a plesiomorphic trait will already have a single origin at the root of the putative tree, so that the effect of parsimony is precisely to provide unique derivations wherever possible. (Secondary plesiomorphies, being a kind of apomorphy, are treated likewise.) Grouping by synapomorphy would thus have to behave like parsimony, but further, the latter applies to actual traits, whereas the logic of true synapomorphies does not. Superficially, the use of the synapomorphy principle in phylogenetic inference seems to be just a consequence of the logical connection between true synapomorphies and genealogies, but it cannot be just that, as the condition of that logic— that the traits are indeed synapomorphies—need not be met. Grouping by putative synapomorphy is instead a consequence of the parsimony criterion.

### ABUNDANCE OF HOMOPLASY

There are two main varieties of the position that use of the parsimony criterion depends crucially on the supposition that homoplasy is rare in evolution. In the first, the observation that requirements for homoplasy are minimized is taken as prima facie evidence that the supposition is needed. In the second, the claim is advanced in conjunction with some more elaborate, often statistical, argument. The conclusion from the first kind of reasoning is quite general, while that from the second is necessarily limited by the premises of the argument employed. If the first kind of criticism were correct, there would be little point to considering arguments of the second sort. I shall thus first point out why the first type of objection rests on a fallacy.

To evaluate the claim that an inference procedure that minimizes something must ipso facto presuppose that the quality minimized is rare, it is useful to consider a common application of statistics. In normal regression analysis, a regression line is calculated from a sample of points so as to minimize residual variation around the line, and the residual variation is then used to estimate the parametric residual variance. Plainly the choice of line has the effect of minimizing the estimate of the residual variance, but one rarely hears this procedure criticized as presupposing that the parametric

residual variance is small. Indeed, it is known from normal statistical theory that the least squares line is the best point estimate of the parametric regression line, whether the residual variance is small or not. The argument that the parsimony criterion must presume rarity of homoplasy just because it minimizes required homoplasy is thus at best incomplete. That reasoning presumes a general connection between minimization and supposition of minimality, but it is now plain that no such general connection exists. Any successful criticism of phylogenetic parsimony would have to include more specific premises.

The same conclusions can readily be reached in a specifically phylogenetic context. Suppose that for three terminal taxa A, B, C, there are 10 putative synapomorphies of A + B and one putative apomorphy shared by B and C. We assume for simplicity of discussion that the characters are independent and all of equal weight, and that attempts to find evidence to support changes in the data have already failed. Parsimony then leads to the preference for ((A, B), C) over alternative groupings. We will be interested in whether abundance of homoplasy leads to preference for some other grouping. If it does not, then the claim that parsimony presupposes rarity of homoplasy is at best not generally true.

It is plain that the grouping (B, C), A) is genealogically correct if the one B + C character is, in fact, a synapomorphy, and that ((A, B), C) is instead correct if the A + B characters are synapomorphies. Truth of the latter grouping does not require, however, that all 10 of the putative synapomorphies of A + B be accurate homologies. If just one of those characters were truly a synapomorphy while all the other characters in the data were in fact parallelisms, the genealogy would necessarily be ((A, B), C). That A and B share a common ancestor unique to them, in other words, does not logically require that every feature shared by A and B was inherited from that ancestor. In the extreme, if all the characters were parallelisms, this would not imply that ((A, B), C) is genealogically false. Under those circumstances the data would simply leave the question of the truth of that (or any other) grouping entirely open.

The relationship between characters and genealogies thus shows a kind of asymmetry. Genealogy ((A, B), C) requires that the B + C character be homoplasious, but requires nothing at all concerning the A + B characters. The genealogy can be true whether the conforming characters are homoplasious or not. One kind of objection to phylogenetic parsimony runs that ad hoc hypotheses are indeed to be minimized, but this does not mean minimizing homoplasies, because a genealogy also requires ad hoc hypotheses of homology concerning the characters that conform to it. It is seen that such is not the case. Only characters conflicting with a genealogy lead to requirements for ad hoc hypotheses, and so the only ad hoc hypotheses needed to defend a genealogy are hypotheses of homoplasy.

The sensitivity of inference by parsimony to rarity of homoplasy is readily deduced from these observations. If homoplasy is indeed rare, it is quite likely with these characters that ((A, B), C) is the correct genealogy. In order for that grouping to be false,

it would be required at least that all 10 of the A + B characters be homoplasious. As these characters are supposed to be independent, the coincidental ocurrence of homoplasy in all 10 should be quite unlikely. Suppose, then, that homoplasy is so abundant that only one of the characters escapes its effects. That one character might equally well be any of the 11 in the data, and if it is any one of the 10 A + B characters the parsimonious grouping is correct. That grouping is thus a much better bet than is ((B, C), A). At the extreme, as has already been seen, if homoplasy is universal, the characters imply nothing about the genealogy. In that case the parsimonious grouping is no better founded than is any other, but then neither is it any worse founded.

It seems that no degree of abundance of homoplasy is by itself sufficient to defend choice of a less parsimonious genealogy over a more parsimonious one. That abundance can diminish only the strength of preference for the parsimonious arrangement; it can never shift the preference to a different scheme. In this the relationship of abundance of homoplasy to choice of genealogical hypothesis is quite like that between residual variance and choice of regression line. Large residual variance expands the confidence interval about the line, or weakens the degree to which the least squares line is to be preferred over nearby lines, but it cannot by itself lead to selection of some other line that fits the data even worse.

## STOCHASTIC MODELS

The supposition of abundance of homoplasy by itself offers no grounds for preferring unparsimonious arrangements, but it is easy enough to arrive at that preference by resorting to other premises. Felsenstein (1973, 1978, 1979) objects to parsimony on statistical grounds. He suggests (as others have) that genealogies ought to be inferred by statistical estimation procedures. In his approach he devises stochastic models of evolution, then applies the principle of maximum likelihood, choosing the genealogy that would assign highest probability to the observed data if the model were true. With the models that he has investigated, it develops that the maximum likelihood tree is most parsimonious when rates of change of characters are very small, under which circumstances the models would also predict very little homoplasy. He concludes from this that parsimony requires rarity of homoplasy. In his 1978 paper he discusses a model according to which both parsimony and clique methods would be certain to yield an incorrect genealogy if a large enough random sample of independent characters were obtained. He contends that maximum likelihood estimation under the same conditions would yield the correct tree, as that estimate possesses the statistical property termed consistency. That last is the logical property that if an indefinitely great number of independent characters were sampled at random from the distribution specified by the model, then the estimate would converge to the parameter of the model, the hypothetical true tree.

Felsenstein does not try to defend his models as realistic. His attitude on their purpose seems to be instead that "if a method behaves poorly in this simple model framework, this calls into question its use on real characters" (1981, p. 184), or perhaps (1978, p. 409), "If phylogenetic inference is to be a science, we must consider its methods guilty until proven innocent." The first is preposterous except on the supposition that reasoning from false premises cannot lead to false conclusions. As for the second: To the extent that these models are intended seriously, they comprise empirical claims on evolution. If science required proof concerning empirical claims in order to draw conclusions, no kind of science would be possible.

Felsenstein nonetheless apparently believed that he had demonstrated that practical application of parsimony requires rarity of homoplasy, but in fact such is hardly the case. The dependence of parsimony on rarity of homoplasy is in Felsenstein's analysis a consequence of his models. These models, as he is well aware, comprise "strong assumptions about the biological situation" (1978, p. 403). If those assumptions do not apply to real cases, then so far as Felsenstein can show, the criticism of parsimony need not apply to real cases either. But, again, Felsenstein does not maintain that the assumptions of his models are realistic. He has not shown that abundance of homoplasy implies preference for unparsimonious genealogies. Instead he has shown at most that if homoplasy were abundant, and if in addition the conditions of his models prevailed in nature, then one should prefer unparsimonious schemes. We have already seen that abundance of homoplasy by itself does not justify departure from parsimony. If Felsenstein's argument offers any reason for that departure, then that reason would have to rely on the supposition that his models apply to nature. An ironic result indeed. The original criticism of parsimony was that it required an unrealistic assumption about nature. It now seems instead that unparsimonious methods require such assumptions, whereas parsimony does not.

Felsenstein's arguments from consistency and maximum likelihood have a related drawback. Consistency is a logical relationship between an estimation method and a probability model. In the hypothetical case imagined by Felsenstein, his method would have obtained the right answer, but whether the method would work in practice depends on whether the model is accurate. If it is not, then the consistency of the estimator under the model implies nothing about the accuracy of the inferred tree. The status of a procedure as a maximum likelihood estimator is also bound to the probability model. If the model is false, the ability of a procedure to find the most likely tree under the model implies nothing of how likely the chosen tree might actually be. Likewise the conclusion that parsimony would arrive at the wrong tree depends on the model, and so the hypothetical analysis implies little about the practical accuracy of parsimony.

One might say, of course, that the model illustrates a potential weakness of parsimony: That criterion will fail if the conditions of the model should happen to be met. And how are we to know that this will not happen? This seems in fact to be the intended substance of Felsenstein's remarks. While admitting that his premises are unrealistic, he rejects realism as a criticism of his attack on parsimony, claiming that an objection based on realism "amounts to a confession of ignorance rather than a validation of the inference method in question" (1978, p. 408). A derivation that implies nothing about reality is not much of an improvement on ignorance, of course. To the extent that Felsenstein has a point, then, it seems to be just that parsimony is invalid because we cannot be certain that it will not lead to errors of inference. But there is nothing distinctive about Felsenstein's model in that regard. One may always concoct fantastic circumstances under which scientific conclusions might prove incorrect. It is hardly necessary to resort to mathematical manipulations in order to produce such fears. One need only imagine that his characters have evolved in just the right way to lead him to a false conclusion. Or, with Descartes, that his perceptions and reason have been systematically and maliciously distorted by a demon. None of these possibilities can be disproved, but it hardly matters. There is likewise no reason for accepting any of them, and so collectively they amount to no more than the abstract possibility that a conclusion might be wrong. No phylogeneticist—or any scientist—would dispute that anyway, and so such "objections" are entirely empty. That thinking provides no means of improving either conclusions or methods, but instead offers, if anything, a rejection of all conclusions that cannot be established with certainty. If that attitude were taken seriously, no scientific conclusions whatever could be drawn.

## EXPLANATORY POWER

A number of authors, myself among

them (Farris, 1973, 1977, 1978), have used statistical models to defend parsimony, using, of course, different models from Felsenstein's. Felsenstein has objected to such derivations on grounds of statistical consistency: as before, parsimonious reconstructions are not consistent under his models. That is no more than an equivocation, as the models differ, and consistency is a relationship between method and model. But I do not mean by this to defend those favorable derivations, for my own models, if perhaps not quite so fantastic as Felsenstein's, are nonetheless like the latter in comprising uncorroborated (and no doubt false) claims on evolution. If reasoning from unsubstantiated suppositions cannot legitimately question parsimony, then neither can it properly bolster that criterion. The modeling approach to phylogenetic inference was wrong from the start, for it rests on the idea that to study phylogeny at all, one must first know in great detail how evolution has proceeded. That cannot very well be the way in which scientific knowledge is obtained. What we know of evolution must have been learned by other means. Those means, I suggest, can be no other than that phylogenetic theories are chosen, just as any scientific theory, for their ability to explain available observation. I shall thus concentrate on evaluating proposed methods of phylogenetic analysis on that basis.

That ad hoc hypotheses are to be avoided whenever possible in scientific investigation is, so far as I am aware, not seriously controversial. That course is explicitly recommended by Popper, for example. No one seems inclined to maintain that ad hoc hypotheses are desirable in themselves; at most they are by-products of conclusions held worthy on other grounds. Nonetheless, I suspect that much of the criticism of the phylogenetic parsimony criterion arises from a failure to appreciate the reasons why ad hoc hypotheses must be avoided. Avoiding them is no less than essential to science itself. Science requires that choice among theories be decided by evidence, and since the effect of an ad hoc hypothesis is precisely to dispose of an observation freely, there could be no effective connection between theory and observation, and the concept of evidence would be meaningless. The requirement that a hypothesis of kinship minimize ad hoc hypotheses of homoplasy is thus no more escapable than the general requirement that any theory should conform to observation; indeed, the one derives from the other.

There are a number of properties commonly held to characterize a theory that gives a satisfactory account of observation. The theory must first of all provide a description of what is known, else it would serve little purpose. As Sober (1975) puts it, theories serve to make experience redundant. But not all descriptions are equally useful. Good theories describe in terms of a coherent framework, so that experience becomes comprehensible; in short they are explanatory. Explanations in turn may be judged on their ability to cover observations with few boundary conditions, that is, with little extrinsic information. Sober has characterized theories satisfying this goal as most informative, or simplest. All these criteria are interrelated in the case of phylogenetic inference, so that they effectively yield a single criterion of analysis. These connections have already been recognized, and I shall summarize them only briefly.

I have elsewhere (Farris, 1979, 1980, 1982) already analyzed the descriptive power of hierarchic schemes. I showed that most parsimonious classifications are descriptively most informative in that they allow character data to be summarized as efficiently as possible. That conclusion has aroused some opposition, as syncretistic taxonomists had been inclined to suppose that grouping according to (possibly weighted) raw similarity gave hierarchies of greatest descriptive power. There seems to be no reason for taking that view seriously, however, as no attempt has been made to derive clustering by raw similarity from the aim of effective description of character information.

In my treatment I found that a hierarchic classification provides an informative or efficient description of the distribution of a feature to the degree that the feature need occur in the diagnoses of few taxa. The utility of efficient descriptions is precisely that they minimize redundancy. As I have observed before (particularly Farris, 1980), the presence of a feature in the diagnosis of a taxon corresponds to the evolutionary interpretation that the feature arose in the stem species of that taxon. There is thus a direct equivalence between the descriptive utility of a phylogenetic taxon and the genealogical explanation of the common possession of features by members of that group. Sober (1975) has stressed the importance of informativeness of theories, and has developed a characterization of informativeness in terms of simplicity. It is

no surprise to find that simplicity is related to parsimony, and Beatty and Fink (1979) have lucidly discussed the connection in terms of Sober's ideas. Sober (1982) has likewise concluded that phylogenetic parsimony correponds to simplicity (efficiency, informativeness) of explanation.

In choosing among theories of relationship on the basis of explanatory power, we wish naturally to identify the genealogy that explains as much of available observation as possible. In general, deciding the relative explanatory power of competing theories can be a complex task, but it is simplified in the present case by the fact that genealogies provide only a single kind of explanation. A genealogy does not explain by itself why one group acquires a new feature while its sister group retains the ancestral trait, nor does it offer any explanation of why seemingly identical features arise independently in distantly related lineages. (Either sort of phenomenon might, of course, be explained by a more complex evolutionary theory.) A genealogy is able to explain observed points of similarity among organisms just when it can account for them as identical by virtue of inheritance from a common ancestor. Any feature shared by organisms is so by reason of common descent, or else it is a homoplasy. The explanatory power of a genealogy is consequently measured by the degree to which it can avoid postulating homoplasies.

It is necessary in applying that last observation to distinguish between homoplasies postulated by the genealogy and those concluded for other reasons. A structure common to two organisms might be thought to be a homoplasy on grounds extrinsic to the genealogy. Such a conclusion would amount to specifying that the structure is not a point of heritable similarity. A genealogy would not explain such a similarity, but that would be no grounds for criticizing the genealogy. Rather, the extrinsic conclusion would make the feature irrelevant to evaluating genealogies by effectively stipulating that there is nothing to be explained. The same would hold true for any trait that is known not to be heritable, such as purely phenotypic variations. The explanatory power of a genealogy is consequently diminished only when the hypothesis of kinship requires ad hoc hypotheses of homoplasy.

By analogy with the abundance of homoplasy argument, it might be objected that seeking a genealogical explanation of similarities is pointless, inasmuch as most similarities are likely to be homoplasies anyway. If homoplasy were universal, that point might well hold. It seems unlikely, however, that homoplasy is universal. It is seldom maintained that segmented appendages have arisen independently in each species of insect. Universality of homoplasy would imply in the extreme that organisms do not generally resemble their parents, a proposition that seems at best contrary to experience. That the character distributions of organisms generally correspond to a hierarchic pattern, furthermore, seems comprehensible only on the view that the character patterns reflect a hierarchy of inheritance. Indeed, the recognized organic hierarchy was one of the chief lines of evidence for Darwin's theory of descent with modification. The idea that homoplasy is abundant is not usually intended in such extreme form, of course. Usually it is meant just to suggest that there is room for doubt concerning whether a shared feature is a homology or a homoplasy. Under those circumstances, however, genealogies retain explanatory power. More to the point, the explanatory power of alternative genealogies is still related to their requirements for homoplasies. Suppose that one genealogy can explain a particular point of similarity in terms of inheritance, while a second hypothesis of kinship cannot do so. If that point of similarity is, in fact, a homoplasy, the similarity is irrelevant to evaluating genealogical hypotheses, as has already been seen. If the similarity is, instead, a homology, then only the first genealogy can explain it. If there is any chance that the similarity is homologous, the first genealogy is to be preferred.

There is nothing unusual in the relationship of genealogical hypotheses to characters: scientific theories are generally chosen to conform to data. But it is seldom possible to guarantee that observations are free of errors, and it is no criticism of a theory if it turns out that some of the observations that conform to it are susceptible to error. If a theory does not conform to some observation, however, then the mere suspicion that the observation might be erroneous is not logically adequate to save the theory. Instead the data must be dismissed outright by recourse to an ad hoc hypothesis. Establishing that an observation is erroneous, on the other hand, simply makes it irrelevant to evaluating the theory. The relationship between the explanatory power of genealogies and their requirements for ad hoc hypotheses is likewise characteristic of theories in general. Any observation relevant to evaluating a theory will either conform—and so be explained—or fail to do so, in which case an ad hoc hypothesis is needed to defend the theory. It is generally true that a theory explains relevant observations to the degree that it can be defended against them without recourse to ad hoc hypotheses.

## INDEPENDENCE OF HYPOTHESES

Identifying the more explanatory of two alternative hypotheses of kinship is accomplished by finding the total of ad hoc hypotheses of homoplasy required by each. Reckoning those totals will generally involve summing over both separate characters and over observed similarities within characters. Only required ad hoc hypotheses diminish the explanatory power of a putative genealogy. It is thus important to ensure that the homoplasies combined in such totals are logically independent, since otherwise their number need not reflect required ad hoc hypotheses. If two characters were logically or functionally related so that homoplasy in one would imply homoplasy in the other, then homoplasy in both would be implied by a single ad hoc hypothesis. The "other" homoplasy does not require a further hypothesis, as it is subsumed by the relationship between the characters. This is the principle underlying such common observations as that only independent lines of evidence should be used in evaluating genealogies, and that there is no point to using both number of tarsal segments and twice

that number as characters. Phylogeneticists seldom attempt to use logically related characters as separate sources of evidence (although an example of this mistake is discussed by Riggins and Farris, 1983), and so it seems unnecessary to discuss this point further here.

A different sort of interdependence among homoplasies may arise in considering similarities within a single character. Suppose that 20 of the terminal taxa considered show a feature X, and that a putative genealogy distributes these taxa into two distantly related groups A and B of 10 terminals each. There are 100 distinct two-taxon comparisons of members of A with members of B, and each of those similarities in X considered in isolation comprises a (pairwise) homoplasy. Those homoplasies do not constitute independent required hypotheses, however. The genealogy does not require that similarities in X within either group be homoplasies; it is consistent with identity by descent of X within each group. If X is identical by descent in any two members of A, and also in any two members of B, then the A-B similarities are all homoplasies if any one of them is. The genealogy thus requires but a single ad hoc hypothesis of homoplasy. Of course the numbers in the groups do not matter; the same conclusion would follow if they were 15 and 5, or 19 and 1.

Similar reasoning can be extended to more complex examples, but the problem can be analyzed more simply. If a genealogy is consistent with a single origin of a feature, then it can explain all similarities in that feature as identical by descent. A point of similarity in a feature is then required to be a homoplasy only when the feature is required to originate more than once on the genealogy. A hypothesis of homoplasy logically independent of others is thus required precisely when a genealogy requires an additional origin of a feature. The number of logically independent ad hoc hypotheses of homoplasy in a feature required by a genealogy is then just one less than the number of times the feature is required to originate independently. (The lack of a structure might, of course, be a feature; "origins" should be interpreted broadly, to include losses.)

## LENGTHS

That last observation reduces to the rule that genealogies with greatest explanatory power are just those that minimize the (possibly weighted) total of required independent origins of known features. There is another way of putting that characterization, in terms of length. Each required origin of a feature can be assigned (although not necessarily uniquely) to a particular branch of a putative tree, and the weighted total of the origins in a branch can be regarded as that branch's length. If such lengths are summed over branches of the tree, the result is the total of required origins, or the length of the tree. Early work on automatic techniques of parsimony analysis (particularly the Wagner method formulation of Kluge and Farris, 1969) used the length conception of parsimony. That formulation has turned out to be technically very useful and has facilitated considerable progress in methods of analysis. (Basic principles are described by Farris, 1970; for some applications

of greatly improved procedures see Mickevich, 1978, 1980; Schuh and Farris, 1981; Mickevich and Farris, 1981). Nonetheless, its use was in a way unfortunate, for the length terminology has probably caused more misunderstanding than has any other single aspect of parsimony methods.

The length measure used by Kluge and Farris is coincidentally a familiar mathematical measure of distance in abstract spaces, the Manhattan metric. Once ideas have been reduced to formulae, it is easy to forget where the formulae came from, and to devise new methods with no logical basis simply by modifying formulae directly. Phylogenetic reconstructions typically infer the features of hypothesized ancestors, so that the length of a branch lying between two nodes of a tree can be regarded as the distance between two points in the space of possible combinations of features. If one notes only that length is to be minimized, then he might just as well seek trees of minimum Euclidean length, or indeed of minimum length in any of the other uncountably many possible measures of abstract distance. But even that does not exhaust the possibilities. Numerical values—lengths of a sort—can be calculated for branches without regard to the possible features of nodes, by fitting the tree directly to a matrix of pairwise distances between terminal taxa (such methods are reviewed by Farris, 1981). Such trees, too, might be selected to minimize "length," and this might be done for any of the huge number of ways of arriving at a matrix of pairwise distances.

The analogy through length has allowed methods such as this to become confused with parsimony analysis, and that confusion has played a role in specious criticisms of phylogenetic methods. Felsenstein (1981)—one of the main proponents of the idea that "parsimony" might mean almost anything—for example, attributes "parsimony" to one such method that had been used by Edwards and Cavalli-Sforza (1964). Rohlf and Sokal (1981) used a procedure for fitting branch lengths to a distance matrix to analyze the data of Schuh and Polhemus (1980), then criticized the parsimony analysis of the latter authors on the grounds that the distance-fitted tree is "shorter." As Schuh and Farris (1981) pointed out, the length that Rohlf and Sokal attribute to their tree is quite meaningless, inasmuch as it is smaller than the number of origins of features required to account for the data (for related discussion see Farris, 1981).

That will serve as a general commentary on this class of methods, which are too numerous in their possibilities to discuss here individually. The lengths arrived at by such calculations are generally incapable of any interpretation in terms of origins of features, and the evaluation of trees by such lengths consequently has nothing to do with the phylogenetic parsimony criterion. What is worse, the trees produced by these methods frequently differ in their grouping from parsimonious genealogies, and to that extent the use of these procedures amounts to throwing away explanatory power.

## PAIRWISE HOMOPLASIES

The situation is somewhat different with some types of comparative data, such as matrices of immunological distance, in which no characters are directly observed. I have emphasized before (Farris, 1981) that the parsimony criterion cannot be directly applied to such cases, and so I shall not consider them here. (The paper just cited offers other bases for evaluating methods of distance analysis.) Some analogies with distance analytic methods, however, can be related to the present discussion.

In fitting a tree to distances, branch lengths are used to determine a matrix of pairwise tree-derived distances between terminal taxa. The derived distance between a pair is just the sum of the lengths of the branches that lie on the path connecting the two taxa on the tree. Evaluation of the fit of the tree to observed distances is based on conformity of the derived to the observed distance values, this being measured by, say, the sum of the unsigned differences between the corresponding elements of the matrices (other measures are discussed by Farris, 1981). Parsimony analyses can also be used to produce derived distances, the patristic differences of Farris (1967). I had earlier (Farris, 1967) termed the departure of patristic from observed differences (pairwise) homoplasies, and from this, as well as by analogy with distance analytic procedures, one might be tempted to evaluate genealogies according to the total of those pairwise homoplasies (such a suggestion has been made by D. Swofford). The drawback of doing so is already clear from earlier discussion: the pairwise homoplasies are not independent.

A more extreme problem of interpretation of pairwise homoplasies arises in some methods for analyzing electrophoretic data. Suppose that each of three terminal taxa A, B, and C is fixed for a different allele at some locus, and that these taxa are related through an unresolved tree with common ancestor X. There are a number of ways of calculating distance between gene frequency distributions (see Farris, 1981). To fix ideas, suppose that the Manhattan distance on frequencies is used. The distance between any two terminals is then 2. The ancestor X might plausibly be assigned frequency 1/3 for each observed allele. In that case the distance between X and any terminal is 4/3, the patristic difference between any two terminals is 8/3, and the corresponding pairwise homoplasy is 2/3. That implies that there is homoplasious similarity between any two terminals, but the conclusion is nonsense, inasmuch as there is no similarity between them at all. The three terminals simply have 3 entirely different conditions of the locus.

The details of that example depend on how gene frequencies are assigned to X, but no assignment can bring all the pairwise "homoplasies" to 0 simultaneously. In part this observation reflects the difficulties inherent in any attempt to utilize distances between gene frequency distributions as evidence in phylogenetic analysis (discussed in further detail by Farris, 1981). Of greater interest for present purposes is what the example reveals about alleles as characters. The algebraic reason for the existence of those spurious homoplasies is that the distance coefficient treats shared 0 frequencies as points of similarity. Two taxa are assessed as similar in that both lack some allele, whereas in fact they simply possess different alleles. It is clear that those shared absences offer no independent assess-

ment of the resemblances among the taxa, as the 0 frequency in any one allele is a necessary consequence of the fixation of any other. This problem then results from treating dependent quantities as if they were independent.

That difficulty is not limited to analysis of frequency data. Mickevich and Johnson (1976) introduced a method in which frequencies are transformed into a two-state coding: any frequency above a cutting point is coded as 1 (presence), any other as 0 (absence). The standard Wagner method is then used to find a tree minimizing required origins of states for the coded data. This procedure obviates many of the difficulties of analyzing frequency data through distance measures, but it still suffers from dependence of variables. Fixation of one allele will necessarily control the codes of others at the same locus. The number of state origins for the coding thus need not indicate the number of independent hypotheses of homoplasy for a genealogy, and this procedure should not then be regarded as a parsimony method. The problem of interdependence can, however, be avoided by choosing a better means of coding. Mickevich and Mitter (1981) and Mitter and Mickevich (1982) have made impressive progress in developing coding methods for analyzing electrophoretic data.

### COVERING ASSUMPTIONS

Inasmuch as the aim is to minimize ad hoc hypotheses, it might seem that one could do better still by posing single hypotheses to cover several separate cases of homoplasy. Any putative genealogy might on that reasoning be defended against any character by concocting some premise to imply that all similarities in that character are homoplasies—or against any set of characters by dismissing evidence in general. I shall refer to such mass dismissals of evidence as covering assumptions.

The danger of using covering assumptions can be readily seen through a consideration of usual scientific practice. Suppose that an experiment is designed to evaluate a theory on the basis of readings from several instruments, and that some of the readings do not conform to the theory. If the nonconforming observations are only a few of the many readings made, the theory may seem to offer a generally satisfactory explanation; it is less so to the degree that such observations are abundant. Even then attempts may be made to salvage the theory. If the offending readings all come from the same instrument and so are logically related, they might be dismissed through the premise that the instrument is defective. (If it is found to be defective, so much the better.) But if no connection can be found among the nonconforming readings, the claim that they are coincidentally erroneous would have to be viewed with suspicion. Even the best theories seldom conform to every relevant observation, and so theories are well founded to the degree that nonconforming observations are rare. If contradictory observations could be dismissed as uninformative without regard to their abundance, the link between theory and observation would be tenuous at best.

Of course this is generally recognized, and attempts to defend theories by doing away with entire masses of evidence are typically rationalized by postulating mechanisms to account for what would otherwise be coincidental departures of observation from expectation. The legitimacy of that procedure depends crucially on validity of the postulates used. If the postulated mechanisms can themselves be corroborated by other sources of evidence, their use to defend the original theory is justified, and indeed they constitute improvements or extensions of the original theory. But if such mechanisms cannot be defended on extrinsic grounds, then they amount to no more than ad hoc excuses for the failure of the theory. Logically (albeit not rhetorically), they have no more force than the flat assertion that all nonconforming observations must be erroneous because the theory is true. Covering assumptions must be forbidden in scientific study, not only because they are ad hoc, but more particularly because they provide false license to dismiss any amount of evidence whatever.

The reason for prohibiting covering assumptions might be encapsulated by the observation that their use would allow theories to be chosen without regard to explanatory power. This effect can be seen directly in phylogenetic application. If 20 terminals share a particular feature, a genealogy consistent with a single origin of that feature explains those similarities fully. A hypothesis of kinship that broke those terminals into two separate groups of 10 would not explain all the similarities among taxa, but it would still explain similarities within those groups. A tree that divided the same terminals into 4 separate groups of 5 would explain still less of observed similarities, but would still retain some explanatory power, while a scheme that required 20 separate origins would leave the observed similarities entirely unexplained. Some ad hoc rationale might be used to combine 3 or 19 logically independent hypotheses of homoplasy into a "single" hypothesis. The possibility of that combination might be interpreted to mean that all these genealogies but the first conform equally well with observation. If such a course were followed, the differences in explanatory power among the last three hypotheses of kinship would play no role in choosing among them.

Almost any method that led to departure from parsimony might be suspected of involving a covering assumption. One might presume that the various length measures discussed before arise from some underlying premises that would amount to assumptions about the nature of evolution. But inasmuch as those premises, supposing that they exist, have never been made explicit, there is no real possibility of evaluating them as theories, and it is more immediately useful to view those methods as resulting simply from misunderstanding of the explanatory relationship between genealogies and characters.

Felsenstein's maximum likelihood methods offer fine examples of reliance on covering assumptions. The stochastic models would—if they were realistic—explain why seemingly independent characters would depart systematically from a parsimonious arrangement, hence would justify preference for unparsimonious schemes. Likewise, that neither Felsenstein nor anyone else maintains that those models can be corroborated makes it clear that in practice that justification

would be entirely specious. But most of these methods have never been advocated for practical application, anyway. Felsenstein's own recent efforts center on likelihood interpretations of procedures that had already been advocated on other grounds, as I shall discuss later. It is of more practical interest to analyze methods that have been proposed more or less seriously.

## IRREVERSIBILITY

Some techniques have been proposed as restricted "parsimony" methods. In these the number of origins of features is minimized, subject to the condition that some kinds of origins be rare or forbidden. Commonly, methods of this sort embody some version of the idea that evolution of individual characters is irreversible. In the method of Camin and Sokal (1965) secondary plesiomorphies are supposed not to occur, and so are excluded from reconstruction, the tree being chosen to minimize parallelisms. In the "Dollo" method of Farris (1977), origins of structures are supposed to be unique— structures once lost cannot be regained—and the tree is chosen to minimize secondary plesiomorphies.

Any of these methods might yield the same genealogy as would be obtained without the restriction, but none of them needs to do so, and in general applying the restriction will increase the number of hypotheses of homoplasy needed to defend the conclusion. Since there is no particular limit to that increase, using the restriction amounts simply to dismissing en mass any evidence that might otherwise seem to vitiate the conclusion. The motivation for doing so seems often to be more a matter of technical convenience than of conviction of the propriety of the restriction. That seems particularly to apply to the Camin-Sokal method, as it was one of the earliest techniques to be implemented as a computer algorithm. The reason for my own (Farris, 1977) development of the Dollo method likewise had little to do with the realism of the assumption. That study was intended primarily to show logical flaws in Le Quesne's (1974) earlier attempt to analyze the same problem.

In a serious study, defending conclusions that depended crucially on use of a restricted method would require defending the restriction itself. I would not claim that the supposition of irreversible character evolution could never be supported by extrinsic evidence. I would suggest, however, that what acceptance that idea has gained has been based mostly on generalizations derived from hypotheses of kinship. The common notion that evolution generally proceeds from many, similar, parts to fewer, differentiated, parts, for example, seems to have been arrived at by induction from putative lineages. If the putative phylogenies used to draw such conclusions had been arrived at by presupposing irreversibility, then the conclusion would have no legitimate empirical support. If the idea of irreversibility is supported at all, then, it must have been derived from analyses that did not depend crucially on its truth. The evidence for a directed evolutionary trend, then, would be that the postulated trend conforms to a pattern of kinship that is in turn supported by other evidence—that is, that itself conforms to other characters. If it were known that evolution is irreversible, application of that knowledge

might lead to genealogical inferences that otherwise might seem unparsimonious. But in fact no such thing is known, and the attempt to apply an empirically supported claim of irreversibility as a criticism of parsimony leads to a peculiar difficulty. Any body of characters might be made to appear consistent with the postulate of irreversibility. It is always technically possible to construct a tree so that all homoplasies take the form of parallelisms. It might seem from this that character information could never challenge the theory. But if the evidence for irreversibility was originally based on character distributions, then it would be quite unwarranted to analyse further cases so as to force them into conformity with irreversibility. The effect of doing so would be precisely to confer on irreversibility the status of an empirical conclusion that cannot be questioned by evidence—a contradiction in terms. In order to avoid that fallacy, it is necessary to allow that character information may support a conclusion of reversal. Whenever a putative reversal offers a more complete (that is, as already seen, more parsimonious) explanation of observed similarities than does a reconstruction enforcing irreversibility, irreversibility must be discounted. (In that particular case trends might still be accepted as rough descriptive generalizations.)

A proponent of irreversibility might nevertheless insist that when an analysis that does not presuppose irreverssibility gives a different result from another that does use that premise, then the conclusion of the former depends crucially on the supposition that reversal is possible. Moreover, as the procedure just outlined will always discount irreversibility when parsimony requires, there is no way of rejecting the possibility of reversal. That possibility might seem, then, to be an ad hoc hypothesis, so that a conclusion of reversal actually requires more ad hoc hypotheses than would be suggested just by counting independent origins. But even if possibility of reversal did constitute an ad hoc hypothesis, it would certainly not be an additional independent hypothesis, for it is entailed by the particular hypothesis of reversal postulated. That observation, in fact, contains the key to the defect of the whole objection. If a particular conclusion of reversal could be legitimately criticized as presupposing the possibility of reversal, then any scientific conclusion whatever could be dismissed as requiring the supposition of its own possibility. The argument outlined is seen in that light to be simply another rationalization for discarding evidence.

It is clear that the reasoning outlined effectively views irreversibility and the possibility of reversal as competing theories. The charge that possibility of reversal cannot be rejected by parsimony analysis would be pertinent only as a criticism of a way of testing an empirical claim. But that view is itself suspicious. Irreversibility is certainly an empirical claim, and, furthermore, it is plainly testable in principle, inasmuch as it prohibits something, namely reversals. The possibility of reversal, on the other hand, can hardly be by itself an empirical claim in the same sense (although the claim that particular reversals have occurred might be), as it does not prohibit anything. One might think that admitting that reversal might occur, if

it is not itself directly an empirical contention, nonetheless implies one, in that using a method that can discard irreversibility for parsimony would necessarily yield conclusions of reversal. But in fact it is quite possible for a parsimonious reconstruction to lack requirements for reversal. (The contrary, of course, is also possible, and is often observed. But that is a consequence of the idea in conjunction with particular observations, not of the idea itself.) While irreversibility and the possibility of reversal seem superficially to be simply alternative theories, then, they are in fact not the same kind of idea. The first is a theory that forbids conclusions that might otherwise seem supported by observation, and, when confronted with such cases, can be saved only by ad hoc supposition. The second is simply an attitude. The possibility that irreversibility (or any theory) is false must be considered in order to test the theory. No kind of empirical science would be possible without such attitudes.

## POLYMORPHISM

Because of their reliance on covering assumptions to justify otherwise unnecessary ad hoc hypotheses of homoplasy, the Camin-Sokal and Dollo techniques should not be regarded as proper parsimony methods, prior usage notwithstanding. The situation may be different, however, for another restricted procedure. In the chromosome inversion model of Farris (1978), it is presumed that each of two alternative inversion types originated uniquely. Inversion types may nonetheless show incongruence with a genealogy through independent fixations from polymorphic ancestral populations, and the tree is chosen to minimize such fixations. The accuracy of the premise of unique origin might, of course, be questioned, but the idea is accepted by specialists on grounds extrinsic to genealogical hypotheses, and I shall not attempt to dispute its validity here. A further observation in this connection, however, seems worthwhile.

As this model presumes a unique origin for each inversion type, it might seem that similarity between organisms would on this premise be due to inheritance regardless of the genealogy postulated, so that the relationship between parsimony and genealogical explanation would no longer hold. The inherited similarity covered by the premise, however, holds only for chromosomes of individuals. Resemblance between populations fixed for the same inversion may still be explained by inheritance, or else the coincidental result of independent fixations. As it is populations that are grouped in postulating a genealogy, it is still possible to compare alternative genealogical hypotheses on explanatory power. There is in fact nothing unusual in this conclusion. It is generally true that features used to arrange taxa are characteristics of populations, rather than of individuals. The observation that deer have antlers is just a contracted way of stating that normal, adult, male deer in breeding condition possess those structures. The females, young, and deformed are not given a separate place in the system by reason of lacking the characteristic. The same principle underlies Hennig's emphasis of the idea that holomorphs rather than specimens, are classified. Mickevich and Mitter

(1981) arrive at the same concept in developing their greatly improved methods for analyzing electrophoretic data. They concentrate on recognizing suites of alleles as features of populations, rather than attempting to use single alleles—traits of individuals—as characters.

## PHENETIC CLUSTERING

Clustering by raw similarity (phenetic clustering) has sometimes been advocated as a means of making genealogical inferences, typically with the justifying assumption that rates of evolutionary change (or divergence) are nearly enough constant so that degree of raw similarity reflects recency of common ancestry. The method is most often used with comparative biochemical data, but it has been recommended for morphological data as well (for example by Colless, 1970).

Constancy of rate is rather a different theory from irreversibility of evolution, but many of the comments made earlier apply here as well. Phenetic clustering might coincidentally produce a parsimonious scheme, but it certainly need not do so, and again there is no limit in principle to the number of otherwise unnecessary requirements for hypotheses of homoplasy that this method might impose. The assumption is certainly an empirical claim, and advocates of the method usually defend it by producing evidence for rate constancy. (Colless is an exception; he shows no inclination to resort to evidence.) That evidence typically takes the form of correlations between observed raw similarities and putative recency of common ancestry. Those last naturally depend on hypotheses of kinship, and this raises the familiar dilemma. If the genealogies used as evidence depended crucially on rate constancy, there would be no evidence. Supposing, then, that they do not, the evidence must consist of agreement between the theory and arrangements that conform to character distributions. Just as before, if the premise of rate constancy is used to justify unparsimonious conclusions, the effect is to consider rate constancy as empirical and irrefutable at once. Likewise parsimony analysis might be accused of presupposing that rates can vary, but discussion of that idea would precisely parallel what has already been said in connection with irreversibility.

A molecular evolutionist is quite happy with the generally good correlation that is observed between raw similarity and putative recency of common ancestry; for him it substantiates the molecular "clock." But as I have emphasized before (Farris, 1981), such correlations are not enough to justify clustering by raw similarity. The correlations reported show considerable scatter. The implication of this, accepting the usual interpretation of the general correlation, is that rates of divergence vary somewhat. Even if it is often true, then, that genealogically most closely related taxa are also mutually most similar, there are evidently exceptions. Those exceptions could not be identified if genealogy were inferred by presupposing that raw similarity reflects kinship. To make accurate inferences in such cases—to discover what the cases are—it is necessary to use a method that can discount raw similarity as indicative of kinship if the data seem to require doing so—if

doing so is required to achieve a more complete explanation of observed features. By analogy with the discussion of irreversibility, the same conclusion would be reached just by requiring that the relationship of raw similarity to kinship be vulnerable to evidence. It seems, then, that a correlation between raw similarity and kinship—even if it often holds—can provide no legitimate grounds for accepting unparsimonious inferences.

I commented before on the distinction between an ad hoc covering assumption and a corroborated improved theory able to account systematically for observations that would otherwise seem coincidental departures from its predecessor. This distinction suggests a further defect in the attempt to defend phenetic clustering on grounds of a correlation between raw similarity and kinship. In a legitimate extension of theory, the old coincidences are not dismissed as such, but explained by the extension. The process, that is, expands explanatory power, rather than discarding it. Suppose that clustering by raw similarity in some case requires otherwise unnecessary hypotheses of homoplasy, and that the conclusion is defended on grounds of a theoretical relationship between raw similarity and recency of common ancestry. If this is not ad hoc, then the theory must offer an explanation of the putative homoplasies. It is far from clear, however, that it can do so. Homoplasies, as already observed, are not explained by the inferred genealogy, from the standpoint of which the shared features that they represent are so only coincidentally. Inasmuch as raw similarities are calculated from features, it seems curious that they could either explain or be explained by a scheme that left the features themselves unaccounted for. In order for a relationship between raw similarity and kinship to explain homoplasies, furthermore, it would seem necessary to suppose that that relationship rests on some real mechanism. That mechanism would have to have the property that organisms would come to possess features in common for reasons other than inheritance, and in just such a way as to maintain the correlation between raw similarity and recency of common ancestry. As no known natural process appears to have this property, it would seem that use of a postulated correlation between raw similarity and kinship to defend clustering by raw similarity rests necessarily on an ad hoc covering assumption.

A related conclusion can be reached by another route. Phenetic clustering ignores considerations of parsimony and so effectively proceeds by freely introducing whatever hypotheses of homoplasy are needed to derive a result conforming to the rate constancy premise. The procedure would be highly questionable on statistical grounds alone, then, if homoplasy were supposed to be rare. The method then requires the assumption that homoplasy is abundant, and indeed its proponents are prominent in criticizing parsimony as requiring rarity of homoplasy. The premise that homoplasy is abundant, however, poses a problem for clustering by raw similarity as well. That method infers recency of common ancestry of two taxa from the fraction of characters in which the two are similar. If homoplasy were rare (and rates constant), that would be superficially reasonable.

Similarity between two lineages would decrease in clocklike fashion as ancestral similarities were lost. But if homoplasy is abundant, many of the similarities between two taxa are likely to be homoplasious, in which case they need indicate nothing about how recently the pair diverged. Two populations having only a remote common origin, and so (if rates were constant) very little homologous similarity, might have many recently acquired homoplasies, and so be judged to be of recent common ancestry. It is easy enough to identify conditions under which inferences might still be valid. If pairwise homoplasies were all the same, or nearly so, homoplasy would not alter the relative degrees of raw similarity among taxa, and then (if rates of change were constant) the method would still work. Phenetic clustering effectively presumes, then, that the variance of pairwise homoplasies is small. Keeping that variance small would be the task of the hypothetical mechanism just discussed.

While phenetic clustering does not consider homoplasies as such, it does select a tree by finding a constant-rate (ultrametric) model that conforms to observed raw similarities as closely as possible. If rates of evolution were constant, homologous similarities would conform to the constant rate model, so that departure from the model would be due to variation in pairwise homoplasies. The phenetic clustering procedure most commonly applied for genealogical inference, UPGMA, has precisely the effect of minimizing the variance of pairwise departures of observed from ultrametric similarities (Farris, 1969b). Phenetic clustering and parsimony analysis are similar, then, in the sense that each minimizes a criterion. But whereas abundance of homoplasy need not imply error by parsimonious inference, large variations in pairwise homoplasies would certainly vitiate the conclusion of phenetic clustering. Phenetic clustering, unlike parsimony, depends crucially on minimality in nature of the quantity that it minimizes. Clustering by raw similarity possesses the very sort of defect that its proponents had incorrectly claimed as a weakness of phylogenetic analysis.

## CLIQUES

Clique methods rely on parsimony to interpret suites of congruent characters, but their trees require homoplasy for characters outside the selected clique, and often the clique tree will be quite unparsimonious for those characters. In practice the excluded characters are often numerous, so that basing the inferred genealogy just on the clique imposes a considerable loss of explanatory power. These methods are then prime suspects for reliance on a covering assumption, but for a long time it was not clear from the clique literature what that assumption was supposed to be.

Le Quesne (1972) offered an approximate method (later made exact by Meacham, 1981) for finding the probability that a suite of characters would all be congruent if features were distributed independently and at random among taxa. He suggested selecting the clique with the lowest such probability, and other proponents also commonly refer to cliques as "least likely." It is possible that this idea is intended as a justification of clique methods. If so, the justifying reasoning amounts to no more than misunderstanding of statistics. If a clique were evaluated just on its probability under a random model, the evaluation would be bound to the model. In that case the covering assumption of cliques would be that characters—being randomly distributed—have no relationship to genealogy. Perhaps it was intended that low probability under a null model would lend credence to an alternative, genealogical interpretation of the clique, but that idea, too, rests on a fallacy. Observing that a clique (or anything else) has low probability under a model might provide statistical grounds for rejecting the model, but it does not by itself offer any basis for choosing any particular alternative hypothesis. Once a model has been rejected, the probabilities it assigns to events necessarily become irrelevant. In this case rejecting the null model is uninformative, as no one interested in making phylogenetic inferences would have taken it seriously anyway. The statistical reason for accepting a new hypothesis is that it assigns much higher probability to observation than does the old. In normal statistics, a large enough difference between sample means serves as grounds for rejecting the hypothesis that the two samples were drawn from populations with the same parametric mean. If an alternative hypothesis is chosen so that it assigns maximum probability to the observed difference, the new theory conforms best with observation. But one hardly proceeds by choosing observations so as to minimize their nominal probability under the original hypothesis, let alone using such observations as the basis for choosing a new theory. Making statistical genealogical inferences from characters that had been used to reject the hypothesis of randomness would likewise require choosing a genealogy that would assign maximum probability to available characters—a maximum likelihood tree. I have already commented on the difficulties of applying that approach in practice, but this case is far worse. No model other than the rejected one of randomness is provided, and so neither are any grounds whatever for accepting the tree from the "least likely" clique as a genealogical inference best conforming to observation. (Felsenstein has made much the same point.)

As none of the ideas just discussed provides any legitimate rationale for clique methods, those procedures must rest on an undisclosed assumption, if indeed they rest on anything at all. It is not difficult to discern what that assumption would have to be. Cliques are usually chosen to comprise as many mutually congruent characters as possible, and any characters that must be discarded to achieve this are simply counted as excluded. If the genealogy corresponding (by parsimony) to the clique is accepted, each of the excluded characters will require at least one hypothesis of homoplasy, but the number required may well vary among those characters. As characters are counted just as excluded or not, the number of hypotheses of homoplasy required by excluded characters plays no role in the analysis: similarities in those characters are dismissed en mass. The covering assumption involved is thus like the archetypical one discussed before. Ad hoc hypotheses of a sort are counted, but the counts do not reflect simple hypotheses of homoplasy. Instead any

and all similarities in each excluded character are discounted by recourse to a "single" covering assumption. Excluding a character amounts to treating all similarities in it as irrelevant to assessing kinship. Those similarities could all be logically irrelevant only if they were all homoplasies. The covering assumption utilized is, then, that excluding a character—concluding that it shows some homoplasy—implies that all points of similarity in that character are homoplasies.

As discussed before, the collective dismissal of similarities in a character would be justified if the multiple required origins of features were not logically independent. It is readily seen, however, that such is not the case. The conclusion that endothermy has evolved independently in mammals and in birds does not imply that each species of bird or mammal has independently achieved that condition. Such being the case, it is likewise clear, from earlier discussion, that use of such a covering assumption leads to loss of explanatory power. As before, a single requirement of homoplasy may leave many of the similarities in a character explained, while a large enough number of required homoplasies will leave the same similarities entirely unexplained. Counting characters as simply excluded or not produces an evaluation oblivious to that distinction.

An attempt might be made to defend clique methods by advancing their covering assumption as an empirical claim on evolution, although of course doing so would raise the same sort of difficulties already discussed for irreversibility and rate constancy. Clique advocates have not tried to take that course—perhaps for fear of inviting ridicule. Unlike superficially tenable premises such as rate constancy, the clique assumption implies a theory that no one would take seriously as a realistic possibility. A different approach to rationalizing cliques has been taken by Felsenstein, though. He has proposed two stochastic models (Felsenstein, 1979, 1981) under which he derives cliques as maximum likelihood estimates of genealogy. Both of these operate, just as would be expected from the clique assumption, by supplying principles that would excuse dismissing characters as units. In the 1979 paper this effect is achieved by introducing the possibility of a carefully selected type of error. Any character incongruent with the accepted tree is characterized as erroneous, and this is taken to mean that the character has been so completely misinterpreted that it is uninformative on genealogy. In the other model, incongruent characters are instead regarded as having changed so frequently in evolution as to be unrelated to genealogical grouping. Felsenstein himself emphasized that his models are inconsistent with the observed frequency of incongruence among characters. Realism thus plays no role in these justifications, which seem aimed instead at defending the clique assumption just by translating it into statistical terminology. Both rationalizations, consequently, have the same faults as the clique assumption itself.

The error idea rests on a misrepresentation of how systematists recognize characters. At one time it was believed that the eyes of octopi and of vertebrates were the same. That was certainly an error, for the two organs differ in both structure and ontogeny. But that the mistake was made does not mean that the sameness attributed to the eyes of rats and of mice is likewise a misrepresentation. Concluding a homoplasy in a feature may well invite renewed inspection and possible reinterpretation. But that reinspection certainly need not lead to dismissing all the agreements between taxa that the original feature had been intended to summarize. While it is reasonable to attribute some homoplasies to errors, it does not follow from this that those errors will turn out to have universal effects. It is seen from this that Felsenstein's use of the idea of wholesale error as a defense of the clique assumption amounts to no more than stating the desired conclusion as a premise. The defect of cliques is just that they treat every conclusion of homoplasy as if it implied universal homoplasy. Felsenstein attributes homoplasies to errors, but bolsters cliques only by supposing that any conclusion of error implies universal error. Neither implication is valid, and so either is merely an ad hoc rationalization for dismissing relevant evidence.

Much the same applies to Felsenstein's second argument. Dismissing an incongruent character on the grounds that it must have changed very frequently clearly depends on discounting the possibility that it changed only a few times. As Felsenstein (1981, p. 183) puts it, the clique method is suitable when "it is known that a few characters have very high rates of change, and the rest very low rates, but it is not known which characters are the ones having high rates." He does not disclose, however, how one comes to know the rate of change of a character without a prior phylogenetic analysis. Nor does he explain how one would apply that undisclosed method to gain the knowledge that his method calls for, without in the process incidentally learning which characters had the high rates. Nor, again, does he offer any pretense of a reason why rates should restrict themselves to be either very high or else very low —or why rapidly changing characters ought to be "few." In the absence of such explanation, it is seen that the covering assumption that one conclusion of homoplasy implies universal homoplasy has once again been "defended" simply by restating it as the entirely equivalent—and equally unsubstantiated—premise that any feature that originates more than once must have done so a very great number of times.

### CONCLUSION

Advocacy of nonphylogenetic methods has consistently been based on the charge that parsimony depends on unrealistic assumptions. That allegation has never been supported by substantial argument. It has been instead motivated by the dependency of other approaches on false suppositions: Proponents of other views have tried to bolster their position through the pretense that no means of phylogenetic analysis can be realistic.

Parsimony analysis is realistic, but not because it makes just the right suppositions on the course of evolution. Rather, it consists exactly of avoiding uncorroborated suppositions whenever possible. To a devotee of supposition, to be sure, parsimony seems to presume very much indeed: that evolution is not irreversible, that rates of evolution are not constant, that all characters do not evolve according to identical stochastic processes, that one conclusion of homoplasy does not imply others. But parsimony does not suppose in advance that those possibilities are false—only that they are not already established. The use of parsimony depends just on the view that the truth of those—and any other—theories of evolution is an open question, subject to empirical investigation.

The dichotomy between parsimony and supposition is just that; parsimony offers no barrier to evolutionary theories as such. Rate constancy—or any other supposition—seems to be in conflict with parsimony in the abstract, as it seems to offer a different basis for making genealogical inferences. But it would conflict with parsimony in application only in conjunction with observation, if maintaining the supposition required discarding a parsimonious—explanatory—interpretation of evidence. In that case, however, the same evidence would serve to question the supposition, which could then be defended only by presupposing its truth, or—entirely equivalently—simply dismissing the evidence. But if parsimonious interpretation of evidence did not refute the supposition, then the latter would become a corroborated theory. Parsimony does not require that no such theories will be corroborated, but offers a means for that corroboration, provided evidence allows it. Unlike prior supposition, empirically supported evolutionary theories can offer no criticism of parsimony, for those theories could have become corroborated just to the extent that they require few dismissals of evidence. The insistence by proponents of suppositions that parsimony is unrealistic, it is then seen, is merely a subterfuge. Ostensibly the objection is to parsimony, but in fact the complaint is that some cherished idea does not conform to evidence.

I return finally to the questions raised at the beginning. Phylogenetic analysis is necessarily based on parsimony, both because it is precisely the criterion that leads to grouping according to putative synapomorphy, and because empirical investigation is impossible without avoiding ad hoc hypotheses. Only synapomorphy provides evidence of kinship, for the attempt to use raw similarity as evidence would necessarily either rest on un-

corroborated—and so nonevidential—supposition, or else could lead to no conclusion conflicting with synapomorphy. And phylogenetic analysis is most certainly empirical, for in applying the parsimony criterion, it chooses among alternative hypotheses of relationship on the basis of nothing other than their explanatory power. Differing as it thus does from all other approaches, phylogenetic systematics alone provides a logical basis for the empirical study of the relationships among organisms.

## REFERENCES

Beatty, J., and W. L. Fink, 1979, review of *Simplicity*. *Syst. Zool.*, 28: 643–651.

Camin, J. H., and R. R. Sokal, 1965, A method for deducing branching sequences in phylogeny. *Evolution*, 19: 311–326.

Cartmill, M., 1981, Hypothesis testing and phylogeny reconstruction. *Zeit. zool. Syst. Evolut.-forsch.*, 19: 73–95.

Colless, D. H., 1970, The phenogram as an estimate of phylogeny. *Syst. Zool.*, 19: 352–362.

Edwards, A.W.F. and L. L. Cavalli-Sforza, 1964, Reconstruction of evolutionary trees. In V. H. Heywood and J. McNeill, eds., *Phenetic and Phylogenetic Classification*. London, Systematics Association, pp. 67–76.

Farris, J. S., 1966, Estimation of conservatism of characters by constancy within biological populations. *Evolution*, 20: 587–591.

——, 1967, The meaning of relationship and taxonomic procedure. *Syst. Zool.*, 16: 44–51.

——, 1969a, On the cophenetic correlation coefficient. *Syst. Zool.*, 18: 279–285.

——, 1969b, A successive approximations approach to character weighting. *Syst. Zool.*, 18: 374–385.

——, 1970, Methods for computing Wagner trees. *Syst. Zool.*, 19: 83–92.

——, 1973, On the use of the parsimony criterion for inferring evolutionary trees. *Syst. Zool.*, 22: 250–256.

——, 1977, Phylogenetic analysis under Dollo's law. *Syst. Zool.*, 26: 77–88.

——, 1978, Inferring phylogenetic trees from chromosome inversion data. *Syst. Zool.*, 27: 275–284.

——, 1979, The information content of the phylogenetic system. *Syst. Zool.*, 28: 483–519.

——, 1980, The efficient diagnoses of the phylogenetic system. *Syst. Zool.*, 29: 386–401.

——, 1981, Distance data in phylogenetic analysis. In *Advances in Cladistics: Proceedings of the First Meeting of the Willi Hennig Society*, ed. V. A. Funk and D. R. Brooks, New York Botanical Garden.

——, 1982, Simplicity and informativeness in systematics and phylogeny. *Syst. Zool.*, 31: 413–444.

Farris, J. S. and A. G. Kluge, 1979, A botanical clique. *Syst. Zool.*, 28: 400–411.

Farris, J. S, A. G. Kluge, and M. F. Mickevich, 1982, Phylogenetic analysis, the monothetic group method, and myobatrachid frogs. *Syst. Zool.*, 31: 317–327.

Felsenstein, J., 1972, Maximum likelihood and minimum steps methods for estimating evolutionary trees from data on discrete characters. *Syst. Zool.*, 22: 240–249.

——, 1978, Cases in which parsimony or compatibility methods will be positively misleading. *Syst. Zool.*, 27: 401–410.

——, 1979, Alternative methods of phylogenetic inference and their interrelationship. *Syst. Zool.*, 28: 49–62.

——, 1981, A likelihood approach to character weighting and what it tells us about parsimony and compatibility. *Biol. Jour. Linn. Soc.*, 16: 183–196.

Funk, V. A., and D. R. Brooks, 1981. National science foundation workshop on the theory and application of cladistic methodology. Organized by T. Duncan and T. Stuessy. University of California, Berkeley, 22–28 March 1981. *Syst. Zool.*, 30: 491–498.

Gaffney, E. S., 1979, An introduction to the logic of phylogeny reconstruction. In J. Cracraft and N. Eldredge, eds., *Phylogenetic Analysis and Paleontology*, New York, Columbia University Press, pp. 79–112.

Hennig, W., 1966, *Phylogenetic Systematics*, Urbana, Ill., University of Illinois Press, 263 pp.

Kluge, A. G., and J. S. Farris, 1969, Quantitative phyletics and the evolution of anurans. *Syst. Zool.*, 18: 1–32.

Le Quesne, W. J., 1969, A method of selection of characters in numerical taxonomy. *Syst. Zool.*, 18: 201–205.

——, 1972, Further studies based on the uniquely derived character concept. *Syst. Zool.*, 21: 281–288.

——, 1974, The uniquely evolved character concept and its cladistic application. *Syst. Zool.*, 23: 513–517.

Meacham, C. A., 1981, A probability measure for character compatibility. *Math. Biosci.*, 57: 1–18.

Mickevich, M. F., 1978, Taxonomic congruence. *Syst. Zool.*, 27: 143–158.

——, 1980, Taxonomic congruence: Rohlf and Sokal's misunderstanding. *Syst. Zool.*, 29: 162–176.

Mickevich, M. F., and M. S. Johnson, 1976, Congruence between morphological and allozyme data in evolutionary inference and character evolution. *Syst. Zool.*, 25: 260–270.

Mickevich, M. F., and C. Mitter, 1981, Treating polymorphic characters in systematics: A phylogenetic treatment of electrophoretic data. In V. A. Funk and D. R. Brooks, eds., *Advances in cladistics: Proceedings of the First Meeting of the Willi Hennig Society*. New York Botanical Garden, pp. 45–60.

Mickevich, M. F., and J. S. Farris, 1981, The implications of congruence in *Menidia*. *Syst. Zool.*, 30: 351–370.

Nei, M., 1972, Genetic distance between populations. *Amer. Nat.*, 106: 283–292.

Riggins, R., and J. S. Farris, 1983, Cladistics and the roots of angiosperms. *Syst. Bot.*, 8: 96–101.

Rohlf, F., and Sokal, R., 1981, Comparing numerical taxonomic studies. *Syst. Zool.*, 30: 459–485.

Schuh, R. T., and J. T. Polhemus, 1980, Analysis of taxonomic congruence among morphological, ecological, and biogeographic data sets for the Leptopodomorpha (Hemiptera). *Syst. Zool.*, 29: 1–26.

Schuh, R. T., and J. S. Farris, 1981, Methods for investigating taxonomic congruence and their application to the Leptopodomorpha. *Syst. Zool.*, 30: 331–351.

Sober, E., 1975, *Simplicity*, London, Oxford University Press, 189 pp.

——, 1982, Parsimony methods in systematics. *Hennig Society II*, New York, Columbia University Press.

Watrous, L. E., and Q. D. Wheeler, 1981, The out-group comparison method of character analysis. *Syst. Zool.*, 30: 1–11.

Wiley, E. O., 1975, Karl R. Popper, systematics, and classification: A reply to Walter Bock and the evolutionary taxonomists. *Syst. Zool.*, 24: 233–243.

——, 1981, *Phylogenetics: The Theory and Practice of Phylogenetic Systematics*. New York, John Wiley and Sons, 439 pp.

# Acknowledgments

## I. Guiding Ideas in Evolutionary Biology

1. "The Structure of Evolutionary Genetics." From chapter 1 of *The Genetic Basis of Evolutionary Change,* Columbia University Press, New York (1974), pp. 3–18. Reprinted by permission of the publisher and the author.

2. "Typological versus Population Thinking." From E. Mayr, *Evolution and the Diversity of Life,* Harvard University Press (1975), pp. 26–29. Reprinted by permission of the publisher and the author.

3. "The Strategy of Model Building in Population Biology." From *American Scientist,* 54, no. 4 (1966), pp. 421–431. Reprinted by permission of the publisher and the author.

## II. Fitness

4. "Darwin's Untimely Burial." From *Ever since Darwin,* New York, W. W. Norton (1977), pp. 39–48. Reprinted by permission of the publisher and the author.

5. "The Propensity Interpretation of Fitness." From *Philosophy of Science,* 46, no. 2 (1979), pp. 263–286. Reprinted by permission of the publisher and the authors.

6. "Adaptation and Evolutionary Theory." From *Studies in the History and Philosophy of Science* (1978), 9, no. 3, pp. 181–206. Reprinted by permission of the publisher and the author.

7. "The Logical Status of Natural Selection and Other Evolutionary Controversies." From M. Bunge (ed.), *The Methodological Unity of Science,* Dordrecht (1973), pp. 84–102. Reprinted by permission of the publisher and the author.

8. "The Supervenience of Biological Concepts." From *Philosophy of Science* (1978), 45, pp. 368–386. Reprinted by permission of the publisher and the author.

## III. The Units of Selection

9. "Caring Groups and Selfish Genes." From *The Panda's Thumb,* New York, W.W. Norton (1980), pp. 85-92. Reprinted by permission of the publisher and the author.

10. "Replicator Selection and the Extended Phenotype." From *Zeitschrift für Tierpsychol.* (1978), 47, pp. 61-76. Reprinted by permission of the publisher and the author.

11. "Reductionistic Research Strategies and Their Biases in the Units of Selection Controversy." From T. Nickles (ed.), *Scientific Discovery,* D. Reidel Publishing (1980), pp. 213-259. Reprinted by permission of the publisher and the author.

12. "Holism, Individualism, and the Units of Selection." From P. Asquith and R. Giere (eds.), *PSA 1980,* Proceedings of the Philosophy of Science Association Meetings, E. Lansing, Michigan (1981). Reprinted by permission of the publisher and the author.

13. "Artifact, Cause, and Genic Selection." From *Philosophy of Science* (1982), 49, no. 2, pp. 157-180. Reprinted by permission of the publisher and the authors.

## IV. Adaptation

14. "Adaptation." From *The Encyclopedia Einaudi,* Milan (1980). Reprinted by permission of the publisher and the author.

15. "The Spandrels of San Marco and the Panglossian Paradigm: A critique of the adaptationist programme." From *Proc. R. Soc. London* (1978), 205, pp. 581-598. Reprinted by permission of the publisher and the authors.

16. "A Critique of Optimization Theory in Evolutionary Biology." From chapter 8 of *Caste and Ecology in the Social Instincts,* Princeton University Press (1978), pp. 292-315. Reprinted by permission of the publisher and the authors.

17. "Optimization Theory in Evolution." From *Annual Review of Ecology and Systematics* (1978), 9, pp. 31-56. Reprinted by permission of the publisher and the author.

## V. Function and Teleology

18. "The Structure of Teleological Explanations." From *The Structure of Science,* Indianapolis: Hackett Publishing Co. (1961), pp. 398-428. Reprinted by permission of the publisher and the author.

19. "Functions." From *Philosophical Review* (1973), 82, pp. 139-168. Reprinted by permission of the publisher and the author.

20. "Wright on Functions." From *Philosophical Review* (1976), 85, pp. 70-86. Reprinted by permission of the publisher and the author.

21. "Functional Analysis." From *Journal of Philosophy* (1975), 72, no. 20, pp. 741-764. Reprinted by permission of the publisher and the author.

## VI. The Reduction of Mendelian Genetics to Molecular Biology

22. "The Standpoint of Organismic Biology." From *The Structure of Science,* Indianapolis: Hackett Publishing Co. (1961), pp. 428-446. Reprinted by permission of the publisher and the author.

23. "Reduction in Biology: Prospects and Problems." From R. S. Cohen et al. (eds.), *PSA 1974,* pp. 613–632; Dordrecht, 1976. Reprinted by permission of the publisher and the author.

24. "Reduction in Genetics." From R. S. Cohen et al. (eds.), *PSA 1974,* pp. 653–670; Dordrecht, 1976. Reprinted by permission of the publisher and the author.

25. "Informal Aspects of Theory Reduction." From R. S. Cohen et al. (eds.), *PSA 1974,* pp. 653–670; Dordrecht, 1976. Reprinted by permission of the publisher and the author.

26. "Reductive Explanation: A Functional Account." From R. S. Cohen et al. (eds.), *PSA 1974,* pp. 671–710; Dordrecht, 1976. Reprinted by permission of the publisher and the author.

27. "Unifying Science Without Reduction." From *Stud. Hist. Phil. Sci.* (1977), 9, 143–162. Reprinted by permission of the publisher and the author.

## VII. The Nature of Species

28. "Species Concepts and Their Applications." From ch. 2 of *Populations, Species, and Evolution,* Cambridge: Harvard University Press (1963). Reprinted by permission of the publisher and the author.

29. "The Biological Species Concept: A Critical Evaluation." From *American Naturalist* (1970), 104, pp. 127–153. Reprinted by permission of the publisher and the author.

30. "Contemporary Systematic Philosophies." From *Annual Review of Ecology and Systematics* (1970), 1, pp. 19–53. Reprinted by permission of the publisher and the author.

31. "Phylogenetic Systematics." From *Annual Review of Entomology* (1965), 10, pp. 97–116. Reprinted by permission of the publisher and the author.

32. "A Matter of Individuality." From *Philosophy of Science* (1978), 45, pp. 335–360. Reprinted by permission of the publisher and the author.

33. "Biological Classification: Toward a Synthesis of Opposing Methodologies." From *Science* (1981), pp. 510–516. Reprinted by permission of the publisher and the author.

34. "Cases in Which Parsimony or Compatibility Methods Will Be Positively Misleading." From *Systematic Zoology* (1978), 27, 401–410. Reprinted by permission of the publisher and the author.

35. "The Logical Basis of Phylogenetic Analysis." From *Hennig Society II,* Columbia University Press (1982). Reprinted by permission of the publisher and the author.

# Index